15th International Workshop on Advanced Computing and Analysis Techniques in Physics Research

(ACAT 2013)

Journal of Physics: Conference Series Volume 523

Beijing, China
16-21 May 2013

ISBN: 978-1-63439-304-1
ISSN: 1742-6588

Printed from e-media with permission by:

Curran Associates, Inc.
57 Morehouse Lane
Red Hook, NY 12571

Some format issues inherent in the e-media version may also appear in this print version.

Printed by Curran Associates, Inc. (2014)

For permission requests, please contact the Institute of Physics
at the address below.

Institute of Physics
Dirac House, Temple Back
Bristol BS1 6BE UK

Phone: 44 1 17 929 7481
Fax: 44 1 17 920 0979

techtracking@iop.org

Additional copies of this publication are available from:

Curran Associates, Inc.
57 Morehouse Lane
Red Hook, NY 12571 USA
Phone: 845-758-0400
Fax: 845-758-2634
Email: curran@proceedings.com
Web: www.proceedings.com

TABLE OF CONTENTS

TRACK 1

COMPUTING TECHNOLOGY FOR PHYSICS RESEARCH

012001 LESSONS LEARNED FROM THE K COMPUTER PROJECT—FROM THE K COMPUTER TO EXASCALE 1
Y Oyanagi

012002 OPPORTUNITIES AND CHOICE IN A NEW VECTOR ERA 7
A Nowak

012003 PERFORMANCE QUALITY MONITORING SYSTEM FOR THE DAYA BAY REACTOR NEUTRINO EXPERIMENT 16
Y B Liu

012004 A CONCURRENT VECTOR-BASED STEERING FRAMEWORK FOR PARTICLE TRANSPORT 22
J Apostolakis, R Brun, F Carminati, A Gheata, S Wenzel

012005 ALICE EXPERT SYSTEM 31
C Ionita, F Carena

012006 EXPERIENCE, USE, AND PERFORMANCE MEASUREMENT OF THE HADOOP FILE SYSTEM IN A TYPICAL NUCLEAR PHYSICS ANALYSIS WORKFLOW 37
E Sangaline, J Lauret

012007 GPUS FOR REAL-TIME PROCESSING IN HEP TRIGGER SYSTEMS 42
R Ammendola, A Biagioni, L Deri, M Fiorini, O Frezza, G Lamanna, F Lo Cicero, A Lonardo, A Messina, M Sozzi, F Pantaleo, Ps Paolucci, D Rossetti, F Simula, L Tosoratto, P Vicini

012008 HIGH PERFORMANCE COMPUTING ACTIVITIES IN HADRON SPECTROSCOPY AT BESIII 50
B Liu

012009 INITIAL EXPLORATIONS OF ARM PROCESSORS FOR SCIENTIFIC COMPUTING 54
D Abdurachmanov, P Elmer, G Eulisse, S Muzaffar

012010 JALIEN – A NEW INTERFACE BETWEEN THE ALIEN JOBS AND THE CENTRAL SERVICES 60
A G Grigoras, C Grigoras, M M Pedreira, P Saiz, S Schreiner

012011 LESS IS MORE. WHY OBERON BEATS MAINSTREAM IN COMPLEX APPLICATIONS 67
F V Tkachov

012012 MANAGING A TIER-2 COMPUTER CENTRE WITH A PRIVATE CLOUD INFRASTRUCTURE 73
S Bagnasco, D Berzano, R Brunetti, S Lusso, S Vallero

012013 ANALYSIS OF PERFORMANCE IMPROVEMENTS FOR HOST AND GPU INTERFACE OF THE APENET+ 3D TORUS NETWORK 80
R Ammendola A, A Biagioni, O Frezza, F Lo Cicero, A Lonardo, P S Paolucci, D Rossetti, F Simula, L Tosoratto, P Vicini

012014 PERFORMANCE OPTIMISATIONS FOR DISTRIBUTED ANALYSIS IN ALICE 88
L Betev, A Gheata, M Gheata, C Grigoras, P Hristov

012015 USE OF CHECKPOINT-RESTART FOR COMPLEX HEP SOFTWARE ON TRADITIONAL ARCHITECTURES AND INTEL MIC 97
K Arya, G Cooperman, A Dotti, P Elmer

012016 STUDY OF CACHE PERFORMANCE IN DISTRIBUTED ENVIRONMENT FOR DATA PROCESSING 105
D Makatun, J Lauret, M Šumbera

012017 TGEOCAD: AN INTERFACE BETWEEN ROOT AND CAD SYSTEMS 113
C Luzzi, F Carminati

012018 THE ATLAS MUON AND TAU TRIGGERS 121
L Dell'Asta

012019 THE EVOLUTION OF THE TRIGGER AND DATA ACQUISITION SYSTEM IN THE ATLAS EXPERIMENT 129
A Krasznahorkay

012020 ATLAS DISTRIBUTED COMPUTING: EXPERIENCE AND EVOLUTION ... 137
A Nairz

012021 A WEB-BASED DEVELOPMENT ENVIRONMENT FOR COLLABORATIVE DATA ANALYSIS ... 148
M Erdmann, R Fischer, C Glaser, D Klingebiel, M Komm, G Müller, M Rieger, J Steggemann, M Urban, T Winchen

TRACK 2

DATA ANALYSIS – ALGORITHMS AND TOOLS

012022 ADVANCES IN TRACKING AND TRIGGER CONCEPTS ... 153
I Kisel

012023 A NEURAL-NETWORK CLUSTERISATION ALGORITHM FOR THE ATLAS SILICON PIXEL DETECTOR ... 164
K J C Leney

012024 ALIGNMENT OF THE CMS SILICON TRACKER ... 173
T Lampén

012025 MAXIMUM LIKELIHOOD RECONSTRUCTION FOR THE DAYA BAY EXPERIMENT ... 181
D Xia

012026 DATA ANALYSIS OF TRACKS OF HEAVY ION PARTICLES IN TIMEPIX DETECTOR ... 186
S Hoang, R Vilalta, L Pinsky, M Kroupa, N Stoffle, J Idarraga

012027 SIMULATION-AIDED OPTIMIZATION OF DETECTOR DESIGN USING PORTABLE REPRESENTATION OF 3D OBJECTS ... 194
J Balewski

012028 THE AUTOMATED MATRIX-ELEMENT REWEIGHTING AND ITS APPLICATIONS AT THE LHC ... 199
A Mertens

012029 MULTIVARIATE POLYNOMIAL REGRESSION REGULARIZED VIA FIT FUNCTION UNCERTAINTY ... 205
P Kövesárki, I C Brock

012030 QUASI-OPTIMAL WEIGHTS: A VERSATILE TOOL OF DATA ANALYSIS ... 214
F Tkachov

012031 ATLAS TRIGGER SIMULATION WITH LEGACY CODE USING VIRTUALIZATION TECHNIQUES ... 221
G Galster, J Stelzer, W Wiedenmann

012032 MADANALYSIS 5: STATUS AND NEW DEVELOPMENTS ... 226
E Conte, B Fuks

012033 DELPHES 3: A MODULAR FRAMEWORK FOR FAST-SIMULATION OF GENERIC COLLIDER EXPERIMENTS ... 234
M Selvaggi

012034 EVENT RECONSTRUCTION AND ANALYSIS IN THE R3BROOT FRAMEWORK ... 239
D Kresan, M Al-Turany, D Bertini, R Karabowicz, A Manafov, A Rybalchenko, F Uhlig

012035 CONCEPTS AND PLANS TOWARDS FAST LARGE SCALE MONTE CARLO PRODUCTION FOR THE ATLAS EXPERIMENT ... 246
E Ritsch

012036 OPTIMIZING ATLAS CODE WITH DIFFERENT PROFILERS ... 255
S Kama, R Seuster, G A Stewart, R A Vitillo

TRACK 3

COMPUTATIONS IN THEORETICAL PHYSICS: TECHNIQUES AND METHODS

012037 MODERN SUMMATION METHODS FOR LOOP INTEGRALS IN QUANTUM FIELD THEORY: THE PACKAGES SIGMA, EVALUATEMULTISUMS AND SUMPRODUCTION ... 262
C Schneider

012038 ANALYTICAL CALCULATION OF HEAVY QUARKONIA PRODUCTION PROCESSES IN COMPUTER ... 279
V V Braguta, A K Likhoded, A V Luchinsky, S V Poslavsky

012039 AUTOMATIC CALCULATION IN QUARKONIUM PHYSICS 287
B Gong, L-P Wan, J-X Wang, H-F Zhang

012040 AUTOMATED COMPUTATION OF SCATTERING AMPLITUDES 300
G Ossola

012041 AUTOMATED ONE-LOOP COMPUTATION IN QUARKONIUM PROCESS WITHIN NRQCD FRAMEWORK 313
F Feng

012042 BCVEGPY AND GENXICC FOR THE HADRONIC PRODUCTION OF THE DOUBLY HEAVY MESONS AND BARYONS 321
X-G Wu

012043 SANC SYSTEM AND ITS APPLICATIONS FOR LHC 328
R Sadykov, A Arbuzov, D Bardin, S Bondarenko, P Christova, L Kalinovskaya, V Kolesnikov, A Sapronov, E Uglov

012044 NEW DEVELOPMENTS IN FEYNRULES 336
A Alloul, N D Christensen, C Degrande, C Duhr, B Fuks

012045 THREE-LOOP BETA-FUNCTIONS AND ANOMALOUS DIMENSIONS IN THE STANDARD MODEL 342
A V Bednyakov, A F Pikelner, V N Velizhanin

012046 DEVELOPMENT OF AN OBJECT ORIENTED LATTICE QCD CODE "BRIDGE++" 350
S Ueda, S Aoki, T Aoyama, K Kanaya, H Matsufuru, S Motoki, Y Namekawa, H Nemura, Y Taniguchi, N Ukita

012047 RECENT DEVELOPMENTS ON FORM 358
T Ueda, J Vermaseren

012048 NUMERICAL MULTI-LOOP CALCULATIONS WITH SECDEC 364
S Borowka, G Heinrich

012049 FIVE-LOOP NUMERICAL EVALUATION OF CRITICAL EXPONENTS OF THE φ^4 THEORY 373
L Ts Adzhemyan, M V Kompaniets

012050 FORMCALC 8: BETTER ALGEBRA AND VECTORIZATION 381
B C Nejad, T Hahn, J-N Lang, E Mirabella

012051 THE BLACKHAT LIBRARY FOR ONE-LOOP AMPLITUDES 388
Z Bern, L J Dixon, F Febres Cordero, S Höche, H Ita, D A Kosower, D Maître, K J Ozeren

012052 DISTRIBUTED AND MULTI-CORE COMPUTATION OF 2-LOOP INTEGRALS 393
E De Doncker, F Yuasa

012053 ELECTROWEAK AND QCD RADIATIVE CORRECTIONS TO DRELL-YAN PROCESS FOR EXPERIMENTS AT THE LARGE HADRON COLLIDER 405
V A Zykunov

012054 PRECISE CALCULATION FOR HEAVY GAUGE BOSON PRODUCTION IN THE LHT MODEL 414
G Lei, Z Ren-You, M Wen-Gan

012055 ON CROSS-SECTION COMPUTATION IN THE BRANE-WORLD MODELS 426
D Kirpichnikov

012056 GOSAM APPLICATIONS FOR AUTOMATED NLO CALCULATIONS 431
G Cullen, H Van Deurzen, N Greiner, G Heinrich, G Luisoni, P Mastrolia, E Mirabella, G Ossola, T Peraro, J Reichel, J Schlenk, J F Von Soden-Fraunhofen, F Tramontano

012057 COMPUTATION OF MULTI-LEG AMPLITUDES WITH NJET 440
S Badger, B Biedermann, P Uwer, V Yundin

012058 AUTOMATIC ONE-LOOP CALCULATIONS WITH SHERPA+OPENLOOPS 448
F Cascioli, S Höche, F Krauss, P Maierhöfer, S Pozzorini, F Siegert

012059 LITERED 1.4: A POWERFUL TOOL FOR REDUCTION OF MULTILOOP INTEGRALS 456
R N Lee

012060 GENERALIZED HARMONIC, CYCLOTOMIC, AND BINOMIAL SUMS, THEIR POLYLOGARITHMS AND SPECIAL NUMBERS 464
J Ablinger, J Blümlein, C Schneider

012061 MULTI-LOOP INTEGRAND REDUCTION WITH COMPUTATIONAL ALGEBRAIC GEOMETRY 474
S Badger, H Frellesvig, Y Zhang

012062 ON APPLICATIONS OF MATHEMATICA PACKAGE "FAPT" IN QCD 482
V Khandramai

012063 ONE LOOP INTEGRATION WITH HYPERGEOMETRIC SERIES BY USING RECURSION RELATIONS 490
N Watanabe, T Kaneko

012064 THE NLO CALCULATIONS OF HEAVY QUARKONIUM PRODUCTION AT B FACTORIES 498
Y-J Li, G-Z Xu, K-Y Liu, Y-J Zhang

DISCUSSION AND OTHERS

012065 FROM BLACK HOLES TO COSMOLOGY : THE UNIVERSE IN THE COMPUTER 507
J-P Luminet
012066 SUMMARY OF THE ACAT ROUND TABLE DISCUSSION: OPEN-SOURCE, KNOWLEDGE SHARING AND SCIENTIFIC COLLABORATION 519
F Carminati, D Perret-Gallix, T Riemann
Author Index

Journal of Physics: Conference Series **523** (2014) 011001 doi:10.1088/1742-6596/523/1/011001

Preface

This volume of *Journal of Physics: Conference Series* is dedicated to scientific contributions presented at the 15th International Workshop on Advanced Computing and Analysis Techniques in Physics Research (ACAT 2013) which took place on 16–21 May 2013 at the Institute of High Energy Physics, Chinese Academy of Sciences, Beijing, China.

The workshop series brings together computer science researchers and practitioners, and researchers from particle physics and related fields to explore and confront the boundaries of computing and of automatic data analysis and theoretical calculation techniques.

This year's edition of the workshop brought together over 120 participants from all over the world. 18 invited speakers presented key topics on the universe in computer, Computing in Earth Sciences, multivariate data analysis, automated computation in Quantum Field Theory as well as computing and data analysis challenges in many fields. Over 70 other talks and posters presented state-of-the-art developments in the areas of the workshop's three tracks: Computing Technologies, Data Analysis Algorithms and Tools, and Computational Techniques in Theoretical Physics. The round table discussions on open-source, knowledge sharing and scientific collaboration stimulate us to think over the issue in the respective areas.

ACAT 2013 was generously sponsored by the Chinese Academy of Sciences (CAS), National Natural Science Foundation of China (NFSC), Brookhaven National Laboratory in the USA (BNL), Peking University (PKU), Theoretical Physics Cernter for Science facilities of CAS (TPCSF-CAS) and Sugon.

We would like to thank all the participants for their scientific contributions and for the en- thusiastic participation in all its activities of the workshop. Further information on ACAT 2013 can be found at http://acat2013.ihep.ac.cn.

Professor Jianxiong Wang
Institute of High Energy Physics,
Chinese Academy of Science

Journal of Physics: Conference Series **523** (2014) 011001 doi:10.1088/1742-6596/523/1/011001

Committees

Local Organizing Committee

- Gang Chen (IHEP, CAS)
- Ying Chen (IHEP, CAS)
- Xuebin Chi (Supercomputer Center, CAS) Co-Chairman
- Tiejun Deng (IHEP) Administrative Secretary
- Yuanning Gao (Tsinghua University)
- Bin Gong (IHEP) Scientific Secretary
- Weidong Li (IHEP, CAS)
- Wengan Ma (University of Science and Technology of China)
- Yajun Mao (Peking University)
- Congfeng Qiao (UCAS)
- Jianxiong Wang (IHEP, CAS) Co-Chairman

Journal of Physics: Conference Series **523** (2014) 011001 doi:10.1088/1742-6596/523/1/011001

Track Conveners and Advisers

- Computing Technology for Physics Research
  - David Britton
  - Gang Chen
  - Jerome Lauret
  - Axel Naumann
- Data Analysis - Algorithms and Tools
  - Pushpalatha Bhat
  - Denis Oliveira Damazio
  - Weidong Li
  - Depei Qian
  - Thomas Speer
- Computations in Theoretical Physics: Techniques and Methods
  - Andrej Arbuzov
  - Gudrun Heinrich
  - Andrei Kataev
  - Wengan Ma
  - Daniel Maitre
  - Tord Riemann
  - Peter Uwer

International Advisory Committe

- Andrej Arbusov (BLTP JINR)
- Pushpalatha Bhat (Fermilab)
- Fawzi Boudjema (LAPTH)
- David Britton (Glasgow University)
- Federico Carminati (CERN)
- Xuebin Chi (Supercomputer Center, CAS)
- Denis Oliveira Damazio (BNL)
- Bruce Denby (U Pierre et Marie Curie)
- Prabhakar S Dhekne (Bhabha Atomic Research Centre)
- Nigel Glover (IPPP-Durham)
- Gudrun Heinrich (MPI-Munich)
- Viacheslav Ilyin (MSU)
- Sverre Jarp (CERN)
- Toshiaki Kaneko (KEK)
- Matthias Kasemann (DESY)
- Andrei Kataev (INR)
- Christian Kiesling (MPI-Munich)
- Yoshimasa Kurihara (KEK)
- Jerome Lauret (BNL)
- Denis Perret-Gallix (IN2P3/CNRS) Chairman

- Harrison B Prosper (Florida State University)
- Depei Qian (Beihang University)
- Fons Rademakers (CERN)
- Sudhir Raniwala (Rajasthan University)
- Tord Riemann (DESY)
- Jose Seixas (Rio de Janeiro Federal University)
- Yoshimitsu Shimizu (KEK)
- Thomas Speer (Brown University)
- Liliana Teodorescu (Brunel Univserity)
- Peter Uwer (Humboldt University)
- Jos Vermaseren (NIKHEF)
- Jianxiong Wang (IHEP, CAS)
- Monique Werlen (LAPTH)
- Zhaoxi Zhang (ITP, CAS)

Sponsors

Journal of Physics: Conference Series **523** (2014) 011002 doi:10.1088/1742-6596/523/1/011002

Peer review statement

All papers published in this volume of *Journal of Physics: Conference Series* have been peer reviewed through processes administered by the proceedings Editors. Reviews were conducted by expert referees to the professional and scientific standards expected of a proceedings journal published by IOP Publishing.

Lessons learned from the K computer project ---from the K computer to Exascale---

Yoshio Oyanagi[1]
Graduate School of System Informatics, Kobe University
Rokko-dai 1-1, Nada-ku, Kobe, 657-8501 Japan

E-mail: oyanagi@people.kobe-u.ac.jp

Abstract. The history of the supercomputers in Japan and the U.S. is briefly summarized and the difference between the two is discussed. The development of the K Computer project in Japan is described as compared to other PetaFlops projects. The difficulties to be solved in the Exascale computer project now being developed are discussed.

1. Brief History of Supercomputers

Since the first half of the 1980s, following the efforts to produce commercial vector computer Cray-1 in the U. S., Japanese computer vendors have been producing a number of powerful supercomputers to solve large scale scientific and engineering problems. The history of supercomputers in Japan before 1999 is described in a previous article[1)]. The trends in Japan as well as in the U.S. can be summarized as follows.

1.1. Premordial age (1970s)

The vector processing was first proposed by Senzig and Smith of IBM[2)]. It was implemented in the IBM2983 Array Processor (1969) as an attached processor to the I/O channel of mainframe computers. Independent vector computers were developed by two vendors: ASC of Texas Instruments (1972) and STAR-100 of Control Data Corporation (1973). These machines, although pioneering, were not commercially successful. The first commercially successful vector supercomputer was Cray-1 of Cray Research Inc. (1976).

At the same time Japanese vendors were also developing vector computers. One year after the Cray-1, Fujitsu shipped FACOM 230-75 APU (1977), which incorporated three data buffers of 256 words for vector operations. It also supported list-directed vector accesses. Hitachi and NEC made an IAP (Integrated Array Processors) for their top-of-the-line mainframe computers. The performance of these computers, however, was modest as compared to the Cray-1.

As for parallel processing, Burroughs Corporation and University of Illinois developed a pioneering parallel computer ILLIAC IV (1973). Burroughs started to develop BSP (Burroughs Scientific Processor) but the project was cancelled in 1980. It is to be noted that the ICL in U.K. developed DAP machine, which consists of one-bit processors (1979).

1.2. Vector age (1980s)

1 To whom any correspondence should be addressed.

Full-fledged vector supercomputers were announced and produced in the 1980's. In the U.S., CDC produced Cyber-203 and Cyber-205 (1981). Cray Research developed Cray XMP-4 (1984), Cray-2 (1985) and Cray YMP (1988). ETA, a subsidiary of CDC, developed ETA-10 (1987). IBM developed 3090 VF (1985). There were also several minisupercomputers, such as Convex C1 (1985) and C2 (1988) or Supertek S1 (1989).

Japanese vendors shipped high performance vector computers: Hitachi's S810 (1983) and S820 (1987); Fujitsu's VP200 (1983) and VP2600 (1989); and NEC's SX-2 (1985) and SC-3 (1990). In contrast to the U.S. vector computers, Japanese machines have the following features:

1) compatibility with the mainframe computer
2) single processor with many vector pipes
3) large main memory
4) large vector registers
5) list-directed vector operation

At the same time, a considerable number of parallel venture companies emerged in the U.S. On the other hand, although there were some parallel computing research activities in Japan, they were mainly in academia and were considered as special-mission processors.

1.3 Vector parallel and commodity parallel (1990's)

Parallelism became the key trend in both vector and scalar machines in the 1990's. In the U.S., Cray Research shipped several parallel vector computers: C90 (1991), T90 (1995) and SV1 (1998). At the same time, several U.S. companies produced parallel computers using commodity processors: TMC's CM-5 (1992), Convex' SPP (1994), IBM's SP1 (1993) and SP2 (1994) and Cray Research's T3D (1993) and T3E (1996) and many others. It is to be noted that the highly parallel machines constructed in the DoE's ASCI Program were also based on commodity processors.

In contrast, the mainstream of Japanese supercomputers was based on parallel vector architecture: Hitachi's S3800 (1993); Fujitsu's NWT (Numerical Wind Tunnel, 1993) and its commercial version VPP500 (1993); and NEC's SX-4 (1995). At the same time, Japanese companies developed parallel machines with scalar processors: Hitachi's cp-pacs (1996), SR2201 (1996) and SR8000 (1998); Fujitsu's AP1000 (1994) and AP3000 (1997); NEC's Cenju-2 (1993), Cenju-3 (1994) and Cenju-4 (1997). Some of those machines were sold as a test bed rather than a production machine. The NWT, SR2201 and cp-pacs occupied top positions in some of the Top500 lists, which started in 1993.

1.4 Massively parallel era (2000's)

Massively parallel machines become the main trend in the 2000's. In the U.S. from the end of the 1990's, the ASCI Program (later called ASC Program) continued to build a number of massively parallel machines based on commodity processors: ASCI Red (1997), ASCI Blue Pacific (1998), ASCI Blue Mountain (1998), ASCI White (2000), ASCI Q (2002) and ASCI Purple (2005). IBM's Blue Gene/L (2004) and Cray's Red Storm (2005) later joined the ASC Program. The DARPA started HPCS (High Productivity Computing Systems) in 2002 and the NSF started the TeraGrid Project in 2001. All the machines developed in those projects were sold as commercial products to universities and laboratories as well as to industry sectors.

The biggest news in Japan in this period is the completion of NEC's Earth Simulator in 2002, which occupied the top position in five consecutive Top500 lists (2002-2004). The emergence of the huge vector supercomputer stimulated the U.S. and doubled the U.S. supercomputer budget. In contrast, Fujitsu left vector business and adopted massively parallel computers based on the Sparc64 and x86 processors.

1.5 Observations

In the last decades of the 20th century, U.S. and Japan were the major producers of supercomputers. There was, however, big difference between the policies of the two countries.

Until late 1990's, Japanese vendors focused on vector machines and users in Japan enjoyed the power of vectorization. Besides hardware, vendors provided very good vectorizing compilers, so that users were in a sense spoiled. In Japan, vendors thought that parallel machines were for specialized purposes (eg. image processing) and users dared not try to harness parallel machines in the 1980's. Although some computer scientists in Japan were interested in building parallel machines, they were not used for real applications in science and technology. Around 2000 Japan was at least ten years late in parallel processing for scientific computing as compared to the U.S.

It is to be noted that practical parallel processing for scientific computing in Japan was started by application users: qcd-pax (1989), NWT (1993), GRAPE series (1989-) and the Earth Simulator (2002).

2. Challenge to PetaFlops

2.1 PetaFlops Conferences

As early as in 1994, when the NWT took No. 1 in the Top500 list with 124 GFlops Linpack performance, a conference aiming at the PetaFlops (10^6 GFlops) was held in the U.S.[3)]. After several workshops, the PETAFLOPS II conference was held in Santa Barbara in 1999[4)].

2.2 Reluctant Japan

At that time, there were no significant activities in Japan towards PetaFlops. The Japanese government established the IT Strategic Headquarter in 2001 and promoted high speed Internet and inexpensive Internet services, but the levelling up of supercomputers was not considered as a national project. They thought that the supercomputers may be constructed according to the needs of each field.

2.3 After the Earth Simulator

Only after the success of the Earth Simulator in 2002 the discussion started on a possible supercomputer in the PetaFlops region. The Information Science and Technology committee in the Mext (Ministry of Education, Culture, Sports, Science and Technology) started discussion on the possible measures to promote computational science and technology from August 2004.

The recommendation of this Mext committee was to promote a national project to construct a leading edge supercomputer. The Mext decided to start this project on July 25, 2005 and in October the Riken (The Institute of Physical and Chemical Research) was selected the developer of the supercomputer. The working group in the Mext identified ten killer applications: life science, astrophysics, space and aeronautics, materials, atomic energy, environment, disaster prevention, fluid dynamics, plasma and industrial design. The observation of the working group was that the multiphysics-multiscale simulation would be important in those fields and a hybrid architecture was appropriate to keep high performance in different types of computing. The author agreed on the first point that multiphysics simulation is important, but it did not mean that a hybrid architecture was suited for such simulation.

2.4 Conceptual design

The original proposal was a hybrid of three parts: scalar part (1 PetaFlops), large scale processing (vector) part (0.5 PetaFlops) and special-purpose processor part (20 PetaFlops). On September 2007, after long discussion, the Riken finalized the conceptual design of the hardware. The system consists of two parts: scalar processor part and vector processor part. The scalar processor was to be built by Fujitsu and the vector part by NEC and Hitachi. The total Linpack performance was to exceed 10 PetaFlops.

The site selection was another big issue. Out of 15 proposals, Riken finally decided to build it in Kobe.

The strategic committee in the Mext identified five strategic fields in SPIRE (Strategic Programs for Innovative Research) on July 22, 2009: 1) Predictive bioscience, medicare and drug design, 2) New material and new energy, 3) Earth environmental prediction for disaster prevention and mitigation, 4) Next-generation manufacturing, 5) Origin and structure of material and universe.

2.6 Outside Japan
PetaFlops was a target in various part of the world. In Los Alamos National Laboratory, the Roadrunner attaind 1.026 PetaFlops in 2008 using the Cell processor as accelerators. It was the first supercomputer which exceeded 1 PetaFlops Linpack performance. In Oak Ridge National Laboratory, the Jaguar attained 1.059 PetaFlops in 2008. It is a homogeneous system like the K Computer.

In China there came several Petascale supercomputers. Nebulea in Shenzhen attained 1.271 PetaFlops with NVIDIA GPU. Tian-he 1A in Tianjin Supercomputer Center got 2.566 PetaFlops in Linpack.

2.7 Government revitalization unit
Unfortunately, NEC and Hitachi retired from the project due to bad economy on May 13, 2009. The Riken, however, decided to continue the joint development with Fujitsu to build 10 PetaFlops machine using scalar processors only on May 14, 2009.

A general election of Japan was held on August 30, 2009 for the lower house of the Diet of Japan and the opposition Democratic Party won the majority. The Democratic Party government introduced a budget screening by the Government Revitalization Unit. On Friday, November 13, 2009, the third working group of the Government Revitalization Unit examined the construction budget of the supercomputer. The budget proposal stressed that the Japanese supercomputer would win the No. 1 when it is completed. The chairperson of the working group, Ms. Lien Fang Murata, asked the Mext staff, "Why should it be No. 1 in the world?" "Is No. 2 not enough?" Another criticism was about the demise of the vector part in the project.

After some discussion the conclusion of the working group was "to freeze the project." This conclusion, although not final, met strong reactions from both academia and industry. Consequently, Government decided to overturn the conclusion in December 2009, with considerable modification of the project. It was no longer required to get the top position.

2.8 Completion
The nickname of the supercomputer "京 (Kei)" or "K" was decided on July 5, 2010 after soliciting public proposals. "京" is a Japanese numeral meaning 10^{16}. In China, however, this Chinese character is not used as a numeral.

The first eight racks were shipped to the Kobe Riken site on September 28, 2010. It took 10 months to install the full 864 racks in Kobe. In the ISC2011 conference in Hamburg in June 2011, the K Computer won the No.1 with the Linpack speed of 8.162 PetaFlops using 80% of the full system. In the SC11 conference in Seattle in November 2011, the K Computer won again the top position with 10.51 PetaFlops using the full system. The K Computer also won the No. 1 positions in four HPC Challenge benchmarks: Global HPL, Global Random Access, EP Stream per system and Global FFT. As a real application "The first principle calculation of a Silicon nanowire of 100,000 atoms on the K Computer" won the Gordon Bell prize.

The open access to the K Computer started in September 2012. Applications should be submitted and evaluated for users to have access to the K Computer. Some of the users are from the industry.

3. Challenge to ExaFlops
According to our experience of the K Computer, 10 PetaFlops is not enough for high precision simulation of complex real world systems. Next target is the ExaFlops (10^{18} Flops), which is no easy target, but there have been various efforts to approach this goal.

3.1 Challenges in the U. S.

As early as in 2004, "The Path to Extreme Supercomputing" conference was held in Santa Fe to discuss the possibility of ExaFlops.

In the SC08 in Austin, the IESP (International Exascale Software Project) started as an international collaboration of US, Japan and Europe. The target was to develop system software such as OS, compiler or middleware by international collaboration. It had meetings in Santa Fe, Paris, Tsukuba, Oxford, Mauii, San Francisco and Kobe.

In Europe, similar project started, the EESI (European Exascale Software Initiative). It also had several workshops.

3.2 Japanese efforts

The Exascale efforts in Japan began with participating in the IESP in 2008. A voluntary group of people organized SDHPC (Strategic Development of High Performance Computing) in 2010 and had ten workshops up to now.

As an official effort, the Mext started two working groups for the Exascale computing in July 2011. They published two reports[5)] in March 2012: 1) Application working group white paper, and 2) HPCI technology roadmap white paper. The latter included architecture, system software, programming and numerical library. The co-design of application and architecture is important. It is proposed to interface application and architecture on a two-dimensional map of relative memory bandwidth (B/Flop) as the x-axis and relative main memory capacity (B/Flops) as the y-axis. We first thought that in the Exascale region each application might require different architecture. We finally found that relatively universal computer can cover considerable area of applications. We may need, however, other architectures for high bandwidth jobs or big data processing.

In February 2012, the Mext set up Next Generation HPCI Working Group (chaired by Oyanagi) for two years. This working group published an interim report[6)] in June 2013. The recommendation is that the government should build one flagship machine to cover relatively wide area of applications. It should be supplemented by other leading machines which have different characteristics. The final report will be published in March 2014.

At the same time, the Mext funded four feasibility study projects for two years, one for application and three for architecture. The application team is to find and estimate a number of social and scientific problems to be solved only by Exascale computing. Three architectures to be considered in the feasibility study are: 1) homogeneous scalar processor, 2) accelerator, and 3) vector processor.

In August 2013, the Mext submitted a budget proposal of 30 million US dolars for the Exascale conceptual design in the 2014 fiscal year (April 2014 – March 2013) to the Ministry of Finance. The total budget for seven years is estimated roughly 1 billion US dollars.

3.3 Technological issues

Very few are thinking of so-called disruptive technologies such as superconducting processor or quantum computing to build Exascale systems. The majority is going to use CMOS semiconductor technology along the future trend. Even using 10 nm or less, there are lots of technological issued to be addressed by our development.

3.3.1 Power wall

From the economic situation, the power consumption of the future Exascale computer should be at the same level of the current K Computer, i.e. around 20 MW. This means we should develop 100 times higher Flops/W than the K Computer. Even considering future progress of semiconductor technology, it is not an easy task. The problem is that more energy is lost in data move than calculations. We should develop algorithms to reduce data moving, not to reduce numerical operations.

3.3.2 Memory wall

The nominal Flops may increase thanks to the large number of transistors in a chip, but the bandwidth between processor and memory is limited due to the number of pins of a chip. There are number of proposals to address this issue. One idea is to increase the bandwidth by optical connection or 3D stacking of chips. Other idea is to put the on-chip memory in the processor. A cache.is a kind of on-chip buffer memory but is not addressable and controlled by the OS. Addressable on-chip memory is proposed but then the programming would be very difficult without clever compilers.

3.3.3 Reliability wall

Since the number of circuit elements is enormous in ExaFlops, the failure rate may be proportional to it. Fault tolerance in hardware, OS, interconnection and software will be crucially important.

3.3.4 Programming wall

The number of processes or threads will be huge and the memory hierarchy deep. It would be very difficult to write programs on such a machine. Computing efficiency, that is the ratio of measured Flops to the peak Flops is important. In the case of the K computer, it is not easy to exceed 10%. Some users stay below 1%. The situation might be worse for Exascale computers.

4. Conclusion

We have seen that due to the success of vector computers in the 1980's in Japan, parallel processing was behind the U.S. and Europe. We note that practical parallel computer in Japan (NWT, cp-pacs, the Earth Simulator etc.) were built in the initiative of application users.

Although the K Computer project met with strong head wind in the early stage, it is completed and is working stably and producing scientific outcomes. Japan is catching up with the U.S. and Europe in parallel computing.

We are planning to build Exascale supercomputers around 2020. Difference applications require different architecture in terms of B/Flop and B/Flops. Since many applications strongly demand such machines, we are making efforts to acquire taxpayers' support to build Exascale machines.

References

[1] Oyanagi Y 1999 *Parallel Computing* **25** 1545 (Further references therein)

[2] Sentig D N and Smith R V 1965 *AFIPS Proc. FJCC* **27** 117

[3] *Proc. Enabling Technologies for Petaflops Computing (Pasadena, U.S.A., 22-24 February 1994)* ed. T Sterling, P Messina and P H Smith (the MIT Press)

[4] *Proc. PETAFLOPS II, Second Conference on Enabling Technologies for Peta(fl)ops Computing (Santa Barbara, U.S.A. 15-19 February 1999)* ed. P Messina, T Sterling and P Smith

[5] http://open-supercomputer.org/wp-content/uploads/2012/03/FutureHPCI-Report.pdf (in Japanese)

[6] http://www.mext.go.jp/b_menu/shingi/chousa/shinkou/028/gaiyou/1337595.htm (in Japanese)

Opportunities and choice in a new vector era

A Nowak[1]
[1] CERN openlab CTO Office, Geneva, Switzerland

E-mail: Andrzej.Nowak@cern.ch

Abstract. This work discusses the significant changes in computing landscape related to the progression of Moore's Law, and the implications on scientific computing. Particular attention is devoted to the High Energy Physics domain (HEP), which has always made good use of threading, but levels of parallelism closer to the hardware were often left underutilized. Findings of the CERN openlab Platform Competence Center are reported in the context of expanding "performance dimensions", and especially the resurgence of vectors. These suggest that data oriented designs are feasible in HEP and have considerable potential for performance improvements on multiple levels, but will rarely trump algorithmic enhancements. Finally, an analysis of upcoming hardware and software technologies identifies heterogeneity as a major challenge for software, which will require more emphasis on scalable, efficient design.

1. Introduction

CERN openlab and its Platform Competence Center spent the past decade investigating mainstream and upcoming platform technologies. During that time, the computing landscape has witnessed many changes, but two are of particular importance. On one end of this decade, intrinsic improvements in single-thread processor efficiency ceased to be sourced from manufacturing process improvements and aggressive frequency scaling, and had to be provided through microarchitectural optimizations such as out of order enhancements, changes in branch predictors, cache optimizations, buffer size increases and so on. High Energy Physics (HEP) software traditionally makes pretty good use of such mechanisms. On the other end of this decade, we find ourselves snowed under with vectors, hardware threads, cores – all various forms of parallelism.

This new era presents challenges as it presents opportunities. The discussion contained in the paper presented addresses these main points.

2. Observations

2.1. Hardware

Worries about computing platform performance often stem from an inefficient use of available resources - which is a generic and persistent challenge. One such challenge in the past was the growing compute-memory gap, illustrated on the classic Hennessy & Patterson plot (Figure 1).

Journal of Physics: Conference Series **523** (2014) 012002 doi:10.1088/1742-6596/523/1/012002

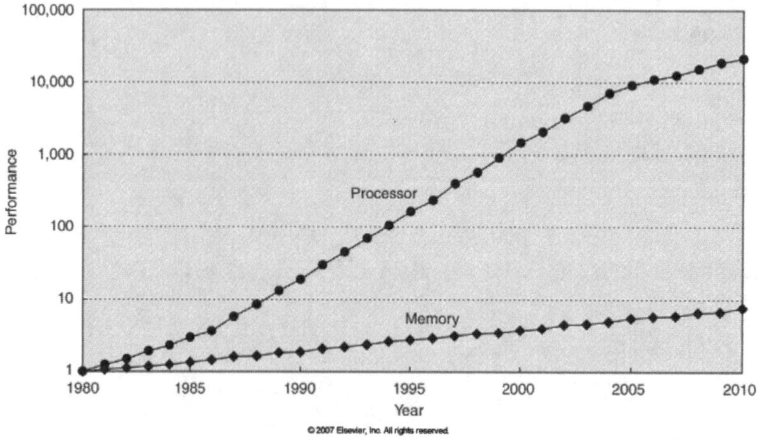

Figure 1: CPU-memory performance gap. Modelled after "Computer Architecture": Hennessy, John L.; Patterson, David A.

In modern computing, it is less common for bottlenecks in a platform to shift in a major way, e.g. from disk to memory bandwidth, and thus even well-tuned workloads are rarely adversely affected by hardware upgrades. A notable exception is the rise of accelerators, which in a "step function" offered difficult to extract compute capabilities, but plenty of bandwidth to go along (Figure 2).

Let us consider the recent history of mainstream compute capabilities. Vector width has quadrupled on x86 since the introduction of MMX (and went much further on GPUs!), hardware threading made a triumphant comeback after a brief appearance in the Pentium 4 family, and the "Haswell" microarchitecture from Intel now features 8 execution ports in lieu of 6 on earlier models. Core count in a chip has grown from 2 in awkward MCP designs to over 60 in the Intel KNC, while multi-socket, multi-node installations are standards in datacenters. All in all, the appearance of a vast land of opportunity created by the simultaneous expansion of multiple performance dimensions has a strong impact on the ratio between the theoretical limits of computing efficiency and the targets actually achieved – as discussed further.

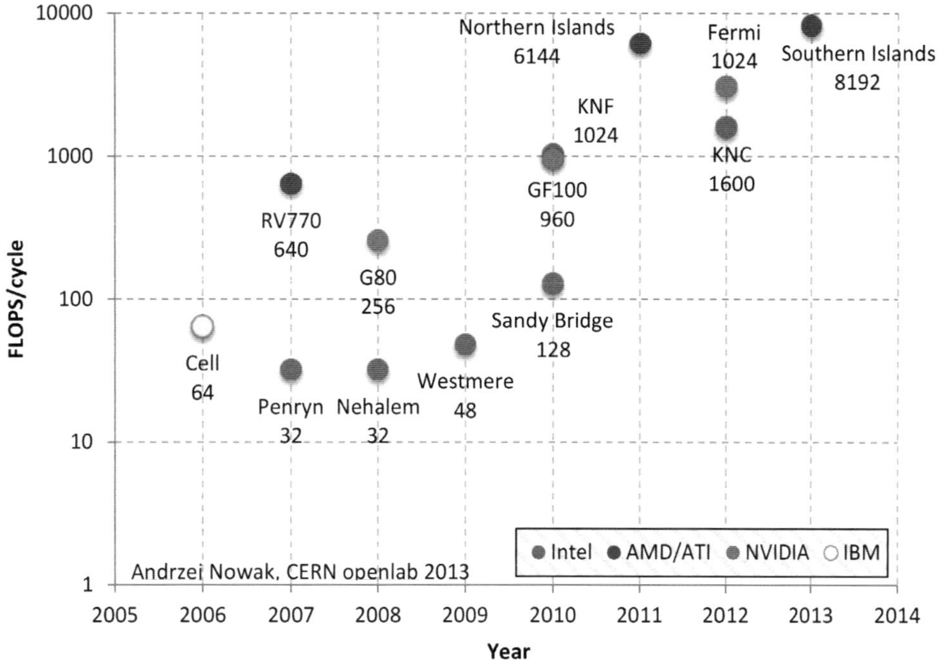

Figure 2: Peak single precision, general purpose FLOPs/cycle in a single chip

This illustrates the need for a shift in programming and optimization focus, in particular in compute-oriented workloads - from generic latency and hotspot-oriented efforts to sweeping redesigns of whole software packages. While it is common to extrapolate Moore's Law and its benefits at least 5 years into the future, industry specialists warn of major movements in connectivity, memory and energy efficiency. For example, the ratio of bytes per FLOP is no longer a compute density metric, but also one that implies an optimal or suboptimal energy balance in a chip.

Table 1: Memory operations and their energy costs in a platform

Operation	Energy cost
L1 access	2 pJ
L2 access	150 pJ
RAM	2 000 pJ
FLOP	Future - ?

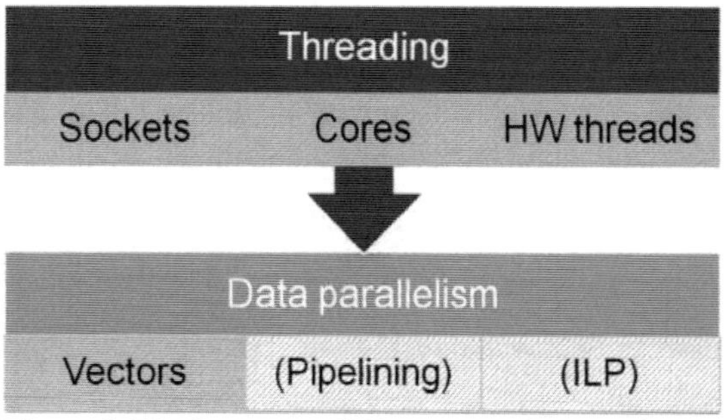

Figure 3: Hardware parallelism mapped onto software

This trend can be noted by observing the widespread community interest in accelerators, vectors and even new architectures – after many years of indivisible x86 reign on our desktops and in our datacenters, numerous experiments with ARM, NVIDIA and even MIPS architectures are underway.

2.2. Software stagnation

On the other side, production HEP software remained the same from the point of view of its capability to take advantage of technical novelties. As discussed in [1], large, code-heavy C++ based frameworks dominate, with two major toolkits, ROOT [2] and Geant4 [3,4], still single-threaded and lacking vectorization (for the most part). It should not come as a surprise that, roughly speaking, the gap in performance between the theoretical limit and optimized code is 10x, and the distance between the performance of scalar, single-threaded HEP code and optimized code is another 10x. These numbers are decomposed into more detailed factors in Table 2 and Table 3, and explained in detail in [1].

Table 2: Multiplicative factors for parallelism in three different software and datacentre scenarios

	SIMD	ILP	HW THREADS	CORES	SOCKETS
THEORY	4	4	1.35	8	4
OPTIMIZED	2.5	1.43	1.25	8	2
HEP	1	0.80	1	6	2

Table 3: Multiplied factors for parallelism in three different software and datacentre scenarios

	SIMD	ILP	HW THREADS	CORES	SOCKETS
THEORY	4	16	21.6	172.8	691.2
OPTIMIZED	2.5	3.57	4.46	35.71	71.43
HEP	1	0.80	0.80	4.80	9.60

We need to program for tomorrow's hardware – today.

The important lesson here touches not only on modularity and adaptability to changes in hardware, both of which must become the daily bread of many key developers, but also on the viabililty, sustainability and performance of the software engineering process in HEP. New challenges linked to heterogeneity, discussed further, will put our methods to an even more stringent test. In 2007, when the many-core era was just around the corner, CERN openlab began its efforts of community evangelization in preparation for the inevitable explosion in datacentre core count. It quickly became apparent that our software would benefit from more forms of hardware parallelism than one, which fortunately still map onto a manageable set on the software side (Figure 3). However, many physicists we spoke to did not have the bandwidth to worry about issues such as hardware underutilization. Now the gap has grown sufficiently large for many to find the resources to address this challenge. For its part, CERN has initiated the Forum on Concurrent Programming Models and Frameworks and is investing in major improvements in Geant4, such as Geant-V and the multi-threaded Geant4 prototype, and ROOT. Another prominent effort is the EU FP7 "ICE-DIP" project, which aims to develop next-generation data acquisition capabilities using new technologies.

Large HEP jobs are deeply affected by object-oriented development, to the extent that top functions in a flat profile consume low single-digit percentages of time or instructions, while it takes hundreds if not thousands of entries to account for the bottom 50% of the profile – which means that even if we get rid of them completely (hardly a possibility), the workload would run at most twice as fast. Such frameworks are almost beyond hope when it comes to the *efficient* use of opportunities such as vectorization, and need a data oriented design approach to benefit from them. It's not all bad, however: prominent examples of success exist as well. For example, the GSI team working on the CBM/ALICE online codes achieved a 120'000x speedup on their code. The interesting part is that only a factor of 120x came from microarchitectural optimizations and parallelization, while a factor of 1000x came from implementing a significantly more optimal algorithm (Figure 4). There is also a good result from the CERN openlab side: work done on the MLFit benchmark allowed it to scale by nearly 2x when moving from an Intel "Westmere" based server to a "Sandy Bridge" one (Figure 5).

	Stage	Description	Time/track	Speedup
		Initial scalar version	12 ms	–
Intel	1	Approximation of the magnetic field	240 μs	50
Intel	2	Optimization of the algorithm	7.2 μs	35
Intel	3	Vectorization	1.6 μs	4.5
Cell	4	Porting to SPE	1.1 μs	1.5
Cell	5	Parallelization on 16 SPEs	0.1 μs	10
		Final simdized version	0.1 μs	120000

10000x faster on any PC (stages 1–3)

Comp. Phys. Comm. 178 (2008) 374-383

Figure 4: Speedups for optimized ALICE/CBM online code, figure copied from I. Kisel et al. [5][6]

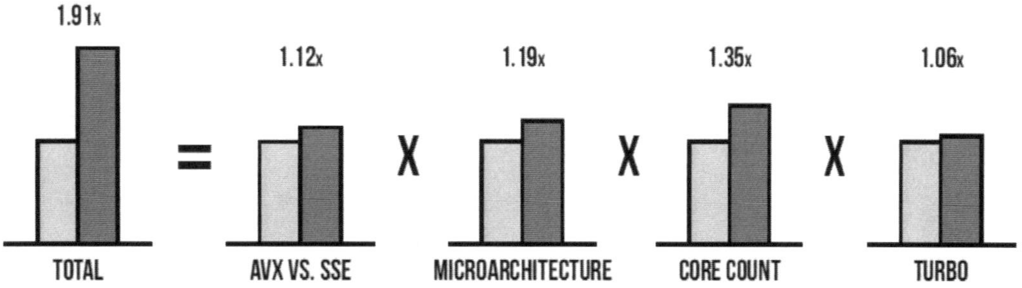

Figure 5: "Westmere" (light blue) vs. "Sandy Bridge" (dark blue) server performance, frequency scaled

In what concerns specific technologies for, say, vectorization, they are differentiated not just by their features, but also by the syntax or notation they employ and their implementation (library, compiler extension, etc). While new syntax options such as Cilk+ are quite appealing, to the extent that they're desired in the C++ standard, there has yet to appear a sufficiently convincing technical implementation that provides convenience and performance at the same time. On the multi-threading front, TBB is being integrated into a prototype version of Geant4, also with notable success.

There is, of course, no guarantee that vector growth will continue, but it is still a growing trend, with AVX3 sporting 512-bit vectors on mainstream Xeon systems, and much wider vector-like units still being deployed in GPUs en-masse.

Finally, even though the case for parallelism on multiple levels might be clear by now, it will still take some work to establish which is the best path to pursue, which technologies are the most promising (and sustainable) ones, and which syntax, notation and programming concept will fit the HEP community.

3. Heterogeneity

One emerging challenge with respect to contemporary computing strategies and software development is heterogeneity. It has been around for a long time at the datacentre level, where for example a Grid user only sees a virtualized interface to his system and does not know which capabilities the CPU assigned to his job will have. At the chip level, AMD has its "Trinity" APU technology, while Intel introduced graphics processors supporting "Sandy Bridge" processors and its successors. Very few (experimental) software packages make use of such hardware. Finally, at the platform level, we see heterogeneity not only at the level of compute capabilities of interconnected devices but also in terms of their architecture, memory configuration, integration and so on. Before such a device is engaged to collaborate with the main CPU, many questions must be asked, also at the software level. How will the device feed on data and communicate? Will it have sufficient memory, performing as expected? Will it meet latency constraints? Will it provide floating point results compatible with the main processor? Should the code run natively on the accelerator or be offloaded when the time is right? The answer to those questions and actions taken will depend on the problem and the structure of next-generation software frameworks. Another platform related question is that of memory – in recent years there has been a trend of stratification. Two levels of caches expanded to three and multi-socket systems provide multiple NUMA domains. A hot topic in this domain is high bandwidth, low latency, potentially non-volatile memory, which could act as a layer between L3 and RAM, or play a role in storage (Figure 6 and Table 4).

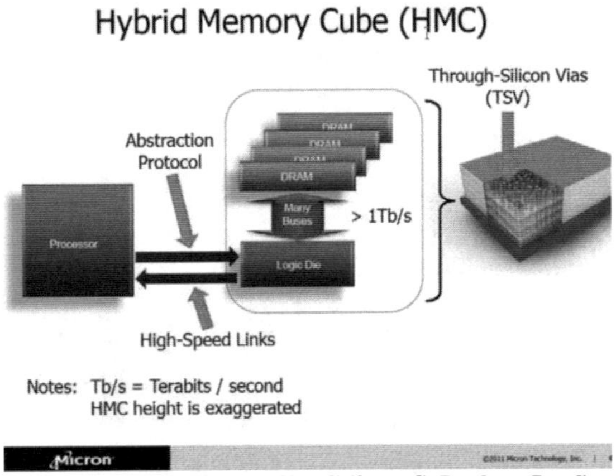

Figure 6: Micron HMC (from S. Borkar, Intel)

	Bandwidth	Energy
DDR3 (Today)	> 10GB/s	> 50 pJ / bit
HMC (Prototype)	> 100 GB/s	< 10 pJ / bit

Table 4: HMC prototype parameters

The software side will feel pressure on two fronts. On one side, compilers and toolsets might begin to diverge, and the OS kernel will have to efficiently arbitrate communication between the host and slot-in devices. Such has been the experience of CERN openlab while working on the Intel MIC project since 2008, where none of the physics benchmarks used or their dependencies were fully ready for cross-platform compilation and usage. The porting times shown in Table 4 illustrate that a small benchmark such as the Trackfitter was easy to port, even though it required a manual reimplementation of several key functions, to match the architecture of the KNF. MLFit, which was written using large standards (OpenMP and MPI) and stripped of ROOT dependencies, was essentially a matter of recompilation and updating the build system. However, the first port of the multi-threaded Geant4 prototype, which relies on a lot of Geant4 source code and has several external dependencies, took more than a month to get right. The column showing the time needed for new (updated) ports demonstrates that in our experience, introducing support for a certain degree of heterogeneity was an investment that paid off and did not require major reviews.

	LOC	1st port time	New ports	Tuning
TF	< 1'000	days	N/A	2 weeks
MLFit	3'000	< 1 day	< 1 day	weeks
MTG4	2'000'000	1 month	< 1 day	< 1 week

Table 5: Approximate KNF/"Larrabee" (MIC) porting and tuning times by HEP benchmark

The low-power ARM architecture is making its 64-bit entry into datacenters worldwide, and those in the HEP community who wish to use it [7][8] will face challenges very similar to those experienced in our work with Intel MIC. For example, compiler support for NEON vector instructions is still quite young, and several other areas in popular Open Source software need attention. As recently as in 2012, it was still a time consuming feat to properly install and support Linux on one of the "desktop" ARM development boards.

On the other software front, it is not yet apparent how to maintain a single source that could be compiled to or offloaded to different backends [9], and whether it should be maintained as "straight" C/C++ code or as targeted kernels, or perhaps even in another form. Several projects tried to bridge this productivity and engineering gap – RapidMind, Intel's Ct, OpenCL, custom Intel compiler extensions and several others – but none of them have managed to provide a portable source AND portable performance at the same time. Some established industry standards come to the rescue, e.g. OpenMP4 now includes accelerator and offload directives that were pre-published with OpenACC, but it will still take some time before the industry comes to a well-defined conclusion.

A particular and painful consequence of symmetric processing is the challenge of numerical convergence [10][11]. In some cases, there is no other choice but to accept that floating point results will diverge between the CPU and an accelerator or co-processor. Vectors might be processed differently, data and operation ordering may change, associativity may vary, compilers will generate different code and "fast" function variants might be used as defaults. In any modern "offload" device (GPUs and Intel MIC included) there exist two or three variants of math functions – those that strictly adhere to standards but are excruciatingly slow, those that explicitly sacrifice accuracy for speed, and those that offer a reasonable compromise by operating on limited ranges of input arguments. Unfortunately for coders, these new options often imply a revision of floating-point in the source – an undertaking that is not concluded overnight on frameworks with 5 million lines of code. Ultimately, such an exercise can be a very useful one, in particular to develop methods that assess the quality of computation results through means other than their bit-for-bit binary compatibility on multiple platforms. That is also why it is important to closely follow efforts on math library optimization, such as [12].

4. Conclusion

As always, the computing landscape is changing rapidly. However, it has become the responsibility of the platform and system user to "seize the day", and to turn all the hardware in a computing platform into an advantage – as opposed to neglecting silicon that has already been paid for. This should be taken into account in the light of general optimization truths which suggest that optimization should start with the algorithm, as that often is where the highest potential for gains resides.

At least now, it would seem that next generation code must make some choices between programmability and performance – neither can be sacrificed and neither can be ignored. Universal, open standards must continue to play a leading role for reasons of sustainability and interoperability. Scaling in multiple hardware and software dimensions seems indispensable – that includes numerous aspects such as parallelization, vectorization, multi-process domain, message passing readiness, "asynchronicity" and lack of coherency. Well-designed algorithms and frameworks will be a major contributor to success in this area. Finally, power efficiency is becoming a practical worry as well – future chips might require an optimal or specific balance of data movement and computation.

There is a long road ahead of the HEP community, one that will surely take many turns and hold many surprises. Without doubt, many mainstream technologies the community uses today have already peaked, and it could well be that we currently find ourselves at a crossroads similar to the one that took us into the PC era [13].

5. Acknowledgements

The dissemination of this work was supported by "ICE-DIP", an EU FP7 Marie Curie grant, #316596.

6. References

[1] Jarp S, Lazzaro A and Nowak A 2012 The future of commodity computing and many-core versus the interests of HEP software *J. Phys. Conf. Ser.* **396** 052058

[2] Brun R and Rademakers F 1997 ROOT — An object oriented data analysis framework *Nucl. Instruments Methods Phys. Res. Sect. Accel. Spectrometers Detect. Assoc. Equip.* **389** 81–6

[3] Agostinelli S, Allison J, Amako K, Apostolakis J, Araujo H, Arce P, Asai M, Axen D, Banerjee S, Barrand G, Behner F, Bellagamba L, Boudreau J, Broglia L, Brunengo A, Burkhardt H, Chauvie S, Chuma J, Chytracek R, Cooperman G, Cosmo G, Degtyarenko P, Dell'Acqua A, Depaola G, Dietrich D, Enami R, Feliciello A, Ferguson C, Fesefeldt H, Folger G, Foppiano F, Forti A, Garelli S, Giani S, Giannitrapani R, Gibin D, Gómez Cadenas J J, González I, Gracia Abril G, Greeniaus G, Greiner W, Grichine V, Grossheim A, Guatelli S, Gumplinger P, Hamatsu R, Hashimoto K, Hasui H, Heikkinen A, Howard A, Ivanchenko V, Johnson A, Jones F W, Kallenbach J, Kanaya N, Kawabata M, Kawabata Y, Kawaguti M, Kelner S, Kent P, Kimura A, Kodama T, Kokoulin R, Kossov M, Kurashige H, Lamanna E, Lampén T, Lara V, Lefebure V, Lei F, Liendl M, Lockman W, Longo F, Magni S, Maire M, Medernach E, Minamimoto K, Mora de Freitas P, Morita Y, Murakami K, Nagamatu M, Nartallo R, Nieminen P, Nishimura T, Ohtsubo K, Okamura M, O'Neale S, Oohata Y, Paech K, Perl J, Pfeiffer A, Pia M G, Ranjard F, Rybin A, Sadilov S, Di Salvo E, Santin G, Sasaki T, et al 2003 Geant4—a simulation toolkit *Nucl. Instruments Methods Phys. Res. Sect. Accel. Spectrometers Detect. Assoc. Equip.* **506** 250–303

[4] Allison J, Amako K, Apostolakis J, Araujo H, Dubois P A, Asai M, Barrand G, Capra R, Chauvie S and Chytracek R 2006 Geant4 developments and applications *Nucl. Sci. Ieee Trans.* **53** 270–8

[5] Kisel I, Kulakov I and Zyzak M Parallel Implementation of the KFParticle Vertexing Package for the CBM and ALICE Experiments *Computing in High Energy and Nuclear Physics 2012*

[6] Kisel I, Kulakov I and Zyzak M Parallel Algorithms for Track Reconstruction in the CBM Experiment Computing in High Energy and Nuclear Physics 2012

[7] Neufeld N Measurements of the LHCb software stack on the ARM architecture

[8] Elmer P Explorations of the viability of ARM and Intel Xeon Phi for Physics Processing

[9] Rohr D, Gorbunov S, Szostak A, Kretz M, Kollegger T, Breitner T and Alt T 2012 ALICE HLT TPC Tracking of Pb-Pb Events on GPUs *J. Phys. Conf. Ser.* **396** 012044

[10] Corden M J Differences in floating-point arithmetic between Intel Xeon processors and the Intel Xeon Phi coprocessor

[11] Corden M J and Kreitzer D 2009 *Consistency of floating-point results using the intel compiler or why doesn't my application always give the same answer* (Technical report, Intel Corporation, Software Solutions Group)

[12] De Dinechin F From CRLibm to Metalibm: assisting the production of high-performance proven floating-point code

[13] Jarp S, Simmins A, Yaari R and Tang H 1995 *PC as physics computer for LHC?*

Performance quality monitoring system for the Daya Bay reactor neutrino experiment

Y B Liu on behalf of the Daya Bay collaboration
Institute of High Energy Physics, Chinese Academy of Sciences, Beijing, China
E-mail: liuyb@ihep.ac.cn

Abstract. The Daya Bay Reactor Neutrino Experiment started running on September 23, 2011. The Performance Quality Monitoring system (PQM) has been developed to monitor the detector performance and data quality. The main feature is its ability to efficiently process multi-data-stream from three experimental halls. The PQM processes raw data files from the Daya Bay data acquisition system, generates and publishes histograms via a graphical web interface by executing the user-defined algorithm modules, and saves the histograms for permanent storage. That the whole process takes only around 40 minutes makes it valuable for the shift crew to monitor the running status of all the sub-detectors and the data quality.

1. Introduction

The Daya Bay experiment is designed to determine precisely the neutrino mixing angle θ_{13} with a sensitivity better than 0.01 in the parameter $\sin^2 2\theta_{13}$ at 90% confidence level. Three experimental halls (EHs) were built to contain eight functionally identical antineutrino detectors (ADs) [1]. Before October 18, 2012, two ADs were located in EH1 and one in EH2 (the near halls). Three ADs were positioned in the far hall, EH3. After the installation of the last two ADs, 2 and 4 ADs are located in each near hall and the far hall, respectively. In each experimental hall there are also muon detectors, including the inner (IWS), outer (OWS) water shields, and a resistive plate chamber (RPC) tracker. With the 6-AD configuration, the Daya Bay collaboration reported the first observation of non-zero $\sin^2 2\theta_{13}$ [2], and has recently updated the result of $\sin^2 2\theta_{13} = 0.089 \pm 0.010(\text{stat.}) \pm 0.005(\text{syst.})$ [3] in a three-neutrino framework.

During physics data taking, events from different experimental halls are separately assembled in raw data format and stored into persistent files on the online disk array by the data acquisition system (DAQ). Event rates of 8-AD configuration for EH1, EH2, EH3 are about 1.2 kHz, 1.0 kHz, and 600 Hz, respectively. The total number of raw data files for each day is about 320, approximately 1 GB volume per file. The online histogram presenter of the DAQ provides the hit-maps for sub-detectors. In addition, high-level histograms are necessary to be produced for monitoring the running status of all sub-detectors and the data quality (DQ).

The Performance Quality Monitoring system (PQM) is developed to utilize the offline software to process multi-data-stream from three experimental halls, generates more user-defined histograms, including some high-level ones containing the calibration and reconstruction information, and thus plays an important role in offline monitoring for the experiment. Even with the total event rate of around 3 kHz, the PQM can handle the whole process around 40 minutes.

In this article, we present the offline computing environment at Daya Bay in Section 2. In Section 3, the PQM design is outlined. The performance and the information display of the PQM are shown in Section 4. We give a short summary in the last section.

2. Offline computing environment at Daya Bay

There are 11 servers in the offline computing environment at Daya Bay, including a file server, a data transfer server, an offline database server, a web server, two PQM servers (one for backup), and the rest of 5 servers forming user farms. The Portable Batch System (PBS) is implemented to allocate computational jobs among the available computing resources.

The file server provides disk array for onsite data processing. To prevent interruptions to the DAQ, the data transfer server is responsible for copying raw data files to the offline file server. The offline database server copies the information that is necessary for offline analysis from the online database to the offline database. The PQM server is responsible for the implementation of the functionality of the PQM for monitoring the detector performance and the DQ. This will be described in detail in Section 3. The web server is used to display user-defined histograms in analysis algorithm modules.

The Daya Bay offline software (NuWa) [1] using Gaudi [4] as the underlying software framework is available in the offline computing environment. In this framework, reconstruction algorithms have been developed to reconstruct the energy and the vertex of the antineutrino event from the charge pattern of the PMTs. One can find the details of the calibration and reconstruction methods in [1]. By default, the latest calibration constants in the offline database are loaded into the reconstruction algorithms, reflecting the best understanding of the detectors.

3. Implementation of the PQM

The PQM is designed to efficiently process all events in all raw data files as soon as possible. It communicates with the offline database to get the latest raw data file list and submit jobs to the PBS to process them in parallel. At the end of each job, some selected histograms are printed for web display. The shift crew can compare the new histograms with standard ones to report the sub-detector status and the DQ.

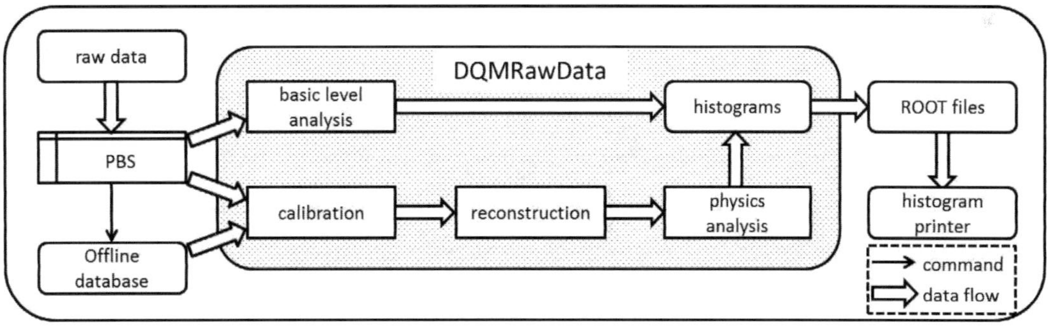

Figure 1. The data flow of the PQM job.

3.1. Data processing chain

To manage the data flow of the PQM as shown in figure 1, a control script in Python language has been developed and runs in a background mode on the PQM server. When the DAQ closes a raw data file, the offline database will be updated with a new record copied from the online database. A copy of the file will be transferred to the file server in less than 5 minutes, assuming a size of (~1 GB). When the data transfer is completed, the status of this file is set as TRANSFERRED in both online and offline database.

The control script queries the offline database at a fixed time interval (10 seconds). If a new record is found and the data status returned by the query shows that the corresponding raw data file is already on the file server, a job for processing the new file is submitted to the PBS. The PQM job uses a recently manually tagged NuWa version to reconstruct data with the latest calibration constants in the offline database, runs analysis algorithms, and fills the results into user-defined histograms. All of the histograms are dumped into a ROOT [5] file at the finalization step of the algorithms, which is then merged with the accumulated ROOT file for the same run. After that, a histogram printer in C++ language is executed taking the merged ROOT file as input to print selected histograms into figures. The control script also checks if the job is finished at a 10-second interval by detecting an empty TXT file created at the end of the job. For finished jobs, the control script transfers the corresponding figures to the PQM disk for web display (see Section 4). When the run ends and all the corresponding raw data files are processed, the control script saves the accumulated ROOT file on the PQM disk for permanent storage.

3.2. Analysis modules in PQM

Currently four modules are developed for the PQM to analyze data and fill user-defined histograms in the NuWa framework: one for histograms of electronics channel information, one for the histograms of calibrated PMT information of ADs and the water shields, one for the histograms of the reconstruction for the ADs, and the other one for the histograms of the RPCs. The histograms displayed by the web interface are shown in table 1. The detailed description of the variables listed in table 1 can be found in [1, 6, 7]. It provides the flexibility to add more algorithm modules as well as to fill more user-defined histograms.

Table 1. The histograms produced by analysis modules in PQM for monitoring the detector performance and the DQ.

Detector Unit	Level	Histograms
PMT	Basic	Mean of ADC, TDC and preADC, RMS of ADC, TDC and preADC, ΔADC, dark noise, dark rate and TDC vs channel ID, hit rates, ADC sum vs. trigger types, ADC sum, number of blocked triggers vs. run time, number of channels, trigger types
	Reconstruction	y vs. x, z vs. radius, energy, energy vs. radius, energy vs. z, event rate vs. radius, event rate vs. z
RPC	Basic	Map of patch rate, map of module rate, map of channel multiplicity, distributions of time between module and system triggers, maps of layer efficiency, distribution of layer efficiency, electronics error types vs. time, maps of channel hit count, channel hit count vs. channel ID, distribution of number of layers per readout, distribution of number of modules per system trigger, maps of layer singles rate, distribution of layer singles rate, module and system trigger rates vs. time.

Since a different configuration of detectors is adopted for EH3, all the analysis modules take the strategy of creating histograms dynamically for each detector, and even for each PMT channel to reduce the output file volume and to improve the processing efficiency. If one event read from the raw data file is the first one for the current detector, a new set of histograms for this detector are created. Otherwise, the event is analyzed and filled into existed histograms.

4. Performance and information display

The PQM is required to provide more information about the running status of all sub-detectors in 3 EHs than the online histogram presenter provided by the DAQ. The data transfer from the DAQ takes less than 5 minutes. The job can be submitted to the PBS about 10 seconds after the data transfer is finished. Running NuWa job modules takes around 30 minutes. In the meanwhile, some user-defined histograms produced by job modules are saved in ROOT files. They are printed and published via the web interface in around 2 minutes. The job failure rate is less than 1% during nomarl data taking. The typical latency of the display of the PQM plots is 40 minutes after a new raw data file is closed by the DAQ. The most time-consuming part is running the calibration and reconstruction algorithms. Since all algorithms are configurable in the PQM job, it is easy to remove them and only show the basic-level information of the detectors to reduce the latency to less than 20 minutes. This special configuration of PQM jobs was usually implemented during the commissioning period, since the onsite experts could see the detector status as soon as possible.

A web interface, as shown in figure 2, running on the web server is utilized to display the selected user-defined histograms. On the main page (see figure 2(a)), two options are provided for users: one is for showing histograms for the current runs of each site, named as 'realtime', the other one is for all the previously processed runs, named as 'history'. For the latter, one can choose the run number in a drop-down menu listing the site ID and run type, i.e. 'Physics', 'FEEDiag' and so on. On the 'realtime' page (see figure 2(b)), site buttons are available for different sites: 'DayaBay', 'LingAo' and 'Far'. The 'realtime' page refreshes at a fixed time interval of 1 minute. For each site, the user can also use detector buttons to view histograms for different sub-detectors. On the 'history' page, there are only detector buttons, because this page is for the run selected by the user, and the specific run is for only one site.

(a) (b)

Figure 2. Web interface for the PQM.

Several example histograms are given here. Figure 3 shows the number of blocked triggers in one event as a function of time. Generally, if a cosmic-ray muon passes through the detector and induces a shower with extremely large energy deposit, overflow of the electronics occurs, and the local trigger and the corresponding data package can be blocked. The blocked trigger number is recorded and read out. If the fraction of the blocked trigger number over the total trigger

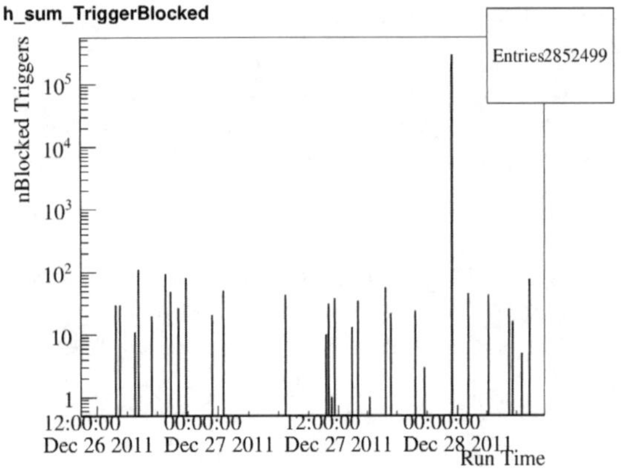

Figure 3. The number of blocked triggers in one event vs. run time.

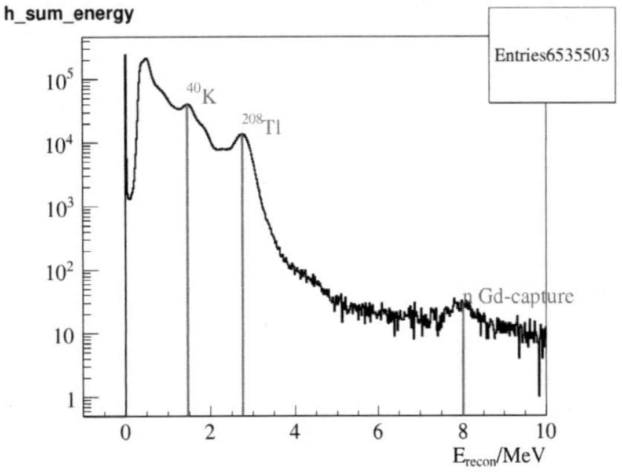

Figure 4. The reconstructed energy distribution of all events for AD1.

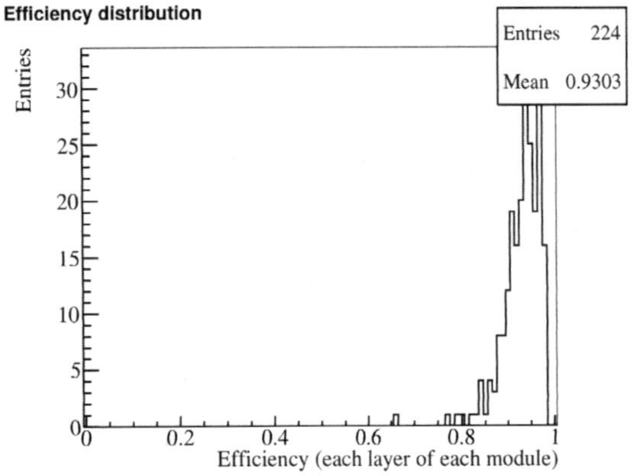

Figure 5. The efficiency of each layer for the RPCs.

number is larger than a pre-defined value, the shift crew should report to the DQ working group (DQWG).

Figure 4 presents the reconstructed energy spectrum for all triggers for an AD. The peaks of ^{40}K (1.46 MeV), ^{208}Tl (2.61 MeV), and the gamma peak from neutron capture on Gd ($\sim$8 MeV) are labeled in fixed positions in the histogram. The red line for ^{208}Tl is set to $\sim$2.75 MeV due to the energy non-linearity which is not corrected in the analysis algorithm. A relative shift between the red lines and the observed peaks indicates problems in the data, e.g. incorrect energy scales read from the offline central database or unexpected behaviour of the PMTs.

The statistics of the efficiency of each RPC layer [6, 7] is displayed in figure 5. The efficiency is calculated in the algorithm of analysis modules for the PQM. The efficiency is a critical quantity to evaluate the performance of the RPCs used as an anti-coincidence detector for muons for the experiments. As shown in the figure, most of RPC layers are working with the efficiency of about 90%. The performance of the RPCs can change due to the gas pressure or other unexpected reasons. The RPC experts and the DQWG are informed if the RPC layer efficiency distribution distorts greatly by comparing with the standard ones.

There are also many other histograms to monitor the detector performance and the DQ. The shift crew can compare them with the corresponding standard figures to check the running status of the sub-detectors and the DQ. In the case that irregularities are observed, the shifter should inform the related experts and report to the DQWG.

A supernova monitoring program is under development to join the Supernova Early Warning System (SNEWS) [8], in which individual neutrino-sensitive experiments send supernova (SN) burst datagram to SNEWS coincidence computer at Brookhaven National Laboratory (BNL). Supernova burst neutrinos are important in studying the dynamics of core collapse and supernova explosion, neutrino properties, and many other interesting studies. The three-hall configuration design at Daya Bay can significantly reduce false SN alarms caused by the spallation background. We will have online and offline triggers for monitoring SN bursts. Given the short latency time of the PQM, the offline trigger for SN bursts will be added to the PQM job.

5. Summary

The onsite data processing is running smoothly. The PQM, which has been developed for the Daya Bay experiment, has the ability to provide user-defined histograms to the shift crew with a latency time of about 40 minutes. It satisfies the requirement for the data taking. By comparing the histograms with the standard ones, potential DQ problem can be found. The PQM will contribute to the SN monitoring program in the Daya Bay experiment in the near future.

Acknowledgments

This work is supported by the National Natural Science Foundation of China (Y2118M005C). I would like to thank Gaosong Li for the contribution to the ReconDataHistogram, Logan Lebanowski and Qingmin Zhang for the contribution to the RpcDataHistogram, and Xiangpan Ji for the contribution to CalibDataHistogram.

References

[1] Daya Bay collaboration 2012 *Nucl. Instrum. Methods* A **685** 78
[2] Daya Bay collaboration 2012 *Phys. Rev. Lett.* **108** 171803
[3] Daya Bay collaboration 2013 *Chinese Physics* C **37** 011001
[4] Gaudi http://cern.ch/gaudi/
[5] ROOT http://root.cern.ch/drupal/
[6] Zhang Q M, He M, Xu J L, Zou J H, Ning Z, Zhao J, Lu H Q, Pěč V, Lebanowski L 2012 *J. Phys.: Conf. Series* **396** 022061
[7] Ning Z et al. 2013 *J. Instrum.* **8** T03007
[8] Wei H Y 2013 *Supernova Trigger in the Daya Bay Reactor Neutrino Experiment* 15th Int. Ws. on Advanced Computing and Analysis Techniques in Physics Research (Beijing, China)

A concurrent vector-based steering framework for particle transport

John Apostolakis, René Brun, Federico Carminati, Andrei Gheata and Sandro Wenzel
European Organization for Nuclear Research (CERN) - Geneva, Switzerland

E-mail: Federico.Carminati@cern.ch

Abstract. High Energy Physics has traditionally been a technology-limited science that has pushed the boundaries of both the detectors collecting the information about the particles and the computing infrastructure processing this information. However, since a few years the increase in computing power comes in the form of increased parallelism at all levels, and High Energy Physics has now to optimise its code to take advantage of the new architectures, including GPUs and hybrid systems. One of the primary targets for optimisation is the particle transport code used to simulate the detector response, as it is largely experiment independent and one of the most demanding applications in terms of CPU resources. The Geant Vector Prototype project aims to explore innovative designs in particle transport aimed at obtaining maximal performance on the new architectures. This paper describes the current status of the project and its future perspectives. In particular we describe how the present design tries to expose the parallelism of the problem at all possible levels, in a design that is aimed at minimising contentions and maximising concurrency, both at the coarse granularity level (threads) and at the micro granularity one (vectorisation, instruction pipelining, multiple instructions per cycle). The future plans and perspectives will also be mentioned.

1. Introduction

High Energy Physics always depended upon the availability of very large computational resources to extract the physics from the collected data and it has been well served by the steady increase in computing power during the last 30 years synthetized by the so-called Moore's law. From the 80's until 2004, this increase came mainly in the form of faster cycle computers, doubling the software performance every 18 months with no need for code refactoring. High Energy Physics made full use of this bonanza and attempts to optimise the offline code to take into account the intrinsic parallelism of the problem being dealt with (independent tracks, events, detectors and so on) and the new architectures appearing on the market (RISC, multicore machines, SIMD capable CPUs, GPU co-processors) were few and of marginal impact, at least in the offline domain.

The situation is now changing due to two concomitant factors. On one side, since the CPU clock frequency reached the 3GHz mark around 2005, fundamental physical limits have blocked its further progress. Computer companies still claim they offer an 18-month doubling of the computing capacity, but this now comes in the form of enhanced parallelism at various levels, which must be skilfully exploited by the application software. In other words, while until 2005 the increase in performance was provided in the form of raw computational speed, now it is offered in terms of "opportunities" that the user has to seize. It goes without saying that this does not come for free, above all because some applications are more prone to be optimised than others.

The second important factor is that, in the next few years, the new or upgraded experiments will see an increase in the amount of data collected of a factor between 10 and 1000, and therefore require at least a corresponding increase of the computational resources, against a background of dwindling funds, where maintaining a constant budget will be already a substantial success.

These factors combined suggest that it is urgent to explore what the opportunities are to improve the performance of our code. The assessment of the current situation can be found in an interesting paper published last year by the OpenLab project [1]. This paper shows that there is, at least in principle, plenty of scope for the improvement of our code.

Several initiatives have been started to improve and optimise existing code (see for instance [2,3,4] and similar publications at CHEP 2012), however for the moment their only coordination is via the CERN Concurrency Forum [5]. The advantage of starting from existing code is that the improvements can be of immediate use to the community. However such initiatives are also, almost by definition, constrained by the features of the existing code: it is the flip side of their short time usefulness.

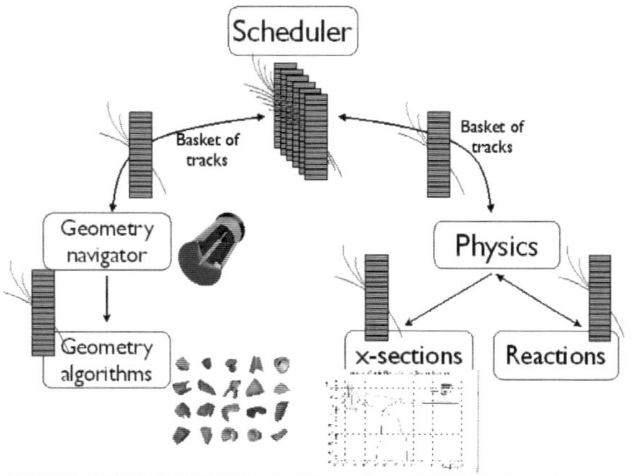

Figure 1. General layout of the Vector Prototype.

We consider it necessary to complement these initiatives with a wide-ranging exploration unhindered by any backward compatibility consideration. For this we have chosen the particle transport application used in detector simulation. The Geant Vector Prototype project (in short Geant-V) started based on a set of preliminary studies performed in 2012. High Energy and Nuclear Physics has a long established tradition of developing particle transport codes to simulate particle detectors both in the design phase and to evaluate the efficiency of particle detection in the data analysis phase ([6,7,8,9,10,11]), and there is substantial expertise in the community on this kind of problems. Monte Carlo simulation is one of the most demanding computing task, due to the slow decrease of the statistical fluctuations around the estimated mean, which is proportional to the inverse of the square root of the number of events simulated. Moreover particle transport simulation is one of the most experiment-independent applications in High Energy Physics. Unlike reconstruction and analysis, where most of the code is specific to a given experimental apparatus, most of the simulation code is generic, and it is expected that any improvement reached in this code could profit all existing and future experiments.

2. The Geant-V Project

The objective of the Geant-V project is to explore different design options to maximise the efficiency of the transport code on modern CPUs and GPUs, in order to obtain the best possible performance in terms of events produced per unit time. For this, parallelism at all levels has to be exposed and exploited.

A particle transport problem has a large intrinsic parallelism, insofar all events are independent and different tracks within the same events are also independent. Exploitation of this parallelism is necessary, but alone not sufficient, as it may improve memory footprint but not increase the number of events per second actually simulated. There is also significant potential for fine grain parallelism in treating low-level operations including geometry and physics. This too must be exposed and exploited, increasing the number of instruction per cycle and exploiting instruction pipelining and vector processing within the CPUs and GPUs.

Present particle transport is essentially scalar and sequential. A particle is followed from its birth to its end, across whatever portion of the detector it traverses. This means that different geometry paths and materials are loaded and then unloaded for each particle history, with a considerable danger of exceeding the capacity of the faster levels of the cache ("trashing" of the cache memory). Particle transport is mostly local, in the sense that the majority of the "transport steps" happen in a small portion of the volumes, typically 50% of the steps in less than 1% of the volumes. This is the result of our preliminary studies that we have presented at CHEP 2012 [12]. We started from this consideration to design a particle transport programme that could exploit this locality by transporting vectors of particles (baskets) that are contained in the same logical volume and are transported in the same material.

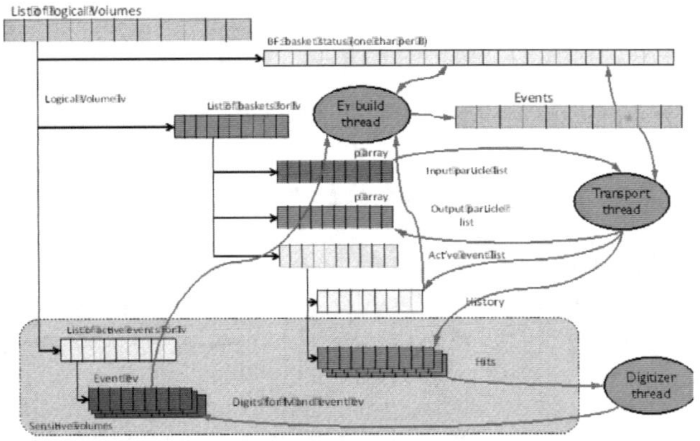

Figure 2. Layout of the scheduler.

In exploring the possibilities to recast particle simulation so that it takes advantage from all performance dimensions and technologies we are concentrating our efforts in two dimensions:

1. Coarse grain parallelism. This exploits the "trivial parallelism" intrinsic in Monte Carlo simulations and leads to reduction of memory footprint on shared memory, multi-core machines. On systems with attached processors (GPUs or Xeon Phis) this allows to have several of them working in parallel, trading the minimum amount of data between them and the main processor.

2. In-core micro-parallelisation or vectorisation. This allows increasing throughput and ultimately is the source of the improved performance. It is expected that the gain in performance derive not only from the parallelism at the instruction level (vectorisation, instruction pipelining and multiple instructions per cycle), but also from a better use of both data and instruction caches due to the enhanced "locality" of the code. This requires a redesign of the data structures and processing strategy that usually also increases locality thereby reducing the cache misses.

In the following we will give a short report of the current situation in the four main areas in which we have factorised our problem: the scheduler, the geometry navigator, the basic geometry routines and the physics. The relation between these elements in the transport process is shown in **Figure 1**. The scheduler has the role to orchestrate the queues of baskets. The input baskets are given to the geometry navigator and to the physics subsystem for handling. The geometry subsystem passes them to the geometry algorithms to determine the distance to the boundaries and to the volumes contained into the logical volume. When on a boundary, the navigator determines the next volume that is to be entered by the particles in the global geometry. The navigator gives back to the scheduler a new basket of particles in various logical volumes that need to be re-dispatched to the logical volume baskets for further processing. The physics subsystem determines the distance to the next interaction for the particles of a basket and, when the interaction is supposed to happen, it generates the secondary particles and again feeds back particles in baskets to the scheduler for further re-dispatching. Each basket is handled by a separate task, while one more task handles the scheduler.

2.1. The scheduler

It is possible to obtain the desired performance improvements only if we manage to transport sets of particles "together" to maximise the locality of the operations. This calls for the efficient handling of lists of particles to be fed both to the geometry and to the physics methods, which should be in turn properly optimised to treat them. A scheduler handles these lists of tracks during transport. The general layout of the scheduler can be seen in **Figure 2**. The general schema of the scheduler is the following. Each logical volume has a list of baskets to which corresponds a list of status flags. Each basket has one input and one output buffers of particles, and a list of active events. For each event in this list there is one or more buffer of hits, if this is a sensitive detector, and a history buffer. The scheduler performs the following actions:

1) A basket which has the status flag "ready for transport" is associated with a transport task (run by a thread) which transports all particles in the input buffer and, as they leave the volume, interact or stop, puts them in the output buffer for further processing. The basket is marked "ready for dispatching".
2) A dispatching task (again associated with a thread) looks for baskets "ready for dispatching" and puts their particles in buffers that are "ready for filling". When a basket is full, it is marked "ready for transport".
3) When the basket "ready for transport" is dispatched, it is marked as "ready for filling".

A detailed analysis of the program flow leads us to believe that locks are minimised in this process flow. We are now carefully designing the handling of the special cases of the beginning and the end of the cycle and of the "flushing" of events to avoid memory inflation.

As we have seen (see [12]), the difference in population of the different logical volumes can be very relevant. During a single event some volumes may never reach a population of particles convenient to handle via the basketised transport. To make sure that this does not happen, our design allows the transport of several events in parallel. This is an important modification of the transport schema now in operation in various Monte Carlo programmes, and it will call for an attentive revision of the scoring strategy of the user code.

When an event is finished, i.e. all its particles have been transported, a separated task will be used to digitise the event by going through all the baskets of the sensitive volumes and collecting the hits for this event, transforming them into digits. We are investigating SIMD optimisations for this task because all hits of a detector have, by definition, a similar structure.

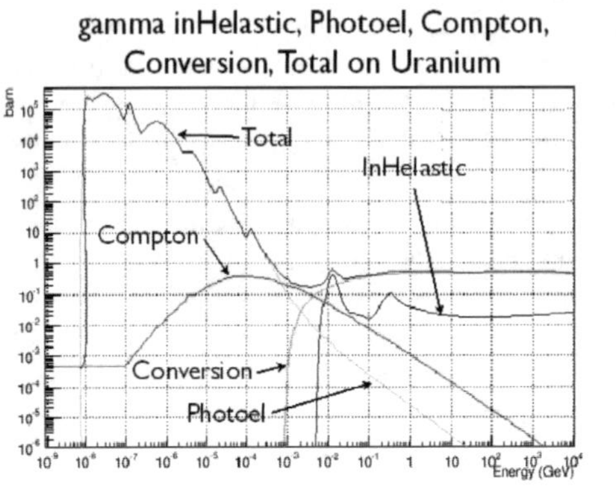

Figure 3. Example of cross section tabulation from Geant4.

2.2. Physics

The plan for the prototype includes optimising physics algorithms, including cross section tables, sampling and generation of secondaries, passing them lists of particles and optimising their SIMD performance. This however has to be regarded as a long-term goal, as the development of a SIMD optimised physics package may take a substantial amount of time. Nevertheless, for the purpose of verifying our design, we needed both a realistic physics shower development and a realistic geometry package. Our problem has very little chance to be linear, so simple scaling from "toy models" is most probably meaningless. We will explain in the following sub-section how we are dealing with the geometry. To be able to simulate a realistic shower development without interfacing with large packages, such as Geant4, we decided to extract from Geant4 both the tabulated cross sections and a sampling of the final states and use them in the prototype. To do so, we have selected the major mechanisms, namely bremsstrahlung, e+ annihilation, Compton scattering, discrete ionisation, elastic scattering, inelastic scattering, pair production, photoelectric effect and capture, and for each particle we have tabulated all Geant4 cross sections for all materials from Hydrogen to Uranium in the range 10 keV to 1 TeV. Using 100 logarithmically spaced bins produces an ~90 MB file. For each energy bin, each reaction, each particle and each element we generate a number of final states that we store in yet another file. When sampling 20 final states per bin, with 100 bins as before, we have a file of ~500 MB (including multiple scattering and energy loss, see later). This means that for a typical detector, which is composed of less than 30 elements, we can store the whole physics in ~300 MB in memory (taking into account Root compression), which is still a reasonable size for an in-memory database.

For continuous processes we follow the same approach. We have tabulated the specific energy loss, the average angle of multiple scattering and the average dispersion for the multiple-scattering angle. All this information is to be used at transport time to generate continuous (i.e. along the step) processes and discrete processes. When a physics process is selected, the set of final states corresponding to the energy closest to the one of the incoming particle is determined, and one of these final states is randomly selected and rotated along the direction of the incoming particle, with a random azimuthal angle. In future we will also rescale the energy of the final state to preserve energy balance. We expect that this give us realistic shower development even if the quality of the physics will be probably very diverse depending on the problem and the range of energy treated. It could however be a nice basis for a possible fast simulation development. In **Figure 3** we show an example of cross section tabulation with all the reactions of a photon on Uranium.

For the moment we consider only natural mixtures of isotopes, but a future development will extend this method to the single isotopes. The cross sections are held in C-like data structures wrapped in lightweight Root classes with minimal use of virtuality, to make it easier to port the code on different co-processors. The extracted cross sections and the final states are held in two Root files. For the moment the methods used to extract cross sections and sample final states have only scalar signatures; the implementation of vector signatures and the SIMD optimization will come next.

2.3. The geometry navigator

In order to have a first version of the prototype capable of tackling complex geometries, we needed to start from a mature navigator and we chose the Root one. The first obstacle we encountered was that the Root navigator was not designed for a multi-threaded context, and the geometry description was carrying stateful navigation information to optimise the search algorithms. The first job was to separate the state describing the current position and direction of a particle flying through the detector from a stateless, read-only description of the geometry in memory. Once this was achieved, we could transport several particles in parallel holding their status in thread-local data, while all threads were using a shared geometry description.

This enabled us to do multi-threaded particle transport with the Root geometry TGeo. It is worth to note that this new version of TGeo is now used in production, as it is in effect a more rationalised and efficient code even for sequential transport and geometry navigation. This code is also a step in the direction of the attached processors like GPUs, insofar the geometry description (and physics tables) can be loaded once and for all in the coprocessor and only the stateful particle data have to be traded.

The next step will be to be able to transport the baskets of particles. For this a "vectorised" version of the navigator will have to be developed. We have started the design phase of this new navigator.

2.4. The geometry algorithms

We have used the basic geometry routines as a playground with which to experiment with in-core micro-parallelism and vectorisation (kernel optimisation). The procedure here is essentially to recast the algorithms so that they would handle not one but several particles at the time. This is achieved via the introduction of a new "vector" signature, and it corresponds to the idea of "moving" the loop from outside the method to its inside. However, apart from the trivial gain via the reduction of the number of calls, this is not nearly enough to exploit the SIMD features of the CPU. In order to use SIMD CPU capabilities, the compiler needs to emit special assembly instructions (e.g. "add" versus "vaddp") to the hardware. For this multiple options exist:

1. Autovectorisation: This consists of exposing the parallelism in the code and letting the compiler exploit the opportunities to replace scalar instructions with vector ones (without code changes). This would of course be the ideal option for portability and maintenance, but in practice it very rarely works without pragmas or heavy re-factoring of the code.
2. Explicit vector-oriented programming via intrinsic instructions. This manually instructs the compiler to use vector instructions. This requires refactoring and code changes to work, and

Journal of Physics: Conference Series **523** (2014) 012004 doi:10.1088/1742-6596/523/1/012004

at the lowest level can be very machine and architecture dependent. Template-based API hiding low-level details like the Vc library [13] can be very helpful. Our experience shows that this approach gives good performance, portability and only little platform dependency. However it does require some code changes and refactoring.

3. Language extensions, such as Intel Cilk Plus Array notation [14] or OpenACC [15]. Our investigation in this area is just beginning and we will not elaborate more on it in this paper.

Given the above, we concentrated on the second approach above. We intend to continue studying the other two approaches, however, in order to come quickly to an "order of magnitude" answer as to whether micro-parallelisation and vectorisation could give us the expected performance benefits. So we started with the optimisation of methods to calculate the distance from inside and outside for three simple geometrical shapes (box, cone and tube) using the Vc library on an Intel Xeon machine. The first results can be shown in **Figure 4** and are indeed quite encouraging. It must be stressed that this is a very preliminary result and there is a lot of work ahead in SIMD-optimizing more complicated shapes, extending to other methods and to the coordinate transformation that uses a good fraction of the CPU time spent in geometry.

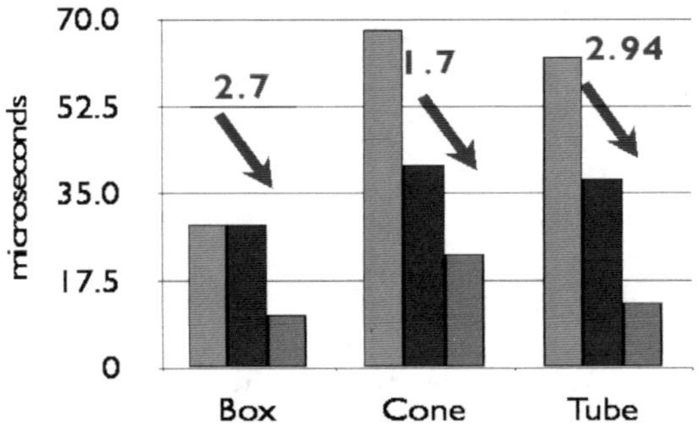

Figure 4. Performance improvement of the SIMD version of the method calculating the distance from a point inside the shape along the flypath for 1024 particles. The first column corresponds to the use of Root 5.34.09, the second an optimised version of Root and the third the SIMD version of the algorithms implemented via the Vc library with the AVX vector instructions. The numbers show the speedup of the SIMD version with respect to the optimised Root.

3. Conclusions

Work toward the realisation of a high performance particle transport prototype optimised for SIMD machines (the Geant Vector Prototype) is progressing steadily. The major components have been identified and designed and their implementation is in an advanced state. Preliminary results on the SIMD optimisation of the geometry routines are promising. The next steps will be the implementation of the Scheduler, the extension of the tests with the optimised SIMD geometry routines and the optimisation of more shapes, the completion of the sampling of the final states of the interactions and the realisation of a vector navigator. This will allow us to assemble a complete application that could do some basic particle transport and perform accurate timings. An important factor will also be the consideration of how our design could take advantage of GPUs.

Another important item that we will have to tackle is the estimation of the overheads implied in the creation and handling of the baskets and in the creation of thread-safe data structure. With the parallelization of the processing, keeping the results will be an additional challenge, involving also work on the random number generator. Inevitably this too will have an overhead that again we have to measure and qualify.

In conclusion we can say that the first steps in the design of a high performance detector simulation application have been taken and they are promising, while a vast and interesting programme of work lies ahead.

Acknowledgements

We would like to thank Pere Mato of the SFT group for useful discussions. We would also like to thank Marilena Bandieramonte from Catania University and Raman Shegal from the Bhabha Atomic Research Centre, Mumbai, India for their collaboration.

Bibliography

1. Jarp S, Lazzaro A and Nowak A 2012 The future of commodity computing and many-core versus the interests of HEP software. *International Conference on Computing in High Energy and Nuclear Physics* **396** pp 52-58 doi:10.1088/1742-6596/396/5/052058
2. Dongx X, Cooperman G, Apostolaks J, Nowak A, Jarp S, Asai M and Brandt D 2012 Creating and Improving Multi-Threaded Geant4 *International Conference on Computing in High Energy and Nuclear Physics* **396** doi:10.1088/1742-6596/396/5/052029
3. Dong X, Cooperman G and Apostolakis J 2010 Multithreaded Geant4: Semi-automatic Transformation into Scalable Thread-Parallel Software *Euro-Par 2010 - Parallel Processing* **6272** pp 287–303
4. Canal P, Elvira D, Hatcher R, Jun S and Mrenna S 2013 A Vision on the Status and Evolution of HEP Physics Software Tools *arXiv:1307.7452v1*
5. Group C 2013 *Forum on Concurrent Programming Models and Frameworks* http://concurrency.web.cern.ch
6. Johnston R 1963 A General Monte Carlo Neutronics Code *LAMS* **2856** Los Alamos Scientific Laboratory, Los Alamos
7. Cashwell E, Neergaard J, Taylor W and Turner G 1972 MCN: A Neutron Monte Carlo Code *Los Alamos Scientific Laboratory Report LA* **4751** Los Alamons National Laboratory, Los Alamos
8. Ford R and Nelson W 1978 The EGS code system – Version 3 *SLAC* **210** Stanford Linear Accelerator Center, Stanford
9. Brun R et al 1993 GEANT Detector Description and Simulation Tool (Version 3.21) *CERN Program Library* **W5013**, CERN, Geneve
10. Battistoni G, Muraro S, Sala P, Cerutti F, Ferrari A, Roesler S, Fassò A and Ranft J 2007 The FLUKA code: Description and benchmarking 2006 *Albrow M and Raja R eds. Hadronic Shower Simulation Workshop 2006*, Fermilab 6-8 September 2006 **896** pp 31-49
11. Allison J et al 2006 Geant4 developments and applications Nuclear Science *IEEE Transactions* **53** 1 pp 270-278
12. Apostolakis J, Brun R, Carminati F and Gheata A 2012 Rethinking particle transport in the many-core era towards GEANT 5 *Computing in High Energy Physics 2012* **396-2**

doi:10.1088/1742-6596/396/2/022014

13. Kretz M and Lindenstruth V 2012 Vc: A C++ library for explicit vectorization. Software: Practice and Experience 42 **11** pp 1409–1430
14. Intel Corporation 2013 *Cilk Plus* https://www.cilkplus.org
15. OpenACC Consortium 2013 *OpenACC Home* http://www.openacc-standard.org

ALICE Expert System

C Ionita[1] and F Carena[2]

[1,2] ALICE DAQ, CERN, Route de Meyrin 385, 1217 Meyrin, Switzerland

E-mail: `costin.ionita@cern.ch`

Abstract. The ALICE experiment at CERN employs a number of human operators (shifters), who have to make sure that the experiment is always in a state compatible with taking Physics data. Given the complexity of the system and the myriad of errors that can arise, this is not always a trivial task. The aim of this paper is to describe an expert system that is capable of assisting human shifters in the ALICE control room. The system diagnoses potential issues and attempts to make smart recommendations for troubleshooting. At its core, a Prolog engine infers whether a Physics or a technical run can be started based on the current state of the underlying sub-systems. A separate C++ component queries certain SMI objects and stores their state as facts in a Prolog knowledge base. By mining the data stored in different system logs, the expert system can also diagnose errors arising during a run. Currently the system is used by the on-call experts for faster response times, but we expect it to be adopted as a standard tool by regular shifters during the next data taking period.

1. Introduction

The ALICE collaboration is studying a state of matter called quark-gluon plasma, which existed just after the Big Bang when the Universe was still extremely hot. In order to achieve this goal, many different detectors have been put in place, each providing a different piece of information to physicists. The control of these detectors is achieved using several "online systems" [1]: Detector Control System (DCS), Data Acquisition (DAQ), Trigger system (TRG), and High Level Trigger (HLT). On top of these, the Experiment Control System (ECS) coordinates the actions performed by all online systems.

Keeping these systems in a state compatible with Physics data taking is not always a trivial task. Alongside hardware or software faults, human error is also a major factor in lowering the data taking efficiency. This comes as a result of the fact that the shifters who monitor the state of these systems are not always expert users. For this reason, shifters often do not know how to handle the inherent complexities and must seek further assistance from an on-call expert, usually resulting in lost data taking time.

After analyzing the most frequent problems occuring in practice, we came to the conclusion that not all problems require advanced expert knowledge. In fact, some of these issues can be diagnosed, and then solved, just by inspecting the state of certain sub-systems. In this paper, we describe the ALICE expert system, which aims to automate the diagnosis of frequently occuring issues and make smart recommendations for troubleshooting.

In order to understand the state of the ECS at any given moment, the expert system queries a number of SMI objects [2], which describe the behaviour of the ECS as a finite state machine. This information is correlated with the information extracted from the configuration databases, and then stored as facts in a Prolog knowledge base. Based on the facts known at a certain time, the system infers whether a Physics or technical run is possible. If a run cannot be started, the system prompts the user with a set of failure reasons, as well as with possible solutions provided by human experts.

Note however that some problems cannot be detected a priori. In this case, the Physics run can actually start, but it will fail at run time. In this scenario, the system looks at the error messages (with the highest severity) stored by the ALICE infoLogger system [3], and attempts to make a recommendation using again human expert knowledge.

2. Architectural overview

Before going into more details, it is important to get a high level overview of the data flow in the ALICE Data Acquisition (DAQ) system. Every collision produces a very large amount of data, recorded as "event fragments" using dedicated electronic equipment. Event fragments are sent from the Front End Read Out (FERO) electronics of the detectors, through optical links, to Local Data Concentrators (LDC), which generate a number of sub-events. Each sub-event is then shipped through a Gigabit Ethernet network to Global Data Concentrators (GDC) where the full events are built. All LDCs and GDCs used in a given run are grouped together under a high level entity called a partition.

The expert system builds its knowledge base using a C++ component that extracts information from multiple sources. First, we have the configuration databases containing vital information, such as the list of detectors included in a partition, whether the GDCs are selected statically or dynamically, and so on. The *DataExtractor* connects to these databases, extracts and parses the raw data, and then stores it in more convenient data structures. At the next stage, the *KBManager* takes the information received from the *DataExtractor*, aggregates it with state information provided by the different SMI objects, and then asserts it as Prolog facts (Fig. 1). The exact SMI objects that have to be queried in this process have been established using human expert knowledge of the ECS.

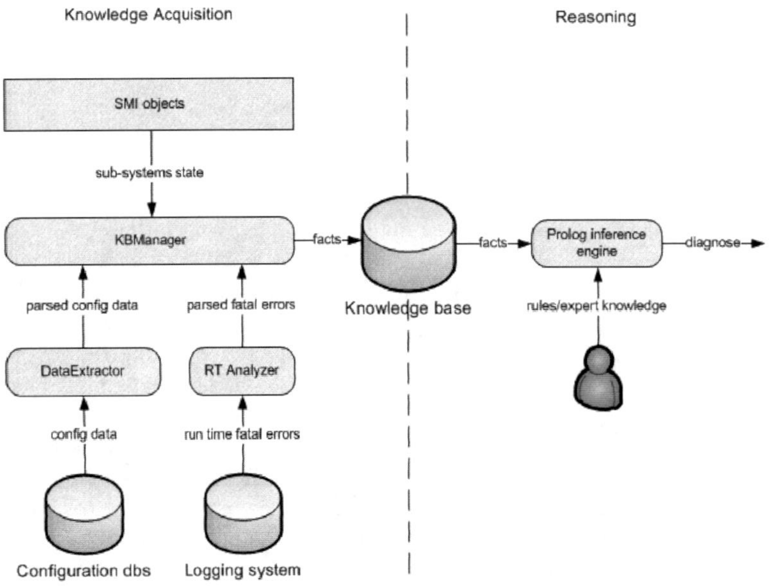

Figure 1. Architectural overview

Another source of information is the infoLogger system, which provides an overview of the errors occuring at run time. The Run Time Analyzer (*RT Analyzer*) connects to this database and attempts to extract the errors with the highest severity, which are usually the root cause of a run interruption. Similar to the *DataExtractor*, these errors are again parsed and put into a more convenient format, while at the next stage, the *KBManager* asserts them as Prolog facts into our knowledge base.

It is important to note that the reasoning is completely decoupled from the knowledge acquisition. Thus, the *KBManager* performs no filtering of the data, as its only role is to build a large pool of facts that can be used for reasoning. This means that the process can be inefficient at times, since some of the asserted facts might be useless. For instance, there is no need to store the information of all possible GDCs if the GDCs are selected statically. Ideally we would only store information about the required GDCs. However, the *KBManager* is not aware whether the GDCs are selected dynamically or statically and it does not attempt to build any decision trees. This is done at a later stage by the inference engine. The reason we have chosen this design is to keep the reasoning contained within the Prolog inference engine, and not spread across other components. If the logic changes in the future, we need only modify the predicate definition in the inference engine, while the knowledge acquisition component will be kept the same.

3. Knowledge representation and reasoning

A generic expert system is made up of three main components, as described in Fig. 2: the production memory, the working memory and the inference engine.

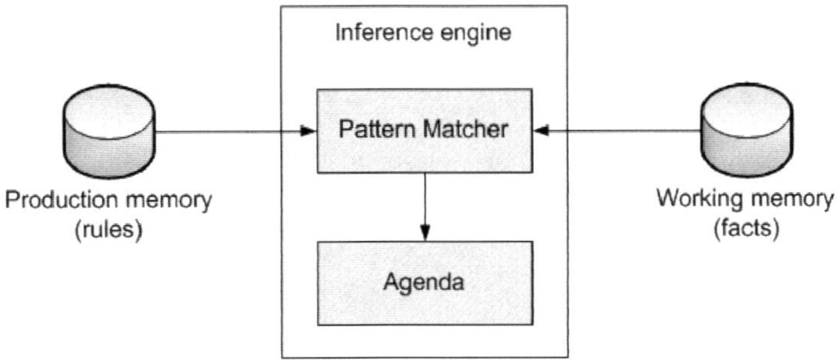

Figure 2. Generic expert system

The working memory contains all the known facts about the world. For instance, if detector *a* is running, then we can represent this information as a predicate *running(a)*. The production memory contains all the rules (expert knowledge) that the system uses to assert new facts or to determine the satisfiability of a goal. In the example below, a control server is available if it is running and not locked:

```
rc_servers_available([H|T]) :-
  check_running(H),
  check_not_locked(H),!,
  rc_servers_available(T).
```

The inference engine can reason about the facts in the knowledge base using either forward chaining or backward chaining. In the first approach, the engine runs through all the rule definitions and triggers rules whose conditions match facts in the knowledge base. These rules add new data to the knowledge base, which can then trigger more rules. The process continues iteratively until no rules are triggered anymore (Fig. 3).

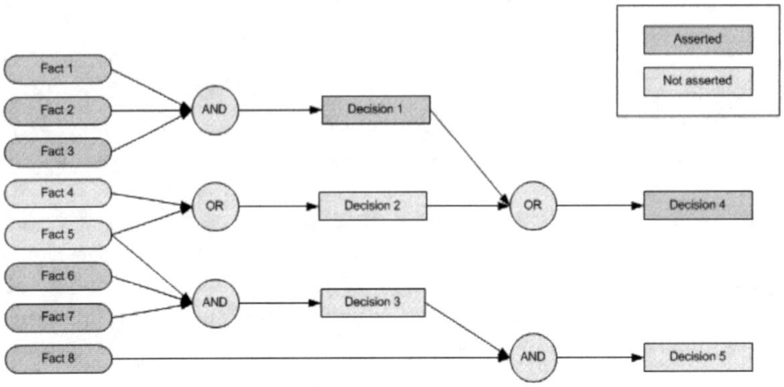

Figure 3. Forward chaining

In backward chaining, the system starts from a goal definition and attempts to satisfy its preconditions. The process continues recursively until a predicate can be evaluated as true or false. These predicates correspond to the leaves of a decision tree (Fig. 4). Note that in general a closed world assumption is required, i.e. if a predicate is not explicitly indicated to be true, then it can be assumed to be false.

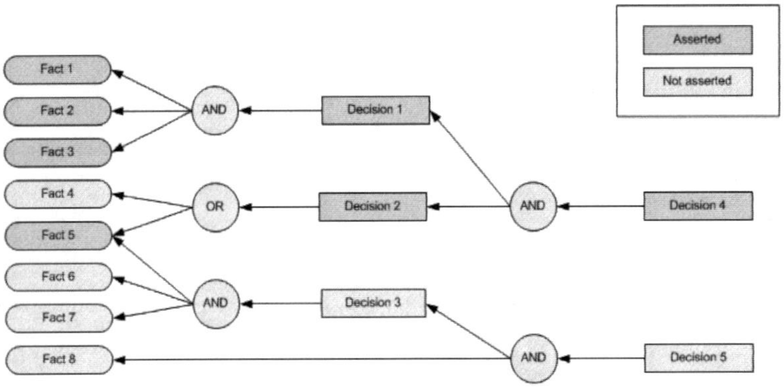

Figure 4. Backward chaining

The main disadvantage of forward chaining is that we cannot establish the reasons why a goal could not be satisfied, which is a critical information when assisting the human shifters. Backward chaining will attempt to backtrack whenever a predicate cannot be satisfied, and therefore, we can use this feature to store all reasons for which a certain action could not be performed. Consequently, Prolog is a natural choice for our inference engine, since it uses backward chaining.

Note, however, that Prolog cannot deal natively with the requirement of providing all reasons for a failed predicate. For this reason, we had to use a workaround, taking advantage of its backtracking capabilities. Each basic fact is wrapped with a top level predicate, that adds a reason to a list whenever the basic fact is not true, as shown below:

```
check_granted(X) :-
  granted(X);  %if true, continue

  failures(F), % if not true, get the list of current failures
  append(F,[X,'not granted'],F2), % append new failure
  retractall(failures(_)),% remove old list
  assert(failures(F2)), fail. % assert new list and return false
```

4. A priori and a posteriori detection

The main role of the system is to assist the human shifters in determining why Physics data taking is not possible in a given system state or configuration. Apart from that, an additional set of actions (pre-requisites for Physics data taking) can also be diagnosed individually. The entire process is fully automatic, and the system is able to recursively drill down on a certain problem, until the most specific diagnosis is found. As shown in Fig. 5, the Physics run cannot start because the FERO is not ready. While in principle this problem can be easily solved by running a MAKE_FERO_READY command, the system continues the analysis and determines that such a command would fail since the DAQ resources are blocked by another entity. Thus, the actual recommendation given by the system is to first release the DAQ resources, which, in turn, allows both the FERO ready and the starting of a Physics run.

Another important observation is that some of the facts cannot be known a priori and the potential problems can only be detected at run time. This is why we added the *RT Analyzer* as a separate component. Unlike a priori diagnosis, run-time analysis is not real-time, as the system needs to wait until the events have been stored by the logging system. Note that this does not interfere in any way with the typical workflow, as no action can be performed anyway until a Physics run comes to a complete halt.

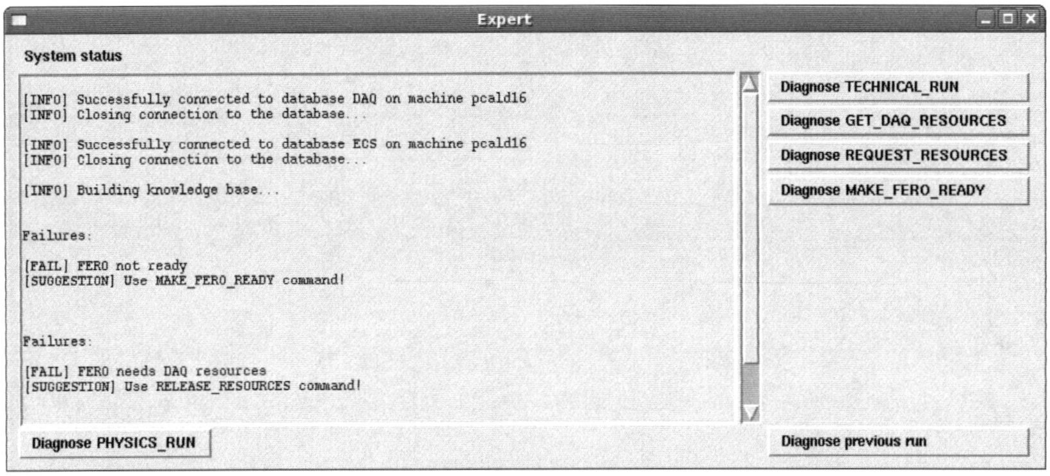

Figure 5. Tcl/Tk GUI

5. Conclusions

In this paper, we have described an expert system capable of assisting human shifters in the ALICE control room. Our system extracts information from the configuration databases and the logging system, and then dynamically builds a Prolog knowledge base. The rules used for inference have been designed using human expert knowledge.

Our next step is to better integrate the expert system with the other components of the ECS. This way, instead of being just an assistant, the system will be able to perform actions on behalf of users, e.g. run scripts, send email alerts etc. Secondly, we are also aiming to improve the user experience, by revamping the current GUI, or perhaps even replacing it with a web application.

References

[1] ALICE Collaboration 2008 *JINST*
[2] Gaspar C. Franek B. 1997 *CHEP*
[3] Chapeland S. et al. 2013 *Proceedings to be published in CHEP*

ACAT2013 IOP Publishing
Journal of Physics: Conference Series **523** (2014) 012006 doi:10.1088/1742-6596/523/1/012006

Experience, use, and performance measurement of the Hadoop File System in a typical nuclear physics analysis workflow

E Sangaline[1] and J Lauret[2]
[1]Physics Department, University of California Davis, Davis, CA 95616-5270, USA
[2]Physics Department, Brookhaven National Laboratory, Upton, NY 11973-5000, USA

E-mail: sangaline@nuclear.ucdavis.edu

Abstract. The quantity of information produced in Nuclear and Particle Physics (NPP) experiments necessitates the transmission and storage of data across diverse collections of computing resources. Robust solutions such as XRootD have been used in NPP, but as the usage of cloud resources grows, the difficulties in the dynamic configuration of these systems become a concern. Hadoop File System (HDFS) exists as a possible cloud storage solution with a proven track record in dynamic environments. Though currently not extensively used in NPP, HDFS is an attractive solution offering both elastic storage and rapid deployment. We will present the performance of HDFS in both canonical I/O tests and for a typical data analysis pattern within the RHIC/STAR experimental framework. These tests explore the scaling with different levels of redundancy and numbers of clients. Additionally, the performance of FUSE and NFS interfaces to HDFS were evaluated as a way to allow existing software to function without modification. Unfortunately, the complicated data structures in NPP are non-trivial to integrate with Hadoop and so many of the benefits of the MapReduce paradigm could not be directly realized. Despite this, our results indicate that using HDFS as a distributed filesystem offers reasonable performance and scalability and that it excels in its ease of configuration and deployment in a cloud environment.

1. Introduction

Most modern NPP experiments currently utilize the ROOT framework for data analysis and binary ROOT files for storage [1]. The ROOT distribution includes XRootD which is a highly scalable and fault-tolerant distributed storage solution designed as an extension of ROOT's built in rootd file server [2]. XRootD is commonly used in conjunction with ROOT and has reliably handled the storage and analysis requirements of a number of experiments over the years. The configuration of XRootD does not, however, lend itself to applications where resources are being dynamically added and removed on short time scales. While exploring options for more flexible scaling of computing resources with virtualization we have found that XRootD is best suited for relatively static allocations of resources. This has motivated us to investigate other possibilities for distributed filesystems that are better suited for dynamic scaling.

Hadoop is a software framework for the distributed storage and processing of large data sets [3]. It is maintained as an open source project by the Apache Software Foundation and implements many of the ideas behind Google's MapReduce and Google File System (GFS) [4]. The underlying technology was designed to achieve robust scalability on commodity hardware and thus has redundancy, fault tolerance, and flexibility in deployment built in as key features. These aspects have played a significant role in Hadoop's adoption in industry where it is used to store and analyse datasets that are large even by NPP standards. For example, the Large Hadron Collider (LHC) produces 15 PB of data per year while Facebook's Hadoop clusters grow by over 150 PB of data per year [5], [6]. The success of Hadoop in handling storage and cluster growth of this magnitude makes it a possible candidate for consideration within NPP applications.

We aim to evaluate the performance of the Hadoop File System (HDFS) with respect to the needs of NPP and our experiences with dynamic resource allocation [7]. This work serves to complement prior investigations of HDFS's applicability in NPP by evaluating additional access methods and further investigating I/O scaling [8], [9].

2. Testing Environment

For the purposes of testing, a cluster of 25 nodes was constructed. A Xen kernel under Scientific Linux 5.9 was installed as the host system and a guest virtual machine was configured on each node and allocated one core of a dual core 1.8 GHz AMD Opteron Processor, 4 GB of RAM, four 2 TB drives, and a 1 Gb/s network interface attached to a 1 Gb/s switch [10]. The virtual nodes were configured to run CentOS 5.9 with the full STAR software environment. Apache Hadoop 1.1.2 was deployed on the virtual cluster using a 64 MB block size as well as MapR's commercial distribution of Hadoop with an M5 evaluation license. MapR required direct access to the four storage drives while for Apache Hadoop the drives were instead configured as a RAID 5 array formatted as ext3.

Using the Hadoop client and accessing the local filesystem directly were used as baselines to assess the relative overheads of Hadoop itself and of the various access methods. It should be noted that a libhdfs plugin is now available in ROOT but that compatibility issues prevent the STAR environment from being used with newer versions of ROOT. Instead, several different methods for mounting HDFS and accessing it through standard POSIX libraries were configured:

- Fuse-DFS. A mountable interface to HDFS based on the Filesystem in Userspace project (FUSE) [11].
- HDFS NFS Proxy. A project which allows HDFS access via an NFS server [12].
- MapR. A proprietary alternative which also allows the filesystem to be mounted via NFS [13].

3. Tests and Results

A variety of tests were run to get a basic comparison of the reliability and access rates for the different methods. The systems were configured to drop caches before and after each test and to include sync times in the performance measurements. The I/O speeds were investigated as a function of file size, replication, and the number of active clients. Unless otherwise noted, all rates refer to sequential file access.

For single client file access it was found that there was little change in I/O rates with the replication, although this does not include the time for full replication to be reached. Local disk write rates saturated at 30 MB/s while the Hadoop client saturated at 24 MB/s and both HDFS NFS Proxy and Fuse-DFS saturated at around 12 MB/s. Write operations would mainly be used for importing data files into a cluster and with the Hadoop client this would bring relatively low overhead. The read rates can be seen in figure 1 where we notice considerable overhead relative to local disk access. It was found that HDFS NFS Proxy had serious stability issues with file sizes above 200 MB.

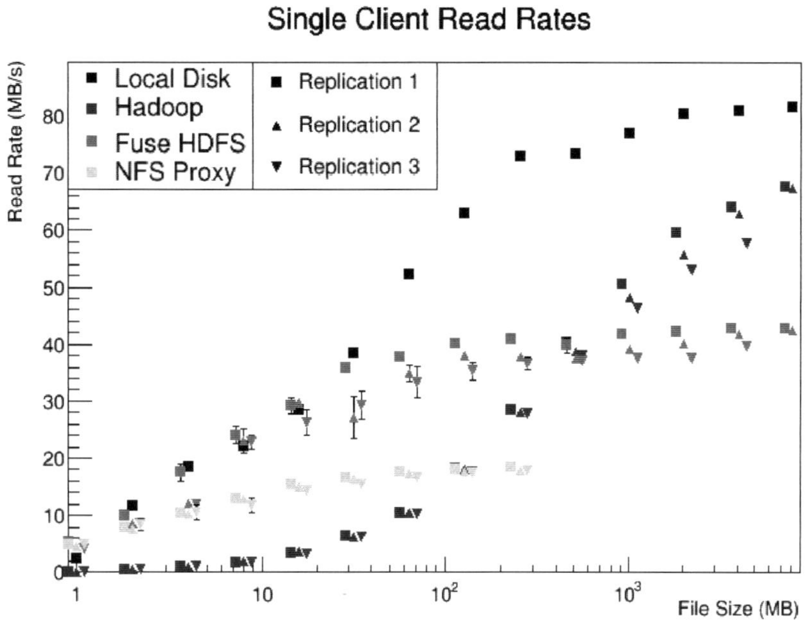

Figure 1. The read rates for various access methods as a function of file size.

The multiple client read rates offerinsight into the scalability in the situation where multiple analyses might need to access the same files. From figure 2 we find that for all interfaces the total read rate scales linearly with the number of clients, even as the number of clients grows considerably larger than the replication. We also find that MapR significantly outperforms the other interfaces for reading.

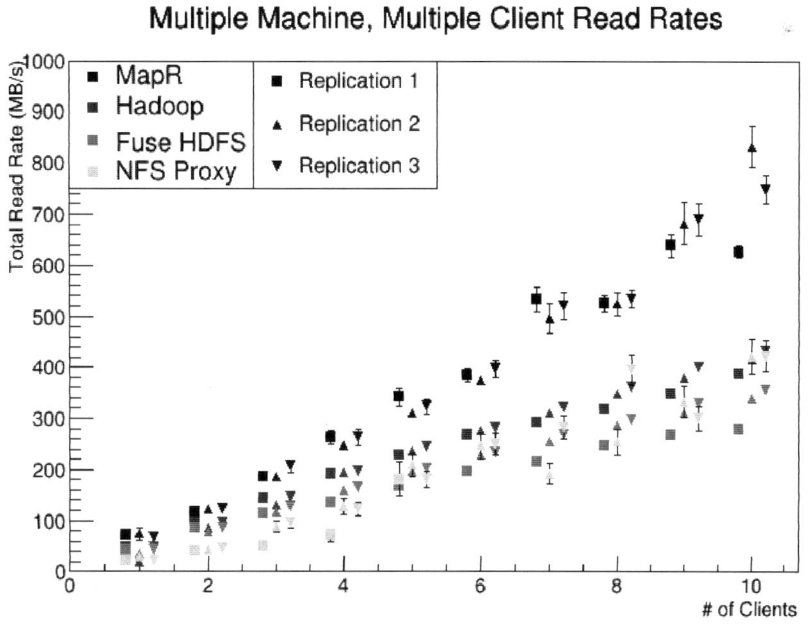

Figure 2. The scaling of read rates as a function of number of clients for a single 1 GB file.

HDFS was designed for sequential access and, because of this, the previously discussed rates are larger than what we would expect for non-sequential reads. To get a more realistic view of what sort of performance we could expect we created 1 GB ROOT files containing Au+Au events and wrote analysis code within the STAR framework to run over them. The code applies basic event quality cuts and fills several histograms, tasks that would be typical of a simple physics analysis. Running these analysis on the Hadoop virtual cluster yielded interesting results which can be found in table 1. Fuse-DFS is comparable in performance to local disk access and significantly outperforms MapR, contrary to the tests with sequential reads.

Table 1. STAR Analysis Rates

Access Method	Read Rate (MB/s)
Fuse-DFS	9.9+/-0.1
MapR	3.1+/-0.1
Local Disk	9.7+/-0.6

4. Conclusions and Outlook

We have performed basic performance testing for various HDFS access methods in the context of a virtualized cluster for NPP analysis. There is further work to be done in terms of understanding performance under more realistic conditions but we can already form some important conclusions.

We were able to quickly and easily configure and deploy Hadoop images on new machines. When switching between MapR and Hadoop we were required to frequently make changes in disk and server configurations, to format HDFS, and to repopulate the cluster with files. Nodes were trivial to deploy and large data files could be transferred into the cluster with little overhead and with replication handled automatically. Qualitatively speaking, this was all relatively straightforward and satisfies the properties we were looking for in something that could be easily deployed on short timescales.

Our multiple client testing demonstrated linear scaling beyond the replication, and in the future we will run more extensive tests in order to understand how saturation will occur. Tests run which performed actual analysis tasks found the analysis rate for Fuse-DFS to be consistent with local disk access and to outperform the proprietary MapR. This result is very promising and offers hope that performance in actual applications might be better than in the sequential tests. Overall, we find that HDFS would likely make a viable distributed filesystem for quick deployment of NPP analysis clusters and hope to continue our studies to better understand performance under saturation conditions.

5. Acknowledgements

We would like to acknowledge and thank the DOE and NSF for funding that supported this work as well as the University of California Davis and Brookhaven National Laboratory for resources that they made available.

References

[1] Moneta L et al. 2008 Recent developments of the ROOT mathematical and statistical software *J. Phys.: Conf. Ser.* **119** 042023
[2] XRootD http://xrootd.slac.stanford.edu/
[3] Apache Hadoop http://hadoop.apache.org/
[4] Ghemawat S, Gobio H and Leung S-T 2003 The Google File System *Proc. 19th ACM Symposium on Operating Systems Principles*

Journal of Physics: Conference Series **523** (2014) 012006 doi:10.1088/1742-6596/523/1/012006

[5] Kanti P 2009 Black Holes at the *LHC Lecture Notes in Physics* **769** 387-423
[6] Under the Hood: Scheduling MapReduce jobs more efficiently with Corona https://www.facebook.com/notes/facebook-engineering/under-the-hood-scheduling-mapreduce-jobs-more-efficiently-with-corona/10151142560538920
[7] Borthakur D 2013 Hdfs architecture http://hadoop.apache.org/docs/r1.2.1/hdfs_design.html
[8] Bockelman B 2009 Using Hadoop as a grid storage element *J. Phys.: Conf. Ser.* **180** 012047
[9] Riahi H *et al.* 2012 Using Hadoop File System and MapReduce in a small/medium Grid site *J. Phys.: Conf. Ser.* **396** 042050
[10] The Xen Project http://www.xenproject.org/
[11] Filesystem in userspace http://fuse.sourceforge.net/
[12] HDFS NFS Proxy https://github.com/cloudera/hdfs-nfs-proxy
[13] The MapR Distribution for Apache Hadoop http://www.mapr.com/Download-document/7-MapR-White-Paper

GPUs for real-time processing in HEP trigger systems

GAP Collaboration

R Ammendola[1], A Biagioni[1], L Deri[2], M Fiorini[3], O Frezza[1], G Lamanna[4], F Lo Cicero[1], A Lonardo[1], A Messina[1,5], M Sozzi[2,4], F Pantaleo[2,4], PS Paolucci[1], D Rossetti[1], F Simula[1], L Tosoratto[1] and P Vicini[1]

[1] INFN Roma,P.le A.Moro,2 - 00185 Roma - Italy
[2] Pisa University, Largo B.Pontecorvo,3 - 56127 Pisa - Italy
[3] Ferrara University, Via Saragat,1 - 44122 Ferrara - Italy
[4] INFN Pisa, laro B.Pontecorvo,3 - 56127 Pisa - Italy
[5] Roma University, P.le A.Moro,2 - 00185 Roma - Italy and CERN - Geneve - Switzerland

E-mail: `Gianluca.Lamanna@cern.ch`

Abstract. We describe a pilot project (GAP - GPU Application Project) for the use of GPUs (Graphics processing units) for online triggering applications in High Energy Physics experiments. Two major trends can be identified in the development of trigger and DAQ systems for particle physics experiments: the massive use of general-purpose commodity systems such as commercial multicore PC farms for data acquisition, and the reduction of trigger levels implemented in hardware, towards a fully software data selection system ("trigger-less"). The innovative approach presented here aims at exploiting the parallel computing power of commercial GPUs to perform fast computations in software not only in high level trigger levels but also in early trigger stages. General-purpose computing on GPUs is emerging as a new paradigm in several fields of science, although so far applications have been tailored to the specific strengths of such devices as accelerators in offline computation. With the steady reduction of GPU latencies, and the increase in link and memory throughputs, the use of such devices for real-time applications in high energy physics data acquisition and trigger systems is becoming relevant. We discuss in detail the use of online parallel computing on GPUs for synchronous low-level triggers with fixed latency. In particular we show preliminary results on a first test in the CERN NA62 experiment. The use of GPUs in high level triggers is also considered, the CERN ATLAS experiment being taken as a case study of possible applications.

1. Introduction

The scientific project described in this paper is based on the use of Graphic Processing Units (GPUs) for scientific computation. In recent years, the use of massively parallel computation is gaining ground in several fields of scientific research, in order to overcome some shortcomings of the present microprocessor technology. Modern super-computers are structured as arrays of several distributed computing units sharing data through dedicated high-bandwidth connections. Even processors in commodity personal computers (PCs) contain several computing cores.

Table 1. Characteristics of computing dedicated GPUs.

Video Card	n. of cores	Processing Power (GFLOPS)	Memory Bandwidth (GB/s)
NVIDIA TESLA C1060 (2009)	240	933	102.4
NVIDIA TESLA C2050 (2011)	448	1288	144
NVIDIA TESLA K20 (2013)	2496	3520	208
NVIDIA GFORCE GTX680 (2012)	1536	3090	192
AMD RADEON 5870 (2010)	1600	2720	153.6
AMD RADEON 7970 (2012)	2048	4096	288

GPUs, developed to speed up graphics-related computing tasks, are naturally organized around a parallel architecture. Because of such architecture, and the fact that most of the chip resources are devoted to computation - differently from a normal processor (CPU) in which a large fraction of those is required for other functions such as caching and handling of peripherals - allow to achieve a large computing power using a limited amount of space and power. GPUs are recently used also in applications very different from those for which graphic processors were originally conceived. So-called GPGPU (General Purpose computing on GPU) is nowadays common in several fields of scientific research requiring large computing power, such as hydrodinamical modeling, complex system simulation, image reconstruction for medical diagnostics. Major GPU vendors are strongly favouring this trend by exposing the raw computing power of graphic processors to users within a general-purpose paradigm. The two most widespread development frameworks and languages to program GPUs are $CUDA^{®}$ by NVIDIA® and OpenCL by the Kronos consortium. While the first one is devoted to the use of NVIDIA® devices, the latter was developed with the support of several vendors who want to allow a common approach to the programming of multi-core devices (both GPUs and CPUs).

Mainly due to the technological advances required by the massive computer game and video editing markets, current video processors provide computing power, in terms of operations per second (FLOPs) which are vastly larger than those offered by CPUs.

Table 1 shows the characteristics of some of the most recent GPUs on the market which are particularly suited to scientific computation. Besides the large number of cores, these processors have a large bandwidth to dedicated on-board memory, which is a key factor to allow a large computing throughput. On the other hand, one of the limitations for the use of GPUs in scientific computation is due to the need to access such devices through a PC bus which - despite the significant bandwidth provided today by the new PCI-express gen 3, up to $\sim 16GB/s$ - is often the most important bottleneck for real-time applications.

2. Real-time triggering in HEP

In High Energy Physics (HEP) experiments, the trigger system has a central role, which often determines the significance and the discovery potential. The trigger system must decide, usually based on limited and partial information, whether the physics event observed in a detector is interesting. The use of such real-time selection allows to optimize the usage of the finite data acquisition bandwidth, and to limit the amount of disk space required to store the data for the subsequent offline analysis stage. Online selection is performed by arranging the trigger system in a cascaded set of computation levels. The first, or the lowest, trigger level is usually implemented in hardware, often based on custom electronics tailored to the experiment and normally distinct from the data acquisition system. Nowadays these systems are often implemented using programmable logic (FPGA), which offer quite some flexibility and reconfiguration capability. In some cases, however, these kind of computations are performed on dedicated VLSI chips developed for the specific project [2], sacrificing flexibility and scalability for speed.

The following, "higher", trigger levels are today commonly implemented in software, using dedicated farms of commodity PCs. The use of commercial technology, computers and networks, allows to have very flexible and scalable trigger systems. The event rate which these trigger levels must handle has been already largely reduced by the upstream hardware trigger levels, allowing for complex decisions, based on more complete event informations, just by increasing the computing resources.

The time required to complete an event selection is not a big issue in higher trigger levels. The large amount of inexpensive memory (RAM) in today's PCs allow to store events for times of the order of seconds, thus relaxing the latency requirements for the algorithms. This is usually not the case for the lowest hardware trigger levels, since the memory buffers in the readout electronic boards, where incoming events are temporary stored before the trigger decision, often limit the maximum latency to tens of microseconds.

A better approach would be therefore one in which all data are brought directly to PCs, eliminating the hardware trigger levels. However, present HEP experiments must handle data rates so large to make such "triggerless" architecture unrealistic from a resource point of view: the bandwidth of general-purpose network systems is often insufficient to bring all data on PCs, and the raw computing power of individual PCs does not match, for complex computations, the required elaboration rates, to keep the number of PCs to affordable and manageable levels.

3. The use of GPUs in low-level triggers

Even if a fully "triggerless" approach is at the moment unrealistic, graphic processors represent a viable alternative to fill the gap between a multi-level trigger system with a hardware lowest level and a system which does not require any real-time hardware processing on reduced event information. Indeed GPUs do provide a large raw computing power on a single device, thus allowing to take complex decisions with a speed which can match significant event rates.

In a standard multi-level trigger architecture GPUs can be easily exploited in the higher software levels: being powerful computing devices, they can boost the capabilities of the processing system, thus allowing more complex computations to be performed without increasing the scale of the system itself. This is the case of the use of GPUs in the software trigger level (LVL2) of the CERN ATLAS experiment, as discussed later on.

The use of GPUs in lowest trigger levels, on the other hand, requires a careful assessment of their real-time performances. A low total processing latency and its stability in time (on the scales of "hard" real-time) are indeed requirements which are not of paramount importance in the applications for which GPUs have been originally developed. As mentioned above, the issue is related to the fact that in common usage of GPUs as graphics co-processors in computers, data is transferred to the GPU - and results are transferred back - through the PCI-express computer bus. Moreover, in order to better exploit the parallel GPU architecture, computing cores must be saturated, thus requiring the computation on a significant number of events after a buffering stage.

Considering a system in which data transfer occurs through a standard ethernet network, figure 1 shows the steps required for GPU computing, in a standard approach. Data from the detector reaches the Network Interface Card (NIC) which copies them periodically on a dedicated area in the PC RAM, from which they are copied to the user space memory where applications can process them. Here events are possibly buffered in order to prepare a sufficient data load for the following stages, and they are copied to GPU memory through the PCI express bus. The host (the PC on which the GPU card is plugged) has the task of launching the GPU kernel, which operates on the data. Computation results can be sent back to the host for further

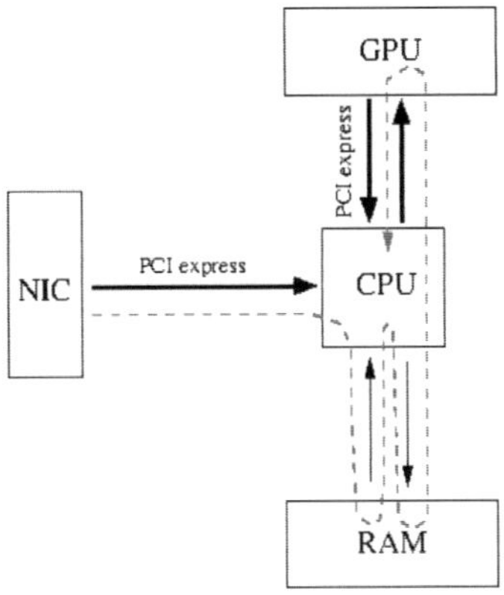

Figure 1. Schematic view of data transport from NIC to GPU

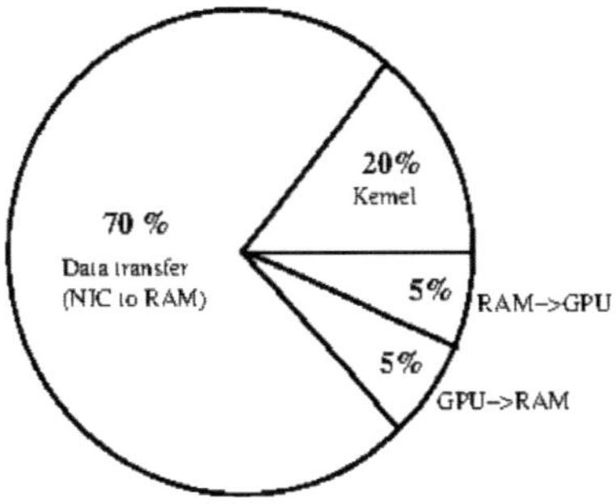

Figure 2. Time budget of data processing on GPU.

processing or distribution to the detectors, to ultimately trigger the reading of the complete data. Figure 2 shows, for a data packet of 1404 bytes, the budget for the processing time for a use case in which the kernel fits the data to a circle which, as discussed later, is a typical example of a computation required to extract significant information from raw detector data. In this example the most important contribution to the total latency is actually due to the data transfer latency from the NIC to the GPU memory. In order to reduce this contribution, in the project described here we studied two approaches: the use of a dedicated NIC device driver with very low latency (PFRING, by NTOP[1]) and the use of a direct data transfer protocol from a custom FPGA-based NIC to the GPU (Nanet); these two approaches will be described in detail in the following. The use of a custom NIC driver allows to perform the low-level trigger processing directly on a standard PC, but the results - in terms of total latency - might be affected by the PC characteristics (mother-board, CPU, etc.) because the data transfer time and its fluctuations are controlled by the host computer. In the second approach, instead, the data transfer is fully controlled by the custom NIC without any participation from the PC, at the price of using a non-standard hardware device.

3.1. Driver DNA-PFRING

Socket transmission on standard drivers does not guarantee the highest packet capture speed, especially if transmission is realized with small-size packets and high rate. As described above, the copy of packets from the NIC to the system buffers and then to the userland, is done by the CPU in two steps. PFRING is a new kind of socket that works in connection with the DNA (Direct NIC Access) driver, both developed by NTOP [1], in order to allow direct copy of packets from the NICs buffers to the memory through DMA (Direct Memory Access). A schematic drawing of the PFRING working principle can be found in Fig.3: using the DNA driver, data transfer is managed by the NICs NPU (Network Processing Unit) processor, through circular buffers that are directly accessible for the applications, and therefore for the copy to the GPUs memory. The scheduler that manages data transfer from the NIC to the GPUs memory has been implemented in a partially preemptive way in parallel streams. This allow to hide data transfer latency by exploiting concurrent copy-execution allows by last generation GPUs. Preliminary tests (Figure 4) show that this scheme improves by a factor higher than 2 the data transfer time and, most importantly, reduces fluctuations to a negligible level.

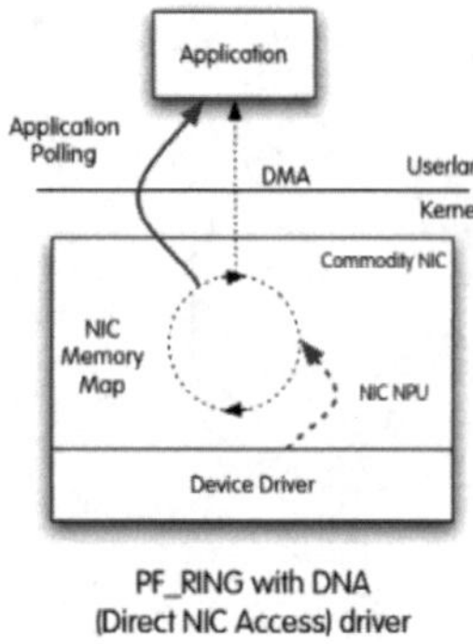

Figure 3. Scheme data processing using DNA driver (*courtesy of NTOP*).

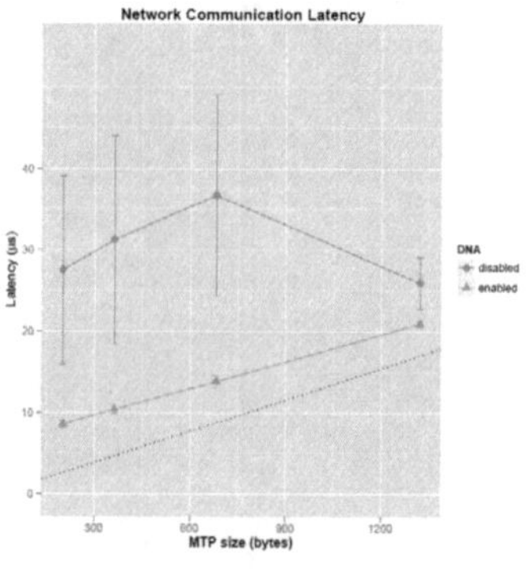

Figure 4. Comparison between standard data transmission and by using PFRING socket.

3.2. Nanet

The NaNet board [4] is a customized version of the APEnet+ [3] NIC designed to be integrated in the GPU-based low level trigger system of the NA62 RICH detector. Adding to the APEnet+ design the logic to manage a standard GbE interface, NaNet is able to exploit the GPUDirect P2P capabilities of NVIDIA Fermi/Kepler GPUs equipping a hosting PC to directly inject into their memory an UDP input data stream from the detector front-end, with rates compatible with the low latency real-time requirements of the trigger system.

In order to render harmless the unavoidable OS jitter effects that usually hinder system response time stability, the main design rule is to partition the system so that the hosting PC CPU can be offloaded from any data communication or computing task, leaving to it only system configuration and GPU kernel launch tasks. Within NaNet, this meant that data communication tasks were entirely offloaded to a dedicated UDP protocol-handling block directly communicating with the P2P logic: this allows a direct (no data coalescing or staging is performed) data transfer with low and predictable latency on the GbE link $\rightarrow$ GPU data path.

The `UDP OFFLOAD` block comes from an open core module [1] built for a Stratix II 2SGX90 development board. Focus of that design is the unburdening of the `Nios II` soft-core microprocessor onboard the Stratix II from UDP packet management duties by a module that collects data coming from the Avalon Streaming Interface (`Avalon-ST`) of the Altera Triple-Speed Ethernet Megacore (`TSE MAC`) and redirects UDP packets along a hardware processing data path. The `Nios II` subsystem executes the InterNiche TCP/IP stack to setup and tear down UDP packet streams which are processed in hardware at the maximum data rate achievable over the GbE network.

Bringing the open core into the NaNet design required some modifications, first of all the hardware code was upgraded to work on the Stratix IV FPGA family; this upgrade made available the improved performances of an FPGA which is two technology steps ahead in respect to the Stratix II.

The synthesis performed on a Stratix IV achieves the target frequency of 200 MHz (in the current APEnet+ implementation, the `Nios II` subsystem operates at the same frequency).

Current NaNet implementation provides a single 32-bits wide channel; it achieves 6.4 Gbps at the present operating frequency, 6 times greater than what is required for a GbE channel.

Data coming from the single channel of the modified `UDP OFFLOAD` are collected by the `NaNet CTRL`. `NaNet CTRL` is a hardware module in charge of managing the GbE flow by encapsulating

[1] NIOS II UDP Offload Example, `http://www.alterawiki.com/wiki/Nios_II_UDP_Offload_Example`

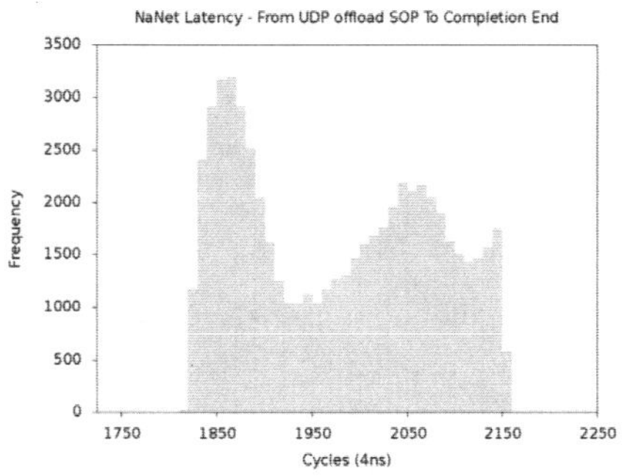

Figure 5. Distribution plot over 60000 samples of a NaNet packet traversal time.

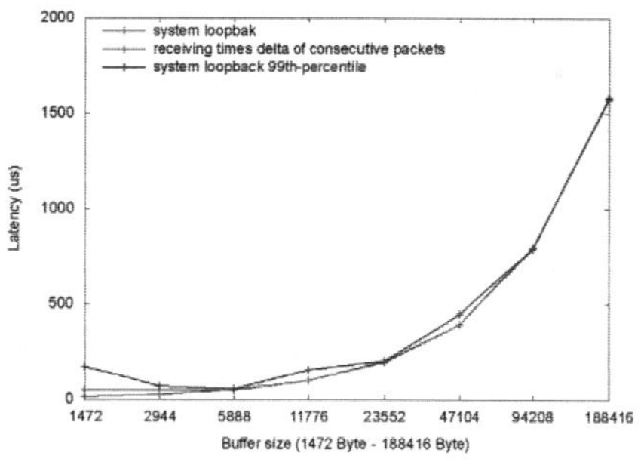

Figure 6. Average and 99-percentile NaNet latency vs. buffer size.

packets in the typical APEnet+ protocol (`Header`, `Payload`, `Footer`).

Benchmarks for latency and bandwidth were carried out. In order to measure the system latency and its fluctuations a "system loopback" configuration was used: connecting one GbE interface of the hosting PC to the NaNet, it was possible able to generate and receive a UDP stream in a single host process, measuring latency as the difference of host processor Time Stamp Counter register at send and receive time of the same UDP packet (see figure 6).

Latency inside the NIC was measured adding 4 cycles counters at different stages of packet processing; their values are stored in a profiling packet footer with a resolution of 4 ns; for a standard 1472 bytes UDP packet, traversal time ranges between 7.3 us and 8.6 us from input of `NaNet CTRL` to the completion signal of the DMA transaction on the PCIe bus (see figure 5).

For the same packet size, saturation of the GbE channel is achieved, with 119.7 MB of sustained bandwidth.

We foresee several improvements on the `NaNet` design:

- Implementing a custom logic dedicated to destination address calculation for the data receiving circular buffer in GPU memory, currently implemented by the `Nios II`, in order to lower latency and its fluctuations, especially at lower receiving buffer sizes.
- Adding a buffering stage to the `NaNet CTRL`, enabling the coalescing of consecutive UDP payload data into a single APEnet+ packet payload, improving bandwidth figure also with small-sized UDP packets.
- Increasing the number of supported GbE channels, to sustain eventually the higher bandwidth demands from the experimental requirements.
- Implementing a 10-GbE data link interface.

3.3. A physics case: NA62

The NA62 particle physics experiment at CERN [6] is considered as a use case for the study of the use of GPUs at the lowest trigger level. The NA62 trigger is organized in three levels: the first level is implemented on the very same FPGA-based boards which perform detector data readout [5], while the following two levels are implemented on PCs. The first hardware level (Level 0, L0) must handle an input event rate of order 10 MHz and, with a rather long maximum latency of 1 ms. It applies a rejection factor around 10, to allow a maximum input rate of 1 MHz to the second trigger level (Level 1, L1) implemented in software. This level, together with the following and last one (Level 2, L2), must reduce the rate to about 10 kHz in order to allow permanent data storage for later offline analysis.

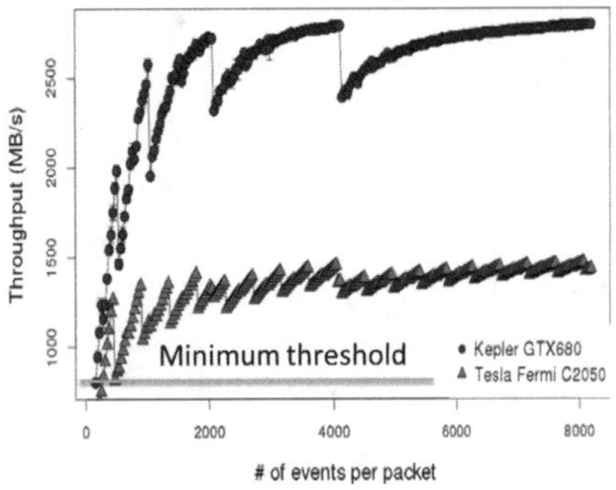

Figure 7. Throughput as a function of number of events for last generation GPUs.

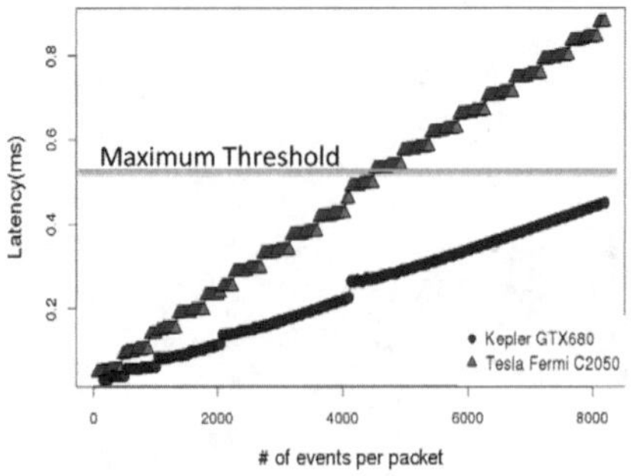

Figure 8. Total latency (including data transfer and computing).

In the standard implementation of L0, the trigger primitives contributing to the building of the final trigger decision are computed on the readout board FPGAs, and are very simple conditions, mostly based on the event hit pattern or multiplicity. The use of processors, such as GPUs as discussed in this paper, would rather allow to build more complex physics-related trigger primitives, such as energy or direction of the final state particles in the detectors.

As a first use case, we studied the possibility of reconstructing, in GPUs, the ring-shaped hit patterns in a RICH Cerenkov detector. Such detector, described in ref.[7], can provide a measurement of the velocity and direction of charged particles (such as muons and pions) which traverse it, thus contributing to the computation of other physical quantities such as the decay vertex of the K^+ and the missing mass. The use of such information allows to implement highly selective trigger algorithms also for other interesting decay modes.

We studied different ring reconstruction algorithms in order to assess which ones are the most suited to a GPU parallel implementation.

As described in ref.[8] the "math" algorithm, based on a simple coordinate transformation of the hits which reduces the problem to a least square procedure, was found to be the best one in terms of computing throughput (for single rings). This algorithm was implemented and tested on different GPUs, such as the NVIDIA Tesla C1060, Tesla C2050 and GeForce GTX680 (in increasing order of processing core generation). The computing performance of the C2050 and GTX680 proved to be a factor 4 and 8 higher than that of the C1060. In figure 7 we show the computing throughput for these devices as a function of the number of events processed in one batch. The effective computing power is seen to increase with the number of events to be processed in one go; the horizontal line shows the requirement related for an online trigger based on the RICH detector in NA62.

Figure 8 shows instead (for NVIDIA Tesla C2050 and GeForce GTX680 devices) the total latency, which includes data transfer times to and from the GPU and the kernel execution time. The significant reduction of the latency for the newer GTX680 GPU is due to the faster data transfer due to the presence of the gen.3 PCI express bus. Also in this case the maximum latency allowed by the NA62 application is seen to be attainable when a reasonable number of events is processed in one batch.

4. Application of GPUs in High Level Triggers

The ATLAS trigger [9] has performed remarkably well so far. The constant increase of the LHC luminosity up $3 \times 10^{34} Hz/cm^2$ and the successive upgrade at $1 \times 10^{35} Hz/cm^2$ requires a constant renovation of the trigger strategy. Currently, a first upgrade is foreseen in 2018 [10], when realtime tacking capabilities will also be available, followed by a complete renovation of the trigger and detector systems in 2022. The ATLAS trigger system is organized in 3 levels. The first-level trigger (LVL1) is built on custom electronics, while the second-level (LVL2) and the event-filter (EF) are implemented in software algorithms ran on commodity PC farms. The LVL2, based on the concept of the Region-of-iterest (RoIs), offers a natural study case for the deployment of GPUs in the realtime environment of a LHC experiment. The LVL2 algorithms are now implemented as approximated solutions of complex primitives such to stay within the time budget. The deployment of GPUs will allow to better exploit the potentiality of the detectors, implementing refined algorithms with higher selection efficiency. Such a result will improve the sensitivity to interesting physics signals and the ability to cope with an higher rate of multiple pp interactions (pileup). A fundamental aspect to exploit the computational performances of the GPUs is the possibility of parallelizing the execution of the algorithms.

5. Conclusions

The use of graphics processing units in scientific computing has become very common in the last years. The GAP Project [11], recently funded, will study the GPUs in real-time HEP trigger systems and for medical imaging. In particular, for the lowest trigger levels, work has to be done in order to reduce the contributions to the total latency due to data transfer from the detectors to the GPU. Two strategies are being pursued at the moment: the first makes use of a special driver that allows direct copy of the data from the NICs buffers avoiding redundant copies; the second one foresees to use a FPGA-based board for establishing a peer-to-peer connection with the GPU. Possible applications of this technique are the low level trigger of the NA62 Experiment at CERN, in particular for the reconstruction of photon rings in the RICH detector. Preliminary results show that current GPUs are suitable for sustaining the rate and minimizing the latency to an acceptable level for the experiment. As possible applications to high trigger, as a study case, we will consider a possible extension of the muon trigger of the ATLAS Experiment at CERN.

Acknowledgments

The GAP project is partially supported by MIUR under grant RBFR12JF2Z "Futuro in ricerca 2012". This work is partially supported by the EU Framework Programme 7 project EURETILE under grant number 247846.

References

[1] *http://www.ntop.org*
[2] Annovi A, Beretta M, Volpi G, Beccherle R, Bossini E, Crescioli F, Dell'Orso M and Giannetti P et al. 2011 *IEEE Nucl. Sci. Symp. Conf. Rec.* 141
[3] Ammendola R, Biagioni A, Frezza O, Lo Cicero F, Lonardo A, Paolucci PS, Rossetti D and Simula F 2012 *J. Phys. Conf. Ser.* 396
[4] Lonardo A et al. 2013 *http://on-demand.gputechconf.com/gtc/2013/presentations/S3286-Low-La tency-RT-Stream-Processing-System.pdf*
[5] Avanzini C et al. 2010 *Nucl. Instrum. Meth. A* 623
[6] Lamanna G 2011 *J. Phys. Conf. Ser.* 335 .
[7] Angelucci B, Anzivino G, Avanzini C, Biino C, Bizzeti A, Bucci F, Cassese A and Cenci P et al. 2010 *Nucl. Instrum. Meth. A* 205
[8] Collazuol G, Lamanna G, Pinzino J and Sozzi MS 2012 *Nucl. Instrum. Meth. A* 662
[9] Atlas Collaboration 2008 *JINST* 3 P08003
[10] Atlas Collaboration 2012 CERN-LHCC-2011-012
[11] *http://web2.infn.it/gap*

High performance computing activities in hadron spectroscopy at BESIII

Beijiang Liu (for BESIII collaboration)
Institute of High Energy Physics, CAS, Beijing, China
E-mail: liubj@ihep.ac.cn

Abstract. BESIII experiment is operated in the τ-charm threshold energy region. It has collected the world's largest data samples of J/ψ, $\psi(3686)$, $\psi(3770)$ and $\psi(4040)$ decays. These data are being used to make a variety of interesting and unique studies of light hadron spectroscopy, precision charmonium physics and high-statistics measurements of D meson decays. As one of the experiments at the high luminosity frontier, data processing at BESIII is computationally very expensive for large data sets. In this presentation, we report two recent progresses in using high performance computing: a Tag-based preselection for data reduction and GPUPWA, a PWA framework harnessing the GPU parallel computing.

1. Introduction

The fundamental theory of the strong interactions (Quantum Chromodynamics, QCD) predicts that there should exist new forms of matter, such as glueballs (pure-gluon objects) and hybrids (qqbar states with explicit gluon) [1]. Experimental search of these predictions and subsequent investigation of their properties would provide validation of and valuable input to the quantitative understanding of QCD. The experimental study of hadron spectroscopy is to systematically map out all the resonances with the determination of their properties like mass, width, spin-parity as well as partial decays widths. The key to success lies in high levels of precision during the measurement and high statistics in the recorded data set complemented with sophisticated analysis methods.

BESIII (Beijing Spectrometer) [2] is a new state-of-the-art 4π detector at the upgraded BEPCII (Beijing Electron and Positron Collider) [3] that operated in the τ-charm threshold energy region between (about) 2 and 4.6 GeV with a designed luminosity of $10\times10^{33}cm^2s^{-1}$. The BESIII experimental program addresses issues in light hadron physics, charmonium spectroscopy and decays, D and Ds meson decays, and numerous topics in QCD and τ physics [4]. To date, BESIII has accumulated data samples corresponding to 1.3 billion J/ψ decays, 0.6 billion $\psi(3686)$ decays and 2.9 fb at the peak of the $\psi(3770)$ resonance, which decays to $D\bar{D}$ meson pairs nearly 100% of the time. These are all world's-largest data samples at these c.m. energies and the J/ψ sample is the first ever to be collected in a high quality detector like BESIII.

Generally, a typical physics analysis can be categorized into three phases as shown in Table 1.

In this presentation, we report two recent progresses of using high performance computing in BESIII physics analysis: a Tag-based pre-selection for data reduction and the GPUPWA, a PWA framework for high statistics partial wave analyses harnessing the GPU parallel computing.

Table 1. Summary of the systematic errors from the event selection.

Phase	Purpose	Task	Bottle neck
Algorithmic Analysis	Reduction: event selection, record useful information	Processing reconstruction data	I/O, CPU
Interactive Analysis	Final selection, plotting, interactive studies	Processing Ntuple/Tree	I/O
Statistical Analysis	Extract physics	Fit	CPU

2. TAG-based preselection

The typical branching fractions for interesting physics channels are of the order of $O(10^{-3})$ [4, 5]. The traditional event-wise accessing of BOSS (Bes Offline Software System) [6] is not effective for the selective accessing with the rate of $O(10^{-3})$. Event tags (TAG) are event-level metadata which support efficient identification and selection of events of interest to a given analysis. TAG content includes, for example, run number, high level trigger decisions, event properties (like visible energy), and particle data (photons, charged tracks, electrons, muons, pions, kaons, protons, *etc*). To facilitate queries for event selection, TAG data can be stored in ROOT files or a relational database. The current TAG size is approximately 0.1 kilobyte per event. The computing cost of pre-selection with TAGs is much less than the pre-selection with the full reconstruction events. TAGs allow jobs to process only those reconstruction events that satisfy a given predicate. In the current implementation, the TAG data for every event is stored in the same reconstruction file as an individual ROOT tree. C++ modules are written using the TAG attributes for event selection. No payload data is retrieved for unselected events and data files containing only unselected events are not accessed. To improve the I/O performance of the event selection jobs, Hadoop and HDFS [7] could be employed for job management and distributed storage. The analysis algorithms are able to access the TAG data and reconstruction data from the local disks.

3. GPUPWA

Extracting resonance properties from experimental data is however far from straightforward; resonances tend to be broad and plentiful, leading to intricate interference patterns. In such an environment, simple fitting of mass spectra is usually not sufficient and a partial wave analysis (PWA) is required to disentangle interference effects and to extract resonance properties. In the cases discussed here, the full kinematic information is used and fitted to a model of the amplitude in a partial wave decomposition. The partial wave amplitude is constructed with an angular part and a dynamical part. The model parameters are determined by an unbinned likelihood fit to the data, while the event-wise efficiency correction is included. In a typical PWA (we use the radiative decay $J/\psi \to \gamma\eta\eta$ [12] as an example), the quasi two-body decay amplitudes (isobar model) in the sequential decay process $J/\psi \to \gamma X, X \to \eta\eta$, are constructed using covariant tensor amplitudes described in Ref. [8]. The basis of the likelihood fitting is that a hypothesized probability density function (PDF) would produce the data set under consideration. The probability to observe the event characterized by the measurement ξ is :

$$P(\xi) = \frac{\omega(\xi)\epsilon(\xi)}{\int d\xi \omega(\xi)\epsilon(\xi)}, \tag{1}$$

where $\epsilon(\xi)$ is the detection efficiency and $\omega(\xi) \equiv \frac{d\sigma}{d\Phi}$ is the differential cross section, and $d\Phi$ is the standard element of phase space.

$$\frac{d\sigma}{d\Phi} = |\sum_{j}^{N_{wave}} \Lambda_j A_j(\xi)|^2, \tag{2}$$

where A_j is the partial wave amplitude with coupling strength determined by a complex coefficient Λ_j. The normalization integral is performed numerically by Monte Carlo techniques. The likelihood for a particular model is

$$\mathcal{L} = \prod_{i=1}^{N_{data}} P(\xi_i). \tag{3}$$

A series of likelihood fits are performed for parameter estimation and model evaluation. In the log likelihood calculation, the likelihood value of background events are given negative weights, and are removed from data since the log likelihood value of data is the sum of signal and background.

Most computing problems in particle physics are trivially parallel, as data consists of independent events. The traditional approach to parallelism is thus to treat different subsets of events on different cores/machines/in different locations and then have a relatively lightweight piece of code perform a synthesis. For PWA, this synthesis (namely the likelihood sum) has to be performed very frequently (for every fit iteration) and thus network latency quickly dominates for distributed architectures. As the calculations per event are relatively simple and exactly the same for all events, it is desirable to have a very large number of (simple) cores available in a single machine.

GPUPWA [9] has been developed as the working framework of BESIII, harnessing GPU parallel acceleration. GPUPWA is now developed with *OpenCL* [10] as described in [11]. The performance is improved by O(100) of magnitude comparing to FORTRAN implementation used in BES II (without multi-threading).

The framework now provides facilities for amplitude calculation, minimization and plotting and is widely used for analyses (e.g. [12, 13]) at BES III. It continues to be developed and is available at [14]. A cluster of 35 computing nodes with 140 GPUs is established to serve the PWA at BESIII.

4. Summary

Now five years from our first collisions, BESIII has established a broad and successful program in charm physics. Recently, in 2012, even larger samples have been accumulated at the J/ψ and $\psi(3686)$; total samples are now about 1.2 billion and 0.35 billion decays, respectively. Furthermore, our 2013 dataset includes more data near 4260 MeV, and also a large sample at the Y(4360). With the excellent performance of the accelerator and detector, more interesting results are expected. High performance computing techniques will play a more and more important role in the physics analysis of high statistics experimental data.

5. Acknowledgements

This presentation is partly supported by Youth Innovation Promotion Association, CAS.

References

[1] Close F E 1988 *Rept. Prog. Phys.* **51** 833
Godfrey S and Napolitano J 1999 *Rev. Mod. Phys.* **71** 1411
Amsler C and Tornqvist N A 2004 *Phys. Rept.* **389** 61
Klempt E and Zaitsev A 2007 *Phys. Rept.* **454**, 1
Crede V and Meyer C A 2009 *Prog. Part. Nucl. Phys.* **63** 74

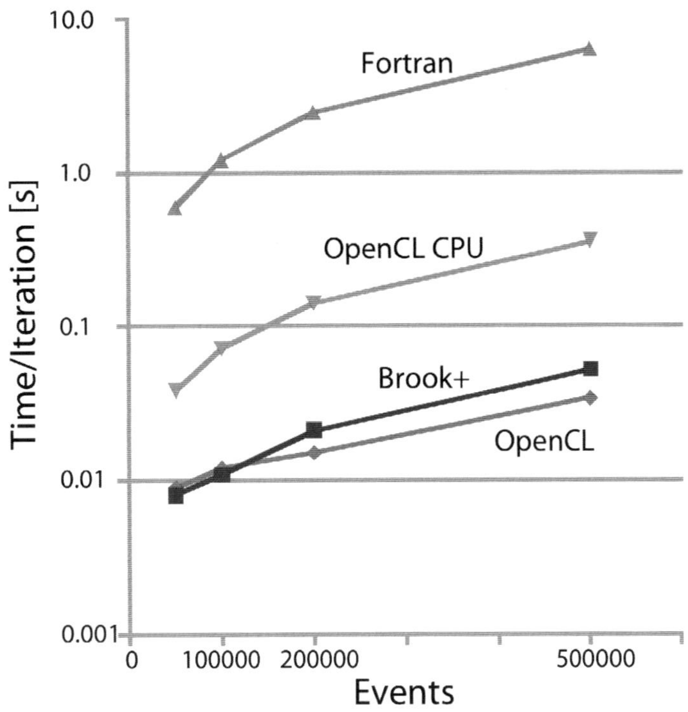

Figure 1. Comparison of the running time of one fit iteration of the benchmark $J/\psi \to \gamma K^+K^-$ analysis in different implementations on the same machine. Both *OpenCL* datasets were obtained using the AMD implementation. The detailed information for the setup is described in Ref. [11].

[2] Ablikim M *et al* (BESIII Collaboration) 2010 *Nucl. Instrum. Meth. A* **614** 345
[3] Bai J Z*et al* (BES Collaboration) 1994 *Nucl. Instrum. Meth. A* **344** 319
Bai J Z*et al* (BES Collaboration) 2001 *Nucl. Instrum. Meth. A* **458** 627
[4] Asner D M, Barnes T, Bian J M, Bigi I I, Brambilla N, Boyko I R, Bytev V and Chao K T *et al* 2009 *Int. J. Mod. Phys. A* **24**, S1
[5] Beringer J *et al* (Particle Data Group) 2012 *Phys. Rev. D* **86** 010001
[6] Li W 2006 *Proc.Int.Conf.Comput.High Energy and Nucl.Phys* 225
[7] `http://hadoop.apache.org/`
[8] Zou B S and Bugg D V 2003 *Eur. Phys. J. A* **16** 537
[9] Berger N Liu B J and Wang J K 2010 *J. Phys. Conf. Ser.* **219**, 042031
[10] `http://www.khronos.org/opencl/`
[11] Berger N 2011 *AIP Conf. Proc.* **1374** 553
[12] Ablikim M *et al* [BESIII Collaboration] 2013 *Phys. Rev. D.* **87** 092009
[13] Ablikim M *et al* [BESIII Collaboration] 2012 *Phys. Rev. Lett.* **108** 112003
[14] GPUPWA package, available at `http://sourceforge.net/projects/gpupwa/`

Initial explorations of ARM processors for scientific computing

David Abdurachmanov[1], Peter Elmer[2], Giulio Eulisse[3], Shahzad Muzaffar[3]

[1] Digital Science and Computing Center, Faculty of Mathematics and Informatics, Vilnius University, Vilnius, Lithuania

[2] Department of Physics, Princeton University, Princeton, NJ 08540, USA

[3] Fermilab, Batavia, IL 60510, USA

E-mail: `Peter.Elmer@cern.ch`

Abstract. Power efficiency is becoming an ever more important metric for both high performance and high throughput computing. Over the course of next decade it is expected that flops/watt will be a major driver for the evolution of computer architecture. Servers with large numbers of ARM processors, already ubiquitous in mobile computing, are a promising alternative to traditional x86-64 computing. We present the results of our initial investigations into the use of ARM processors for scientific computing applications. In particular we report the results from our work with a current generation ARMv7 development board to explore ARM-specific issues regarding the software development environment, operating system, performance benchmarks and issues for porting High Energy Physics software.

1. Introduction

The computing requirements for high energy physics (HEP) projects like the Large Hadron Collider (LHC) [1] at the European Laboratory for Particle Physics (CERN) in Geneva, Switzerland are larger than can be met with resources deployed in a single computing center. This has led to the construction of a global distributed computing system known as the Worldwide LHC Computing Grid (WLCG) [2], which brings together resources from nearly 160 computer centers in 35 countries. Computing at this scale has been used, for example, by the CMS [3] and ATLAS [4] experiments for the discovery of the Higgs boson [5, 6]. To achieve this and other results the CMS experiment, for example, typically used during 2012 a processing capacity between 80,000 and 100,000 x86-64 cores from the WLCG. Further discoveries are possible in the next decade as the LHC moves to its design energy and increases the machine luminosity. However, increases in dataset sizes by 2-3 orders of magnitude (and commensurate processing capacity) will eventually be required to realize the full potential of this scientific instrument. The scale and longevity of the LHC computing require continual R&D into new technologies which may be relevant in the coming years. In this paper we report on our investigations into one such technology, low power ARM processors, for scientific computing.

2. Processor Architectures and Power Efficiency

The construction of the WLCG was greatly facilitated by the convergence around the year 2000 on commodity x86 hardware and the standardized use of Linux as the operating system for scientific computing clusters. Even if multiple generations of x86 hardware (and hardware from both Intel and AMD) are provided in the various computer centers, this was a far simpler situation than the typical mix of proprietary UNIX operating systems and processors.

Until around 2005, a combination of increased instruction level parallelism and (in particular) processor clock frequency increases insured that performance gains expected from Moore's Law would be seen by single sequential applications running on a single processor. The combination of Linux, commodity x86 processors and Moore's law gains for sequential applications made for a simple software environment.

Since around 2005, however, processors have hit scaling limits, largely driven by overall power consumption [7]. The first large change in commercial processor products as a result of these limits was the introduction of "multicore" CPUs, with more than one functional processor on a chip. At the same time clock frequencies ceased to increase with each processor generation and indeed were often reduced relative to the peak. The result of this was that one could no longer expect that single, sequential applications would run faster on newer processors. However in the first approximation, the individual cores in the multicore CPUs appeared more or less like the single standalone processors used previously. Most large scientific applications (HPC/parallel or high throughput) run in any case on clusters and the additional cores are often simply scheduled as if they were additional nodes in the cluster. This allows overall throughput to continue to scale even if that of a single application does not. It has several disadvantages, though, in that a number of things that would have been roughly constant over subsequent purchasing generations in a given cluster (with a more or less fixed number of rack slots, say) now grow with each generation of machines in the computer center. This includes the total memory required in each box, the number of open files and/or database connections, increasing number of independent (and incoherent) I/O streams, the number of jobs handled by batch schedulers, etc. The specifics vary from application to application, but potential difficulties in continually scaling these system parameters puts some pressure on applications to make code changes in response, for example by introducing thread-level parallelism where it did not previously exist.

There is moreover a more general expectation that the limit of power consumption on future Moore's Law scaling will lead to more profound changes going forward. In particular, the power hungry x86-64 "large" cores of today will likely be replaced by simpler and less power hungry "small" cores with a greater emphasis on aggregate throughput performance per watt, rather than just raw performance. This has rekindled interest in solutions that would lead back to a more heterogeneous computing environment.

3. ARM Processors

A strong contender for this evolving low power (high performance/watt) server market is the ARM processor [8] due its nearly complete dominance in the low power mobile market for smartphones and tablets, which has also seen dramatic growth since around 2005. The size of the mobile market, and its traditional focus on low power, has led to interest in using these processors also in a server environment. As such ARM-based server products such as Boston Viridis [9] are starting to appear.

The ARM processor has a long history [10] dating back to Acorn Computers and the early days of personal computers. It is a RISC processor and the current generation (ARMv7), used in most high-end mobile devices and the new server products, is a 32bit processor. We are interested in the "A" series of general purpose "Application" processors. (ARM also produces "R" and "M" series designs for use in real-time and embedded microcontroller environments, respectively.) A 64bit version of the ARM processor (ARMv8) has also been designed and is expected to appear in server products from fall 2013 or early 2014. Intel has also announced the development of products (Silvermont) aimed at a low power market. However, much like the current mobile market, it isn't positioned to dominate the low power server market as it has dominated the commodity processor market in the past.

As the ARM processors are general purpose and run Linux, only a standard port of the CMS software is required, similar to what was done, for example, to port the CMS software from 32bit (ia32) to 64bit (x86-64). Such a port is reasonably straightforward relative to the changes required to use other high performance per watt solutions (e.g. GPGPU's, which require actual software rewrites), thus the effort required for these initial investigations was also relatively modest.

4. Test Setup

For the tests described in this paper we used a low-cost development board, the ODROID-U2 [11]. The processor on the board is an Exynos 4412 Prime, a System-on-Chip (SoC) produced by Samsung for use in mobile devices. It is a quad-core Cortex A9 ARMv7 processor operating at 1.7GHz with 2GB of LP-DDR2 memory. The processor also contains an ARM Mali-400 quad-core GPU accelerator, although that was not used for the work described in this paper. The board has eMMC and microSD slots, two USB 2.0 ports and 10/100Mbps Ethernet with an RJ-45 port. Power is provided a 5V DC power adaptor.

The cost of the board alone was $89 and with the relevant accessories (cables, a cooling fan, a 64GB eMMC storage module, etc.) the total cost was $233. This extremely modest cost permitted us to do meaningful initial investigations without investing in a full-fledged server.

All build tests were done using a 500GB 3.5″ ATA disk connected via USB. Runtime tests were done with output written to the eMMC storage.

For the Linux operating system on the board we used Fedora 18 ARM Remix with kernel version 3.0.75 (provided by Hardkernel, the vendor for the ODROID-U2 board) due to its similarities to Scientific Linux CERN (SLC). It is fully hard float capable and uses the floating point unit on the SoC. The kernel was reconfigured to enable swap devices/files, which is required for CMSSW compilation. A 4GB swap file was used in our build environment.

In order to compare results from the ARM board we also used two typical x86-64 servers currently deployed at CERN. The first is a dual quad-core Intel Xeon L5520 @ 2.27 GHz (Nehalem) with 24GB of memory. The second is dual hexa-core Intel Xeon E5-2630 @ 2.00GHz (Sandy Bridge) with 64GB of memory. Both machines were equipped with a large local disk for output and used software installed on an afs filesystem at CERN. These machines were purchased about three years apart and very roughly represent the range of x86-64 hardware being operated at the time of our ARM tests.

5. Experimental Results

5.1. The CMSSW software stack

The software written by the CMS collaboration itself (CMSSW) consists of approximately 3.6M source lines of code (SLOC), as measured by the SLOCCount tool [12]. The entire software stack includes also 125 "external" packages, including HEP software packages like ROOT [13] (~1.7M SLOC), Geant4 [14] (~1.2M SLOC) and many general open source packages: GCC, boost, Qt, Python, etc.

5.2. Build platform for ARM tests

For the compilation and linking of this large set of software we considered three options: compilation directly on the ODROID-U2 board itself, cross compilation for the ODROID-U2 board from an x86-64 host and compilation for the ODROID-U2 board from an emulation environment such as QEMU [16] running on an x86-64 host

Our experience with ARM emulation with QEMU prior to purchasing the ODROID-U2 led us to believe that it was not yet quite mature enough to provide a stable build environment sufficient for our needs. Even though it was clear that the small ODROID-U2 development board is much less powerful than most standard x86-64 servers, we decided to attempt compilation of the full stack directly on the ARM board. This was motivated by the idea that should we eventually adopt ARM as a standard production architecture, we would probably aim for direct compilation on ARM servers.

5.3. Issues arising during the ARM port

During the port to ARMv7 we had to resolve a number of issues in the software and build recipes:

Oracle: We did not have Oracle libraries for the ARMv7 architecture. By construction, however, no standard Grid-capable CMS applications depend on Oracle. Thus this affected only a small number of CMSSW packages used primarily for writing calibrations into the Oracle database at CERN. CMS applications which read calibrations do not interact directly with Oracle, but instead access the calibrations via the FroNTier web service [17], with no direct dependency on Oracle.

Configuration: There were a number of minor configuration issues, for example:

- The `-m32` and `-m64` options do not work. (On x86-64 CMS had made a complete transition to 64bit a couple of years prior to this work.)
- In a number of places there were x86-based assumptions leading to attempts to configure for x86-64 SSE and AVX

Memory use: Compilation of some translation units (primarily generated ROOT dictionaries) exhausted the virtual memory address space. Here the solution was simply to refactor the dictionaries.

ROOT Cintex: A patch was needed for the Cintex trampoline in ROOT to support the ARM architecture. This was submitted to the ROOT developers.

Signedness: x86-64 and ARM treat the signedness of char/bit-fields differently, Intel is signed and ARM is unsigned by default. This was dealt with on ARM by imposing the use of the compiler options `-fsigned-char` and `-fsigned-bitfields`, along with a few small code modifications to fix non-portable code.

Dictionary generation and I/O: There were several bugs in the ROOT I/O infrastructure, as well as non cross-platform types (that crept in after CMS transitioned to 64bit on x86-64 and stopped regularly producing 32bit builds), that at time of these tests prevented us from properly reading and writing ROOT files.

5.4. Build times

With changes to the build recipes resulting from fixing these issues we were able to build all of the standard CMSSW externals and 99% of the CMSSW code. The CMSSW code which did not build was the small subset requiring Oracle. This demonstrates the advantages of relying primarily on open source software and, when closed source software cannot be avoided, carefully restricting the code which can depend on the closed source libraries.

After making the changes described above, we achieved the following results for the total build times directly on the ODROID-U2 board:

- ~4h - compilation of a "bootstrap" kit consisting of the GCC compiler (version 4.8.0) and a small set of packages (rpm, apt, zlib, ncurses, nspr, sqlite, etc.) that we use for packaging and distributing the results of our builds
- ~12h - compilation of all of the 125 external packages not included in the "bootstrap" kit
- ~25.5h - compilation of the CMSSW code itself as well as a set of generated ROOT dictionaries

These are quite reasonable results. Taking into account that the externals do not change frequently, these results are already very close to consistent with an eventual "nightly" integration build where we compile the very latest versions of all of the CMSSW code, but reuse pre-existing builds of the externals.

5.5. Run time tests

For a run time test and benchmark we used an actual CMS application from the build described above rather than a synthetic benchmark. This application performs a Monte Carlo simulation of 8 TeV LHC Minimum bias events using Pythia8 [18] (event generation) followed by simulation with Geant4 [14]. Due to the problem mentioned above with generation of the dictionaries used for data input/output, output was turned off. The data output however has little effect on the total CPU time (and thus these benchmarks) as the CPU cost is heavily dominated by the Geant4 simulation. The application itself is single-threaded (sequential) and thus ran on a single core at a time. To simplify testing we ran only a single job at a time on each machine with the aim of measuring the single core performance. Multiple tests were performed of sufficient length to estimate properly an average per event time and dedicated tests were run to subtract off job startup times. A proper validation of the application output was complicated by the lack of an output file, but checks done by enabling printout indicated consistency between ARM and x86-64. The results for performance (events simulated per minute per core) are shown in Table 1.

Table 1. Results of run time tests

Type	Cores	Power (TDP)	Events /minute /core	Events /minute /Watt
Exynos4412 Prime @ 1.704 GHz	4	4W	1.14	1.14
dual Xeon L5520 @2.27GHz	2×4	120W	3.50	0.23
dual Xeon E5-2630L @2.0GHz	2×6	120W	3.33	0.33

In order to calculate values for the performance per watt, it would be a bit misleading to compare the total power used by real (x86-64) servers with a small development board. To get a better estimate, which more directly compares the processors themselves, we used the "thermal design power" (TDP) numbers for the processors themselves. Here TDP numbers for the two Intel Xeon x86-64 processors were taken from their website [15]. For the ODROID-U2 we were not able to find specific TDP numbers, but based on our own measurements we have estimated the TDP-equivalent fully loaded power at about 4W. Using these values we have calculated an equivalent performance (in events per minute) per watt in the last column of Table 1. A clear advantage in terms of performance per watt is seen for the ARM processor.

6. Conclusions

We have done a port of the entire CMS software stack, including 3.6M SLOC of CMS-written code in CMSSW and 125 external support packages, to the ARMv7-based ODROID-U2 development board. We chose to build directly on the development board itself and measured build times consistent with operating a "nightly" build of the CMS software. We report performance and performance per (TDP) watt numbers both for the ARMv7 board and for two typical x86-64 servers at CERN. On the basis of these results we conclude that ARM-based low power servers, if they succeed in the market, show great potential for use with typical HEP high throughput computing applications.

Acknowledgment

This work was partially supported by the National Science Foundation, under Cooperative Agreement PHY-1120138, and by the U.S. Department of Energy.

References

[1] Evans L and Bryant P 2008 LHC Machine *JINST* **3** S08001
[2] Bird I 2011 Computing for the Large Hadron Collider *Annual Review of Nuclear and Particle Science* **61** 99-118
[3] Chatrchyan S et al (CMS Collaboration) 2008 The CMS experiment at the CERN LHC *JINST* **3** S08004
[4] Aad G et al (Atlas Collaboration) 2008 The Atlas Experiment at the CERN Large Hadron Collider *JINST* **3** S08003
[5] Chatrchyan S et al (CMS Collaboration) 2012 Observation of a new boson at a mass of 125 GeV with the CMS experiment at the LHC *Phys.Lett.* **B716** 30-61
[6] Aad G et al (Atlas Collaboration) 2012 Observation of a new particle in the search for the Standard Model Higgs boson with the ATLAS detector at the LHC *Phys.Lett.* **B716** 1-29
[7] Fuller S H and Millet L I (Editors) 2011 *The Future of Computing Performance: Game Over or Next Level?* The National Academies Press.
[8] `http://www.arm.com/products/processors/index.php`
[9] `http://www.boston.co.uk/solutions/viridis/default.aspx`
[10] `http://www.arm.com/about/company-profile/milestones.php`
[11] Puechased via the Hardkernel website: `http://www.hardkernel.com`
[12] Source code and documentation available at `http://www.dwheeler.com/sloccount/`
[13] `http://root.cern.ch`
[14] Agostinelli S et al 2003 Geant4 - a simulation toolkit *Nuclear Instruments and Methods in Physics Research* **A 506** 250-303
[15] `http://ark.intel.com/products/40201/` and `http://ark.intel.com/products/64586/`
[16] `http://www.qemu.org`
[17] Blumenfeld B, Dykstra D, Lueking L and Wicklund E 2008 CMS Conditions Data Access using FroNTier *J. Phys.: Conf. Ser.* **119** 072007
[18] Sjostrand T, Mrenna S and Skands P 2008 A Brief Introduction to PYTHIA 8.1 *Comput. Phys. Commun.* **178** 852-867

JAliEn - A new interface between the AliEn jobs and the central services

A.G. Grigoras[1], C. Grigoras[1], M.M. Pedreira[1], P. Saiz[1], S. Schreiner[2]

[1] CERN, 1211 Geneva, Switzerland

[2] Technische Universität Darmstadt, Hochschulstraße 10, 64289 Darmstadt, Germany

E-mail: alina.grabriela.grigoras@cern.ch, costin.grigoras@cern.ch, miguel.martinez.pedreira@cern.ch, pablo.saiz@cern.ch, steffen.schreiner@cased.de

Abstract. Since the ALICE experiment began data taking in early 2010, the amount of end user jobs on the AliEn Grid has increased significantly. Presently 1/3 of the 40K CPU cores available to ALICE are occupied by jobs submitted by about 400 distinct users, individually or in organized analysis trains. The overall stability of the AliEn middleware has been excellent throughout the 3 years of running, but the massive amount of end-user analysis and its specific requirements and load has revealed few components which can be improved. One of them is the interface between users and central AliEn services (catalogue, job submission system) which we are currently re-implementing in Java. The interface provides persistent connection with enhanced data and job submission authenticity. In this paper we will describe the architecture of the new interface, the ROOT binding which enables the use of a single interface in addition to the standard UNIX-like access shell and the new security-related features.

1. Introduction

The ALICE [1] experiment uses AliEn (ALIce ENvironment) [2] middleware to leverage heterogeneous computing resources and provide users with a unified view of the system. It allows users to execute jobs wherever there are free resources matching the job requirements and transparently access existing data or upload job results to any of the 70 disk and tape storage elements.

The system currently has a set of databases at its core, holding the details of all experiment files, all jobs to be executed and the scheduled data transfers. All interactions with the databases go through a set of central services that enforce authentication, authorization and access to the stored objects and also to the physical files stored on the distributed storage infrastructure.

Current AliEn is assembled from about 200 packages written in Perl, C and C++, with two distinct channels of communicating to the central services: via web services or through xrootd-based communication channels. Similarly there are two distinct command line user interfaces, a full Perl implementation for site administration and a mix of bash scripting and C/C++ code for normal users.

In order to reduce the complexity of this system and streamline the operation and maintenance of this environment we have started to rewrite the core functionality in Java, profiting from the built-in serialization, easy to use fully TLS [5]-based communication, object caching and other powerful features. The result of these efforts is the jAliEn[3] Grid middleware project. Below are the details of the implementation so far and the plans for the future development needed in order to fully replace the current communication channels.

2. System architecture

The entire system is now a single Java package that maps each logical entity from the current AliEn implementation to a corresponding serializable object encapsulating the details (LFN, GUID, PFN, SE, JDL, Job, principals, quotas etc). To retrieve such objects an API based again on serializable objects extending a base Request class is offered to the users, similar to a Java Remote Method Invocation [4]. These requests encapsulate requester identity which is checked on the authoritative entities. Furthermore, some of the requests can implement a Cacheable interface, allowing them to be cached in memory of the various components and returned immediately to the requester instead of being forwarded for resolving to the authoritative entities. This is only done for long-lasting objects like SE details or GUID replica locations.

2.1. Overall layout

Services layout follows closely the current AliEn services layout:

- JCentral: a central component that handles the direct access to the core databases
- JSite: a connection multiplexing and caching service running on trusted site machines
- JBox: end-user (or job) service handling client-side authentication and upstream connections

In Figure 1 below the current interaction model in AliEn is presented in comparison with the Java implementation.

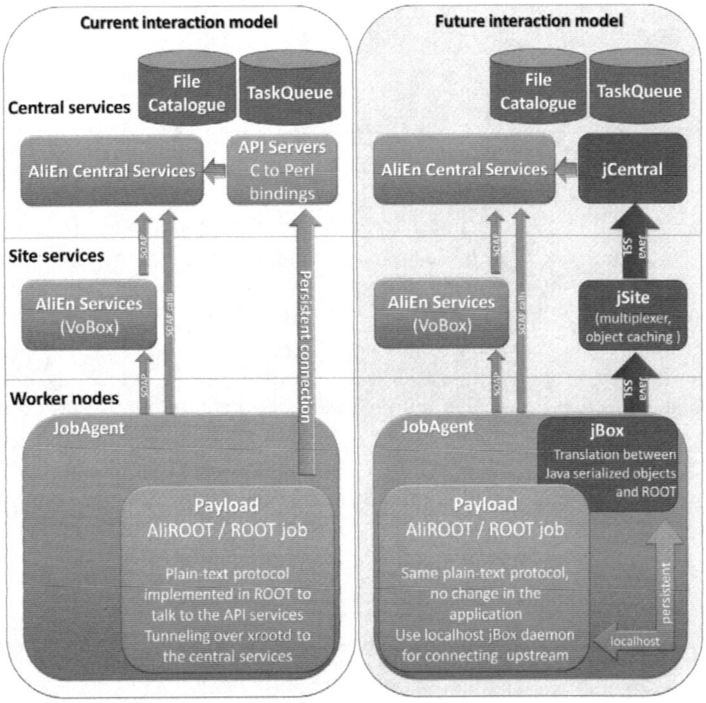

Figure 1. Current and future interaction models in AliEn

Communications of jAliEn components are exclusively based on mutually authenticated and encrypted connections based on the TLS [5] protocol while using the Bouncy Castle Crypto API [6]. The authentication is established using the Grid user's X.509 [7] certificate and X.509 host certificates on service side.

2.2. JCentral

This component is responsible for handling the direct database connections and being the authoritative source to resolve any API requests. In practice the service listens on a single TCP port and expects TLS [5] connections to be established.

The AliEn principle is established based on the DN of the certificate used to connect to the service. This identity is then tagged to any Request object received on the ObjectStream. Any JCentral instance is able to answer any received Request, allowing the service to be horizontally scaled on as many central machines as needed. This is different from the current AliEn services where different capabilities are served by different services listening on different TCP ports, also with different security constraints and server technologies.

The aim was to have a minimalistic API and so far there are 25 distinct Request object types that cover all aspects of interacting with the central databases: getting logical and physical file details, performing list and find operations, manipulating catalogue entities, managing jobs and getting trace information and so on. As such it is up to the client to split a more complex request in a suite of such atomic operations in order to perform the entire command. For example a command to submit a job based on the specifications in a logical file (JDL) present in the catalogue would be translated in series of requests to the central services to locate the physical copies of the file, to get the access tokens to them, to actually retrieve the content, parse it on the client side and interpret the requirements, do the same for other files referenced from the JDL and finally prepare the entry that is to be injected in the central task queue. This delegates an important part of the processing to the entity that requests the service, while currently all these operations are executed centrally by the AliEn services and it frequently happens for them to accumulate a backlog of requests to process since the processing depends on external entities.

2.3. JSite

In order to avoid establishing persistent connections from each entity in the environment (which currently would be more than 50000 concurrent jobs and some hundreds of users, but at least a 4x increase has to be foreseen) this entity is used to multiplex requests on a single upstream connection. The same TLS [5] mechanism is used to decorate the requests, with the difference that the upstream connection will trust that the identity was verified by this layer and will not overwrite it any more. This implies that the services will run in a trusted environment, similarly to how the current AliEn site services are running, on site-local machines dedicated to the experiment.

In addition to forwarding the Request objects upstream and dispatching the replies to the corresponding channel this service is also able to hold Cacheable-implementing objects in memory and returns the cached reply directly to the client, as long as the entry is still valid. The time to live is set by the authoritative service (JCentral). All services that see this object will cache it in memory, so that if a similar request is generated within the lifetime of the reply it can be resolved directly in the memory of the client.

While this layer is designed to be cascaded indefinitely, in practice a single layer of services distributed on the site VOBox-es is enough to handle the site-generated message traffic (since even the largest sites currently run less than 10000 jobs to one VOBox). This leaves the option to scale the services either horizontally (several service instances on large sites) or vertically (for example with regional concentrators).

2.4. *JBox*

This service runs as a daemon on the end-user machine or in the environment of each job. It handles the authentication upstream based on the full X.509 certificate. This is again different from the current procedure in AliEn, requiring Grid users to periodically create a password-less X.509 proxy certificate [8], which conceptually represent temporary certificates as powerful as the main certificate itself. It will also require one-time X.509 certificates to be handed to each job for authentication as well, while currently jobs authenticate with a shared secret which is a unique token associated in the central task queue to every job ID.

JBox acts like ssh-agent, running as a daemon on the user machine and creating a temporary file in which connection details of the current instance are stored (local TCP port, shared secret). The file is protected with regular permissions that only allow the same user to read its content (0400). JBox will only establish upstream connections (to JSite or directly to JCentral) when it receives a command to execute. Idle connections are automatically closed after a timeout in order to reduce the load on the upstream services. A configurable timeout can be set on the authentication itself, to force the user to resupply the key password before being able to continue interacting with the environment. This improves the security over the current system by not having any unprotected certificate or proxy certificate stored persistently on the client's file system at any moment.

JBox listens on a local-machine TCP port only (the first available one in the 10100 and 10200 range). Along with the TCP port the process ID of the current daemon is written in the same temporary file. Subsequent attempts to start a service for the same account verify if the specified service is still running and exit immediately in that case as a protection against wrong commands.

This component is also responsible for parsing the commands received from the user prompt and executing the intended operations on the client side. For this it makes use of dynamic class loading in Java to instantiate a class with a name derived from the invoked command and pass it the arguments received from the command line. This way it is very easy to add new commands either in the core or by each user in its private installation. Each of these command classes implements a help function that exposes all expected or accepted arguments.

From the local host this component can accept any number of connections expecting a very simple text-based protocol. The actual front-end can then be anything from a plain terminal (like telnet) to a custom-made shell or (on a slightly different implementation) ROOT with its TAliEnGrid [9] wrapper. After passing the shared secret, further lines are interpreted as full commands to be executed and parsed as described above. Commands are processed sequentially on the same connection, but concurrent connections are not blocking each other so different application or shell instances can operate independently.

Included in the package, a custom shell component called Jsh provides the expected command line editing, history and tab-completion of commands. Jsh verifies upon startup if a JBox service is available and otherwise triggers the according service to start in background, while interacting with the user to supply the certificate's password in order to allow the initialization of the upstream connection within the service.

3. Standalone Grid middleware for certified Grid jobs

Beyond the integration of jAliEn in AliEn as a client component for Grid users and jobs, jAliEn was envisioned as a complete future Grid middleware, based on the AliEn schema of databases and directory services. In [10][11], these capabilities of jAliEn providing certified Grid jobs and files based on a new pilot job mechanism using cascaded digital signatures were presented.

4. Performance

As a reference point for the mechanism's scalability, the presented communication framework was benchmarked based on LFN queries to the file catalogue, equivalent to the information necessary to request for a simple directory listing. The testing involved the full stack of mutually authenticated and encrypted communication and object serialization.

The performed test cases were established using three test machines: one machine (Intel Core 2 Duo CPU 2.80GHz, 8GB RAM, solid-state drive) providing all central Grid service instances, including databases and a directory service, establishing a fully independent lab test environment. And two machines as Grid clients '*clientA*' (Intel i5 CPU 2.50GHz, 8GB RAM, solid-state drive) and '*clientB*' (Intel Core 2 Duo CPU 1.86GHz, 4GB RAM, solid-state drive). In three test cases, both machines executed 20 concurrent threads, each querying for all LFNs within a certain directory in a loop, each thread simply issued the same queries one after another, for duration of exactly 10 minutes. The queried directory contained 20 file entries (LFNs), therefore a client received all information existent within the catalogue concerning a certain LFN entry (e.g. parent and child node IDs, owner and permissions, file checksum, date of creation, size, physical replicas etc) as a whole. For each directory request the wall time was taken, while ignoring any processing or displaying of the respective information on client side and without any caching mechanisms in place since the purpose was to load the central services as much as possible.

Figure 2 below shows the results of the first test case with 20 concurrent threads per machine, based on 2 operating system processes with 10 internal threads each, all of which are requesting the respective LFNs in loop. In Figures 3 and 4 the same logic is used to stress the central services from two machines, on a larger scale. Figure 3 shows the response time with 300 concurrent threads per machine respectively 500 concurrent threads in Figure 4. In both cases there were 50 threads per client process instance, so 6 and respective 10 processes as client instances. In all test cases, the connection to the central service is established once per process, accordingly all internal threads of a process share one pre-established connection. This scenario mimics the JSite component, since the component has a single uplink and aggregates requests from locally executed Grid user interfaces or jobs.

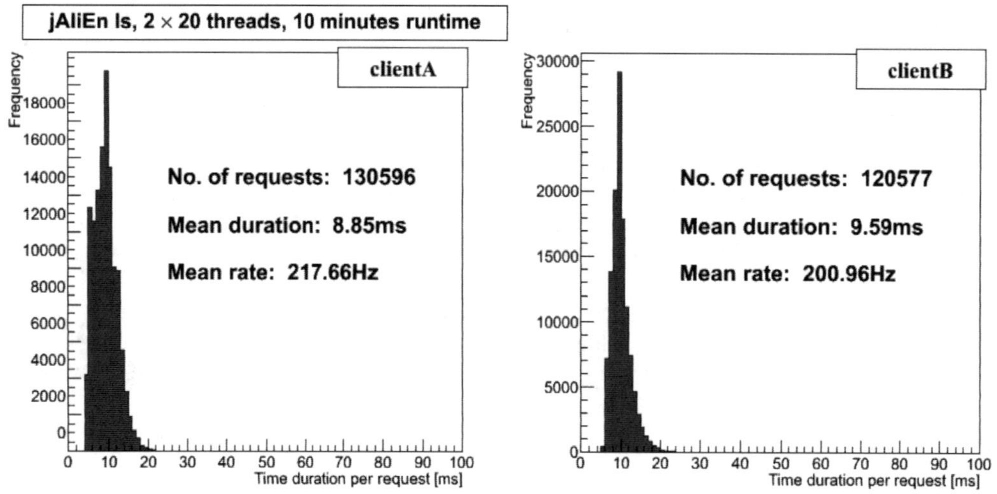

Figure 2. Two machines with each 20 concurrent requests for directory entries in loop.

Concerning the first case with 20 concurrent request loops per machine, the two clients were each able to receive more than 120k request responses successfully with a mean time of less than 10 milliseconds.

In the second test case with 300 requesting threads, the mean time for a request was approx. 27 milliseconds and the bulk of requests stayed below 90 milliseconds. In the third test case with 500 threads, the average request time was approx. 44 milliseconds and the bulk of requests were executed in less than 150 milliseconds.

In all three cases the jAliEn central service instance was able to serve requests at a rate of more than 400Hz (418.62Hz, 436.59Hz and 431.91Hz, which are the combined frequencies from the two clients that were run simultaneously in each test). As a reference point for comparison: the busiest AliEn central service API instance serves on average requests at approx. 250Hz and up to 400Hz during a peak load period with more than 50k concurrently active Grid jobs.

Despite the test setup using standard non-high-performance hardware and the fact that all relevant central service instances in the test environment were running on one single system, it was still able to match the performance of the production servers (HP DL380 G8 servers with 24 cores each).

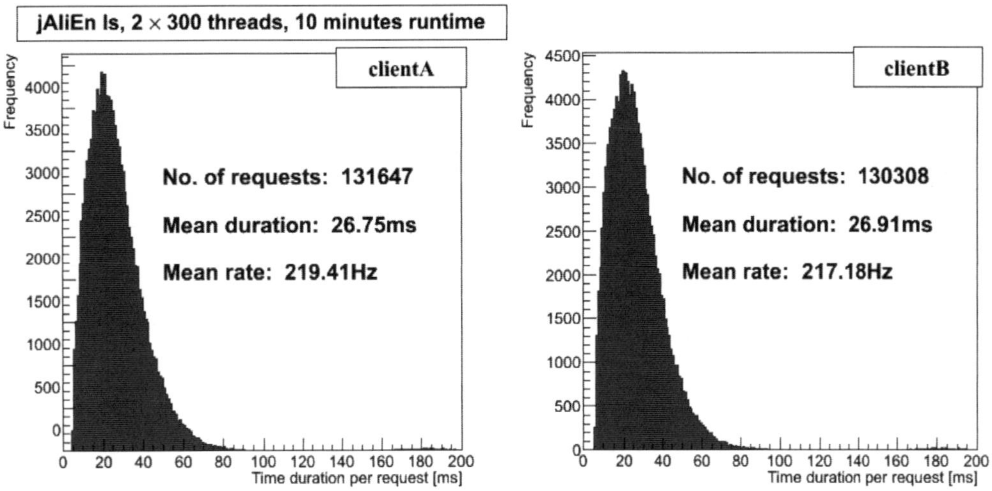

Figure 3. Two machines with each 300 concurrent requests for directory entries in loop.

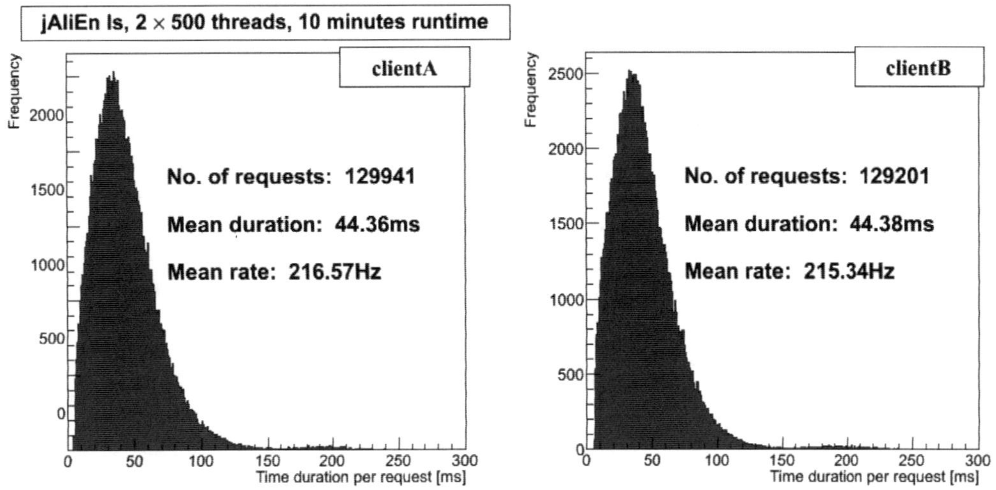

Figure 4. Two machines with each 500 concurrent requests for directory entries in loop.

5. Summary

In this paper we have presented the architecture, main features and performance tests of the new implementation of communication channels in jAliEn. The new middleware approach written in Java using serialized objects is highly efficient and will allow the system to run more jobs in the future.

Due to the utilization of only mutually authenticated and fully encrypted communication while providing single sign-on functionality based on only X.509 certificates (and thereby setting aside X.509 proxy certificates) significant security improvements concerning Grid client and service systems as well as Grid credentials can be achieved. Finally, jAliEn combines these security features with a high-level object-driven Grid communication protocol and a huge potential in terms of performance and scalability.

6. References

[1] ALICE - A Large Ion Collider Experiment. http://aliceinfo.cern.ch/ .

[2] AliEn. http://alien2.cern.ch/ .

[3] jAliEn. http://jalien.cern.ch/ .

[4] Oracle Corporation. Java Remote Method Invocation – Distributed Computing for Java http://www.oracle.com/technetwork/java/javase/tech/index-jsp-138781.html .

[5] T. Dierks and E. Rescorla. The Transport Layer Security (TLS) Protocol Version 1.2. RFC 5246 (Proposed Standard), August 2008.

[6] The Legion of the Bouncy Castle. Bouncy Castle API. http://www.bouncycastle.org/ .

[7] D. Cooper, S. Santesson, S. Farrell, S. Boeyen, R. Housley, and W. Polk. Internet X.509 Public Key Infrastructure Certificate and Certificate Revocation List (CRL) Profile. RFC 5280 (Proposed Standard), May 2008.

[8] S. Tuecke, V. Welch, D. Engert, L. Pearlman, and M. Thompson. Internet X.509 Public Key Infrastructure (PKI) Proxy Certificate Profile. RFC 3820 (Proposed Standard), June 2004.

[9] Andreas Joachim Peters. AliEn Grid User Reference Guide V1.0. http://project-arda-dev.web.cern.ch/project-arda-dev/alice/apiservice/ .

[10] Steffen Schreiner, Costin Grigoras, Alina Grigoras, Latchezar Betev and Johannes Buchmann. A security architecture for the ALICE Grid Services. Proceedings of Science (PoS), The International Symposium on Grids and Clouds, ISGC2012:027, 2012.

[11] Steffen Schreiner, Costin Grigoras, Maarten Litmaath, Latchezar Betev and Johannes Buchmann. Certified Grid Job Submission in the ALICE Grid Services. Open Access Journal of Physics: Conference Series (JPCS), vol. 396, part 3, 032096, 2012.

Less is more. Why Oberon beats mainstream in complex applications

F V Tkachov

Institute for Nuclear Research, Russian Academy of Sciences, Moscow 117312 Russia

E-mail: ftkachov@inr.ac.ru

Abstract. The second part of the two-part account of the enabling technologies behind a successful completion of the Troitsk-nu-mass experiment, summarizes an almost 20 years' experience of using Oberon/Component Pascal in complex applications ranging from the cutting-edge computer algebra (b-quark decays etc.) to experimental data processing (neutrino mass etc.) to exploratory algorithm design work (the optimal jet finder etc.) to systematic computer science education (the international project Informatika-21).

1. Introduction

The currently best ν mass bound was published by the Troitsk-ν-mass experiment [1]. This happened after the experiment had been plagued by the so-called "Troitsk anomaly" for about 10 years. The anomaly melted away after a reanalysis that was made possible by two enabling technologies: the statistical method of quasioptimal weights for parameter estimation (see a companion talk [2]) and the Oberon technologies for software development reviewed in the present talk.

In regard of software, everything could in theory have been done with fortran or even C++ but at a much higher cost. A proper *technique* is about minimizing effort while attaining the best result; such a technique is rooted in how well one understands the basic principles of what one is doing. The success of the reanalysis proves soundness of the adopted approach towards software development.

In fact, a complete reanalysis of the Troitsk-ν-mass data [1] was not planned. The initial purpose was only to implement a new statistical method [2] and to play with real data with a view to clarify the treatment of the Poisson background as a test of the new method. The software development tool was chosen to be a derivative of the Oberon System [3] known as the BlackBox Component Builder [4], [5], [6]. It was known from a previous experience with the BEAR computer algebra engine [7] to be easy to use, efficient, safe, flexible and producing a clean compiled code (a combination of features shared by all Oberons). The resulting piece of software proved — as a result of adhering to the principles of the underlying system [3], [8] — to be easy to operate, safely extensible, adaptable and efficient enough that the testing of the new method grew into a complete reanalysis of the data. The successful completion of [1] within the severe manpower and other limitations that the Troitsk-ν-mass team operated under, is proof of the sound rationality behind the minimalistic Oberon as opposed to the prodigal complexity of the mainstream tools.

The software used in [1] was designed to be generic enough so that it could be used as a supplemental material for my university programming course. The course is part of the educational project «Informatika-21» [9]) that promulgates scientific rationality for the professional and general (physicists', linguists' etc.) IT education in the form of a unified backbone of algorithmics/programming courses spanning students of ages 11-20 and based on a single perfectly designed programming platform — Oberon. In the capacity of coordinator of the «Informatika-21» project, I enjoyed the privilege of discussing the matters of software development in general and programming education in particular, with a considerable number of excellent experts (rocket scientists, compiler optimization gurus, UAV designers, school teachers, university professors, etc.; cf. [10]). As a result, my views of the subject have sharpened compared with the earlier talk [5].

2. The IT bubble of excessive complexity

I would like to discuss software development from the point of view of a "non-professional" programmer, i.e. one for whom computer programming is not a full-time job but part of application problem solving. That using a computer inevitably leads to a full-blown programming in complex applications is seen from the fact that all significant software packages develop script languages. Such languages are usually poorly designed, inefficient and unreliable. They also limit access to the low level programming and thus do not allow one to utilize the full power of the computer. The resulting two-tier programming model (an "efficient" system level with the notoriously unsafe C/C++ and a somewhat less unreliable but inefficient application level with a multitude of special purpose languages) represents an unnecessarily complicated environment that prevents non-professional programmers from exploiting their knowledge of the application problem domain to the full. There is, as a result, a vast grey area where programming problems are excessively hard for non-professionals to tackle with the complicated mainstream tools, yet a recruitment of a professional programmer cannot be justified.

The surging complexity of the IT industry tools is visualized in the following figure (based on one borrowed from [12])

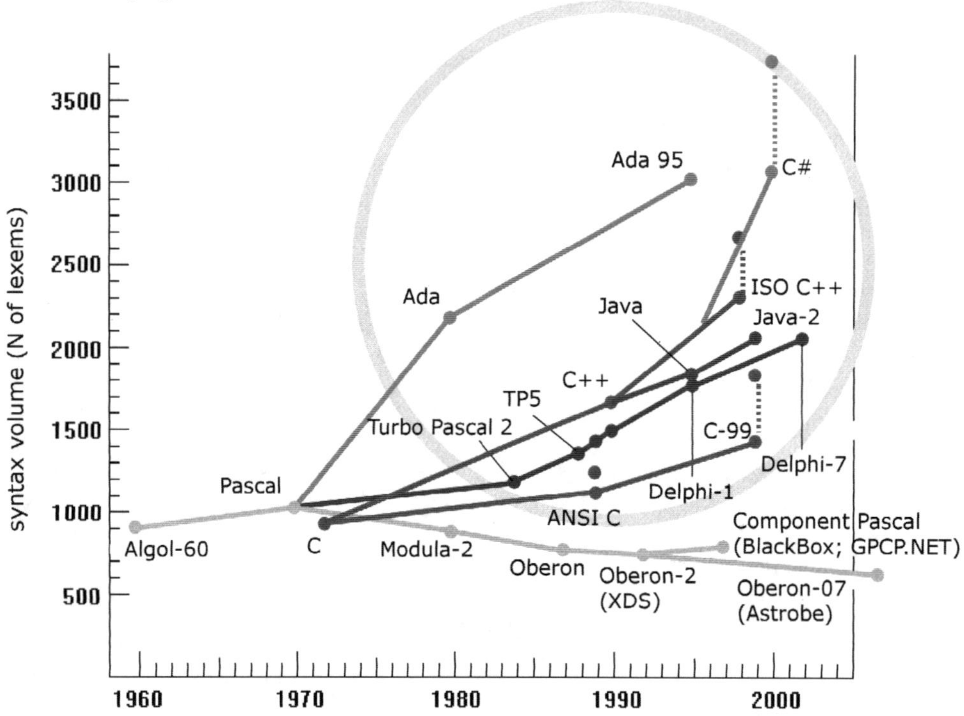

Figure 1. The syntax volume of some programming languages.

The figure is not intended to compare the complexity of various programming languages but only to convey an idea of the IT industry's growing bubble of complexity. The figure is complemented by the so-called Wirth's Law stating that the software gets slower faster than the hardware gets faster.

The only line that goes down in the figure is the one corresponding to the languages designed by Niklaus Wirth and his school at ETH, Zürich, who for many years have been combatting the complexity of programming and programming languages. But let us discuss the bubble first; it is important to understand its nature because, quite obviously, some fundamental forces are at play.

3. Understanding the IT complexity bubble

There are two causes of the excessive IT complexity bubble visualized in Figure 1.

1. The first cause is the **combinatorial nature of normal intellect**. (The adjective normal is chosen to correlate with the normal science of Thomas Kuhn [13].) This was observed in a multitude of experiments conducted by primatologists and zoopsychologists starting from Wolfgang Koeler in the early XX c. This means that the basic mechanism of intellect is to find a (finite) combination of known actions with the objects within view in order to achieve a desired goal. If a combination is not found whereas the goal remains in view, one gets restless and roams around, which brings new objects into the scope of one's attention thus increasing chances of finding the solving combination. Thus, the normal intellect tends to pile up things and objects, to add patches and props to bring the prospective solution closer to the target — a phenomenon well seen e.g. by observing how kids do their programs. They are reluctant to go back to their program in order to simplify it once it seems to do what is required. A habit of cleaning up the mess — which may require a critical ability in order to separate what is important and to be left from what is not and to be discarded — has to be learned either by painful experiences (the Kalashnikov Principle: Excessive complexity is vulnerability) or via a proper training. The parallels with Figure 1 are quite obvious, with an additional complication that the cleaning up of the mess in that case may mean redesigning the language. Such a redesign the industry — together with their customers — are reluctant to do for stated reasons of compatibility, even if a redesign would in the long run be beneficial to all.

A related point worth making here to explain the steep rise of the curves in Figure 1, is that with the IT, the normal/combinatorial intellect has for the first time in the history of Humanity been largely freed from the restrictions of the resistance of materials.

2. The other, social reason is that the excessive complexity creates what economists call **asymmetry of information**. This serves as a source of an extra economic and social rent derived by IT experts (cf. the Nobel Prize for economics awarded to Akerlof, Spence and Stiglitz in 2001 for their analysis of the market of used cars with "the bad driving out the good" in the market). It is not, for example, uncommon to observe a youngster who proudly pontificates about software development and programming educaton in virtue of having simply acquired an amount of knowledge about obscure features of an overly complex mainstream tool. The youngster's less computer savvy companions may indeed start to regard and treat him as a true expert. This example is actually a paradigmatic one for a good part of the IT industry.

Alchemy preceded chemistry and astrology preceded astronomy. In the absence of an established IT education system based on a scientific foundation, there is no reason why the fast expanding IT sphere should be any different. Indeed, the bubble visualized in Figure 1 testifies to the fact that the IT industry passes through a parascientific stage characterized by a paradoxical if implicit assumption that the growing complexity of programming languages and tools might help one solve the key problem of software development — the one of controlling the complexity of software.

4. What is Oberon/Component Pascal

Oberon remains the only systematic research project to have successfully identified a minimal rational core of technologies for efficient software development. It perhaps deserves to be called anti-language due to its contrast with the typically bloated mainstream languages.

Oberon descends from the old Pascal of 1970 via Modula-2 (in fact, Modula-2 might have been described as Pascal-80, and Oberon, as Pascal-86). However, one should beware of myths and wrong associations because Oberon is a fundamentally more powerful language than either Pascal or Modula-2 because of its automatic garbage collection. That feature, however, does not prevent it from being a fast compiled language: the garbage collection is only invoked when dynamic memory allocation is used. Oberon is also a fundamentally safer language (no segviols, ever) because it extends the strict static type safety mechanism to dynamic records.

There is a number of Oberon dialects (cf. the fact that the exact position of an optimum may be hard to pinpoint precisely due to its instability with respect to minor changes of the optimum criteria): the original ETH Oberon [3], Active Oberon [14], Oberon-2 (available as XDS Oberon [15]), Component Pascal (a 1997 refinement of Oberon-2 from Oberon microsystems, Inc. available with the BlackBox Component Builder [4] as well as the Gardens Point Component Pascal compiler for Microsoft.NET [16]), Oberon-07 (with a 2011 revision available from [17]). Most of the above are available in open source. All these languages are very small and for the purposes of the present talk will be referred to as Oberon/Component Pascal.

— Oberon/Component Pascal can be described as a "vanishing" imperative programming language with everything that can be put away into libraries excluded from the language; the remaining language core is designed with utmost care, with unlimited, safe and efficient extensibility in mind.

— The language is very small, its syntax rules fit on one page, and the language report is only about twenty pages long plus or minus a few pages depending on the dialect.

— A very fast one-pass compiler produces a clean compiled code with no compiler-generated overhead which normally requires an optimization to get rid of as is the case e.g. with C++, (Floating point optimization is available via an external tool [15].)

— The language is highly readable in the Pascal tradition, and carefully designed to be robust against typing errors.

— It is statically type safe, including dynamic records.

— It features independently compiled modules that serve as units of information hiding, are dynamically linked and (un)loaded; as a result, the programming system gives one an interactive feel despite being based on a true compiler.

— It is object-oriented via a very efficient basic mechanism of single record extension.

— It is automatically garbage collected, which together with record extensibility makes it a synthetic language with flexibility comparable to that of functional languages (as proved e.g. by the BEAR computer algebra engine [5], [7]).

A popular argument in the discussions of programming languages is that functional languages are higher level than imperative ones. However, farther from hardware does not necessaritly mean closer to human, as is easily seen when one tries to teach kids. In fact, high level application problems are synthetic in nature, combining elements of imperative, functional and markovian (substitutions-based) paradigms. The hardware, however, is imperative, so a synthetic language based on the imperative paradigm, such as Oberon, offers a clear advantage in regard of efficient use of the hardware.

A superior readability and semantic transparency of Oberon are also relevant for the concept of open source. Indeed, access to program sources is essential for its adapation to changing requirements. The most adaptable is the code that is the most accessible. The most accessible code is one written in the simplest language. Ergo, the Oberon code is more open than the C++ code in proportion to how Oberon is simpler and more readable than C++.

5. Why Oberon technologies rather than just a programming language

Every tool — even as simple as a hammer — requires a proper technique to use it efficiently, i.e. to achieve the best results with minimal effort and maximal safety. Tools and techniques come together, and a proper balance is key. (If one is not aware of one's technique, it is probably best described as chiropractic.)

This observation is highly relevant for a complicated tool such as programming language. To make the most of Oberon the language one should use a proper development environment and possess a set of basic algorithmic skills (see below).

An Oberon development environment supports the *text-as-interface* paradigm: any text document can serve as a command prompt, and any text fragment can be designated as program input (similar to, but much more flexible than the command line parameters). Then text documents can serve as flexible, customizable menus, and simultaneously as storage for various sets of parameter values — a feature of great usefulness in applications involving parameter fitting, optimizations, etc.

Lastly, instead of learning defects and peculiarities of a big language, one invests effort into basic algorithmics [18], into learning Dijkstra's method to construct complex loops [19], and basic architectural patterns (Carrier-Rider, separation of interface from implementation, the message bus, etc.).

6. Bridging efficiency and flexibility

A popular conference topic in the 80s was the so-called numerical-analytical interface. The fundamental problem here is that numerical applications that are usually handled with fortran etc. tend to require flexibility of dynamic data structures as more adaptive intellect is built into the algorithms. Such a flexibility, however, used to be only available via the functional languages. On the other hand, large-scale computer algebra tasks, usually handled via dynamic languages, may benefit greatly from compiled routines. This is illustrated by the following figure:

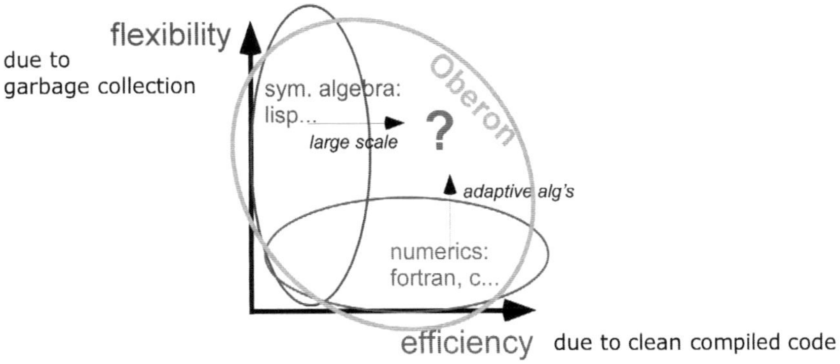

Figure 2. The flexibility *vs* efficiency dilemma.

One small, simple and reliable Oberon covers both bases, and does what is usually achieved via complex (and inherently less reliable) combinations such as C++ and python, or fortran and Mathematica, etc.

The success of Java at the expense of C++ in the late 90s proved the latter's vulnerability due to its complexity — in a complete agreement with the above mentioned Kalashnikov Principle. Yet the high-energy physics community chose C++ as a successor to fortran. Note that by 1992, the Oberon System that never had a segviol had been ported to a dozen platforms, an optimizing XDS compiler was already available, and in 1993 the full Oberon functionality was made available in the popular operating systems. In view of these facts and after 20 years' experience of software development, it is hard not to regard the adoption of C++ as a gross failure of the HEP community as a ***scientific*** community.

Journal of Physics: Conference Series **523** (2014) 012011 doi:10.1088/1742-6596/523/1/012011

7. Conclusions

A central problem of software development is controlling the complexity of software. The problem is partly rooted in the way the human brain operates and is fostered by the asymmetry of information in the absence of a proper programming education system. **Containing the gratuitous growth of complexity must be a permanent concern whenever the information technologies are involved.** Such a containment starts with a proper choice of the programming language and related tools. There is abundant and various experience to support the notion that a minimalistic language designed with utmost care is key to efficient development of reliable software. Such a minimalistic language is bound to resemble Oberon modulo insignificant syntactic variations that are more likely to be defects of design than anything else (cf. the recent Google's Go [20]). Such defects would be least tolerated by non-professional programmers and in the educational context.

References

[1] Aseev V N et al. 2011 *Phys. Rev. D* **84** 112003 (arXiv:1108.5034).

[2] Tkachov F V 2013 *Quasi-optimal weights: a versatile tool of data analysis.* Talk at ACAT 2013, Beijing.

[3] Wirth N and Gutknecht J 1992 *Project Oberon.* Addison-Wesley.

[4] Oberon microsystems, Inc. http://www.oberon.ch/blackbox.html

[5] Tkachov F V 2001 *From novel mathematics to efficient algorithms. Do we have proper SD foundation to build future?* Talk at CPP2001, Japan (arXiv:hep-ph/0202033).

[6] *Workshop Oberon Day at CERN* 2004 http://www.inr.ac.ru/~blackbox/Oberon.Day/

[7] Blokland I et al. 2004 *Phys. Rev. Lett.* **93** 062001 (arXiv:hep-ph/0403221);
Blokland I et al. 2005 *Phys. Rev. D* **71** 054004 (arXiv:hep-ph/0503039);
Blokland I et al. 2005 *Phys. Rev. D* **72** 033014 (arXiv:hep-ph/0506055);
Czarnecki A, Slusarczyk M, Tkachov F V 2006 *Phys. Rev. Lett.* **96** 171803 (arXiv:hep-ph/0511004).

[8] Szyperski C (with Gruntz D and Murer S) 2002 *Component Software – Beyond Object-Oriented Programming* Addison-Wesley / ACM Press.

[9] International educational project «Informatika-21» since 2002 http://www.inr.ac.ru/~info21

[10] *Workshop Oberon Day* 2011 http://www.oberonday2011.ethz.ch/

[11] *Educational project «Informatika-21»* (since 2002) http://www.inr.ac.ru/~info21/ (in Russian)

[12] Sverdlov S Z 2007 *Programming Languages and Compilation Methods* Piter Press (in Russian).

[13] Kuhn Th S 1962 *The Structure of Scientific Revolutions* University of Chicago Press.

[14] Prof. Jurg Gutknecht's group at ETH Zurich: http://nativesystems.inf.ethz.ch/

[15] XDS Oberon (optimizing) http://www.excelsior-usa.com/xdsx86.html

[16] Gardens Point Component Pascal: http://gpcp.codeplex.com/

[17] Oberon-07 for embedded applications: http://www.astrobe.com/

[18] Wirth N 2012 *Algorithms and data structures. Version for Oberon* http://www.inf.ethz.ch/personal/wirth/books/AlgorithmE1/AD2012.pdf

[19] Dijkstra E W 1976 *A Discipline of Programming* Prentice-Hall.

[20] http://golang.org

Managing a tier-2 computer centre with a private cloud infrastructure

Stefano Bagnasco[1], Dario Berzano[1,2,3], Riccardo Brunetti[1,4] , Stefano Lusso[1] and Sara Vallero[1,2]

[1]Istituto Nazionale di Fisica Nucleare, Via Pietro Giuria 1, 10125 Torino, Italy
[2]Dipartimento di Fisica, Università degli Studi di Torino, Via Pietro Giuria 1, 10125 Torino, Italy
[3]CERN - European Organization for Nuclear Research, CH-1211 Geneva 23, Switzerland
[4]Now at Colt Engine S.r.l., Pianezza, Italy

E-mail: stefano.bagnasco@to.infn.it

Abstract. In a typical scientific computing centre, several applications coexist and share a single physical infrastructure. An underlying Private Cloud infrastructure eases the management and maintenance of such heterogeneous applications (such as multipurpose or application-specific batch farms, Grid sites, interactive data analysis facilities and others), allowing dynamic allocation resources to any application. Furthermore, the maintenance of large deployments of complex and rapidly evolving middleware and application software is eased by the use of virtual images and contextualization techniques. Such infrastructures are being deployed in some large centres (see e.g. the CERN Agile Infrastructure project), but with several open-source tools reaching maturity this is becoming viable also for smaller sites. In this contribution we describe the Private Cloud infrastructure at the INFN-Torino Computer Centre, that hosts a full-fledged WLCG Tier-2 centre, an Interactive Analysis Facility for the ALICE experiment at the CERN LHC and several smaller scientific computing applications. The private cloud building blocks include the OpenNebula software stack, the GlusterFS filesystem and the OpenWRT Linux distribution (used for network virtualization); a future integration into a federated higher-level infrastructure is made possible by exposing commonly used APIs like EC2 and OCCI.

1. Introduction

The Computer Centre at INFN Torino hosts a medium-sized Scientific Computing infrastructure that caters to the needs of a number of applications, the largest ones being a WLCG Tier-2 Computing Centre [1] for the ALICE experiment at the CERN LHC [2] and a PROOF-based Parallel Interactive Analysis Facility [3] for the same experiment. The centre is planned to grow in the near future by integrating more use cases, including among others a Tier-2 computing centre for the BES-III Experiment at IHEP (Beijing) and a small computational farm for the local Theoretical Physics group.

However, while the amount of resources and the variety of applications is steadily increasing, manpower unfortunately is not. It is thus becoming almost mandatory to consolidate such resources to achieve scalability and economies-of-scale not only by centralizing procurement but also and mainly by separating application management, that can be delegated to experienced users, from infrastructure management that can not. In this sense, Computing Centres need to become providers of computing and storage resources, not (only) of high level services.

Beyond being a trend and a buzzword in the IT world, the Cloud Computing paradigm actually promises to be a way to raise the level of abstraction and implement such separation. By converting our computing farm into a Cloud infrastructure according to the Infrastructure-as-a-Service (IaaS) paradigm, we were successful in reducing the amount of manpower needed by management tasks. Processes such as the dynamical reallocation of resources across the applications, the maintenance of large deployments of complex and rapidly evolving middleware and application software, are eased by the use of virtual images and contextualization techniques. At the same time the flexibility of the infrastructure was increased, allowing us to easily provide on-demand computing resources to a number of smaller applications and short-term requirements.

Furthermore, a number of larger-scale Cloud Computing project are starting in Italy and Europe to deploy large-scale distributed infrastructures: a local Cloud, provided its implementation is designed for interoperability, follows some existing standard, and exposes widely-used interfaces, should allow us to immediately take part in such activities.

Given these considerations, and taking into account the need not only to maintain but also to develop the infrastructure with minimum manpower and effort, the system was designed along three guidelines:

- Provide a production service to users as soon as possible
- Favour manageability and flexibility over performance
- Ensure interoperability with existing and upcoming infrastructures

This translated into a number of design and implementation choices that will be described in the following sections. We choose only stable and widely used tools and components, such as the OpenNebula cloud stack and the GlusterFS filesystem, and try to develop in-house as few pieces as possible. Furthermore, an agile development cycle was adopted, with resources given to the users as soon as possible and features and functionalities introduced only when they become needed by the applications.

2. The components

The middleware stack of choice is OpenNebula [4], an open-source software suite aimed at deploying, managing and monitoring Infrastructure-as-a-Service clusters widely used both in industry and in the research community. Even though most of the more recent Cloud projects use OpenStack [5], at the time we made the original choice the latter was judged not to be mature enough for a production environment. Furthermore, OpenNebula is arguably more suited for the task at hand (an advanced approach to data centre virtualization, with IaaS as a reference paradigm), whereas other stacks are more aimed at the full-fledged deployment of AWS-like infrastructures [6].

OpenNebula has many attractive features:

- it has a modular and easily customizable architecture, mainly based on shell and ruby scripts;
- it provides most of the needed features and functionalities, with the notable exception, as of version 3.8, of an integrated tool for network-on-demand, which prompted the developments described in section 5. ;
- it exposes, even if with different levels of maturity, industry-standard interfaces such as OCCI and EC2, along with the native OCA;
- it includes SunStone, a functional web application for most of the management tasks.

The versions used for the work described in this paper were 3.6 and later 3.8. Wide use is made of contextualization techniques, which allow us to maintain only a small number of very generic , very simple Operating System images, which are then set up and configured ("contextualized") only at boot time. We are currently using the standard contextualization tools natively provided by OpenNebula [7] and maintaining a number of shell scripts, but work is ongoing to adopt CloudInit [15], a more advanced tool recently adopted by many Cloud Computing projects.

Also for the backend storage (as opposed to *data* storage, which holds data to be accessed by the virtual workers, *backend* storage is used mainly to provide shared space to VMs and to distribute images across the physical hosts) a widely used tool was chosen, whose robustness and scalability has been proven by several very large deployments in the scientific and high-performance computing.

GlusterFS [8] is an open-source filesystem for network-attached storage currently developed and maintained by RedHat. It provides horizontal scalability by aggregating storage servers ("bricks") over the network into one large parallel file system with no master host: all synchronizations are peer-to-peer and clients access data directly from the node hosting it. While reasonably easy to set up in the simpler configurations, it is at the same time flexible enough to cater to different needs with a single tool; in particular, GlusterFS can mimic some RAID functionalities, like striping or mirroring, at the filesystem level so that a filesystem can be optimized for performance or reliability. This feature was exploited as described in section 4.

OpenNebula, while implementing tools for the definition and management of virtual networks at the cloud controller and hypervisor level, at the time lacked a tool for their implementation in the virtual infrastructure. We currently deploy Virtual Routers as small virtual machines based on OpenWRT [14], a lightweight Linux distribution designed for embedded systems; they provide internal and external connectivity to the virtual machines of a given application, alongside with tools for network configuration and management; details will be given in section 5.

3. A tale of two clusters

The components of most Scientific Computing systems can be very roughly divided into three categories: general services, worker nodes, and storage. General services are needed to orchestrate the computation, provide access to the cluster, manage and monitor the system or provide specific services for the relevant use case (a typical example is an LRMS head node); worker nodes are the number crunchers that perform the actual calculation (being it data analysis, simulation, numerical computation etc.); storage services provide access to input data to be analysed and, if needed, storage space for output. The latter are distinguished from the storage system needed by the Cloud Infrastructure to operate (e.g. for distributing the virtual machine images across hosts); storage services for data are beyond the scope of this paper and below we will only briefly describe one type of data storage we use.

"Servers" that provide general services and "Workers" that provide computational power have different requirements. In order to efficiently accommodate the two classes, the infrastructure comprises two groups of hosts ("Clusters" in OpenNebula terminology), the "Services Cluster" and the "Worker Clusters"; "Server-class" virtual machines will be instantiated on the Services Cluster, and "Worker-class" ones on the Workers clusters.

The services are often critical for the functionality of the whole system, so they need resilient, redundant hardware and some form of High Availability may be desirable. Some services require inbound connectivity (at a bare minimum, for granting access to the system), whereas the local disk performance requirements may not be exceedingly tight. The hosts in this cluster are built on server-class hardware with redundant RAID disks and power supplies. They are dual-homed with interfaces in both the public network and the workers' private network (private and public IPs) with both in- and out-bound connectivity to provide outside access to private machines. Conversely, workers that do data analysis need a high-throughput access to data storage, usually don't need a public IP since data to be processed can be ether on local disk or on a local server reachable through the private network. Losing a Worker means only losing a small fraction of computational power, so they can use less redundant (and less expensive) hardware. The workers are typically dual-twin servers that share a redundant power supply among four motherboards, but have good network connection with the storage servers; the infrastructure currently includes about 40 such hosts.

Any generic application (a "virtual farm" in the following) on our infrastructure can be viewed as a number of virtual machines in the Workers cluster managed by a number of services that run in virtual machines in the Services cluster. At the bare minimum, the only service needed is a Virtual Router to provide network services and access from the outside to a user's virtual Worker Nodes.

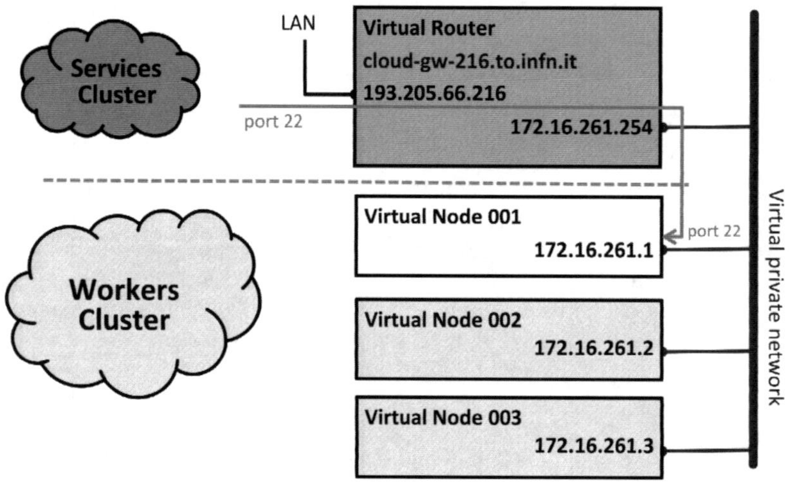

Figure 1. Example deployment of a Virtual Farm on the two clusters. VMs in the Workers Cluster are isolated from other VMs running on the infrastructure and connected to the external network through the Virtual Router. Higher-level services can be run on the Worker cluster head node (e.g. a batch queue manager) or, if they are more critical or need outbound connectivity, on a separate VM on the Services cluster.

The two clusters have been provided with different types of backend storage ("Datastores" in OpenNebula terminology) in order to optimize the performances and satisfy the above requirements, as described in section 3.

4. Storage backend architecture and GlusterFS

The backend storage is used to distribute virtual images, in our case either RAW or QCOW2, to the physical hosts. As mentioned above, the Services and the Workers clusters have different requirements, most of which are fulfilled by Gluster filesystems served by two fileservers equipped with redundant hardware (RAID disks and power supplies) and 10Gbps Ethernet interfaces. An external redundant FibreChannel controller provides the physical storage units.

The virtual images repository resides on a simple Gluster filesystem that currently comprises a single brick, but that can be scaled out by simply adding more bricks should more capacity be needed. All hosts mount this filesystem, from which the image needs to be copied by the relevant OpenNebula Transfer Manager to the datastore on which the running machine images will reside before it boots. Both the datastores and the copy mechanisms are different for the Services and Workers clusters.

As mentioned, services may be crucial to the smooth running of the whole site, so one wants them to be resilient, i.e. to be able to tolerate with little or no interruption of service the failure of a hardware or software component. To this aim, all hosts in the Services cluster share a Gluster filesystem on which the image is copied just before the virtual machine boots. The filesystem is replicated across the two fileservers to ensure availability in case of the failure of one, with the Gluster's self-healing feature enabled, and is shared across all physical hosts to allow for live migration of virtual machines from one host to another. Even if we currently don't implement automatic High Availability mechanisms, this at the very least allows us to e.g. put offline a physical host for maintenance without service interruption, at the cost of a latency in the startup of the virtual machine while the image is copied from the repository to the Shared Datastore; this is generally deemed tolerable since the restart of a service is a rare event. Also the small price to be paid in terms of performance from having a host accessing its running image over the network is justified by the higher stability of the system.

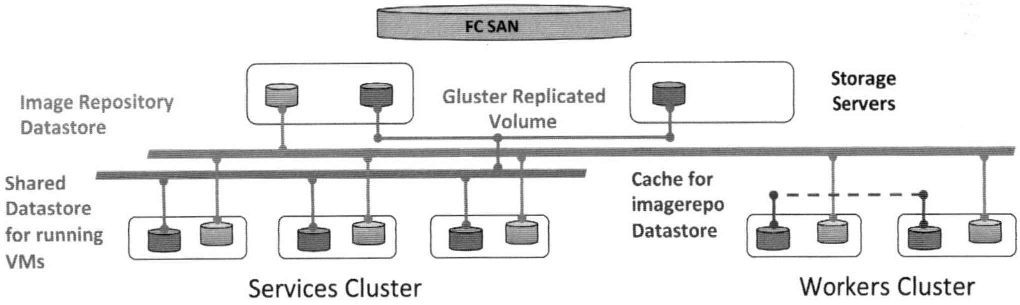

Figure 1. Architecture of the Backend Storage; Data Storage is not included.

On the other hand, dynamic allocation of resource entails the need for the fast startup of large numbers of virtual machines, which would be impossible if the images needed to be copied every time. Furthermore, the computing capacity of the site depends heavily on the performance of worker nodes, so network latency towards the running image is to be avoided. Thus, hosts in the Workers cluster share no filesystem; their running images are cached asynchronously on the local disk of the host by a custom torrent-like tool (built upon scpWave [9] and rsync [10]) and managed by a custom-written Transfer Manager, so that they are always already there when a virtual machine needs to boot. Of course this implies that a little care is required in the management of the images in order not to saturate the local disks that comprise this Cached Datastore.

For the sake of completeness we just mention that also one of the applications running on the cloud infrastructure, the PROOF-based Virtual Analysis Facility [15], takes advantage of a performance-optimized 50TB Gluster filesystem also for data access.

5. Virtual Farm networking and the V-Router

As already mentioned, OpenNebula was lacking a tool to build and manage on-demand virtual networks to provide virtual farms with an isolated, flexible and secure network environment. In particular, for example, a tool was needed to provide access from the outside to private virtual machines.

In our infrastructure, all physical hosts share a common private network, while the ones in the Services cluster also have a second interface on a public network. Then, each application has its own private fully-featured class-C network built on top of the physical network. At level 2, these networks are isolated from each other using appropriate ebtables [11] rules defined on the hypervisor network bridge, to which obviously the users have no access.

At level 3, each private network (except the larger one for the WLCG Tier-2, which is managed directly on the core switch of the infrastructure) is managed by a Virtual Router: a small virtual machine (one physical CPU, 150 MB memory) running OpenWRT, a lightweight Linux distribution originally designed for embedded systems. OpenWRT provides tools for network configuration and management such as NAT and DHCP, firewalling and port forwarding along with a web interface for their configuration and monitoring. The Virtual Routers run on the services clusters, both for robustness and to allow them to be accessed from the outside.

In the standard configuration, port 22 on the public network of the Virtual Router is forwarded to port 22 on the private network of one of the nodes in the virtual farms, thus providing ssh access to manage the nodes; communication between nodes in the same virtual farms is open, whereas of course they have no way to access the private networks of other clusters.

6. Conclusions and outlook

Currently the Torino Cloud hosts two main applications: a fully-fledged WLCG Tier-2 site for the ALICE experiment, in which all Grid services except the Storage Elements run on the Services cluster and the Worker Nodes on the Workers cluster; and a PROOF-based Virtual Analysis Facility for the same experiment, which has been described in detail elsewhere [15]. This last application, based upon CernVM [11], PoD [13] and PROOF, is currently the one that better exploits the IaaS features of the infrastructure. Alongside those, a number of smaller use cases have been successfully tested and more are planned to join in the next future, including a Tier-2 centre for the BES-III experiment.

The infrastructure has now been in production for nearly two years; the tools of choice proved themselves satisfyingly stable and robust. This, together with the flexibility and manageability of the IaaS model, helped to reduce the management load on the Data Centre staff: as a trivial example, the frequent updates of the Grid middleware are now much faster and easier, as is reallocation of resources across different applications.

There is of course still much room for improvement, and we are planning several updates to the system, both to improve the stability of the infrastructure and to provide new functionalities.

For example, we are exploring the opportunities given by the CernVM "ecosystem", in which the ongoing development of PROOF and the Virtual Analysis Facility is tightly entangled, to provide higher-level Platform-as-a-Service tools to experiments. We are also working on the Amazon EC2 interface of OpenNebula to ensure interoperability with other similar infrastructures and we plan to participate in upcoming projects aimed at developing higher-level federated cloud infrastructures. Moreover, we are moving the first steps towards providing self-service systems for advanced users to deploy their virtual application farm in as simple as possible way.

Acknowledgements
The present work is partially funded under contract 20108T4XTM of "Programmi di Ricerca Scientifica di Rilevante Interesse Nazionale" (Italy).
The authors want to acknowledge the support by the ALICE and BES-III communities, and in particular by their Italian components.

References
[1] http://wlcg.web.cern.ch/
[2] http://aliceinfo.cern.ch/
[3] http://aaf.cern.ch/
[4] http://www.opennebula.org
[5] http://www.openstack.org/
[6] See for example http://blog.opennebula.org/?p=4042
[7] http://opennebula.org/documentation:archives:rel3.8:context_overview
[8] http://www.gluster.org
[9] https://code.google.com/p/scp-wave/
[10] http://rsync.samba.org/
[11] http://ebtables.sourceforge.net/
[12] http://cernvm.cern.ch/portal/
[13] Malzacher P and Manafov A 2010 *J. Phys.: Conf. Ser.* **219** 072009
[14] https://openwrt.org/
[15] https://help.ubuntu.com/community/CloudInit
[16] Berzano D *et al.* 2012 *J. Phys.: Conf. Ser.* **368** 012019
[17] http://aws.amazon.com/ec2/

Journal of Physics: Conference Series **523** (2014) 012013 doi:10.1088/1742-6596/523/1/012013

Analysis of performance improvements for host and GPU interface of the APENet+ 3D Torus network.

R Ammendola A[1], A Biagioni[2], O Frezza[2], F Lo Cicero[2], A Lonardo[2], P S Paolucci[2], D Rossetti[2], F Simula[2], L Tosoratto[2] and P Vicini[2]

[1]INFN Roma II, Via della Ricerca Scientifica 1 - 00133 Roma, Italy
[2]INFN Roma I, P.le Aldo Moro 2 - 00185 Roma, Italy

E-mail: piero.vicini@roma1.infn.it

Abstract. APEnet+ is an INFN (Italian Institute for Nuclear Physics) project aiming to develop a custom 3-Dimensional torus interconnect network optimized for hybrid clusters CPU-GPU dedicated to High Performance scientific Computing. The APEnet+ interconnect fabric is built on a FPGA-based PCI-express board with 6 bi-directional off-board links showing 34 Gbps of raw bandwidth per direction, and leverages upon peer-to-peer capabilities of Fermi and Kepler-class NVIDIA GPUs to obtain real zero-copy, GPU-to-GPU low latency transfers. The minimization of APEnet+ transfer latency is achieved through the adoption of RDMA protocol implemented in FPGA with specialized hardware blocks tightly coupled with embedded micro-processor. This architecture provides a high performance low latency offload engine for both trasmit and receive side of data transactions: preliminary results are encouraging, showing 50% of bandwidth increase for large packet size transfers. In this paper we describe the APEnet+ architecture, detailing the hardware implementation and discuss the impact of such RDMA specialized hardware on host interface latency and bandwidth.

1. Introduction

Scaling towards exaFLOPS systems in HPC requires to select an high performances interconnection network solution able to also tackle with requirements of low power consumption, high efficiency and resilience. Most of the newest HPC platforms, as testified by TOP500 [1] and the Graph500 that enlist world's current most powerful computing systems, are mainly large clusters interconnected by a switch-less, multi-dimensional toroidal mesh. This solution has been proven effective to ensure lower costs in terms of power and equipment compared with a fat-tree one with comparable interconnect technology. Moreover scaling such mesh is easier, even while maintaing comparable or better communication latency for computing algorithms with a good degree of locality.

At the same time, TOP500 shows that GPU accelerators are gaining market in the supercomputing arena.

On this path we designed and deployed a PCIe network board for tera-scale clusters, the APEnet+ card [2], based on state-of-the-art FPGA. With its six fully bidirectional, point-to-point links, the APEnet+ board can effortlessly turn an assembly of off-the-shelf many-core x86_64 systems into a cluster with a 3D-torus network topology; an overview of the board architecture is given in Sec. 2.

The Network Interface is also able to directly access GPUs memory leveraging upon the peer-to-peer (P2P) capabilites of Fermi- and Kepler-class NVIDIA GPUs. This is a first-of-its-kind feature for a non-NVIDIA device that allows unstaged off-board GPU-to-GPU transfers with unprecedented low latency, as described in Sec. 2.1.

The overall architecture of APEnet+ has been evolutionarily improved by adding or re-engineering its parts in order to relieve or bypass the bottlenecks. The set of performance figures that this evolving design allowed to achieve are comparable with those of top offerings from the commercial world for the same PCIe generation; Sec. 3 describes more in-depth the areas where most of the work was performed.

Thanks to the reconfigurability of the FPGA platform and the modularity of our architecture, it has been possible to reuse APEnet+ in different contexts, for example as a low-latency interface between a read-out board and a GPU farm in the data acquisition system of the low level trigger of a high-energy-physics experiment (NaNet project)[3, 4].

Short term future roadmap foresees the delivery of a new release of the APEnet+ card relying on 28nm FPGAs and PCIe Gen3 bus. A glimpse on this work is given in Sec. 4 together with some conclusions.

2. Architecture Overview

APEnet+ is a point-to-point, low-latency, 3D-torus network controller integrated in a PCIe Gen2 board based on Altera Stratix IV FPGA developed at "Istituto Nazionale di Fisica Nucleare" (INFN).

The APEnet+ network architecture has, at its core, the *Distributed Network Processor* (DNP), which acts as an off-loading network engine for the computing node, performing inter-node data transfers.

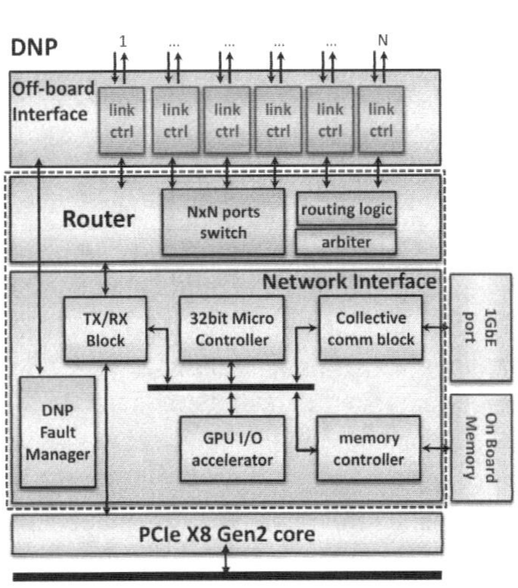

Figure 1. Overview of APEnet+. The DNP is the core of the architecture - composed by the Off-Board Interface, Router and Network Interface macro blocks - implemented on the FPGA. The system interfaces to the host through the PCIe bus.

The DNP has been developed as a parametric Intellectual Property library, meaning that some fundamental architectural features can be chosen among the available ones, others can be customized at design-time and new ones can easily be plugged in.

A highly modular design has been employed separating the architectural *Core* from the *Off-board Interface* block, connected by a customizable number of ports as shown in Fig. 1.

The Off-Board Interface manages the node-to-node communication flow over links equipped with a bi-directional Serializer/Deserializer, error checking and DC-balancing; it allows point-to-point, full-duplex connections for each node.

The Core block structure is split into a *Router* component, responsible for data routing and dispatch, and a *Network interface*, the packet injection/processing logic comprising host interface, Tx/Rx logic and μC.

The Router establishes dynamic links among the ports of the cross-bar switch, managing conflicts on shared resources.

The Network Interface block has basically two main tasks: on the transmit data path, it gathers data coming in from the PCIe port and forwards them to the relevant destination ports; on the receiving side, it provides hardware support for the Remote Direct Memory Access (RDMA) protocol, allowing remote data transfer over the network without involvement of the CPU of the remote node.

The GPU I/O Accelerator in Fig.1 allows to interface to the GPU for direct memory access, as described in Sec. 2.1.

APEnet+ features has been proved on the field by deploying the 3D toroidal network for the QUonG HPC platform at INFN Roma, Italy. QUonG is an INFN initiative targeted at developing a computing system dedicated to Lattice QCD computations; it is a massively parallel computing platform leveraging on commodity multi-core processors coupled with last generation GPUs as computing nodes interconnected by a point-to-point, high performance, low-latency 3D torus network, which is particularly suited to the transmission patterns of the set of algorithms LQCD belongs to. QUonG installation at the moment counts 16 nodes, each one equipped with 2 M2075 NVidia GPU and one APEnet+ board, and shows an aggregated peak performance of ~32 TFLOPS [5].

2.1. Peer-to-Peer GPU memory access

A peculiar feature of APEnet+ can be exploited when the cluster nodes are equipped with Fermi- and Kepler-class NVIDIA GPUs: APEnet+ is the first non-NVIDIA device able to directly access their memory leveraging upon their peer-to-peer (P2P) capabilites. In this way Remote GPU-to-GPU data transfers are possible without staging and involving the CPU, resulting in a very low transfer latency. Figure 2 gives a pictorial view of the data flow on the TX side from the GPU memory through the Network Interface when the P2P is enabled, compared to what happens using a device that does not provide P2P access to the GPU memory as in Fig. 3. Without the P2P a copy is necessary from the GPU memory to the CPU memory to allow the Network Interface to dispatch the message. The same happens on the receiving side.

All things considered this APEnet+ feature allows to obtain a gain in reduced GPU-to-GPU transfer latency as shown in Fig. 4. A thorough description of APEnet+ P2P implementation can be found in [6] together with an example of an accelerated application; we remark that off-board transfer latency in the GPU-to-GPU case is competitive with current academic research in this field [7, 8].

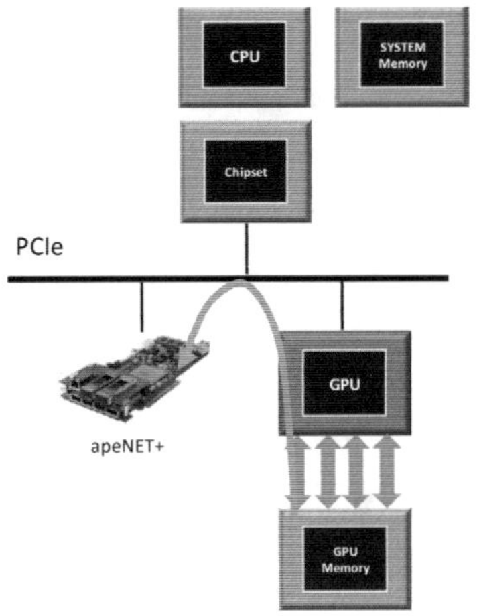

Figure 2. APEnet+'s GPU TX flow exploiting P2P.

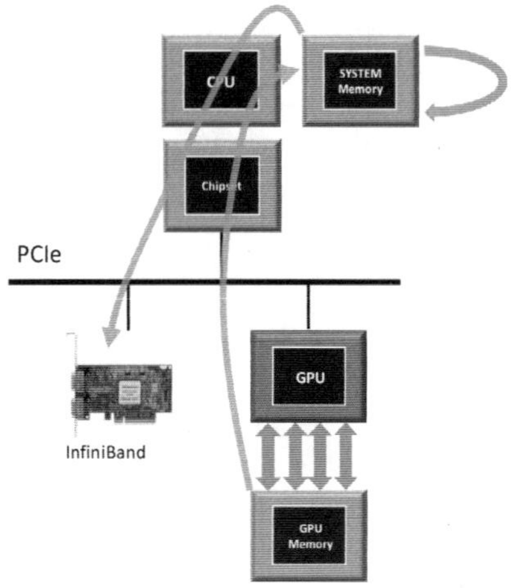

Figure 3. Traditional GPU TX data flow without the use of P2P: a copy GPU-to-CPU memory is necessary to complete the data send.

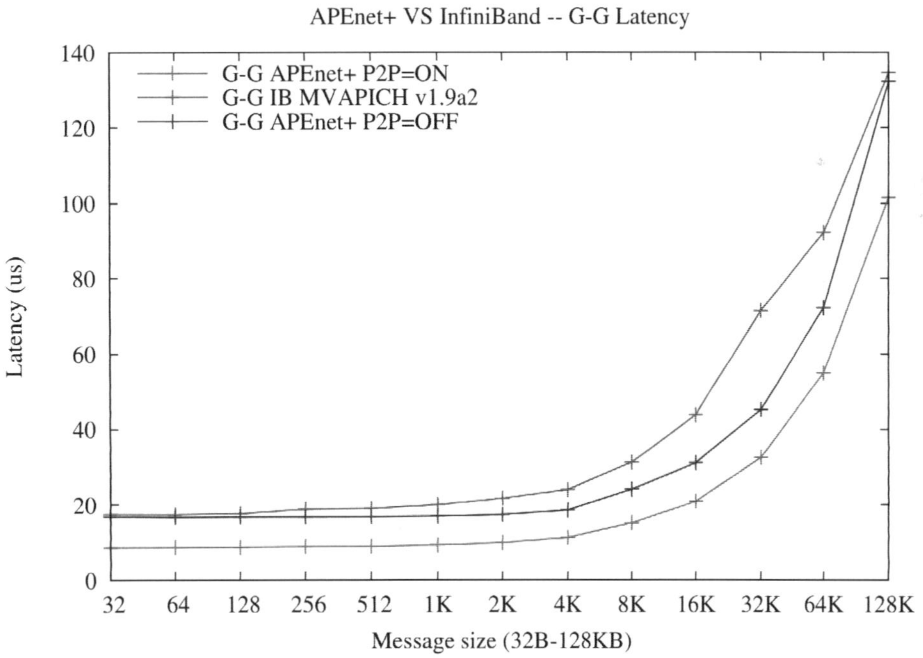

Figure 4. Plot showing the effect of APEnet+ P2P on GPU-to-GPU data transfer latency. Results are also compared with latencies of an Infiniband interconnect of the same PCIe generation, obtained using MVAPICH v1.9a2.

Journal of Physics: Conference Series **523** (2014) 012013 doi:10.1088/1742-6596/523/1/012013

3. New hardware blocks for performance improvements

The use of FPGAs made possible for the development of APEnet+ a tight design cycle and modular and reconfigurable architecture. In this way we could evolve the architecture by working around the critical areas as shown by benchmarks, and remove or reduce the performance bottlenecks.

In the following section we describe some of this work performed on key areas of the IP.

3.1. Tx speed-up: Double DMA channel

The *Tx Block*, depicted in Fig. 5, is the DNP logic block in charge of loading data from CPU memory.

The host data buffer transmission is completely handled by the APEnet+ Kernel driver: PCIe data read transactions (Tx) are issued after the information contained in command-packets, each one composed by four 128bit-words: *Header*, *Footer*, *Cmd0* and *Cmd1*. The Tx Block uses the first two command words (Header and Footer) to encapsulate data coming from the host memory, creating data-packets that will be sent to the network, while Cmd0 and Cmd1 contain information used respectively to program the PCIe DMA data read – *i.e.* packet size and address – and to communicate the end of the transaction. Even commands are read in DMA mode, so the Host communicates to the reading logic the number N of incoming command packets and notifies when they are ready. Command-packets are placed in the host memory in a circular buffer; the read logic read and consume them through the *Multi Packet Instantiator*, possibly gathering more than one command in a single PCIe DMA read transaction.

In the first implementation these DMA read operations on the PCIe were completely sequential. In order to reduce the latency between two consecutive reading processes, we introduced a mechanism able to switch between two DMA channels (DMA0 and DMA1) popping alternatively new commands from the *Tx Cmd Fifo*.

Fig. 6 compares the read bandwidth measured using one or two DMA channel implementation, showing the bandwidth improvement equal to a gain $\sim 40\%$ on the TX side bandwidth.

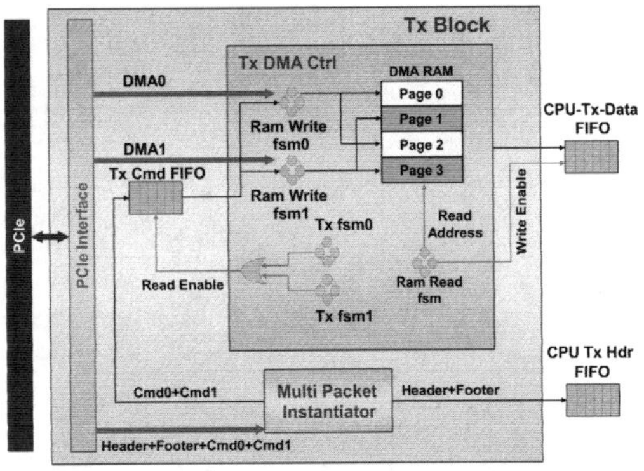

Figure 5. APEnet+'s CPU TX flow.

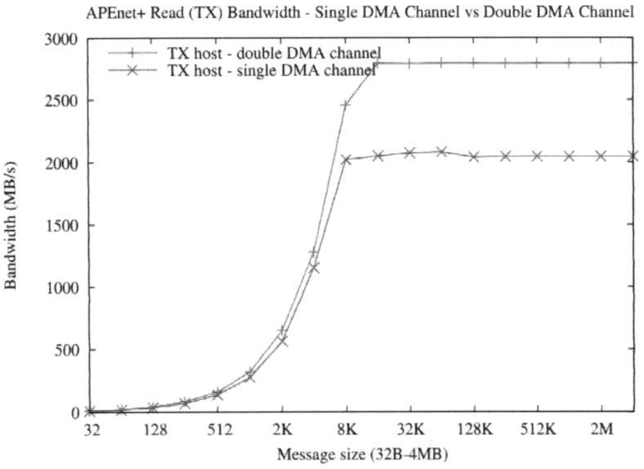

Figure 6. APEnet+ Tx single vs. double DMA results.

3.2. Rx speed-up: on-board memory management

The RDMA protocol implementation requires that the virtual memory management task has to be performed by the APEnet+ hardware on the receive side. Each packet header contains the virtual address of the destination buffer and once the packet reaches the remote destination, a virtual to physical address translation is needed to execute the DMA write in the host memory. So, the impact of this highly demanding task has to be limited to lower the latency of packet data transfer.

In the first implementation we delegated the complete task of memory management to the soft-core Altera Nios2 μC clocked at 200 MHz. At run-time, a table with user registered buffers and pages is loaded into the μC on-chip RAM by the APEnet+ device driver. A queue of incoming message requests is built upon packet arrivals: for each packet in queue the registration of the destination buffer is checked, and buffer information is gathered, including the buffer decomposition into pages and V2P address translation.

Such a μC implementation has shown a severe performance penalty with an average of 600 clock cycles to perform a single virtual address processing. To speed up the execution of these tasks we designed a specialized Translation Lookaside Buffer (TLB), similar to a modern CPU's Memory Management Units. A TLB is an associative cache where a limited amount of entries can be stored in order to perform memory management tasks, significantly reducing Nios2 processing time in case of a cached translation — a 'hit' — or forwarding the operation to the μC in case of a 'miss'. In Fig. 7 is depicted how the TLB interacts with the rest of the design. The current hardware TLB implementation takes 31 clock cycles (logic is running at 250 MHz) to perform the complete task of buffer look-up and address translation.

With a simple inter-node bandwidth test it is possible to measure the advantage given by

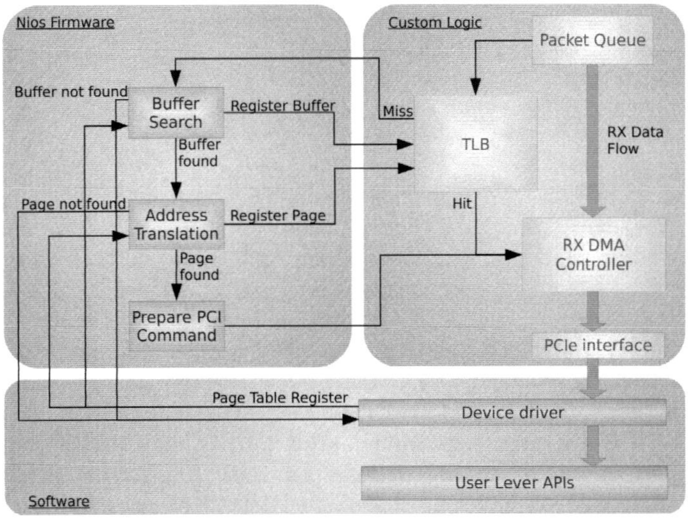

Figure 7. Schematic representation of memory management tasks. Three areas are interested: Software, with Page Table Registration; μC Firmware, interacting with the Device Driver for Registration and error handling (Buffer or Page not found events); Custom Logic, with the TLB in charge of packet processing and instructing the DMA engine logic.

the use the TLB. Results are shown in Fig. 8 for a TLB with 32 cacheable entries, where a comparison is made with the same test with the TLB disabled. Both Host- and GPU-bound data transfer are plotted, showing a performance benefit for message size up to 128 kB in the Host case and to 2 MB in the GPU case, due to the limited number of entries in this test (32) and to the memory paging size (4 kB for host memory, 64 kB for GPU memory).

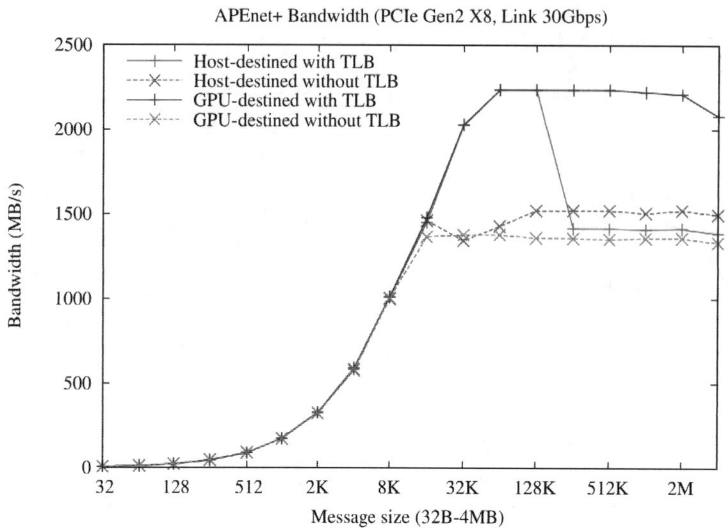

Figure 8. Bandwidth comparison in inter-node communication with (continuous lines) and without TLB (dashed lines). Use of TLB allows to reach a maximum bandwith of $\sim 2240MB/s$, instead of $\sim 1520MB/s$ in the host case and $\sim 1380MB/s$ in the GPU case when the TLB is disabled. The performance penalty for host-bound message size larger than 128 kB can be overcome by building a larger TLB.

4. Conclusion

APEnet+ is our custom Network Interface that can be used to implement high performance, low latency multidimensional toroidal networks for HPC clusters. A custom NIC for the QUonG cluster would have been impossibly lengthy and costly to build without the resources offered by FPGAs, which allow a tighter design cycle (prototyping, benchmarking and re-engineering for faster logic or bug fixes) and the fast development of a more reliable product. In this paper we described a couple of the iteration on this design cycle, that definitely helped to make the APEnet+ card the first P2P-enabled non-NVIDIA device, guaranteeing off-board GPU-to-GPU data transfers with with unprecedented low latency performances.

At the moment we are working to accommodate in our design the new features of next-gen FPGA: the all-round improvement coming from the 28 nm technology push, a PCIe Gen3 interface with upgraded encoding (from 8b/10b to 128b/130b) for doubled host bandwidth, acceleration from 8.5 Gb/s to 14.1 Gb/s for the transceivers, a more performing μC and an embedded 10GbE interface. While the 10GbE interface is mostly of interest for the evolution of the NaNet board [4], the resulting increased throughput will greatly benefit a new revision of the APEnet+ card, together with more resources to investigate application-specific acceleration methods or routing strategies.

Acknowledgments

This work was partially supported by the EU Framework Programme 7 project EURETILE under grant number 247846.

References

[1] *http://www.top500.org*

[2] Ammendola R, Biagioni A, Frezza O, Lo Cicero F, Lonardo A, Paolucci P.S, Rossetti D, Simula F, Tosoratto L and P Vicini 2012 *J. Phys. Conf. Ser.* **396** 042059
[3] Lonardo A *et al.* "Building a Low-Latency Real-time GPU-based stream processing system", *http://on-demand.gputechconf.com/gtc/2013/presentations/S3286-Low-Latency-RT-Stream-Processing-System.pdf*
[4] Lamanna G *et al.* 2013 to appear in *Proceedings of 15th International Workshop on Advanced Computing and Analysis techniques in Physics (ACAT)*
[5] Ammendola R, Biagioni A, Frezza O, Lo Cicero F, Lonardo A, Paolucci P S, Rossetti D, Simula F, Tosoratto L and Vicini P 2011 *Application Accelerators in High-Performance Computing (SAAHPC), 2011 Symposium* 113–122
[6] Ammendola R, Bernaschi M, Biagioni A, Bisson M, Fatica M, Frezza O, Lo Cicero F, Lonardo A, Mastrostefano E, Paolucci P S, Rossetti D, Simula F, Tosoratto L and Vicini P 2013 to appear in *Proceedings of 2013 IEEE International Symposium on Parallel and Distributed Processing Workshops (IPDPSW)*
[7] Hanawa T, Kodama Y, Boku T and M Sato 2013 "Interconnection network for tightly coupled accelerators architecture", in *2013 IEEE 21st Annual Symposium on High-Performance Interconnects (HOTI), San Jose, CA, U.S.A.* 79–82
[8] Bittner R, Ruf E and Forin A 2013 *Cluster Computing* 1–10 ISSN 138–7857 (*http://dx.doi.org/10.1007/s10586-013-0280-9*)

Performance optimisations for distributed analysis in ALICE

L Betev[1], A Gheata[1], M Gheata[2], C Grigoras[1] and P Hristov[1]
[1] European Organization for Nuclear Research (CERN) - Geneva, Switzerland
[2] Institute for Space Sciences – Bucharest, Romania

E-mail: andrei.gheata@cern.ch

Abstract. Performance is a critical issue in a production system accommodating hundreds of analysis users. Compared to a local session, distributed analysis is exposed to services and network latencies, remote data access and heterogeneous computing infrastructure, creating a more complex performance and efficiency optimization matrix. During the last 2 years, ALICE analysis shifted from a fast development phase to the more mature and stable code. At the same time, the frameworks and tools for deployment, monitoring and management of large productions have evolved considerably too. The ALICE Grid production system is currently used by a fair share of organized and individual user analysis, consuming up to 30% or the available resources and ranging from fully I/O-bound analysis code to CPU intensive correlations or resonances studies. While the intrinsic analysis performance is unlikely to improve by a large factor during the LHC long shutdown (LS1), the overall efficiency of the system has still to be improved by an important factor to satisfy the analysis needs. We have instrumented all analysis jobs with "sensors" collecting comprehensive monitoring information on the job running conditions and performance in order to identify bottlenecks in the data processing flow. This data are collected by the *MonALISa*-based ALICE Grid monitoring system and are used to steer and improve the job submission and management policy, to identify operational problems in real time and to perform automatic corrective actions. In parallel with an upgrade of our production system we are aiming for low level improvements related to data format, data management and merging of results to allow for a better performing ALICE analysis.

1. Introduction

The ALICE analysis workflow requires processing of large data sets with complex event structure in a distributed environment. This justifies sharing the input data between analysis tasks using a 'read once - process many times' policy. Such approach requires de-serializing the full event information to satisfy the large diversity of analysis needs, putting a non-negligible burden on the I/O. From a data processing perspective, the individual tasks are optimized for data locality in a quasi-democratic distributed system. This implies that jobs are being scheduled such that most pull data via LAN, not excluding however a small share of remote files accessed using *xrootd* [1]. The remote reading can pose a serious limitation for the less CPU-intensive analyses, which can become I/O bound. The behaviour becomes pathological in case of individual user analysis, leading to a lower overall ALICE analysis jobs efficiency. This paper describes the recent developments carried out in the core offline team to optimize the ALICE distributed analysis and identify and redress as much as possible the sources of inefficiency. The reason for this work is triggered by today's analysis needs in ALICE, projected to year 2018 through the requirements of the Run3 of LHC.

We have attempted as a first step to quantify the impact of the various sources of inefficiency to our existing data flow. This analysis is quite general and applicable for other distributed analysis systems as it allows understanding what elements can be optimized and to which extent. As a second step we have instrumented the analysis jobs with real-time monitoring, allowing identifying throughput limitations depending on resource configuration and its state, for example storage server load, site configuration or analysis job type. Correlating this information with expected values collected in a test facility allows taking corrective actions, described in the following chapters. The "hunt" for efficiency in analysis reveals both strong and weak features in our analysis model and gives a general approach for analysis optimization in a distributed environment.

2. Figures from today and goals for tomorrow

During the LHC *Run1*, ALICE was producing in average about *4 PBytes* of data suitable for analysis per month. In the same month, the data was processed about five times using about one third of the total available ALICE Grid [2] resources. During Run3, the RAW data reconstruction will be done on a dedicated online facility - the O^2 [3]. By assuming a constant Grid capacity growth of about 20% per year and its fair share between analysis and simulation we estimate an increase in analysis power by a factor of 5 by 2018. On the other hand, during Run3, the event rate is expected to increase by a factor of about 50, which are several factors above the projected processing over acquisition ratio. This problem is applicable for all LHC experiments and calls for detailed reviews of the analysis workflows to improve the overall efficiency.

To collect the necessary data for the efficiency studies, we decided to implement "sensors" into the analysis framework, which send information to a centralized monitoring system. The first objective of the analysis of the collected monitoring data was to identify bottlenecks in the workflow, allowing implementing different methods of data access and throughput optimization. This implies reducing the time to solution and at the same time increasing the efficiency by reducing the overheads in different stages of the distributed analysis workflow. In addition a simplification of the event structure and making low-level optimizations like caching or vectorisation is necessary.

ALICE uses from the beginning of the data taking a common analysis framework, described in detail in previous papers [4,5]. This brings many advantages to users and allows for more efficient organized analysis. Some of these advantages are:

- Having common interfaces hiding data diversity and backend complexity;
- Freedom to run user analysis on all resources in a transparent way;
- Uniform analysis model and style covering the full workflow;
- Powerful job control and traceability;
- Sharing input data for many analysis in a train-like fashion, with an increased quota for organized analysis

A downside of the approach was to have quite "generous" event information in the general analysis data formats to accommodate any type of analysis, leading to a bigger event size and reducing the analysis throughput and efficiency. The organized "train" model [5] also implies a larger impact of individual errors on the turnaround time and overall efficiency in distributed analysis. We also observed bad usage patterns of the framework in user analysis, which trigger in many cases inefficient use of resources.

3. Low level view on analysis efficiency

Trying to quantify efficiency in all phases of an analysis job, we end-up with a simple classification shown in figure 1. The very first stage is reading the event data from either a network stream or the local disk. The time spent for this I/O operation is proportional to the event size and with one over the streaming throughput, having as overhead the rate of I/O operations performed times the latency (or round trip time). While a fast network or a local SSD can help fetching the data faster into memory, it is very important to minimize the rate of I/O requests, especially for WAN streams. For ALICE analysis, the data streaming is steered by the ROOT framework [6] and the way to achieve this is to activate tree caching. This will be discussed based on a simple example in this section. In figure 1, the reading and writing repeats after each event to fetch the next event and possibly to write some tree output.

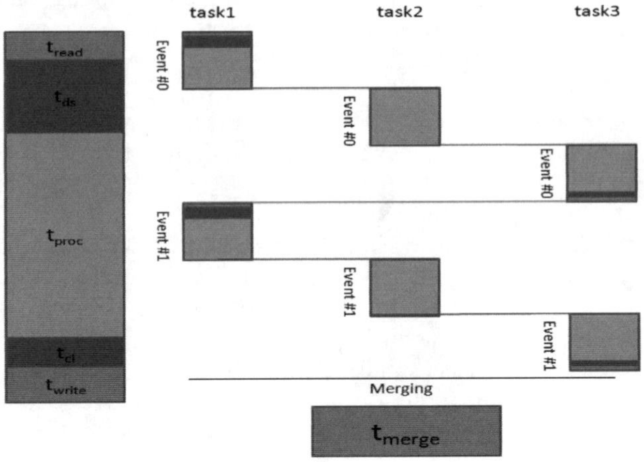

Figure 1. Processing phases for ALICE analysis trains. The total time spent per event is shared by I/O (t_{read}+t_{write}), de-serialization of the buffers into memory objects (t_{ds}), processing (t_{proc}) and event clean-up (t_{cl}). These phases repeat sequentially for each processed event in case asynchronous tree prefetching is not activated in ROOT. The merging time for the analysis outputs of different jobs has to be accounted in addition, as well as the specific distributed processing overheads.

In a second stage, the event is de-serialized by ROOT from memory buffers. While this operation is accounted as CPU time by the OS, it represents in fact hidden I/O and has to be considered as such in the overall budget. Again, this time depends on the event size, but also on its complexity (i.e. number of branches that have to be de-serialized). In case of ALICE AOD format, we have recently measured that simply flattening the event structure into a single level *ntuple* while keeping the same information as in the original data structure (which uses composition in an object-oriented approach) reduces by a factor of two the deserialization time and highly improves the *ROOT* file compression factor. Table 1 presents some simple measurements for an empty analysis train, which is just reading all events in a file in different conditions. This illustrates the throughput limits to which the analysis can become I/O bound due to either reading latency or deserialization.

Table 1. Throughput limits when reading ALICE AOD data in different conditions. While in the local case the limit comes from de-serializing data, for WAN reading without caching the I/O requests is typically bound by the round trip time. A heavy load on the disk servers can additionally worsen the situation. The CPU cycles contributing to the apparent efficiency are just used for deserialization and put an upper limit to the achievable throughput.

Device throughput and access time		**Job throughput**	**Apparent job efficiency**
Local spinning:	50 MB/s, 13 ms	5.93 MB/s	86.5%
Local SSD:	266 MB/s, 0.2 ms	6.83 MB/s	94.1%
WAN:	7.4 MB/s, 63 ms RTT		
No load on the remote disk server		0.46 MB/s	13.4%
Heavy load on the disk server		0.083MB/s	2.5%

While reading the ALICE AOD from the local disk, the tests reveal an I/O bound behaviour. Most of the job time in this case is taken by de-serialisation, which limits the throughput to a maximum of about 7 MB/sec for the test machine. To be noted that this limit depends linearly on the CPU speed of the machine, fact observed directly on the I/O bound jobs being run at different computing centres.

The third phase of processing is the actual task execution, which needs to be maximized with respect to the I/O. This can be done in ALICE analysis by chaining many analyses in a single train. While this can achieve the goal of reading once and processing many times, it introduces extra memory needs and pushes the time to analysis completion. A compromise is reached by combining I/O and CPU intensive tasks in a train, thus optimizing the overall CPU/IO performance.

Some additional analysis overheads consist in writing the output data and cleaning-up of the event transient data structures (like list of tracks) in between reading events. In the ALICE case this represents less than 1% of the total job time and will not be discussed here. What represents a real bottleneck in getting the final results for distributed analysis is the merging of all partial results produced due to the splitting of the input data to several sub-jobs. The current solution adopted in ALICE analysis and possible further improvements of this algorithm will be discussed in the following section.

We haven't discussed the impact of the job and data management on the distributed analysis performance. This does not affect the job efficiency, but can increase significantly the time to completion. Different types of overheads can be triggered by unevenness of the sub-job completion time, delays in the jobs submission due to resource load or failed job resubmissions, thus generating long completion tails. The merging procedure introduces the necessity of additional synchronization steps as described in the following section.

Finally, analysis efficiency is also a function of the analysis algorithm quality. Monitoring performance on train-by-train basis provides information on the code quality too. We have observed also efficiency improvements in case of using more compact data sets like analysis object data, or when reading only partially the input event information. To spot the sources of analysis inefficiency, we have to consider all these details and measure the effect of each on the overall budget.

4. Overview of operations 2012/2013

Considering the overall Grid activity for this period, the approximate share of resources used by the main ALICE distributed computation task was 60% simulation, 10% raw data reconstruction and 30% data analysis. From the 30%, the share between organized and individual user analysis was evenly split. The time profile of the individual user analysis efficiency with more than 200 users is shown in figure 2.

The average efficiency of 26% for analysis jobs in 2013 is contributing to the total ALICE job efficiency with a weight of 20%. This averages among user test and production jobs, most often non-optimized. These numbers are also influenced by the fact that most ALICE physics working group's analyses are now executed in organized central trains.

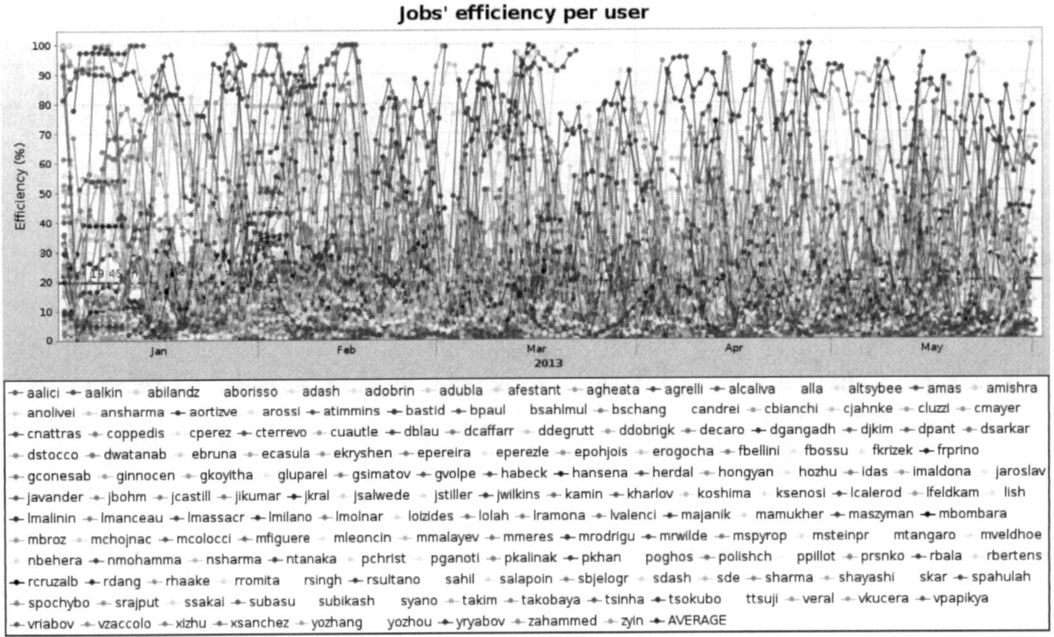

Figure 2. Time profile of individual user analysis efficiency for 5 months of operation in 2013. The average job efficiency was less than 20% for this period, spanning all possible values. This shows that such analysis mode is far from being optimal.

The benefits of running organized analysis come mostly from the fact that many analyses share the same data, which gives a large I/O reduction compared to running them individually. In addition, the CPU efficiency behaves like the best of the components. One of the downsides of organized analysis is the need to test thoroughly the performance of each individual component. This problem was addressed by a new LEGO-like system, allowing to assemble large analysis trains from individual components in an automated way [7]. The system performs individual component checks before any GRID job submission, preventing running any faulty analysis code. In addition, the LEGO frontend is based on a easy to use web interface and allows both users to register their analysis and interesting datasets, and administrators to group and schedule analysis trains according the working group priorities.

Figure 3 presents the typical interface and components of the LEGO framework. To be noted that this framework brought several improvements in both performance and code stability.

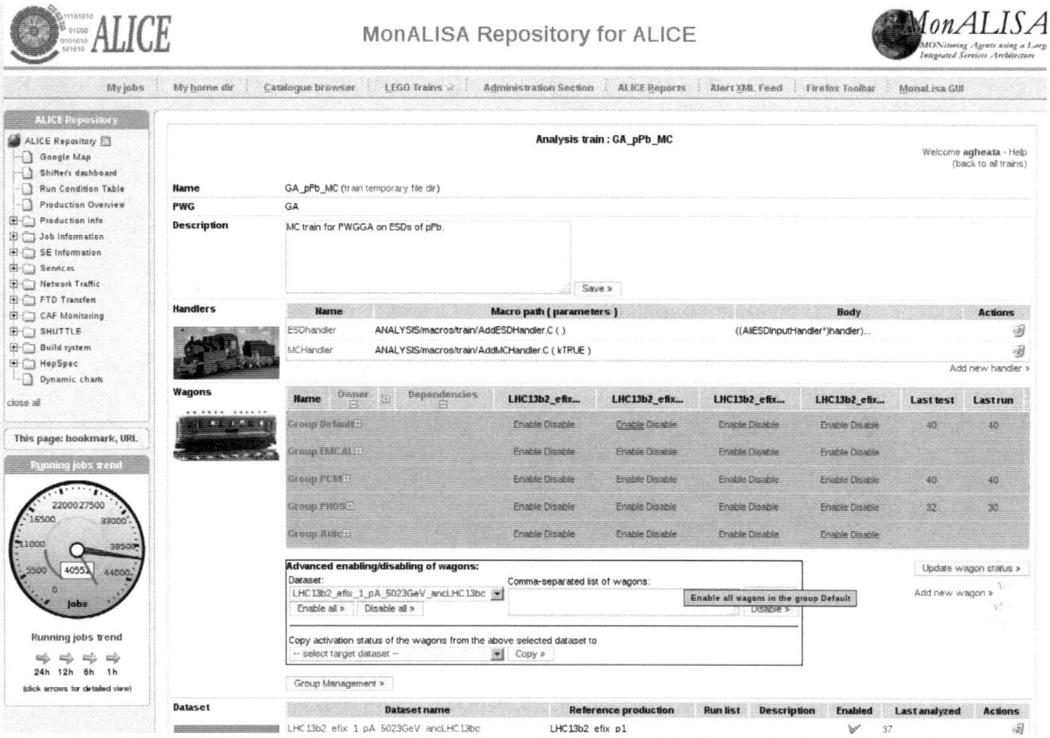

Figure 3. The ALICE LEGO framework interface, allowing registering analysis modules, datasets and composing trains. Each train is tested component by component on a dedicated facility, and then executed in the ALICE GRID. Every stage in this process is monitored using the MonALISA system [8].

The LEGO framework allowed for a detailed monitoring of the performance, both for sub-job efficiency and global execution of all involved stages: testing, job scheduling, execution and merging. During train operation, we have observed large inefficiency spikes for a large set of jobs that otherwise were behaving correctly in local tests. This behaviour was correlated to the fact that the job splitting algorithm managed only partially to assign local files to analysis jobs, so the faulty behaviour was assigned to the fact that few ROOT files were accessed through WAN. Indeed, ROOT issues a large number of read requests and not caching them leads to extremely expensive overheads related to latency to the remote disk server. Adding tree-caching support to the framework improved with a factor of more than two the overall analysis jobs efficiency, without having to modify the dispatching model.

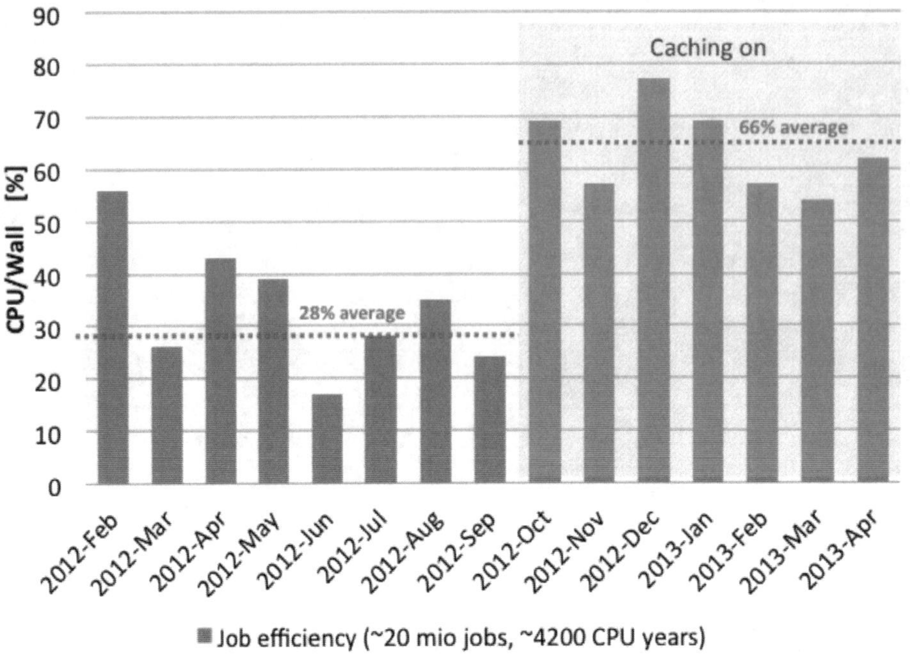

Figure 4. Overall efficiency improvement for ALICE organized analysis due to the introduction of *TTreeCache* support.

The current situation is that despite introducing tree caching, the individual user analysis efficiency is still low. To improve the overall Grid efficiency, moving ½ of the individual user jobs to organized analysis would result in an immediate efficiency gain of 7%. Another observation is that the time to solution is almost 3 times faster for compact analysis formats for I/O bound analysis, which encourages the migration to AOD-based analysis. Our analysis led to classifying the main actions to perform to alleviate the I/O problem, which is quite important for ALICE analysis:

- Caching I/O queries and prefetching. Recent measurements indicate extra gains of 10-15% when activating data prefetching in ROOT.
- Reducing the analyzed event size, having to assess the price of cutting down generality and multiplying the number of datasets.
- Selective disabling branches can alleviate I/O cost by a factor of 5, at the price of reducing the size of the analysis trains to just few modules.
- Encouraging custom filtering and skimming procedures in organized analysis can substantially reduce the input information requirements in particular cases. We distinguish the case of rare signal analysis requiring a small fraction of input events, or analysis requesting a small fraction of the available event information.

5. Instrumenting analysis performance

We have implemented low-level "sensors" in the analysis manager to collect detailed performance information at job level. The sensors collect timing information per analysis session, measuring the throughput and efficiency for every input file of the job. In addition, they collect information on the data sources, worker node identity or timing per analysis module within a train. The information is being read, processed and summarized by the monitoring system, allowing us to detect pathologies in the data flow, study data type effects and investigate possible flaws in train components. A typical summary output of a single train processing is given in Figure 5.

Site activity									
Site	Job eff. ▲	HepSpec06	All files	Local files	Remote files	CERN EOS	NDGF DCACHE	SNIC DCACHE	FZK SE
CERN 12190 jobs (60.94%)	79.08%	10.25	69829 files 3.574 MB/s	69776 (99.92%) 3.59 MB/s	53 (0.076%) 0.562 MB/s	**69776 (99.92%)** **3.59 MB/s**	7 (0.01%) 1.643 MB/s	5 (0.007%) 1.543 MB/s	2 (0.003%) 2.823 MB/s
BOLOGNA 3 jobs (0.015%)	73.31%	8.841	18 files 3.175 MB/s	12 (66.67%) 3.898 MB/s	6 (33.33%) 2.374 MB/s	6 (33.33%) 2.374 MB/s			
FZK 854 jobs (4.269%)	65.57%	9.837	4587 files 2.376 MB/s	4025 (87.75%) 3.399 MB/s	562 (12.25%) 0.782 MB/s	258 (5.625%) 0.952 MB/s	275 (5.995%) 0.626 MB/s	13 (0.283%) 0.828 MB/s	**4025 (87.75%)** **3.399 MB/s**
GSI 398 jobs (1.99%)	10.95%	9.926	2179 files 0.473 MB/s	1124 (51.58%) 3.394 MB/s	1055 (48.42%) 0.219 MB/s		651 (29.88%) 0.245 MB/s	338 (15.51%) 0.184 MB/s	4 (0.184%) 1.31 MB/s
BRATISLAVA 191 jobs (0.955%)	10.22%	9.918	1000 files 0.455 MB/s	261 (26.1%) 1.011 MB/s	739 (73.9%) 0.378 MB/s	117 (11.7%) 0.467 MB/s	397 (39.7%) 0.346 MB/s	149 (14.9%) 0.39 MB/s	66 (6.6%) 0.454 MB/s
SUT 8 jobs (0.04%)	10.19%	7.641	42 files 0.355 MB/s		42 (100%) 0.355 MB/s		14 (33.33%) 0.343 MB/s	28 (66.67%) 0.361 MB/s	
YEREVAN 1 jobs (0.005%)	9.963%	10.81	4 files 0.372 MB/s		4 (100%) 0.372 MB/s		3 (75%) 0.388 MB/s		

Figure 5. Job efficiency and throughput are highly dependent on the data locality (one can check the percentages of "local files" and "remote files"). "Pathologies" can be easily monitored with this tool, pointing to transient operational problems, excessive load of resources or site misconfiguration problems.

5.1. Merging strategy

Merging the analysis jobs outputs is a general problem for analysis and can introduce large overheads. Having a single merging job per master job does not scale in execution time and memory requirements in the general case. We have adopted a different solution to alleviate the linear scaling with the number of files and to profit by the intrinsic parallelism offered by a distributed system. We organize merging in stages, with a limited number of files per job in a given stage. In this process we profit from the standard scheduler policies by providing a standard collection of output files to be merged at the end of the analysis jobs. The staged merging exhibits a log(N) scalability, where N is the number of stages (maximum 5). In Figure 6 we present job profiles provided by MonALISA system, which include the merging stages. Using this merging policy, we observe that the overhead is much less than the total job time.

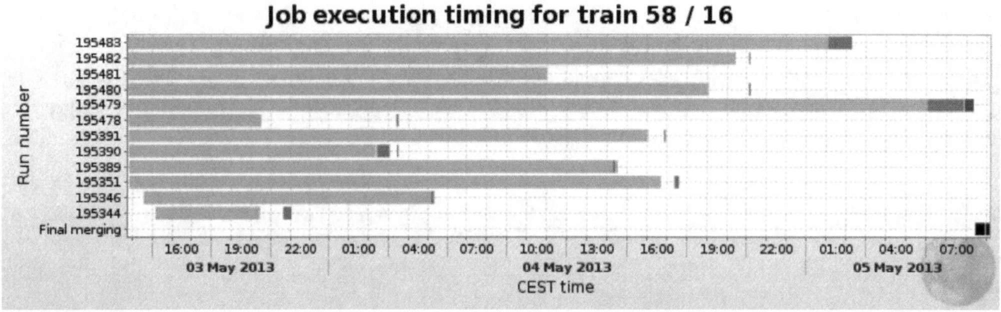

Figure 6. Typical ALICE analysis job time profile. The green bars represent the data processing time, the blue bars represent the merging stages per data set (in this case a run), while the final merging stage is the black bar starting as soon as the last merging job per run is completed.

6. Conclusions

The paper describes the current efforts to improve the performance of the ALICE distributed analysis jobs. The investigation method is quite general in its principles and aims to understand the limitations and possible bottlenecks of the current framework and computational fabric. We are able to monitor job efficiency information and throughput in relation with several factors: the input file location, input data type or execution site. The analysis of the collected monitoring data allowed for finding solutions to improve the overall efficiency in ALICE organized analysis by a large factor, and represent invaluable lessons for the evolution our analysis data formats and distributed processing patterns. Such studies are triggered by the future computing requirements in the LHC Run3, to be addressed within the ALICE O^2 project [3]. Many of these requirements are common for all LHC experiments. The evolution of the ALICE analysis framework toward a better performing system is continuing and the progress will be reported in future communications.

7. References

[1] The XROOTD project, xrootd.org
[2] Bagnasco S *et al* 2008 *J.Phys.:Conf.Ser.* **119** 062012
[3] The ALICE O^2 project, alicematters.web.cern.ch/?q=ALICE_02project
[4] Gheata A 2008 ALICE analysis framework *Proc. XII Int. Workshop on Adv. Comp. and Analysis Techniques in Phys. Research (Erice) PoS(ACAT08)* p028
[5] Gheata A 2012 Making distributed ALICE analysis simple using the GRID plug-in *J. Phys.: Conf. Ser.* **368** 012014
[6] The ROOT project, root.cern.ch
[7] The ALICE LEGO train system, https://alimonitor.cern.ch/trains/
[8] Legrand I *et al* 2009 *Computing Physics Communications* **180** Issue 12 December 2009 pp 2472-2498

Use of checkpoint-restart for complex HEP software on traditional architectures and Intel MIC

Kapil Arya[1], Gene Cooperman[1], Andrea Dotti[2], Peter Elmer[3]

[1] College of Computer and Information Science, Northeastern University, Boston, MA, USA
[2] SLAC National Accelerator Laboratory, Menlo Park, CA 94025, USA
[3] Department of Physics, Princeton University, Princeton, NJ 08540, USA

E-mail: `Peter.Elmer@cern.ch`

Abstract. Process checkpoint-restart is a technology with great potential for use in HEP workflows. Use cases include debugging, reducing the startup time of applications both in offline batch jobs and the High Level Trigger, permitting job preemption in environments where spare CPU cycles are being used opportunistically and efficient scheduling of a mix of multicore and single-threaded jobs. We report on tests of checkpoint-restart technology using CMS software, Geant4-MT (multi-threaded Geant4), and the DMTCP (Distributed Multithreaded Checkpointing) package. We analyze both single- and multi-threaded applications and test on both standard Intel x86 architectures and on Intel MIC. The tests with multi-threaded applications on Intel MIC are used to consider scalability and performance. These are considered an indicator of what the future may hold for many-core computing.

1. Introduction

The computing requirements for high energy physics (HEP) projects like the Large Hadron Collider (LHC) [1] at the European Laboratory for Particle Physics (CERN) in Geneva, Switzerland are larger than can be met with resources deployed in a single computing center. This has led to the construction of a global distributing computing system known as the Worldwide LHC Computing Grid (WLCG) [2], which brings together resources from nearly 160 computer centers in 35 countries. Computing at this scale has been used, for example, by the CMS [3] and Atlas [4] experiments for the discovery of the Higgs Boson [5, 6]. To achieve this and other results the CMS experiment, for example, typically used during 2012 a processing capacity between 80,000 and 100,000 x86-64 cores from the WLCG. Further discoveries are possible in the next decade as the LHC moves to its design energy and increases the machine luminosity. However, increases in dataset sizes by 2-3 orders of magnitude (and commensurate processing capacity) will eventually be required to realize the full potential of this scientific instrument.

Building the software to run on and operate such a computing system is a major challenge. The distributed nature of the system implies that ownership and control of the resources is also distributed, and thus the resources are by necessity heterogeneous in nature. This heterogeneity appears both in terms of specific x86 hardware generations and in patch levels of the deployed Linux operating systems. As these resources are at times shared with other projects custom modifications of systems for HEP-specific or experiment-specific reasons are in general not possible. The very large number of CPU hours used also introduces significant reliability requirements on the software. The software itself is non-trivial: each experiment typically is dependent on many millions of lines of code, written for the most part in C++ with contributions from up to a thousand physicists. Given the upgrade and evolution plans for the LHC, these software projects, begun in the late 1990's, will likely need to evolve with computing technology through the 2020's.

Such an environment however provides significant opportunities for innovation. In this paper we examine one interesting technology, *process checkpoint-restart*, which has great potential for use in HEP workflows in such a system. We first describe the specific use cases of interest and the requirements to make the technology useful in the HEP environment. We then provide some benchmarks for the use of checkpoint-restart with the CMS software on today's x86-64 processors. And finally, we examine aspects of this technology which are of interest given the possible evolution of processor technologies and resource availability in the coming years. We look in particular at the use of checkpointing with architectures like Intel's Xeon Phi, a member of Intel's MIC (many integrated core) architecture.

2. Process Checkpoint-Restart

In many circumstances it is desirable to "checkpoint" the state of a UNIX process, or a set of processes, to disk with the possibility of restarting it at a later time.

2.1. Use Cases

There are a number of interesting use cases for this functionality:

Debugging: The very large number of jobs and CPU hours required for HEP computing makes high reliability of the software quite important. The distributed and heterogeneous nature of the computing system however makes debugging problems somewhat difficult, as the first step to resolving most code behavior problems is being able to reproduce the problem. While a traditional "core" file may provide information about the process state after a crash has happened, it doesn't allow one to step through the program to see the behavior leading to the crash. In the case where the crash happens after a job has run for many hours, reproducing a problem by rerunning it from the beginning can also be quite expensive. If however a job were to checkpoint its state from time to time, it would be possible to use the last checkpointed state before a crash to reproduce and replay the problem quickly.

Avoiding CPU-intensive initialization steps: Many applications are constructed such that they have CPU-intensive initialization steps. Examples include in-memory geometry construction from a simplified geometry description, physics cross section table calculations, etc. Typically this is done to allow generality of software implementation for multiple possible job configurations: it is easier to calculate quantities derived from job configurations on the fly at the beginning of each job than to store those quantities for a possibly infinite number of potential job configurations. HEP workflows are however constructed such that a particular job configuration may be used for a very large number of jobs, where the geometry or physics process configuration are the same and only random number seeds or input files change from job to job. The result is that the same calculation is done in every single job instance in a particular workflow. In most cases where jobs run for a long time this job initialization time is negligible relative to the total running time.

However two cases exist where job initialization time can be problematic. First, very short duration jobs can sometimes be required for other software reasons or for operational reasons related to resource availability or input dataset structure. In this case the overhead from long startup initialization times can be a significant fraction of the CPU utilization.

Second, as will be described later, there are strong reasons to consider the use of multi-threaded applications in the future. In the case where the startup initialization itself cannot be easily parallelized and will be executed sequentially on a single core, the initialization itself may effectively idle a large number of other cores eventually needed for the event processing. In this case, a single instance of the job can be run and checkpointed just after the initialization phase. That checkpoint can then be used to restart a much larger number of instances of the application in batch, with only minor reconfiguration to set input files and/or random number seeds. The instances running in batch can thus avoid the startup CPU cost.

Allowing preemption during opportunistic resource use: From time to time, it is possible to "opportunistically" use computing resources which belong to some other organization when the other organization does not have enough work to keep the processors fully utilized for some period of time. In some of these cases the period of time for which the resources are made available may not be well defined in advance and the resources may need to be handed back to their owner in an unscheduled fashion. In this case it is useful to be able to "preempt" running opportunistic jobs, checkpoint their state to disk and restart them when opportunistic use is again possible.

Interactive "workspaces": In interactive programs, such as event displays and analysis tools, the user provides inputs which lead to particular state of the program at a given time. Being able to save that state out, for example before going home for the day, and restart later is often desirable.

Very long running jobs: In situations where jobs must run for an extremely long time, sometimes days or weeks, they can be sensitive to hardware or infrastructure failures or interfere with required site maintenance. In these cases it can be quite useful to checkpoint periodically the program to avoid losing and needing to repeat the calculations from scratch after such failures.

Managing "tails" for multi-threaded applications: Several HEP experiments are moving in the direction of multi-threaded frameworks, which (initially) process events on different threads. As the CPU time per event can vary significantly (with long "tails" to the distribution) at the end of the job, one thread may still be processing an event which takes a long time while the other threads/cores are idled. One possibility for managing such situations would be to checkpoint the job with a single active thread and restart a number of such jobs at a later time together, to keep the full set of CPU cores active.

2.2. Possible solutions

A rudimentary "checkpoint-restart" can sometimes be achieved in an application-specific fashion. For example a typical HEP event processing framework can be constructed to perform a simple "checkpoint" by flushing completed output events to its output file after every N input events have been processed. In addition it is necessary to write some sort of "metadata" to track any other relevant internal state needed to restart the job, e.g. how far the job had progressed through its input events, the state of random number generators, etc. In this example a "restart" would then be performed by restarting the framework and passing it information to allow it to reconfigure itself to match the state it was in at the time of the output checkpoint. This requires however the addition and maintenance of dedicated code, both in the framework itself and externally in the workflow management system. In some cases, where third party libraries are used which also maintain state, it can be quite complex to truly restore the same state. If the state is encapsulated within the code of the library, for software engineering reasons, it can also be impossible.

A much better, and more general, solution is true process-level checkpointing. This is a technology which has existed for a long time, especially in High Performance Computing (HPC) and batch systems, however often the particular implementations are tied to specific environments. Thus the technology has not seen general use in HEP high throughput distributed computing. In this paper we examine the use of a transparent, user-level checkpointing package for distributed applications called Distributed MultiThreaded CheckPointing (DMTCP) [7]. The features of DMTCP make it more appropriate for deployment in the HEP distributed computing environment.

3. Distributed MultiThreaded CheckPointing (DMTCP)

DMTCP (Distributed MultiThreaded CheckPointing) is free, open source software (`http://dmtcp.sourceforge.net`, LGPL license). The DMTCP project traces its roots to late 2004. A key feature of DMTCP for use in the heterogeneous HEP computing environment is that it works in user space, with no kernel-level modifications required. As such it is works with a wide range of Linux kernel versions. It also works with multi-threaded applications and compression of output checkpoint files is possible.

Its usage can be as simple as:

```
dmtcp_launch ./myapp arg1 ...
dmtcp_command --checkpoint     [ from a second terminal window ]
dmtcp_restart ckpt_myapp_*.dmtcp
```

DMTCP is also "contagious". If a process begins under DMTCP control, then any child processes will also be under DMTCP control, and any remote processes (spawned through "ssh") will also be under DMTCP control. At the time of checkpoint, a script, `dmtcp_restart_script.sh`, is written, and the script can restart all processes across all nodes for the given computation.

The newly released DMTCP version 2.0 (as of Oct. 3, 2013), supports DMTCP plugins to flexibly adapt to external conditions. For example, the DMTCP plugin interface permits application-initiated checkpoints, as well as application-delayed checkpoints during critical operations. Alternatively, the `--interval` flag of `dmtp_launch` permits automatic periodic checkpointing.

DMTCP plugins make it easier to use event hooks to detach such external resources as a database prior to checkpoint, and to reconnect during restart. While DMTCP will save and restore the file offset of open files, event hooks make available an alternative of cleanly closing valuable files during checkpoint, and re-opening them during restart. In another example, DMTCP virtualizes network addresses to enable transparent migration to a new cluster. Finally, if a large region of memory is not actively used at the time of checkpoint, then the size of the checkpoint image can be considerably reduced. An event hook allows the application to write zeroes into the inactive memory at checkpoint time, and DMTCP will then replace the zeroes by zero-fill-on-demand pages (empty pages to be recreated on demand).

Among the contributed plugins for DMTCP is the rm (resource manager) plugin, which supports use of DMTCP within the Torque and SLURM batch queues. Plugin support for additional batch queues is planned. Similarly, future support for the Intel SCIF network is planned, allowing one to checkpoint a computation over a network of Intel MIC CPUs. The SCIF plugin will be based on the existing contributed plugin to support checkpointing over InfiniBand. Other contributed plugins support checkpointing a network of virtual machines. Virtual machines ease the job of deploying complex software.

4. Experimental results with x86-64

To investigate the characteristics of DMTCP with HEP software, we chose to make tests using the CMS simulation application. The machine used to do the test was a dual quad-core Intel Xeon L5520 operating at 2.27GHz, with 24GB of memory. Tests were done with checkpoint files written to a local disk as well as the normal job output files. The test job had no input event file. However it reads conditions via a web squid from a remote database. DMTCP version 1.2.7 and CMSSW version 6_2_0, compiled with GCC version 4.7.2, were used for the tests. For simplicity the test job used with checkpointing was the only CPU-intensive process running on the machine at the time the tests were performed. The tests were done using the external DMTCP coordinator trigger to produce a checkpoint, rather than the API.

The CMS application generated simple Minimum Bias events and simulated them with Geant4 [8]. The particular test job itself has a 2 minute initialization phase and then takes an average time per event of ~13 seconds. The memory footprint is ~1GB VSIZE (750MB RSS). A typical uncompressed checkpoint takes ~1.5s and the resulting size on disk of the checkpoint file was ~750MB. When triggering checkpointing with the compression on, ~10s was required. The checkpoint image was however significantly smaller, at only ~220MB. In both cases no problems were seen in restarting the application from the checkpoint files.

5. Processor Architectures

The construction of the WLCG was greatly facilitated by the convergence around the year 2000 on commodity x86 hardware and the standardized use of Linux as the operating system for scientific computing clusters. Even if multiple generations of x86 hardware (and hardware from both Intel and AMD) are provided in the various computer centers, this was a far simpler situation than the previous typical mix of proprietary UNIX operating systems and processors. Until around 2005, a combination of increased instruction level parallelism and (in particular) processor clock frequency increases insured that performance gains expected from Moore's Law would be seen by single sequential applications running on a single processor. The combination of Linux, commodity x86 processors and Moore's Law gains for sequential applications made for a simple software environment.

However since around 2005 processors have hit scaling limits, largely driven by overall power consumption [9]. The first large change in commercial processor products as a result of these limits was the introduction of "multicore" CPUs, with more than one functional processor on a chip. At the same time clock frequencies ceased to increase with each processor generation and indeed were often reduced relative to the peak. The result of this was one could no longer expect that single, sequential applications would run faster on newer processors. However in the first approximation, the individual cores in the multicore CPUs appeared more or less like the single standalone processors used previously. Most large scientific applications (HPC/parallel or high throughput) run in any case on clusters and the additional cores are often simply scheduled as if they were additional nodes in the cluster. This allows overall throughput to continue to scale even if that of a single application does not. It has several disadvantages, though, in that a number of things that would have been roughly constant over subsequent purchasing generations in a given cluster (with a more or less fixed number of rack slots, say) now grow with each generation of machines in the computer center. This includes the total memory required in each box, the number of open files and/or database connections, increasing number of independent (and incoherent) I/O streams, the number of jobs handled by batch schedulers, etc. The specifics vary from application to application, but potential difficulties in continually scaling these system parameters puts some pressure on applications to make code changes in response, for example by introducing thread-level parallelism where it did not previously exist.

There is moreover a more general expectation that the limit of power consumption on future Moore's Law scaling will lead to more profound changes going forward. In particular, the power hungry x86-64 "large" cores of today will likely be replaced by simpler and less power hungry "small" cores. One example of such a technology is the Intel MIC architecture, as implemented in the Intel Xeon Phi coprocessor card.

6. Experimental Results with Xeon Phi

To test the use of checkpointing on the Xeon Phi, we used a beta version of Geant4 version 10 which provides support for event-based multi-threaded applications. We did not use the full CMS simulation for this, but instead a simpler benchmark application (FullCMS) which uses the actual CMS geometry imported from GDML file. The experimental setup was a standard Intel Xeon box with 32 logical cores, equipped with an Intel Xeon Phi 5120P coprocessor card. In our test the application is started and checkpoints are triggered in different moments. Then we have restarted the application from the checkpoint file and we have verified that the application resumes correctly from the saved state. This condition is verified checking that the final output of the simulation equals the original one. A comparison of random number engine status at the end of the job. Since in a Geant4 application there is a very large use of random numbers (billions of calls to the engine in typical application), if the status of the application from the checkpoint file does not match exactly the original one the sequence of the random number calls will be different, producing a different final state of the random number engine.

A typical Geant4 application with multi-thread support consists of a sequential part in which the geometry of the experimental setup is built in memory and physics processes are initialized (e.g. material-dependent cross sections are calculated). Threads are then spawned, initialized and they start to simulate events independently (see Figure 1). To reduce the total memory footprint the most memory consuming objects are shared between threads. The need for synchronization between threads (locks, barriers) is minimized since only read-only objects are shared.

We have performed tests to verify the correct behavior of DMTCP for two of the Use Cases described in Section 2.1.

In the first case we have instrumented the Geant4 application code with a call to DMTCP to trigger a checkpoint file at the end of the initialization phase (Figure 2). On the Intel Xeon Phi accelerator the initialization takes about 5 minutes. The checkpointing itself takes about 1 minute (the application working directory, physically located on the host, was mounted by the coprocessor through NFS) and the resulting checkpoint image file is 1.4GB (uncompressed). Restarting from the checkpoint image file takes less than 10s. Functionally this appears to work as expected. The resulting checkpoint image file can thus be distributed to other nodes and the simulation process "cloned" without the startup cost, simply by resetting the random number seeds.

For the second test we have emulated the use case of some threads being slower than other in producing results. This can happen if one or more threads is simulating more complex-than-average events. To control such behavior we have modified our application in a way that half of the threads were responsible of simulating simple and fast events (low energy single particles) while the second half of threads was responsible for longer ones (ten times higher energy). In current Geant4 multi-threaded mode the application will wait for all threads before terminating thus leaving half of the MIC cores unused. We have instrumented the application code to verify when the number of active threads drops below a given value, in such a case a checkpoint is triggered (Figure 3). Also in this case we have verified that the application was behaving as expected.

It is important to note that we did not have to modify Geant4 "kernel" code to enable checkpointing, all code modifications were done at the application level. For the first test (checkpointing at first event) we have provided feedback to Geant4 developers that have introduced a new user-hook, not present in the initial design of the code, that allows for the execution of (optional) user code just before the first event is simulated but after all threads have been fully initialized (this guarantees that the checkpointing is performed in a reproducible state of the application). Both tests show that checkpointing can be used to increase the efficiency of resource usage also on accelerator technologies where the minimization of the time spent in sequential fractions of the code or with only few threads active is fundamental to efficiently use the hardware resources.

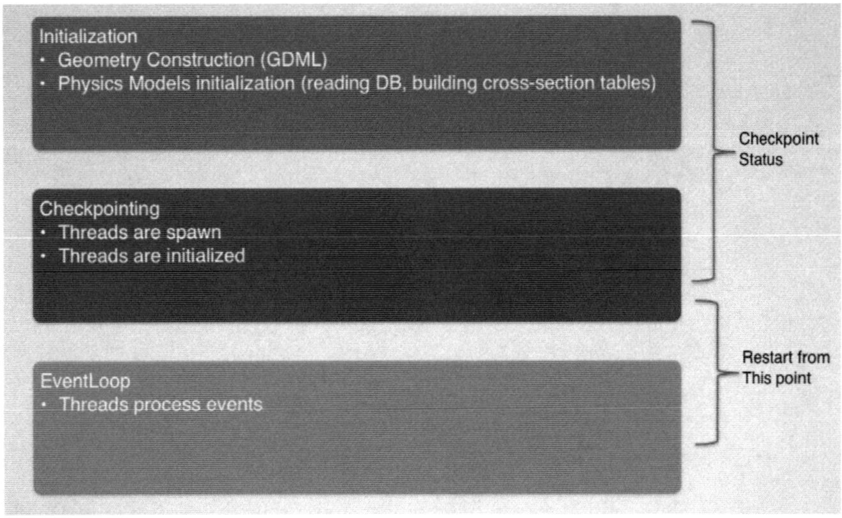

Figure 1. Startup structure of a typical Geant4 simulation job

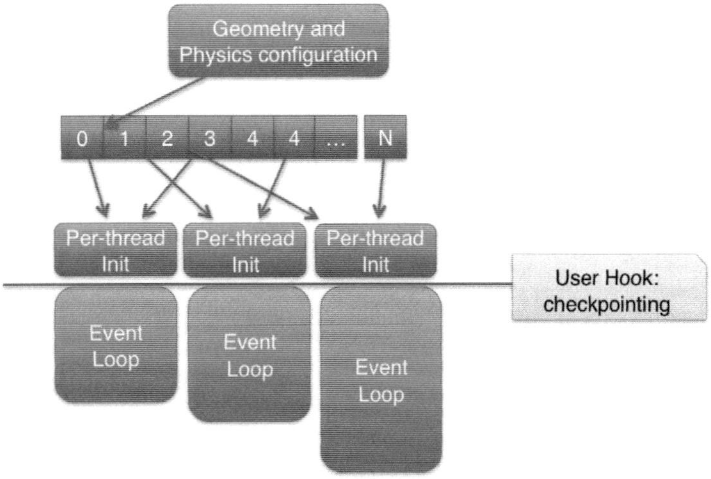

Figure 2. User hook for checkpointing to avoid startup CPU cost

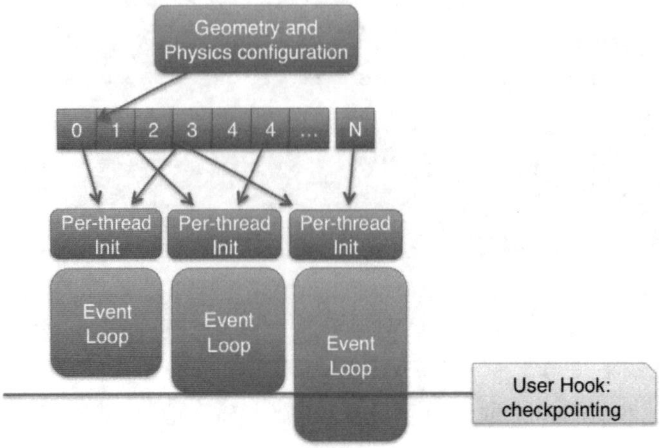

Figure 3. User hook for checkpointing to avoid "tails" CPU cost

7. Conclusions

We have made investigations into the use of the checkpoint-restart technology DMTCP with HEP applications from CMS and Geant4. We have reported on the performance seen, both on a traditional x86-64 architecture and on Intel's Xeon Phi, for situations relevant for a number of interesting use cases for HEP computing. We believe that the results obtained are very encouraging and demonstrate the viability of the use of this technology in the HEP environment.

Acknowledgments

This work was partially supported by the National Science Foundation under Grant OCI-0960978 (Arya, Cooperman) and Cooperative Agreement PHY-1120138 (Elmer) as well as by the U.S. Department of Energy (Dotti).

References

[1] Evans L and Bryant P 2008 LHC Machine *JINST* **3** S08001
[2] Bird I 2011 Computing for the Large Hadron Collider *Annual Review of Nuclear and Particle Science* **61** 99-118
[3] Chatrchyan S et al (CMS Collaboration) 2008 The CMS experiment at the CERN LHC *JINST* **3** S08004
[4] Aad G et al (Atlas Collaboration) 2008 The Atlas Experiment at the CERN Large Hadron Collider *JINST* **3** S08003
[5] Chatrchyan S et al (CMS Collaboration) 2012 Observation of a new boson at a mass of 125 GeV with the CMS experiment at the LHC *Phys.Lett.* **B716** 30-61
[6] Aad G et al (Atlas Collaboration) 2012 Observation of a new particle in the search for the Standard Model Higgs boson with the ATLAS detector at the LHC *Phys.Lett.* **B716** 1-29
[7] Ansel J, Arya K and Cooperman G 2009 DMTCP: Transparent Checkpointing for Cluster Computations and the Desktop *Proc. IEEE Intl. Parallel and Distributed Processing Symposium* (Rome)
[8] Agostinelli S et al 2003 Geant4 - a simulation toolkit *Nuclear Instruments and Methods in Physics Research* **A 506** 250-303
[9] Fuller S H and Millet L I (Editors) 2011 *The Future of Computing Performance: Game Over or Next Level?* The National Academies Press.

Study of cache performance in distributed environment for data processing

Dzmitry Makatun[1,3], Jérôme Lauret[2] and Michal Šumbera[3]
[1]Faculty of Nuclear Physics and Physical Engineering, Czech Technical University in Prague
[2]STAR, Brookhaven National Laboratory, USA
[3]Nuclear Physics Institute, Academy of Sciences, Czech Republic

E-mail: makatun@rcf.rhic.bnl.gov

Abstract. Processing data in distributed environment has found its application in many fields of science (Nuclear and Particle Physics (NPP), astronomy, biology to name only those). Efficiently transferring data between sites is an essential part of such processing. The implementation of caching strategies in data transfer software and tools, such as the Reasoner for Intelligent File Transfer (RIFT) being developed in the STAR collaboration, can significantly decrease network load and waiting time by reusing the knowledge of data provenance as well as data placed in transfer cache to further expand on the availability of sources for files and data-sets. Though, a great variety of caching algorithms is known, a study is needed to evaluate which one can deliver the best performance in data access considering the realistic demand patterns. Records of access to the complete data-sets of NPP experiments were analyzed and used as input for computer simulations. Series of simulations were done in order to estimate the possible cache hits and cache hits per byte for known caching algorithms. The simulations were done for cache of different sizes within interval 0.001 - 90% of complete data-set and low-watermark within 0-90%. Records of data access were taken from several experiments and within different time intervals in order to validate the results. In this paper, we will discuss the different data caching strategies from canonical algorithms to hybrid cache strategies, present the results of our simulations for the diverse algorithms, debate and identify the choice for the best algorithm in the context of Physics Data analysis in NPP. While the results of those studies have been implemented in RIFT, they can also be used when setting up cache in any other computational work-flow (Cloud processing for example) or managing data storages with partial replicas of the entire data-set.

1. Introduction

Efficient usage of available cache space is important for transferring and accessing data in computational grids. Though, a great variety of caching algorithms is known, a study is needed to evaluate which one can deliver the best performance in data access considering the realistic demand patterns.

Cache cleaning algorithms can be applied to keep in the cache of data-transfer tools files that may be reused. The size of those cache is small (several percent of the entire dataset) and the clean up has to take place regularly to make space for further transfers. Another task can, for example, be to delete a part of local data replica if no longer in use or requested. The problem posed by cache cleanup is to select and delete files which are the least likely to be used again. An investigation to find the most appropriate algorithm is required.

In this study, all the caching algorithm were implemented following the concept known as "water-marking". Water-marking is an approach where thresholds are set for the cache cleanup starts and stops. It considers the current disk space occupied by the data in cache and the high-mark and the low-mark for cache size are externally set up and can be adjusted as needed. When the used cache size exceeds the high-mark, the cache clean-up starts, and files are deleted until the used cache size gets below the low-mark. The time interval between clean-ups depends on combination of high/low marks, cache size and data-flow. With watermarking concept more computational demanding algorithms can be implemented as the cleanup may be independent of the data transfers.

2. Access patterns

Several data access patterns were extracted from log files of data management systems at sites of HEP/NPP experiments in order to simulate caching. Three different access patterns were used as input for our simulations:

STAR1: the pattern was extracted from Xrootd [8] logs taken from the STAR experiment's Tier-0 site of RHIC Computing Facility at Brookhaven National Laboratory (RCF@BNL), it consist of records made during a 3 months period (June-August 2012) of all datasets available in STAR.

STAR2: extracted from the same data management system, but of records made during a 7 months period (August 2012 - February 2013) under similar conditions.

GOLIAS farm is part of regional computing center for particle physics at the Institute of Physics (FZU) in Prague, and is part of a Tier-2 site for the CERN/ATLAS experiment. The facility also performs data processing for another experiment - AUGER, which makes less than 1% of the total requests. The pattern was extracted from DPM [9] logs for a 3 months period (November 2012 - February 2013).

The usage of access patterns corresponding to different time periods and experiments helps to verify the results of our simulations. As input of our simulations, we tried to focus on a few characteristic access patter. The key parameters we came up with for the three access patterns are given in Table 1. Both STAR access patterns have similar parameters. It is noteworthy to mention that the first one was taken right before the Quark Matter 2012 conference and the second one, right after. This is important as the access to data is intensified before a conference and without verification, it would be doubtful if our findings would be stable across time. The number of files requested only once during the period, is less than 10% in both cases.

The FZU/GOLIAS access pattern is taken from another experiment with different data-storage structure, DPM is used here within a Tier-2 data access context (user analysis). This access pattern is much less uniform and differs from the other two: the size of files is not explicitly limited and can reach 18 GB, the number of requests for a file varies from 1 up to 94260, with an average 5, while it varies from 1 to 203 for the STAR patterns. 44% of files were requested only once

Our analysis is not sensitive to the particular Data Management system, (XRootD or DPM), further we explain differences in the access patterns with specifics of experiment and role of the cluster in the system (tier level).

3. Cache simulation

The efficiency of caching can be estimated by two quantities, the cache hits H and cache hits per megabyte of data H_d from here on referred to as *cache data hits*:

$$H = \frac{N_{cache}}{N_{req} - N_{set}} \tag{1}$$

Table 1. Summary of three user access patterns used as input for simulations. The selection of two time sequence in STAR and one from a different experiment aims at verifying stability of our result and findings.

		STAR1	STAR2	GOLIAS
Time period	months	3	7	3
Number of requests	$\times 10^6$	33	52	21
Data transferred	PB	50	80	10
Maximal number of requests for one file	–	192	203	94260
Average number of requests per file	–	19	15	5
Number of unique files	$\times 10^6$	1.8	1.7	3.8
Total size of dataset	PB	1.45	2	1
Maximal file size	GB	5.3	5.3	18
Average file size	GB	0.8	1	0.3

$$H_d = \frac{S_{cache}}{S_{req} - S_{set}} \tag{2}$$

where N_{req} is the total number of requests, S_{req} -the total amount of transferred data in bytes, N_{set} -the number of unique files witch were requested at least ones during the considered period, S_{set} - the total size of those unique files in bytes, N_{cache} - the number of files transferred from cache, S_{cache} - the amount of data transferred from cache in bytes.

By maximizing the cache hits H one reduces the number of files transferred from external sources and thus reduces the overall make-span due to transfer startup overhead for each file. By maximizing the cache data hits H_d one reduces the network load, since more data is reused from the local cache.

If the access pattern is completely random, the expected cache hit and cache data hits would be equal to $cache\ size/storage\ size$, so it can be useful to compare the actual cache performance to this estimation.

Altogether 27 different caching algorithms were simulated. But the majority of studied algorithms did not bring any improvements over the simplest one (FIFO). Only the algorithms that appeared to be the most efficient are discussed in this paper:

- **First-In-First-Out (FIFO)**: evicts files in the same order they entered the cache. Performance of this trivial algorithm provide a good comparison benchmark against more sophisticated ones which can demand significant computational resources.

○ **Least-Recently-Used (LRU)**: evicts the set of files which were not used for the longest period of time.

● **Least-Frequently-Used (LFU)**: evicts the set of files which were requested less times since they entered the cache.

★ **Most Size (MS)**: evicts the set of files which have the largest size.

\+ **Adaptive Replacement Cache (ARC)**[5]: splits cached files into two lists: L1 - files with $access\ count = 1$, and L2 - files with $access\ count > 1$. LRU is then applied to both list, the self adjustable parameter $p = cache\ hits\ in\ L1/cache\ hits\ in\ L2$ defines the number of cached files in each list. The general idea is to invest more into the list which delivers more hits.

$*$ **Least Value based on Caching Time (LVCT)**[4]: Deletes files with the smallest value of the Utility Function:

$$UtilityFunction = \frac{1}{CachingTime \times FileSize} \quad (3)$$

where **Caching Time** of a file F is total size of all files accessed after the last request for the file F.

$\bigtriangledown$ **Improved-Least Value based on Caching Time (ILVCT)**[3]: Deletes files with the smallest value of the Utility Function:

$$UtilityFunction = \frac{1}{NumberOfAccessedFiles \times CachingTime \times FileSize} \quad (4)$$

where **Caching Time** is the same as for LVCT and **Number Of Accessed Files** is a number of files requested after the last request for selected file.

4. Results

Three series of simulations with three access patterns were performed for each algorithm (90 simulations in total for each algorithm):

- 10 simulations with cache size 1-90 % of storage with fixed low-mark 75% and high-mark 95%. Those simulations aim to understand what would happen if we have large storage cache. Those cases are aligned with a DPM and Xrootd use where most (if not all) the dataset(s) are in the system.
- 10 simulations with cache size 1.2 - 0.0025% of storage with fixed low-mark 75% and high-mark 85%. This high-mark was selected in order to leave enough margin (15%) in case when a large file is requested at the moment when the cache is almost full. We used those simulations to understand the behavior of cache cleanup if the cache size is by several orders less than the dataset size. This is in fact a most common case for transfer cache on data transfer nodes.
- 10 simulations with fixed cache size 10% of storage,fixed high-mark 95% and variable low mark within 0-90%. We performed those simulations to better understand the effect of data retention on cache (delete the least in hope of re-use).

In order to compare one algorithm against another an average improvement can be calculated in a following way:

$$Average\ improvement = \frac{\sum_{i=1}^{n} \frac{value2_i - value1_i}{value1_i}}{n}, \quad (5)$$

where n is the total amount of simulations with equal parameters for both algorithms, i is the number of the simulation,$value1$ - cache hits or cache data hits of a reference algorithm (FIFO), $value2$ - cache hits or cache data hits of a compared algorithm.

Table 2 contains the results of comparison of all the algorithms represented in this paper against FIFO. 60 values for each algorithm were taken from results of simulation series 1 and 2 in order to calculate the average improvement. According our results, the LFU algorithm does not bring any improvement over FIFO due to its well known flaw - it accumulates files which were popular for a short period of time, and those files prevent newer ones from staying in cache. The ARC algorithm was developed as an improvement to LRU, and not surprisingly, it outperforms LRU by ~5% in cache hits and ~7% in cache data hits. Therefore, LFU and LRU algorithms could be excluded from the further analysis in our case studies.

Table 2. Average improvement of algorithms over FIFO.

Algorithm	cache hits	cache data hits
MS	**116** %	-20 %
LRU	8 %	5 %
LFU	-27 %	-18 %
ARC	13 %	**11** %
LVCT	**86** %	2 %
ILVCT	28 %	2 %

The graphical detailed results of simulations for all 3 series are given at Figures 1-3. The performance of FIFO and 3 algorithms appeared to be the most efficient (MS, ACR and LVCT) is presented at the plots.

Difference between Tier-2 and Tier-0 access patterns leads to distinct cache performance. Only the data dedicated for the ongoing analysis is placed at the Tier-2 site, while at the Tier-0 site all the experimental data is stored. As a result – the access pattern at the Tier-2 site has stronger access locality. STAR1 and STAR2 access patterns correspond to Tier-0 site and GOLIAS to a Tier-2 site. Thus, any particular algorithm at the plots delivers higher cache hits and cache data hits for GOLIAS access pattern than for STAR1 and STAR2.

The behavior of algorithms is similar within each dataset that is, their respective performance ordering is the same. This observation implies that if one of the algorithm appears to be most efficient for one of the datasets it is also the most efficient for the other datasets. This statement is also true for the rest of simulated algorithms not present on our figure. Though the communities represented by the STAR and GOLIAS access patterns are somewhat similar, this result is slightly surprising as our case studies represent two time sequence within the same usage and totally uncorrelated experiments. It would be interesting to study those algorithms in a different experimental context (outside the HEP/NPP communities) but this study is outside the scope of our paper.

The MS algorithm has shown outstanding cache hits, but the lowest cache data hits. At the same time the LVCT has cache hits comparable to the MS while cache data hits are 2% improved over the FIFO. This algorithm could be an optimal when the cache hits is the target optimization parameter. The ARC algorithm has shown the highest cache data hits for studied access patterns.

The dependence of algorithms performance on low mark is presented at Figure 3. With higher low mark the number of clean-ups increases and that is why the difference between algorithms becomes more notable. Performance of efficient algorithms (FIFO, LRU, ARC and LVCT) increases steadily with the low mark. One should be careful when setting up a cache low mark at a particular site, since a higher low mark can increase cache performance significantly, but at the same time it can result in running cache clean-ups too often, consuming significant computational resources (and potentially increasing the chance to interfere with data transfers hence, degrading transfer performances if delete/writes/read overlap).

5. Conclusion

Performance of cache algorithms implemented with watermarking concept was simulated for a wide range of cache sizes and low marks. Three access patterns from Tier-0 and Tier-2 sites of 2 different experiments were used as input for simulations. Regardless of the cache size, Tier-level and specificity of experiment the LVCT and ARC appeared to be the most efficient caching algorithms for the communities we investigated. While we found the stability of relative

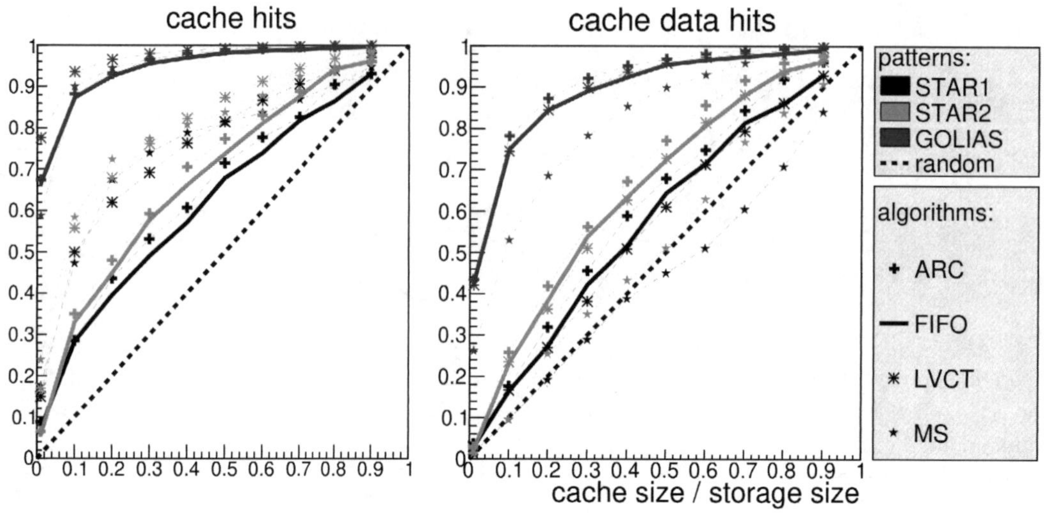

Figure 1. Results of simulation. Performance of algorithms for cache of larger size can be compared. For all of the simulations on this plot the following parameters were fixed: low mark = 0.75, high mark = 0.95

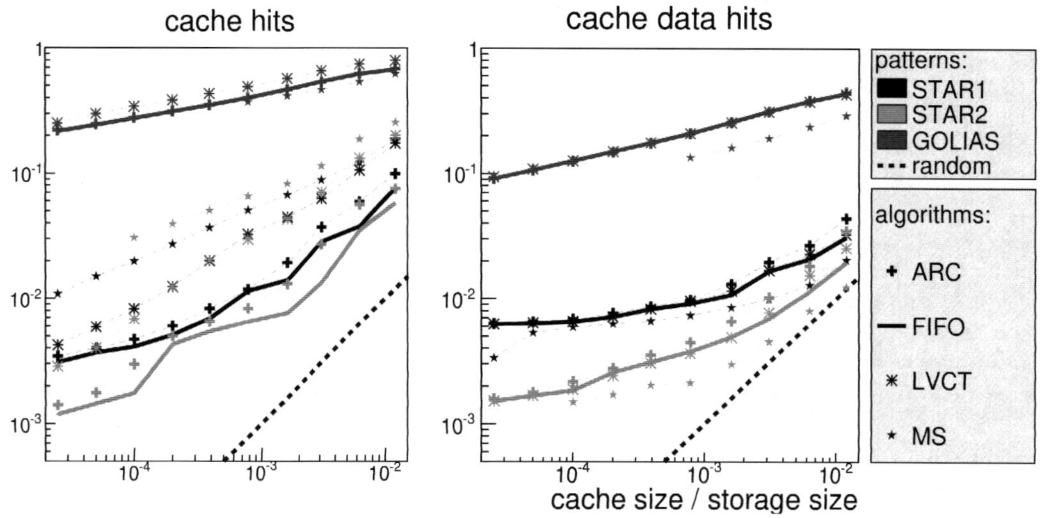

Figure 2. Results of simulation. Performance of algorithms for cache of smaller size can be compared. For all of the simulations on this plot the following parameters were fixed: low mark = 0.75, high mark = 0.85

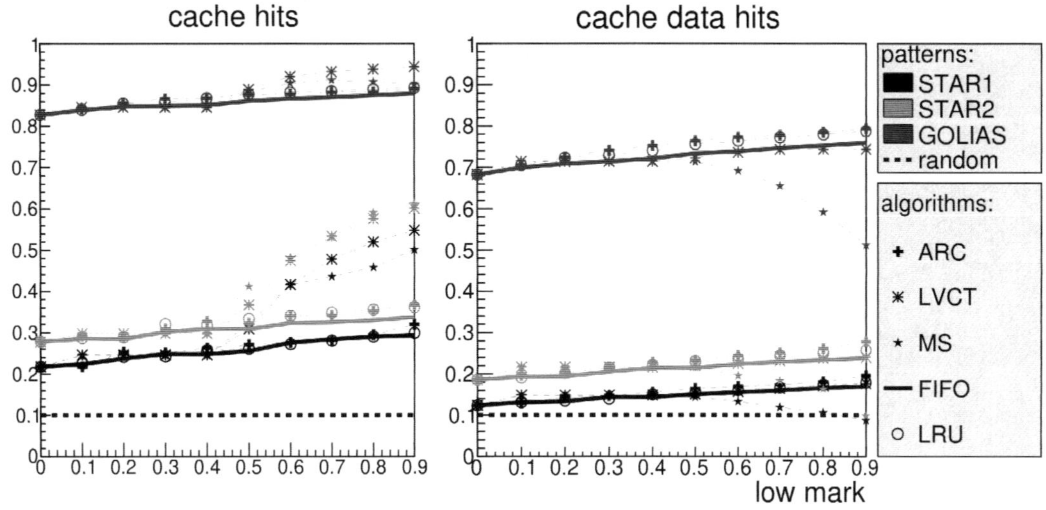

Figure 3. Results of simulation: dependence of cache performance on low mark. For all of the simulations on this plot the following parameters were fixed: cache size / storage size = 0.1, high mark = 0.95

algorithms' performance surprising at first, we attribute this result to an access pattern which is intrinsically similar in nature. An extension of this work could be the investigation of this result in a different experiment context which is a work beyond our initial goal. LVCT and ARC could certainly be safely implemented in tools such as RIFT.

- If the goal is to minimize makespan due to a transfer startup overhead the LVCT algorithm should be selected.
- If the goal is to minimize the network load the ARC algorithm is an option.

Acknowledgments

The work has been supported by the grant 13-02841S of the Czech Science Foundation (GACR) and the US Department of Energy. The support of the Visegrad Fund is gratefully acknowledged. Furthermore the authors would like to acknowledge with much appreciation the help of the staff of GOLIAS computer farm for kindly providing them with the log files and access to the facility for running simulations.

References

[1] Makatun D, Lauret J and Sumbera M 2012 Distributed Data processing in high-energy physics *Proc. of PhD students workshop at FNSPN CVUT* (Prague) ISBN 978-80-01-05138-2
[2] Zerola M, Lauret J, Bartak R and Sumbera M 2012 One click dataset transfer: toward efficient coupling of distributed storage resources and CPUs *J. Phys.: Conf. Ser.* **368**
[3] Achara J P, Rathore A, Gupta V K and Kashyap A 2010 An improvement in LVCT cache replacement policy for data grid *Proc. of the 13th Int. Workshop on Advanced Computing and Analysis Techniques in Physics Research* (Jaipur) p 44
[4] Song Jiang and Xiaodong Zhang 2003 Efficient Distributed Disk Caching in Data Grid Management *Proc. of the IEEE Int. Conf. on Cluster Computing (CLUSTER03)* (Hong Kong) pp 446-51
[5] Megiddo, Nimrod and Modha D S 2004 Outperforming LRU with an adaptive replacement cache algorithm *Computer* **37** 58-65
[6] Fast Data Transfer *Project web-site: http://monalisa.cern.ch/FDT/*
[7] High Performance Storage System *Project web-site: http://www.hpss-collaboration.org/*

[8] Xrootd *Project web-site: http://xrootd.slac.stanford.edu/*
[9] Disk Pool Manager *Project web-site:* https://svnweb.cern.ch/trac/lcgdm/wiki/Dpm

TGeoCad: an Interface between ROOT and CAD Systems

C Luzzi [1] [2] and F Carminati [2]

[1] University of Ferrara, Via Giuseppe Saragat, 1, 44020 FE, Italy
[2] CERN, 1211 Geneva 23, Switzerland

E-mail: `cluzzi@cern.ch`

Abstract. In the simulation of High Energy Physics experiment a very high precision in the description of the detector geometry is essential to achieve the required performances.
The physicists in charge of Monte Carlo Simulation of the detector need to collaborate efficiently with the engineers working at the mechanical design of the detector. Often, this collaboration is made hard by the usage of different and incompatible software. ROOT is an object-oriented C++ framework used by physicists for storing, analyzing and simulating data produced by the high-energy physics experiments while CAD (Computer-Aided Design) software is used for mechanical design in the engineering field.
The necessity to improve the level of communication between physicists and engineers led to the implementation of an interface between the ROOT geometrical modeler used by the virtual Monte Carlo simulation software and the CAD systems.
In this paper we describe the design and implementation of the TGeoCad Interface that has been developed to enable the use of ROOT geometrical models in several CAD systems. To achieve this goal, the ROOT geometry description is converted into STEP file format (ISO 10303), which can be imported and used by many CAD systems.

1. Introduction

The incompatibility between the ROOT [1] geometrical modeler and the CAD systems comes from the different representation of the geometrical shapes.

In a CAD system, the shapes are treated as *boundary represented objects* (BREPS), a BREPS is defined by the description of its boundaries. Boundary representation models are based on two elements: topology and geometry. The topology gives information about the relationships between the main topological items, i.e. faces (bounded portion of a surface), edges (bounded piece of a curve) and vertices (points). The geometry describes the integration of these elements in space. Each solid is placed by a translation/rotation matrix in an assembly of many solids and the different assemblies form a complete hierarchy of geometrical objects.

On the other side, in ROOT the Constructive Solid Geometry method is adopted according to which the solids are defined directly as three-dimensional primitives and created out of a set of simple shapes like cuboids, cylinders, prisms, pyramids, spheres. Complex shapes are created by using boolean operation between simple shapes. Each shape can then be placed by translation/rotation matrices in assemblies. For more information about solid modelling see [2].

Another substantial difference between ROOT and CAD systems is the hierarchy of assemblies.

In a CAD system the basic unit is a solid part that can have an arbitrary number of nested assemblies.

A ROOT geometry is composed by a number of volumes. The largest one is used as a container for all other volumes in the detector geometry. Each volume created is placed inside another one, called the mother volume. A volume is composed by its shape and all its physical characteristics as the material of the volume, whether it contains any sensitive detector elements, the magnetic field, etc [3].

The main task of the TGeoCad interface is to analyse the ROOT shape representation, extract all the information about the solid and reproduce it in a BREPS format using the OpenCASCADE Technology [4] libraries which provide also the possibility to write BREPS assemblies in a STEP file.

2. The Open CASCADE Technology

The Open CASCADE Technology (OCCT) is an open source software development platform written in C++. It has been designed for rapid production of sophisticated domain-specific design applications.

OCCT allows to develop application dealing with two or three-dimensional (2D or 3D) geometric modeling in general-purpose or specialized CAD systems, manufacturing or analysis applications, simulation applications, or illustration tools.

The OCCT libraries are grouped into six modules, those used by the TGeoCad interface are:

- *Foundation Classes:* it is used for the description of elementary geometric shapes (points, vectors, lines, circles, conics, planes), for the positioning of geometries in space or on a plane by means of an axis or a coordinate system and for the definition of geometric transformations such as translation, rotation and symmetries;
- *Modeling Data:* it allows building pure topological data structures defining relationships between simple geometric entities and giving the possibility to assign to a shape an orientation and a location. Using the abstract topological data structures each shape can be divided into various topological components as shown in Fig.1:

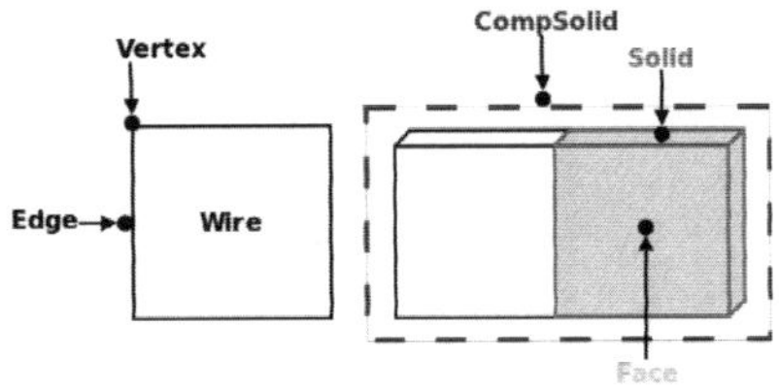

Figure 1. Topological data structures

- *Modeling Algorithm:* it provides geometric and topological algorithms for creating vertices, edges, faces and solids. It gives the possibility to directly build primitive objects such as boxes, wedges and rotational objects. Moreover, sweeping and Boolean operations are also provided.
- *Extended data exchange (XDE) module:* it allows writing applications based on the Open CASCADE Technology in the IGES and/or STEP (AP203, AP214) file formats.
- *Open Cascade Application Framework (OCAF):* it is a Rapid Application Development (RAD) framework used for specifying and organizing application data according to which data structure are organized by mean of a reference model. The reference can point to any

type of data in the application and it is implemented in the form of labels. Application data such as a shape itself or a shape location are attached to these labels as attributes. The OCAF application is used to manage an XDE document which is a container for the reference keys composed by a set of labels organized in a tree structure. Each label has a tag expressed as an integer value and it is identified by a string build by concatenation of tags from the root of the tree, e.g. [0:1:2] (see Fig.2).

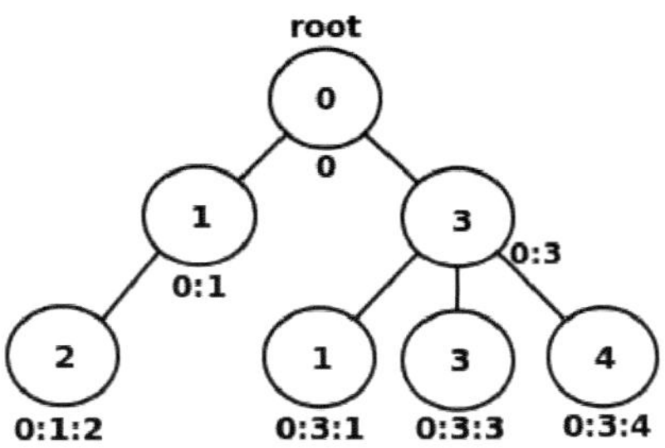

Figure 2. The labels tree

3. The STEP Standard

STEP stands for STandard for the Exchange of Product model data and it is the ISO 10303 standard intended to handle a wide range of product-related (electronic, electro-mechanical, mechanical, architectural, process plant, furniture, etc) data covering the entire life-cycle of a product (design, analysis, planning, manufacture).

The STEP format was designed to meet the need of having a common set of CAD/CAM tools within most companies using a neutral format for the file exchange in order to reduce the number of needed translators.

The components of the STEP standard are divided into several series of parts. Each part series contains one or more types of ISO 10303 parts. The OpenCascade technology gives the possibility to create files according to the STEP parts AP203 and AP214.

The AP 203 specifies the application protocol for configuration controlled three-dimensional design. It is related to the transfer of product shape models, assembly structure, and configuration control information such as part versioning and release status. Using AP 203 assemblies of positioned and oriented part models can be exchanged.

The AP 214 specifies core data for automotive mechanical design processes. It is an extension of AP203 and it also manages manufacturing tools, tolerance data, numerical control data, data for kinematic simulation, 2D drawings, and product data management information.

The STEP format is a neutral format well supported by almost any 3D CAD software. It provides a mechanism to describe product data independent from any particular system. For more information about the STEP standard see [5].

4. Details of the conversion

TGeoCad has different way to reproduce a ROOT shape in the OpenCASCADE platform:

- Shapes created step by step starting from points such as box, parallelepiped, trapezoid etc.
  - For example the ROOT TGeoTrd1 is translated creating edges from points, wire from edges, faces (planar surfaces) from wires, shells from faces and solid from shells;

- Shapes created using OCCT capabilities for solid primitives creation and boolean operations such as tube, cone, sphere, etc.
  - As shown in the Fig.3, the TGeoCone is created using the OCCT libraries which giving the height and radius information allow to create an inner and outer cones and to substract the inner cone from the outer cone.

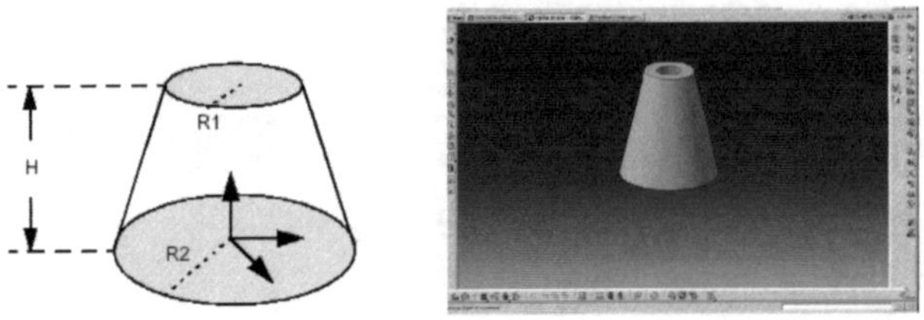

Figure 3. a. OCCT cone representation; b. A TGeoCone translated in a STEP file and loaded in CATIA.

- Shapes created by using modeling algorithm (extrusions, revolutions, lofts) applied to basic geometries such as the hyperboloid;
- TGeoCompositeShape created using OCCT Boolean operations between two or more shapes.

The conversion method is selected according to the type of each shape. When it is possible we use directly the OCCT libraries for solid primitives, sometimes combined with the OCCT modeling algorithms. When there is not a suitable OCCT library for the creation of a specific solid we create it starting from points.

5. Structure of the interface

Three main classes compose the TGeoCad interface:

- *TGeoToStep*: a TGeoObject based class that takes the TGeoManager pointer and creates the geometry step file;
- *TGeoToOCC*: implements all the methods shown in the Table n.1 to translate each ROOT shape in the corresponding OCCT shape;

Table 1. Methods implemented to traslate ROOT shapes in OCCT shapes.

ROOT	TGeoToOCC	ROOT	TGeoToOCC
TGeoBBox(..)	OCC_Box(..)	TGeoTrd2(..)	OCC_Trd(..)
TGeoSphere(..)	OCC_Sphere(..)	TGeoTubeSeg(..)	OCC_Tube(..)
TGeoArb8(..)	OCC_Arb8(..)	TGeoCtub(..)	OCC_Cuttub(..)
TGeoConeSeg(..)	OCC_Cones(..)	TGeoTube(..)	OCC_TubeSeg(..)
TGeoCone(..)	OCC_Cones(..)	TGeoPcon(..)	OCC_Pcon(..)
TGeoPara(..)	OCC_ParaTrap(..)	TGeoTorus(..)	OCC_Torus(..)
TGeoGtra(..)	OCC_ParaTrap(..)	TGeoPgon(..)	OCC_Pgon(..)
TGeoTrd1(..)	OCC_Trd(..)	TGeoEltu(..)	OCC_Eltu(..)
TGeoHype(..)	OCC_Hype(..)	TGeoXtru(..)	OCC_Xtru(..)
TGeoComposite(..)	OCC_Composite(..)		

- *TOCCToStep*: reproduces the ROOT geometry tree (mother-children relationship) in the XDE document and writes it to the STEP file:

– As first step an OCAF application is created in order to manage the XDE document, afterwards shapes are created and memorized on the XDE document without reporting any information about the relathionship between shapes.
As shown in Fig.4, starting from the top of the ROOT geometry tree the *OCCShapeCreation(..)* method translates each ROOT shape in the OCCT version using all the methods shown in the Table n.1.
For each translated shape a new label is created and memorized into the XDE document. The correspondence shape-label is also stored in a map of volumes and labels in order to keep memory of the shapes already translated.
Therefore, a check is done for each shape taken from the root file. If the shape-label correspondence is present in the map it means that the shape has already been translated then the label is copied from the map to the XDE document using the up-to-date location of the current root shape. If there is no correspondence present in the map, the shape needs to be translated.

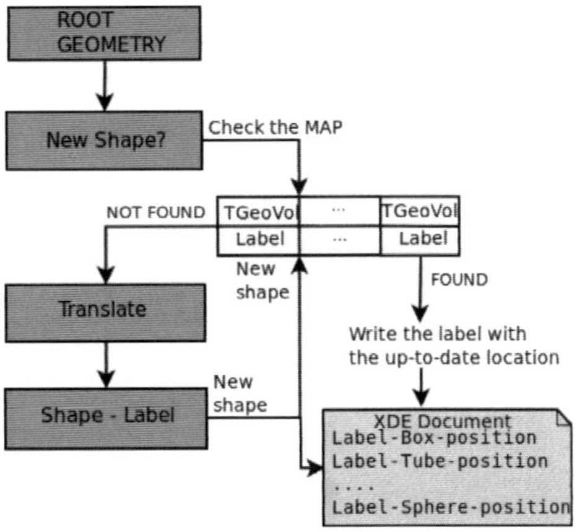

Figure 4. The OCCShapeCreation method

– As second step, Fig. 5 shows the *OCCTreeCreation(..)* method which starts to browse the geometry and for each node from the end to the top of the ROOT physical tree the mother and daughter label references are found in the map. In addition the daughter location matrix is obtained from the root file. These parameters are used to calculate the position of the daughter with respect to the mother. The connection between the daughter label and the mother label is then created, resulting in a new label, which is added to the XDE document.
The OCCT shapes are added, step by step, to an assembly entity where each of its

Journal of Physics: Conference Series **523** (2014) 012017 doi:10.1088/1742-6596/523/1/012017

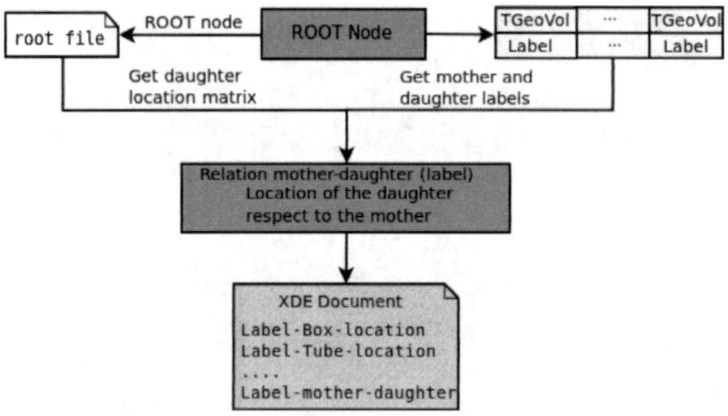

Figure 5. The OCCTreeCreation method

sub-shapes define a sub label therefore each node of the assembly refers to its subshapes.

– At last, the XDE document containing the whole geometry is written to the STEP file. An example of STEP file is shown in the Fig. 6

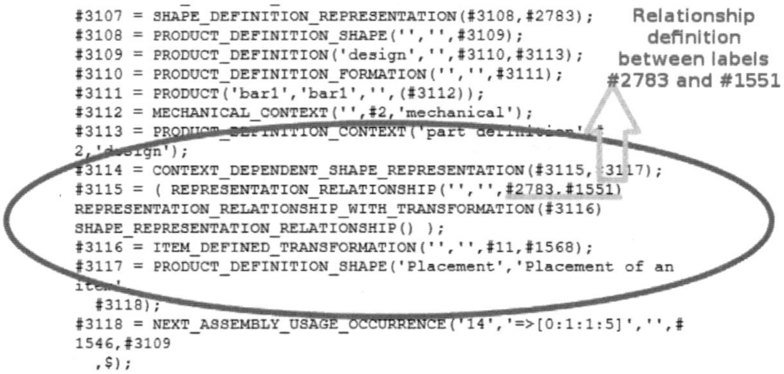

Figure 6. Relationship between two shapes reproduced in the STEP file.

Fig. 7 shows the whole architecture of the TGeoCad interface.

6. Technical details

In order to have the TGeoCad interface available in ROOT, OpenCascade must be installed and ROOT needs to be compiled using the configuration options:

```
./configure --enable-geocad;
--with-occ-incdir:  location of OpenCascade inc files
--with-occ-libdir:  location of OpenCascade lib files
```

The TGeoCad has to be used from ROOT by loading a geometry in memory, creating an object of the TGeoToStep class and calling the method CreateGeometry(). A STEP file called geometry.stp containing the geometry will be created on the current directory.

```
root[0] gSystem->Load("libGeoCad.so");
root[1] .x roottest.C
```

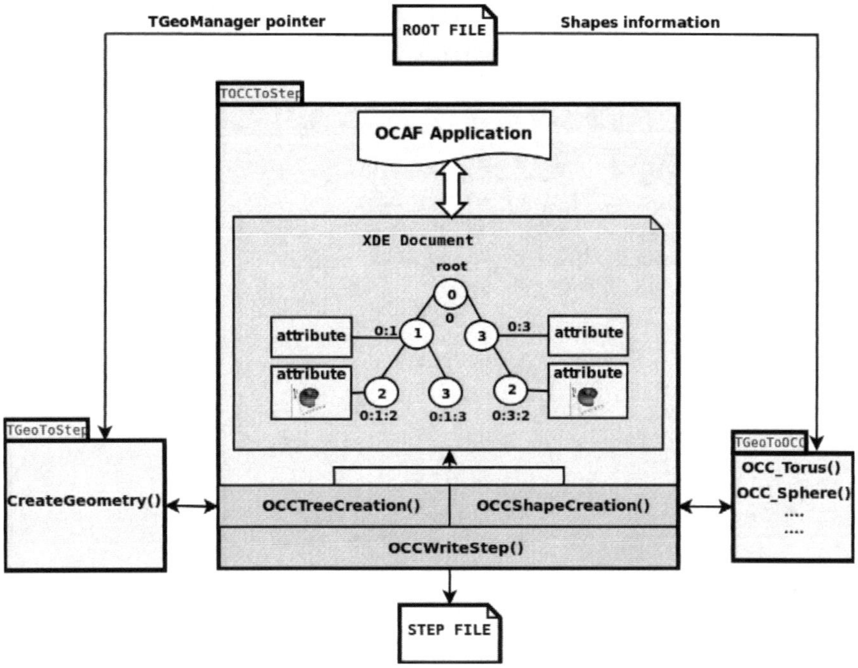

Figure 7. The TGeoCad architecture

```
root[2] TGeoToStep *myStep = new TGeoToStep (gGeoManager);
root[3] myStep->CreateGeometry();
```

7. Conclusion

The table 2 reports the number of volumes and nodes of the ROOT geometry, the dimensions of the ROOT files and of the related STEP files, as well as the execution time of each conversion.

Table 2. Information related to the conversion of several geometries.

Nr. Volumes	Nr. Nodes	ROOT file	STEP File	Execution time
13	485	5.3 KB	143.1 KB	0.18 s
90	184	19.7 KB	1.4 MB	792.53 s
645	646	36 KB	10.3 MB	10.27 s
936	936	13.1 KB	15.6 MB	16.22 s
7143	29046966	328.2 KB	13.9 MB	6303.58 s

The execution time is not directly proportional to the number of volumes or to the dimension of the root file. We can assert that, in the majority of the cases, the execution time depends on the number of boolean operations required to reproduce the geometry. These boolean operations are time consuming in OpenCascade.
The dimension of the STEP file depends on the size of the ROOT file and on how many times a same node is repeated in the geometry. Every time that a node is repeated in the ROOT geometry, the implementation of the shape does not take place but only the reference to the object is re-written in the file by modifying the location.

Journal of Physics: Conference Series **523** (2014) 012017 doi:10.1088/1742-6596/523/1/012017

Acknowledgements
We would like to thank Prof. Eleonora Luppi, Dr. Latchezar Betev, Dr. Rene Brun, Dr. Andrei Gheata and the whole ROOT team for their support and advices throughout this work.

References
[1] *An Object-Oriented Data Analysis Framework* http://root.cern.ch
[2] Freeman W. H. & Co. 1988 *Introduction to Solid Modeling* New York NY USA ISBN:0-88175-108-1
[3] *The ROOT geometry package* http://root.cern.ch/download/doc/18Geometry.pdf
[4] *The OpenCASCADE Technology* http://www.opencascade.org/
[5] *ISO 10303* http://www.iso.org

The ATLAS muon and tau triggers

L Dell'Asta[1] on behalf of the ATLAS Collaboration
[1] Boston University, US
E-mail: dellasta@cern.ch

Abstract. The ATLAS experiment at CERN's Large Hadron Collider (LHC) deploys a three-level processing scheme for the trigger system. The development of fast trigger algorithms and the design of topological selections are the main challenges to allow for a large program of physics analyses. In the following, two of the ATLAS trigger signatures are described: the muon and the tau triggers. The structure of the three levels of these two trigger signatures are explained in detail as well as their performance during the first three years of operation.

1. Introduction

ATLAS is one of the two multipurpose experiments at the LHC. Physics goals include the search for the Higgs boson and new physics beyond the Standard Model. The ATLAS detector consists of an inner tracking detector surrounded by a superconducting solenoid which provides a 2 T magnetic field, electromagnetic and hadronic calorimeters and a muon spectrometer with a toroidal magnetic field. More details can be found in Ref. [1].

In 2012, the LHC provided pp collisions at 20 MHz, i.e. every 50 ns. Of the incoming rate, ATLAS recorded approximately 400 Hz for physics analyses. There are various reasons not to record every collision. Firstly, the processes that are of interest to physicists occur at rates of 10, 1 or < 0.1 Hz, which corresponds to a tiny fraction of the total events produced. For example, a process like $H \rightarrow ZZ \rightarrow \mu\mu\mu\mu$ occurs once in 10^{13} pp interactions (at 14 TeV). Secondly, computing resources are limited. It takes $\sim$10 s to perform the offline processing of one event, so even at just 400 Hz, around 4000 CPUs are needed. Moreover, one event in raw format takes $\sim$1.5 MB, and almost the same size is needed in addition after its reconstruction. With this data size, up to 4300 TB a year per experiment are needed.

The trigger system takes care of selecting, among the 20 MHz of pp collisions, the 400 Hz that will be then analyzed by physicists. The triggers are based on identifying combinations of candidate physics objects (*signatures*), e.g. muons and taus. The trigger system is configured via a trigger *menu* which defines trigger *chains*. A sequence of reconstruction and selection steps for specific objects in the trigger system is specified by a trigger chain which is often referred to as a trigger. Trigger signatures and trigger menus are driven by physics goals. Just to give two examples: the Higgs boson can decay as $H \rightarrow ZZ \rightarrow \mu\mu\mu\mu$. Therefore a typical signature of the production of the Higgs boson would be the presence of four muons. It is thus important to have a trigger system which can select efficiently events with many muons. Another example is the decay of the Higgs boson in tau pairs, $H \rightarrow \tau_{had}\tau_{lep}$. In this case the signature is the presence of an hadronic tau and a lepton, e.g. a muon. A combined trigger is therefore useful.

Interesting events as the ones just described are hidden in the very busy LHC environment. Collisions at the LHC are characterized by high pile-up (i.e. multiple interactions per crossing).

In Table 1 the peak instantaneous luminosities during the first three years of operation are shown. In 2012, when a peak instantaneous luminosity of $7.73 \times 10^{33}\text{cm}^{-2}\text{s}^{-1}$ was reached, events with 35 interactions were recorded, as shown in Fig. 1. When the number of multiple interactions is so high, events are also characterized by a high charged track multiplicity. In Fig. 2 the event display of a $Z \to \mu\mu$ event candidate is shown. The interesting event is overlaid with 25 additional interactions. In such a busy environment several trigger levels are needed to be able to exploit tracking information at the trigger selection.

Table 1. Peak instantaneous luminosities and recorded integrated luminosities during the first three years of ATLAS data taking.

Year	Center of mass energy	Peak Instantaneous Luminosity	Recorded Integrated Luminosity
2010	7 TeV	$2.1 \times 10^{32}\text{cm}^{-2}\text{s}^{-1}$	45.0 pb^{-1}
2011	7 TeV	$3.65 \times 10^{33}\text{cm}^{-2}\text{s}^{-1}$	5.25 fb^{-1}
2012	8 TeV	$7.73 \times 10^{33}\text{cm}^{-2}\text{s}^{-1}$	21.7 fb^{-1}

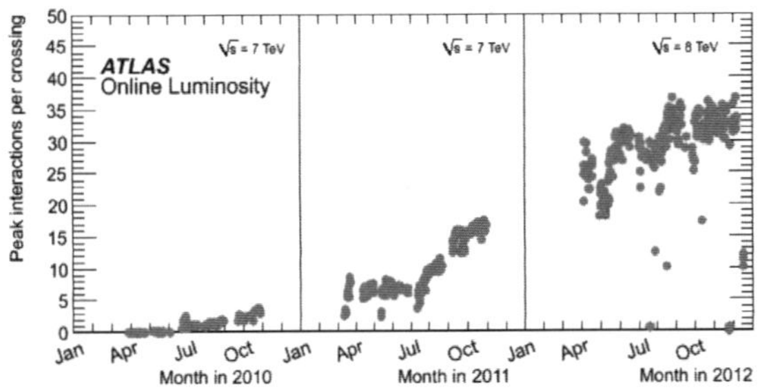

Figure 1. The maximum mean number of events per beam crossing versus day during the *pp* runs of 2010, 2011 and 2012 [2].

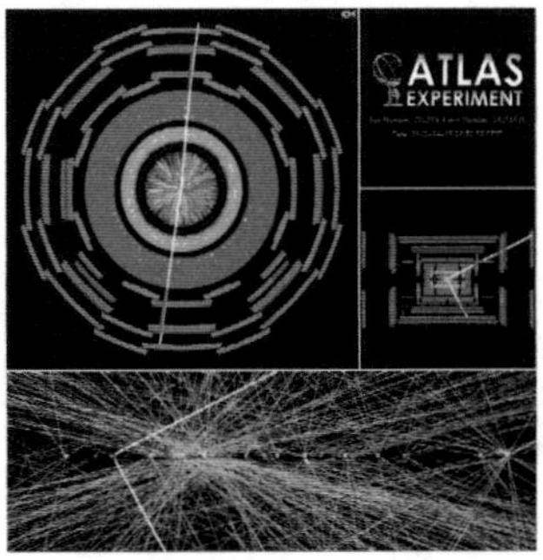

Figure 2. A $Z \to \mu\mu$ candidate event together with 25 additional pile-up reconstructed vertices [3].

The ATLAS trigger system is divided into three levels. At each level, the rate is significantly reduced and more detailed information can be used. The rate reduction factor between the incoming rate and the first level of the trigger is ~300, while the two following levels have reduction factors of ~10.

The first level of the trigger system is called Level 1 (L1) and it is hardware-based. It analyzes coarse granularity data from calorimeter and muon detectors separately and identifies Regions-of-Interest (RoI), detector regions to be further analyzed by the High Level Trigger (HLT). After the L1, there is the HLT which is subdivided into two separate levels that are software-based. The Level 2 (L2) accesses the full granularity data within the RoI, which corresponds approximately to 2% of the total event size. At this stage, detector information like tracking is combined. Topological requirements can also be implemented at L2. The Event Filter (EF) uses algorithms similar to offline for object reconstruction. The EF exploits the seed from the L2 and uses the full event data.

In the following, two of the ATLAS trigger signatures are described in detail: the muon and tau triggers.

2. Muon trigger

Muons played a fundamental role in the discovery of the Higgs boson. The discovery announced on the 4th July 2012 (see Ref. [4]) was partly based on the search for the Higgs boson decaying into a Z boson pair, with each Z boson decaying into two muons.

Muons are characterized by the presence of a track in the Muon Spectrometer (MS) and a track in the Inner Detector (ID). ATLAS has specific detectors devoted to triggering these leptons [5].

The Level 1 muon trigger makes use of different detector technologies in different regions of the detector. A quarter-section of the muon system is shown in Fig. 3. The barrel region ($|\eta| <$ 1.05) is instrumented with Resistive Plate Chambers (RPC), while the endcap regions (1.05 $< |\eta| <$ 2.4) are instrumented with Thin Gap Chambers (TGC). The geometric coverage of the L1 trigger in the endcap regions is ~99% and ~80% in the barrel region. The limited geometric coverage in the barrel region is due to a crack at around $\eta = 0$, space used by services of the ID and the calorimeters, the feet and rib support structures of the ATLAS detector and two small elevators at the bottom of the muon spectrometer.

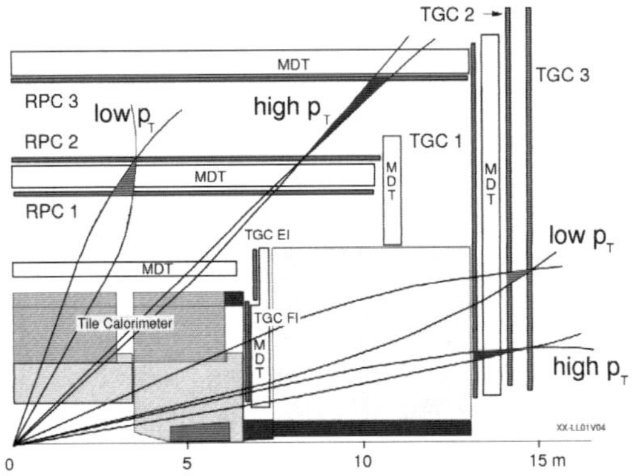

Figure 3. Quarter-section of the muon system in the $z - y$ plane [5].

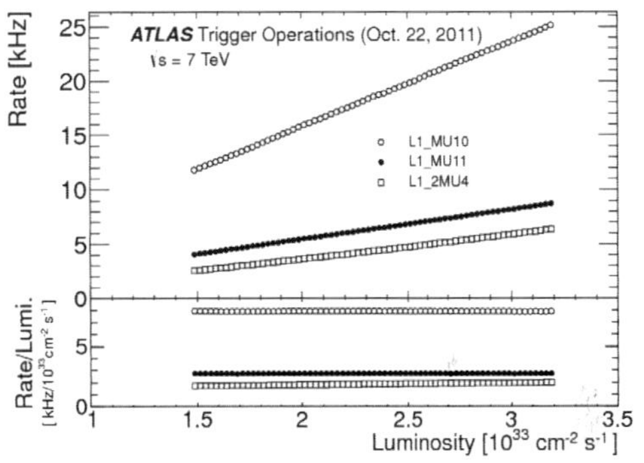

Figure 4. Level 1 muon trigger rates versus the instantaneous luminosity in 2011 [5].

At Level 1, muon candidates are identified by custom-built hardware that forms a coincidence of hits in layers of trigger chambers. The hit pattern along the muon trajectory is used to estimate the transverse momentum, p_T , of the muon. Six thresholds were used during 2012: three for high p_T (11, 15 and 20 GeV) and three for low p_T (4, 6 and 10 GeV). L1 triggers are usually named L1_MUX, where X is the p_T threshold in GeV. It is important to note that the low p_T thresholds require the coincidence of three layers of muon chambers both in the barrel and in the endcap regions, while the high p_T thresholds require two station coincidence in the barrel and three in the endcaps. A L1 muon trigger signal carries the p_T information of the muon and the position information of the RoI.

In Fig. 4 the rates for three different L1 triggers as a function of the instantaneous luminosity are shown. It can be seen that L1 rates scale linearly with the instantaneous luminosity, making them robust against pile-up.

Figures 5 and 6 show the L1_MU11 trigger efficiency with respect to offline in the barrel and in the endcap regions respectively. L1_MU11 was one of the main L1 muon triggers used in analyses in 2011. These efficiencies were obtained with a tag-and-probe analysis of $Z \rightarrow \mu\mu$

events and they include also the geometric acceptance of the detectors. In the barrel region the efficiencies are lower than in the endcap regions, ~72.5% and ~93.5% respectively. The high p_T thresholds have 5-10% lower efficiency than the low p_T thresholds in the barrel region: this is due to the fact that the latter require the coincidence of only two layers of muon chambers.

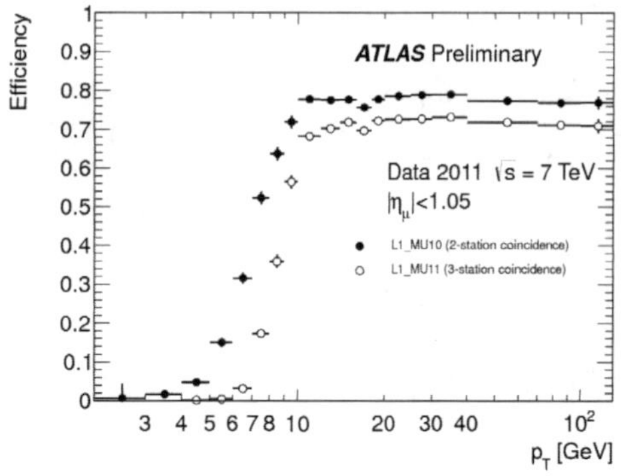

Figure 5. L1 trigger efficiency with respect to isolated offline combined muons for the barrel region as a function of the offline combined muon transverse momentum. The efficiencies include geometric acceptance of the detectors [5].

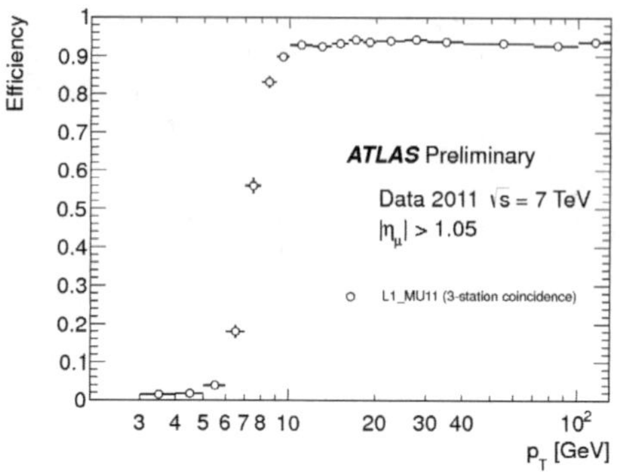

Figure 6. L1 trigger efficiency with respect to isolated offline combined muons for the endcap regions as a function of the offline combined muon transverse momentum. The efficiencies include geometric acceptance of the detectors [5].

At Level 2, the candidate from L1 is refined by using the precision data from the Monitored Drift Tubes (MDT) chambers. The L2 muon standalone algorithm (SA) constructs a track using the MS data within the RoI defined by the L1 seed. First, a pattern recognition algorithm selects hits from the MDT inside the RoI. Second, a track fit is performed using the MDT hits, and a p_T measurement is assigned from Look Up Tables (LUTs). Reconstructed tracks in the ID are then combined with the tracks found by the L2 muon SA algorithm by a fast track combination algorithm (CB) to refine the track parameter resolution. Additionally, the isolation algorithm starts from the result of the combined algorithm and incorporates tracking and calorimetric information to find isolated muon candidates (i.e. muons with low near-by activity).

It is essential, for good trigger performance, that the muon track parameters are reconstructed with enough accuracy, i.e. that the track parameters measured online are as close as possible to the ones measured offline, after reconstruction. To evaluate the goodness of the L2 measurements, the resolution of the track parameters is checked. The residuals between the L2 and offline muon track parameters ($1/p_T$, η and ϕ) are checked in bins of p_T. The widths of the residual distributions are extracted in each bin with a Gaussian fit. The resolution of the inverse p_T track parameter as a function of the offline muon p_T in the barrel is shown in Fig. 7. As expected, the CB algorithm has a better resolution.

Two types of reconstruction algorithms are available at Event Filter. The OutsideIn algorithm starts from the RoI identified by L1 and L2, reconstructing segments and tracks using information from the trigger and precision chambers. The track is then extrapolated back to the beam line to determine the track parameters at the interaction point, thus forming a muon candidate using the MS information only, resulting in the EF standalone trigger. Similar to what is performed by the L2 algorithms, the muon candidate is combined with an ID track to form an EF muon combined trigger. The InsideOut algorithm starts instead from the ID tracks and extrapolates them to the MS region. Due to the extremely busy environment of the ID, the

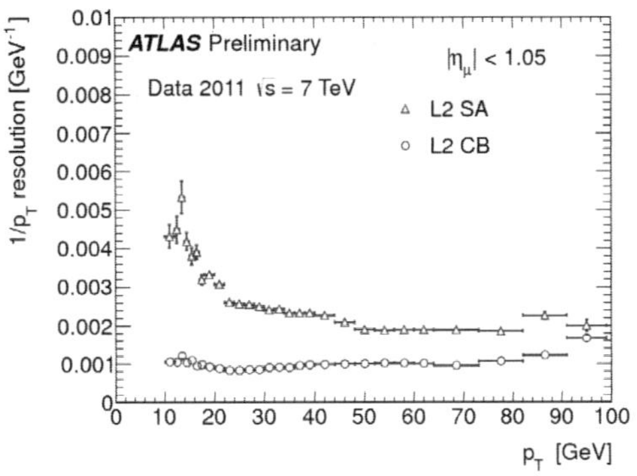

Figure 7. Resolution of the inverse p_T track parameter of the L2 SA and L2 CB algorithms as a function of the offline muon p_T in the barrel region [5].

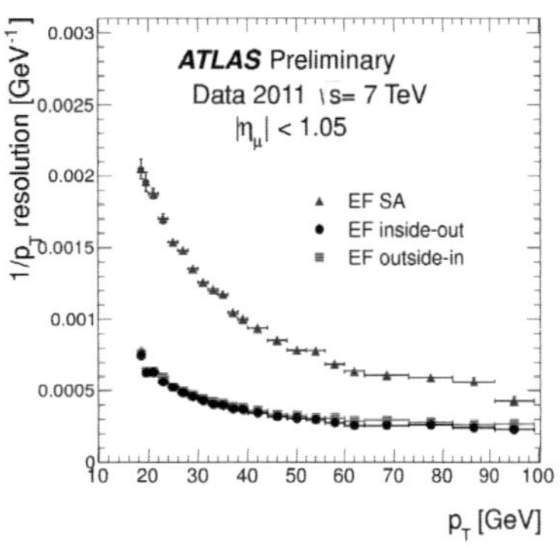

Figure 8. Resolution of the inverse p_T track parameter of the EF SA and EF InsideOut and OutsideIn CB algorithms as a function of the offline muon p_T in the barrel region [5].

InsideOut algorithm is slower, as shown in Fig. 9. The complementary strategies employed by these two algorithms minimize the risk of losing events at the online selection.

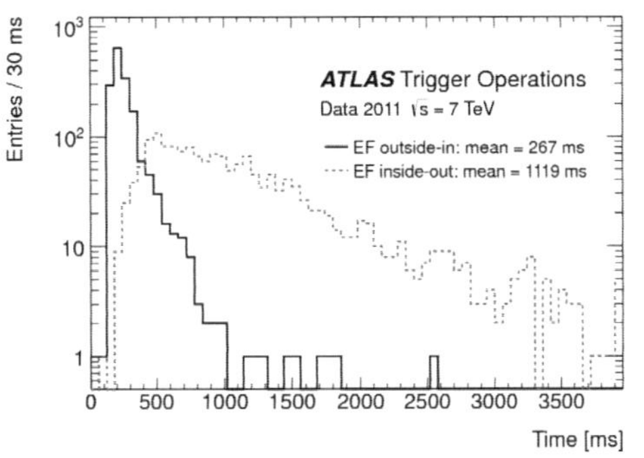

Figure 9. Execution times of the EF OutsideIn (solid line) and InsideOut (dashed line) muon triggers [5].

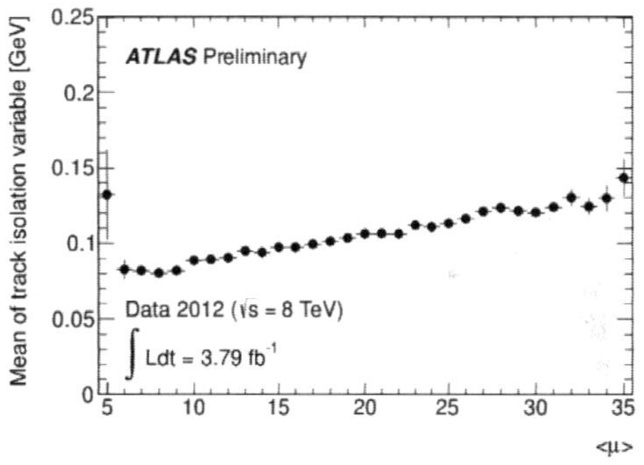

Figure 10. Track isolation variable as a function of number of interactions per bunch crossing [6].

The EF InsideOut and OutsideIn algorithms were running separately during 2011 data taking. To save computing time, in 2012 the two algorithms were merged in a single chain, running the OutsideIn algorithm first and then, if that failed, the InsideOut one.

In addition, isolation algorithms are run at the HLT. These provide a reduction in the rate of the muon trigger while keeping p_T thresholds low. The most critical aspect for isolation algorithms is pile-up robustness. The ATLAS muon trigger has two different types of isolation algorithms. One algorithm is based on information from calorimeters: it uses the energy deposits in the electromagnetic and hadronic calorimeters. The other algorithm uses tracking information: it is based on tracks around the muon candidate. Track based isolation was used in 2012 data taking and it was proved to be pile-up robust (see Fig. 10).

3. Tau trigger

Taus play an important role in the search of the Higgs boson as well as supersymmetric and exotic particles. Taus decay hadronically 65% of the time. Therefore, jets from QCD processes are an overwhelming background to hadronic taus.

Taus can be distinguished from QCD jets: taus are characterized by low track multiplicity and form a narrow, well collimated jet. Tau jets are also isolated: there is no activity around the narrow cone that contains the tau-candidate decay products. These features are exploited by the tau trigger [7].

The Level 1 tau trigger uses electromagnetic (EM) and hadronic (HAD) calorimeter trigger towers with granularity $\Delta\eta \times \Delta\phi = 0.1\times0.1$. Hadronic tau decay modes are identified by the following features (shown in Fig. 11): the sum of energy in 2×1 pairs of EM towers, the energy in 2×2 HAD towers behind the EM cluster, the energy in a 4×4 isolation ring around the 2×2 core region. The core region is defined as the two-by-two trigger tower region of $\Delta\eta \times \Delta\phi = 0.2 \times 0.2$. The isolation region is defined as a four-by-four trigger tower region minus a two-by-two core region. In 2012, five different $E_{\rm T}$ thresholds were used: 8, 11, 15, 20, 40 GeV. To keep both the tau trigger rates under control and tau thresholds low, isolation requirements were applied. The absolute $E_{\rm T}$ in the EM isolation region was required to be smaller than 4 GeV.

In Fig. 12 different L1 tau trigger rates as a function of the instantaneous luminosity are shown.

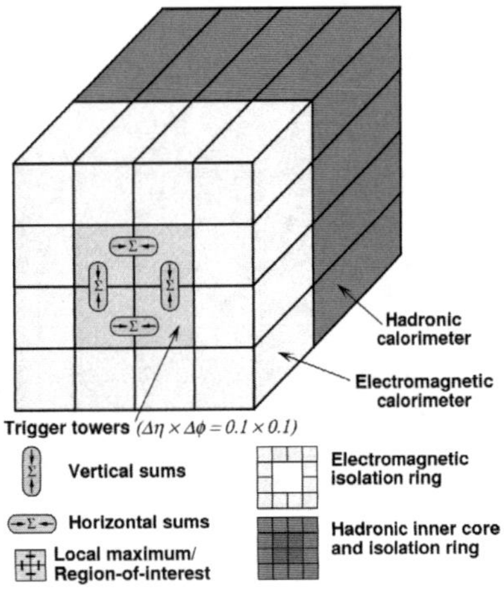

Figure 11. Building blocks of the tau algorithm with the sums to be compared to programmable thresholds.

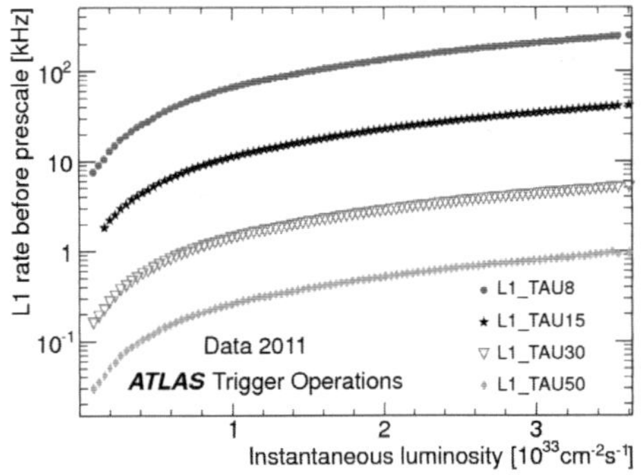

Figure 12. Level 1 tau trigger rates versus the instantaneous luminosity in 2011 [7].

The Level 2 tau trigger uses both calorimetric and tracking information. The calorimeter selection is applied to the RoI. The L2 uses the full granularity information from all layers of the calorimeters within a region of $\Delta\eta \times \Delta\phi = 0.8 \times 0.8$. It refines the position of the RoI and obtains the total $E_{\rm T}$ and shape variables, used to identify hadronically decaying taus. One of these variables is the EM radius, defined as $R_{\rm EM} = \frac{\sum E\Delta R}{\sum E}$, where the sum extends over all cells in the first three layers of the EM calorimeter. In 2012, in order to improve the algorithm pile-up robustness, the cone used to compute both $E_{\rm T}$ and shape variables changed from 0.4 (used earlier) to 0.2. The effect of this change on the EM radius is shown in Fig. 13. The trigger efficiency as a function of the average number of interactions per bunch crossing is flatter when tightening the cone size.

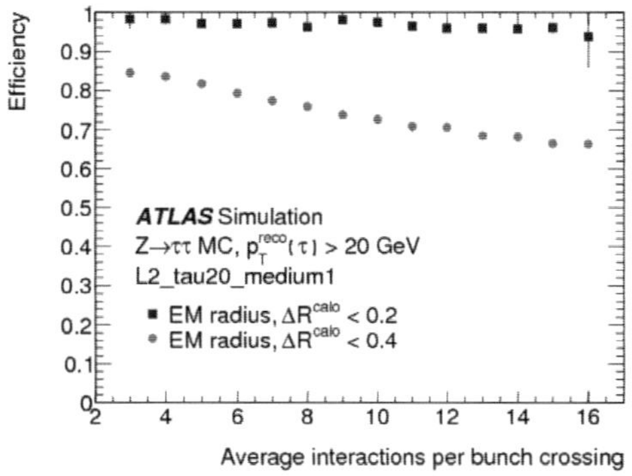

Figure 13. L2 tau trigger efficiencies using the L2 EM radius calculated in cones of $\Delta R < 0.2$ and 0.4 as a function of the average number of interactions per bunch crossing [8].

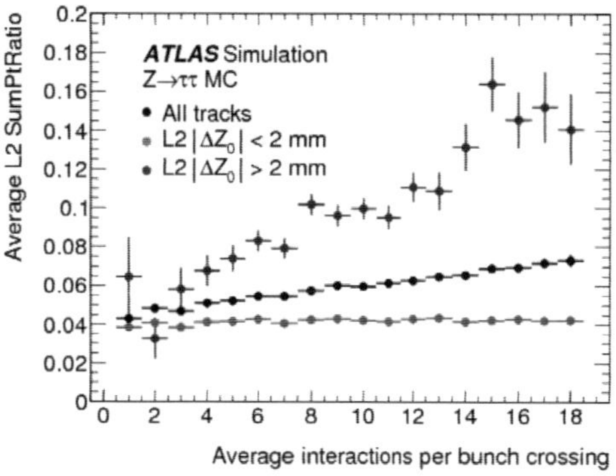

Figure 14. Average L2 SumPtRatio as a function of the average number of interactions per bunch crossing for different track selections [8].

Tracking in the tau RoIs uses fast custom algorithms based on combinatorial pattern recognition followed by a fast Kalman filter track fit. The tracking efficiency is good and comparable to the one found at the Event Filter. Track counting and track-based isolation use tracks found in core and isolation regions of radii 0.1 and 0.3 respectively. One of the tracking variables used to identify taus at L2 is the SumPtRatio, which is the ratio of the scalar sum of the p_T of tracks in the isolation region. In 2012, the track selection criteria was changed to improve pile-up robustness: only tracks with an impact parameter compatible ($|\Delta z| < 2$ mm) with the leading track are used. The effect of this improvement is shown in Fig. 14, where the average of the SumPtRatio variable as a function of the average number of interactions per bunch crossing is shown.

At the Event Filter, a more accurate tau reconstruction is performed. In 2011 the hadronic tau identification was cut-based. Cuts were parameterized as a function of E_T , number of tracks associated to the tau jet, and track and cluster shape variables. The definition of these variables was the same as the one used by the offline reconstruction. In 2012, to make the online tau identification as similar as possible as in offline, an identification based on multivariate analyses was introduced, using the TMVA Boosted Decision Tree (BDT) (see Ref. [9]). A *medium* identification criteria was chosen, with efficiencies of 85% and 80% for one- and multi-prong taus.

With the new algorithm implementations at L2 and EF, the dependence on pile-up was substantially reduced. Figures 15 and 16 show the efficiency of L1, L2 and EF as a function of the number of reconstructed vertices in 2011 and 2012 data respectively. The efficiency was measured using a tag-and-probe analysis of $Z \to \tau\tau \to \tau_{had} + \mu+$ neutrinos events.

4. Conclusions

In this paper the ATLAS muon and tau triggers have been presented. The ATLAS detector has a three level trigger system, which reduces the incoming rate of 20 MHz LHC collisions to ~400Hz of recorded data. Trigger signatures and trigger menus are driven by the physics goals, e.g. SM precision measurements and Higgs boson search.

Dedicated muon and tau triggers are available in ATLAS. These triggers were used to select events used in analyses that brought the discovery of a particle compatible with the SM Higgs boson. The description of both triggers and their performance have been shown. The muon

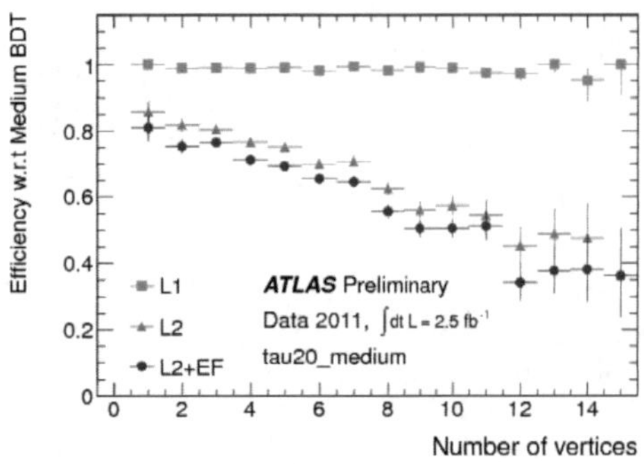

Figure 15. Tau trigger efficiency with respect to offline identified tau candidates as a function of number of vertices measured in 2011 data [8].

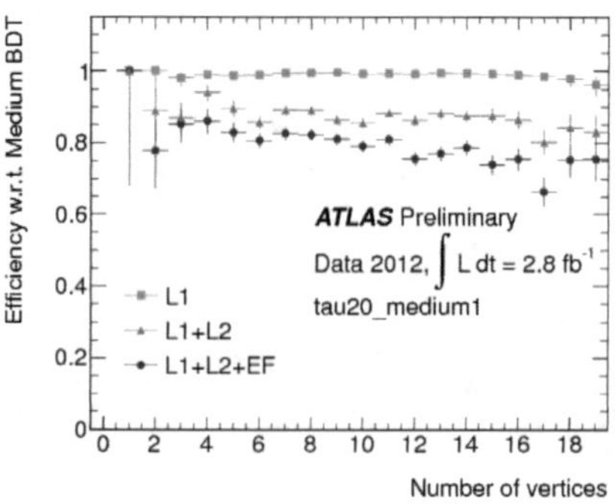

Figure 16. Tau trigger efficiency with respect to offline identified tau candidates as a function of number of vertices measured in 2012 data [8].

trigger combines different detector information and track parameters are measured online with good precision. The trigger efficiency (including the geometric acceptance of the detectors) is approximately 72.5% in the barrel and 93.5% in the endcaps. These efficiensies have been proved to be pile-up robust. The tau trigger is essential for $H \to \tau\tau$ searches. Different algorithms are available in the trigger to help to separate taus from QCD jets. This separation is mainy based both on tracking and calorimeter shape variables. The algorithms used variables that have improved pile-up robustness during the 2012 data taking.

References

[1] The ATLAS Collaboration 2008 The ATLAS Experiment at the CERN Large Hadron Collider *JINST* **3** S08003.

[2] ATLAS Luminosity Public Results web page `https://twiki.cern.ch/twiki/bin/view/AtlasPublic/LuminosityPublicResults`

[3] ATLAS Event Display web page `https://twiki.cern.ch/twiki/bin/view/AtlasPublic/EventDisplayStandAlone`

[4] The ATLAS Collaboration 2012 Observation of a new particle in the search for the Standard Model Higgs boson with the ATLAS detector at the LHC *Phys. Lett.* B **716** 1-29

[5] The ATLAS Collaboration 2012 Performance of the ATLAS muon trigger in 2011 ATLAS-CONF-2012-099 `https://atlas.web.cern.ch/Atlas/GROUPS/PHYSICS/CONFNOTES/ATLAS-CONF-2012-099/`

[6] ATLAS Muon Trigger Public Results web page `https://twiki.cern.ch/twiki/bin/view/AtlasPublic/MuonTriggerPublicResults`

[7] The ATLAS Collaboration 2013 Performance of the ATLAS tau trigger in 2011 ATLAS-CONF-2013-006 `https://atlas.web.cern.ch/Atlas/GROUPS/PHYSICS/CONFNOTES/ATLAS-CONF-2013-006/`

[8] ATLAS Tau Trigger Public Results web page `https://twiki.cern.ch/twiki/bin/view/AtlasPublic/TauTriggerPublicResults`

[9] Hoecker A, Speckmayer P, Stelzer J, Therhaag J, von Toerne E and Voss H 2007 TMVA: Toolkit for Multivariate Data Analysis *PoS ACAT* **040**

The evolution of the Trigger and Data Acquisition System in the ATLAS experiment

A Krasznahorkay[1,2] on behalf of the ATLAS Collaboration
New York University, NY, USA
E-mail: `Attila.Krasznahorkay@cern.ch`

Abstract. The ATLAS experiment, aimed at recording the results of LHC proton-proton collisions, is upgrading its Trigger and Data Acquisition (TDAQ) system during the current LHC first long shutdown. The purpose of the upgrade is to add robustness and flexibility to the selection and the conveyance of the physics data, simplify the maintenance of the infrastructure, exploit new technologies and, overall, make ATLAS data-taking capable of dealing with increasing event rates. The TDAQ system used to date is organised in a three-level selection scheme, including a hardware-based first-level trigger and second- and third-level triggers implemented as separate software systems distributed on separate, commodity hardware nodes. While this architecture was successfully operated well beyond the original design goals, the accumulated experience stimulated interest to explore possible evolutions. We will also be upgrading the hardware of the TDAQ system by introducing new elements to it. For the high-level trigger, the current plan is to deploy a single homogeneous system, which merges the execution of the second and third trigger levels, still separated, on a unique hardware node. Prototyping efforts already demonstrated many benefits to the simplified design. In this paper we report on the design and the development status of this new system.

1. Introduction

The ATLAS experiment [1] is one of the major experiments of the Large Hadron Collider (LHC) [2]. It is a general purpose experiment, aiming at studying the Standard Model Higgs Boson, and looking for physics beyond the Standard Model of particle physics. It consists of various semi-conductor based charged particle tracker detectors (the Inner Detector), a liquid-argon based electromagnetic calorimeter, and a hadronic calorimeter made of steel and scintillator tiles. The outside of the detector is covered by a muon spectrometer for identifying muons leaving the detector.

The detector provides many millions of read-out channels, able to capture data every 25 nanosecods. This can not all be kept for further data analysis of course. The Trigger & Data Acquisition (TDAQ) system is needed to collect information from all parts of the detector, and select on average 400 events per second[3] to be keep for further processing.

The TDAQ system performed very well during LHC's Run 1 (2009-2012), in many aspects well beyond its design values. It allowed ATLAS to collect events relevant for physics analyses with a high efficiency and sustainable event recording rates. However it is clear that to do

[1] Currently at: European Laboratory for Particle Physics (CERN), Geneva, Switzerland
[2] On leave from: Institute of Nuclear Physics of the Hugnarian Academy of Sciences (ATOMKI), Hungary
[3] Average number for the 2012 data taking period.

the same for Run 2 (2015-2018), which will provide the LHC experiments with much more demanding conditions, the TDAQ system has to be improved to keep its high efficiency and low event-recording rate.

This document summarizes some of the hardware and software developments that are happening during LHC's Long Shutdown 1 (LS1, 2013-2014), and how they will impact ATLAS's data taking in LHC's Run 2.

1.1. The ATLAS TDAQ System in Run 1

ATLAS uses a 3-level trigger system, organised as follows:

- **Level-1** (LVL1) composed of specially designed, dedicated hardware components making decisions about the events based on simple reconstruction methods of information just from the calorimeters and the muon spectrometer.
- **Level-2** (LVL2) composed mainly of dedicated trigger algorithms running on commodity PCs, reconstructing physical objects primarily in small regions of the dector (Region of Interest, RoI), seeded by LVL1.
- **EventFilter** (EF) running mainly offline reconstruction code, having possible access to the full event information.

Level-2 and the EventFilter are referred to collectively as the High Level Trigger (HLT).

Figure 1 shows the logical structure of the TDAQ system, citing the original design parameters of the system, and the operating parameters that were in effect by the end of LHC's Run 1.

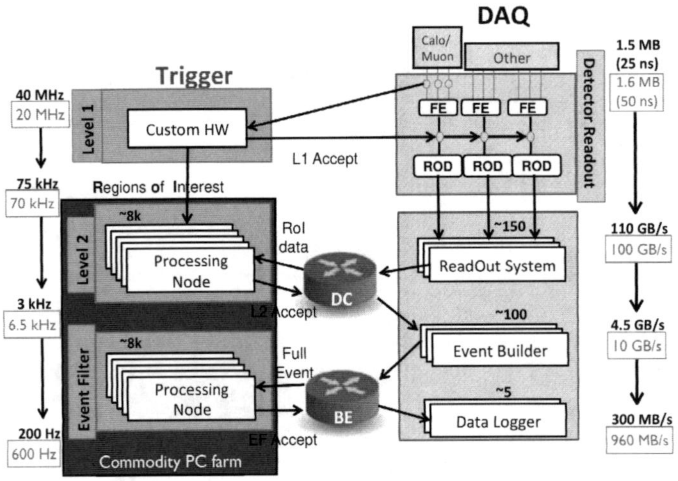

Figure 1. Schematic view of the ATLAS TDAQ system as used in Run 1. The (unframed) labels in black show the design parameters of the system, the (framed) labels in red show the parameters achieved at the end of Run 1.

The data acquisition network was partitioned into multiple segments, as shown in Figure 2. Two separate Data Collection (DC) networks were transporting Read Out System (ROS) data fragments to the LVL2 trigger processes running on the computing nodes (XPU). The two networks were coupled by the Event Building (SFI) nodes that collected all the fragments of the events accepted by LVL2, and sent the full events onward to the EF nodes. The events accepted by the EF, and were written to permanent storage via the Sub-Farm Output (SFO) nodes.

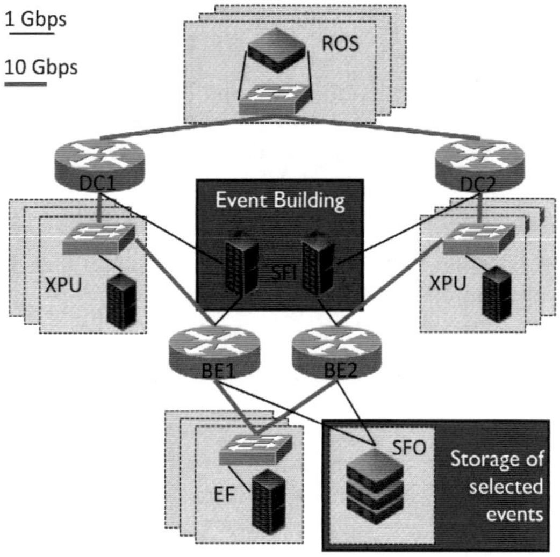

Figure 2. Schematic view of the ATLAS DAQ network as used in Run 1.

1.2. Changes for LHC's Run 2

LHC's Long Shutdown 1 is used to increase both the energy and the intensity of the proton-proton collisions in the machine. Table 1 summarizes the most important changes in the parameters of the machine.

Table 1. Characteristic properties of LHC's Run 1, and the expected properties of the machine for Run 2. The pile-up is expressed in average number of interactions per bunch crossing.

Period	E_{cm} [TeV]	# of bunches	Bunch separation [ns]	Peak lumi. [$cm^{-2}s^{-1}$]	Pile-up [#/BC]	Event size [MB]
Run 1	8	1380	50	$\sim 7 \times 10^{33}$	~ 35	~ 1.5
Run 2	~ 13	1380 - 2700	25 - 50	$1.5 - 2 \times 10^{34}$	up to 80	1.7 - 2.4

According to current expectations the machine will provide proton-proton collisions at 13-14 TeV, with several times the intensity of that in Run 1, with up to an expected 80 proton-proton interactions per bunch crossing. To achieve our physics goals, which include continuing the searches for new physics in electroweak final states, and performing precision measurements on the newly discovered Higgs boson, we will need to:

- Speed up the HLT code, to compensate for increased processing time due to high pileup and allow more use of slower offline reconstruction algorithms.
- Make the trigger and offline analysis selections more similar to each other, increasing the trigger selection efficiency, and simplifying analyses a little bit.
- Introduce more specialised, more selective triggers, to help reduce trigger rates.

The code optimisation is critical, as observations indicate a steeper than linear CPU time response as a function of the average number of pile-up vertices in the events. This means that with the current code we would not be able to fit into the resource limits set by the HLT farm.

Physics analyses will also need to use more specialised triggers than in Run 1, when it was still possible to afford selecting events with relatively simple requirements. As an example, in Run 2 it is expected that even the LVL1 trigger looking for 20 GeV muons will give an acceptance rate $\gg$10 kHz. Since the 100 kHz LVL1 readout rate is a hard limitation in the system, the trigger configuration will need a strategy combining higher thresholds, isolations and single muons in coincidence with some other condition(s) for selecting events for LVL2.

2. Hardware Upgrades

This section summarizes the hardware upgrades being made to ATLAS during LS1 that have an impact on the TDAQ system.

2.1. The Level-1 Topological Trigger

The Level-1 trigger system in Run 1 was only able to look for single objects (like jets and muons), or require simple combinations of such single objects to be present in the event. To extend the functionality of the LVL1 trigger, a new hardware component will be added to the system, the LVL1 Topological Trigger [3].

It will receive detailed information (such as energy and direction) about the candidate objects found by the LVL1 calorimeter and muon triggers. This information will then be processed by dedicated algorithms implemented in its FPGAs, looking for signatures such as:

- A muon close to a jet.
- An object with a certain transverse mass.
- An event with an effective mass above a given criterium.

The topological trigger will only be able to send the limited information of which signatures were found to the HLT due to the 100 kHz readout rate for this information. This is a complication for the HLT which would use the knowledge of which L1 object combination passed the topological criteria to seed the RoI-guided HLT reconstruction. The solution is to simulate the L1 topological trigger hardware: the FPGA code will also be implemented in C++, validated against the hardware response, and run online in the HLT.

2.2. Fast TracKing (FTK)

Another dedicated hardware component is now being developed to help with track finding and reconstruction in the HLT. Called the Fast Tracker ([4]), it will receive all the semi-conductor data from the inner detector charged particle tracker at each LVL1 accept signal, with up to 100 kHz rate. The hardware then finds and reconstructs charged track candidates using pattern matching on specially pre-processed data in an associative memory. This way the hardware provides tracking information for the whole detector very quickly, without the use of HLT processing time or readout bandwidth.

The information can then be used in the HLT in a number of ways:

- Calculating the isolation of identified leptons from other charged particles.
- Finding the primary vertices of the event using all the reconstructed tracks.
- Seeding the HLT track reconstruction to reconstruct charged tracks with high precision in a large area of the detector.
- The reconstructed tracks may be used directly in the b-tagging and tau reconstruction algorithms to improve their performance with the additional tracking information.

There is currently both a very active code development program underway to make use of the FTK tracks in the HLT reconstruction, and a large set of trigger performance studies going on to evaluate the performance of using FTK tracks as part of the object/event selection.

According to the current schedule, the FTK system will cover the barrel system by July 2015, and the entire detector by the end of 2015.

2.3. Simplified HLT & DAQ Achitecture

While the system architecture described in Section 1.1 served ATLAS well in Run 1, technology advancements allow it to be simplified for Run 2.

With the rolling replacement of core network routers during LS1, the simpler design shown in Figure 3 has been possible so that it still provides greater bandwidth to meet the needs of increased event data sizes and access patterns expected with the high pileup conditions of Run 2.

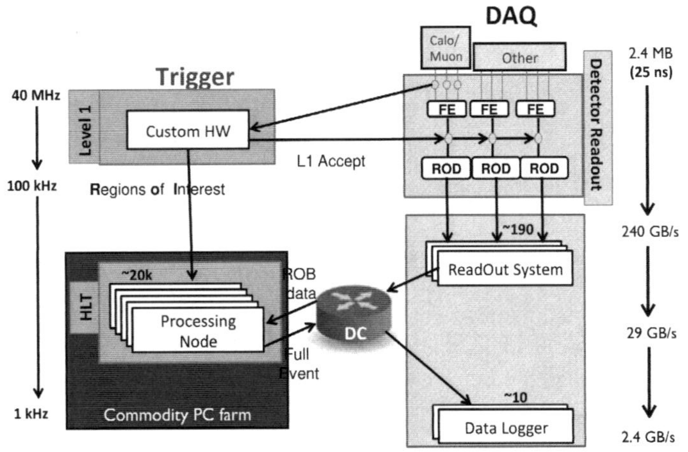

Figure 3. Schematic view of the ATLAS TDAQ system planned to be used in Run 2.

This has also made is possible to merge the L2, EF and event building functionality into a single HLT node. This is advantageous for balancing HLT resources and allows for a much simpler software design too, described in Section 3.1.

Figure 4 shows how the system will use a single network – with multiple switches – for transporting the event fragments to the HLT nodes, as the HLT nodes will now take care of both the LVL2 and EF decision making and the event building as well. The accepted, fully built events will be sent on the same network to the same type of SFO nodes that were used in the previous system as well.

This network design is easily scalable, in case the HLT farm needs to be expanded during Run 2.

3. Software Upgrades

ATLAS set out to simplify, and at the same time improve its trigger software significantly during LS1. There are more development projects than could be listed here (for instance updating the trigger Event Data Model, improving the trigger analysis tools, speeding up and improving the efficiencies of algorithms, etc.). For this reason this section just focuses on the two most important developments happening in the core software of the ATLAS trigger.

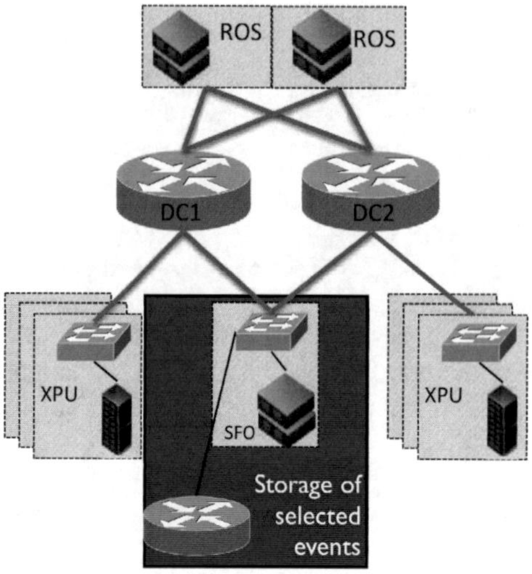

Figure 4. Schematic view of the ATLAS DAQ network as it will be used at the beginning of Run 2.

3.1. The Merged High Level Trigger

Having two High Level Trigger levels (LVL2 and EF) proved to give us very good flexibility during Run 1 to construct trigger configurations that could be used even at the highest LHC intensities. But it became evident by the end of the running period, that the system should be simplified somewhat in order to ease day-to-day operations of the data acquisition.

Some of the imperfections of the old system were:

- *The rigidity of the network configuration.* Computing resources needed to be assigned to either LVL2 or the EF, and re-configuring the network was a non-trivial task. This is addressed by the changes decribed in Section 2.3.
- *Calculation of the same quantities at LVL2 and the EF.* Because of the limitation of how much data could be sent from LVL2 to the EF, some reconstruction that were performed at LVL2, needed to be repeated in the EF.
- *Duplications in network transfers.* The information requested by LVL2 algorithms needed to be requested once more by the event building node in order to build the full event, increasing the request rate hitting the Read-Out System nodes by $\sim$10%.

The simplified software design, made possible by the merged HLT, is shown in Figure 5.

The new system will still separate algorithms running either at the LVL2 or the EF stage. But the algorithms themselves will be run on a single computing node, inside a single process. At the same time, the event building will also run on the same node. The separate LVL2 Processing Unit (L2PU) and EF Processing Task (PT) applications will be merged into a single HLT Processing Unit (HLTPU) application. The tasks of the Data Flow Manager (DFM), the Sub-Farm Input (SFI) and Event Filter Dataflow (EFD) components will be taken over by the Data Collection Manager (DCM).

The defining difference between the algorithms at that point will be that LVL2 algorithms will be executed before the full event would be built, while EF algorithms would have the ability to access the full event information, just like in the previous implementation.

The main improvements expected from the new system are the following:

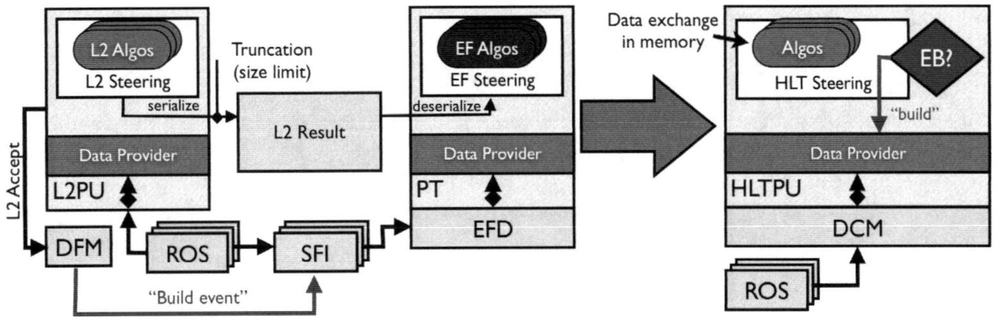

Figure 5. Schematic view of the structure of the HLT code as used in Run 1 shown on the left, and the schema under development for Run 2 is shown on the right. The components are explained in the referring text.

- As all the algorithms will be running inside the same process, they will be free to exchange as much information between each other as needed.
- The data collection system on the processing units will be capable of caching event fragments, so that once event building is requested for a given event, the HLT node will only need to collect the data not requested by one of the LVL2 algorithms already.
- The event building point will be much more flexible in the new system. It should be possible to find the optimal event building point inside any trigger configuration in order to optimise the network traffic, queue/buffer lengths and CPU time.

The ability to exchange much more information between algorithms than before allows us to explore new reconstruction strategies. One example is how the Inner Detector track reconstruction code is being re-designed. In the new system it will no longer be necessary to repeat some parts of the track finding at both levels. The system is being re-designed ([5]) to combine fast LVL2-like data unpacking and pattern recognition with subsequent refinements based on offline reconstruction tools.

3.2. Multi-Processing in the HLT

The clear trend in computing is to move toward computers with more and more CPU cores. Up to now we could take advantage of the multi-core CPUs by running multiple independent HLT processes on them. Unfortunately in the many-core era this strategy will not scale due to memory constraints.

HLT processes need about 1.5 GB of memory to run in our current system. This means that on an N-core system we need to start N HLT processes, and we need to provide N×1.5 GB memory to them. With current servers being available with 48 CPU cores, and future servers on the horizon with O(100) cores, this setup will soon not be able to provide enough memory for the processes.

The problem is tackled by taking advantage of the fact that much of the data held in memory by the individual processes is the same, for example the magnetic field description and detector conditions data. The unique data needed by the processes is 300 MB/process, which is much more realistic to have available even on a many-core system.

Using the Linux kernel's copy-on-write feature, it is possible to start a single process that is forked into multiple processes after initialisation, allowing the running of many HLT processes in parallel, with much reduced memory requirements compared to starting them individually. This setup has been demonstrated successfully in recent tests, making it ATLAS's baseline model for running HLT processes in Run 2.

4. Summary

The ATLAS TDAQ system worked exceptionally well during LHC's Run 1. It allowed us to efficiently select the events needed for all the physics analyses and the discovery of the Higgs boson. With the almost 2 year first long shutdown of the LHC underway to upgrade the energy and intensity capabilities of the accelerator, we are in the process of making improvements to the system so that it will be able to serve us just as efficiently during LHC's Run 2.

There are many developments happening to simplify and improve both the hardware and the software design of the TDAQ system. These will allow us to have even more flexibility in developing and configuring how we select proton collisions that are interesting to physics analyses.

All the developments are well underway, with full scale tests of the system expected in 2013 and 2014.

References

[1] ATLAS Collaboration 2008 *Journal of Instrumentation* **3** S08003
[2] Evans L and Bryant P 2008 *Journal of Instrumentation* **3** S08001
[3] Simioni E *et al.* 2012 Topological and Central Trigger Processor for 2014 LHC luminosities Tech. Rep. ATL-DAQ-PROC-2012-041 https://cds.cern.ch/record/1459209 CERN Geneva
[4] Shochet M *et al.* 2013 Fast TracKer (FTK) Technical Design Report Tech. Rep. CERN-LHCC-2013-007 ATLAS-TDR-021 https://cds.cern.ch/record/1552953 CERN Geneva
[5] ATLAS Collaboration 2013 Studies for the development of the Inner Detector trigger algorithms at ATLAS Tech. Rep. ATL-DAQ-PUB-2013-002 https://cds.cern.ch/record/1602918 CERN Geneva

ATLAS distributed computing: experience and evolution

A Nairz[1], on behalf of the ATLAS Collaboration
[1] CERN, CH–1211 Geneva 23, Switzerland
E-mail: armin.nairzcern.ch

Abstract. The ATLAS experiment has just concluded its first running period which commenced in 2010. After two years of remarkable performance from the LHC and ATLAS, the experiment has accumulated more than 25 fb^{-1} of data. The total volume of beam and simulated data products exceeds 100 PB distributed across more than 150 computing centres around the world, managed by the experiment's distributed data management system. These sites have provided up to 150,000 computing cores to ATLAS's global production and analysis processing system, enabling a rich physics programme including the discovery of the Higgs-like boson in 2012. The wealth of accumulated experience in global data-intensive computing at this massive scale, and the considerably more challenging requirements of LHC computing from 2015 when the LHC resumes operation, are driving a comprehensive design and development cycle to prepare a revised computing model together with data processing and management systems able to meet the demands of higher trigger rates, energies and event complexities. An essential requirement will be the efficient utilisation of current and future processor technologies as well as a broad range of computing platforms, including supercomputing and cloud resources. We will report on experience gained thus far and our progress in preparing ATLAS computing for the future.

1. Introduction

The first extended operations period of the Large Hadron Collider (LHC) from 2010 to early 2013 (the so-called "Run-1") was extremely successful for the ATLAS experiment [1]. In total, ATLAS recorded an integrated luminosity of about 27 fb^{-1} of pp collision data (5.3 fb^{-1} at a centre-of-mass energy of $\sqrt{s}$ = 7 TeV in 2010–2011, 21.7 fb^{-1} at $\sqrt{s}$ = 8 TeV in 2012), corresponding to 7.6 billion events and a data volume of 7.4 PB, as illustrated in figures 1 and 2. The greatest physical success was undoubtedly the discovery of a Higgs-like boson, as announced in July 2012.

For about one month per annual run period, the LHC was colliding heavy ions (Pb-Pb in 2010–2011, p-Pb in 2013); during those periods ATLAS recorded 167 pb^{-1} of Pb-Pb and 30 nb^{-1} of p-Pb collision data, amounting in total to 720 million events and a data volume of 740 TB. Based on this data sample, ATLAS has been pursuing a rich, fruitful research programme on exploring the properties of the quark-gluon plasma.

The ATLAS distributed computing system, the subject of this paper, was also working extremely well during Run-1 [2],[3]. It has proven an indispensable component of the ATLAS experiment and contributed substantially to its overall success.

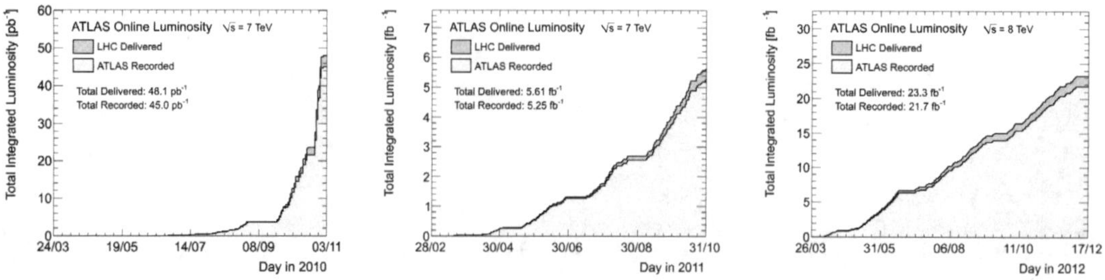

Figure 1. Integrated luminosity of *pp* collisions data delivered by the LHC (green histograms) and recorded by ATLAS (yellow histograms) in 2010 (left), 2011 (centre), and 2012 (right).

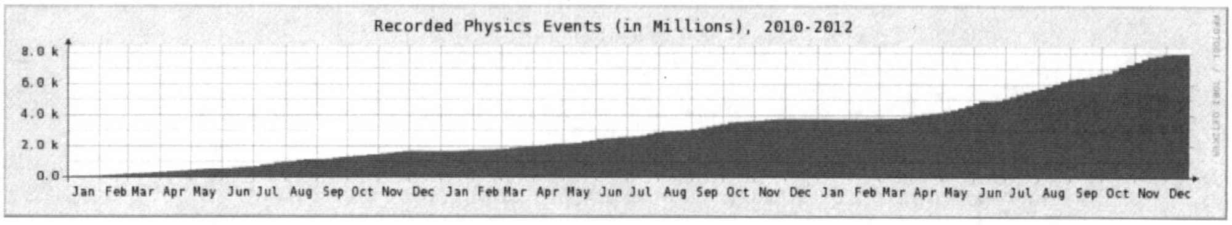

Figure 2. Number of *pp* collision events for physics analysis ("physics events") recorded by ATLAS in the period 2010–2012.

This paper is organised in two parts. The first part introduces the ATLAS computing system (section 2) and describes its main constituents, Tier-0 (section 3), distributed data processing and data management (section 4), together with their respective performances and achievements during Run-1. The second part (section 5) discusses the challenges ATLAS computing has to face for the next LHC operation period ("Run-2"), and gives a short overview of the strategies adopted and projects launched to address them, in order to get prepared for the restart of data taking in 2015.

2. Overview of the ATLAS computing system

The ATLAS Distributed Computing (ADC) system manages the worldwide data processing, Monte Carlo simulation production and user analysis activities, together with their associated data accesses and transfers. There are over 100 computing centres (sites) participating in the ADC project; the available, managed resources (as of May 2013) amount to up to 150k computing cores, about 70 PB of total disk space, and about 50 PB of tape-resident data.

2.1. The ATLAS computing model

The sites participating in ADC are part of the more comprehensive LHC Computing Grid (LCG) [4] and correspondingly organised in a hierarchical ("tiered") structure, according to their roles, available infrastructure and resources:

- the Tier-0 centre at CERN, which is mainly responsible for recording raw data onto tape, carrying out calibration processing and first-pass physics data processing;
- ten Tier-1 centres (i.e. large computing centres distributed across the world and well-connected with the Tier-0), whose tasks are the storage of a replica of the raw data on tape and data reprocessing;
- several dozens of Tier-2 centres (i.e. regional computing centres providing disk and CPU resources and associated to their respective Tier-1 centres), which are responsible for running group and end-user analysis jobs.

Monte Carlo simulation production is run both on Tier-1s and Tier-2s, wherever possible. Data are distributed and replicated over the Grid, both for redundancy (i.e. to secure data with replicas) and for increasing the accessibility (i.e. to create more replicas for frequently used, popular data).

Starting from raw data, as recorded by the ATLAS detector, processing at the Tier-0 and on the Grid produces a set of derived data products, serving dedicated purposes:

- ESD (Event Summary Data): output objects of reconstruction
- AOD (Analysis Object Data): an event representation (ATLAS-wide format) with reduced information for physics analysis
- DPD (Derived Physics Data): a data representation for end-user analysis. DPDs are produced for working groups and individual end-users and could have a group-specific format. Examples of DPD formats are dESD (derived from ESD, for performance groups), dAOD (derived from AOD, for physics groups) or NTUP (ROOT n-tuples, both for physics groups and end-users).
- TAG: a format holding event-level metadata (event tags; i.e. short event summaries primarily used for event selection)

A schematic view of the raw and derived data types, together with the associated workflows to produce them, is shown in figure 3.

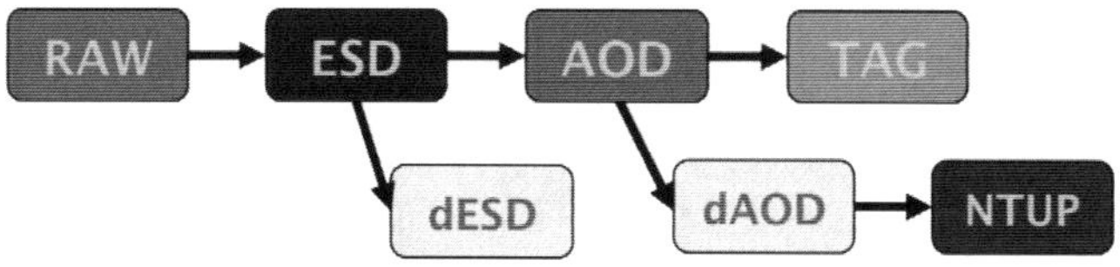

Figure 3. Schematic view of ATLAS raw and derived data types and associated workflows.

More details about the ATLAS computing model, as originally laid out in 2005–2008, can be found in references [5],[6].

2.2. Evolution of the ATLAS computing model

During Run-1 it turned out necessary to adjust the original computing model (as sketched in section 2.1) to new use cases, changing run conditions and technology advances [7],[8].

In response to unexpected usage patterns for the individual data types, data distribution policies were revised and adjusted; e.g. it turned out that users preferred the variety of DPD formats over AODs for their analyses, thus the number of AOD replicas over the Grid could be reduced. In addition, a more dynamic data placement strategy according to data popularity was introduced.

Also run and data-taking conditions changed in the course of Run-1: upgrades in the ATLAS data-acquisition infrastructure allowed an increase of the event recording rate (from originally foreseen 200 Hz to eventual 400–500 Hz), and advances in the LHC led to higher luminosity, resulting in higher pile-up and bigger event sizes. Through introduction of offline raw-data compression at Tier-0, the raw-data volume could be significantly reduced (almost by a factor two), thus saving tape space and allowing for the longer-term placement of raw data on Tier-1 disks to facilitate the access. Since ESD event sizes and their total data volume on Tier-1 disk grew too big, it was decided to keep bulk ESDs on disk only for a limited lifetime (typically a few months) and delete them afterwards.

As the network connectivity between sites had been improving significantly over the years, the original static and hierarchical organisation of Grid sites [9] and the data transfer policy between them was revisited. Based on the measured network performance, ADC moved gradually from a "tree-like" topology with Tier-1/Tier-2 association to a "mesh" of any Tier-$*$ combination,

Journal of Physics: Conference Series **523** (2014) 012020 doi:10.1088/1742-6596/523/1/012020

allowing direct data transfers between all kinds of sites. This resulted in faster transfers, a more efficient use of the computing infrastructure with less load on the system, and eventually more end-user convenience.

3. Tier-0 processing

The ATLAS Tier-0 system (see [10] for a more detailed system overview) was running very reliably, stably and successfully during the first LHC data-taking period.

First-pass data processing at Tier-0 was keeping up with the increasing LHC performance. Dedicated computing resources, both storage and CPU, got adjusted and extended every year to meet the requirements. In 2012 Tier-0 resources comprised more than 2 PB of disk (with tape back-end) and about 3k computing cores of the CERN batch system. The CPU capacity could be increased on demand by up to 4.5k cores through flexible usage of additional, shared public batch resources ("spill-over" mechanism). Figure 4 shows, for the 2012 data-taking period, monitoring plots of the physics raw data rate, as recorded by the ATLAS data-acquisition system and transferred to the Tier-0 (with peaks of 1.3 GB/s), and of the number of concurrently running Tier-0 jobs. (Note that for most of the period Tier-0 was running about 6k parallel jobs. A peak activity of about 7.5k parallel jobs is clearly visible in June 2012, an intensive processing period preceding the discovery of the Higgs-like boson.)

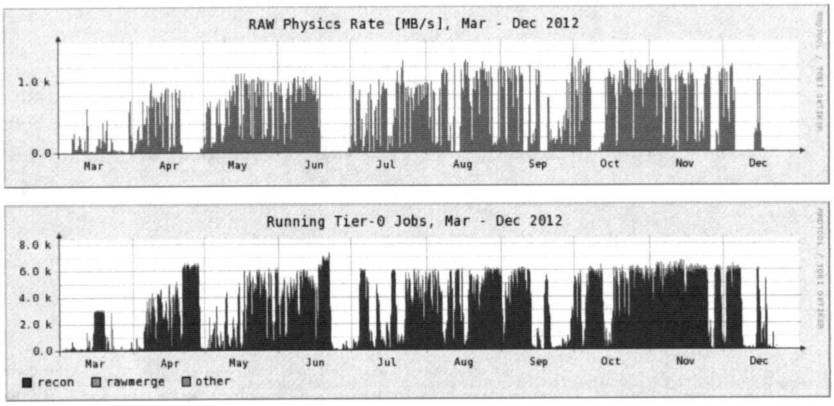

Figure 4. Recorded ATLAS physics raw data rate (top) and concurrently running Tier-0 reconstruction jobs (bottom) for the 2012 data-taking period.

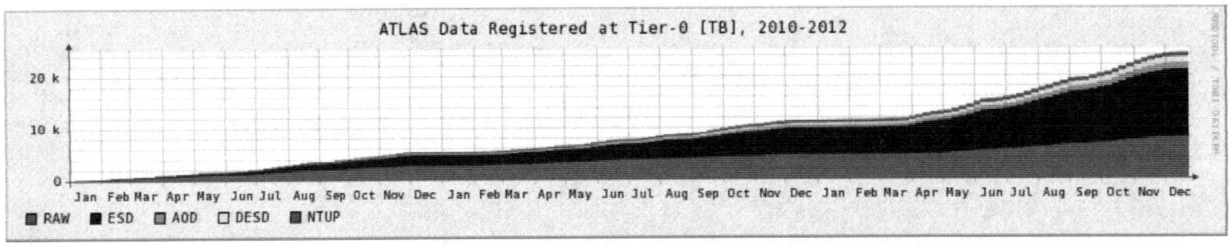

Figure 5. Accumulated total volume of ATLAS raw and derived data types processed by Tier-0 and registered with the DDM system in the period 2010–2012.

Tier-0 processing was organised in two separate steps: in a first step, the so-called "calibration loop", dedicated smaller data streams (most notably the "express stream") were processed with high priority, and the resulting output was provided to detector and data-quality monitoring groups for deriving calibration and alignment information and assessing the data quality. After a delay of usually 48 hours, the second step – processing of the bulk physics data – was launched, already using up-to-date calibration. This resulted in high-quality data reconstruction already from first-pass processing, such that most 2012 data produced by Tier-0 could be used directly in physics analysis, without need of prior reprocessing.

Figure 5 shows the accumulated data volume processed by Tier-0 and registered with the ATLAS distributed data management (DDM) system, thus prepared for the export to the Grid and made available to the users.

4. Distributed computing

The ATLAS distributed computing (ADC) system and infrastructure were working very well and successfully during Run-1 [2],[3].

The ATLAS production system is responsible for managing data processing and running user analysis jobs on the Grid; it is based on PanDA (see [11] for more details) as workload management system. During peak activity in 2012, the system was running up to 150k parallel jobs (figure 6); the usual average job throughput (i.e. the number of successfully completed jobs) per day amounts to about one million jobs, with peaks of up to 1.7 million (figure 7).

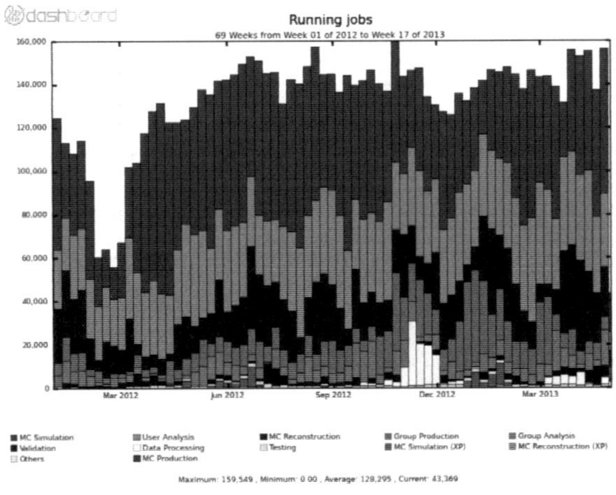

Figure 6. Number of concurrently running ATLAS Grid jobs per activity for the period from January 2012 to April 2013.

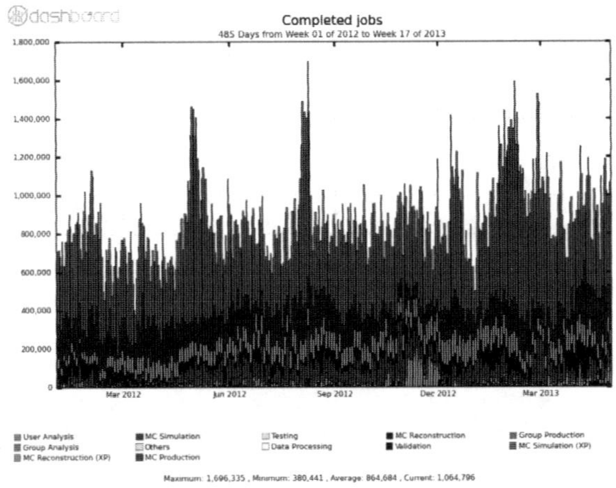

Figure 7. Number of successfully completed ATLAS Grid jobs per day and activity for the period from January 2012 to April 2013.

The majority of jobs run on the Grid are short end-user analysis jobs. Submission of analysis jobs to PanDA takes place through a powerful and simple command-line interface. User-friendly and equally easy-to-use data management tools allow users to have the job output either transferred to their "home" site on the Grid (both automatically and manually), or downloaded to off-Grid computers.

The ATLAS distributed data management (DDM) system (see [12],[13] for more details) manages the data distribution to and between Grid sites. The system successfully completes about one million file transfers per day, amounting to a total daily volume of 400–800 TB (figure 8); those numbers include transfer of the input/output of both production and user-analysis jobs.

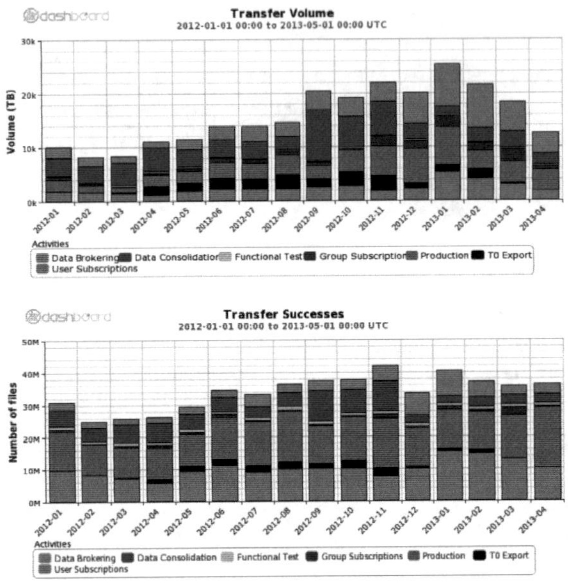

Figure 8. Monthly data volume (top) and number of files (bottom) transferred by the ATLAS DDM system over the Grid for the period from January 2012 to April 2013, for different activities.

As illustrated through figures 6–8 (examples only!), ADC is endowed with a comprehensive monitoring suite [14], which is essential for day-to-day operations, provides the tools for shift services, or allows to extract overview and resource usage summaries.

5. Future challenges: Run-2 and beyond

Regardless of the success and achievements of ATLAS distributed computing during Run-1, conditions for Run-2 (starting in 2015) will be much more demanding, both in terms of external run conditions (LHC, ATLAS data acquisition) and available computing resources.

Higher LHC centre-of-mass energy (13–14 TeV) and higher luminosity will result in higher event multiplicities and more pile-up (i.e. number of *pp* interactions per bunch crossing, or overlaid events). In 2012 the mean pile-up, $\langle\mu\rangle$, was at about $\langle\mu\rangle = 20.7$, requiring about 20–30 seconds wall-time at the Tier-0 to fully reconstruct an event. In Run-2 one has to expect values of $\langle\mu\rangle = 40$ in the most "favourable" circumstances (luminosity levelling at the LHC), but the pile-up could easily be higher. Using the same reconstruction software as in 2012, the projected reconstruction time per event, for a centre-of-mass energy of 14 TeV and $\langle\mu\rangle = 40$, would be at about 70 seconds (figure 9), i.e. more than twice as much as in 2012.

There are also hardware and software upgrades planned in the ATLAS online trigger and data-acquisition system, in order to achieve higher data-taking rates; the target figure is 1 kHz (sustained), i.e. about twice the rate as in 2012.

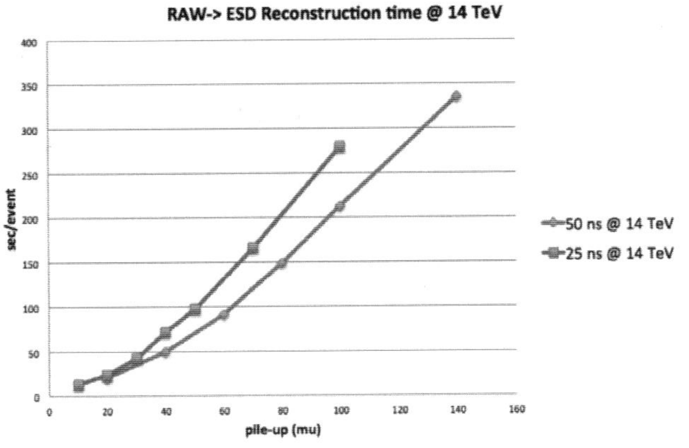

Figure 9. Projected reconstruction time for a *pp* collision event at a centre-of-mass energy of 14 TeV, dependent on the pile-up, for two LHC operations scenarios: a bunch-crossing rate of 40 MHz (25 ns bunch spacing; red graph) and of 20 MHz (50 ns bunch spacing; blue graph), respectively.

To account for the changes in LHC run and ATLAS data-taking conditions, more computing resources will be needed everywhere: CPU, storage (due to more data and bigger event sizes), network, etc. Realistically, however, due to a stretched financial situation, ATLAS cannot expect an increase in the computing budget; facing a "flat budget" scenario, an increase in Grid resources can only be achieved through technological progress. Therefore, in general, the goal will be to better utilise available resources and explore new "affordable" resources wherever possible.

To address those issues and increase the overall efficiency of the system, the ATLAS software and computing community has launched an ambitious upgrade and development programme, which will be introduced briefly and non-exhaustively in the following sections.

5.1. New DDM system (Rucio)

As mentioned in section 4, the current ATLAS DDM system (DQ2) [12],[13] has been working very well and lived up to (and even beyond) expectations. Currently (May 2013) the system administers about 140 PB of ATLAS data on more than 150 Grid sites, and the total data volume typically increases per year by more than 40 PB in more than 100 million files; see figure 10 as an illustration.

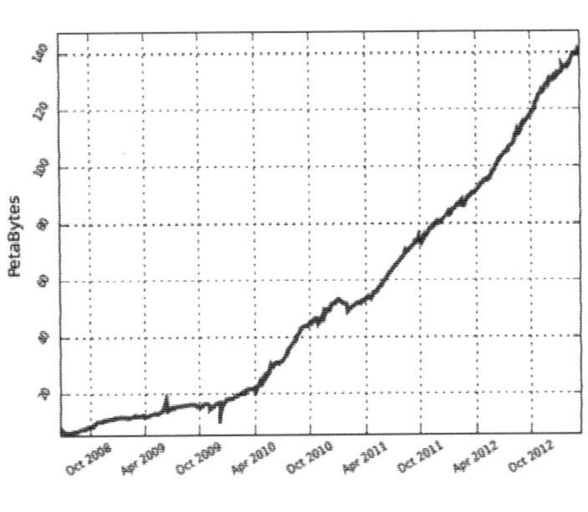

Figure 10. Total volume of data managed by DDM in the period from July 2008 to March 2013. Note the sharp increase since April 2010, the beginning of the 2010 data-taking period.

Nevertheless, projections towards Run-2 conditions indicate that DQ2 will not continue to scale. There are matter-of-principle extensibility issues, due to its design, together with changes to the computing model and the Grid middleware, that are difficult to accommodate.

The ATLAS DDM developers' team has therefore embarked on the project of a fundamental redesign and rewrite of the DDM software (project name *Rucio*) [15]. The next-generation DDM system will address the above-mentioned scalability issues. It will have new concepts for optimisation of storage occupancy, data distribution and network utilisation, and will provide better means of accounting and space quota management. Where appropriate, Rucio will be exploiting new technologies; e.g. it envisages the usage of noSQL database technologies (like Hadoop) and foresees the support for new protocols beyond *gridftp* (like *xroot* and *http*). Rucio development is planned to be concluded by the end of 2013; a fully functional new system should then be ready to go in production in early 2014.

5.2. New Grid production system (ProdSys2)

As mentioned in section 4, the existing Grid data processing system [11] has been working very well. There are, however, similar scalability concerns as with DQ2 (section 5.1), together with difficulties to implement new features and requirements in the existing design.

Again, a fundamental redesign and rewrite campaign for a new Grid production system (project name *ProdSys2*) is in progress [16]. The new system will address the above-mentioned scalability concerns. It will provide better support for new or different workflows, and will offer means of more flexible and dynamic job definitions and job scheduling, e.g. support for multi-core node scheduling and dynamic job splitting. ProdSys2 is planned to be tightly integrated with the "Event Service", a newly conceived service being developed right now, which will enable data processing at a much finer granularity (at event level, instead of the currently only possible file level). Wherever beneficial for the efficacy of the ProdSys2 system, new Rucio concepts (e.g. file-based data management, usage of transfer rules) are planned to be exploited. Similar to Rucio, ProdSys2 is planned to be deployed and to replace the existing system in early 2014.

5.3. Concurrency

Much effort, in the whole community, is presently invested into making better and more efficient use of modern multi-core processors. The issue has also been in the focus of the ACAT2013 workshop (see e.g. [17]).

The ATLAS software and computing community is following several tracks to address the issue. Firstly, effort is going into optimising the existing code [18], parallelisation of algorithms and evaluation of new, customised software libraries. Secondly, work is ongoing on a new concurrent framework, which is planned to support features like full threading and event-level/algorithm-level parallelism. In this area, ATLAS is greatly benefitting from the collaboration with other CERN teams. And thirdly, ATLAS is pursuing the development, testing and deployment of *AthenaMP* [19] (*Athena*: the ATLAS core software package, *MP*: MultiProcess), a framework featuring basic parallelism through processing the event loop by spawning parallel worker processes. AthenaMP also has the advantage of reducing the memory footprint of a job through memory sharing ("copy-on-write" feature) between workers.

AthenaMP is already advanced and will be commissioned and moved to production in the course of 2014. Some operational challenges on the Grid still have to be addressed and resolved [20], such as the efficient matching of jobs to resources (which involves issues like whole-node scheduling, the deployment of multi-core queues, or the optimal usage of sites offering serial and multi-core queues). Flexible input/output file management in AthenaMP is also mandatory and needs to be refined, in particular once the new Event Service (section 5.2) has become operational.

5.4. Cloud computing

In the following two sections ADC initiatives will be introduced whose common goal is the extension of the CPU capacity through the usage of "opportunistic resources". One possibility is the usage of free (e.g. academic) clouds, or alternatively, "cheap" commercial clouds as temporary ("opportunistic") extensions of the ATLAS CPU resources [21]. Cloud computing also has been intensively discussed during the ACAT2013 workshop (see e.g. [22]).

The utilisation of statically allocated virtual machines (VMs) in a cloud has already been successfully demonstrated in the context of ATLAS distributed data processing. Examples are the adaptation of the ATLAS high-level trigger farm for running simulation production during the period of the long LHC shutdown, or several cloud-based PanDA queues that have been running production jobs. In that context, CERN OpenStack as framework and top-level infrastructure has been commissioned and is being used, offering additional 800 computing cores to ATLAS since October 2012.

In addition, ADC is already actively exploring public-private partnerships in cloud computing. This is still an R&D project, but ATLAS has managed to get access to such resources on a scale of several thousand cores, and to run its jobs over sustained periods of weeks to months. As a pioneering activity, ATLAS actively participated in the Helix-Nebula project; recently new projects were started in collaboration with commercial cloud providers, like Amazon (EC2) or Google (GCE).

There are still many open issues to address and resolve, such as the efficient usage of cloud storage, or the way transfers to/from clouds (e.g. for user analysis jobs) are handled. Overall, however, cloud computing is a quickly evolving field offering interesting and attractive opportunities.

5.5. Usage of HPCs

Another possibility of utilising opportunistic resources (see section 5.4) is gaining access to cycles on high-performance computing (HPC) facilities. HPCs are frequently operated by academic or research institutions (or funded by the same agencies that also fund HEP research), which often facilitates the access for members of the HEP community to those facilities. HPCs are particularly attractive because of their huge capacity: equivalents of several hundred thousands of computing cores in a single HPC could easily surpass the capacity of the LCG as a whole, so gaining access to even a small fraction of an HPC may have a substantial impact.

ADC has managed to run first proof-of-principle tests on several HPCs in the US and Europe, e.g. Intrepid (Argonne, USA), SuperMUC (Munich, Germany) and MOGON (Mainz, Germany).

Several open issues regarding the efficient usage of HPCs still have to be resolved. Examples are whole-node scheduling on HPCs, or the question of how to run jobs in an environment where there is no worker-node disk and no outbound I/O. In the context of job scheduling, one has to be able to get fast onto those facilities once they become available; to get fast off them once they disappear; jobs and the job management system have to be robust against disappearing resources; and one has to be able to fill the HPCs with fine-grained workflows. All those issues will be addressed by the new ProdSys2 (section 5.2).

5.6. Federated storage

Using "storage federations" (i.e. clusters of well-connected sites sharing their storage and appearing as an entity to the outside) constitute a means of optimising storage resources, for example through reduced data transfers due to data placement. The concept heavily relies on the good connectivity between sites, and has been made possible by the fast evolution of the network infrastructure in recent years; thus, in a way, the network may become a managed resource as well. Storage federations have also received much attention during the ACAT2013 workshop (see e.g. [22]).

The traditional ATLAS computing model is based on the "DataGrid" concept, where jobs go to the data (and jobs access the data through the local network, LAN) and data (usually whole data sets) may be replicated across sites for better accessibility. Storage federations, on the other hand, provide new access modes and redundancy: jobs now access the data on shared storage resources through the wide-area network (WAN); as (e.g. analysis) jobs may not need all the input files of a whole data set, it may be sufficient to transfer only a part of the data set across sites, or the possibilities of file or event-level caching could be exploited; and, by creating a common name-space across all storage sites, an easy-to-use and homogeneous way of accessing data could be established.

ADC has started evaluating federated storage based on a system of XRootD "redirectors", which turned out to be a working solution [23]. Feasibility studies have been positively concluded after a "full dress rehearsal" exercise carried out in spring 2013, involving sites in the US and Europe. The plan is to include more sites (e.g. also in Asia) in future tests.

6. Conclusions

ATLAS distributed computing has been working extremely successfully during Run-1 in all aspects of large-scale data processing, data management and user analysis.

To get ATLAS distributed computing prepared for the challenging start-up of the LHC in 2015, many interesting and promising ideas and projects are being pursued. The overall goal is to use the scarce computing resources as efficiently as possible, and to open up new resources in an innovative way (like clouds or HPCs) wherever feasible. The projects make good progress, but there is still a lot of work ahead in the years 2013–2014.

Acknowledgements

The author wishes to thank in particular the following ATLAS colleagues for providing valuable information and material: I. Ueda, S. Campana, K. De, R. Gardner, V. Garonne, B. Kersevan, A. Klimentov, T. Lecompte, T. Maeno, S. Panitkin, M. Potekhin, H. Sakamoto, R. Seuster, I. Vukotic, R. Walker and T. Wenaus.

References

[1] The ATLAS Collaboration 2008 The ATLAS Experiment at the CERN Large Hadron Collider *J. Inst.* **3** S08003
[2] Ueda I (for the ATLAS Collaboration) 2012 ATLAS Operations: Experience and Evolution in the Data Taking Era *J. Phys.: Conf. Ser.* **331** 072034
[3] Ueda I (for the ATLAS Collaboration) 2012 The Present and Future Challenges of Distributed Computing in the ATLAS Experiment *Proc. of Science* PoS(ICHEP2012)500
[4] Eck C et al. 2005 *LHC computing Grid: Technical Design Report* CERN-LHCC-2005-024 (Geneva: CERN)
[5] The ATLAS Collaboration 2005 *ATLAS Computing: Technical Design Report* CERN-LHCC-2005-022 (Geneva: CERN)
[6] Jones R and Barberis D 2008 The ATLAS computing model *J. Phys.: Conf. Ser.* **119** 072020
[7] Jones R W L and Barberis D 2010 The evolution of the ATLAS computing model *J. Phys.: Conf. Ser.* **219** 072037
[8] Campana S, Barreiro Megino F, Jézéquel S, Negri G, Serfon C and Ueda I 2012 Evolving ATLAS Computing For Today's Networks *J. Phys.: Conf. Ser.* **396** 032019
[9] The MONARC Project, http://MONARC.web.cern.ch/MONARC/
[10] Elsing M, Goossens L, Nairz A and Negri G 2010 *J. Phys.: Conf. Ser.* **219** 072011
[11] Maeno T et al. (for the ATLAS Collaboration) 2012 Overview of ATLAS PanDA Workload Management *J. Phys.: Conf. Ser.* **331** 072024
[12] Branco M, Cameron D, Gaidioz B, Garonne V, Koblitz B, Lassnig M, Rocha R, Salgado P and Wenaus T 2008 Managing ATLAS data on a petabyte-scale with DQ2 *J. Phys.: Conf. Ser.* **119** 062017
[13] Stewart G A et al. (for the ATLAS Collaboration) 2012 Advances in service and operations for ATLAS data management *J. Phys.: Conf. Ser.* **368** 012005
[14] Schovancová J (for the ATLAS Collaboration) 2012 ATLAS Distributed Computing Monitoring tools after full 2 years of LHC data taking *J. Phys.: Conf. Ser.* **396** 032095
[15] Garonne V et al. 2012 The ATLAS Distributed Data Management Project: Past and Future *J. Phys.: Conf. Ser.* **396** 032045

[16] Maeno T et al. (for the ATLAS Collaboration) 2012 Evolution of the ATLAS PanDA Production and Distributed Analysis System *J. Phys.: Conf. Ser.* **396** 032071
[17] Nowak A 2013 Opportunities and choice in a new vector era *(these proceedings)*
[18] Kama S 2013 Optimizing the ATLAS code with different profilers *(these proceedings)*
[19] Van Gemmeren P, Binet S, Calafiura P, Lavrijsen W, Malon D and Tsulaia V (for the ATLAS Collaboration) 2012 I/O Strategies for Multicore Processing in ATLAS *J. Phys.: Conf. Ser.* **396** 022054
[20] Crooks D et al. (for the ATLAS Collaboration) Multi-core job submission and grid resource scheduling for ATLAS AthenaMP *J. Phys.: Conf. Ser.* **396** 032115
[21] Barreiro Megino F H et al. 2012 Exploiting Virtualization and Cloud Computing in ATLAS *J. Phys.: Conf. Ser.* **396** 032011
[22] Bird I 2013 Grid AND Cloud evolution from now onward *(these proceedings)*
[23] Bauerdick L et al. 2012 Using Xrootd to Federate Regional Storage *J. Phys.: Conf. Ser.* **396** 042009

A Web-Based Development Environment for Collaborative Data Analysis

M Erdmann, R Fischer, C Glaser, D Klingebiel, M Komm[a], G Müller, M Rieger*, J Steggemann[b], M Urban and T Winchen

III. Physics Institute A, RWTH Aachen University, Aachen, Germany

[a] now at Centre for Cosmology, Particle Physics and Phenomenology, Université catholique de Louvain, Louvain-la-Neuve, Belgium

[b] now at CERN, European Organization for Nuclear Research, Geneva, Switzerland

* corresponding author, `rieger@physik.rwth-aachen.de`

E-mail: `vispa@lists.rwth-aachen.de`

Abstract. Visual Physics Analysis (VISPA) is a web-based development environment addressing high energy and astroparticle physics. It covers the entire analysis spectrum from the design and validation phase to the execution of analyses and the visualization of results. VISPA provides a graphical steering of the analysis flow, which consists of self-written, re-usable Python and C++ modules for more demanding tasks. All common operating systems are supported since a standard internet browser is the only software requirement for users. Even access via mobile and touch-compatible devices is possible. In this contribution, we present the most recent developments of our web application concerning technical, state-of-the-art approaches as well as practical experiences. One of the key features is the use of workspaces, i.e. user-configurable connections to remote machines supplying resources and local file access. Thereby, workspaces enable the management of data, computing resources (e.g. remote clusters or computing grids), and additional software either centralized or individually. We further report on the results of an application with more than 100 third-year students using VISPA for their regular particle physics exercises during the winter term 2012/13. Besides the ambition to support and simplify the development cycle of physics analyses, new use cases such as fast, location-independent status queries, the validation of results, and the ability to share analyses within worldwide collaborations with a single click become conceivable.

1. Introduction

The demand for distributed resources and global access to data experienced a rapid growth in the past years. In fact, both, the way we use instrumental technology and the requirements we claim on hardware and software currently undergo a change of paradigms. The development work of scientists is no longer subject to limitations such as locally restricted availability of computing power and data access. Instead, the change of infrastructure becomes the focus of attention in order to fulfill individual needs concerning place and time independence. This is especially achieved by the progress made in the field of networking technology.

The consequent advantages of network (and internet) communication can be applied to scientific workflows. We present a new approach that is based on the desktop version of the Visual

Physics Analysis (VISPA) project, which has been successfully used in various high-energy and astroparticle physics data analyses [1]. The new concept implements the VISPA platform for the application within a standard web-browser. The underlying server-client-approach is based on central software, data, and resource deployment as well as time- and location-independent accessibility. It also provides the opportunity of cooperative analysis creation. Therefore, it can be used as a platform for various scientific applications, e.g. for novel blended learning concepts in a wide scope of physical teaching.

2. The Visual Physics Analysis project

The Visual Physics Analysis (VISPA) project provides a development environment for physics analyses covering the entire analysis spectrum from the design and validation phase to the execution of analyses and the visualization of results. It utilizes the C++ Physics eXtension Library (PXL) [2] as underlying analysis framework.

Analysis that are based on PXL are subdivided into logically separated modules, i.e. blocks of code that are designated to carry out well defined tasks. In addition to predefined I/O modules, PXL provides the possibility to include reusable, user-defined modules, written either in C++ or Python [3]. Connections between these modules describe the data-flow of the analysis, which is subject to an event-by-event approach.

VISPA provides a graphical representation of analyses. It not only reflects their structure, i.e. the sequence of modules and the connections between them, but it also allows for graphical steering. Therefore, the web-based implementation of VISPA includes a variety of extensions, aimed to cover a wide range of functionality.

3. Technical implementation

3.1. Server-client approach

The VISPA platform follows a server-client approach that is typical for modern web applications. A web browser (client) sends a specific request to a web server via the Hypertext Transfer Protocol (HTTP). The server processes the request to create a response that is sent back to the client. There, it can be parsed to either display new content, usually via HTML markup [4], or to update already existing content using the AJAX concept [5].

The server is built on the basis of the Python web framework CherryPy [6]. This allows for a convenient integration of various scientific software packages written in Python, such as PXL, ROOT [7], and NumPy [8]. The user management is realized using a SQL database that stores login information for all registered users. The SQL dialect, e.g. MySQL, PostgreSQL, or SQLite, can be altered as the object-relational mapping (ORM) framework SQLAlchemy [9] is incorporated. The content of HTTP responses is created and cached using the Mako templating library [10].

Although the server-client system is intended to operate on separate computers via network communication, the possibility to install the VISPA platform locally on a single computer is preserved. In this operation mode, the user does not need to register to the platform, which accordingly is accessible only from the local machine.

3.2. Workspaces

The implementation of workspaces is a new key ingredient of the VISPA platform. Scientific analyses often depend to a great extent on the availability of computing power and data access. Hence, in some scientific environments a single machine may not be appropriate for acting as both, web server and processing unit at the same time. For the purpose of splitting the architecture into logically and physically separated units, all user-related computational tasks are sourced out into workspaces.

Every machine running Python that is accessible via SSH over the network can act as a workspace, regardless of whether it is, e.g., a mobile phone, a desktop machine, or a computing cluster. Access is granted if the client is able to connect to the workspace using password or public key authentication. At the beginning of a connection, all necessary files are copied over to the workspace where a receiving process is started. The communication between this process and the VISPA server is subject to the JSON-RPC 2.0 specification [11]. This way, the entire host system of the workspace can be utilized including file access, permissions settings, and CPU usage.

In this model, the VISPA server itself takes on the role of an intermediate node: it serves requests and delivers files related to the VISPA platform on the one hand, and handles connections as well as distributes computational tasks to workspaces on the other hand. A generic overview of this approach is shown in figure 1.

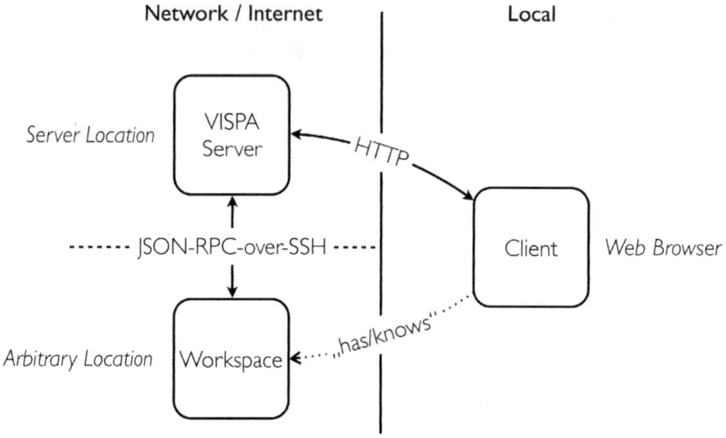

Figure 1. Schematic overview of the workspace approach.

3.3. User interface and extensions

The graphical user interface (GUI) is based on the HTML5 markup language and the JavaScript scripting language, which are both supported by common web browsers. Advanced interaction components and compatibility for mobile devices are implemented using the DOJO framework [12] and the jQuery UI framework [13]. This software foundation is provided by the web server and initially transfered to the web browser on startup. It is then utilized by both, the VISPA platform itself and its extensions. A selection of extensions is illustrated in figure 2.

The *Analysis Designer* provides the GUI for developing physics analyses based on the PXL toolkit. Modules and connections between them can be created and/or changed by drag & drop gestures in order to steer the analysis flow.

The *File Browser* implements the possibility to browse through personal and public files that are available on the currently selected workspace. It provides common features like, e.g., up- and download, copy, and paste functionality as well as the possibility to select and open files in other extensions.

In high energy physics, a data file may contain sophisticated information of multiple events. Files of the format `pxlio`, introduced by PXL, can be inspected event-by-event with the *PXL Browser*. The properties of the constituents can be obtained by clicking on their graphical representations.

For the purpose of directly inspecting, creating, and steering analysis code within the web browser, a *Code Editor* is incorporated based on the Ace Editor project [14].

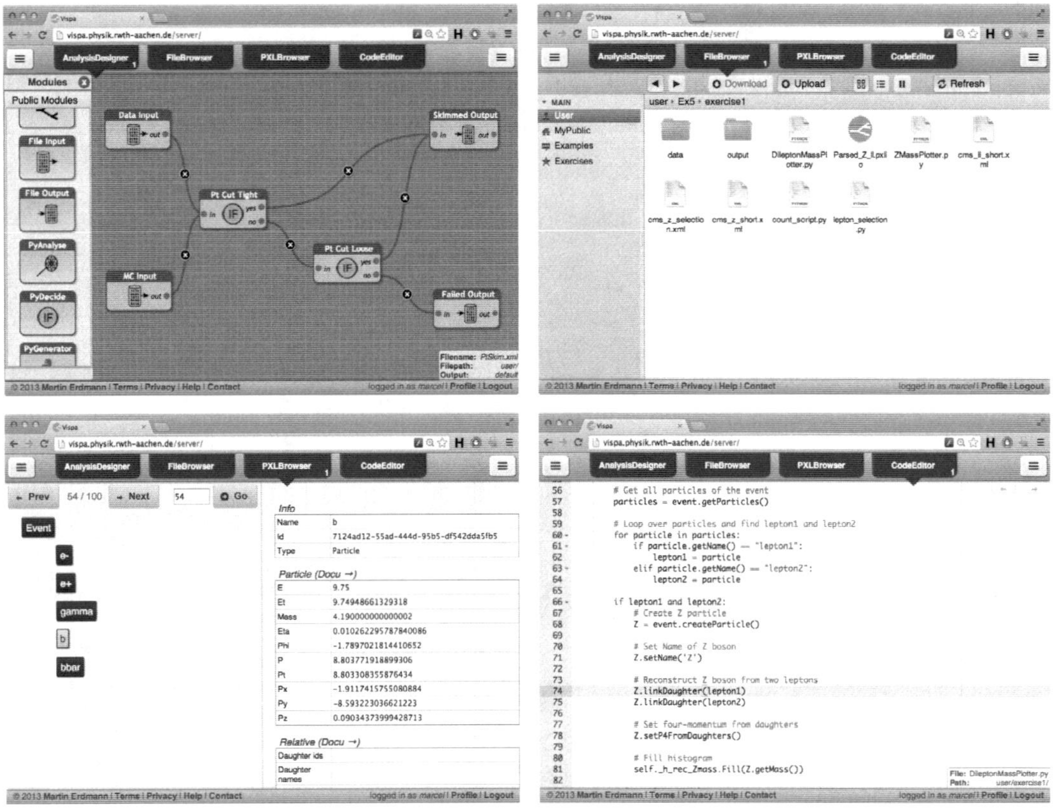

Figure 2. Extensions of the web-based implementation of VISPA (from top-left to bottom-right): Analysis Designer, File Browser, PXL Browser, and Code Editor.

4. A field test with third year students

The practical use of the VISPA platform was tested in the scope of the course "Particle Physics and Astrophysics" held during the winter term 2012/13 with 3 hours of lectures per week and exercises due every 2 weeks. The course consisted of more than 100 third year physics students. 25 % of the points of every exercise could be achieved by successfully working on data analysis tasks using VISPA. A total of 50 % of all exercise points was mandatory to receive the permission to participate in the final exam.

The server setup was specially dedicated to the application with students. An instance of the VISPA platform was implemented on a Linux server with 10 worker nodes in accordance with a security concept created in collaboration with the computing center of the RWTH Aachen University. It turned out that even at peak times, memory and CPU occupancy were sufficient to supply each student request with adequate computing power.

In order to assess the success of the application, a survey was started to obtain feedback about the acceptance and learning benefit of the VISPA platform. 63 students participated the survey that comprised both, predefined questions and the possibility for free text comments. Valuable comments on the workflow were received. The majority concerned the number of interactions, i.e. the number of mouse clicks, needed to perform an iteration of the data analysis exercise. The result of the overall rating for the VISPA platform is shown in figure 3, yielding a positive assessment of the learning project, which is visible on the right side of the distribution.

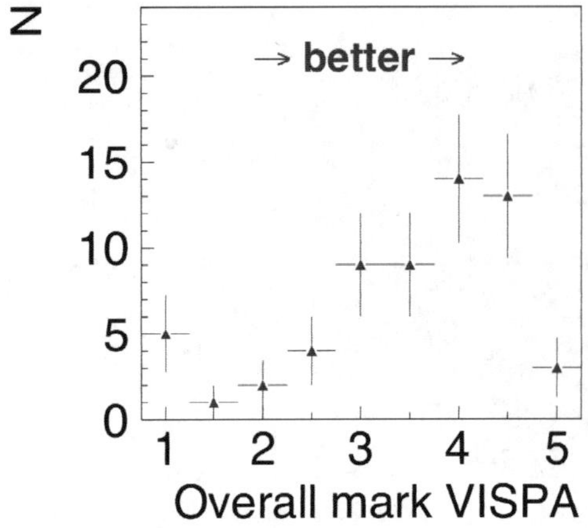

Figure 3. The student's overall rating of the VISPA learning concept on a scale with the highest value being the best mark.

5. Conclusions

We have presented a web-based development environment addressing workflows in high energy and astroparticle physics. It enables the development and execution of physics data analysis in a standard web-browser. Workspaces are a key ingredient of the platform. They allow for remote access to data, software, and resources on arbitrary machines that are accessible via SSH. The potential of the presented web platform is comprehensive. For example, scientists are able to collaborate on analysis concepts and algorithms through the web interface. A new level of transparency may be offered in publications by providing web access to the corresponding analysis. Students may get access to analysis examples exploring public experiment data with professional tools that are standard in todays physics analyses.

Acknowledgments

We wish to thank the organizers of the ACAT2013 conference for their kind support. This work is supported by the Ministerium für Wissenschaft und Forschung, Nordrhein-Westfalen, the Bundesministerium für Bildung und Forschung (BMBF), and the Helmholz Alliance Physics at the Terascale. M Rieger and C Glaser gratefully thank for support by the Deutsche Forschungsgemeinschaft (DFG) and T Winchen gratefully acknowledges funding by the Friedrich-Ebert-Stiftung.

References

[1] Bretz H-P et al. (2012) *A Development Environment for Visual Physics Analysis* JINST **7** T08005 doi:10.1088/1748-0221/7/08/T08005 arXiv:1205.4912v2 `http://vispa.physik.rwth-aachen.de`
[2] Erdmann M et al. (2012) *Physics eXtension Library (PXL)* `http://vispa.physik.rwth-aachen.de/PXL`
[3] *Python Programming Language* `http://www.python.org`
[4] World Wide Web Consortium *HTML5 Specification* `http://www.w3.org/TR/html5`
[5] Holdener A T (2008) *Ajax - The Definitive Guide* (O'Reilly Media, Sebastopol, CA, USA)
[6] The CherryPy Team *CherryPy - A Minimalist Python Web Framework* `http://www.cherrypy.org`
[7] Brun R and Rademakers F (1996) *Nucl. Inst. & Meth. in Phys. Res. A 398 81-86* `http://root.cern.ch`
[8] NumPy Developers `http://www.numpy.org/`
[9] SQLAlchemy Authors and Contributors `http://www.sqlalchemy.org`
[10] Bayer M *Mako Templates for Python* `http://www.makotemplates.org/`
[11] JSON-RPC 2.0 Specification `http://www.jsonrpc.org/specification`
[12] The Dojo Foundation `http://dojotoolkit.org`
[13] The jQuery Foundation `http://jqueryui.com`
[14] Cloud9 IDE and the Mozilla Foundation `http://ace.c9.io`

Journal of Physics: Conference Series **523** (2014) 012022 doi:10.1088/1742-6596/523/1/012022

Advances in tracking and trigger concepts

Ivan Kisel

Goethe University, 60323 Frankfurt am Main, Germany
FIAS Frankfurt Institute for Advanced Studies, 60438 Frankfurt am Main, Germany
GSI Helmholtz Center for Ion Research, 64291 Darmstadt, Germany

Abstract. Increasing beam intensities and input data rates require to rethink the traditional approaches in trigger concepts. At the same time the advanced many-core computer architectures providing new dimensions in programming require to rework the standard methods or to develop new methods of track reconstruction in order to efficiently use parallelism of the computer hardware. As a results a new tendency appears to replace the standard (usually implemented in FPGA) hardware triggers by clusters of computers running software reconstruction and selection algorithms. In addition that makes possible unification of the off-line and on-line data processing and analysis in one software package running on a heterogeneous computer farm.

1. Introduction

Several large international research centers are focused on high-energy physics. Among them are: CERN (Geneva, Switzerland) with ATLAS, CMS, LHCb and ALICE experiments; BNL (Upton, USA) with PHENIX and STAR heavy-ion experiments; and FAIR/GSI (Darmstadt, Germany) with CBM and PANDA future experiments. Focusing on different aspects of high-energy physics and, therefore, having different structures of their detector setups, e.g. of collider or fixed-target types, the experiments confront very similar problems in the processing of experimental data.

In addition to physics governed challenges, a new challenge is brought with modern many-core computer architectures. It requires not only to use new parallel programming languages and frameworks, not only to understand the computer hardware, but in many cases to develop new methods, suitable for parallel data streams, and redesign the traditional methods, which were optimal in the past, now become inefficient running on many-cores.

2. Methods of event reconstruction

Usually the most complicated part of the data processing is the event reconstruction, which consists of four stages: (1) track finding, (2) track fitting, (3) particle identification and (4) search for short-lived particles [1]. To cope with the very large combinatorial procedure of combining one- or two-dimensional hits into five-dimensional particle trajectories, the following methods are used at the track finding stage: conformal mapping, Hough transformation, track following and cellular automaton.

The conformal mapping method (Fig. 1) images the measurement space (x, y) onto the conformal space (u, v) with the transformation:

$$u = x/(x^2 + y^2); v = y/(x^2 + y^2). \quad (1)$$

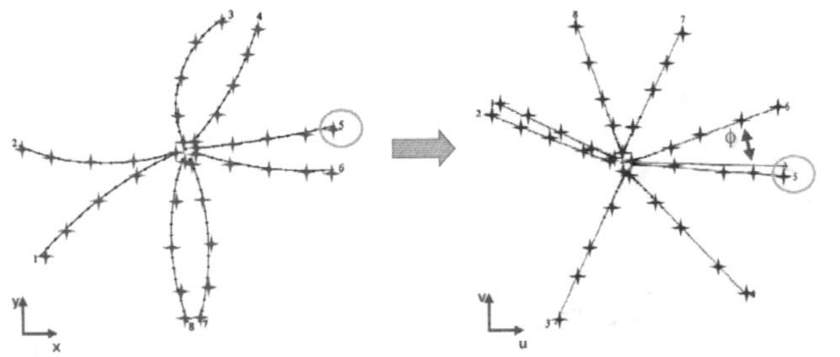

Figure 1. An illustrative representation of the conformal mapping method for particle trajectories in a cylindrical detector with homogeneous magnetic field.

Transforming a circle, which passes through the origin of the coordinate system, into a straight line, the method brings an impressive visual simplification of the problem. This makes finding of particle trajectories in a homogeneous magnetic field extremely simple. Each step of the method is easy to implement in hardware, like in an FPGA. For instance, histogramming angles of the measurements will immediately produce peaks, which correspond to the particle trajectories. Some complications of the method are: one needs to know the interaction point in advance; appropriate for uniform magnetic fields only; provides only rough track parameters estimation; suitable for finding primary tracks, but not secondaries; does not group measurements into track candidates; needs a significant amount of memory for multidimensional histogramming. Thus, the conformal mapping method is useful for implementation in hardware for simple event and trigger topologies.

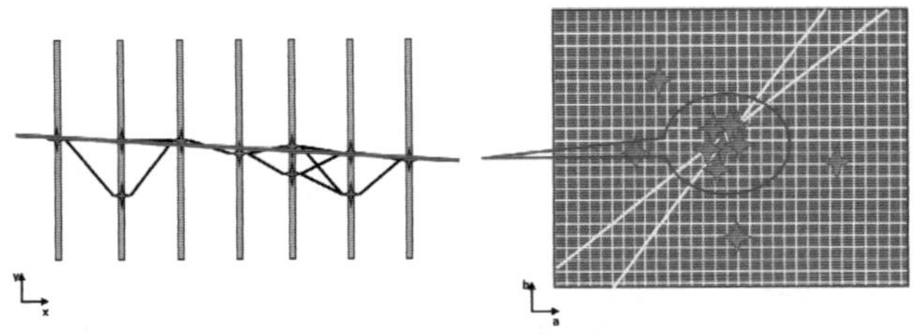

Figure 2. An illustrative representation of the Hough transformation method for a straight line trajectory.

The Hough transformation method (Fig. 2) images the measurement space (x, y) onto the parameter space (a, b) with the transformation:

$$y = a \cdot x + b; \rightarrow b = -x \cdot a + y. \tag{2}$$

Now each measurement (x, y) will determine a straight line in the parameter space. An intersection point of two lines will determine a set of track parameters (a, b). Due to non-precise measurements and effects of multiple scattering, the straight lines will not intersect each other in the same point, but produce a cluster of intersection points, that is easy to localize by histogramming. A situation becomes more complicated when noise hits are present in the detector or when several or many tracks are registered. In this case to avoid unnecessary overfilling of the histogram one can work with pairs of measurements, thus filling the histogram directly with intersection points. The Hough transformation method is a generalization of the histogramming method and is easy to implement in hardware. Some complications of the method: needs a global track model; appropriate for uniform magnetic fields only; does not include effects of multiple scattering; provides track parameters estimations without errors estimations; does not make competition between track candidates; needs a significant amount of memory for multidimensional histogramming. As a result, similar to the conformal mapping method, the Hough transformation method is useful for implementation in hardware for simple event and trigger topologies.

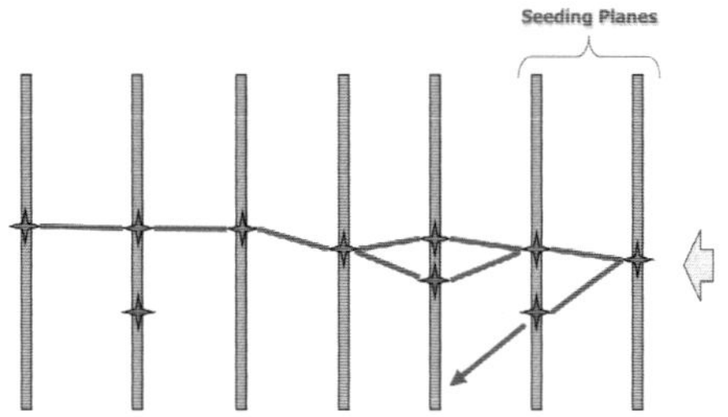

Figure 3. An illustrative representation of the track following method.

The track following method (Fig. 3) implements an intuitive approach to reproduce a particle way, i.e. to move from one measurement to another. One starts from the last detector plane (probably with the least hit density), takes a measurement and assumes that it belongs to a track. Taking another measurement on the next to the last detector plane, one gets a shortest possible track segment, called *track seed*. Further propagation of the track seed from one detector plane to another collects measurements into a track group. If several measurements are found on a detector plane within the propagation region, one creates several *branches* and follows them as separate track candidates. If a detector plane has no measurement within the region of propagation, the current track candidate is deleted, and the procedure is repeated with the next seed (or the next track candidate within the branch). During the track following one can apply the Kalman filter track fitting procedure in order to get simultaneously the estimations of the track parameters and the covariance matrix. Traditionally, development of a new experiment starts with an ideal Monte-Carlo track finder and a realistic Kalman filter track fitter, then it is easy to implement the realistic track finder as track following based on the Kalman filter, thus combining track finding and track fitting. In addition, for a developer it is psychologically easy to accept the hit-by-hit track finding approach. Some complications of the method: it is intrinsically based on a single track assumption; needs the seeding stage; efficiency is limited by the detector efficiency and the seeding efficiency; works at the hit level and forces random access to the hits; after discarding a track candidate can often repeat the same calculations; no global competition between track candidates; limited by the exponential grow of the combinatorial enumeration at high track densities. The track following method is extensively employed in the LHC experiments.

The cellular automaton method (Fig. 4), which can be regarded as a local version of the Hopfield neural network, also creates short track segments, but, contrary to the track following method, it does it between all pairs of detector planes (stage 1). Then it does not work anymore with hits, but with the created segments. Applying a track model (straight line on the illustrative figure) it introduces relations between the segments on neighboring pairs of detector planes in case they have a common hit and a small kink angle. Then it counts consecutively the neighbor segments with their possible position on a track candidate (stage 2). After the tree structure appeared, starting from the last segments with the largest position counters it collects the neighboring segments into track candidates (stage 3). At the last stage the track candidates are fitted with the Kalman filter and selected according to the χ^2-value. Local operations with data bring us to a parallel implementation of the algorithm. The implementation is staged from hits to segments and to track candidates, thus step-by-step collecting the tracking information on the fly. A second order polynomial behavior is observed at high track densities. The track competition stage is naturally included in the algorithm. Some complications of the method:

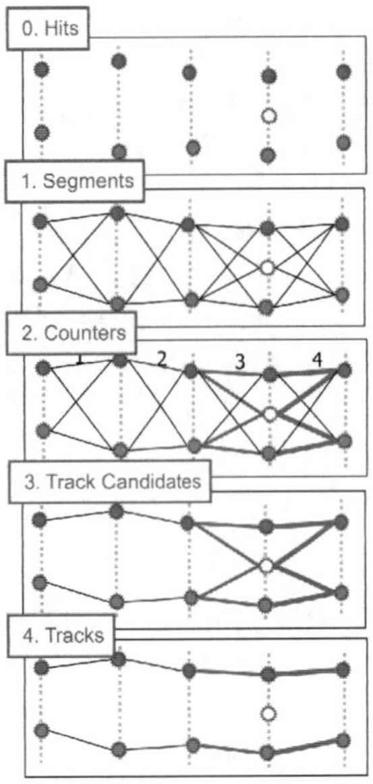

Figure 4. An illustrative representation of the cellular automaton method.

staged algorithm need synchronization after each stage; parallel implementation of the algorithm requires from a developer to have an expert skill in parallel computer architectures, which come into the market only now. The cellular automaton method is successfully used in several heavy-ion experiments.

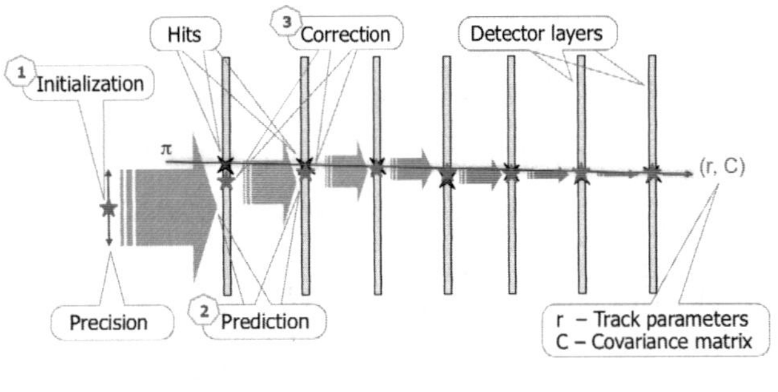

Figure 5. An illustration of the Kalman filter based track fitting algorithm.

Found tracks are then fitted using the Kalman filter method [1]. The Kalman filter (1) starts with an arbitrary initial approximation (Fig. 5); (2) adds hits one after the other; (3) refines the state vector $\vec{r}$ of track parameters and gets the optimal values of the track parameters after the last hit. There is a set of fitting procedures based on the Kalamn filter. The simplest one, described above and called track fitter, one obtains the state vector $\vec{r}$ of track parameters and the covariance matrix C in an outer (first or last) measurement. The Kalman smoother provides the state vector and the covariance matrix in an inner point of the trajectory. For some tracks of high importance for physics analysis one can apply the deterministic annealing filter (DAF), which is based on the Kalman filter and iteratively reduces weights of noise of wrongly associated hits in order to improve the quality of the track parameters estimation.

3. Many-core HPC: cores, threads and vectors

Nowadays, the complexity of event topologies lead to a necessity of using complicated reconstruction algorithms already on-line in the high-level triggers (HLT). In addition, in order to cope with high interaction rates one needs to equip an HLT farm with modern high-performance computing (HPC) hardware (Fig. 6). These are many-core central processing units (CPU) with dozens of cores and graphics processing units (GPU) with thousands of arithmetic units. For efficient use of the many-core hardware one has to implement the algorithms in a parallel manner.

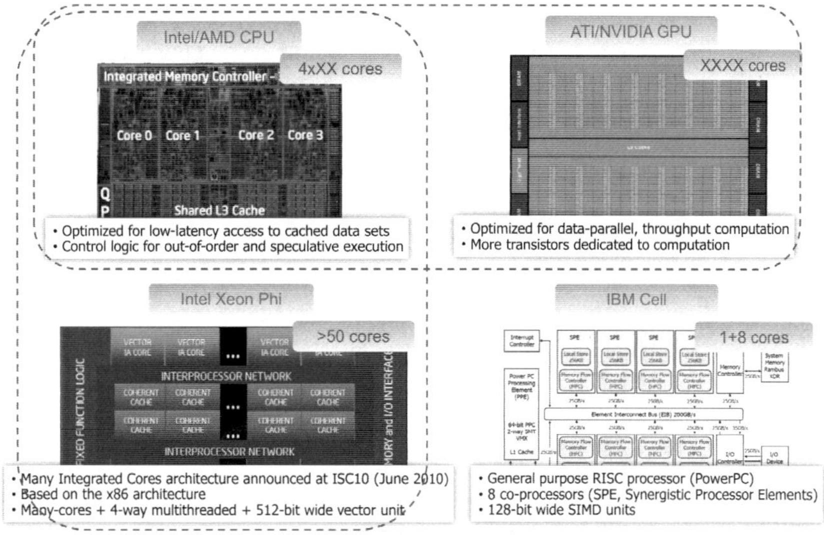

Figure 6. Future HPC systems are heterogeneous.

To illustrate the complexity of the HPC hardware, let us consider a single work-node of an HLT computer farm, a server equipped with CPUs only. Typically it has 2 to 4 sockets with 8 cores each. In case of Intel CPUs, each core can run in parallel 2 hardware threads (processes), that increases the calculation speed by about 30%. The arithmetic units of CPUs operate with vector registers, which contain 4 (SSE), 8 (AVX) or 16 (MIC) data elements. Vectors realize the SIMD (Single Instruction, Multiple Data) paradigm, that means they apply an operation to a vector as a whole, giving a speed-up factor of 4/8/16 with respect to the same operation, but with a scalar. In total, a pure hardware potential speed-up factor of a host is:

$$f = 4 \text{ sockets} \cdot 8 \text{ cores} \cdot 1.3 \text{ threads} \cdot 8 \text{ SIMD} \approx 300, \tag{3}$$

which is already equivalent to a moderate computer farm with scalar single-core CPUs.

Porting an algorithm to a parallel hardware requires usually reworking of the algorithm from the ground up. It includes numerical optimization, memory optimization and parallelization, that brings another 10 to 100 or even up to 1000 speed-up factor.

Thus, the HPC architecture makes possible use of complicated reconstruction and selection algorithms already at the HLT level. In addition, on-line and off-line data reconstruction, selection and analysis become very similar or even identical, that provides enormous flexibility and compactness of a computer farm.

4. Parallel programming

Parallel programming is, first of all, a parallel formulation of an algorithm, and only then its implementation in a parallel language. The hardware provides us two levels of parallelization: a task level parallelism working with cores and threads, and a data level parallelism working with SIMD vectors.

At the task level parallelism one localizes independent parts of the algorithms and run them in parallel on different cores or threads with or with our synchronization between the processes. Implemented it can be using, for instance, the ITBB or OpenMP frameworks.

If the algorithm allows to organize parallel streams of data, which are processed in the same way, like fit of several tracks, these parts can be SIMDized and run without using an extra hardware, but on vectors within the same threads. For that one can use the auto-vectorization, which is provided by the compilers. Unfortunately, this brings typically (and unpredictably) about 20% of speed-up, which is almost negligible comparing to the potential factor 4/8/16. In order to reach the maximum (depending on the data level parallelism of the algorithm), one can program using the SIMD extensions directly or using the SIMD header files or the much advanced Vc library.

The SIMD header files overload the SIMD instructions inlining the basic arithmetic and logic functions, that makes the code compact and easy readable. An illustrative example of a simple code for calculation of a polynomial function of the first order, which is written using SSE instructions, is:

```
__m128 y = _mm_add_ps(_mm_mul_ps(a,x),b);
```

The same function, but implemented using the SIMD header file, recovers the scalar-like form:

```
fvec y = a*x + b;
```

with overloading

```
friend fvec operator+( const fvec &a,
                       const fvec &b ) {
  return _mm_add_ps(a,b); }
friend fvec operator*( const fvec &a,
                       const fvec &b ) {
  return _mm_mul_ps(a,b); }
```

in the SIMD header file.

The header files provide a simple and flexible SIMD implementation with respect to different CPU architectures. It keeps the program code in a scalar-like form, while the CPU specific SIMD extensions are hidden in the header files, which can be chosen automatically depending on the CPU type. Also a header file with the true scalar implementation exists, that allows to make the conventional debugging and testing of the code.

The OpenCL standard provides a higher level of parallel programing by writing a universal code, which can be run on different types of CPU and GPU processing units. Thus, it provides a portable and efficient access to heterogeneous computer platforms. The OpenCL standard supports both vectorization and parallelization between cores of CPUs and GPUs. The vectorized code in OpenCL looks similar to the previous example:

```
float4 y = a*x + b.
```

5. How to parallelize reconstruction in running experiments

Parallelization is a very non-trivial task, but parallelization of the algorithms in running experiments is a real challenge, to which we have to respond within the next years. In addition to all hardware and software complications described above, the experiment has to run stably and predictably, as this is of vital importance for the physics programme. Different experiments have chosen different strategies for parallelization of the reconstruction software (Table 1), that

ATLAS	Modify the existing code
CMS	Implement a new search algorithm
ALICE	Merge the on-line and off-line codes
STAR	Use an existing algorithm as seed finder
CBM	Develop a new code from scratch

Table 1. Strategies of parallelization in different experiments.

are the most appropriate to provide them a smooth transition to an efficient use of the many-core computer architectures [2].

ATLAS will modify the existing code by numerical and computational optimization, multi-threading and auto-vectorization. The HLT reconstruction software will be gradually integrated with the off-line code.

CMS plans to overcome the current tracking limitations of the combinatorial track following by the parallel implementation the Hough transformation algorithm.

ALICE will use the tracks, found by the cellular automaton track finder in HLT, as seeds for the off-line track following algorithm based on the Kalman filter.

STAR is decided to use the cellular automaton track finder as seed finder for the track following algorithm based on the Kalman filter in off-line and to replace in HLT the Hough transformation algorithm by the same cellular automaton track finder, thus providing mutually reproducible results in on-line and off-line data analysis.

CBM, as a future experiment, profits from the possibility to design and develop a new code from scratch, investigating different approaches. The experiment assumes full identity both in software and hardware between the off-line and on-line reconstruction, selection and analysis of data. Current track finding in CBM is based on the cellular automaton approach.

6. CBM: First Level Event Selection

To demonstrate some typical and specific features of the tracking and trigger concepts, let us consider the current status of the First Level Event Selection (FLES) [3] of the CBM experiment (see a simulated central Au-Au collision on Fig. 7). It is similar to HLT in other experiments, but since in CBM there is no previous stages of data selection, it has a different name in order to emphasize its first (and, actually, the last) level of selection. Thus, FLES combines all traditional stages of triggering.

The FLES package consists of several modules (the block diagram is shown on Fig. 8): track finder, track fitter, particle finder and physics selection. As an input the FLES package receives a simplified geometry of the tracking detectors and the hits, which are created by the charged particles crossing the detectors. Tracks of the charged particles are reconstructed by the Cellular Automaton (CA) track finder [4] using to the registered hits. The Kalman filter (KF) based track fit [5] is used for precise estimation of the track parameters. The short-lived particles, which decay before the tracking detectors, can be reconstructed via their decay products only. The KF particle finder, which is based on the KFParticle package [6] is used in order to find and reconstruct the parameters of short-lived particles by combining already found tracks of the long-lived charged particles. The KF particle finder also selects particle-candidates from a large number of random combinations. In addition, a quality check module is implemented, that allows to monitor and control the reconstruction process at all stages. The FLES package is platform and operating system independent.

The FLES package in the CBM experiment will be performed for the on-line selection and the off-line analysis on a dedicated many-core CPU/GPU farm. The farm is currently estimated to have a compute power equivalent to 60 000 modern CPU cores. The FLES algorithms have

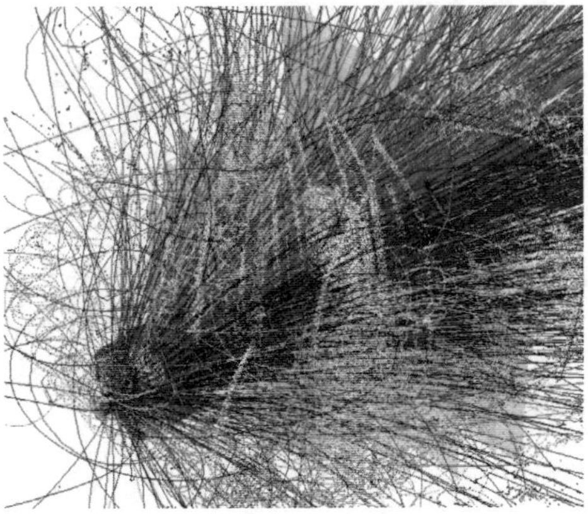

Figure 7. A simulated central Au-Au collision at 25 *A*GeV energy in the CBM experiment with about 1000 charged particles (different colors correspond to different types of particles).

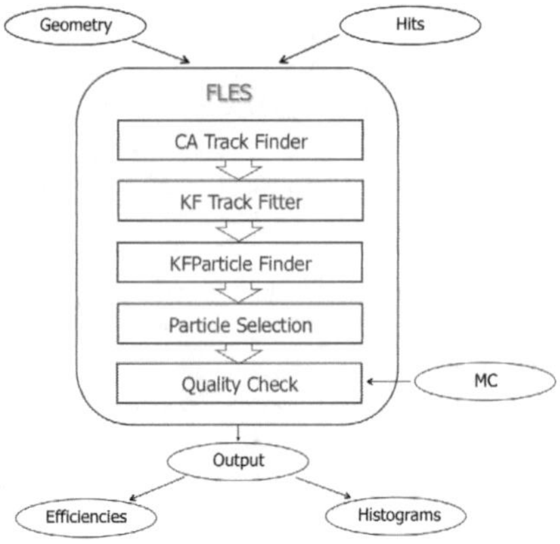

Figure 8. Block diagram of the FLES package.

to be therefore local and parallel with respect to data and thus require a fundamental redesign of the traditional approaches to event data processing in order to use the full potential of modern and future many-core CPU/GPU architectures. Massive hardware parallelization has to be adequately reflected in mathematical and computational optimization of the algorithms. In order to reflect the specific features of the CBM experiment, the track segments in the track finder are created from hits in each three consecutive detector stations to reconstruct the particle momentum. To recover a possible detector inefficiency, hits in the track segments can be also separated by one inefficient station. The track finding procedure is organized in several iterations to make the reconstruction fast and reliable in presence of a high track density: at the first iteration only high-momentum primary tracks are reconstructed, at the second one — low-momentum primary tracks, and then — all other tracks. After each iteration all used hits are removed from further consideration, thus significantly reducing the combinatorics, which is extremely high due to the use of double-sided strip detectors.

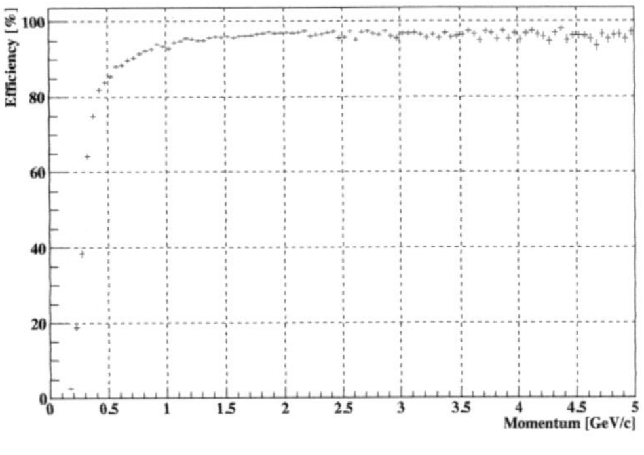

Figure 9. Efficiency of the track reconstruction for minimum bias Au-Au collisions at 25 *A*GeV.

For evaluation purposes a reconstructed track is assigned to a generated particle, if at least 70% of its hits have been caused by this particle. A generated particle is regarded as found, if it has been assigned to at least one reconstructed track. If the particle is found more than once, all additionally reconstructed tracks are regarded as clones. A reconstructed track is called a ghost, if it is not assigned to any generated particle according to the 70% criterion.

Efficiency of the track reconstruction for minimum bias Au-Au UrQMD simulated collisions at 25 AGeV is presented on Fig. 9. In addition the track reconstruction efficiencies for different sets of tracks and ratios of clones (double found) and ghost (wrong) tracks are shown in Table 2. The tests have been performed on a server with Intel Xeon E7-4860 CPUs.

	Efficiency, %	
	mbias	central
All tracks	88.5	88.3
Primary high-p tracks	97.1	96.2
Primary low-p tracks	90.4	90.7
Secondary high-p tracks	81.2	81.4
Secondary low-p tracks	51.1	50.6
Clone level	0.2	0.2
Ghost level	0.7	1.5
Reconstructed tracks/event	120	591
Time/event/core	8.2 ms	57 ms

Table 2. Track reconstruction efficiency for minimum bias and central events

The Kalman filter is intensively used in the combinatorial part of the CA track finder, therefore its fast implementation on modern CPU/GPU computer systems and stability in single precision are crucial. Starting from the idea of using the SIMD unit of modern processors, the KF track fitting algorithm was examined aiming to increase the speed of the CA track finder as a part of the FLES event reconstruction. After the detailed optimization of the memory utilization and the numerical analysis, the KF algorithm had been vectorized [5]. To optimize the memory usage, a magnetic field approximation is used for particle propagation instead of the full magnetic field map, which takes about 70 MB and therefore does not fit into the CPU cache memory. The magnetic field is approximated with a polynomial function of fifth order at each detector station. During the fit of a track the field behavior between the stations is approximated with a parabola taking field values at the three closest measurements along the track. To stabilize the fit, an initial approximation of the track parameters is done by the least square estimator assuming a one-component magnetic field. The first measurement is processed in a special way, which increases the numerical stability of the method in single precision: the equations were simplified analytically using a special form of the initial covariance matrix. The track propagation in the non-homogeneous magnetic field is done by an analytic formula, which is based on the Taylor expansion [7]. The analytic formula allows to obtain the same track fit quality as the standard fourth order Runge-Kutta method, while being 40% faster. Operator overloading has been used to keep flexibility of the algorithm with respect to different CPU/GPU architectures. All these changes have increased the processing speed of the SIMD KF track fit algorithm down to 1 μs per track. This is an improvement by a factor 10000 with respect to the original scalar version of the algorithm [5].

The CBM experiment is an experiment with a forward geometry along Z-axis and, therefore, has a typical set of tracks parameters: x and y track coordinates at a reference z-plane, $t_x = \tan\theta_x$ and $t_y = \tan\theta_y$ are the track slopes in the XZ- and YZ-planes, q/p is an inverse

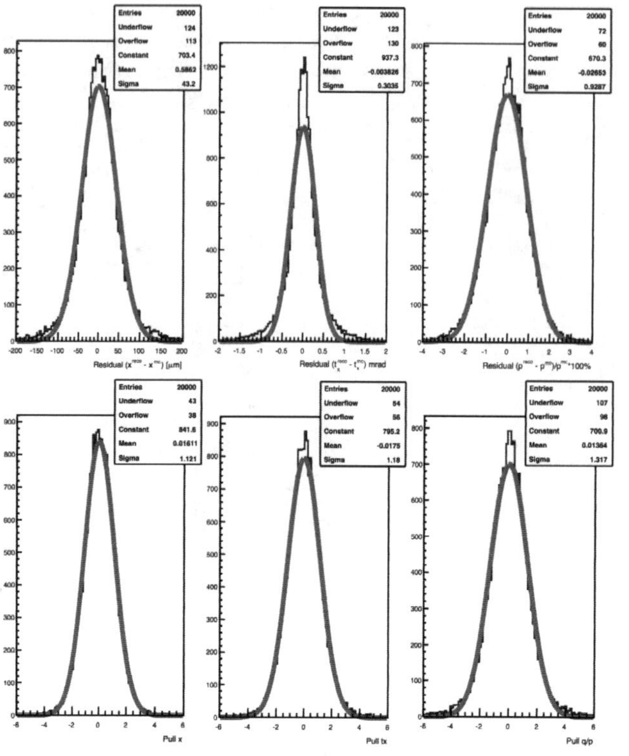

Figure 10. Residuals and pulls distributions of the x (43.2 μm, 1.12) , t_x (0.30 mrad, 1.18) and q/p (0.93 %, 1.32) track parameters.

particle momentum, signed according to the charge of a particle.

Residuals of the track parameters are determined as a difference between the reconstructed parameters and their true Monte-Carlo values. The normalized residuals (pulls) are determined as the residuals normalized by the estimated errors of the track parameters. In the ideal case these should be unbiased and Gaussian distributed with width of 1.0. Thus the pull distributions provide a measure of the track fit quality.

The residuals and the pulls for all track parameters are calculated at the first hit of each track. The distributions for the x, t_x and q/p parameters together with their Gaussian fits are shown on Figure 10 (the results for y and t_y are similar). All distributions are not biased with pulls widths close to 1.0 indicating correctness of the fitting procedure. The slight deviations from 1.0 are caused by several assumptions made in the fitting procedure, mainly in the part of the detector material treatment. The q/p pull is the widest being the most sensitive to these simplifications.

The high track finding efficiency and the track fit quality are crucial, especially for reconstruction of the short-lived particles, which are of the particular interest for the CBM experiment. The reconstruction efficiency of short-lived particles depends quadratically on the daughter track reconstruction efficiency in case of two-particle decays. The situation becomes more sensitive for decays with three daughters and for decay chains. The level of a combinatorial background for short-lived particles depends strongly on the track fit quality. The correct estimation of the errors on the track parameters improves distinguishing between the signal and the background particle candidates, and thus to suppress the background. The ghost (wrong) tracks usually have large errors on the track parameters and therefore are easily combined with other tracks into short-lived particle candidates, thus a low level of ghost tracks is also important to keep the combinatorial background low. As a result, the high track reconstruction efficiency and the low level of the combinatorial background improve significantly the event reconstruction and selection by the FLES package.

The search for the short-lived particles is done in one go, thus minimizing access to the memory. Together with the optimization and vectorization of the code this allows to achieve a high speed even in a presence of a huge combinatorics. The speed of the KF particle finder per core on a server with Intel Xeon E7-4860 CPUs is 1.4 ms per minimum bias Au-Au collision and 10.5 ms per central Au-Au collision at 25 AGeV.

The FLES package has been parallelized with ITBB implementing the event-level parallelism by executing one thread per logical core. Reconstruction of 1000 minimum bias Au-Au UrQMD events at 25 AGeV has been processed in each thread. In order to minimize the effect of the operating system each thread is fixed to a certain core using the pthread functionality provided by the C++ standard library. Fig. 11 shows a strong scalability for all many-core systems achieving the reconstruction speed of 1700 events per second on the 80-cores server.

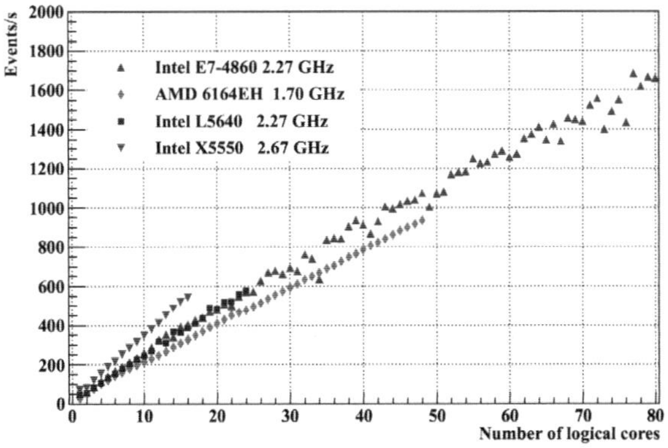

Figure 11. Scalability of the FLES package on many-core servers.

7. Summary

Modern and future many-core computer architectures open new perspectives in tracking and trigger concepts. At the same time, efficient use of the hardware requires much higher computing experience and consolidation of efforts of different experiments.

References

[1] Frühwirth R et al 2000 Data Analysis Techniques for High-Energy Physics. Second Edition, Cambridge Univ. Press

[2] Fourth International Workshop for Future Challenges in Tracking and Trigger Concepts, CERN, 28-30 November 2012

[3] Kisel I, Kulakov I and Zyzak M 2013 Standalone First Level Event Selection package for the CBM experiment *IEEE Trans. Nucl. Sci.*, vol. 60, no. 5, pp. 3703–3708

[4] Kisel I 2005 Event reconstruction in the CBM experiment *Nucl. Instr. and Meth.*, vol. A566, pp. 85–88

[5] Gorbunov S, Kebschull U, Kisel I, Lindenstruth V and Müller W.F.J 2008 Fast SIMDized Kalman filter based track fit *Comp. Phys. Comm.*, vol. 178, pp. 374–383

[6] Gorbunov S and Kisel I 2007 Reconstruction of decayed particles based on the Kalman filter CBM-SOFT-note-2007-003, GSI, Darmstadt

[7] Gorbunov S and Kisel I 2006 Analytic formula for track extrapolation in non-homogeneous magnetic field *Nucl. Instr. and Meth.*, vol. A559, pp. 148–152

A Neural-Network Clusterisation Algorithm for the ATLAS Silicon Pixel Detector

K.J.C. Leney on behalf of the ATLAS Collaboration
University of the Witwatersrand, Johannesburg, South Africa
E-mail: katharine.leney@cern.ch

Abstract. A novel technique using a set of artificial neural networks to identify and split merged measurements created by multiple charged particles in the ATLAS pixel detector is presented. Such merged measurements are a common feature of boosted physics objects such as tau leptons or strongly energetic jets where particles are highly collimated. The neural networks are trained using Monte Carlo samples produced with a detailed detector simulation. The performance of the splitting technique is quantified using LHC data collected by the ATLAS detector and Monte Carlo simulation. The number of shared hits per track is significantly reduced, particularly in boosted systems, which increases the reconstruction efficiency and quality. The improved position and error estimates of the measurements lead to a sizable improvement of the track and vertex resolution.

1. The ATLAS Detector

The ATLAS detector [1] is a multi-purpose detector with a forward-backward symmetric cylindrical geometry and nearly 4π coverage in solid angle. ATLAS uses a right-handed coordinate system with its origin at the nominal interaction point (IP) and the z-axis along the beam pipe. The x-axis points from the IP to the centre of the LHC ring, and the y-axis points upward. Cylindrical coordinates (r, ϕ) are used in the transverse (x, y) plane, with ϕ being the azimuthal angle around the beam pipe and r being the distance to the z-axis. The three major sub-components of ATLAS are the tracking detector, the calorimeter and the muon spectrometer.

2. The ATLAS Pixel Detector

The silicon pixel detector [2] is the innermost layer of the tracking detector and provides precision measurements of the positions of charged particles. It is crucial for the reconstruction of primary vertices which give an estimate of the number of proton-proton interactions per bunch crossing, and for the identification of long-lived particles via the reconstruction of secondary vertices. The detector consists of 1744 identical modules, each containing 46080 silicon pixel sensors. Pixels are 250 μm thick, and nominally measure 50 μm $(r - \phi)$ $\times$ 400 μm (z), although there is a small fraction of longer pixels. A total of 80.8 million read-out channels are provided.

3. Pixel Clustering

Particles traversing the detector typically deposit charge in more than one pixel. The amount of charge deposited is estimated using the time-over-threshold method. Clusters are formed by grouping together pixels with deposited charge that have a common edge or a common corner, as illustrated in figure 1 (a). Starting from the geometrical centre of the cluster, defined as (x_{centre}, y_{centre}) in the local reference frame $x - y$ of the sensor surface, the position of the particle at the point where it crosses the detector layer is computed from the signal heights inside the cluster of pixels:

$$x_{cs} = x_{\text{centre}} + \Delta_x \cdot \left(\Omega_x - \frac{1}{2}\right)$$

$$y_{cs} = y_{\text{centre}} + \Delta_y \cdot \left(\Omega_y - \frac{1}{2}\right),$$

where the parameters $\Omega_{x(y)}$ are defined as:

$$\Omega_{x(y)} = \frac{q_{\text{last row (column)}}}{q_{\text{first row (column)}} + q_{\text{last row (column)}}},$$

and q is the deposited charge. The variables Δx and Δy are calibrated from data, using the linear dependence of the distance between the extrapolated track and the cluster centre on the $\Omega_{x(y)}$ parameters. They are parametrised as a function of the incidence angle in ϕ (θ) of the track with respect to the pixel module and the number of pixels within the clusters in the x (y) axis direction.

Although the standard clustering provides excellent resolution for most clusters, it is inadequate for dense environments with multiple charged particles where charge is deposited in neighbouring pixels and clusters are shared between particles, as is illustrated in figure 1 (b). Objects such as highly energetic jets and multi-prong hadronic decays of tau leptons are most affected. Jets with transverse momentum > 1 TeV will typically produce merged clusters. Unless the shared clusters can be resolved between particles then the track parameters will be mis-estimated. In such cases it is therefore desirable to split the clusters between tracks.

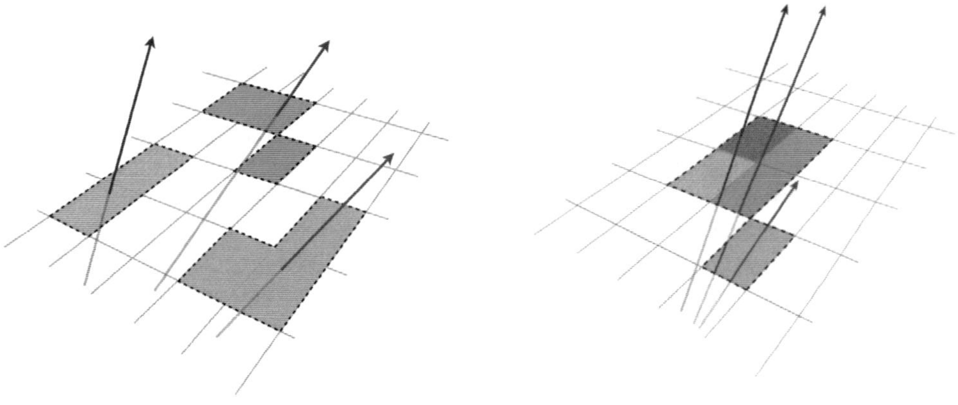

(a) (b)

Figure 1. Pixels with deposited charge are grouped into clusters if they have a common edge or a common corner. In most cases the clusters are well separated and the resolution is excellent (a), but in dense environments with multiple charged particles, charge is deposited in neighbours and clusters are shared (b).

4. Properties of Tracks in Dense Environments

The characteristics of tracks in dense environments, such as in the core of jets, are studied as a function of jet transverse momentum (p_T^{jet}) and of the track distance (ΔR) from the core of the jet for $40\,\text{GeV} < p_T^{jet} < 1200\,\text{GeV}$ using proton-proton collisions with centre-of-mass energy $\sqrt{s} = 8$ TeV recorded with the ATLAS detector during 2012.

The sample corresponds to integrated luminosity of $20.7\,\text{fb}^{-1}$. Monte Carlo samples are modelled using PYTHIA8 [3]. Jets are reconstructed from topological clusters using the anti-k_t algorithm [4] with a distance parameter $R = 0.4$ and required to have $0 < |\eta^{jet}| < 1.2$. Events are selected using a trigger-dependent minimum p_T^{jet} threshold which was imposed on all jets in the event. This threshold is chosen to ensure that the efficiency of a jet passing the trigger is $> 99\%$.

Events are required to have at least one primary vertex reconstructed using Inner Detector tracks. The vertex with the largest $\Sigma\left(p_T^{track}\right)^2$ is identified as the hard-scattering vertex. Tracks are required to pass the following selection:

- Transverse momentum (p_T) $> 0.5\,\text{GeV}$
- One or more hits in the pixel detector.
- Six or more hits in the semi-conductor tracker.
- Transverse impact parameter (d_0) $< 1.5\,\text{mm}$
- $z_0 \sin\theta < 1.5\,\text{mm}$

where d_0 and z_0 are measured with respect to the hard-scattering vertex. No cut on the track χ^2 is applied. Tracks are associated with jets using a simple geometric algorithm. If the distance in $\eta - \phi$ space, $\Delta R_{track-jet}$ between the track and the jet is less than the distance parameter used in the jet reconstruction (R = 0.4), the track is assigned to the jet. Track parameters are evaluated at the vertex perigee and are not extrapolated to the calorimeter.

The mean number of tracks and the mean value of the Σp_T per $\Delta R_{track-jet}$ bin are shown in figure 2. Agreement between data and simulation is good for both distributions for the full range of p_T^{jet} although for the higher bins of p_T^{jet}, jets in data are slightly broader than in simulation. Tracks become more collimated for the more energetic jets, which are observed to have more tracks in the core of the jet.

Distributions of the average number of hits in the pixel detector associated with a track and the average number of shared pixel hits are shown in figure 3. In cases where a track is extrapolated to pass through a silicon module which was not functioning, the 'dead' module is included in the hit count. Tracks in the core of the jet are observed to have fewer hits associated to them, and to share more hits with their neighbouring tracks.

5. A Neural Network Pixel Clustering Algorithm

Neural networks are powerful tools for pattern recognition problems and can handle non-linear correlations between input variables. They are attractive solutions for problems with many degrees-of-freedom, and inputs can be given different weights in the hidden layers of the neural network to finally determine the output. Neural networks are a good choice for a pixel clustering algorithm since many cluster properties are almost meaningless when considered alone (e.g. the charge deposited in a single pixel), but when cluster properties are combined and put into context (e.g. once the charge deposits on adjacent pixels are known) then the information can be combined and used for successful pattern recognition.

The ATLAS pixel clustering uses feed-forward multi-layer perceptron neural networks to compute:

- The number of particles per cluster.

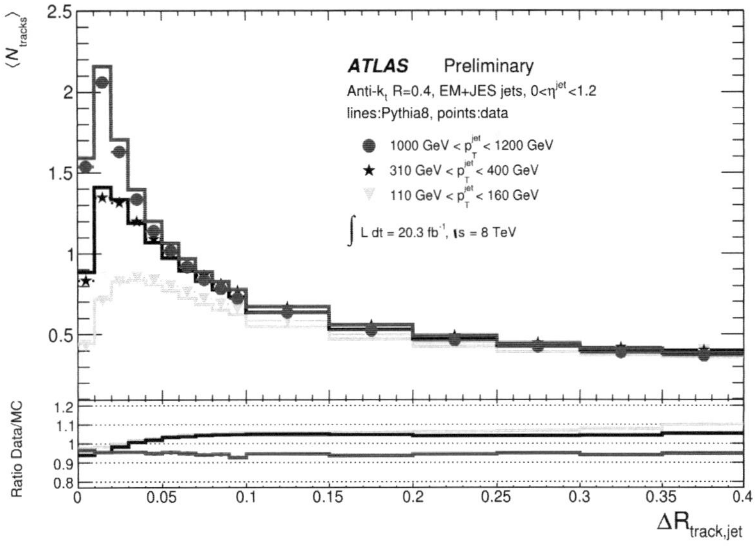

(a)

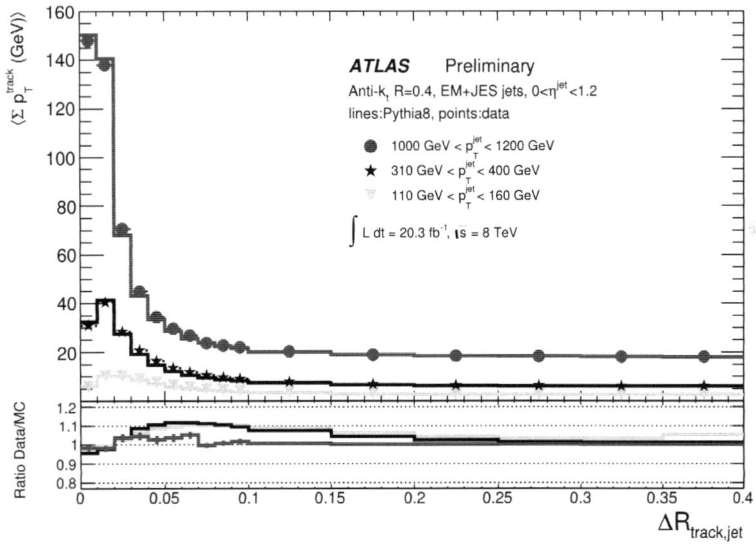

(b)

Figure 2. (a) Mean number of tracks per bin and (b) mean value of $\Sigma p_T^{\text{track}}$ per bin as a function of ΔR, the distance between the track and the jet, for bins in $p_{\text{T}}^{\text{jet}}$ for data and Monte Carlo.

- The position (and uncertainty) in two dimensions of the impact point of the particle(s) on the pixel sensor.

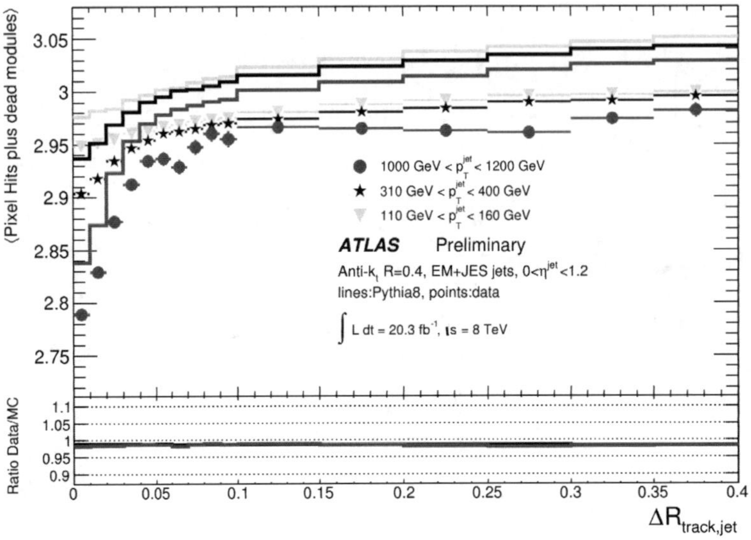

(a)

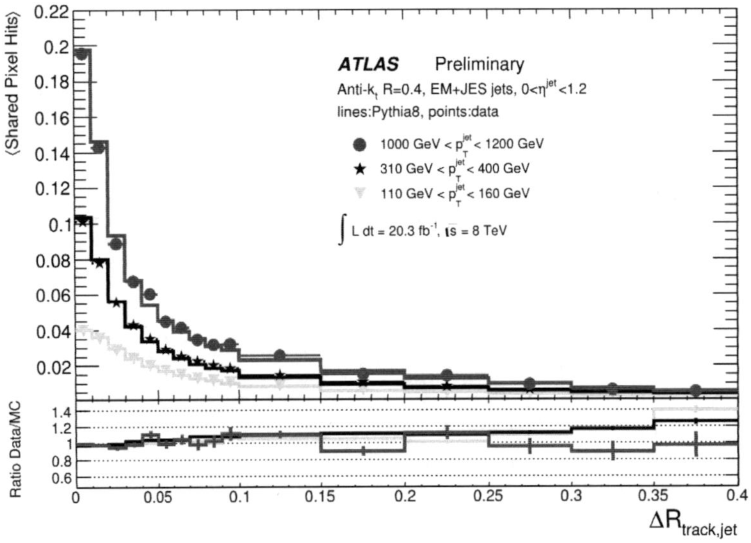

(b)

Figure 3. (a) Mean number of pixel hits and (b) mean number of shared pixel hits as a function of $\Delta R_{track-jet}$, the distance between the track and the jet, for bins in p_T^{jet} for data and Monte Carlo.

The neural network for computing the number of particles per cluster has 60 input nodes, consisting of:

- 7x7 pixel matrix of the collected charge on each pixel.
- Vector of the longitudinal sizes of the pixels in the matrix.
- Direction of the candidate charged particles traversing the cluster.

Particles are initially assumed to have been produced at the estimated centre of the luminous region of the beam-beam interaction. This is estimated using the primary vertices of collisions produced prior to the current interaction. The incidence angle of the particles with respect to the module is calculated by a straight line connecting the beam spot to the cluster position. Once track candidates have been identified, the approach is repeated using this more refined measurement of the particle direction. This is possible because the ATLAS track reconstruction strategy allows measurements to be re-calibrated after they have been assigned to a track candidate. Three output nodes are used, and can be interpreted as the posterior Bayes probabilities for a cluster to correspond to one, two or three charged particles, thus providing an event classification.

An additional set of neural networks, configured for interpolation, are used to estimate the cluster position and the probability density function for the residual of the impact point. Different neural networks are used depending on the estimated number of charged particles traversing the cluster, and trained on the true number of particles in simulation. Additional neural networks are used to calculate the residual in the transverse and longitudinal directions. All networks use the same input variables as the classification network. A total of ten neural networks are needed to assess up to three sub-clusters.

The networks are trained on simulations of pair-produced top-quarks and highly energetic di-jet events, which are divided into test samples and training samples. The number of training patterns is ensured to exceed the number of network parameters by at least a factor of 1000.

6. Performance of the Neural Network Clustering Algorithm

The performance of the splitting technique is quantified using LHC data with $\sqrt{s}$ =7 TeV collected by the ATLAS detector in 2011, and Monte Carlo simulation. Figure 4 illustrates the improved classification and position resolution achieved when using the neural network cluster splitter. The single cluster is resolved into two sub-clusters and the positions of the impact points are accurately resolved.

Figure 5 shows the mean number of pixel clusters split by the pixel cluster splitting neural network as a function of $\Delta R_{\text{track-jet}}$ for different p_T^{jet} bins. Clusters are split at a higher rate in the core of the jet, where the track density is higher (see figure 2 (a)). Slightly better performance is observed in simulation than data. This is likely to be due to the fact that the neural network was trained on simulation.

Tracks are allowed to share a pixel cluster with another track, only if the cluster has not already been split and the neural network output is compatible with a possible merged cluster. The most noticeable improvements are in the innermost layer of the pixel detector (the b-layer) where the particle density is highest. Figure 6 shows the number of shared b-layer hits when comparing the neural network clustering approach to the standard clustering. Ambiguities are shown to be reduced by an order of magnitude when using the neural network.

A dramatic improvement in the cluster resolution (track-to-measurement residual) in the $r\phi$-direction is observed when using the neural network clustering, as shown for 4-pixel clusters in figure 7. The non-linear treatment of charge resolution allows a single peak to be recovered in track-to-hit residuals. Improved cluster resolution leads to improved track parameters, and a 15% improvement in the longitudinal impact parameter is observed when using the neural

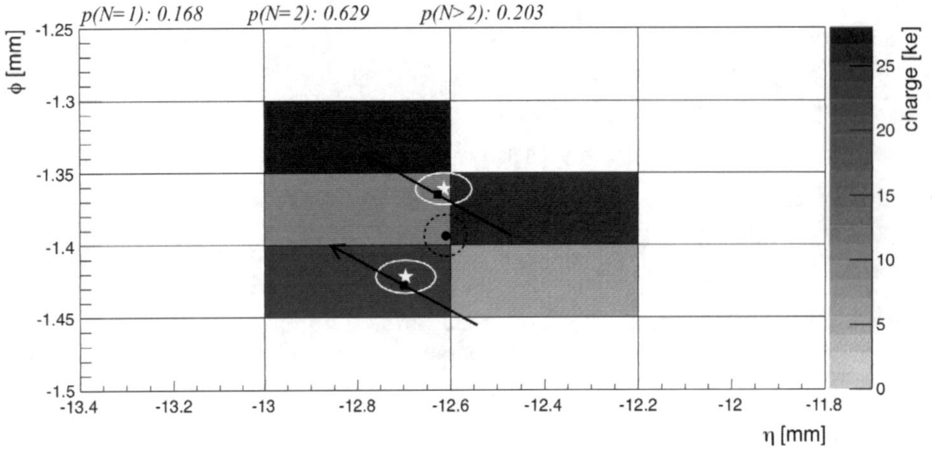

Figure 4. Example of merged clusters created by two particles in Monte Carlo simulation. The arrows show the true directions of the particles, and the square black dots their intersection with the detector layer. The black dot and the dotted black circle show the non-split cluster position and residual respectively. The white stars and circles show the estimated cluster positions and residuals after splitting. The values of $p\,(N = i)$ denote the probabilities for the cluster to be created by i particles, as estimated by the neural network.

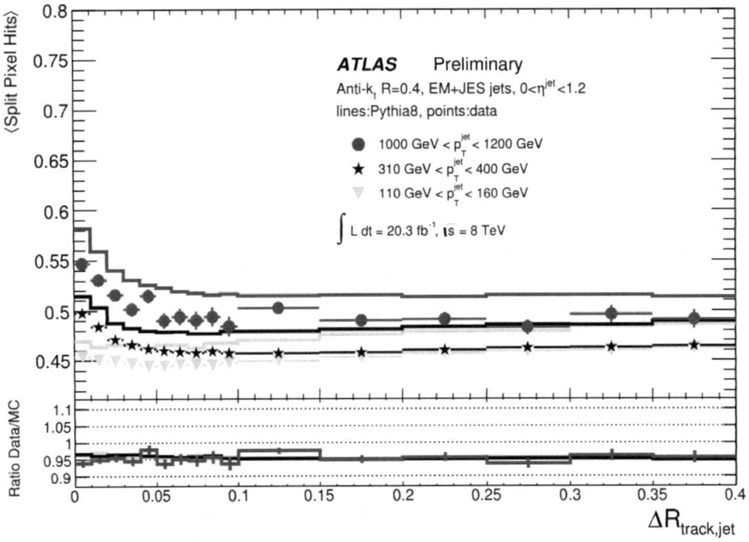

Figure 5. Mean number of pixel clusters split by the neural network pixel cluster splitting algorithm as a function of $\Delta R_{\mathrm{track-jet}}$.

network over the standard clustering. This is illustrated in figure 8. The impact parameter is a key variable for identifying long-lived particles (e.g. heavy flavour quarks).

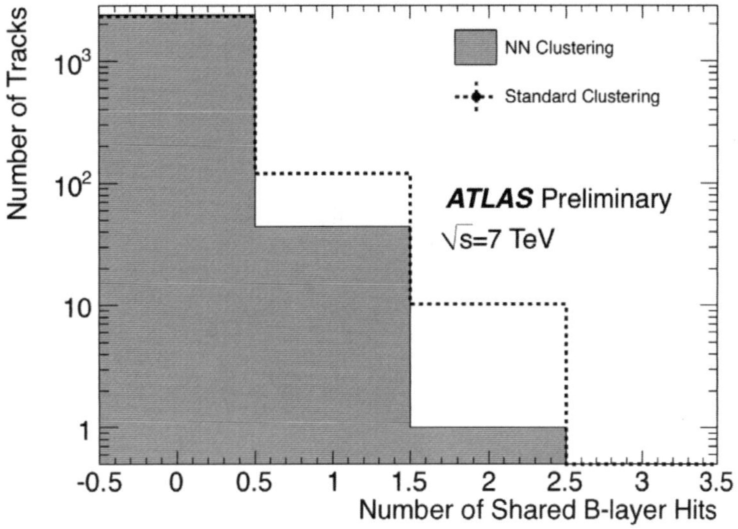

Figure 6. The number of shared hits in the innermost Pixel layer, reconstructed with and without the neural network clustering algorithm [5].

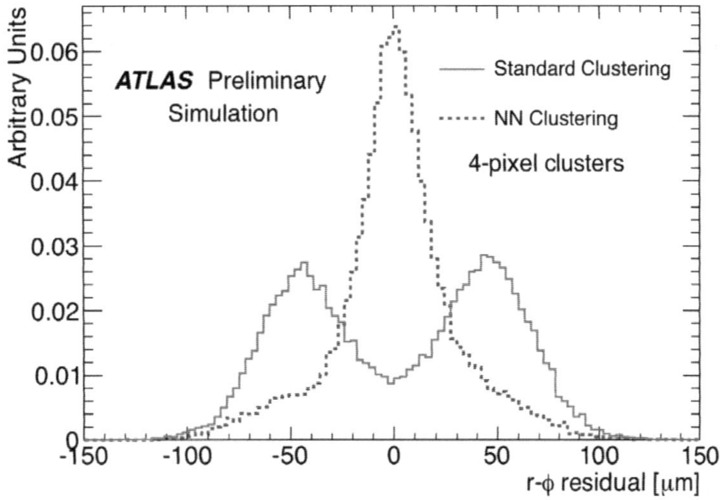

Figure 7. The cluster residual in the $r\phi$-direction for 4-pixel clusters in simulation, reconstructed with and without the neural network clustering algorithm [5].

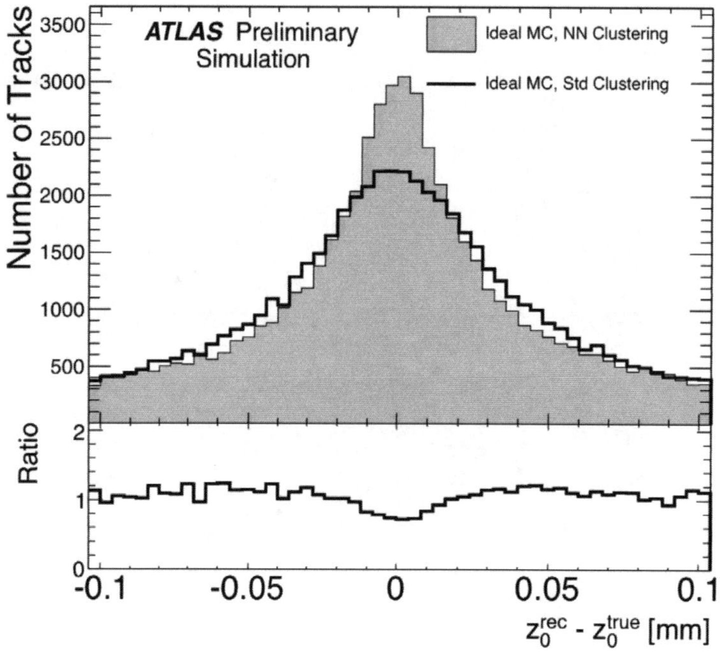

Figure 8. Longitudinal impact parameter resolution in simulation, reconstructed with and without the neural network clustering algorithm [5].

7. Conclusions

A neural network approach to clustering in the ATLAS pixel detector has been used to improve detector performance and make full use of the detector design potential. By taking into account all correlations, clusters created by multiple charged particles can be identified and split. This leads to sizeable improvements in track measurements, particularly in dense environments such as in jet cores and hadronic tau decays. The non-linear treatment of charge collection improves the impact parameter resolution, even for isolated tracks, allowing significantly improvements to the identification of long-lived particles. This improved two-particle separation will become increasingly important during future LHC upgrades, as the number of proton-proton collisions per bunch crossing (and therefore the particle density inside the detector) increases.

[1] ATLAS Collaboration *The ATLAS experiment at the CERN Large Hadron Collider* JINST **3** S08003 (2008)
[2] Aad G et al. *ATLAS pixel detector electronics and sensors* JINST **3** P07007 (2008)
[3] Sjöstrand T, Mrenna S and Skands P *A Brief Introduction to PYTHIA 8.1* CPC **178**:852-867 (2008)
[4] Cacciari M, Salam G P and Soyez G *The anti-k_t jet clustering algorithm* JHEP **4** 063 (2008)
[5] ATLAS Collaboration *Neural Network cluster reconstruction in the pixel detector* ATL-COM-PHYS-2012-476 http://atlas.web.cern.ch/Atlas/GROUPS/PHYSICS/IDTRACKING/PublicPlots/ATL-COM-PHYS-2012-476.

ACAT2013 IOP Publishing
Journal of Physics: Conference Series **523** (2014) 012024 doi:10.1088/1742-6596/523/1/012024

Alignment of the CMS silicon tracker

Tapio Lampén (on behalf of the CMS Collaboration)
Helsinki Institute of Physics, P.O. Box 64, Fin-00014 Helsinki, Finland
E-mail: tapio.lampen@cern.ch

Abstract. The CMS all-silicon tracker consists of 16 588 modules, embedded in a solenoidal magnet providing a field of B = 3.8 T. The targeted performance requires that the software alignment tools determine the module positions with a precision of a few micrometers. Ultimate local precision is reached by the determination of sensor curvatures, challenging the algorithms to determine about 200 000 parameters simultaneously. The main remaining challenge for alignment are the global distortions that systematically bias the track parameters and thus physics measurements. They are controlled by adding further information into the alignment work-flow, e.g. the mass of decaying resonances or track data taken with B = 0 T. To make use of the latter and also to integrate the determination of the Lorentz angle into the alignment procedure, the alignment framework has been extended to treat position sensitive calibration parameters. This is relevant since due to the increased LHC luminosity in 2012, the Lorentz angle exhibits time dependence. Cooling failures and ramping of the magnet can induce movements of large detector sub-structures. These movements are now detected in the CMS prompt calibration loop to make the corrections available for the reconstruction of the data for physics analysis. The geometries are finally carefully validated. The monitored quantities include the basic track quantities for tracks from both collisions and cosmic ray muons and physics observables.

1. Introduction

The central feature of the Compact Muon Solenoid (CMS) apparatus is a superconducting solenoid of 6 m internal diameter. Within the superconducting solenoid volume are a silicon pixel and strip tracker, a lead tungstate crystal electromagnetic calorimeter (ECAL), and a brass/scintillator hadron calorimeter (HCAL). Muons are measured in gas-ionization detectors embedded in the steel return yoke outside the solenoid. Extensive forward calorimetry complements the coverage provided by the barrel and endcap detectors.

CMS uses a right-handed coordinate system, with the origin at the nominal interaction point, the x axis pointing to the centre of the LHC, the y axis pointing up (perpendicular to the LHC plane), and the z axis along the anticlockwise-beam direction. The polar angle θ is measured from the positive z axis and the azimuthal angle ϕ is measured in the x-y plane. Muons are measured in the pseudorapidity range $|\eta| < 2.4$, with detection planes made using three technologies: drift tubes, cathode strip chambers, and resistive plate chambers. Matching muons to tracks measured in the silicon tracker results in a transverse momentum resolution between 1 and 5%, for p_T values up to 1 TeV.

The inner tracker measures charged particles within the pseudorapidity range $|\eta| < 2.5$. It consists of 1 440 silicon pixel and 15 148 silicon strip detector modules and is located in the 3.8 T field of the superconducting solenoid. It provides an impact parameter resolution of $\sim 15\,\mu$m and a transverse momentum p_T resolution of about 1.5% for 100 GeV particles.

The CMS apparatus has an overall length of 22 m, a diameter of 15 m, and weighs 14 000 t. A more detailed description can be found in [1].

To fully exploit the single hit resolution of $9\,\mu$m for pixel and of 23–60 μm for strip sensors, the positions of the sensors must be known to a precision of a few micrometers. This can be best achieved with track-based alignment algorithms. For the CMS tracker these have been applied using tracks from cosmic ray events [2], adding tracks from first LHC collisions [3], and from data taking periods of 2010–2012 [4, 5]. The alignment process has improved during the years, and new features have been added.

The track-based alignment algorithms aim to minimize the χ^2 function, in which the sum of squares of track-hit residuals is collected over a large number of tracks and sensors. The minimization is carried out with respect to selected degrees of freedom (DoF) of each sensor corresponding to its translation, rotation and shape [5]. However, for those sensors which are sensitive only in one coordinate, it is not meaningful to use all these DoFs.

Details of the minimization process can be found in [6, 7] and [8] for the Millepede II and H.I.P. approaches, both used in CMS.

This article gives an overview of the track-based alignment process of the CMS silicon tracker and presents the main improvements and results in 2012, followed by a summary.

2. Alignment in CMS in 2012

The **full-scale alignment** of the CMS silicon tracker consists of alignment of all individual sensors for all their reasonable degrees of freedom (DoF)[1]: 9 DoFs for pixel modules and 8 DoFs for strip sensors (sensitive only in one coordinate). In addition, larger groups of sensors corresponding to physical structures of CMS can be aligned for 9 DoFs. Redundant DoFs are avoided with use of **hierarchy constraints** [2]. This allows good control for time-dependent effects, which typically influence structures larger than individual sensors, for which the alignment compared to nearby sensors is rather stable in time. Therefore, when full-scale alignment parameters are defined with a large dataset, the alignment parameters of large structures can be allowed to vary with time.

The alignment of **high-level structures** is an approach in which frames of modules, layers or even subdetectors (such as the pixel detector) are aligned instead of their components. This approach requires less tracks and is faster to carry out, and is fully sufficient, when a full-scale alignment has previously been carried out, and when there is no reason to suspect deterioration in the sensor-level alignment parameters.

In 2012, new alignments were applied twice in the prompt reconstruction. First, at the beginning of the year, a high-level structure alignment was performed with cosmic ray muon data recorded while there were no collisions provided by the LHC. Later a high-level structure alignment was again carried out together with a pixel module-level alignment taking into account an update in the Lorentz angle calibrations.

In the re-reconstruction of the 2012 data, new alignment parameters were used three times. The main motivation was improvements achieved in the Lorentz angle calibrations.

3. Improvements in Alignment in 2012

Important improvements in the alignment process have been introduced during the past years, especially starting from 2010 and 2011 when abundant amount of high-quality tracks with high momentum became available.

[1] The possible degrees of freedom for alignment are three translational DoFs, three rotational DoFs, and three DoFs corresponding to Legendre polynomials parametrising the deviations from an ideal plane [5].

The following improvements [5] were introduced in 2011, and were thus ready to be used in the data taking of 2012:

- parametrization of sensor shapes;
- consideration of effects of multiple scattering in the track model used in determination of alignment parameters;
- for $Z^0 \rightarrow \mu^+\mu^-$ events: use of the mass of the Z^0 boson as a constraint (in the form of a penalty function to the χ^2 function), together with a vertex constraint enforcing the muons to emerge from a common vertex (described in more detail in [5]); and
- monitoring of time dependence of large structures of the pixel detector.

3.1. Parametrization of sensor shapes

The possibility to model the shape of the sensors with second-order Legendre polynomials has been introduced to the alignment process [5]. Also, the angles and offsets between two daisy-chained sensors in the modules in the outer tracker can be corrected in the alignment.

During construction, the planarity tolerance for the strip sensors was 100 μm [1]. This kind of deviation would create a bias in the measurement direction u of the sensor, depending on the angle ψ of the track with respect to the normal of the sensors as $\Delta u = \Delta w * \tan\psi$, where Δw is the deviation from the ideal plane at the impact point. This is depicted in figure 1. With the same equation, the Δw and thus the shape can be scanned over the sensor surface.

The bias resulting from non-planar sensors is especially large for sensors of the innermost layer of the pixel barrel. At the edge of the 66 mm wide modules, this bias can be of the order of 100 μm.

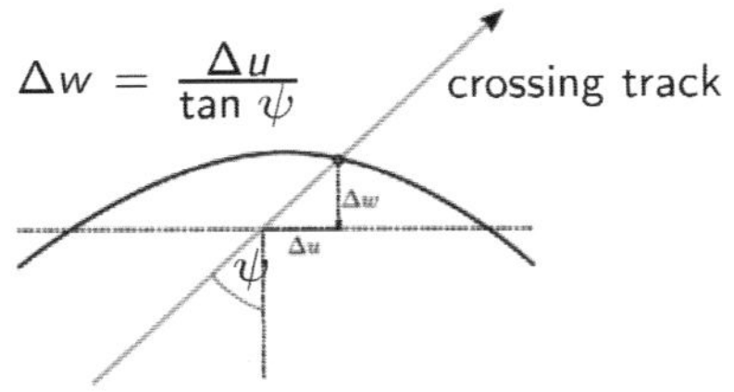

Figure 1. Illustration of a bias Δu in the measurement direction caused by the non-planar shape of a sensor. The sensor differs from the ideal plane by Δw at the crossing point. The angle between the track and the normal of the ideal sensor surface is ψ. The bow is largely exaggerated.

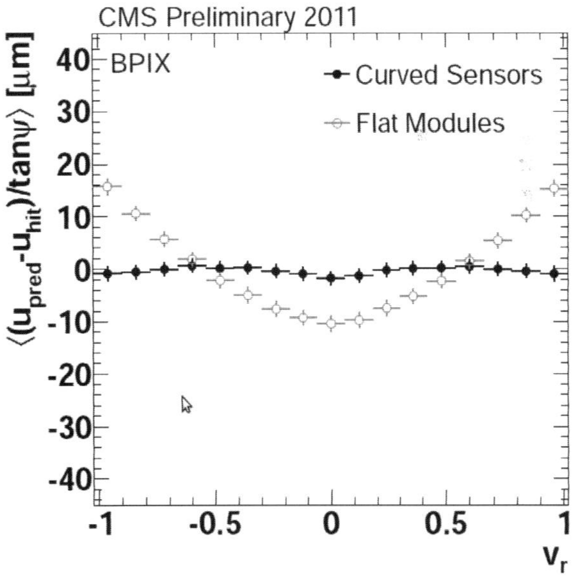

Figure 2. Sensor bow (Δw) as function of the normalized measurement direction v_r, where values of ± 1 correspond to the edges of the sensor.

When the sensor shape parameters are taken into account in hit reconstruction, the related bias is practically corrected. This is illustrated in figure 2, which presents Δw over the normalized relative local track coordinate v_r (corresponding to the global z direction) averaged over pixel barrel sensors. The purple open circles ("Flat Modules") correspond to a situation in which no sensor shape parameters are applied, and average deviation of the order of 10–20 μm can be seen. The black filled circles ("Curved sensors") correspond to the situation when these parameters are used, and the deviations are below 5 μm of absolute value.

3.2. Weak modes and mass constraints

Alignment with the minimization of the χ^2 of track-hit residuals is an efficient way to correct for the uncorrelated sensor misalignments and in this way to improve the precision of measurements and physics variables. A more difficult type of misalignments are, however, those correlated distortions, which do not affect or affect very little the χ^2 function (the "weak modes"). A set of nine basic weak modes and their implication on physics analyses is studied in [4].

Weak modes can be constrained in alignment by using tracks corresponding to multiple topologies (collision tracks with vertices from a wide area, tracks from cosmic rays, etc.). In addition, mass constraints for well-known resonances can be used. For instance, use of cosmic ray tracks is an efficient way to constrain the **telescope weak mode** ($\Delta z \sim r$), which creates a bias in the measured η value of the track.

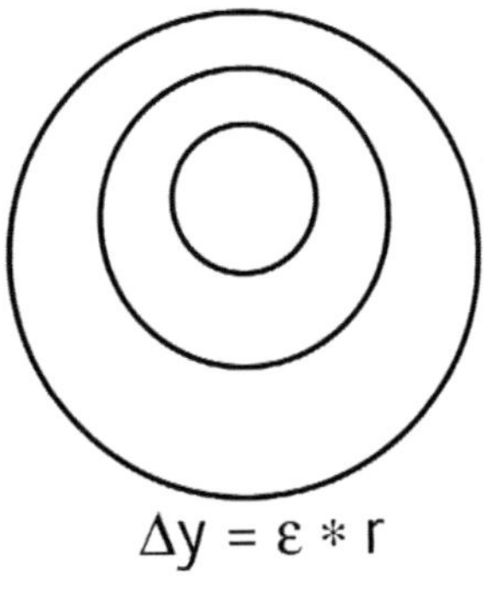

Figure 3. Illustration of the Sagitta weak mode. Individual sensors are not shown, only the layers they are attached to. In this correlated distortion, layers are shifted vertically (Δy) by an amount proportional to their radial coordinate r.

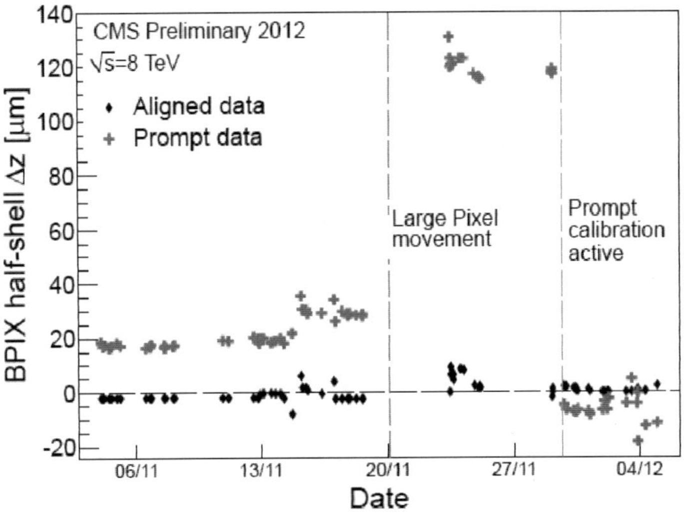

Figure 4. Relative misalignment Δz of the two pixel half barrels along the beam line Δz as function of date in November-December 2012. A 100 μm movement appears on November 22nd, and is corrected with the PCL a week later.

In 2011, a variation of the Z^0 mass as function of the ϕ angle of the positively charged muon from $Z^0 \to \mu^+\mu^-$ decays was observed. This **ϕ-dependent curvature bias** was suspected to be caused by a **sagitta weak mode**, in which sensors are displaced as $\Delta y \sim r$, as shown in figure 3. This systematic distortion can bias the track curvature, which is inversely proportional to the transverse momentum: $\kappa \sim 1/p_T$. It was studied with $Z^0 \to \mu^+\mu^-$ events, which can be used to reveal this bias in the following way. The fitted[2] invariant mass of Z^0 is presented as

[2] The invariant mass distribution is fitted with a wide fit in the range 75–105 GeV/c^2 and the Z width is set to the PDF value of 2.495 GeV/c^2. In the fit, a Breit-Wigner function convoluted with a Crystal ball function is used (it models finite track resolution and radiative tail) summed with an exponential background.

function of the muon direction, separating μ^+ and μ^-. A sinusoidal dependency can reveal the existence of a sagitta weak mode. The result is shown in figure 5 with respect to the ϕ angle, and in figure 6 with respect to the η angle [3]. In 2012, the ϕ dependence of the curvature bias has decreased as a result of applying substantial weights to the $Z^0 \to \mu^+\mu^-$ events in alignment.

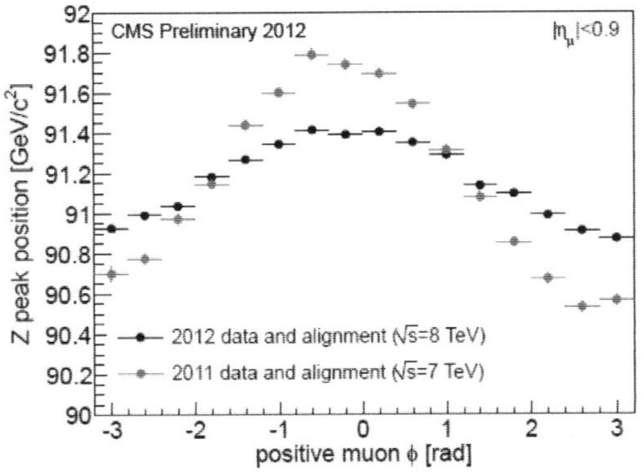

Figure 5. Reconstructed mass for $Z^0 \to \mu^+\mu^-$ decays as a function of ϕ of the μ^+. Results for 2011 data and alignment are shown in red, and those for 2012 in black.

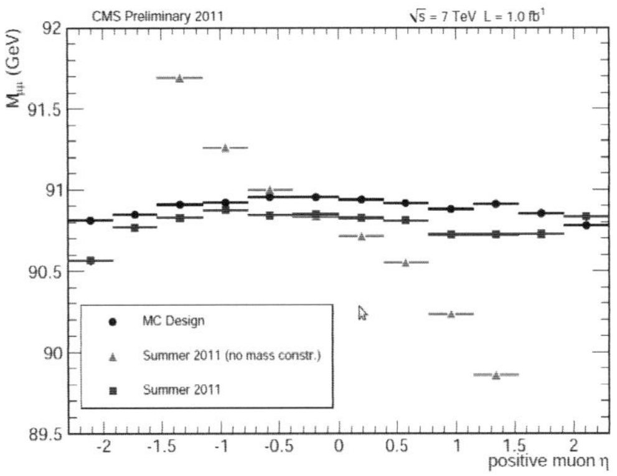

Figure 6. Reconstructed mass for $Z^0 \to \mu^+\mu^-$ decays as a function of η of the μ^+. Results with perfect alignment and simulated events are shown with black circles. Results with and without the mass-constrained Z events used in alignment are shown with blue squares and red triangles, respectively, for data from 2011.

3.3. Prompt calibration loop

In CMS, the **prompt calibration loop** (PCL) is a data stream, in which some measured data is immediately reconstructed for purposes of data quality monitoring, alignment and calibration. With these data, the offline conditions database can be updated so that the actual physics data stream from CMS can be reconstructed after a short, intentional delay with up-to-date conditions.

At the end of 2012, the PCL was able to calculate 6 alignment parameters for the pixel half barrels. It could provide feedback well within the time limit of 48 hours, after which data from the same run was reconstructed and made available to physics analyses.

During the last month of proton-proton collisions of 2012, the PCL was running for monitoring purposes, and thus not used in updating alignment parameters. A major relative movement of the pixel half-shells along the z coordinate was detected on November 22nd. The PCL was activated on November 30th to recover this movement, which later was diagnosed to be caused by a cooling failure. These movements are shown in figure 4, which depicts with red crosses the relative z-shift of the half-shells with the track-vertex residuals as a function of date. The black diamonds refer to the situation when this movement has been corrected for later in re-reconstruction.

[3] This result does not illustrate CMS muon reconstruction and calibration performance. Additional momentum calibration is applied in addition when necessary.

Journal of Physics: Conference Series **523** (2014) 012024 doi:10.1088/1742-6596/523/1/012024

3.4. Lorentz angle calibration

A possibility to calibrate the Lorentz angle has also been implemented in the Millepede II alignment procedure in 2012. As shown in figure 7, in presence of a magnetic field, the Lorentz force deflects drifting signal charge by the Lorentz angle. This effect causes shift of the measured hit positions in the local u direction by $\Delta u = \tan(\theta_{LA}) * d/2$, where θ_{LA} is the Lorentz angle, and d is the width of the sensor.

To calibrate the Lorentz angle (LA), tracks recorded both with magnetic field on and off are needed to disentangle LA effects from genuine misalignments. The full 2012 data with 60 million tracks were used: tracks of isolated muons, $Z^0 \rightarrow \mu^+\mu^-$ events, cosmic ray muons (at both 0 T and 3.8 T), low p_T tracks and collision tracks recorded at 0 T. The Lorentz angle was calibrated for the pixel detector with a granularity of 65 periods of time (of 2012). In addition, 24 spatial parameters were used: three corresponding to the cylindrical layers of the pixel barrel surrounding the interaction point (at radii of 4.4, 7.3 and 10.2 cm [1]), and 8 corresponding to rings in each layer (sensors on one layer with the same z coordinate). This leads to 1560 additional parameters for the alignment procedure.

The Lorentz angle development for all layers (on ring 4) of the pixel barrel is shown in figure 8 as function of integrated luminosity collected by CMS during 2012. The development for layers 1,2 and 3 are shown in figures 9, 10, and 11, respectively. The following observations can be made:

- evolution of the Lorentz angle is consistent in all rings within each layer;
- there is a significant offset of the Lorentz angle values for negative and positive halves of the pixel barrel along Z (rings 1–4 compared to rings 5–8);
- behaviour of the Lorentz angle seems to be different for different layers; it is however suspected that this is caused by different radiation doses, and can follow the same behaviour delayed with time; and
- the largest change of the Lorentz angle (layer 1, ring 4) through whole 2012 is equivalent to the shift of the modules by up to $4\,\mu m$.

The offset seen between Lorentz angles in rings 1–4 and 5–8 is suspected to be caused by a difference in their grounding, which generates a difference in the bias voltage of the sensors. This hypothesis will be verified when data taking restarts.

In 2012, the effect caused by the time-dependent changes of the Lorentz angle is only of the size of a few μm, but this issue will certainly be relevant in 2015, when increased LHC luminosity and the intense radiation will generate time-dependency in the Lorentz angle calibration.

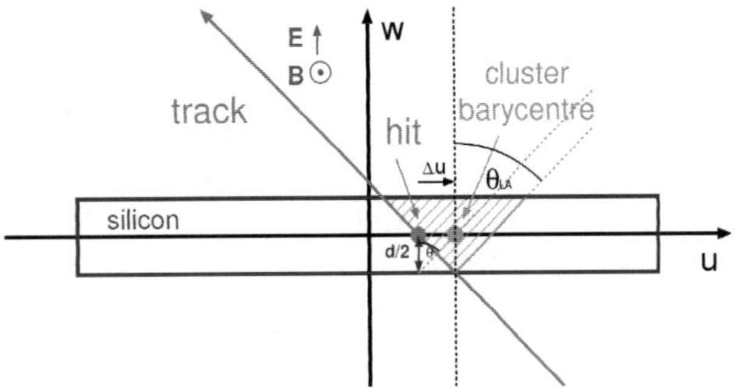

Figure 7. Illustration of the Lorentz angle. Due to the magnetic field, charge carriers in the silicon drift to the direction indicated by orange lines, with an angle with the electric field lines called the Lorentz angle. As a result, the cluster barycentre is reconstructed in a biased position, unless the Lorentz effect is properly accounted for.

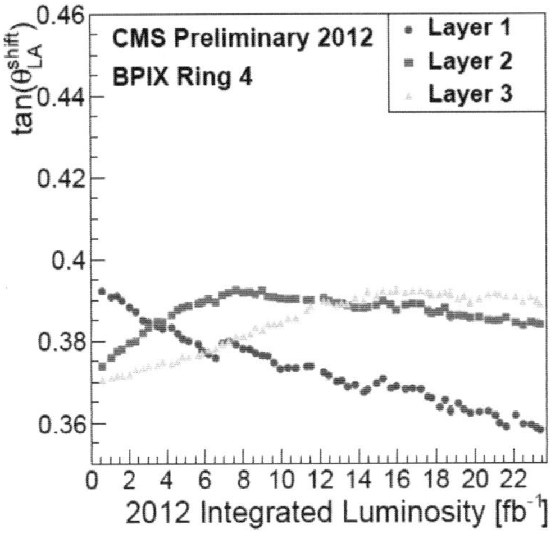

Figure 8. Evolution of the Lorentz angle in 2012 for innermost sensors (ring 4) of all three pixel barrel layers.

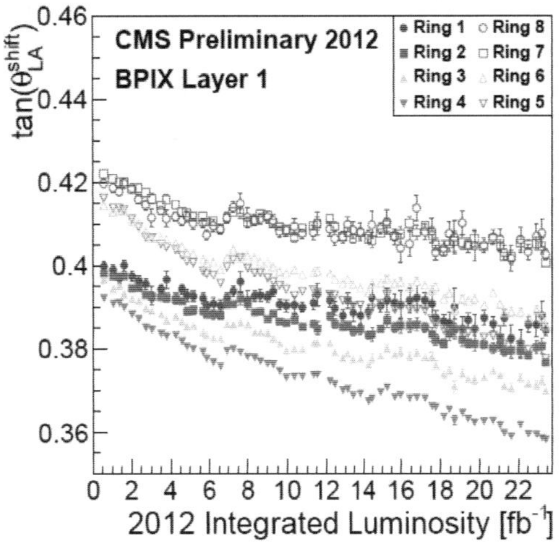

Figure 9. Evolution of the Lorentz angle in 2012 for 8 rings of layer 1 of the pixel barrel.

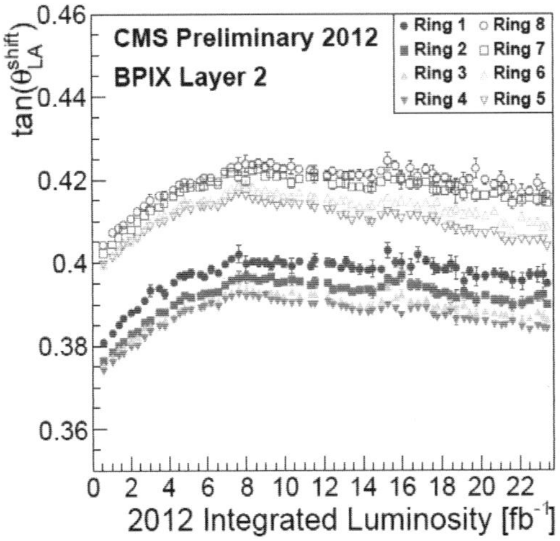

Figure 10. Evolution of the Lorentz angle in 2012 for 8 rings of layer 2 of the pixel barrel.

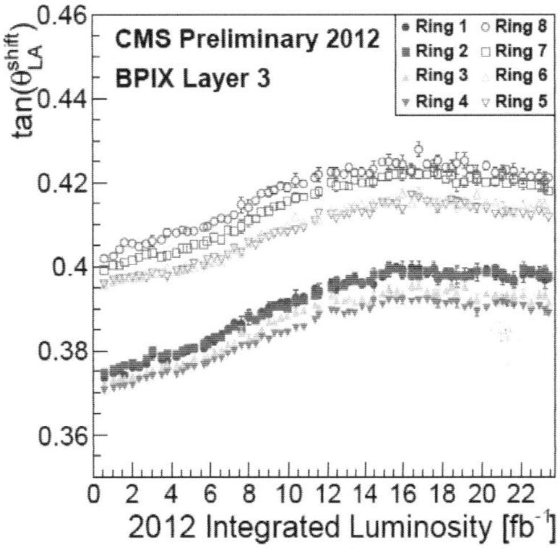

Figure 11. Evolution of the Lorentz angle in 2012 for 8 rings of layer 3 of the pixel barrel.

4. Summary

Alignment of the CMS silicon tracker has been routinely performed with the Millepede II algorithm calibrating 200 000 alignment parameters. The alignment work-flow provides quick response to data taking with a run-by-run alignment of large structures. The improvements in 2012 consist of: calibration of sensor shape parameters, alignment calibration operational in the prompt calibration loop, reduction of amplitude of the ϕ dependent curvature bias with specially treated and weighted $Z^0 \to \mu^+\mu^-$ events, the alignment framework extended to treat also calibration parameters, and Lorentz angle calibration integrated into alignment.

Most of these improvements were in full use already in 2012. When LHC operation begins again in 2015, these improvements will be needed to handle the challenges of the new regime of high luminosity, where for instance time-dependent changes of the Lorentz angle can be expected.

Acknowledgments

The CMS Tracker Alignment group, the Helmholtz Alliance "Physics at the Terascale", the Magnus Ehrnrooth Foundation, and the Waldemar von Frenckell Foundation are acknowledged.

References

[1] CMS Collaboration 2008 *JINST* **3** S08004
[2] CMS Collaboration 2010 *JINST* **5** T03009
[3] Lampén T 2011 Alignment of the CMS silicon tracker, *J. Phys.: Conf. Series* **331** 032043
[4] Draeger J 2010 The alignment of the CMS Silicon Tracker, PoS(ICHEP 2010) **504**
[5] Flucke G 2012 Alignment of the CMS silicon tracker, *J. Phys.: Conf. Series* **368** 012036
[6] Blobel V 2006 Software alignment for tracking detectors, NIM **A566** pp 5-13
[7] https://www.wiki.terascale.de/index.php/Millepede_II
[8] Karimäki V, Heikkinen A, Lampén T and Lindén T 2003 Sensor alignment by tracks, ePrint **physics/0306034** (CMS-CR-2003-022)
[9] Lampén T 2011 CMS silicon strip tracker calibration work-flow and tools , *J. Phys.: Conf. Series* **331** 032044

Maximum likelihood reconstruction for the Daya Bay Experiment

Dongmei Xia (On behalf of the Daya Bay Collaboration)
Institute of High Energy Physics, Beijing, 100049, China

xiadm@ihep.ac.cn

Abstract. The Daya Bay reactor neutrino experiment is designed to precisely determine the neutrino mixing angle θ_{13} . In this paper, we present an algorithm using the maximum likelihood (ML) method to reconstruct the vertex and energy of events in the anti-neutrino detector, based on a simplified optical model describing light propagation. The key parameters of the optical model are calibrated with ^{60}Co source, by comparing the predicted charges of the PMTs with the observed charges. With the optimized parameters, the ML reconstruction provides a uniform energy reconstruction, and a vertex reconstruction with small bias along radial direction.

1. Introduction to the optical model

The optical model is based on the structure of the antineutrino detector (AD). The AD of the Daya Bay experiment has three nested cylindrical volumes separated by concentric acrylic vessels [1]. The innermost volume holds 20 tons of Gd-liquid sintilltor[2] as the antineutrino target. The middle volume is filled with 21 tons of liquild scitillator (LS) which is the gamma catcher. There are 192 8-inch PMTs mounted on eight ladders installed along the circumference and within the mineral oil volume, which is the outer volume of AD. Two reflective panels with a film of Enhanced Specular Reflected (ESR) are placed at the top and bottom of the outer volume to increase the photon-statistics and improve the uniformity of the energy response. Three automated calibration units (ACU-A,ACU-B, ACU-C) are mounted at the top of AD. Each ACU contains a LED as well as two scaled capsules with the radioactive source that can be lowered individually into the Gd-LS along either the centreline or inner edge, or in the LS.

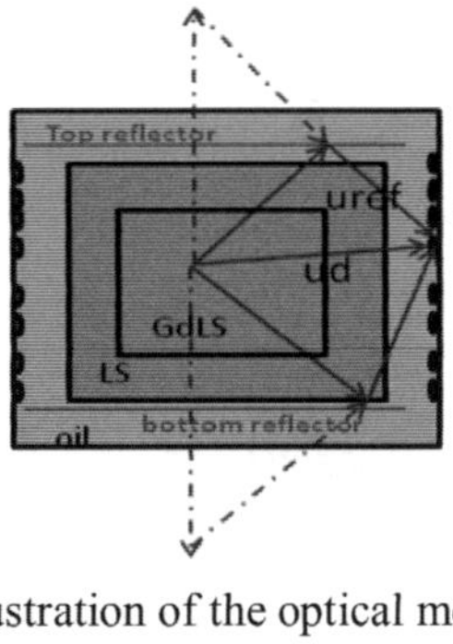

Fig. 1: An illustration of the optical model of the ML reconstruction in AD.

The optical model [3] based on the structure of the AD used in the ML reconstruction is illustrated in figure.1. The predicted charge on each photomultiplier (PMT) is the sum of the photo-electrons (PE) produced by the direct light (μ_d) and the photo-electrons produced by the reflected light (μ_r). The definition of μ_d and μ_r are in equation 1, where Φ is a normalization parameter; $f(\cos\theta_d)$ is the PMT angular response curve; θ_d is the angle between the PMT normal direction and the vector from PMT position pointing to source position; R_d is the distance between PMT and radioactive source; λ_a is the average attenuation length of the liquid scintillator ; ε^{QE} is the PMT relative efficiency; μ_r is the sum of the charges produced by mirror sources, and the subscripts j, k denotes respectively that light is the j-th order reflected by bottom reflector and the k-th order reflected by the top reflector; ε_t ε_b are the reflectivity for the top and bottom reflector, respectively; R_{jk} is the distance from the mirror source to the PMT; θ_{jk} is the angle between the PMT normal direction and the vector from the PMT position pointing to the mirror source position. A sum of μ_d and μ_r gives the total expected charge.

$$\begin{aligned} \mu_d &= \Phi \cdot \frac{f(\cos\theta_d)}{R_d^2} \cdot \exp\left(-R_d/\lambda_a\right) \cdot \varepsilon^{QE} \\ \mu_r &= \Phi \cdot \sum_{j,k} \left(\varepsilon_b^j\right) \cdot \left(\varepsilon_t^k\right) \cdot \frac{f(\cos\theta_{j,k})}{R_{j,k}^2} \cdot \exp\left(-R_{j,k}/\lambda_a\right) \cdot \varepsilon^{QE} \end{aligned} \tag{1}$$

In this optical model, the key parameters that need to be determined from calibration data are as follows:

- The average attenuation length λ_a.
- PMT angular response curve $f(\cos\theta_d)$.
- The top and bottom reflectivity ε_t , ε_b .
- The PMT relative efficiency ε^{QE} .

2. Calibration of key parameters

The calibration of the key parameters is performed by requiring the expected charge distribution on the PMTs to agree with that observed in data. We use the 2.5MeV γ emitted by ^{60}Co for calibration, except for PMT relative efficiency. The PMT relative efficiency is determined by counting the relative occupancy, when the calibration source is at ACUA. A low energy calibration source ^{68}Ge is chosen for the calibrating. With this calibration source, at most one PE is obtained by the PMT.

A χ^2 function is built to calibrate the attenuation length λ_a、 the reflectivity ε_t ε_b and the PMT angular response curve $f(\cos\theta_d)$ simultaneously, as shown in equation (2).

$$\chi^2 = \sum_{j}^{Num} \left(\sum_{i}^{192} \frac{\left(\overline{n_{ij}} - \mu_{ij}\right)^2}{\overline{n_{ij}}} \right) \tag{2}$$

In equation (2), $\overline{n_{ij}}$ is the average observed charge, μ_{ij} is the expected charge and is a function about the parameters to be calibrated $\mu_{ij}\left(\lambda_a, \varepsilon_t, \varepsilon_b, f(\cos\theta)\right)$, 192 is the number of PMTs, *Num* is the total number of calibration sources that locate at different position in the AD, $f(\cos\theta_d)$ is parameterized as $f(\cos\theta) = p_0 + p_1\cos\theta + p_2\cos^2\theta$. Optimal parameters are determined by minimizing χ^2.

Figure 2 shows the ratio of observed charge to expected charge as a function of the incident angle θ and the distance from the PMT to the radioactive source. θ is the angle between the PMT normal direction and the vector from PMT position pointing to source position. The performance of the optical model with the optimized parameters is much better than that with the initial parameters as figure 2 indicates. The initial parameters were measured independently before the assembling of AD.

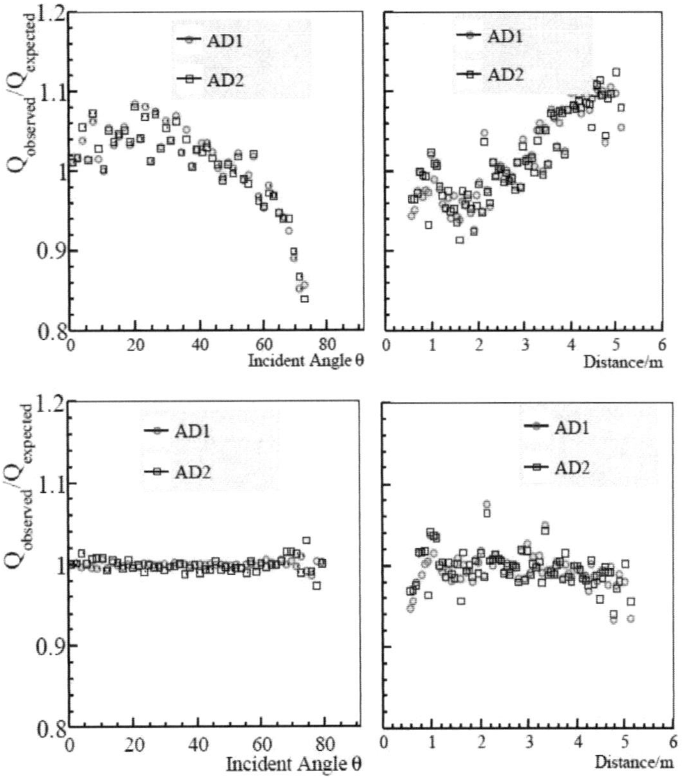

Fig 2: Top: The ratio of observed charge to the expected charge as a function of θ and distance for the old parameters; Bottom: The ratio of observed charge to expected charge as a function of θ and distance for the optimized parameters.

3. Performance of the ML reconstruction

Energy and vertex of events are reconstructed by minimizing the joint likelihood. The joint likelihood is a function of the observing charge pattern and the predicted charge pattern [3]. The predicted charge pattern is calculated with energy E and vertex $\vec{x}$. In this paper, the performance of the ML reconstruction is studied by using the Am-C neutron source; the source was deployed in the detector along various vertical axes and radial directions.

3.1. Energy reconstruction

The accuracy of the energy reconstruction is investigated by comparing the peaks of the reconstructed energy and the true energy of neutron capture events. The energy peaks are determined by fitting the energy spectrum with Double Crystal Ball function as shown in figure 3.

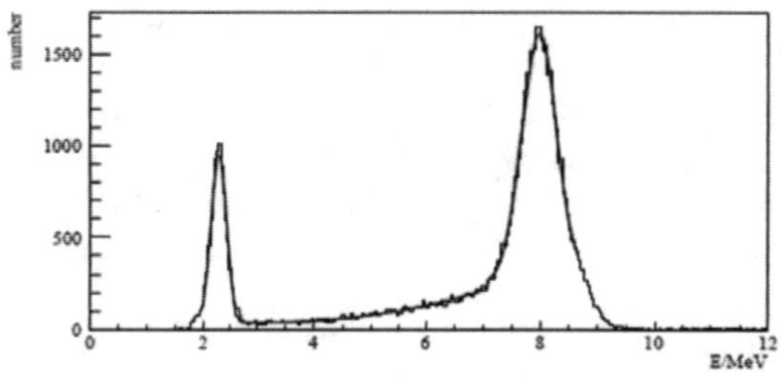

Fig.3.: Energy distribution of the Am-C neutron events.

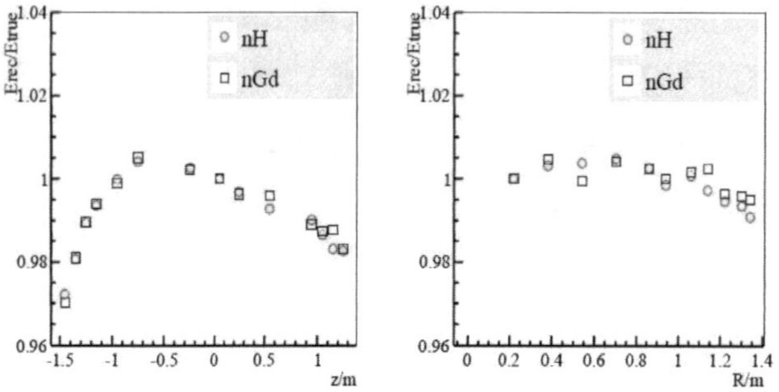

Fig.4.:The vertical (left) and radial (right) energy non-uniformity after ML reconstruction.

From figure 4, the non-uniformity along vertical direction of the energy reconstruction is within 4% for the neutron capture on Hydrogen (nH) events and the neutron capture on Gadolinium (nGd) events. The reconstructed energy is very uniform along the radial direction as shown in figure 4 for neutron source.

3.2. Vertex reconstruction

The bias of the vertex reconstruction is defined as the mean value of differences between the reconstructed vertex and the true vertex in both vertical and radial directions.

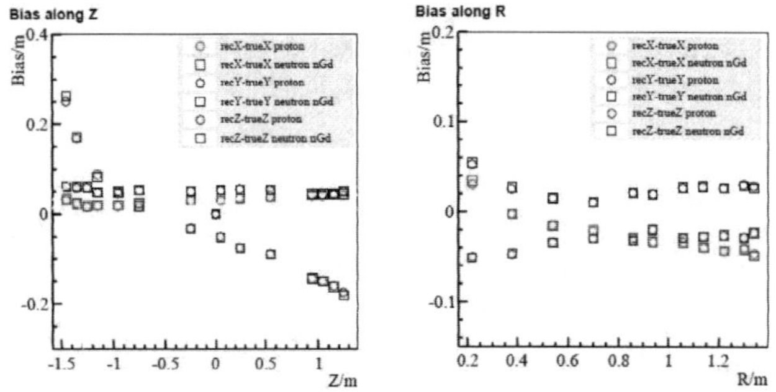

Fig.5.: Bias of the vertex reconstruction along vertical and radial direction.

The bias increases when events are close to the top and bottom reflectors. The maximum Z bias is about 20cm and the bias in X,Y is within 10cm along vertical direction. The bias is within 5cm along radial direction, as shown in figure5.

4. Conclusions

Using the calibration data of ^{60}Co and ^{68}Ge , we optimized the key parameters of the optical model of the ML reconstruction for the Daya Bay experiment. The optical model is improved, and the predicted charge is more consistent to the observed charge. With the optimized parameters, the ML provide a uniform energy reconstruction and a vertex reconstruction without bias along radial direction. The non-uniformity of the energy reconstruction and the bias of the vertex reconstruction along vertical direction are within 4% and 25cm respectively.

References

[1] An F P *et al.* (Daya Bay collabration), Nucl.Instr.Meth.A, (2012),**685**: 78-97.
[2] Ding Y Y *et al.*, Nucl.Instr.Meth.A, (2008), **584**: 238.
[3] Wen L J *et al.*, Nucl.Instr.Meth.A, (2011), **629**: 296-302.

Data Analysis of Tracks of Heavy Ion Particles in Timepix Detector

S Hoang[1], R Vilalta[1], L Pinsky[2], M Kroupa[2], N Stoffle[2,3] and J Idarraga[2]

[1] Department of Computer Science, University of Houston, Houston, TX 77204, USA
[2] Department of Physics, University of Houston, Houston, TX 77204, USA
[3] Lockheed Martin, 1300 Hercules, Houston TX 77058, USA

E-mail: {smhoang, rvilalta, pinsky, mkroupa, nstoffle}@uh.edu, idarraga@cern.ch

Abstract. In this paper, we describe some of the computational challenges that need to be addressed when developing active Space Radiation Monitors and Dosimeters using the Timepix detectors developed by the Medipix2 Collaboration at CERN. Measurement of the Linear Energy Transfer (LET), the source and velocity of incident ionizing radiation, are of initial interest when developing such operational devices because they provide the capability to calculate the Dose-equivalent, and to characterize the radiation field for the design of radiation protective devices. In order to facilitate the LET measurement, we first propose a new method for calculating azimuth direction and polar angle of individual tracks of penetrating charged particles based on the pixel clusters they produce. We then describe an energy compensation method for heavy ion tracks suffering from saturation and plasma effects. Finally, we identify interactions within the detector that need to be excluded from the total effective Dose-Equivalent assessment. We make use of data taken at the HIMAC (Heavy Ion Medical Accelerator Center) facility in Chiba, Japan and NSRL (NASA Space Radiation Laboratory) at the Brookhaven National Laboratory in New York, USA for evaluation purposes.

1. Introduction

The rapid development in semiconductor detector technology at CERN has provided the capability to develop active personal dosimeters and radiation area monitors for use in space radiation environments [1]. The work reported here is based on the Timepix chip developed by the Medipix2 Collaboration at CERN, which when coupled with a Si sensor, can function as an active nuclear emulsion, allowing the visualization of the individual tracks created as the different incident particles traverse the detector [2]. The Timepix chip provides the capability of measuring the energy deposited by each incident particle that traverses the sensor layer. Together with the capability for online readout, this detector opens the door to a completely new generation of active Space Radiation Dosimeters and area monitors.

The Timepix chip is a CMOS-based pixel ASIC made of 256 x 256 pixels, each 55μm square, where the readout electronics for each pixel is embedded within the footprint of that pixel. The chip is attached to an overlying sensor layer using the Flip-Chip® bump-bonding technology. In this particular application, bulk n-type Si with p-type implants at each pixel interface is employed to enable the reverse biasing of the entire sensor layer, allowing the depletion of the free charge carriers from the sensitive volume to reduce noise. Pixel information is read-out in a frame that corresponds to a 256 x 256 pixel image containing the digitized charge information collected during the acquisition time window. Based on the Timepix outputs, the measurement of Linear Energy Transfer (LET) [3], as well as the source and velocity of incident ionizing radiation [4][8], are attainable, and that information provides the capability to calculate the Dose-equivalent and to characterize the radiation field for future radiation protection purposes. The calculation of Dose-equivalent from the raw Absorbed Dose, which is simply the energy deposited per unit mass, requires the track-by-track multiplication of Absorbed Dose value by a Quality Factor that is a function of the LET of the particle producing the track. That Quality Factor can be as high as 30 for high LET tracks [5].

For a Timepix-based detector deployed in space, a penetrating energetic charged particle may be incident on the detector layer isotropically. While it is possible to determine the axis of the particle's trajectory through the sensor layer, there is an ambiguity in terms of the actual physical particle's direction of motion along that axis. By default, we assume that the particle direction is incident from above the sensor layer. There may be mechanisms, in some instances, to resolve the ambiguity, but we have not taken on that challenge yet. Calculation of the two angles --azimuth and polar angle-- described in Section 2 is a critical step in determining the LET, Dose-equivalent, charge and velocity of the particles.

Ionization along the primary particle track causes the creation of electrons and holes, which then drift under the influence of the applied bias voltage. As the carriers drift, they may interact with opposing sign charge carriers, which may result in recombination. This effect is enhanced at the highest densities of electrons and holes (the so-called "plasma-effect") [6]. The estimate of the amount of energy deposited based on the charge collected at the input to the Timepix is thus reduced when this occurs. This effect occurs to the greatest extent in the regions of highest ion density in the sensor, and when the bias voltage field direction is parallel to the track direction. Recombination may cause a reduced amount of energy in the measurement. However, the major losses are introduced by the non-linear response of the in-pixel front-end pre-amplifier stage and the effect of the front-end over-voltage protection diodes. These effects will clearly affect the uncertainty in LET calculations. A reconstruction technique described in Section 3 will be proposed to compensate for the energy loss due to these effects.

As the charged particles traverse the sensor, they may occasionally experience nuclear interactions within the sensor material. These interactions are normally rare and as such, negligible because of the small relative interaction cross section of the nucleus. However, because protons are by far the most common particles in the incident flux, the interactions of incident protons with detector material can produce short-range heavily ionizing nuclear fragments often enough, and these events can mimic genuine incident primary heavy ions that look like high LET particles. These misidentified heavy ions then in turn can contribute significant errors to the Dose-equivalent estimates due to their potentially large Quality Factors. Since our goal is the characterization of the incident radiation field, such interactions are of interest only insofar as we need to know what the primary incident particle was. Therefore, discrimination between these interactions and the tracks of genuine incident heavy ions need to be addressed in order to exclude them from the total effective Dose-Equivalent assessment. The interaction recognition will be shown in Section 4. Conclusion and future work will be described in Section 5.

2. Angle calculation method

When an event occurs, electrons and holes liberated along the particle's path drift in the electric field of the depletion bias voltage, diffusing both laterally and horizontally with respect to the field. The diffused charge will generally spread out over multiple pixels creating a cluster of pixels associated with a single event. This spreading follows a characteristic pattern such that the shape and energy intensity at pixels have precise information embedded within them about the azimuth direction and the angle of incidence with respect to the normal direction.

2.1. Azimuth direction determination

The shape of a cluster resembles a continent (using a geographic analogy) comprising a mountain in the center that is surrounded by peninsulas. Close study of the structure of clusters enabled us to identify three main components: a core part at the center of each cluster, the region immediately surrounding the core, that has been termed the "skirt", and an δ-ray part which comes from the higher energy recoil electrons liberated from the ionization process [3][8](Figure 1). The core contains most of the pixels with high energy, as it is the area where the detector collects most of the charges produced along the traversed path. The skirt contains lower energy pixels, where charge densities are low. The core and the skirt parts constitute the primary track, or continent. The δ-ray part can be considered as peninsulas connecting to the continent. The δ-ray part also contains low energy levels that provide estimates on the velocity or energy of the δ-rays embedded in their length and total energy. The δ-rays begin deep within the core and emerge through the skirt, where they can typically be distinguished clearly. The maximum possible δ-ray energy is a kinematic function of the velocity of the primary ion.

The core part can be used to calculate the azimuth direction, or slope of the track projected onto the plane of the detector. High-energy pixels provide more information about the direction than the low-energy pixels. Thus, instead of using a standard linear least squares fitting, we propose a weighted least square linear fit to find an equation of a line that best describes x-y data pairs of pixels in the core part. The energy at each pixel will be utilized as the weighing factor for this technique.

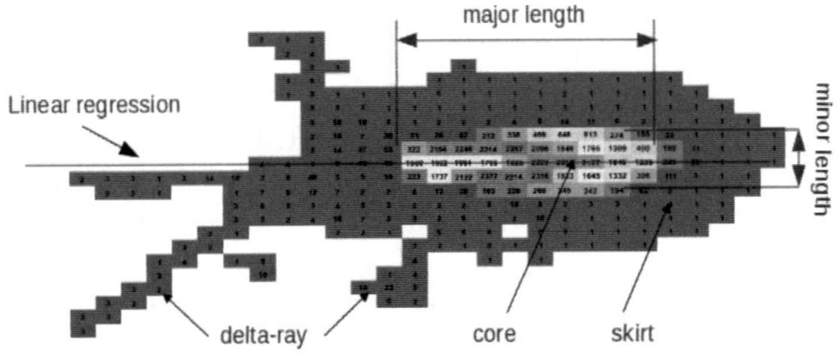

Figure 1. Cluster components

For computational reduction, we apply a threshold cut using a mean energy value of the cluster to get only pixels belonging to the core for model fitting. This threshold value will therefore exclude δ-ray and skirt pixels from data. In order to fit a line $y=\beta_0+\beta_1 x$ to pixels in the core of cluster, we minimized the function of weighted root mean square error as follows:

$$E = \sum e_i [y_i - (\beta_0 + \beta_1 x_i)]^2$$

where e_i is the energy of the pixel, β_0, β_1 are the intercept and the slope of the regression line. x_i, y_i are the coordinates of corresponding pixels in the detector coordinate. E is the error function. By solving this optimization problem, we can calculate the intercept and the slope of the regression line of the cluster as follows:

$$\beta_0 = \frac{\Sigma e_j x_j \cdot \Sigma x_i y_i e_i - \Sigma x^2{}_i e_i \cdot \Sigma e_j y_j}{\Sigma e_j x_j \cdot \Sigma e_i x_i - \Sigma x^2{}_i e_i \cdot \Sigma e_j} \qquad \beta_1 = \frac{\Sigma e_j y_j \cdot \Sigma x_i e_i - \Sigma x_i y_i e_i \cdot \Sigma e_j}{\Sigma e_j x_j \cdot \Sigma e_i x_i - \Sigma x^2{}_i e_i \cdot \Sigma e_j}$$

2.2. Polar angle calculation method

One of the biggest challenges in the determination of the uncertainty in the estimate of the polar angle is that the type of a charged particle at the point of penetrating Timepix chip is unknown. Each type creates a distinct shape and energy on the projected plane of the detector. The δ-ray, skirt, and bias-voltage variation also affect the incident angle resolution. Also, the high-charge non-linearity effects need to be taken into account when they occur with the most heavily ionizing particles. We need to find a general pattern for all types of particles in order to come up with a method for accurately determining the angles of incidence.

We extract the effective *major* track length on the X-axis, and *minor* track width on the Y-axis as described in [3]. We then apply a new formula that combines the track length and width to obtain the best estimate of the angle. Note that using only the major length for angular estimation is not appropriate, as this length will vary considerably according to different applied bias-voltage and different unknown sources of particles.

$$\tan(\alpha) = \frac{D}{T}\left(major - \frac{minor}{2major - \mathrm{minor}} minor \right)$$

In this equation, T=300 (μm) is the detector layer thickness, and D=55 (μm) is the distance between two pixels in the detector. The other constants are assigned by finding a reasonable fit to data for the angles and do not change as much for different configurations of the detector. Note that these assigned constants must satisfy the asymptotic principles in which $tan(\alpha)$ reaches ∞ as *major* reaches ∞ and $tan(\alpha)=0$ as *major*=0. An additional requirement for the vertical track formulation is $tan(\alpha)=0$ as *major=minor*. This formula ensures that the increasing or decreasing of both track lengths by bias-voltage variation effect does not have a big impact on the final angle resolution. Also, this formula can be applied for all types of heavy ion particles that are unknown at this point. D and T parameters can be changed according to a specific design of the Timepix chip and the associated sensor.

3. Energy compensation

3.1. Compensation method

The energy values are suppressed for the pixels in the middle of the clusters where the density of the charges and the resulting input voltages are very high (Figure 2). If we assume that a nominal cluster has the cone-shaped structure that was built by the more-or-less symmetrical accumulation of charges, the appearance of volcano would have to be filled in at a minimum. The slope of the exterior to the summit crater varies depending on the density of the charges. This summit crater needs to be compensated first by flattening out the top of cluster as shown in Figure 3. (Note that this step will not be necessary in the next generation of this technology, known as the Timepix 3 as it has a modified front end pre-amplifier design so that the cluster will have a flat top response for large charges and voltages at the input).

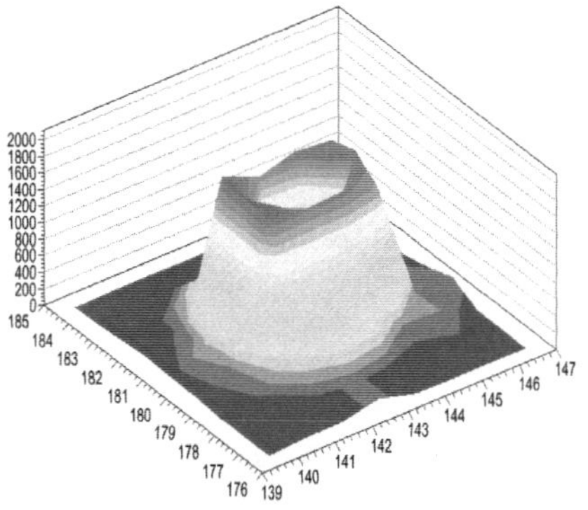

Figure 2. Volcano effect

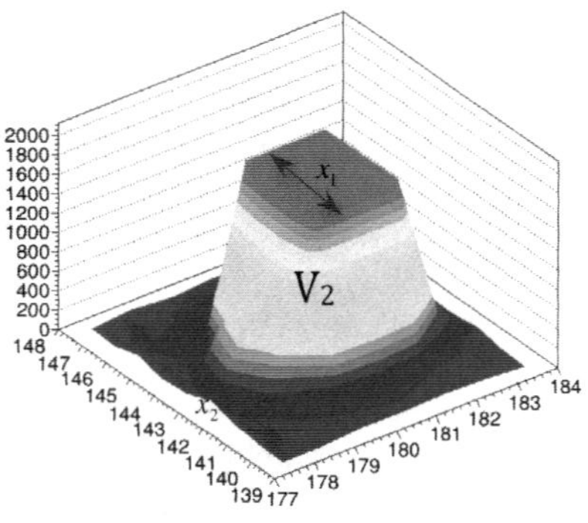

Figure 3. Energy compensation

We assume the total energy of the original cluster without suffering from volcano effect is V. After filling energy loss from the summit crater, we obtain total energy V_2. The remaining energy loss is the missing energy on the top of cluster, which is assumed as V_1. The purpose of the compensation process is to calculate V ($V=V_1+V_2$), given V_2 and the structure of cluster as in Figure 3. This structure allows calculations of minor projected track lengths x_1, x_2 at different the height of cluster according to different energy thresholds. We can calculate V in following equation.

$$V = V_1 + V_2$$

$$\frac{V_1}{V} = \frac{S_1 h_1}{S_2 h} = \frac{x_1 y_1 h_1}{x_2 y_2 h} = \left(\frac{x_1}{x_2}\right)^3$$

$$\frac{V - V_1}{V} = \frac{x_2^3 - x_1^3}{x_2^3}$$

$$\frac{V_2}{V} = \frac{x_2^3 - x_1^3}{x_2^3}$$

$$V = \frac{V_2 x_2^3}{x_2^3 - x_1^3}$$

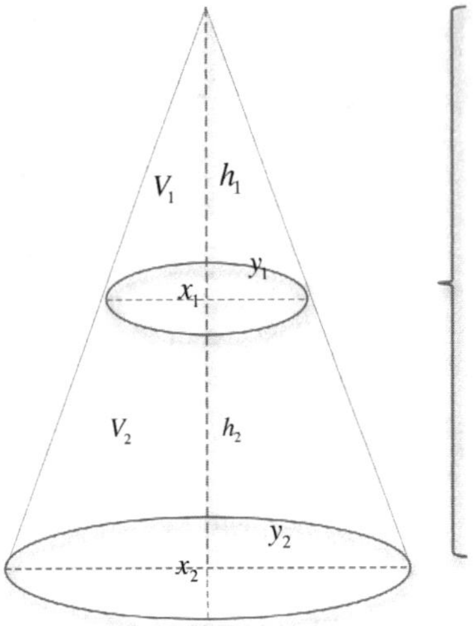

In the formula above, y_1, y_2 are major projected track lengths at different energy thresholds. h_1, h_2 are the height of V_1 and V_2, S_1 and S_2 are the projected area of V_1 and V_2 respectively. $h=h_1+h_2$ is the height of the original cluster after compensation. Note that these formulas are only valid with the assumption that a projected shape of cluster has ellipse-shaped or symmetric quadrilateral. In fact, experimental results show that the assumption is reasonable to yield compensated LET accurately to within 10% at most. A disadvantage of this compensation method is that it is very sensitive to the variation of applied bias voltage, which is affecting the diffusion of charges and cluster structure.

3.2. Incident angle and energy compensation evaluation

We perform an analysis to determine incident angle, LET values for the tracks and evaluate the compensation method. Table 1 shows the LET estimation using angular resolutions described above, combined with total energy deposited from energy calibration method as described in [3]. The LETs generated from different angles are expected to be the same because the relatively small thickness of the sensor does not typically induce a significant variation in heavy ion LET (due to the reduction in the kinetic energy of the incident particle). In this table, the calculated LET values of the particles from H (50, 100MeV), He (100, 180MeV), C (300MeV), O (430MeV), N (290MeV) are very close to the ones calculated from SRIM simulation even for different angles. The high charge non-linear effects do not occur with these particles.

Due to the high charge non-linear effects, which may come in part from the recombination of high densities of electrons and holes, the estimate of the amount of energy deposited is reduced for particles at the end of the table (Si-400MeV, Ar-290MeV, Fe-400MeV), although the angles are well estimated. The red numbers in this table represent raw calculated LET values without compensation for the non-linear effects. The black numbers of the same cell show results after making a correction using the method described in Section 3.1. The corrected mean values of these are close to what is expected. The RMS representing the spread of statistical LET distribution is still big due to the saturation effect. As noted above, the new version Timepix 3 will solve this problem by removing the saturation effect and by providing a flat response region on the top of the volcano.

Table 1. Angle and LET resolution

	SRIM	0 degree		30 degree		60 degree		75 degree	
	Kev/um	Angle	LET	Angle	LET	Angle	LET	Angle	LET
H-50	**2.3**	**8.4**±6.9	**2.26**±0.27	**28**±3.5	**2.27**±0.28	**59.7**±1.7	**2.27**±0.25	**74.2**±1.56	**2.35**±0.24
H-100	**1.36**	**0.6**±1.3	**1.4**±0.3	**30.2**±3.8	**1.3**±0.3	**59.7**±1.2	**1.26**±0.3	**74.5**±1.3	**1.3**±0.3
He-100	**5.42**	**2.1**±6.8	**5.5**±0.75	**30.2**±3.7	**5.67**±0.54	**60.1**±1.3	**5.7**±0.6	**74.7**±1.5	**6.0**±0.54
He-180	**3.65**	**4.7**±7.2	**3.65**±0.5	**29.3**±3.5	**3.8**±0.6	**60.3**±1.4	**3.7**±0.5	**74.6**±1.04	**3.76**±0.4
C-300	**23.0**			**29.7**±4.3	**25.4**±3.6	**60.9**±3.6	**26.6**±4.1		
C-400	**20.0**	**2.1**±5.2	**20.01**±2.8			**60.4**±1.2	**20.6**±2.2		
O-430	**35.0**	**4.3**±6.2	**32.2**±7.7	**29.7**±3.6	**38.8**±5.4	**60.1**±1.3	**39.6**±4.8	**74.7**±0.6	**41**±4.4
N-290	**32.0**	**4.3**±6.6	**31.0**±7.6	**29.1**±3.6	**29.1**±2.8	**60.5**±1.1	**31.9**±2.7	**75.2**±0.6	**33**±2.8
Si-400	**110.0**	**2.6**±3.5	**106.0**±18.8 57.2±5.1	**25.5**±4.6	**113**±24.2 65.5±6.1	59.5±1.8	**125**±17.0 74.9±4.9		
Ar-290	**225**	**4.9**±7.6	**219.2**±18.9 88.15±8.3	**20**±6.4	**237**±19.5 97.6±6.6	**59**±1.57	**218.5**±15.8 102±5.98		
Fe-400	**377.0**	**3.8**±4.5	**374.0**±40.4 126.3±8.2	**24.1**±5.2	**380**±49.0 137±9.5	**61.6**±1.1	**328**±40.5 134.7±12.1	**75.9**±1.5	**340**±23.8 141.8±11.2

4. Interaction recognition

In order to identify the variety of patterns of clusters created from interactions, we conducted experiments with a large sample of proton energies. From that data we can separate high LET clusters that are definitely products of proton interactions. We have been able to identify different types of interactions associated with following patterns:

- **Interaction with two of more clear peaks.** A proton interacts with a Si nucleus producing two or more heavy ions connecting via lower energy pixels. This pattern can be easily recognized by using a suitable energy threshold cut and calculating number of connecting components from the remaining pixels.

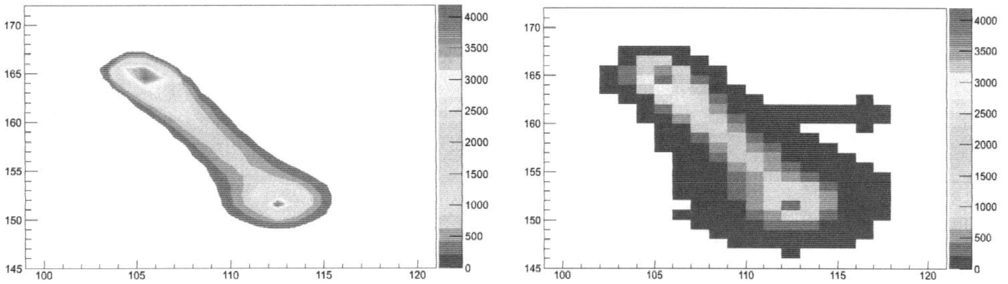

Figure 4. Interaction with 2 primary fragments

- **Interaction that creates a stopping particle**: This interaction transmutes the Si nucleus into a slow particle that ranges immediately or shortly at the vicinity of interaction position. As a result, the forming cluster contains characteristics similar to products of captured neutron with round shape, special morphology, or stopping particle with Bragg curve structure.

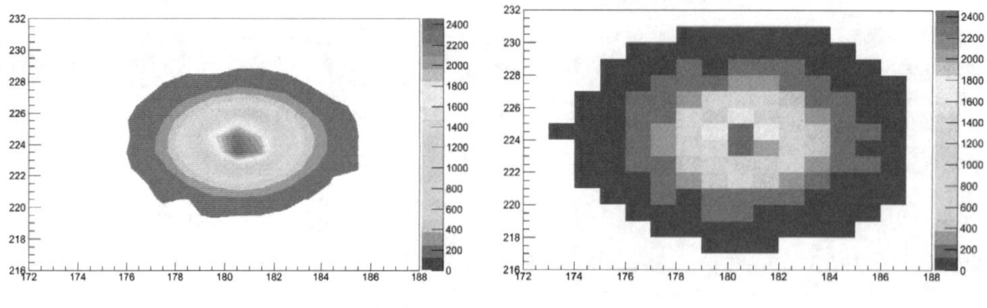

Figure 5. Interaction with stopping

- **Interaction with visible fragments surrounding primary particle**: Unlike the first case, which produces two or more primary fragments, this interaction creates one primary core of cluster and secondary fragments that are clearly different from the skirt of a genuine cluster. We use distance transform [7] to determine levels of pixels and the core of the primary. Calculating the average distance between pixels level 3 in the core and remaining pixels level 2 will help to recognize pixels of fragments. In a genuine cluster, pixels of levels 2 must be the immediate neighbors of pixels level 3 in the core, but this is not the case for these interacting clusters.

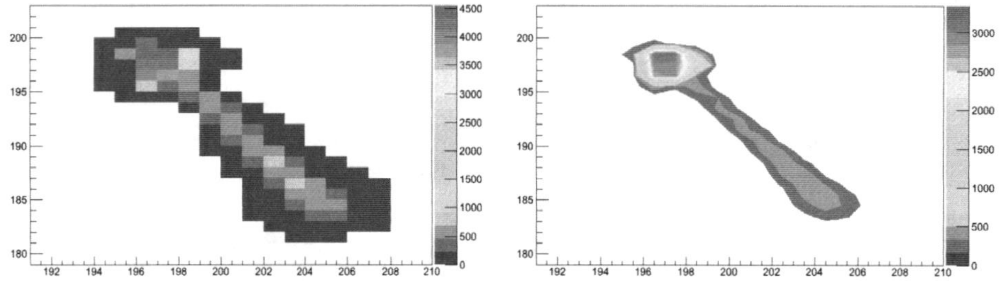

Figure 6. Interaction with one primary and one surrounding secondary

- **Interaction with fragments similar to delta rays**: Similar to the third case, the cluster consists of one primary fragment surrounded by many other fragments. However, the surrounded fragments that mimic the δ-rays are not easy to recognize. We also use the distance transform to isolate the core and calculate the number of inner pixels in fragments. The number of inner pixels will help us determine whether secondary particles are δ-rays or interaction fragments.

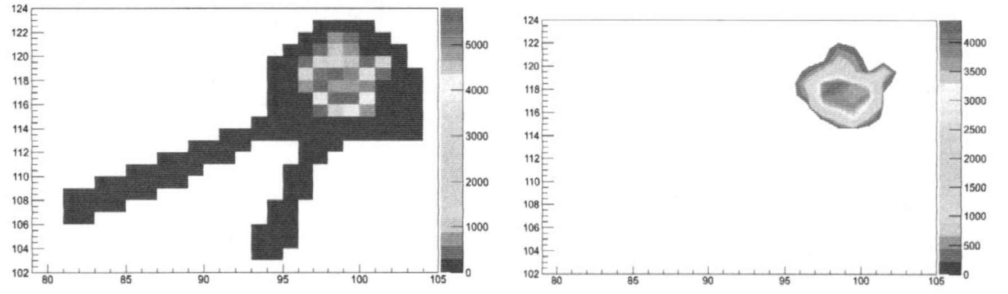

Figure 7. Interaction with one primary and fragments similar to delta-ray

5. Conclusion and future work

In this paper, we proposed a new method for calculating azimuth direction and polar angle of individual tracks of penetrating charged particles, which is a critical step to estimate LET and Dose-equivalent. We then described an energy compensation method to make a correction of LET for heavy ion tracks suffering from plasma effects. Finally, we identified interactions to exclude them from the total effective Dose-Equivalent assessment. Future work will focus on data analysis of delta-rays, which reflect the energy of primary particles, and extract relevant features for identification of ion charge and velocity.

Acknowledgments

The authors would like to express thanks the contributions of the Medipix2 collaboration, HIMAC, and the financial support of NASA Johnson Space Center. We would like to thank the following people in particular: A. Empl, S. Pospisil, Z. Vykydal, D. Turecek, E. Semones, K. Lee, N. Zapp, A. Bahadori, H. Kitamura, N. Yasuda, and Y. Uchihori for their substantial contribution to this research project.

References

[1] Pinsky L, Empl A, Hoang S, Stoffle N, Jakubek J, Vykydal Z, Turecek D, Pospisil S, Kitamura H, Ploc O, Uchihori Y, Yasuda N, Amberboy C, Hauss J, Lee K, Semones E, Zapp N, Parker R and Cooke D 2012 Preparing for the First Medipix Detectors in Space. *Pro. of the IEEE Aerospace Conf.*, Big Sky, Montana

[2] Llopart A, Ballabriga R, Campbell M, Tlustos L and Wong W 2007 Timepix, a 65k programmable pixel readout chip for arrival time, energy and/or photon counting measurements. *Nucl. Inst. And Meth. Phys. Res.*, A **581**, 485-494

[3] Hoang S, Pinsky L and Vilalta R 2012. LET Estimation for Heavy Ion Particles based on a Timepix-based Si Detector. *Int. Conf. on Comp. in High Energy and Nucl. Phys. (CHEP-12)*, New York, USA

[4] Pinsky L, Empl A, Gutierrez A, Jakubek J, Kitamura H, Miller J, Leroy C, Stoffle N, Pospisil S, Uchihori Y, Yasuda N and Zeitlin C 2011 Penetrating heavy ion charge and velocity discrimination with a TimePix-based Si detector (for space radiation applications). *Nucl. Instr. and Meth.* A **633** 190-193

[5] National Council on Radiation Protection 2003 Information Needed to Make Radiation Protection Recommendations for Space Missions Beyond Low-Earth Orbit, *NCRP Report No. 153*

[6] Williams R and Lawson E 1974 The plasma effect in silicon semiconductor radiation detectors. *Nucl. Instr. and Meth.* **120** 261-268

[7] Fabbri R, Costa L, Torelli J and Bruno O 2008 2D Euclidean distance transform algorithms: A comparative survey. *ACM Comp. Survey* **40**

[8] Stoffle N, Pinsky L, Hoang S, Idarraga J, Kroupa M, Jakubek J, Turecek D and Pospisil S 2009 Initial results on charge and velocity discrimination for heavy ions using silicon-Timepix detectors, *JINST 7* C **1**

Simulation-aided optimization of detector design using portable representation of 3D objects

Jan Balewski

Massachusetts Institute of Technology, Cambridge, Massachusetts 02139, USA

E-mail: balewski@MIT.EDU

Abstract. Use of the Standard Tessellation Language (STL) for automatic transport of CAD [1] geometry into Geant is presented. The hybrid approach of combining Geant native and STL objects is preferred. The tradeoffs between the CPU cost of the simulation and the accuracy of tessellation are discussed.

1. Introduction

The design of a new complex detector for a physics experiment typically requires multiple iterations between two teams: the physicists modeling detector properties using simulation package like Geant and the engineers assembling the detector model in CAD to ensure it can be built. The transfer of information from CAD to Geant as well as the double coding of the frequently changing geometry require a sizable investment of resources.

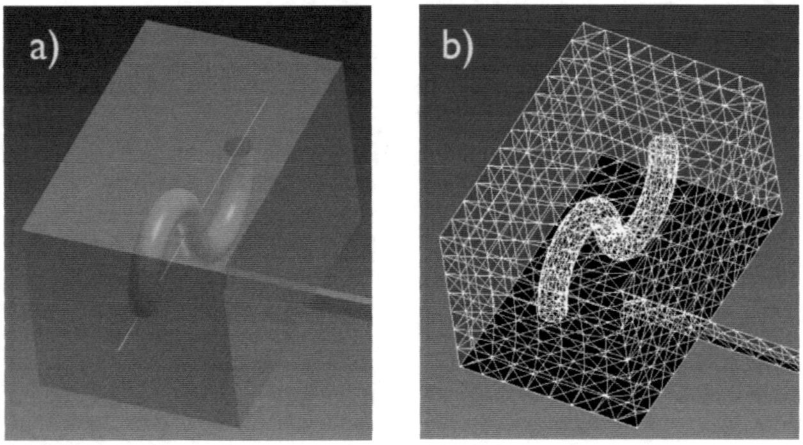

Figure 1. A geometry defined in CAD (a) can be transported to Geant as a mesh of triangles using STL format (b).

[1] CAD is understood as generic name for AutoCad, SolidWorks, or similar tool.

The advent of portable representation of 3D objects via Standard Tessellation Language (STL) file format significantly simplifies transfer of geometry information and reduces the evaluation time of the evolving detector configurations. In section 2 we will describe the basic properties of the STL format, section 3 discuses example of use of STL in Geant simulations.

2. Features of Standard Tessellation Language

The STL file format allows for a convenient way to transport a detailed geometry from a CAD program (see fig. 1a) in to Geant in the form of triangular mesh (see fig. 1b). Most of CAD-like programs have the capability to export 3D objects in to the STL format.

As a prerequisite, one needs to compile Geant with the CADMesh library [1], which in turn requires the ASSIMP library [2]. Next, it suffices to add the following 2 lines to your code to import to Geant the STL file for the object named 'mySphere'

```
CADMesh * mesh = new CADMesh("mySphere.stl", STL,...);
G4VSolid  * mySolid = mesh->TessellatedMesh();
```

From that point one can use the `mySolid` pointer as if it was created by the Geant-native:

```
G4Solid *G4Orb( name="mySphere", rMax=...);
```

Unfortunately, one can't export from CAD the information about the material used to build 'mySphere' and it needs to be re-defined again in Geant.

Although the position and orientation of an object can be passed from CAD to Geant it is advisable to not use this option. CAD uses the absolute reference frame for any object exported as STL. In Geant, as illustrated in fig. 2, we often define a nested geometry of a complex object using nested local coordinate frames. Also, one often places multiple copies of the same Geant object at different absolute locations. To preserve those features of Geant it is more convenient to export from CAD as STL only one element of each type, read it into Geant, then clone it, and re-position as needed.

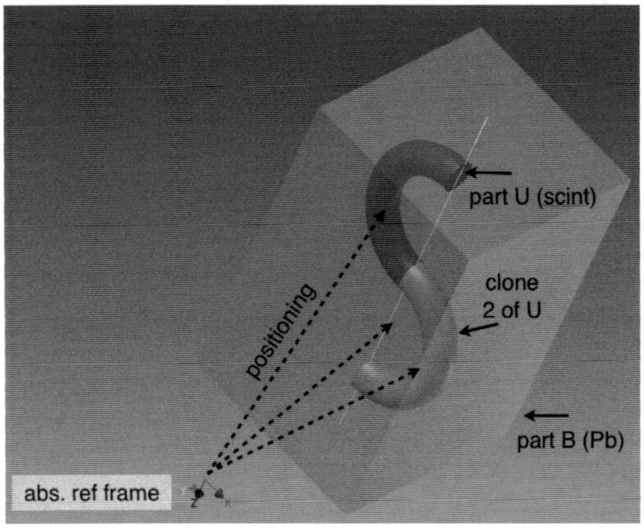

Figure 2. Building a complex object (S-shape) in Geant by replicating and translating the elementary object (U-shape).

There are few tradeoffs in using the STL format. CAD allows user to choose the granulation of the mesh. Smaller triangles will represent a curved surface with higher fidelity. But at the expense of more CPU time spent in simulations, because Geant will need to use smaller steps to resolve boundaries between the volumes. Additionally, the size of the STL file will grow for a smaller mesh step, leading to the longer initialization time of the STL-based geometry encoded in Geant. Fig. 3a shows the CAD ideal model of a half-torus, 4cm in diameter. Fig. 3b illustrates, that if we generously required the maximum deviation between the STL and the ideal shape to not exceed 60μm the STL file will have the size of 51kB and will contain about 1k facets. However, if the highest fidelity of 1μm is requested then the same half-torus would use 2.5MB of the disk space, because now it consist of 50k facets, as shown in fig. 3d.

It is up to the user to decide for which part of the geometry the high fidelity of STL representation is necessary. E.g. to simulate a response of a silicon tracker the 1μm tolerance for STL geometry may be needed, but to study the background produced by a 4m long iron magnet yoke it may suffice to be accurate up to 1mm.

A more saddle issue is a too crude STL mesh 'cuts corners' while approximating series of nested curved objects (think about modeling of an onion). If different layers of such 'onion' were designed and exported from the CAD separately they may partially overlap if re-assembled in Geant. Although Geant probes volumes overlaps using random thrown points it may not detect if two volumes overlap by just 1e-6 fraction of the total volume. But when you run a large scale simulations some particles will get to the ambiguous region and given job may get stuck, run forever. To prevent it, it is advisable to add an additional clearance space between the layers of the onion, make those gaps few times larger than the intended STL mesh accuracy selected while exporting form CAD.

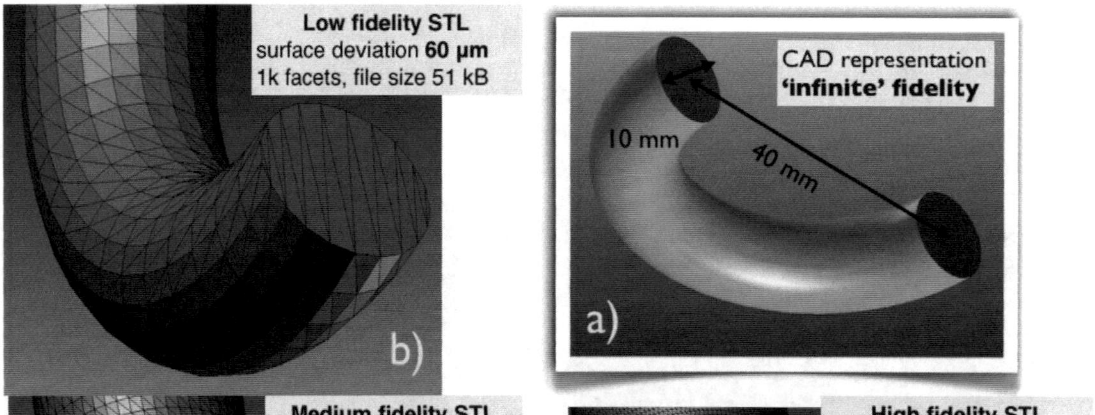

Figure 3. The STL file size and number of facets strongly depend on the required fidelity. (a) is ideal CAD geometry, (b)-(d) show consequences of decreasing maximal deviation of STL from 60μ to 1μm.

2.1. *Tessellation precision vs. CPU cost*

To study the CPU cost of the Geant simulations we have used a simplified implementation of the electromagnetic calorimeter consisting of many S-shaped scintillating fibers embedded in a block of lead, as shown fig. 4a. In the simulation this calorimeter was exposed to 400 positrons, we measured the total CPU time. The graph in fig. 4b shows how the CPU usage depends on the accuracy of the STL representation, and on the energy of the positron. We concluded, that for the fixed positron energy the CPU time increases by 60% if we replace the ideal, Geant native, torus by its crudest STL model. Another factor of 2 in CPU cost is acquired if we require the highest fidelity matching of geometry. As seen from the graph, the pattern is energy independent [2].

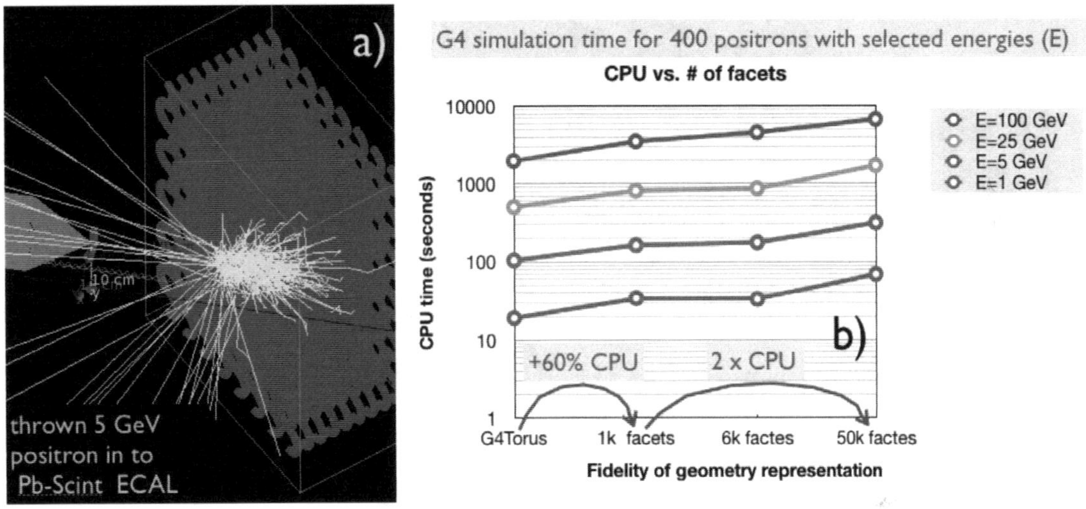

Figure 4. (a) The Geant model of a plastic-lead EM calorimeter used for (b) evaluation of the CPU cost of simulations. Different STL resolutions are labeled on the x-axis, results for different energies of the incident particle are shown with different colors.

3. Hybrid geometry model in Geant

The most practical approach of parallel building of the Geant and CAD models for a large detector is to use the hybrid approach.

Fig. 5 shows an example of the CAD detector model consisting of a multi-disc detector embedded in a phi-symmetric magnet. The parts labeled in blue have simple shapes and it is trivial to implement them in Geant using its native volumes. The more complex parts are labeled in pink. It took a lot of engineering time for the CAD design to adjust the angles, dimensions, make the whole detector assembly fit together. For such case, it is more practical to export the complex parts from CAD using STL, because they will fit again in Geant instantly. The precise location, orientation angles can be read from CAD projections and used while re-assembling the STL volumes in Geant - this saves time too.

[2] One can assume that number of secondary particles generated and tracked in the simulation of the EM cascade remains approximately constant for the fixed energy of the incident particle and does not depend on the granulation of STL mesh.

To conclude, use of Standard Tessellation Language for automatic transport of CAD geometry to Geant is a very advantageous. However, a careful evaluation of costs and benefits of using the STL format for each detector part is advisable. Finally, note it takes some practice to streamline the process of updates of STL geometry.

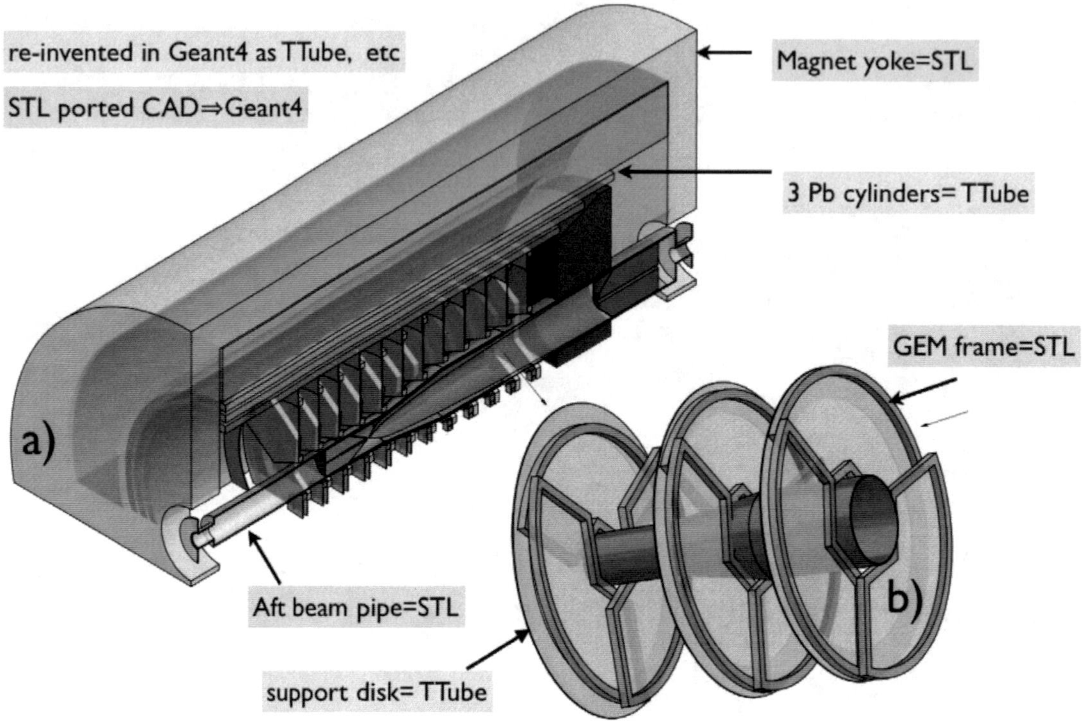

Figure 5. Illustration of a complex geometry of a detector (a). Only more complex parts (b) are exported from CAD to Geant using STL file format.

References

[1] CADMesh library: http://cadmesh.googlecode.com/files/cadmesh-v0.9.tar.bz2
[2] ASSIMP library: http://sourceforge.net/projects/assimp/files/assimp-3.0

The automated Matrix-Element reweighting and its applications at the LHC

Alexandre Mertens
Centre for Cosmology, Particle Physics and Phenomenology (CP^3),
Université Catholique de Louvain,
Chemin du Cyclotron 2, B-1348 Louvain-la-Neuve, Belgium
E-mail: alexandre.mertens@uclouvain.be

Abstract. Given a sample of experimental events and a set of theoretical models, the matrix element method (MEM) is a procedure to select the most plausible model that governs the production of these events. From a theoretical point of view, it is probably the most powerful multi-variate analysis technique since it maximally uses the information contained in the Feynman amplitudes. This technique is now widely known since it has been used for the precision top mass measurement at Tevatron, for example.

The MadWeight software is presented. MadWeight is a phase-space generator designed for the automated numerical estimation of matrix elements based on MadGraph amplitudes. With the modern computing resources, it allows the large-scale deployment of the MEM technique on high-statistics data and simulated samples.

Several applications of the method at LHC are discussed, including the measurement of the spin and parity of the recently discovered boson, signal-to-background discrimination, full differential spectrum estimation and other promising applications.

1. Introduction
The LHC has now delivered 5 and 20 fb^{-1} of data with 7 and 8 TeV in center of mass respectively. The first long shutdown, which will last until 2015, provides an opportunity to upgrade and develop new analyses. In this context, the Matrix Element Method (MEM) is a multivariate analysis (MVA) method in particle physics that takes into account all kinematic information of an experimental event. Using the final state 4-vectors, the MEM computes the posterior probability that a reconstructed event is produced under a given hypothesis. In comparison with other MVA techniques (NN, BDT), MEM is certainly the most theoretically motivated, since it fully exploits the process Feynman amplitude.

This method has been widely used and tested, especially for measurement in the top sector at the Tevatron and for Higgs searches/measurement at the LHC. An automation of this method has been implemented and dubbed *MadWeight* [1].

In the following, the theoretical context of the leading order MEM computation and MadWeight will be first presented before reviewing few applications (b-charge association, differential matrix element method, background rejection and hypothesis testing).

2. The methods

The matrix element method

The probability to observe a reconstructed event x in the theoretical context α is given by:

$$P(x|\alpha) = \int dy P_{\alpha}(y) W(x,y) \quad (1)$$

Where we have defined:

α :	theoretical informations (signal hypothesis).
x :	experimental informations.
$P_{\alpha}(y)$:	The probability to have a partonic configuration y in the theoretical context α.
$W(x,y)$:	The transfer function represent the evolution from partonic configuration y to a reconstructed event x.

The previous expression of $P_{\alpha}(y)$ can be re-written in terms of squared Feynman amplitude $|\mathcal{M}_{\alpha}|^2(y)$ and the parton density functions ($f_1(q_1)$ et $f_2(q_2)$):

$$P(x|\alpha) = \frac{1}{\sigma_{\alpha}\epsilon_{\alpha}} \int d\phi(y) dq_1 dq_2 f_1(q_1) f_2(q_2) |\mathcal{M}_{\alpha}|^2(y) W(x,y) \quad . \quad (2)$$

Since $P(x|\alpha)$ is strictly a probability, the full set of statistical methods and tools on the market can be used.

As example, given a data sample of N events, one can define a likelihood build upon the MEM probability associated with these events.

$$\mathcal{L}(\alpha) = f(N) \prod_{i=1}^{N} P(x_i|\alpha) \quad (3)$$

where $f(N)$ is a renormalization factor.

ISR correction

When considering only leading order matrix element, two methods are usually used to take into account the ISR corrections. The most common approach is to apply a Lorentz transformation to boost the system back into a frame where the transverse momentum of the ISR is null. As illustration, let's consider a reconstructed event with two b-jets and two electrons (see Figure 1). Among others, two processes can lead to this final state: $Zb\bar{b}$ associated production and $t\bar{t}$.

Since the extra radiations are often soft and not fully reconstructed, it is more convenient to derive them from the reconstructed objects belonging to the LO process. If the process contains genuine missing transverse energy ($\not{E}_T$), such as $t\bar{t}$, the $\not{E}_T$ has to be included in the sum. Note that this procedure will not give the expected result in the presence of final state radiations.

- $P(x|Zb\bar{b} \to l^+l^-b\bar{b})$: $p_{ISR} = -(p_{b1} + p_{b2} + p_{e1} + p_{e2})$
- $P(x|t\bar{t} \to l^+l^-b\bar{b}\nu\bar{\nu})$: $p_{ISR} = -(p_{b1} + p_{b2} + p_{e1} + p_{e2}) - \not{E}_T$

The second method is to use LO calculation that contains an extra radiation (i.e. $P(x|Zb\bar{b}j)$), other extra-radiations are then taken into account using the first described method. This method provides a good modeling of the extra hard radiation from both initial and final state but can be time consuming.

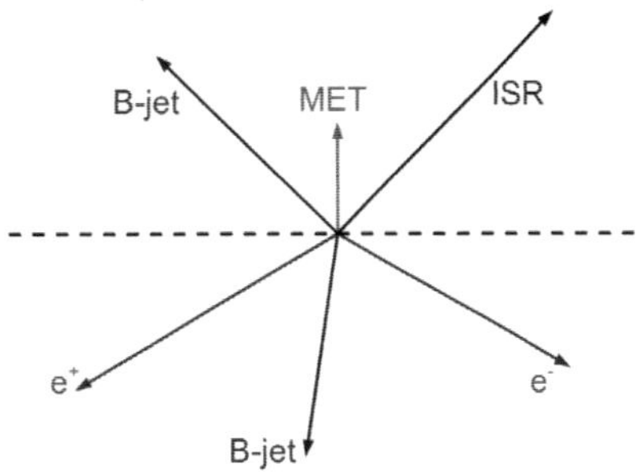

Figure 1. experimental event in which two b-jets and two opposite-sign electrons have been reconstructed.

MadWeight and automation of the MEM
The calculation of the probability entails an intricate numerical integration. A generic module dubbed `MadWeight` has been designed to perform this phase-space integration in an efficient way. The program is based on MadGraph/MadEvent [4] and can be used to test any process implemented in MadGraph. By default, the output probability is the average over all possible parton-jet assignements. MadWeight can also deal with any shape of transfer function provided by the user.

3. Applications at the LHC

3.1. b-Charge association

A method to measure the CP violation with semi-leptonically decayed $t\bar{t}$ events at the LHC has been provided in [2]. The CP violation asymmetries can be defined as mixing or as direct according to the following decay modes:

- **mixing**: $N(b \to \bar{b} \to \ell^+) - N(\bar{b} \to b \to \ell^-)$
- **direct**: $N(b \to \ell^-) - N(\bar{b} \to \ell^+)$

In order to extract the direct and mixing CP violation, one can find out which are b and $\bar{b}$ initially by using MadWeight. The method has been dubbed *b-charge association.* For each event, probabilities P_1 and P_2 corresponding to the two possible choices of the b-charge association can be computed. One can define $W = |(P_1 - P_2)/(P_1 + P_2)|$, which is a variable quantifying the quality of the association (figure 2, left). Selecting events with a high W value will provide a sample with correctly assigned events. For instance one can achieve $\leq 10\%$ mis-association rate with $\approx 70\%$ signal efficiency (figure 2, right). In addition, we observe on left of the figure 2 that this variable can also be used to reject the background.

3.2. Differential Matric Element Method (DMEM) and differential cross-section measurement

We can also use the MadWeight program to estimate the probability density function of arbitrary event kinematic variable. As example, the invariant mass of a $t\bar{t}$ event where both top decays leptonically.

$$\frac{\partial \mathcal{P}}{\partial m}|_{m=m_0} = \frac{1}{\sigma} \int d\phi(y) dq_1 dq_2 f_1(q_1) f_2(q_2) |\mathcal{M}_\alpha|^2(y) W(x,y) \times \delta(m - m_0) \quad (4)$$

For each event, a $\frac{1}{\mathcal{P}} \frac{\partial \mathcal{P}}{\partial m_{tt}}$ curve is computed. Then the sum of the curves provides $\frac{\partial \sigma}{\partial m_{tt}}$. A prospect developed for the LHC was the search for new resonant productions of $t\bar{t}$ pairs decaying

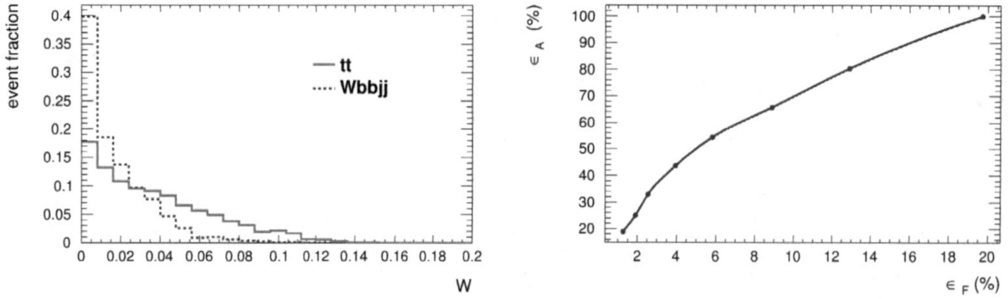

Figure 2. On the left: W variable distribution for $t\bar{t}$and $Wb\bar{b}jj$ events. On the right: Efficiency of the b-charge association as a function of the mis-association probability.

leptonically [9]. The injection of signal event, corresponding to a Z' (spin-1 particle) with a mass of 1 TeV, shows the good discrimination in the left of the figure 3. The possibility to measure the spin of the resonance using the angular variable $\cos(\theta^*_{tt})$ is illustrated in [9].

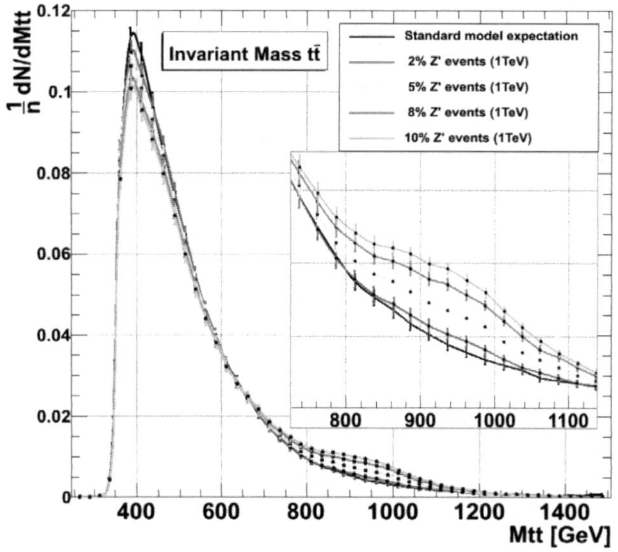

Figure 3. Reconstructed $t\bar{t}$-invariant mass for different ratios of SM and Z'.

3.3. Background discrimination and hypothesis testing

In order to illustrate the use of the MEM in case of background discrimination, let's look at the study of the Higgs boson decaying into four leptons. Benefiting from a very low background and a good mass resolution, the $4l$ final state provides the best spin/parity measurements. Several papers already treat this subject using MEM, see for example [3]. In this section, we will consider only the main background $pp \to ZZ$ and then neglect $pp \to Z + X$ and other backgrounds with a small contribution.

The events with the different spin and parity hypothesis and the background have been generated with MadGraph5 [4], using the "Higgs Characterization" model [5]. We then used Pythia6 [6] and Delphes3 [7] for the detector simulation.

Leptons are selected if the reconstructed transverse momentum is higher than 7 GeV/c and $|\eta|$ smaller than 2.4. An event passes the first selection if four leptons are reconstructed with at least one satisfying $p_T \geq 20$ and a second lepton with $p_T \geq 10$. The two Z-candidates, z_1 and z_2 are then reconstructed with $m_{z_1} \geq m_{z_2}$. A second set of selection is then applied, requiring:

$$50 \leq m_{z_1} \leq 120 \qquad 12 \leq m_{z_2} \leq 120 \tag{5}$$

Finally, a Higgs candidate is reconstructed by summing the four reconstructed lepton 4-vectors. A third selection is then applied, selecting events with a m_{4l} close to the observed excess:

$$110 \leq m_{4l} \leq 160 \qquad . \tag{6}$$

In order to increase the purity of the sample, a simple kinematic discriminant (KD) is defined as $P_{0^+}/(P_{0^+} + P_{ZZ})$. Typically, the background has a small value of KD, while the signal populates a region close to 1 (see Fig. 4, left).

The hypothesis testing is illustrated here by the characterization of the boson at 126 GeV. We will use an extended Likelihood defined as :

$$\mathcal{L}_k = e^{-n_s - n_b} \prod_i \left(n_s \times P_s^k(x_i) + n_b \times P_b(x_i) \right) \tag{7}$$

where k is represents the different spin-parity signal hypothesis for the new resonance (0^+, 0^-, 2^+, etc...). Here we will focus on $\mathcal{L}_{0^+}$ and $\mathcal{L}_{0^-}$. Assuming ≈ 25 fb^{-1}, and considering only the ZZ background. We expect a separation of $\approx 3\sigma$ while neglecting the systematic errors (see Fig. 4, right).

While similar to the CMS approach for the 4 leptons topology [10], this method can be easily extended to channels with neutrinos or jets. Recently searches as complex as $t\bar{t}h$ were treated with the same method [8].

4. Conclusion

The leading order formulation of the matrix element method has been presented as well as the MadWeight program, an automation of this method. Four types of analysis were discussed, namely the b-charge association, the search for new physics using differential matrix element, methods for background discrimination and hypothesis testing. We showed here that the MadWeight program has opened doors to an easy implementation of advanced application of methods. The MEM may therefore acquire a more important role in various type of analysis, ranging from simple four leptons final states to more complex $t\bar{t}h$ decays.

Acknowledgments

First, I would like to thanks the ACAT2013 workshop organisors for giving the opportunity to present this work. I wish to thank Christophe Delaere, Arnaud Pin, Pierre Artoisenet and Michele Selvaggi for their help and feedback. We acknowledge the support from the FNRS and the IAP Program, BELSPO VII/37. This work is partly supported by the IISN convention 4.4503.12.

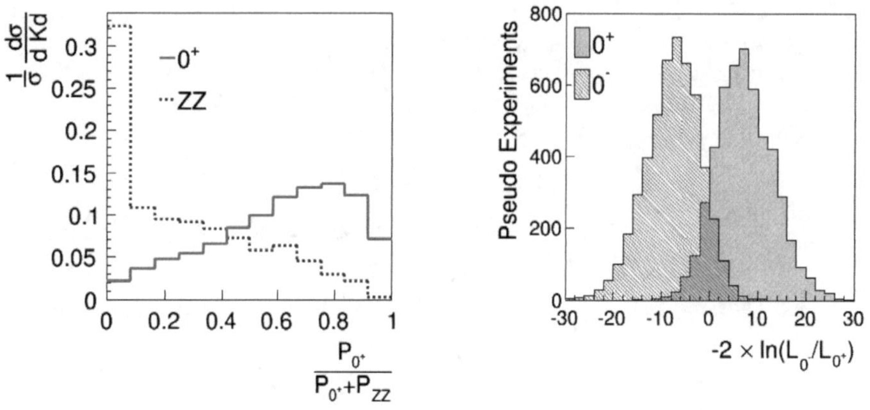

Figure 4. Left: Kinematic discriminant variable for Standard model Higgs boson (red solid line) and for the Background (dashed blue line). On the right: the dashed blue distribution correspond to extended likelihood value with pseudo-experiment considering the background and a odd parity signal The yellow distribution correspond to the background and the Standard model higgs signal.

References

[1] Artoisenet P, Lemaitre V, Maltoni F and Mattelaer O, 2010 *JHEP* **1012**, 068, [arXiv:1007.3300 [hep-ph]].
[2] Gedalia O, Isidori G, Maltoni F, Perez G, Selvaggi M and Soreq Y, 2013 *Phys. Rev. Lett.* **110**, 232002 [arXiv:1212.4611 [hep-ph]].
[3] Bolognesi S *et al*, 2012 *Phys. Rev. D* **86**, 095031, [arXiv:1208.4018 [hep-ph]].
[4] Alwall J et al, 2011 *JHEP* **1106** 128 [arXiv:1106.0522 [hep-ph]].
[5] Artoisenet P *et al*, 2013 *JHEP* **1311** 043 [arXiv:1306.6464 [hep-ph]].
[6] Sjostrand T, Mrenna S and Skands P Z, 2006 *JHEP* **0605** 026 [hep-ph/0603175].
[7] de Favereau J *et al*, 2013, arXiv:1307.6346 [hep-ex].
[8] Artoisenet P, de Aquino P, Maltoni F and Mattelaer O, 2013, arXiv:1304.6414 [hep-ph].
[9] Pin A and Mattelaer O, 2012 *Nuovo Cim. C* **035N3**, 229.
[10] Chatrchyan S *et al.* [CMS Collaboration], 2013 *Phys. Rev. Lett.* **110** 081803 [arXiv:1212.6639 [hep-ex]].

Multivariate polynomial regression regularized via fit function uncertainty

Péter Kövesárki[1] **and Ian C Brock**[2]

[1]University of Wroclaw, Poland

[2]University of Bonn, Germany

E-mail: Peter.Koevesarki@cern.ch

Abstract. This article describes a multivariate polynomial regression method where the uncertainty of the input parameters are approximated with Gaussian distributions, derived from the central limit theorem for large weighted sums, directly from the training sample. The estimated uncertainties can be propagated into the optimal fit function, as an alternative to the statistical bootstrap method. This uncertainty can be propagated further into a loss function like quantity, with which it is possible to calculate the expected loss function, and allows to select the optimal polynomial degree with statistical significance. Combined with simple phase space splitting methods, it is possible to model most features of the training data even with low degree polynomials or constants.

1. Introduction

Regression methods are frequently used in particle physics, usually to quantify a continuous curve or surface that simplifies a statistical sample. Typical examples are the calibration curves for certain detector responses and neural networks trained to identify particle. The mathematical goal is finding a $f : x \to y$ map between the x input space to the y target space in such a way that $f(x)$ predicts the $\mathbb{E}(y|x)$ conditional expectation value with statistical certainty. The least squares algorithm is known to converge to the conditional mean, given it is a finite number and the parametric f function is in the family that contains the solution. This latter information is not always given and one must chose a function family general enough to cover unexpected features. Such a function family are the logistic functions and the radial base functions and generally the *kernels*. For these one usually has to determine an ideal degree of freedom for the fit, namely the number of base functions to be used in order to avoid overtraining of the data and so avoiding picking up non-significant features from the statistical fluctuation. Although these are straightforward procedures, it is computationally intensive to find the global minimum of the sum of squares for the fit. A usually unexploited feature of the least squares method is that the global minimum can be exactly determined for kernels with fixed position in the x space, because the amplitude that minimizes the sum of squares can be calculated with a linear equation, without numerical optimization.

With given $k_i(x), i \in 1..n_d$ kernels and a_i amplitudes the sum of squares for the data points $\{x_j, y_j\}, j \in \{1..N\}$ will take the form

$$E_{\chi^2} = \frac{1}{N}\sum_j \left(y_j - \sum_i a_i k_i(x_j)\right)^2 = \langle y^2 \rangle - \sum_i 2a_i \langle y k_i(x) \rangle + \sum_{il} a_i a_l \langle k_i(x) k_l(x) \rangle \,, \quad (1)$$

where the angled brackets $\langle\rangle$ indicate averaging over the sample. The *loss function* E_{χ^2} in eq. (1) can be minimized in respect of the f_i amplitudes with

$$f_i = \sum_l G_{il}^{-1} h_l \quad (2)$$

by using the matrix $G_{il} = \langle k_i(x) k_l(x) \rangle$ and the vector $h_i = \langle y k_i \rangle$. The G_{il} matrix is symmetric and has $\frac{1}{2}n_d(n_d - 1)$ parameters. A possible way to decrease the number of parameters is to use kernels which are power series $k_i(x) = k_1^i(x)$, resulting in a Hankel-type matrix $G_{il} = \langle k_1(x)^{i+l} \rangle$. A simple power series kernel might be based on the monomials, $k_1(x) = x, k_i(x) = x^i$, resulting in polynomial fitting. Another advantage of fitting a polynomial with a fixed degree instead of Gaussian or sigmoid kernels is that polynomials are not sensitive to the shift of features in the data, in other words they are translation invariant.

2. Uncertainty and covariance of large weighted random sums

The advantage of using the matrix formalism in eq. (1) is that the original training data $\{x_j, y_j\}$ is compressed into the h_i vector and the G_{il} matrix, which can be a great reduction in the number of input parameters. These input parameters are themselves random variables having a certain distribution that in principle could be derived from the generating distribution of the $\{x_j, y_j\}$ sample and the number of measurements N. Due to the central limit theorem, the generating distribution itself does not need to be known, however it has to fulfill certain criteria. Probably the most widely known of the central limit theorems is the one stating that if the generating distribution of the $X_i, i \in \{1..N\}$ random variables have the finite mean $\bar{X}$ and variance σ_X, then the distribution of the variable $s = \frac{1}{\sigma_X}(\sum_i X_i/N - \bar{X})$ converges to the normal distribution. With small modifications this theorem is applicable to h_i and G_{ij}.

The approximation of the covariance matrix of the $p_m = (h_1, ..., h_{n_d}, g_1, ..., g_{2n_d}), m \in \{1, ..., 3n_d\}$ input parameters for polynomial regression is the following. In a general formalism, every data point with index j is a triplet, consisting of an input value x_j, a target value y_j and a weight w_j. With $b \in \{0, 1\}$, $k_0\{1, ..., n_d\}$, $k_1 \in \{1, ..., 2n_d\}$ the input parameter with pseudo index $m = k_b + bn_d$ is calculated as

$$p_m = \frac{1}{\sum_j w_j} \sum_j w_j y_j^b x_j^{k_b} \,.$$

The product $a_j = w_j y_j^b x_j^{k_b}$ for each i index can be treated as a compound random variable. The $1/\sum_j w_j$ was not introduced into this new variable, because that would ruin the independence between the variables. To estimate the probability distribution of p_m the following variables are defined. Let $\langle aw \rangle = \frac{1}{N}\sum_j w_j a_j$ be the weighted average of a_j. This is approximately a Gaussian variable with mean $\langle aw \rangle$ and variance $\sigma^2_{\langle aw \rangle} = \frac{1}{N(N-1)}\sum_j (a_j w_j - \langle aw \rangle)^2$. Define the average weight similarly as $\langle w \rangle = \frac{1}{N}\sum_j w_j$, which is also distributed as a Gaussian with mean $\langle w \rangle$ and variance $\sigma^2_{\langle w \rangle} = \frac{1}{N(N-1)}\sum_j (w_j - \langle w \rangle)^2$. These variables are indeed correlated, and their covariance is $\text{Cov}\,(\langle aw \rangle, \langle w \rangle) = \frac{1}{N(N-1)}\sum_j (a_j w_j - \langle aw \rangle)(w_j - \langle w \rangle)$. The above definitions show, that p_m is a ratio of two Gaussian variables. Its expectation value is p_m, while its variance can be approximated with error propagation. Assuming that $\sigma_{\langle w \rangle}/\langle w \rangle \gg 0$, the approximation of the variance of p_m is

$$\sigma_{p_m}^2 = \begin{pmatrix} \frac{\partial p_m}{\partial \langle aw \rangle} \\ \frac{\partial p_m}{\partial \langle w \rangle} \end{pmatrix}^{\mathrm{T}} \begin{pmatrix} \sigma_{\langle aw \rangle}^2 & \mathrm{Cov}\,(\langle aw \rangle, \langle w \rangle) \\ \mathrm{Cov}\,(\langle aw \rangle, \langle w \rangle) & \sigma_{\langle w \rangle}^2 \end{pmatrix} \begin{pmatrix} \frac{\partial p_m}{\partial \langle aw \rangle} \\ \frac{\partial p_m}{\partial \langle w \rangle} \end{pmatrix}$$

$$= \frac{1}{\langle w \rangle^2}\sigma_{\langle aw \rangle}^2 + \frac{\langle aw \rangle^2}{\langle w \rangle^4}\sigma_{\langle w \rangle}^2 - 2\frac{\langle aw \rangle}{\langle w \rangle^3}\mathrm{Cov}\,(\langle aw \rangle, \langle w \rangle) \tag{3}$$

$$= \frac{1}{\langle w \rangle^2}\sum_j \frac{w_j^2(\langle aw \rangle - a_j)^2}{N(N-1)}\,. \tag{4}$$

It can be seen that eq. (4) gives back the known formula for the standard deviation of $\langle a \rangle$ when all weights are $w_j = 1$, which is also true when the weights are independent of the distribution of a. A similar derivation shows that the covariance between variables p_{m_1} and p_{m_2}, with pseudo-indices $m_1 = k_{b_1} + b_1 n_d$ and $m_2 = k_{b_2} + b_2 n_d$ can be calculated as

$$\mathrm{Cov}\,(p_{m_1}, p_{m_2}) = \frac{1}{\langle w \rangle^2}\sum_j \frac{w_j^2(\langle wy^{b_1}x^{k_{b_1}} \rangle - y_j^{b_1}x_j^{k_{b_1}})(\langle wy^{b_2}x^{k_{b_2}} \rangle - y_j^{b_2}x_j^{k_{b_2}})}{N(N-1)}\,. \tag{5}$$

The covariance and the variance estimation can be generalized to non-monomial kernels, by replacing x^k in the above equations with the given kernel.

It must be noted that traditionally the uncertainty estimates on the fit parameters have different formulas. In most cases the sample $\{x_j, y_j\}$ is augmented with the uncertainty of the target, the conditional variance in the y direction, $\sigma_{y_j}^2$, which can be considered as prior knowledge. In that case the data points in the formation of the estimation of expectation values receive a $\sigma_{y_j}^{-2}$ weight and a $c = \sum_j \sigma_{y_j}^{-2}$ normalization factor. This choice of weight comes from the principle that the different measurements should be combined in a way that minimizes the uncertainty of the result, in this case the estimation of the expectation values. An example could be the $F(x) = a + bx$ least squares regression with a, b unknowns on the $\{x_j, y_j, \sigma_{y_j}^2\}$ sample. The input parameters to this fit are

$$c = \sum_j \frac{1}{\sigma_{y_j}^2}\,, h_0 = \langle y \rangle = \frac{1}{c}\sum_j \frac{y_j}{\sigma_{y_j}^2}\,, h_1 = \langle yx \rangle = \frac{1}{c}\sum_j \frac{y_j x_j}{\sigma_{y_j}^2}\,,$$

$$g_0 = \langle x^0 \rangle = 1\,, g_1 = \langle x \rangle = \frac{1}{c}\sum_j \frac{x_j}{\sigma_{y_j}^2}\,, g_2 = \langle x^2 \rangle = \frac{1}{c}\sum_j \frac{x_j^2}{\sigma_{y_j}^2}\,.$$

The optimal fit parameters are

$$\begin{pmatrix} a \\ b \end{pmatrix} = \begin{pmatrix} g_0 & g_1 \\ g_1 & g_2 \end{pmatrix}^{-1} \begin{pmatrix} h_0 \\ h_1 \end{pmatrix}.$$

As a and b are linear functions of y_j, it is easy to calculate the expectation values needed for the covariance matrix:

$$\mathrm{Cov}\,(a, b) = \mathbb{E}(ab) - \mathbb{E}(a)\mathbb{E}(b) = \frac{1}{c}\sum_j \frac{\partial a}{\partial y_j}\frac{\partial a}{\partial y_j}\sigma_{y_i}^2 = \langle x \rangle = g_1\,. \tag{6}$$

Similarly, $\mathrm{Cov}\,(a, a) = \sigma_a^2 = g_0$ and $\mathrm{Cov}\,(b, b) = \sigma_b^2 = g_2$. This covariance matrix indeed differs from the one derived in eq. (5). The origin of the difference relies in the prior information that was built into the equations. In the case of eq. (5) the weights were provided with the data points, while in the case of eq. (6) the σ_{yi}^2 was given – knowledge of the uncertainty on the $\mathbb{E}(y|x)$

conditional mean. The weights in the former case may come from Monte Carlo integration techniques or from weighted sample separation and it is thought to be fundamentally fixed, while in the latter case it is derived from the principle of optimal data combination. Though it is unclear whether the two methods could be combined, the former method is thought to be superior as it was designed to approximate the $\sigma^2_{y(x)}$ from the sample itself, and also takes into account the uncertainty in the sampling of the input space x. Furthermore, it handles negative and zero weights correctly.

3. Fit function uncertainty

With the knowledge of the uncertainty of the input parameters $p_m = (h_1, ..., h_{n_d}, g_1, ..., g_{2n_d}), m \in \{1, ..., 3n_d\}$, one can estimate the uncertainty of the fitted kernel amplitudes using linear error propagation in eq. (2). The first derivatives of $f_j = \sum_l G^{-1}_{il} h_l, G_{il} = g_{i+l}$ are

$$\frac{\partial f_i}{\partial \langle h_l \rangle} = G^{-1}_{il} ,$$

$$\frac{\partial f_i}{\partial \langle g_o \rangle} = -\sum_{lmn} G^{-1}_{im} \frac{\partial G_{mn}}{\partial g_o} G^{-1}_{nl} h_l .$$

Together with the previously calculated covariance matrix, the uncertainty of the fit function $F(x) = \sum_i f_i x^i$ at a given x point is

$$\sigma^2_{F(x)} = \sum_{nm} \frac{\partial \sum_i f_i x^i}{\partial p_m} \mathrm{Cov}\left(p_m, p_n\right) \frac{\partial \sum_l f_l x^l}{\partial p_n} .$$

The uncertainty of the fitted function does not necessarily cover the true $\mathbb{E}(y|x)$ conditional mean. That only happens if the fit function is general enough to describe all the features of the sample. The meaning of this uncertainty is deeply routed in the central limit theorem. When the central limit theorem was applied to the p_m input parameters, only the fact that the distribution of certain sums can be modeled with a Gaussian distribution came from the theorem, the width and the mean of this Gaussian came from a maximum likelihood fit. This is typically interpreted as a posterior distribution for the true mean, but it can also be interpreted as a model fitted to the sample and predicting where the sum may converge with additional data points. The same can be said about the Gaussian uncertainty of the fit function. It tells us the likelihood where the fit function with the same degrees of freedom would converge with additional data, but not the position of the conditional mean. This is why some methodology is needed to compare fit functions with different degrees of freedom and to find out which method describes better the sample.

4. The uncertainty of the loss function

In the case of polynomial fitting one has to determine the degree of the polynomial that is still statistically meaningful. Using too many degrees of freedom in a fit can result in overfitting or eventually in the interpolation of the data points. In the latter case the E_{χ^2} loss function simply reaches its absolute minimum, zero. However, the uncertainty of the fitted function increases with the number of degrees of freedom and this can be exploited in order to select significant features only, though one has to keep in mind that the Gaussian approximation of the distribution of the p_m input parameters has a limitation. First, the Gaussian approximation is only true if the number of input points N is large enough. Second, the uncertainty of the estimated covariance matrix may also increase to be comparable with the covariance matrix itself if the number of degrees of freedom in the fit is comparable to the number of sample points.

The naïve way of comparing the optimized $F^{\rm opt}_{d_1}(x) = \sum_{ik}^{d_1} h_i G^{-1}_{d_1,ik} x^k$ with degrees of freedom d_1 to $F^{\rm opt}_{d_2}(x)$ with degrees of freedom d_2 would be by calculating their loss functions $E_{d_1,\chi^2} = \frac{1}{N}\sum_j (y_i - F^{\rm opt}_{d_1}(x))^2$ and E_{d_2,χ^2} and checking whether their difference is significantly different from zero. This procedure does not work, as the approximate distribution of E_{d_1,χ^2} is not a good measure of fit quality after the $F^{\rm opt}_{d_1}(x)$ is substituted. This can be observed on the example where the x space is thought to be non-random and only the y coordinates of the sample points can vary. After the substitution of $F^{\rm opt}_{d}(x)$ into E_{d,χ^2} the calculation can be simplified to

$$\begin{aligned} E_{d,\chi^2} &= \langle y^2 \rangle - 2\sum_i F^{\rm opt}_{d,i} h_i + \sum_{ik} F^{\rm opt}_{d,i} G_{ik} F^{\rm opt}_{d,k} \\ &= \langle y^2 \rangle - \sum_{ik} h_i G^{-1}_{ik} h_j \, , \end{aligned}$$

which has an approximate expectation value

$$\begin{aligned} \mathbb{E}(E_{d,\chi^2}) &= \langle y^2 \rangle - \mathbb{E}(\sum_{ik} h_i G^{-1}_{ik} h_j) \\ &= \langle y^2 \rangle - \sum_{ik} h_i G^{-1}_{ik} h_j - \sum_{ik} G_{ik} {\rm Cov}\,(h_i, h_k) \, . \end{aligned}$$

It contains a bias term compared to E_{d,χ^2} and allows $\mathbb{E}(E_{d,\chi^2})$ to be lower than E_{d,χ^2}. This bias means that the approximate distribution of E_{d,χ^2} is not related to the goodness of fit anymore, but to the possible E_{d,χ^2} minima.

Ideally, the best measure of fitness would be a distance-like variable between the fit function and the real $\mathbb{E}(y|x)$ conditional expectation value. A good approximation to that is of course the loss function applied to $F^{\rm opt}_{d}(x)$, but one must differentiate it from the previously described E_{d,χ^2}. In this picture, one must treat the sample as an approximation to $\mathbb{E}(y|x)$, and not as a random variable. Let us call it the cross validation loss function, where only the $F^{\rm opt}_{d}(x)$ is a random variable:

$$E^+_{d,\chi^2} = \underbrace{\langle y^2 \rangle}_{\rm fixed} - 2\sum_i \underbrace{F^{\rm opt}_{d,i}}_{\rm varied} \underbrace{h_i}_{\rm fixed} + \sum_{ik} \underbrace{F^{\rm opt}_{d,i}}_{\rm varied} \underbrace{G_{ik}}_{\rm fixed} \underbrace{F^{\rm opt}_{d,k}}_{\rm varied} \, .$$

As the first derivative of $\frac{\partial E^+_{d,\chi^2}}{\partial p'_m} = 0$ at $p'_m = \mathbb{E}(p_m)$, the expectation value of E^+_{d,χ^2} can be approximated through its second derivative as

$$\begin{aligned} \mathbb{E}(E^+_{d,\chi^2}) &= \langle y^2 \rangle - 2\sum_i h_i \mathbb{E}(F^{\rm opt}_{d,i}) + \sum_{ik} G_{ik} \mathbb{E}(F^{\rm opt}_{d,i} F^{\rm opt}_{d,k}) \\ &= E^+_{d,\chi^2} + \frac{1}{2}\sum_{lm} \frac{\partial^2 E^+_{d,\chi^2}}{\partial p_m \partial p_l} {\rm Cov}\,(p_m, p_l) \, , \end{aligned} \tag{7}$$

where the second derivative of E^+_{d,χ^2} was taken at the expectation values of p_m, and can be expressed as a block matrix

$$\frac{\partial^2 E^+_{d,\chi^2}}{\partial p_m \partial p_l} = \begin{pmatrix} 2G^{-1}_{lm} & -4\sum_{kno} G^{-1}_{lk} \frac{\partial G_{kn}}{\partial p_m} G^{-1}_{no} h_o \\ -4\sum_{kno} G^{-1}_{mk} \frac{\partial G_{kn}}{\partial p_l} G^{-1}_{no} h_o & 6\sum_{knowqg} h_k G^{-1}_{ko} \frac{\partial G_{ow}}{\partial p_l} G^{-1}_{wg} \frac{\partial G_{gn}}{\partial p_m} G^{-1}_{nq} h_q \end{pmatrix} .$$

Since E^+_{d,χ^2} is a quadratic formula and only the fit function is varied, the expectation value cannot be lower than the minimum taken at $F^{\rm opt}_{d}(x)$ by construction. In most cases the second degree Taylor polynomial approximation in eq. (7) in the p_m variables is enough as it is quadratic in the h_i and the dependence on the G^{-1}_{ik} is usually weak for large samples. The explanation is, that G_{ik} depends on the sampling in x which does not affect the $\mathbb{E}(y|x)$ conditional expectation value in the large sample size limit. Therefore for large sample sizes and approximately error-free G_{ik} matrices $\frac{\partial^2 E^+_{d,\chi^2}}{\partial p_m \partial p_l}$ can be approximated with the its upper left sub-matrix, that represents the derivatives of h_l and h_m. In this approximation E^+_{d,χ^2} is a χ^2 variable with d degrees of freedom, given that the covariance matrix $\mathrm{Cov}\,(h_i, h_k)$ is non-singular.

In the full calculation of the standard deviation of E^+_{d,χ^2} one should include the variable $\langle y^2 \rangle$ and its correlation to the other input parameters p_m. Nevertheless, the $\langle y^2 \rangle$ term disappears when a loss function difference is calculated for two different fit functions, and it can be shown that the variance of the loss function differences can be calculated solely from the variance of the individual E^+_{d,χ^2} terms, neglecting the contributions from $\langle y^2 \rangle$. This type of variance is twice the square of the previously described bias term in the expectation value in eq. (7), as expected for a χ^2 distribution:

$$\sigma^2_{d,E^+} = 2\left(\frac{1}{2}\sum_{lm}\frac{\partial^2 E^+_{d,\chi^2}}{\partial p_m \partial p_l}\mathrm{Cov}\,(p_m, p_l)\right)^2 .$$

Similar to the behavior of χ^2 differences, the expectation value of the $E^+_{d2,\chi^2} - E^+_{d1,\chi^2}$ difference is the difference of their expectation values. The same can be said about the variance of $E^+_{d2,\chi^2} - E^+_{d1,\chi^2}$, as it can be calculated from the square root of the individual E^+_{d2,χ^2} variations:

$$\sigma_{d_1,d_2,\Delta E^+} = |\sigma_{d_2,E^+} - \sigma_{d_1,E^+}| .$$

This additional rule makes it extremely simple to find the optimal degrees of freedom d, as there is no need to calculate covariance matrices belonging to the different fit functions. This makes the significance of a set of monomial kernels not only relatively true, but absolute. To optimize the degree of freedom d with a certain significance level, it is enough to minimize $\mathbb{E}(E^+_{d,\chi^2}) + s\sigma_{d,E^+}$, where s can be arbitrarily chosen. Due to the characteristics of the χ^2 distribution, optimizing for the minimal expected E^+_{d,χ^2} results in selecting features, a set of kernels that are at least 50% significant.

5. Determining the right polynomial order

Selecting the right modeling function can be understood as a repeated hypothesis testing. One must decide *a priori* about the null hypothesis and the series of hypotheses to test. In case of fitting a polynomial function, the ordering seems trivial, going from a constant to higher polynomial degree. However, with multivariate input x_μ, a polynomial can be defined with a different degree belonging to each μ index. In this case, one may still choose a common degree, as the optimizing method described in the previous section allows the comparison of fit functions differing in multiple degrees of freedom. A common polynomial order is special in the sense that it is closed under the group of rotations and translations, treating the different μ directions equally.

However, it is not possible to compare the loss functions of an infinite number of polynomials. Not only because it is impracticable, but also because one must stop before the numerical errors exceed the estimated uncertainties in $E^+_{\chi^2}$. It is not straightforward to estimate these numerical errors, but the rule of thumb is, that it increases with both the number of polynomial degrees and the number of input dimensions in the multivariate case. The polynomial degree necessitates to calculate high powers of the input parameters, which both appears in the training and in

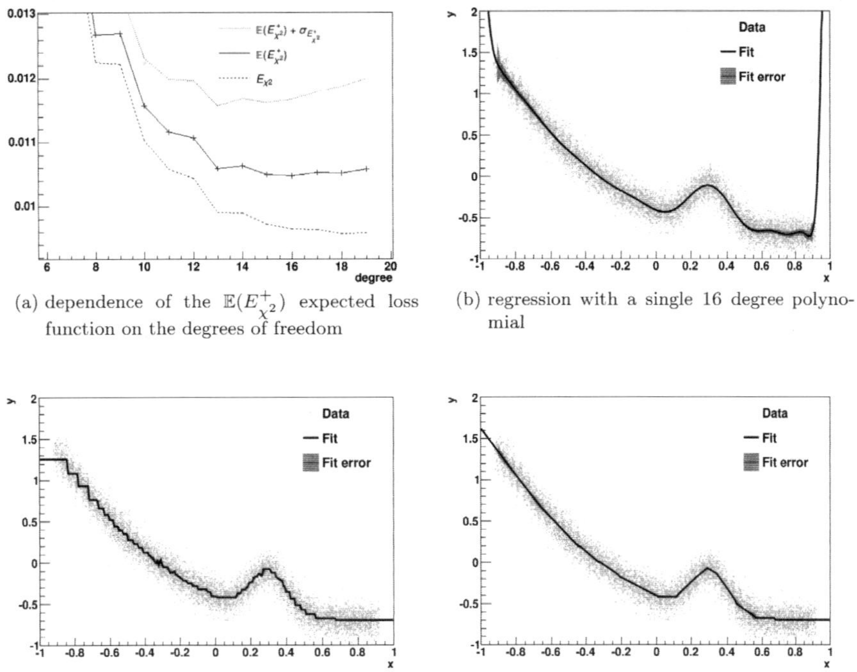

(a) dependence of the $\mathbb{E}(E^+_{\chi^2})$ expected loss function on the degrees of freedom

(b) regression with a single 16 degree polynomial

(c) regression with constants

(d) regression with linear functions

Figure 1: Example of an univariate regression. The training sample is a uniform distribution in $x \in (-0.9, 0.9)$ and a Gaussian smearing in the y direction for every x. The $\mathbb{E}(E^+_{\chi^2})$ expected loss function in fig. 1a has a minimum for the 16 degree polynomial in the examined range. The curve below that belongs to the E_{χ^2} loss function evaluated with the optimal polynomial, and by definition it can only decrease with additional degrees of freedom. The top curve is one standard deviation above the expected loss function. Figure 1b shows the 16 degree optimal polynomial with the largest uncertainties at the two boundaries of the sample. Figure 1c and 1d shows the result of the regression with the splitting procedure described in the text, with allowed maximal polynomial degrees $n_{\text{max}} = 0$ and 1 accordingly. Both of these regressions functions are within one sigma of the conditional $\mathrm{P}(y|x)$ distribution and approximate the $\mathbb{E}(y|x)$ conditional expectation value without picking up statistical noise.

the function evaluation phase. A double precision number can be thought of as a sixteen digit decimal number, and though its nth power is expected to have a relative error of only $n \cdot 10^{-16}$, the numerous subtractions and multiplications needed for the linear equation solver can easily blow this up. In the univariate case, the numerical errors seems to become significant at the polynomial degree around 20 for double precision and 24...30 for 128bit long double precision. In the case of d input dimensions the size of the G matrix grows rapidly with the number of degrees used, since the polynomial coefficients of the $F(x)$ fit function are d-degree symmetric tensors, with $\binom{n+d-1}{n}$ free parameters. All of the free parameters in $F(x)$ contribute to the size of the G matrix. For 20 input dimensions a third degree polynomial has nearly 2000 free parameters, resulting in G matrix size of 2000×2000. Although solving a linear equation with this only takes a few seconds on a modern-day computer, this also means that more than a million instructions are needed to express each unknown of the $F(x)$ polynomial, resulting in large numerical errors.

(a) original

(b) remodeled

Figure 2: Figure (a) shows a photo of a Chinese terra-cotta soldier, whose intensity map was remodeled in fig. (b) in small patches with polynomials. The photo was treated as a random sample, meaning prior knowledge of the uncertainty of the pixel intensities were neglected. The region boundaries were determined with the algorithm described in text, based on the principal axes of the input distribution. The regions were modeled with a two dimensional linear polynomial, with a second layer of univariate regression upon it, to simulate a sigmoid-like behavior. Over- and undershoots from the displayable range were rounded to the maximum and minimum intensities accordingly. A photo instead of a real 3D distribution was chosen here in order to demonstrate that this simple fitting and splitting method can detect the significant features in the data while smoothing the small details, as one would require for a regression method on a statistical sample.

6. Minimizing numerical errors

To overcome the problem of the high polynomial degrees and the large matrices, one can split the sample into many smaller phase spaces, which may require smaller polynomial degrees. For this one must decide on a maximum number of degrees $n_{\max}$ of the polynomial function which is allowed in regression, and $n_{\max} + k$, $k > 1$ for which the expected loss function is scanned. If the optimal degree of the expected loss function is larger than $n_{\max}$, one can apply a predefined algorithm that splits the input phase space. One such algorithm for the univariate case simply splits the input phase space at the x mean, as demonstrated on fig. 1. This requires practically no additional computation, since the $\langle x \rangle$ was already calculated for the regression. The splitting and fitting can be repeated until the full sample is regressed. A possible extension of this approach in the multivariate case finds the multivariate mean $\langle x_\mu \rangle$ first, then splits the sample at this point parallel to the principal axis which is given by the eigenvector with the largest eigenvalue of the $\langle x_\mu x_\nu \rangle - \langle x_\mu \rangle \langle x_\nu \rangle$ matrix; see fig. 2. These methods have the advantage that they place the cut boundaries within the distribution, so the regression on these phase spaces are less likely to produce degenerate solutions. A seemingly more optimal splitting method would be finding the place where the fitted polynomial with $n_{\max} + 1$ degrees have the largest derivative, since this is a hard place to model with an $n_{\max}$ degree polynomial. However, it is non-trivial to define and find this boundary in the multivariate case, and this boundary typically appear nearby the tails of the x distribution, where the fitted function has the largest uncertainty.

7. Conclusions

The presented method is capable of modeling multivariate statistical data with polynomials by detecting the significant features in the data. It is a fast and robust method, as most calculations are computationally very simple and does not require numerical optimization. Similarly to the statistical bootstrap method, the uncertainty of the regression function can be determined from the training sample, but in this case analytically. In combination with a phase space splitting method, it can be extended to fit very complex data, still maintaining numerical stability.

References

[1] Kövesárki P 2012 Polynomial expansion of the binary classification function (*Preprint* `http://arxiv.org/abs/1203.5647`)
[2] Kövesárki P 2012 *Multivariate methods and the search for single top-quark production in association with a W boson in ATLAS* Ph.D. thesis University of Bonn URL `http://hss.ulb.uni-bonn.de/2013/3188/3188.htm`
[3] Metzger W J 2010 *Statistical Methods in Data Analysis* (Radboud Universiteit Nijmegen, The Netherlands) ISBN HEN-343 URL `http://www.hef.ru.nl/~wes/stat_course/statist.pdf`
[4] The European Mathematical Society *Encyclopedia of Mathematics* 1 URL `http://www.encyclopediaofmath.org/index.php?title=Hankel_matrix&oldid=23850`
[5] Bishop C M 2006 *Pattern Recognition and Machine Learning* (Secaucus, NJ, USA: Springer-Verlag New York, Inc.) ISBN 0387310738 iSBN 0-387-31073-8
[6] Ripley B D 1996 *Pattern Recognition and Neural Networks* (Cambridge University Press) ISBN 0521460867 iSBN 978-0521-71770-0
[7] Scargle J D, Norris J P, Jackson B and Chiang J 2013 *The Astrophysical Journal* **764** 167

ACAT2013 IOP Publishing
Journal of Physics: Conference Series **523** (2014) 012030 doi:10.1088/1742-6596/523/1/012030

Quasi-optimal weights: a versatile tool of data analysis

Fyodor Tkachov

Institute for Nuclear Research, Russian Academy of Sciences, Moscow 117312 Russia

E-mail: ftkachov@inr.ac.ru

Abstract. In this first part of the two-part account of the enabling technologies behind a successful completion of the Troitsk-nu-mass experiment, the parameter estimation method of quasioptimal weights is reviewed. In regard of statistical quality, it is on a par with the maximal likelihood method. but exceeds the latter in analytical transparency, flexibility, scope of applicability and numerical robustness. It also couples perfectly with the optimal jet definition and thus provides a comprehensive framework for data processing in particle physics and beyond.

1. Introduction

The currently best ν mass bound was published by the Troitsk-ν-mass experiment [1]. This happened after the experiment had been plagued by the so-called "Troitsk anomaly" for about 10 years. The anomaly went away after a reanalysis that was made possible by two enabling technologies: the statistical method of quasioptimal weights for parameter estimation [2] and the Oberon technologies for software development. The latter are discussed in a companion talk [3]. The present talk reviews the former.

It should be noted that the two technologies per se did not solve any specific physics problem, but provided a flexible and efficient framework to experiment with and implement specific improvements of the analysis in a real-life context under real-life limitations. Everything could in theory have been done with maximal likelihood and fortran or even C++, but at a cost that proved prohibitive in that context. A proper *technique* is about minimizing effort while attaining the best result, it is rooted in how well one understands the basic principles of what one works with.

2. The method

A basic problem of mathematical statistics is, given a parameterized distribution $\pi_M(\mathbf{P})$ and a random sample $\{\mathbf{P}_i\}_{i=1...N}$, to estimate the unknown value M_0 of the parameter M. One seeks to obtain an estimate M_* for the parameter and a standard error estimate σ_* (or the corresponding confidence intervals).

The popular least squares method does not provide a full control; in fact, this project was prompted in the Fall of 2005 by a doubt expressed by the leader of the Troitsk-ν-mass experiment, the late Vladimir Lobashev as to whether the Poisson-distributed background was treated correctly by least squares.

The simple and transparent method of moments is universally regarded as inferior and mostly neglected in data processing applications. Among the few exceptions are [4] and [5]. Both turn out to be special cases of the method of quasi-optimal weights.

The maximal likelihood method, if applicable, yields the best possible estimate, obtained by maximizing the likelihood function,

$$\sum_i \ln \pi_M(\mathbf{P}_i) \le \sum_i \ln \pi_{M_{\mathrm{FFRC}}}(\mathbf{P}_i). \tag{1}$$

The estimate corresponds to the fundamental Fisher-Frechet-Rao-Cramer bound:

$$\sigma^2_{\mathrm{FFRC}} = \frac{1}{N}\left[\text{Fisher's information}\right]^{-1} \tag{2}$$

Note that the maximum can be found by solving

$$\frac{\partial}{\partial M}\sum_i \ln \pi_M(\mathbf{P}_i) = 0 \tag{3}$$

But what if $\pi_M(\mathbf{P})$ is unknown as a formula? This is often the case in high energy physics where only a Monte Carlo event generator may be available but not an explicit expression for the probability distribution (this is due to a very high dimensionality of the event space). On top of that, HEP data processing involves a heavy dose of chiropractic (event selection, jet algorithms, etc.).

Consider the method of generalized moments. For any weight ("generalized moment") $f(\mathbf{P})$, its mean value is a function of the unknown parameter

$$\langle f\rangle = \int d\mathbf{P}\, f(\mathbf{P})\,\pi_M(\mathbf{P}) = F(M) \tag{4}$$

On the experimental side, one can estimate this mean as follows

$$\langle f\rangle_{\exp} = \frac{1}{N}\sum_i f(\mathbf{P}_i) = F_* \tag{5}$$

One then estimates the parameter,

$$M_* = F^{-1}(F_*) \tag{6}$$

This is easily visualized:

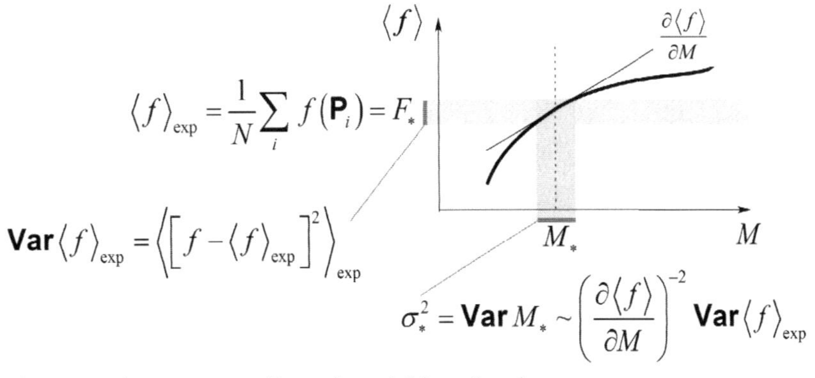

Note that both error estimates are easily evaluated. Note also that

$$\mathbf{Var}\langle f\rangle_{\exp} \sim \frac{1}{N}\mathbf{Var}\langle f\rangle, \quad \mathbf{Var}\langle f\rangle = \left\langle\left[f - \langle f\rangle\right]^2\right\rangle \tag{7}$$

Originally Pearson (1894) dealt with a one-dimensional event space and used:

$$f(\mathbf{P}) = \mathbf{P}^n \tag{8}$$

which yielded suboptimal estimates compared with ML.

However, the method is analytically transparent unlike either of the alternatives, and there are deep mathematical reasons to consider weights. Indeed, an alternative to the set-theoretic notion of a general function is one based on weighted values rather than point values of the function. In this alternative scheme the basic notions are choosen to be functions/mappings (with compositions, morphisms, categories etc.) rather than (sub)sets (with the corresponding set operations such as intersections etc.). An accessible review of these matters is given in [6]. It should be emphasized that such a functional view of the mathematical analysis is as comprehensive as the set-theoretic one. But the heuristics of the two schemes are entirely different: the weighted values are an artificial device in the set-theoretic view whereas in the functional view they are a natural starting point.

In the conventional view, a function is a correspondence $x \to g(x)$. This, however, is fully meaningful only for continuous $g(x)$. There is no natural way to define the value at a discontinuity.

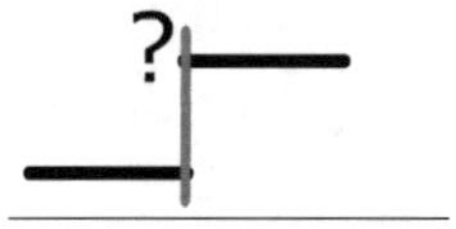

(9)

In that regard a more satisfactory way is to define a function via its weighted values

$$f \to \langle g\, f \rangle \quad \left[\!\left[= \int \mathrm{d}x\, g(x)\, f(x) \right]\!\right] \tag{10}$$

where f are the so-called test functions usually chosen to be smooth (or at least continuous):

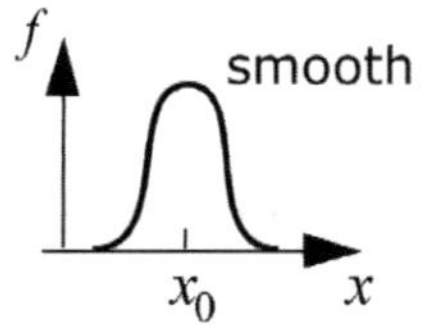

(11)

Such a definition allows one to recover the function values at the points of continuity but the problem of values at discontinuities is bypassed altogether.

In particular — and immediately relevant to approximation methods — instead of comparing point values, one now compares weighted/smeared values. This directly leads to the powerful Bubnov-Galerkin method very widely used in engineering and other numerical applications (see e.g. [7]).

The functional viewpoint singles out the method of generalized weights as a fundamental starting point of any thinking about parameter estimation. So, let us look at it more closely, putting aside the mantra of its inferiority.

One has a simple and explicit expression for the error estimate:

$$\sigma_*^2 \sim \frac{1}{N} \left\{ \left(\frac{\partial \langle f \rangle}{\partial M} \right)^{-2} \left\langle \left[f - \langle f \rangle \right]^2 \right\rangle \right\} \tag{12}$$

Let us minimize it. From the minimum equation

$$\frac{\delta}{\delta f(\mathbf{P})} \{\ldots\} = 0 \tag{13}$$

one easily finds

$$f_{\text{opt}}(\mathbf{P}) = \frac{\partial}{\partial M} \ln \pi_M(\mathbf{P}) \tag{14}$$

This point turns out to be a true minimum, with Fisher's information as the minimum value. This exercise was performed in [8].

However, the optimal weight depends on M, and its actual value M_0 is unknown. Then an iterative procedure naturally suggests itself:

$$M_i \to f_{\text{opt},M_i}(\mathbf{P}) \to M_{i+1} \tag{15}$$

The problem of choosing the initial value M_1 is not specific to this method; it is implicit in any method based on optimizations such as ML and least squares.

Note that each M_i is a correct statistical estimate per se, only the corresponding errors may be suboptimal. Note also that the convergence of the sequence M_i is not an issue here since it is the variance that the iterative procedure is intended to minimize.

The non-optimal weights $f_{\text{opt},M_i}(\mathbf{P})$ are close to the optimal one according to how M_i is close to M_0, whence the name of the method.

Given that the minimum of the variance in the space of weights is quadratic, there is a freedom in the choice of quasi-optimal weights, and the method is rather less demanding in regard of the knowledge of the underlying probability density than the ML method.

That the iterative procedure is equivalent to ML is seen from the fact that

$$\langle f_{\text{opt}} \rangle = 0 . \tag{16}$$

(This assumes that the optimal weight and the averaging are taken at the same M.)

The iterative procedure is similar to the optimizations involved in both the ML and least squares methods. It appears that the method of quasi-optimal weights reveals and efficiently exploits an analytical structure that have been there all along but remained eclipsed by the glory of the maximal likelihood. (One might also say that it fell victim to the personal rivalry between Sir Robert Fisher and Karl Pearson.)

Example 1

Some special cases of the quasi-optimal weighting have been known for a while. The case of linear dependence of the probability distribution on M has been known since 1992 [4]

$$\pi(\mathbf{P}) = \pi_0(\mathbf{P}) + g\pi_1(\mathbf{P}), \quad f_{\text{opt}}(\mathbf{P}) = \frac{\pi_1(\mathbf{P})}{\pi(\mathbf{P})} \tag{17}$$

This has been used in a number of weak signal searches.

This is closely related to the following criterion for hypothesis testing which is locally most powerful near $g = 0$: $\pi_1(\mathbf{P})/\pi_0(\mathbf{P})$.

Another special case was constructed by Jorg Pretz within the context of the g-2 experiments [5].

Example 2

Consider the Cauchy/Breit-Wigner distribution. It has no mean yet the quasi-optimal weights work just fine to determine M:

$$\pi(\mathbf{P}) \propto \frac{1}{(M-\mathbf{P})^2 + \Gamma^2}, \quad f_{M,\text{opt}}(\mathbf{P}) \propto \frac{M-P}{(M-\mathbf{P})^2 + \Gamma^2} \tag{18}$$

In this simple example one can see the various options for an approximate choice of the quasi-optimal weight which, it should be emphasized, need not be chosen to exactly coincide with the analytical formula:

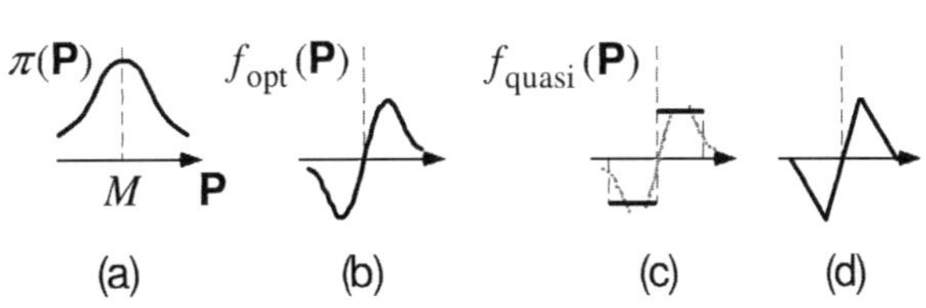

(19)

Any of these shapes would yield a valid estimate for M.

Example 3
Compare Poisson and normal distributions. This concerns situations where one deals with a Poisson distributions but implicitly assumes normality of all distributions in the actual data processing.

Consider the normal distribution

$$\pi_\mu(\mathbf{P}) \propto \exp\left[-(P-\mu)^2 / 2\sigma^2\right] \tag{20}$$

The two optimal weights for the two parameters are as follows:

$$f_{\text{opt},\,\mu} \propto (P-\mu), \quad f_{\text{opt},\,\sigma} \propto (P-\mu)^2 \tag{21}$$

Now consider the Poisson distribution with an integer n in place of the real P:

$$\pi_\mu(n) = e^{-\mu}\frac{\mu^n}{n!}, \quad f_{\text{opt},\,\mu}(n) = \left(\frac{n}{\mu} - 1\right) \tag{22}$$

Using methods devised for the normal distribution to estimate the Poisson parameter μ is similar to working with the distribution

$$\pi_\mu(n) \propto \exp\left[-(n-\mu)^2 / 2\mu^2\right] \tag{23}$$

and the corresponding family of weights is

$$f_{\text{opt},\,\mu}(n) \propto \left(\frac{n}{\mu} - 1\right) + \frac{1}{2}\left(\frac{n}{\mu} - 1\right)^2 \tag{24}$$

The difference from the purely Poisson weight is indicative of a bias introduced thereby. Numerically, for $\mu \sim 0.15$–1.5, the effect is 10–15%, which is large enough to be of concern.

Example 4
Consider how the continuous weights compare with event selection cuts widely used in HEP experiments. Consider a one-dimensional event space [0,1] and the linear probability density:

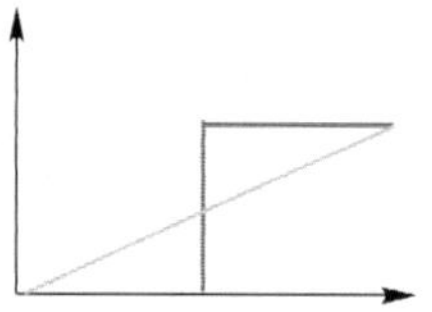

(25)

Consider the variances of the linear weight and the shown cut relative to the linear probability distribution. The ratio of the two variances is 3, and the ratio of the corresponding sigmas is 1.71. In other words, replacing the cut with a continuous weight in this case converts a 3σ signal into a 5σ one.

Example 5
Our last example is the observation of the top quark in all-hadrons channel by D0 as described in [10]. A continuous weight ("an average jet count') was devised empirically and proved to be "particularly powerful" in a multivariate analysis with thirteen variables.

Even if a formula such as $f_{\text{opt}}(\mathbf{P}) = \frac{\partial}{\partial M}\ln \pi_M(\mathbf{P})$ remains somewhat of a theoretical fiction in the case of multijet event space with a Monte Carlo event generator, an explicit formula is always a great force as it sheds light on what otherwise is a trial-and-error search.

3. The case of non-uniform event sample
The method can be extended to the case of non-uniform samples [2].

A set of values U_i of a control parameter is chosen, and for each value a measurement P_i is taken distributed according to a probability distribution $\pi_{i,\theta}(P)$. All these are assumed to depend on the unknown parameter to be estimated accroding to a known formula. If all P_i for different U_i are independent then the situation reduces to the simple case by defining a composite event $\otimes_i P_i$ and the corresponding probability density $\otimes_i \pi_{i,\theta}(P_i)$. The optimal weight then is:

$$\varphi_{\text{opt}}(\{P_i\}) = \sum_i \frac{\partial}{\partial\theta} \ln \pi_{i,\theta}(P_i) = \sum_i \varphi_{\text{opt}_i}(P_i) \tag{26}$$

with the explicit formula for variance:

$$\mathbf{Var}\,\varphi_{\text{opt}} = \sum_i \left[\varphi_{\text{opt}_i}(P_i) - \langle \varphi_{\text{opt}_i} \rangle^{\text{th}}_{\theta_{\text{exp}}} \right]^2 \tag{27}$$

All this is straightforward to code.

4. Notes on usage in the Troitsk-ν-mass experiment

A number of Monte Carlo tests showed that the method is indeed asymptotically equivalent to ML (which fact is quite obvious from the analytical formulae of the method).

Programming implementation of the method is quite straightforward, although some flexibility of the implementation language is desirable (actually used was a minimalistic strictly type-safe language Oberon/Component Pascal, a direct descendant of Pascal, see the companion talk for details [3]).

Numerically, one has to solve a system of equations, which is a simpler task than an optimum search (for a fast Newton-type algorithm, one only needs first derivatives of the mapping involved in the former case, whereas the optimum search requires second derivatives).

The method is less sensitive than, say, MINUIT to the "narrow valleys" situation (although this difficulty cannot be eliminated altogether), as well as to non-smoothness of the functions involved.

One can easily accomodate knowledge about the nature of the probability distributions involved (Poisson etc.).

With a poor statistics, the resulting multidimensional systems of equations may not have a solution; in practice this was not a problem (mismatches that were encountered were small more than enough not to affect the results) but some (theoretical) research may be called for. Similarly, a Student-type correction factors for confidence intervals may be required in the poor statistics case (this problem, however, is a general one and not limited to the quasi-optimal weights; in this regard the present method is neither better nor worse than everything else).

The same conceptual framework earlier yielded the so-called Optimal Jet Definition [11] which was the first to demonstrate an *O(N)* behavior with respect to the number of particles in the final state [12]. The OJD remains the only jet definition derived from first principles in a systematical scientific fashion, without ad hoc constructions. The fact that it meshes well with quasi-optimal weights bodes well for the complicated experimental situations such as the abovementioned one encountered by D0 in the all-jets channel.

Lastly, a fundamental nature of the argument behind OJD is underscored by the fact that a very similar derivation yielded an efficient multi-dimensional density modelling algorithm [13].

To summarize, a general and flexible method of quasi-optimal weights (together with a closely related optimal jet definition, the general multi-dimensional density modelling algorithm, etc.) offers a rather comprehensive unified framework to handle a wide array of HEP data processing problems in a simplified and systematic manner and to help to streamline the motley of (sometimes ad hoc) methods currently in use.

References

[1] Aseev V N et al. 2011 *Phys. Rev. D* **84** 112003 (*Preprint* arXiv:1108.5034).
[2] Tkachov F V 2006 *Transcending The Least Squares* (arXiv:physics/0604127).
[3] Tkachov F V 2013 *Less is more. Why Oberon beats mainstream in complex applications.* Talk

at ACAT 2013, Beijing.
[4] Atwood D and Sony A 1992 *Phys. Rev. D* **45** 2405.
[5] Pretz J 1998 *Reducing the statistical error on the g–2 frequency using a weighting procedure.* Muon g–2 Note No. 326.
[6] Goldblatt R 2006 *Topoi: The Categorial Analysis of Logic*. Dover.
[7] Vilotte J-P 2013 *Data-intensive High Performance Computing and Analysis in Solid Earth Sciences.* Talk at ACAT 2013, Beijing.
[8] Tkachov F V 2002 *Part. Nucl. Lett.* **111** 28-35 (*Preprint* arXiv:physics/0001019).
[9] Bhat P C, Prosper H, Snyder S S 1998 *Int. J. Mod. Phys. A* **13** 5113-5218 (*Preprint* arXiv:hep-ex/9809011).
[10] Tkachov F V, 2002 *Int. J. Mod. Phys. A* **17**, 2783-2884 (arXiv:hep-ph/9901444).
[11] Grigoriev D Yu, Jankowski E, Tkachov F V 2003 *Phys. Rev. Lett.* **91** 061801 (*Preprint* arXiv:hep-ph/0301185).
[12] Tkachov F V 2004 *Nucl. Instrum. Meth. A* **534** 274-278 (*Preprint* arXiv:physics/0401012).

ATLAS trigger simulation with legacy code using virtualization techniques

Gorm Galster[1], Joerg Stelzer[2] and Werner Wiedenmann[3]

[1] University of Copenhagen, DK
[2] Michigan State University, US
[3] University of Wisconsin, US

E-mail: gorm.galster@cern.ch, joerg.stelzer@cern.ch, werner.wiedenmann@cern.ch

Abstract. Several scenarios, both present and future, require re-simulation of the trigger response in the ATLAS experiment at the LHC. While software for the detector response simulation and event reconstruction is allowed to change and improve, the trigger response simulation has to reflect the conditions at which data was taken. This poses a maintenance and data preservation problem. Several strategies have been considered and a proof-of-concept model using virtualization has been developed. While the virtualization with CernVM elegantly solves several aspects of the data preservation problem, the limitations of current methods for contextualization of the virtual machine as well as incompatibilities in the currently used data format introduces new challenges. In this proceeding these challenges, their current solutions and the proof of concept model for precise trigger simulation are discussed.

1. Introduction

Each 50 ns bunches of protons are brought to collide in ATLAS[1], one of the four large detectors experiments at the LHC ring at CERN. At the end of LHC run 1, an average of 35 protons collided per bunch crossing, each generating a spray of particles which are detected by the different sub-detectors of ATLAS.

In order to provide real time data reduction a sophisticated trigger system decide which collision events are stored, effectively reducing the data throughput to hundreds of gigabyte per second. In consequence, all stored data are inherently watermarked by the trigger configuration and trigger algorithms used during data taking. An in-depth analysis of the data thus also becomes a study of the trigger. Having a strategy for simulation of the trigger response is of importance for continued usability of the recorded data.

1.1. ATLAS simulation chain

Physics analyses in high energy physics require that real data are accompanied by simulated Monte Carlo (MC) data. During ATLAS data taking, MC data are produced alongside the real data taking using the same software as is used for online selection and offline reconstruction.

The ATLAS simulation chain is shown in Figure 1. First the detector response to the simulated physics events is simulated. Based on the detector response the trigger simulation adds the trigger decision record to the event before the simulated events are reconstructed and stored.

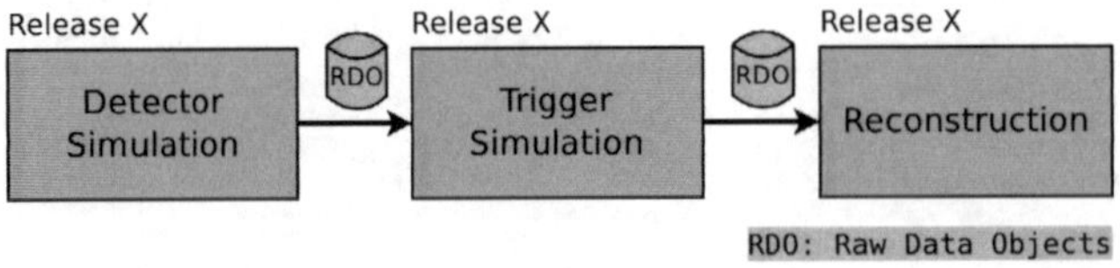

Figure 1. The existing MC production chain. The boxes depict the execution of the respective simulation steps: detector simulation, trigger response simulation and the reconstruction step. The RDO (Raw Data Objects) is the exchange format. The Release X is to denote that the same software version is used for all three steps.

2. Motivation

Several scenarios will require a (re-)simulation of the trigger response:

- An improved description/understanding of the detector response.
- Improved algorithms and methods for offline reconstruction.
- A desire to increase the size of MC samples in future studies.
- The introduction of new event generators.

2.1. Data format challenges

While detector response simulation and event reconstruction should be done with the newest software, the version used for trigger response simulation needs to match that used for data taking. Mismatch between the software versions poses a number of challenges. The most direct challenges are those relating to data format compatibility:

- The trigger response simulation needs to read detector data produced by a new detector simulation. This either implies forward compatibility of the format and content produced by the detector simulation or a possibility to convert the detector response to an older format readable by the old trigger simulation.
- The reconstruction needs to read the old trigger response record. This either implies a conversion step or backward compatibility in the format for the trigger response record such that the reconstruction algorithm can read and understand the produced old trigger response header.

2.2. The environment stability

Besides the challenges related to data formats there are those related to the time scale envisaged for ATLAS. Over a time span of more than a decade it is expected that hardware architectures, operating systems, core components and compilers change, making it impossible to run the old code as-is.

It would thus be necessary to port trigger selection algorithms which, potentially, are not even used any more to the new software releases and certify with every new software release that their selection response remains unchanged. In addition one also has to guarantee that the legacy infrastructure for these trigger algorithms works with the new operating systems, changes in software infrastructure, changes in the compiler and computer hardware developments. Further, knowledge of the working of unused components would have to be preserved so that it is possible to port them continuously to new releases alongside the actual trigger code. This poses a maintenance problem.

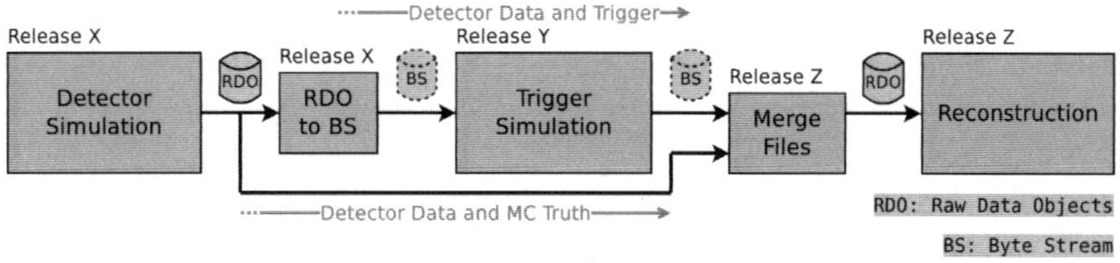

Figure 2. The proposed changes to the existing simulation chain. This chain requires two additional steps and uses an intermediate byte stream (BS) file. The first steps converts the detector specific data of the RDO into BS. After the trigger response record has been added, the RDO is infused with the trigger response record. The resulting RDO is passed on to the reconstruction step.

2.3. Objective and scope

The goal of this study is to investigate viable strategies for a precise trigger simulation reflecting past data taking conditions. The precision requirement can only be realized by the use of the same software as was used during data taking. The proposed strategy should preferably require minimal change to existing infrastructure and additional maintenance even when future developments in hardware and software becomes a reality.

3. Resolving data formats incompatibilities

3.1. The RDO format

The currently used format, in the MC production, as indicated in Figure 1, is the Raw Data Objects (RDO) file format, which is a container format based on ROOT[2] technologies. The RDO files contain as payload structured serialized data objects. The format has the disadvantage with respect to the previously mentioned challenges, that the structure and the content is rapidly evolving in time in response to the demand of more precise detector response simulation and/or addition of new components of the ATLAS detector. While compatibility for reading older files are usually provided, it is not guaranteed, and the complex nature of the format makes it difficult to provide converters for forwards compatibility. Attempts to make such conversions, to allow only a year old trigger simulation to read a modern RDO, were unsuccessful.

3.2. The raw detector format

Another, more native data format exists: the raw detector data format, byte stream[3]. The byte stream format is a chunk-based data format with support for versioning of both the format itself and the data content.

It is already a requirement from ATLAS that the format is kept backwards compatible so that data from any data taking period remain readable. Due to the tight coupling between the byte stream format and thef detector readout, the format is expected to evolve slowly. Further, while not necessarily maintained or kept operational, the code required for the sub-detectors to write old byte stream data already exists. Some sub-detectors of ATLAS already support configuration of the byte stream payload format. Further again, data objects corresponding to detector data are serializable to byte stream (as well as RDO) in order to properly simulate the conditions during online data taking.

The downside of using the byte stream format for simulation is that it inherently only deals with the detector data: MC truth information and other necessary (meta)data are not storable.

3.3. Altering the simulation chain
The trigger response simulation only needs the detector related data of the RDO as input. It adds the trigger response leaving detector data intact. A modified MC production chain, using an intermediate byte stream file, has been developed as shown in Figure 2. Two additional steps are added to the simulation chain. A step that extracts the detector data from the RDO and stores it to the intermediate byte stream file. And the step, after the trigger simulation, which adds the trigger response record to the byte stream file. The trigger response record is read from the byte stream file and added to the RDO. The final RDO file then serves as input to the reconstruction step. These extra steps were implemented within the framework of ATLAS and work has been done to allow for a seamless integration with the existing MC production code. A proof-of-concept setup was rigged and it was shown that the modified simulation chain was viable for all eight tested software versions.

4. Virtualization and data preservation
Virtualization technology solves the before mentioned challenges related to time evolution by allowing effective simulation of older hardware and thus enables a complete encapsulation of operating systems and software.

The use of virtualization, on the other hand, has its down sides: it inherently introduces an overhead. Further, a virtual machine (VM) is for all practical purposes a machine, and not a simple script. Introduction of additional steps running inside of the virtual machine is a major change to the ATLAS MC simulation chain. This implies a major change to the simulation. In addition to the old version of the trigger software, the full machine definitions need to be maintained and kept functional.

Both aspects were addressed in the study using CERN's own VM project, CernVM[4], and its associated technologies.

4.1. The CernVM project
The CernVM philosophy is to have a minimal disk image, which through methods of contextualization, can be configured for use with any version of the experiments software. Disk images are of the order of hundreds of megabytes instead of the several gigabytes of conventional VM disk images. The contextualization information are text files of no more than a few kB. The contextualization and disk inflation then happens when the machine is started.

In order to provide the experimental software, the CernVM images ship with its own HTTP based, locally cache-able, strictly versioned, read-only file system, cvmfs[5]. ATLAS already publishes its software releases and conditions data onto cvmfs file servers.

4.2. Proof of concept implementation
Using libvirt[6], an API for platform virtualization management, with KVM (Linux Kernel-based Virtual Machine)[7] a contextualization was developed building on top of the HEPPIX contextualization method supported by CernVM. The contextualization method, which is mainly intended for grid sites to adjust the raw image to local site policies, was used to tamper enough with the systems initialization process as to make a script that would:

- create and start a fresh CernVM instance,
- contextualize it appropriately,
- have it setup ATLAS release software from cvmfs,
- execute a script with user specified parameters, i.e. the trigger simulation,
- ship out results and log files over xrootd[2],
- and finally dispose of the machine when done.

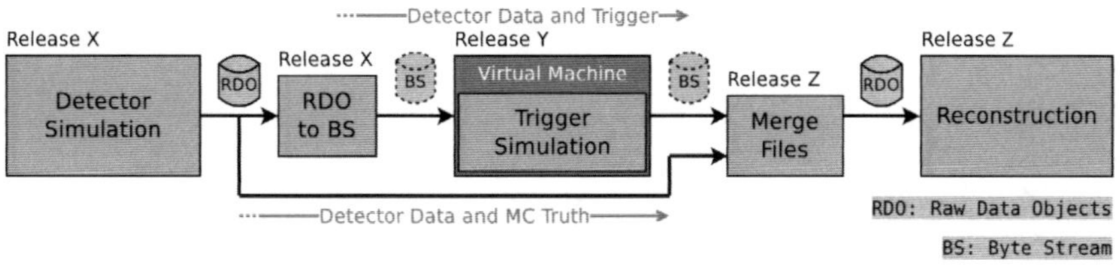

Figure 3. The modified simulation chain where the trigger simulation has been encapsulated in a virtual machine. This setup was used for the proof-of-concept model.

The methods used are not suitable for production since, in the production environment, the software is already being executed on virtual machines. As the chosen infrastructure can not be steered dynamically in the ways described here, this method may lead to a highly undesirable "Russian Doll" where a new VM is created inside a VM. Providing a better solution for dynamic VM allocation would require major changes to existing infrastructure. While not suitable for production the method allowed for rapid integration of virtualization technologies into the existing simulation chain.

At the time of writing it is possible to run the simulation chain depicted in Figure 3 using the proof of concept model on a physical computer. Nevertheless, in order to benefit fully from the CernVM project, better contextualization methods are needed. Requirements for workable methods for setting up a specific ATLAS software release at contextualization as well as elegant ways of providing certificates and credentials for accessing storage services and databases have been fed back to the CernVM project. Further investigation of how to integrate virtualization as seamlessly as possible into the MC production is also needed.

5. Conclusions

It has been demonstrated how to simulate the trigger precisely by using older software versions for the trigger simulation while maintaining modern software versions for detector simulation and reconstruction.

This was achieved by using a more stable data format tightly coupled to the detector hardware. The proposed model introduces only moderate maintenance effort.

Virtualization solutions seem tractable even when software and hardware change radically. The CernVM project already fulfills most of the requirements but more work is needed before production scalability can be assessed.

References

[1] ATLAS Collaboration 2008 The ATLAS experiment at the CERN large hadron collider *JINST* **3** S08003
[2] CERN ROOT Team 2013 ROOT An Object-Oriented Data Analysis Framework `http://root.cern.ch/download/doc/ROOTUsersGuideA4.pdf`
[3] Anjos A d, Beck H P, Gorini B, Vandelli W 2011 The raw event format in the ATLAS Trigger & DAQ *EDMS* 445840
[4] Segal B *et al* 2010 LHC Cloud Computing with CernVM *PoS* **ACAT(2010)** 004
[5] Blomer J *et al* 2011 Distributing LHC application software and conditions databases using the CernVM file system *J. Phys.: Conf. Series* **331** 042003
[6] Red Hat 2013 libvirt: The virtualization API `http://libvirt.org/`
[7] Red Hat 2013 KVM: Kernel Based Virtual Machine `http://linux-kvm.org/`

ACAT2013 IOP Publishing
Journal of Physics: Conference Series **523** (2014) 012032 doi:10.1088/1742-6596/523/1/012032

MadAnalysis 5: status and new developments

Eric Conte[1] **and Benjamin Fuks**[2,3]

(1) Groupe de Recherche de Physique des Hautes Énergies (GRPHE), Université de Haute-Alsace, IUT Colmar, 34 rue du Grillenbreit BP 50568, 68008 Colmar Cedex, France
(2) Theory Division, Physics Department, CERN, CH-1211 Geneva 23, Switzerland
(3) Institut Pluridisciplinaire Hubert Curien/Département Recherches Subatomiques, Université de Strasbourg/CNRS-IN2P3, 23 Rue du Loess, F-67037 Strasbourg, France

E-mail: eric.conte@iphc.cnrs.fr, benjamin.fuks@iphc.cnrs.fr

Abstract. MadAnalysis 5 is a new Python/C++ package facilitating phenomenological analyses that can be performed in the framework of Monte Carlo simulations of collisions to be produced in high-energy physics experiments. It allows one, by means of a user-friendly interpreter, to perform professional physics analyses in a very simple way. Starting from event samples as generated by any Monte Carlo event generator, large classes of selections can be implemented through intuitive commands, many standard kinematical distributions can be automatically represented by histograms and all results are eventually gathered into detailed Html and LaTeX reports. In this work, we briefly report on the latest developments of the code, focusing on the interface to the FastJet program dedicated to jet reconstruction.

1. Introduction

The TeV energy regime is currently being explored at the Large Hadron Collider, at CERN, and searches for new phenomena are a central topic of the associated experimental program. However, as a consequence of the absence of hints for new physics, the states predicted by any theory beyond the Standard Model (BSM) are bound to lie at higher and higher scales. While experimental analyses mostly focus on popular BSM theories and on simplified extensions of the Standard Model (SM), (re)interpretations of the results in any specific theoretical framework and phenomenological investigations of new types of signatures play key roles in the improvement of our understanding of the TeV scale. To this aim, matrix-element-based Monte Carlo event generators are generally employed, the results being matched to parton showering, hadronization and detector simulation algorithms. Phenomenological analyses can then be performed at several levels, from a purely partonic description of the simulated event final states to a more complex description including hadronization and possibly a fast simulation of the detector response. The MadAnalysis 5 package [1] offers a unique framework to perform such analyses, in an easy and efficient way. While about to be fully embedded within the Monte Carlo event generators MadGraph 5 [2] and aMC@NLO [3], MadAnalysis 5 can also be used as a standalone program and outside the MadGraph environment.

In this paper, we describe some of the novel features of the version 1.1.8 of the code and provide an update of the original user guide. After a brief overview of MadAnalysis 5 in Section 2, we illustrate how to use it to design a phenomenological analysis in Section 3. Section 4 is dedicated to a recent module added the code, related to jet reconstruction and allowing one to verify the merging procedures usually employed to combine event samples describing the same physics process but with different final state jet multiplicities. Our conclusions are presented in Section 5.

2. Overview

2.1. MadAnalysis 5 *in a nutshell*

MadAnalysis 5 is a framework designed to facilitate the implementation of phenomenological studies based on Monte Carlo simulations of collisions describing both new physics signal and Standard Model background processes. It allows one, on the one hand, to investigate the characteristics of a particular process, and, on the other hand, to carry out an event selection aiming to extract a given signal from the background. It supports event files encoded under the StdHep [4], HepMc [5], Lhe [6, 7], Lhco [8] and Delphes Root [9] formats.

Properties of the event final states can be probed by means of a small set of intuitive Python commands (see below) acting on particles modeled by user-defined labels and several observables. They can then be represented by histograms whose layout can be tuned according to the preferences of the user. In the same way, event selection criteria can be implemented via dedicated commands and equally applied to a full event (*i.e.*, to select or reject it under a given condition) or to a particle (*i.e.*, to include it or not in the analysis under a specific condition). MadAnalysis 5 can also further compute signal over background ratios whose definition can be adjusted by the user.

Flexible, MadAnalysis 5 also offers sophisticated functionalities allowing one to, *e.g.*, combine particle four-momenta (sums, differences, *etc.*) before the computation of any observable, use the event history (mother-to-daughter relations linking a particle to its decay products) or identify a given particle according to its hardness. All the results are collected into reports available under the Html, PostScript (associated with latex) or Pdf (associated with pdflatex) formats. Interfaced to the FastJet package [10], MadAnalysis 5 also includes several jet-clustering algorithms that can be used for object reconstruction, including or not a simplified detector simulation and several (mis)identification efficiencies.

2.2. *Python console*

Once downloaded from the Internet (see Ref. [11]) and unpacked, the MadAnalysis 5 package does not require any compilation and can be started by issuing

```
bin/ma5
```

from the directory where it has been stored. By default, MadAnalysis 5 is configured for the analysis of events at the parton level, although events at the hadron or at the reconstructed level can be analyzed by launching the program with the `-H` and `-R` arguments, respectively.

The user interacts with MadAnalysis 5 through a console implemented in the Python language that invites him/her to define his/her analysis through a set of commands that can also possibly be collected into a script. The MadAnalysis 5 metalanguage is designed to be simple and concise, as illustrated in the examples presented in the next sections, and the user can be helped in two ways. First, in-line help can be employed by issuing

```
ma5> help
```

in the console, which displays the list of available commands. Second, auto-completion has been implemented and can be used via the *tab* key.

2.3. *Structure of the program*

Particle labels can be either directly defined in the Python console or imported from a Universal FeynRules Output (UFO) model containing names and labels for all the particles of a given model [12]. Additionally, the list of event samples to analyze has to be specified.

MADANALYSIS 5 then takes care of automatically exporting the analysis under the form of a C++ code to be linked to a library called SAMPLEANALYZER, shipped with the package, and to any other optional tools such as FASTJET or DELPHES according to the needs of the user. The C++ program is further compiled and executed, and the results are encapsulated into an XML structure stored in a so-called SAF (standing for SAMPLEANALYZER Format) file. The PYTHON console is eventually automatically taking care of rendering the information human-readable, under the form of POSTSCRIPT, PDF and HTML reports including the defined histograms and the computed selection efficiencies. The full code structure is summarized on Figure 1.

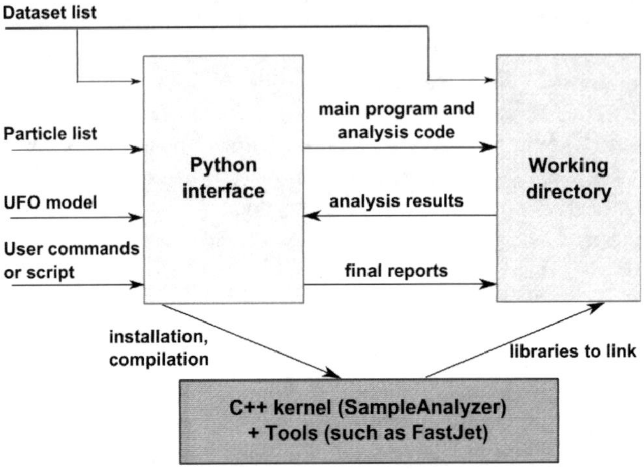

Figure 1. Logical structure of MADANALYSIS 5

On different footings, users can also directly implement their analysis in the C++ language, bypassing the PYTHON console and using all built-in functions available within the SAMPLEANALYZER library. This way of using MADANALYSIS 5 is called the *expert mode* of running, in contrast to the *normal mode* using the PYTHON console, and allows one to implement any analysis which a user can dream of, the only limitation being his/her imagination and programming skills.

2.4. Requirements

When started, MADANALYSIS 5 firstly checks the presence on the system of required dependencies and secondly verifies which of optional libraries are available. The list of the mandatory packages to run properly the program is the following :

- PYTHON 2.6 or a more recent version (although not a 3.x version),
- the GNU GCC compiler together with the associated Standard Template Libraries,
- the ROOT package version 5.27 or more recent [13].

Some specific functionalities of MADANALYSIS 5 may require the use of other external packages and are thus deactivated if necessary. Those consist of

- the LATEX and PDFLATEX compilers (analysis reports in the POSTSCRIPT and PDF formats, respectively),
- the ZLIB package (reading and writing compressed files),
- the FASTJET package (to switch on the module dedicated to jet reconstruction, see Section 4),

- the DELPHES package (reading ROOT files as outputted by DELPHES 3).
- a few representative event samples useful to learn how to use MADANALYSIS 5.

With the exception of the LATEX and PDFLATEX compilers (that must be installed independently by the user), each optional package that MADANALYSIS 5 may require can be installed automatically by using the command `install` from the PYTHON console. For instance, typing in the commands

```
ma5> install gzip
ma5> install delphes
ma5> install fastjet
ma5> install samples
```

leads to the installation of the ZLIB libraries, the FASTJET and DELPHES 3 programs and some Monte Carlo event samples, respectively.

3. Designing an analysis with MadAnalysis 5

In this section, we briefly describe the way to perform a simple analysis with MADANALYSIS 5. More advanced functionalities as well as the expert mode of the code are not discussed and we refer to the manual for more details [1].

3.1. Basic concepts

First, the event samples that must be read and analyzed are specified through the command `import`. By default, all imported samples are collected within a single dataset called `defaultset`, although the user can define as many datasets as necessary by using the command `import` together with the keyword `as`. In general, it is recommended to have one single dataset for each physics process. The wildcard characters `*` and `?` can also be used to import several files simultaneously.

Second, in order to treat the different particle species all along the analysis, MADANALYSIS 5 employs labels referring either to one given type of particle or to several types of particles (the term of multiparticle is used in this case). When MADANALYSIS 5 starts, collections of standard labels for the SM and Minimal Supersymmetric Standard Model particles are loaded. The user has then the possibility to define his/her own labels by the command `define`.

Third, histograms representing either properties of the full event (*e.g.*, the transverse missing energy or the transverse hadronic energy) or properties of a given particle assumed to be part of the event final state content (*e.g.*, its transverse momentum or its multiplicity) can be created by using the command `plot`, followed by the name of the relevant observable (see the manual for more details on the syntax [1]).

Finally, the list of observables that can be represented by histograms can also be used to apply selection requirements either on the events themselves, or to select the final state objects to be considered in the analysis. To this aim, the user can employ two complementary commands, `reject` and `select`, preceding a criterion.

3.2. Illustrative example

We provide, for the sake of the example, a toy analysis based on the events samples provided by installing the optional package SAMPLES. The analysis commands are given by

```
(1) ma5> import  samples/zz.lhe.gz  as  mySignal
(2) ma5> import  samples/ttbar_sl_?.lhe.gz  as  myBkg
(3) ma5> set  myBkg.type  =  background
(4) ma5> define  mu  =  mu+  mu-
(5) ma5> plot  MET
```

```
(6) ma5> plot PT(mu) [logY]
(7) ma5> reject  (mu)  PT  <  20
(8) ma5> select  80  <  M  (mu+  mu-)  <  100
(9) ma5> plot  M(mu+  mu-)
```

On the first line, a dataset called `mySignal` is created and linked to the event file `zz.lhe.gz`. On the second line, one collects the two samples `ttbar_sl_1.lhe.gz` and `ttbar_sl_2.lhe.gz` into a unique dataset named `myBkg`, which is further defined as containing background events on the third line. In contrast, the type of the `mySignal` dataset is left to its default value, *i.e.*, the related events being considered as signal events.

A new multiparticle label is defined on the fourth line by making use of the command `define` and is associated with both muons and antimuons. This label is then used in the rest of the analysis to define selections and histograms. Two histograms are hence defined on the fifth and sixth lines of the analysis, the first one representing the missing transverse energy distribution extracted from each dataset and the second one describing the transverse momentum spectrum obtained after considering all muons and antimuons of each event. By the commands of the seventh and eighth lines, we demand to consider, in the rest of the analysis, all (anti)muons with a transverse momentum smaller than 20 GeV and to keep events containing at least one muon-antimuon pair with an invariant mass included in the range $[80, 100]$ GeV. A final histogram is created on the last line of the analysis and allows one to study the effect of the previous selections on the invariant mass of all muon-antimuon pairs that can be formed in each event.

3.3. Submitting and executing the analysis

The user can execute his/her analysis by typing in, in the PYTHON console, the command

```
submit
```

Behind the scenes, a C++ code corresponding to the implemented analysis is generated, compiled and linked to the SAMPLEANALYZER library. The executable is then run over all datasets and the results are displayed via the different reports. The HTML report can always be open directly from the PYTHON interface by means of the command

```
open
```

A representative histogram that can be found in the report associated with the analysis above is shown in Figure 2.

4. Interface to FastJet package

In order to enable the functionalities of MADANALYSIS 5 related to jet clustering, the FASTJET package has to be installed on the system of the user, either externally to MADANALYSIS 5 or within the program (see Section 2.4).

4.1. Reconstructing jets in hadron-level events

Starting from an event file at the hadron level (in general encoded in the STDHEP or HEPMC file format), MADANALYSIS 5 offers the user a straightforward way to gather the hundreds of hadrons usually present in the event final states into jets by means of any of the algorithms included in the FASTJET library. After this procedure, the events are described in terms of jets, electrons, muons, taus, photons and missing transverse energy. While leptonically-decaying taus are accounted for in terms of their decay products (electrons or muons and missing energy), only hadronically-decaying taus are considered as a distinct physics objects (and defined as taus).

In order to enable jet clustering within MADANALYSIS 5 (assumed to be started in the reconstructed-level mode), it is sufficient to specify the algorithm to be used as the value of the `main.clustering.algorithm` variable. The allowed choices are `kt` (longitudinally-invariant

Dataset	Integral	Entries / events	Mean	RMS	%Underflow	%Overflow
mysignal	256	2.02	48.3857	32.11	0.0	0.0
mybkg	75706	0.493	49.6728	34.55	0.0	0.0

Statistics table

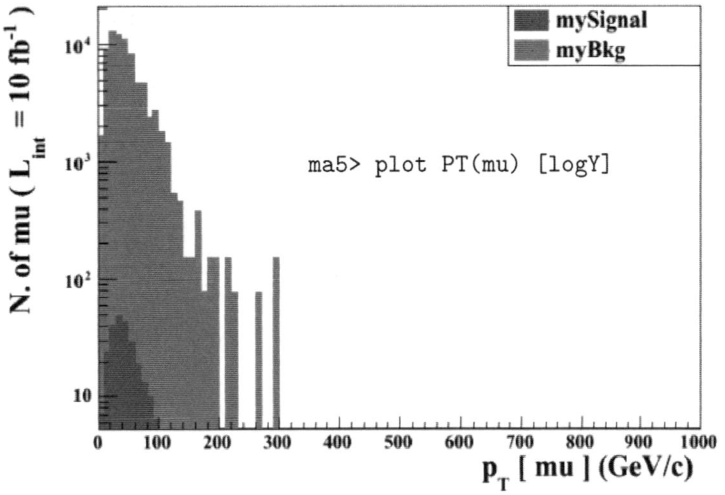

Figure 2. Histogram representing the transverse momentum spectra of all (anti)muons for the two datasets denoted by `mySignal` and `myBkg`. The command associated with the histogram generation is superimposed to the figure.

k_T algorithm [14, 15]), `cambridge` (the Cambridge/Aachen algorithm [16, 17]), `antikt` (the anti-k_T algorithm [18]), `genkt` (the generalized k_T algorithm [10]), `siscone` (the SisCone algorithm [19]), `cdfmidpoint` (the CDF MidPoint algorithm [20]), `cdfalgo` (the CDF JetClu algorithm [21]) and `gridjet` (the GridJet algorithm [10]). For instance, employing the anti-k_T algorithm can be enforced by typing

```
ma5> set main.clustering.algo = antikt
```

As soon as a given jet algorithm is selected, new options depending on the algorithm parameters become available. In the (illustrative) example of the anti-k_T algorithm, the variables `main.clustering.radius` and `main.clustering.ptmin`, respectively representing the jet-radius parameter and the jet transverse-momentum threshold, can be set by the user.

We now turn to the description of all other options coming with the jet clustering module of MadAnalysis 5 and that are related to the identification of the physics objects. After the application of the algorithm, each particle is uniquely identified and cannot contribute to several collections of physics objects. A remark on leptons is however in order here. By default, the jet algorithm distinguishes leptons issued from the hard process from those originating from hadron decays. While the first category of leptons are reconstructed as leptons and included in the relevant collection, leptons of the second category are feeding the jet algorithm and therefore included as parts of the reconstructed jets. This default behavior can be modified by setting the variable `main.clustering.exclusive` to `false`. This forces the jet algorithm to include all visible particles for jet reconstruction, so that a given lepton can consequently be part of the relevant lepton collection and also contribute to a specific jet.

MadAnalysis also includes a simple algorithm dedicated to the b-tagging of the reconstructed jets, so that a jet is identified as originating from a b-quark when it can be matched to a b (anti)quark, *i.e.*, when it lies within a cone of radius R

around one of the parton-level b quarks. The size of the cone (R) is specified as the value of the `main.clustering.bjet_id.matching_dr` variable, while setting the parameter `main.clustering.bjet_id.exclusive` to `true` indicates the code that one cannot match several jets to one given b quark. By default, the b-tagging algorithm is perfect, although (mis)identification efficiencies can also be specified. For example, typing the three commands

```
ma5> set main.clustering.bjet_id.efficiency = 0.60
ma5> set main.clustering.bjet_id.misid_cjet = 0.10
ma5> set main.clustering.bjet_id.misid_ljet = 0.05
```

first sets the b-tagging efficiency to 60%. Next, the two other commands teach the b-tagging algorithm that c-quark-initiated jets and other lighter jets are tagged as b-jets in respectively 10% and 5% of the cases. If more advanced b-tagging methods are required by the user, such as implementing a tagging algorithm depending on the jet transverse momentum, he/she is recommended to use instead a program simulating the entire detector response such as DELPHES [9]. Furthermore, similar options are available for hadronically-decaying taus.

4.2. Storing jet-clustered events

Events obtained after jet clustering can be stored in a file whose name is specified by the user as the value of the variable `main.outputfile`. For instance,

```
ma5> set main.outputfile = "myOutput.lhe"
```

allows one to save the reconstructed events in a file named `myOutput.lhe`. The file format is directly specified by the file extension, the suitable choices being either `lhe` (LHE format) or `lhco` (LHCO format). Compressed files can also be used if the optional ZLIB libraries are installed, the related file extensions being in this case `lhe.gz` and `lhco.gz`. The files created by MADANALYSIS 5 are however not strictly following the LHE and LHCO conventions originally defined in Refs. [6, 7, 8].

- Some variables associated with an LHCO event, such as the calorimeter energy deposits, are ignored by MADANALYSIS 5.
- In the case of LHE events, initial state particles are recorded with a status code set to -1, final state particles with a status code equal to 1 and intermediate particles involved in the hard process with a status code set to 3. Other intermediate particles, such as those generated by parton showering, are omitted from the output file. Each reconstructed physics object, *i.e.*, each of the light jets, b-tagged jets, electrons, muons, hadronically-decaying taus, photons as well as the missing transverse energy, is identified with a specific code inspired by the Particle Data Group (PDG) particle numbering scheme, chosen to be 21, ± 5, ± 11, ± 13, ± 15, 22 and 12, respectively. Finally, the link between an intermediate particle of the hard process and the induced final state objects is saved by means of the associated mother-to-daughter relations as they can be encoded in the LHE format.

We refer to the header of the outputted files for more details.

4.3. Validating multiparton matrix-element merging

In order to achieve an accurate description of the jet properties of a given event sample, two complementary approaches are usually combined. Matrix elements are employed for the description of the hard jets and parton showering for the one of the soft and collinear jets. Consequently, a double counting of the QCD emission is possible as radiation can now emerge from two sources. Several merging procedures have been subsequently developed in the last decades. MADANALYSIS 5 offers a way, based on its interface with FASTJET, to validate the resulting *merged* samples. This relies on a check of the smoothness of the differential jet rate

(DJR) distributions, the DJR observables strongly depending on the merging parameters and procedure as they describe the distribution of the scale at which the jet multiplicity of an event final state switches from N to $N+1$. The DJR histograms can be included in the MadAnalysis 5 reports by typing

```
ma5> set main.merging.check = true
```

5. Summary

MadAnalysis 5 is a unique, flexible and user-friendly framework allowing for professional analyses of simulations of collisions to be produced in high-energy physics experiments. It can be used equivalently for the investigation of parton-level, hadron-level or reconstructed events and is therefore able to read many input formats. The program comes with two modes of running, the so-called normal and expert modes. In the first case, the user is invited to define his/her analysis by typing command lines in a Python console. In the second case, he/she can directly implement his/her analysis in C++, using a large class of predefined functions included in the core of the package.

In this paper, we have emphasized new functionalities related to an interface with the FastJet program and which include the possibility of using different jet algorithms to cluster hadrons into jets, when analyzing hadron-level events. In addition, extra features have been developed and allow to mimic the response of an ideal detector and to output the reconstructed events into a simplified Lhco or Lhe format. Moreover, MadAnalysis 5 is now also capable to generate DJR distributions allowing for the validation of the jet merging procedure traditionally employed in state-of-the-art phenomenological (and also in some experimental) analyses.

Acknowledgments

The authors are grateful to the ACAT organizers for the very nice conference. They also thank A. Alloul for his numerous checks of the code and for his help during the development and the validation of the later releases. This work has been supported by the French ANR 12 JS05 002 01 BATS@LHC and by the Theory-LHC France initiative of the CNRS/IN2P3.

References

[1] Conte E, Fuks B and Serret G 2013 *Comput.Phys.Commun.* **184** 222–256 (*Preprint* 1206.1599)
[2] Alwall J, Herquet M, Maltoni F, Mattelaer O and Stelzer T 2011 *JHEP* **1106** 128 (*Preprint* 1106.0522)
[3] Alwall J, Frederix R, Frixione S, Hirschi V, Maltoni F, Mattelaer O, Torrielli P and Zaro M *in preparation*
[4] http://cepa.fnal.gov/psm/stdhep/c++/
[5] Dobbs M and Hansen J B 2001 *Computer Physics Communications* **134** 41 – 46
[6] Boos E, Dobbs M, Giele W, Hinchliffe I, Huston J *et al.* 2001 (*Preprint* hep-ph/0109068)
[7] Alwall J, Ballestrero A, Bartalini P, Belov S, Boos E *et al.* 2007 *Comput.Phys.Commun.* **176** 300–304 (*Preprint* hep-ph/0609017)
[8] http://www.jthaler.net/olympicswiki/
[9] de Favereau J, Delaere C, Demin P, Giammanco A, Lemaitre V *et al.* 2013 (*Preprint* 1307.6346)
[10] Cacciari M, Salam G P and Soyez G 2012 *Eur.Phys.J.* C **72** 1896 (*Preprint* 1111.6097)
[11] https://launchpad.net/madanalysis5
[12] Degrande C, Duhr C, Fuks B, Grellscheid D, Mattelaer O *et al.* 2012 *Comput.Phys.Commun.* **183** 1201–1214 (*Preprint* 1108.2040)
[13] Brun R and Rademakers F 1997 *Nucl.Instrum.Meth.* A **389** 81–86
[14] Catani S, Dokshitzer Y L, Seymour M and Webber B 1993 *Nucl.Phys.* B **406** 187–224
[15] Ellis S D and Soper D E 1993 *Phys. Rev. D* **48**(7) 3160–3166
[16] Dokshitzer Y, Leder G, Moretti S and Webber B 1997 *Journal of High Energy Physics* **1997** 001
[17] Wobisch M and Wengler T 1998 (*Preprint* hep-ph/9907280)
[18] Cacciari M, Salam G P and Soyez G 2008 *JHEP* **0804** 063 (*Preprint* 0802.1189)
[19] Salam G P and Soyez G 2007 *JHEP* **0705** 086 (*Preprint* 0704.0292)
[20] Blazey G C, Dittmann J R, Ellis S D, Elvira V D, Frame K *et al.* 2000 47–77 (*Preprint* hep-ex/0005012)
[21] Abe F *et al.* (CDF Collaboration) 1992 *Phys.Rev.* D **45** 1448–1458

DELPHES 3: A modular framework for fast-simulation of generic collider experiments

Michele Selvaggi
Centre for Cosmology, Particle Physics and Phenomenology (CP3),
Université Catholique de Louvain,
Chemin du Cyclotron 2, B-1348 Louvain-la-Neuve, Belgium
E-mail: michele.selvaggi@uclouvain.be

Abstract. The new version of the DELPHES C++ fast-simulation framework is presented. The tool is written in C++ and is interfaced with the most common Monte Carlo file formats (LHEF, HepMC, STDHEP). Its purpose is the simulation of a multipurpose detector response, which includes a track propagation system embedded in a magnetic field, electromagnetic and hadronic calorimeters, and a muon system. The new modular version allows to easily produce the collections that are needed for later analysis, from low level objects such as tracks and calorimeter deposits up to high level collections such as isolated electrons, jets, taus, and missing energy.

1. Introduction

Sophisticated detectors are designed in order to detect and precisely measure particles originating from high energy collisions. For most phenomenological studies, a complete detector description is often not needed and a simplified approach based on the parametrisation of the detector response is in general good enough. In 2009, the DELPHES framework was designed to achieve such a goal [1].

DELPHES takes as input the most common event generator output data-formats and performs a fast and realistic simulation of a general purpose collider detector. To do so, long-lived particles emerging from the hard scattering are propagated to the calorimeters within a uniform magnetic field. Photons and charged leptons final observables are computed by smearing the initial momenta according to the resolution of the relevant sub-detectors. High-level reconstructed quantities such as jets and missing energy can be computed either starting from simple calorimeter deposits, or with the so-called particle-flow algorithm.

With respect to its predecessor [1], the present DELPHES version includes a more accurate particle-flow description allowing to combine and optimally use the information of all sub-detectors. This approach is particularly suitable to the treatment of pile-up, which has also been included in DELPHES 3.0. From a technical perspective, the code structure is now fully modular, providing a greater flexibility to the user. Finally, an event display has been added to the DELPHES package.

2. Software implementation

2.1. Data-Flow

DELPHES is a modular framework written in C++ and is based on the ROOT analysis framework [2]. It accesses data from different file formats HEPMC [3], STDHEP [4] and the LesHouches event format (LHEF) [5]). Event files coming from external Monte-Carlo generators are first processed by a Reader. The Reader converts stable particles into a collection of universal objects. This collection is then processed by a series of modules. The jet clustering procedure is performed by means of the FastJet package [6]. Finally, DELPHES allows to store and analyze events in a ROOT tree format [2]. The modular system allows to configure and schedule modules via a configuration file, add modules, change data-flow, alter output information.

2.2. Technical Performance

The main motivation for a tool like DELPHES is to minimize the resources needed on top of those used for event generation: small memory footprint, efficient usage of CPU and small file size.

Figure 1 illustrates how well DELPHES achieves these goals. Memory usage does not exceed a few hundreds of megabytes and remains constant after the initial memory allocation. Processing time can be as low as a few milliseconds and is expected to follow the scaling law of the underlying jet reconstruction algorithm.

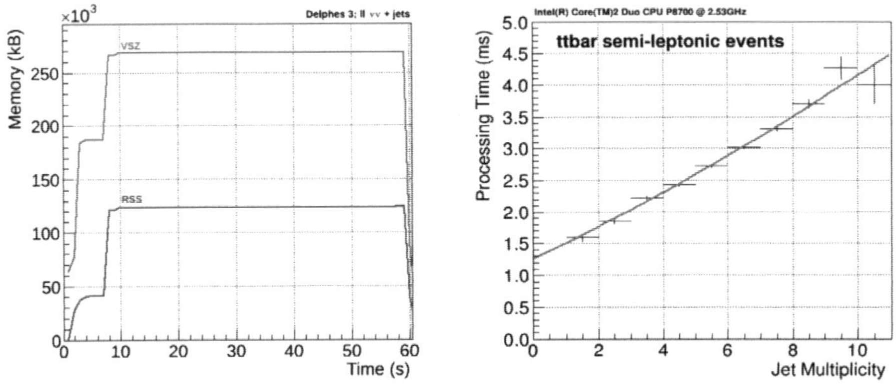

Figure 1. Left: memory usage as function of time after the program start. Right: processing time per event as function of jet multiplicity on inclusive $t\bar{t}$ events.

3. New features

3.1. Particle-Flow

The philosophy of the particle-flow approach is to maximally make use of the information provided by the various sub-detectors for reconstructing the event. This modus operandi is adopted by several collaborations (see for example [7]) but intrinsically depends on the specificity of the experimental device. In DELPHES, we opted for a simplified approach based on the tracking system and the calorimeters for implementing the particle-flow event reconstruction. If the momentum resolution of the tracking system is higher than the energy resolution of calorimeters, it can be convenient to use the tracking information within the tracker acceptance for estimating the charged particles momenta. The particle-flow algorithm produces two collections of 4-vectors — particle-flow tracks and particle-flow towers — that can serve later as input for reconstructing jets and missing transverse energy with a higher resolution.

The comparison of the performance achieved with DELPHES fast-simulation tool against the CMS experiment full-simulation [7] is shown in Figure 2. Jets were clustered with the help of the anti-k_T algorithm [8] and the events were generated with MADGRAPH5 [9]. The comparison is shown for quantities that have been reconstructed using calorimeter information only, and using the particle-flow approach. The resolutions obtained with DELPHES and the CMS full-simulation show a good agreement.

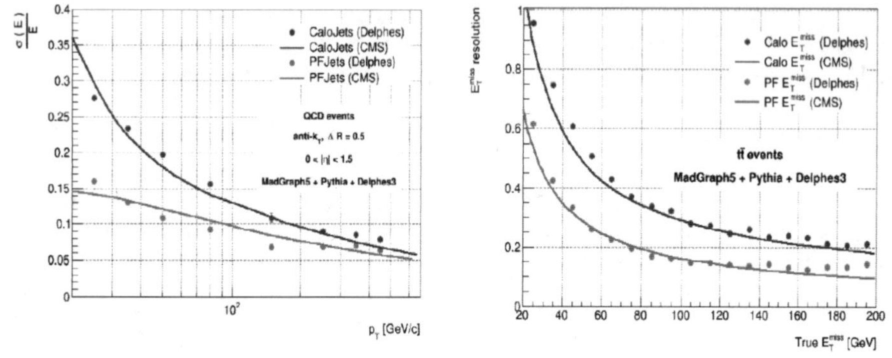

Figure 2. Comparison of the performance achieved with DELPHES fast-simulation tool against the CMS experiment full-simulation. Left: Jet energy resolution as a function of the jet p_T in QCD events. Right: Missing transverse energy resolution as a function a of the true event missing transverse energy in top-pair events.

3.2. Pile-up simulation

At the LHC, several collisions per bunch-crossing will occur in high luminosity conditions. Such additional *pile-up events*, can have the effect of deteriorating the detector response. In DELPHES, pile-up interactions are extracted from a pre-generated low-Q^2 QCD sample. These minimum-bias interactions are randomly placed along the beam axis according to some longitudinal spread that can be set by the user (see Figure 3). The actual number of pile-up interactions per bunch-crossing is randomly extracted from a Poisson distribution.

Pile-up directly affects the performance of jets, E_T^{miss} and isolation. In real experiments, pile-up interactions are identified by means of vertex reconstruction. If such interactions occur far enough from the hard interaction, a precise vertexing algorithm is able to detect them. Combining vertexing and tracking information allows to identify contaminating charged particles from pile-up. On the other hand, since neutral particles do not produce tracks, neutral pile-up contamination can only be estimated *on average*. In DELPHES, pile-up subtraction can be done in two steps:

3.2.1. Charged pile-up subtraction We assume that vertices corresponding to pile-up interactions occurring at a distance $|z| > Z_{vtx}$ can be reconstructed. The parameter Z_{vtx} is obviously related to the spatial vertex resolution of the detector. We assume that pile-up interactions occurring at $|z| < Z_{vtx}$ cannot be disentangled from those originating from the high-Q^2 process. Therefore every particle-flow track originating from such vertices cannot be subtracted from the event, while every particle-flow track originating from a vertex positioned at $|z| > Z_{vtx}$ can be removed from the event.

3.2.2. Residual pile-up subtraction Other techniques are needed in order to extract and remove residual contributions: these include particles that are too close to the hard interaction vertex to be identified as pile-up products with tracking information, charged particles that failed track reconstruction and neutral particles. In DELPHES we have opted for the Jet Area method [10, 11]. This approach, widely used in present collider experiments, allows to extract an average contamination density ρ on an event-by-event basis. The residual pile-up density ρ, can be used to correct observables that are sensitive to the residual contamination, mainly the jet energies and the isolation.

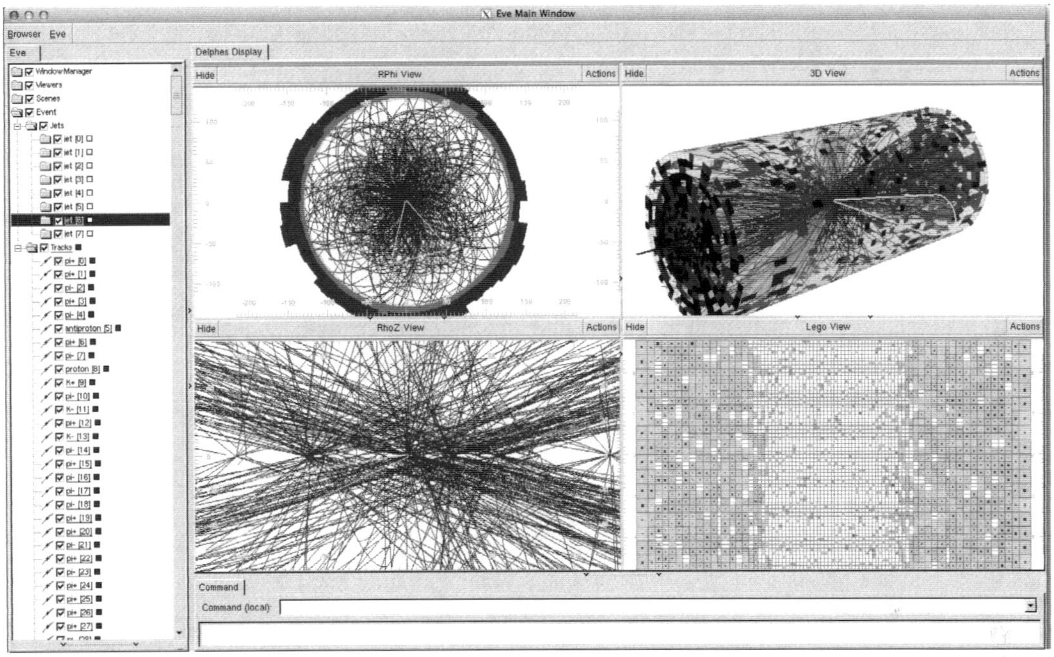

Figure 3. QCD event with 50 pile-up interactions shown with the DELPHES event display based on the ROOTEVE libraries [12]. Transverse view (top left), longitudinal view (bottom left), 3D view (top right), (η,ϕ) view (bottom right).

4. Conclusion

The version 3.0 of DELPHES, a framework designed to perform a fast and realistic simulation of a general purpose collider experiment, was presented. New features such as the particle-flow event reconstruction, and pile-up simulation have been included and validated.

Acknowledgments

First we wish to thank the ACAT 2013 workshop organizers for giving the opportunity to present the new version of DELPHES. We would like to thank Séverine Ovyn and Xavier Rouby for their invaluable contribution to the early DELPHES development. We wish to thank the DELPHES users for continuously providing feedback and deep insights. We acknowledge the support from the FNRS and the IAP Program, BELSPO VII/37. This work is partly supported by the IISN convention 4.4503.13.

Journal of Physics: Conference Series **523** (2014) 012033 doi:10.1088/1742-6596/523/1/012033

References

[1] Ovyn S, Rouby X and Lemaitre V 2009 DELPHES, a framework for fast simulation of a generic collider experiment *Preprint* hep-ph/09032225

[2] Antcheva I *et al* 2009 A C++ framework for petabyte data storage, statistical analysis and visualization *Comput. Phys. Commun.* **180** 2499

[3] Dobbs M *et al* 2001 *Comput. Phys. Commun.* **134** 41

[4] Garren L and Lebrun P 2006 StdHep User Manual `http://cepa.fnal.gov/psm/stdhep/`

[5] Alwall J *et al* 2007 A Standard format for Les Houches event files *Comput. Phys. Commun.* **176** 300

[6] Cacciari M, Salam G P and Soyez G 2012 FastJet User Manual *Eur. Phys. J.* C **72** 1896

[7] CMS Collaboration 2009 Particle-Flow Event Reconstruction in CMS and Performance for Jets, Taus, and MET CMS-PAS-PFT-09-001

[8] Cacciari M, Salam G P and Soyez G 2008 The Anti-k(t) jet clustering algorithm *JHEP* **0804** 063

[9] Alwall J, Herquet M, Maltoni F, Mattelaer O and Stelzer T 2011 MadGraph 5 : Going Beyond *JHEP* **1106** 128

[10] Cacciari M, Salam G P and Soyez G 2008 The Catchment Area of Jets *JHEP* **0804** 005

[11] Cacciari M, Salam G P and Soyez G 2008 Pileup subtraction using jet areas *Phys. Lett.* B **659** 119

[12] Tadel M 2008 Raw-data display and visual reconstruction validation in ALICE *J. Phys. Conf. Ser.* **119** 032036

Journal of Physics: Conference Series **523** (2014) 012034 doi:10.1088/1742-6596/523/1/012034

Event Reconstruction and Analysis in the R3BRoot Framework

Dmytro Kresan, Mohammad Al-Turany, Denis Bertini, Radosław Karabowicz, Anar Manafov, Alexey Rybalchenko and Florian Uhlig

GSI Helmholtzzentrum für Schwerionenforschung mbH, Planckstraße 1, D-64291, Darmstadt, Germany

E-mail: D.Kresan@gsi.de

Abstract. The R^3B experiment (Reaction studies with Relativistic Radioactive Beams) will be built within the future FAIR / GSI (Facility for Antiproton and Ion Research) in Darmstadt, Germany. The international collaboration R^3B has a scientific program devoted to the physics of stable and radioactive beams at energies between 150 MeV and 1.5 GeV per nucleon. In preparation for the experiment, the R3BRoot software framework is under development, it deliver detector simulation, reconstruction and data analysis. The basic functionalities of the framework are handled by the FairRoot framework which is used also by the other FAIR experiments (CBM, PANDA, ASYEOS, etc) while the R^3B detector specifics and reconstruction code are implemented inside R3BRoot. In this contribution first results of data analysis from the detector prototype test in November 2012 will be reported, moreover, comparison of the tracker performance versus experimental data, will be presented.

1. Introduction

The future R^3B (Reactions with Rare Radioactive Beams) experiment is one of the scientific pillars of the FAIR facility to be built in Darmstadt, Germany. It is organised as an international collaboration including about 50 institutes from all over the world. Series of experiments will perform reaction studies with exotic nuclei far off the stability region with the focus on nuclear structure and dynamics. Also astrophysical aspects and technical applications are considered as part of the scientific program.

The NeuLAND (New Large Area Neutron Detector) will be used in the R^3B setup for detection of fast neutrons, in particular for the classification of multi-neutron events ($\leq$6) with precise identification of a first interaction and determination of neutron momenta (in the range from 50 - 1000 MeV) by measurement of position and time of flight. For the summary on the detector concept of NeuLAND see [1, 2].

The aim of the work presented here was to achieve a high performance of the reconstruction algorithm with respect to efficiency, purity and resolution of the identification of such physics observable as, e.g., the so-called tetra-neutron (4 neutrons with a narrow distribution in space and time). The test beam experiment, performed at GSI in November 2012, has delivered first data taken with the full-size NeuLAND prototype. These results were compared to the simulation, leading to improvements in the reconstruction algorithm.

2. Simulation environment

The R^3B experiment is currently in the preparation phase, in which the detector prototypes are being tested. Thus the simulations of the detector performance and comparison versus experimental data is of a high priority. The R3BRoot software package is a branch of the FairRoot framework developed at GSI starting from 2004, which supports all future FAIR experiments with the simulation, reconstruction and data analysis infrastructure ([3]). It is based on ROOT ([4]) and VMC (Virtual Monte Carlo) interface, allowing easy switching between Geant3 [5] and Geant4 [6] transport engines. The task mechanism, inherited from ROOT, provides the possibility for user-friendly steering macros. The framework consists out of 1050 source files, containing 135000 lines of C++ code.

Figure 1 shows the visualisation of several R^3B modules, top-right corner - NeuLAND detector for the neutron measurement. Other detectors from left to right: CALIFA (gamma spectroscopy) with the target inside, super conducting GLAD magnet, Drift Chambers, and ToF wall.

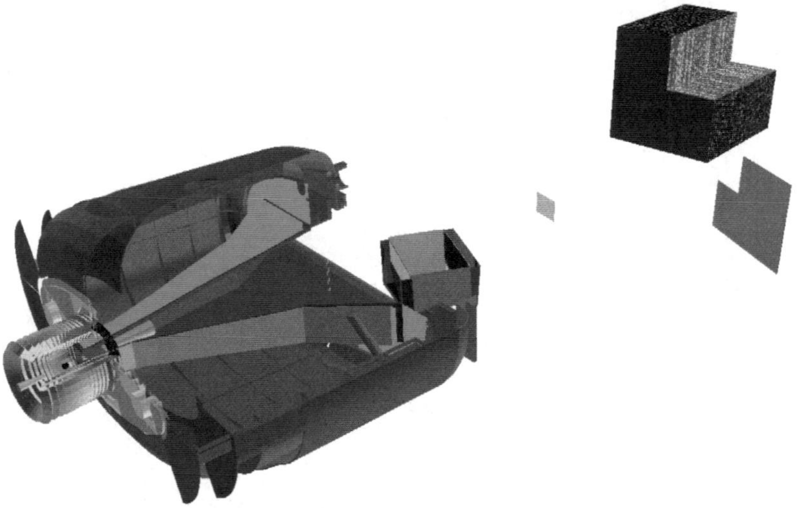

Figure 1. Visualisation of the R^3B detector geometry implemented in R3BRoot.

2.1. NeuLAND geometry

The neutron detector in R^3B will be constructed out of 3000 plastic scintillator bars (read-out from both ends with a single-channel photo-multiplier), organised in 30 double-planes (horizontal and vertical directions iteratively) implementing in this way a fully-active scintillator concept. The dimensions are 2.5 x 2.5 meters with a depth of 3 m. Depending on the specific requirements of the experiment with respect to acceptance and resolution, NeuLAND will be placed at different distances downstream of the target from 14 to 35 m. As input we have used reaction products of ^{132}Sn on hydrogen with stripping of 1 to 4 neutrons having relative energy of 100 and 500 keV. The transport of particles and light in the detector volume by Geant3 (using the GEANT-CALOR interface for neutron transport) was extended by applying light quenching. In the reconstruction we have considered 150 ps time resolution (experimental result from single bar tests), light attenuation and time decay.

3. Simulation results

The reconstruction procedure implies two steps: first the hit producer (digitiser) and second the neutron tracking. The hit producer task is taking Monte Carlo Points (results of transport calculations) as input and creates digis (the same data level as in the real experiment), applying light attenuation in the detector material and a light read-out from the two far ends of a scintillator bar by photomultipliers. The created digis contain information about time, charge and position and serve as input for the second task of the reconstruction algorithm.

3.1. Calorimetric properties

Before finding the first hits of the incident neutrons, the decision on the event class (amount of neutrons) has to be taken. It is done, by applying a set of pre-calibrated 2-D cuts on the values of the number of clusters (groups of digis representing the track of a charged particle, which are created by analysing digi pattern in an event with respect to position and time) and total energy deposit in the detector volume (see Figure 2).

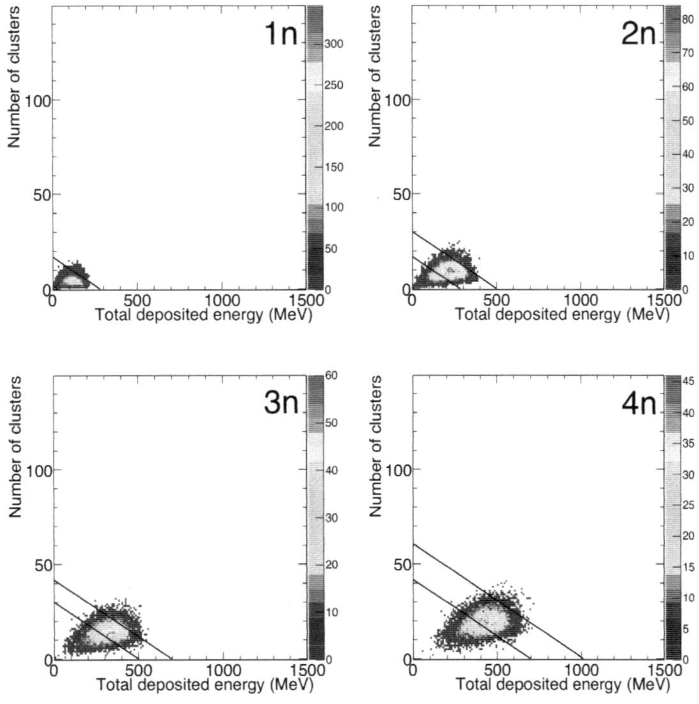

Figure 2. Distribution of number of clusters versus total deposited energy.

Simulation results in an anti-correlation between N_{CL} (vertical axis) and E_{TOT} (horisontal axis), which allows deriving number of impact neutrons with higher efficiency and purity than in case of 1-D cut on total energy only (efficiency for 4n case was increased from 40 to 60%, and contamination by 3n was dropped from 30 to 10%).

3.2. Neutron identification matrices

Table 1 lists the values for the multi-neutron recognition probabilities in %. Columns show generated number of neutrons, rows - detected by the reconstruction explained above. Diagonal elements in the table stand for efficiency, non-diagonal - for misidentification probabilities. Achieved values are in the region of the design goal and are typical for an astrophysical experiment.

These results were obtained for ^{132}Sn at 600 AMeV beam energy with 500 keV relative energy and NeuLAND detector placed at 14 m downstream the target.

Table 1. Neutron identification matrix for beam energy of 600 AMeV, and E_{rel}=500 keV, see subsection 3.2.

	1n	2n	3n	4n
1n	**92**	22	2	0
2n	2	**71**	32	7
3n	0	6	**55**	32
4n	0	0	10	**57**

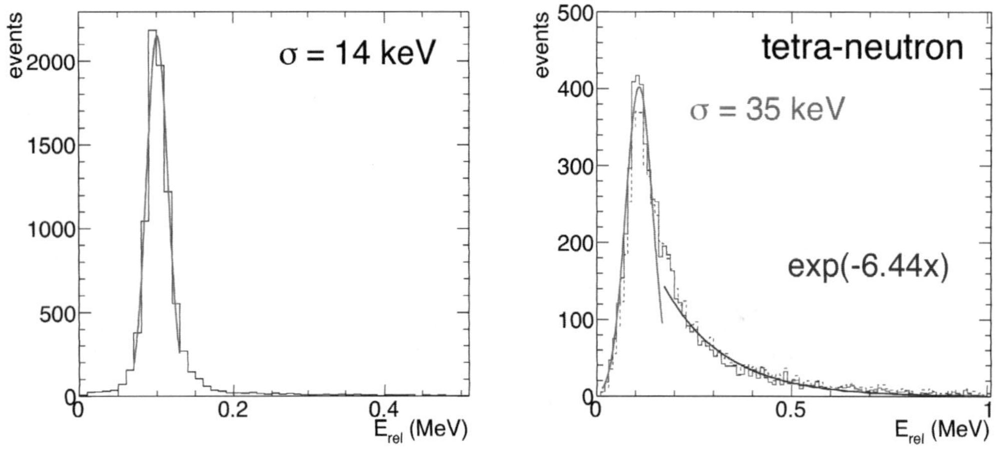

Figure 3. Astro-physics: 1 incident neutron. Tetra-neutron: 4 incident neutrons (beam energy 600 AMeV, E_{rel}=100 keV).

3.3. Reconstruction algorithm

After the number of incident neutrons is determined as explained above, the tracking procedure starts. First of all, the cluster pattern of an event is analysed with respect to the kinematical topology of elastic scattering of neutrons on hydrogen. The secondary hits of elastically scattered neutrons are eliminated from the further analysis. The remaining clusters are sorted according to the values of absolute deviation of potential neutron velocity from the beam velocity (smaller deviation has higher priority) and to the energy deposit in the cluster (higher energy deposit has higher priority). Additional requirement on the sorting of clusters, is that the very first cluster in time stays first. The primary interactions of the known number of incident neutrons are then taken from the list of sorted clusters.

3.4. Relative energy spectra

The reconstruction algorithm is finalised by calculating the neutron momenta (using distance to the target and ToF) and relative energy. Figure 3 shows in this way reconstructed spectrum for 2 different physics cases. The design goal resolution of better than 20 keV is achieved. We have concluded from the detailed analysis of single neutron momentum resolution, that the tail towards higher values on the right plot stems from the higher probability of misidentification in case of 4 incoming neutrons.

4. Prototype test experiment

The full-size NeuLAND prototype detector was tested at GSI in November 2012 with a deuteron beam at various energies from 200 AMeV - 1,5 AGeV. The aim of the experiment was to test the whole read-out chain in the integrated system with other detectors, estimate the total efficiency of neutron measurements, and to obtain the value of time resolution. In addition, one can study quasi free scattering of deuteron on hydrogen by analysing data also from Crystal ball and Silicon trackers. This work is currently ongoing.

Figure 4 shows the technical drawing of the experimental setup in cave "C" at GSI. In the middle - Crystal Ball, the Silicon Tracker and the Target, further downstream - the ALADIN magnet and the NeuLAND prototype.

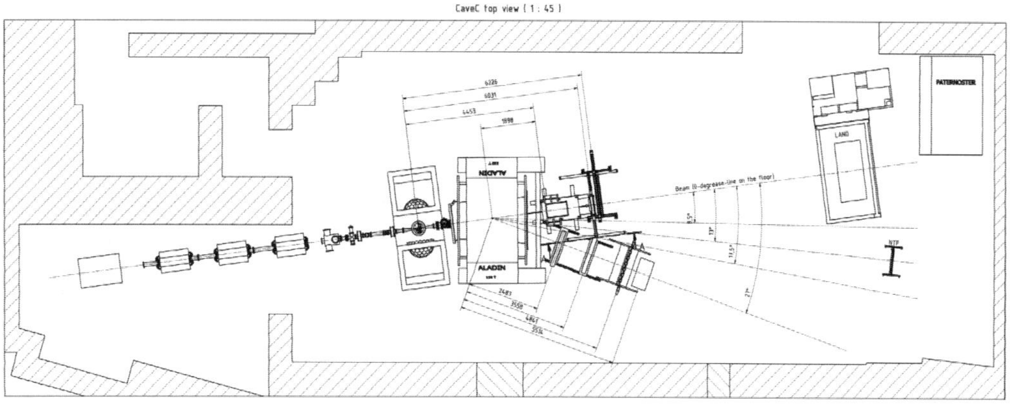

Figure 4. Sketch of the prototype test setup.

The detector was built out of 146 modules with 15 single planes only in vertical direction, total depth is 1,5 m. In this work we focus on the first-level comparison of the simulation results and the experimental data, namely single channel response and total energy deposit.

Journal of Physics: Conference Series **523** (2014) 012034 doi:10.1088/1742-6596/523/1/012034

The old LAND data and R3BRoot simulation results have been conducted which helped in understanding important detector effects. In order to reproduce the measured energy deposit per paddle, one has additionally to introduce individual low charge thresholds, resolution and such effect as photo-multiplier saturation.

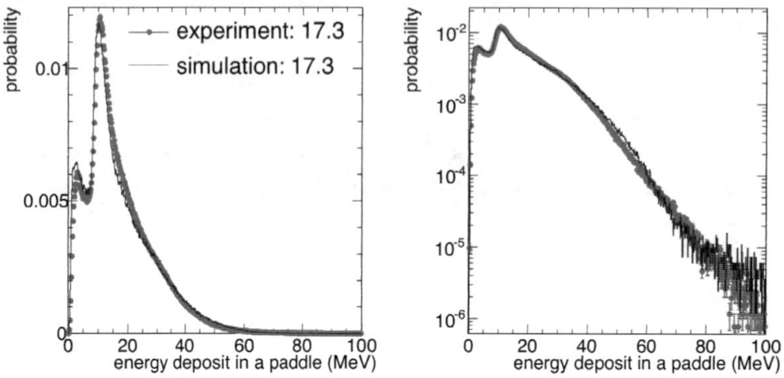

Figure 5. Energy per paddle distribution. Points - experimental data, histogram - simulation.

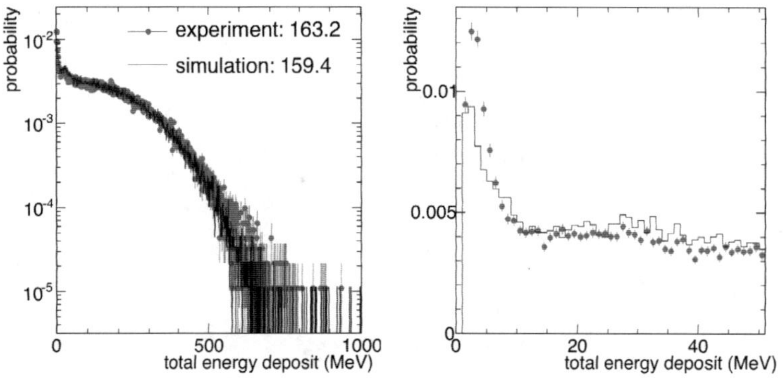

Figure 6. Total energy deposited in the NeuLAND prototype. Points - experimental data, histogram - simulation.

The comparison is presented in Figure 5 for single paddle spectra, and in Figure 6 for the energy deposit in entire detector volume. We conclude very good agreement in the shape, the mean values, and the most probable values of the distribution.

5. Summary

The future R^3B experiment at FAIR is fully supported with the simulation software by the R3BRoot framework and work is ongoing to provide the tools for data analysis and online monitoring. With respect to the NeuLAND detector, the performance of the ToF spectrometer was significantly improved. These studies and results were included in the Technical Design Report (TDR) for the subsystem, which was approved beginning 2013. In addition, the TDR for the subsystem CALIFA (gamma spectrometer), which is based on R3BRoot simulation results, was also approved by FAIR.

The main direction of the algorithm development within the R3BRoot framework is the possibility to use the same code for both simulation and future data analysis (offline and online). That is why the issue with the verification and validation of the reconstruction task is of a big importance. We conclude overall good agreement of simulated and measured detector performance in the particular case of the NeuLAND prototype. With respect to the timing of the algorithm, we observe most of the calculation time spent on input / output of data, being in the order of $\approx$100 ms per event, i.e. processing speed of 5.7 events/s for the simulation and 22 events/s for the analysis (on 2.8 GHz Intel Xeon CPU). An additional constraint on the size of transferred data objects should improve the timing performance. Taking into account maximum beam intensity of the R3B experiment 10^3 - 10^5 beam particles per second and moderate reaction products multiplicity, we conclude that the online data reconstruction with deployment on a cluster with $\approx$ 100 cores will be possible.

References

[1] Boretzky K et al. 2011 *GSI Sci. Rep.* p 174
[2] Boretzky K et al. 2012 *GSI Sci. Rep.* p 190
[3] Al-Turany et al. 2012 *J. Phys.: Conf. Ser.* **396** 022001
[4] Brun R and Rademakers F 1997 *Nucl. Inst. and Meth. in Phys. Res.* A **389** pp 81-86
[5] Brun R Hansroul M and Lassalle J C 1982 *CERN-DD-EE-84-1*
[6] Agostinelli S et al. 2003 *Nucl. Inst. and Meth. in Phys. Res.* A **506** pp 250-303

ACAT2013 IOP Publishing
Journal of Physics: Conference Series **523** (2014) 012035 doi:10.1088/1742-6596/523/1/012035

Concepts and Plans towards fast large scale Monte Carlo production for the ATLAS Experiment

E Ritsch[1]
on behalf of the ATLAS Collaboration
[1] CERN, Geneva 23, Switzerland
E-mail: elmar.ritsch@cern.ch

Abstract. The huge success of the physics program of the ATLAS experiment at the Large Hadron Collider (LHC) during Run 1 relies upon a great number of simulated Monte Carlo events. This Monte Carlo production takes the biggest part of the computing resources being in use by ATLAS as of now. In this document we describe the plans to overcome the computing resource limitations for large scale Monte Carlo production in the ATLAS Experiment for Run 2, and beyond. A number of fast detector simulation, digitization and reconstruction techniques are being discussed, based upon a new flexible detector simulation framework. To optimally benefit from these developments, a redesigned ATLAS MC production chain is presented at the end of this document.

1. Introduction

The current ATLAS Monte Carlo production chain (further described in section 2) has proven very successful over the past years of data taking with the ATLAS experiment [1]. However, the limitations of a CPU intensive, high-accuracy simulation approach become more and more apparent as the demand for larger number of simulated event samples grows rapidly with every year of LHC operation. Therefore, various fast simulation approaches are developed in ATLAS (section 3). In a recent development, fast simulation methods have been bundled in an Integrated Simulation Framework (ISF). The ISF allows to balance accuracy with CPU consumption on a particle level, by selecting particles for accurate simulation only where needed and using fast simulation methods where feasible. Once very fast detector simulation is applied, digitization and reconstruction become the dominating time consumers in the MC production chain. Therefore also fast digitization (section 4) and fast reconstruction (section 5) algorithms are being developed. The fast simulation, digitization and reconstruction methods discussed in this document are based on simplified models and algorithms, thus showing greater computing performance compared to conventional simulation, digitization and reconstruction methods. The production of very fast MC samples will ultimately require to restructure the MC production chain as a whole (see section 6).

2. Current Monte Carlo Production Chain and its Limitations

In order to provide simulated Monte Carlo event samples, the ATLAS collaboration applies a well defined chain of processing steps and conversions between them. Each step produces a different output format. At the end of this chain, both simulated and recorded data are made

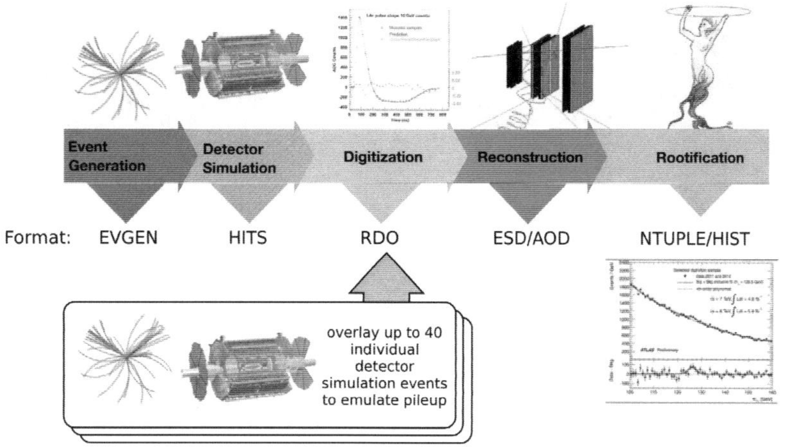

Figure 1. The Monte Carlo production chain as it is currently being used by the ATLAS collaboration. Each step produces an output file which is the input for the next step. In the digitization step, in-time and out-of-time pileup detector effects are overlaid with the simulated signal event.

to share a common data format which allows for direct comparison. The ATLAS Monte Carlo production chain [2], as it is currently in use, is best described by five main steps, which are also depicted in figure 1:

Event Generation concerns the simulation of truth-level particle collision processes which are stored in the EVGEN output format.

Detector Simulation computes the sensitive detector energy deposits (HITS) caused by final state particles taken from EVGEN files. Commonly used detector simulators are Geant4 [3, 4] and an ATLAS fast shower simulation method.

Digitization and Pile-up convert sensitive detector HITS into raw data objects (RDO) i.e. the simulation equivalent of the detector response. During digitization, in-time and out-of-time pileup effects are merged into the signal event. Apart from differences in the file format, the data content of simulated RDO objects is equivalent to bytestream (BS) data recorded from the detector.

Reconstruction finds and identifies physics objects which are stored as event summary data (ESD) and more condensed analysis object data (AOD). With respect to the reconstructed physics objects, the ESD and AOD file contents is equivalent between simulated and recorded data.

Conversion of AOD into ROOT n-tuples is applied to allow fast ROOT[5]-only access to the simulation results.

Figures 2 and 3 show the distribution of CPU and disk usage on the ATLAS Grid sites among the most prominent usage categories, demonstrating that the Monte Carlo production is the biggest relative consumer of Grid resources in ATLAS. MC detector simulation accounts for more than 50 percent of total ATLAS Grid CPU utilization. The main cause for this is the very time consuming Geant4 full detector simulation, of which 2.7 billion collision events were produced in the year 2012. For comparison, 1.8 billion fast shower simulation events were produced in the same year.

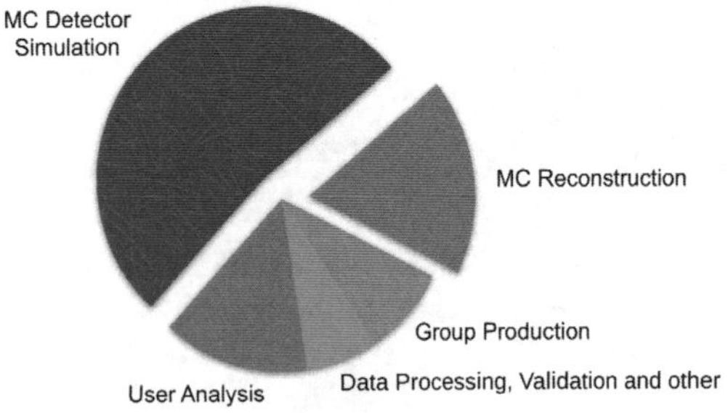

Figure 2. The relative CPU utilization for the most prominent usage categories averaged over all ATLAS Grid sites in the year 2012. In terms of CPU utilization, ATLAS data processing plays a minor role compared to MC production.

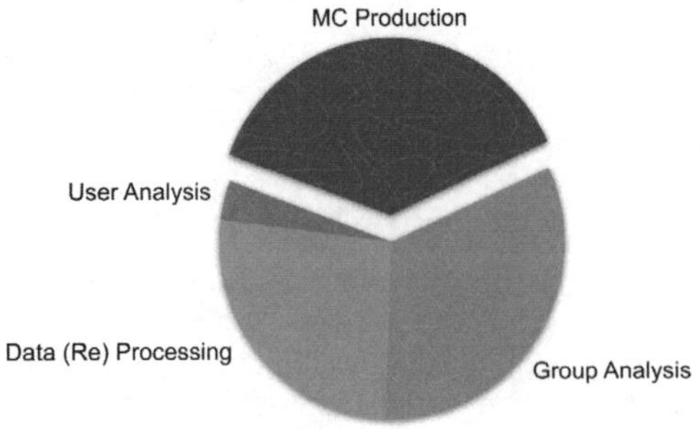

Figure 3. The relative disk utilization for the most prominent usage categories averaged over all ATLAS Grid sites in the year 2012.

3. Fast Detector Simulation

A number of developments are undertaken to speed up ATLAS detector simulation. The possible detector simulation approaches are described qualitatively in figure 4. The following full and fast simulation technologies are use in ATLAS:

Geant4 detector simulation is the most detailed and most accurate ATLAS detector simulation available. Each single particle interaction with the detector material is computed. Secondaries stemming from such interactions are treated in the same way as primary particles coming from the event generator. This approach is therefore also the most time consuming simulation method.

The frozen showers library [6] replaces parts of the full Geant4 simulation by a pre-computed particle shower evolution. By default, ATLAS uses frozen shower libraries for the forward electromagnetic calorimeter simulation together with Geant4 simulation for the rest of the detector.

Fatras [7] is a fast ATLAS tracker simulation. Fatras is based on the reconstruction geometry [9] that describes the detector with a set of layers, while the particle transport is performed by the ATLAS reconstruction extrapolation engine [8]. The particle interaction processes with the detector material are modeled by simplified and fast algorithms.

FastCaloSim [10] is a parameterized ATLAS calorimeter response simulation. FastCaloSim is used in conjunction with Geant4 to form the ATLFASTII simulation flavor. In this configuration, Geant4 is used for particle simulation in the ATLAS inner detector and for muon simulation in all sub-detectors, while FastCaloSim is used to compute calorimeter energy deposits for all particles but muons. FastCaloSim's parametrization is stored in pre-computed lookup tables that are generated from high-statistics single particle samples using Geant4. Being a parameterized simulation, FastCaloSim implements tuning parameters which are used to tune the simulation output against recorded data. This allows the fast simulation to reproduce recorded data in some cases more closely than Geant4 full simulation.

ATLFAST [11, 12] was a parametric ATLAS detector simulation approach which had its main use during the design phase of the detector. The simulation relied on smearing truth-level input in order to directly produce physics object output. Event though ATLFAST was phased out due to the high level of accuracy needed for physics studies with recorded data,

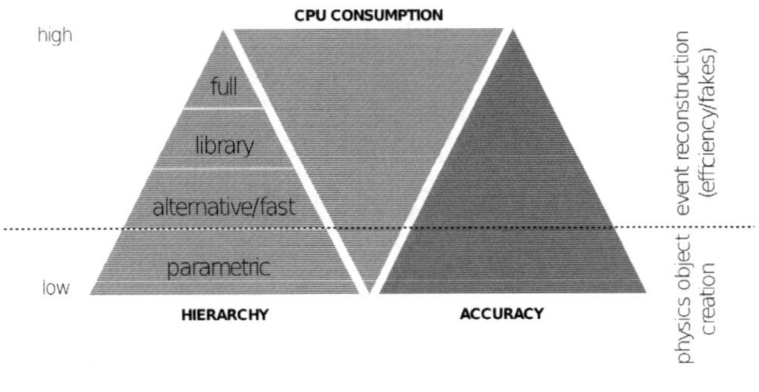

Figure 4. This simulation hierarchy shows a qualitative comparison of different detector simulation methods. Finding the optimal method is a compromise between simulation speed and accuracy. The great algorithmic differences between the methods generally result in a number of independent simulator implementations for one particular detector.

parametric simulation still plays a role for detector upgrade studies.

3.1. The Integrated Simulation Framework

Due to the increasing number of fast ATLAS detector simulation flavors, a common detector simulation framework was developed. The Integrated Simulation Framework allows to combine various simulators (full and fast) for different particles within each event. This way fully accurate simulators can be chosen for particles of interest for a particular physics analysis (e.g. signal particles) and fast simulators can be selected for the rest of the event (e.g. underlying event). This flexibility allows for an optimal balance between speed and accuracy of the ATLAS detector simulation. In addition to ISF setups which are targeted at specific physics use cases, the framework can also be configured to provide full Geant4 or ATLFASTII ATLAS detector simulation. Thus, the ISF will become the single entry point for all ATLAS detector simulation.

Figure 5 gives a qualitative description of the flexibility offered by the ISF. In this example setup, electrons and muons are the particles to be simulated with the highest accuracy, which is why these will be sent through the Geant4 detector simulation. Since particles around electrons could impact the electron measurement, all particles in the proximity of electrons will also be simulated with Geant4. The rest of the event is not required to be simulated with the highest possible accuracy in this example, therefore fast simulators are used for the remaining particles. This mix of simulators provides high accuracy for the signal particle simulation and high speed simulation for the global event variables.

The flexibility of the ISF also allows for partial event simulation. In this case, only selected parts of the event are simulated (e.g. the hard scattering process without the underlying event) and all other particles are excluded from detector simulation. Studies with simulated di-photon Higgs boson decays have shown that significant speed improvements are possible in this mode of ISF operation (figure 7). In addition to the gain in speed, the simulation output file size will be also reduced due to a lower number of simulated energy depositions in sensitive detector elements. Global event variables resulting from partial event simulation are clearly unusable. Thus, partial event simulation may be used where only the properties of the actually simulated particles (e.g invariant di-photon mass) are of interest in the particular physics analysis.

The core ISF design (figure 6) is fully integrated into the ATLAS Athena framework [13]. The **SimulationKernel** is responsible for the particle loop, sending out the particles for simulation to various simulators. These simulators in turn might be creating secondary particles which will then be handed back into the ISF framework. This is done via the **ParticleBroker**, which

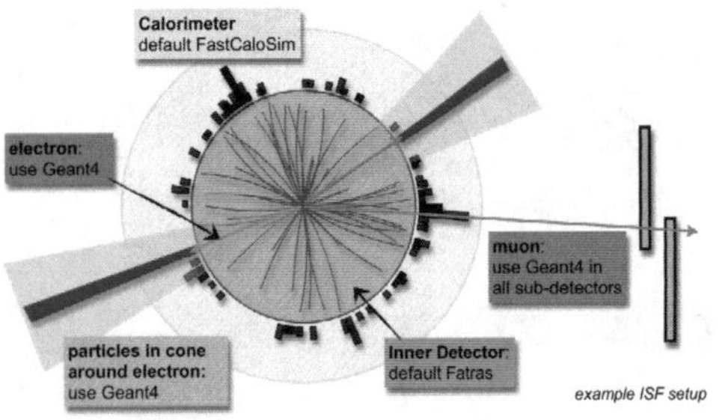

Figure 5. The flexibility of the Integrated Simulation Framework allows to assign individual particles within one event to different simulators. This allows to combine accurate particle simulation where required and fast particle simulation where sufficient for subsequent physics studies.

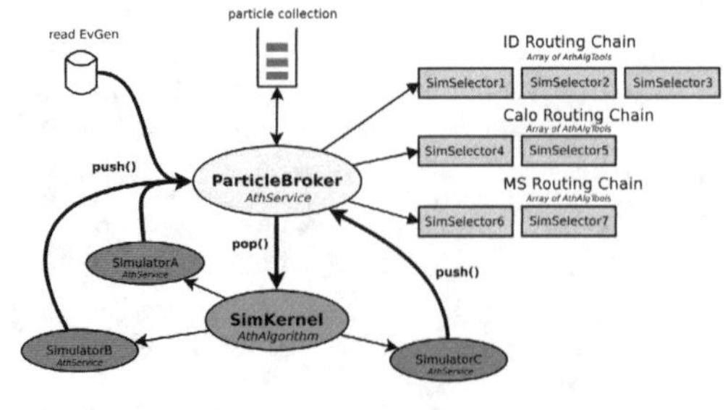

Figure 6. The core ISF design involves a SimulationKernel sending particles to the different ATLAS detector simulators. These simulators return secondary particles back to the ISF via the ParticleBroker. The ParticleBroker uses SimulationSelectors to determine the next simulator for each new particle.

ISF Simulation Setup	Speedup	Accuracy
Full Geant4	1	best possible
Geant4 with FastCaloSim	~25	approximated calorimeter
Fatras with FastCaloSim	~750	all subdetectors approximated
Fatras with FastCaloSim only simulating particels inside cones around photons	~3000	all subdetectors approximated + partial event simulated

Figure 7. Comparing different ISF detector simulation setups shows that very significant speedups with respect to full Geant4 detector simulation are achievable.

is also responsible for making the simulator choice for each new particle. In order to do so, the ParticleBroker uses chains of **SimulationSelectors**, which have bool-type return values. The SimulationSelectors signal the ParticleBroker whether or not they would select the given particle for simulation. The first SimulationSelector in a chain that accepts the given particle, decides on the simulator to be used for this particle. Each SimulationSelector is associated with one simulator only and each sub-detector has an individual chain of SimulationSelectors. This routing algorithm provides a well defined simulator choice for each particle in the ISF particle collection.

4. Fast Digitization

With fast detector simulation methods to be used in MC production soon, efforts are undertaken to also speed up the digitization step. During the digitization, the previously simulated energy deposits inside sensitive detector elements are converted into the detector readout format by simulating the detector response. Traditionally this has been done with very high accuracy in order to mimic the actual detector response in great detail. However, in the context of fast MC production, these methods are now re-visited in order to achieve a further speed-up in the MC

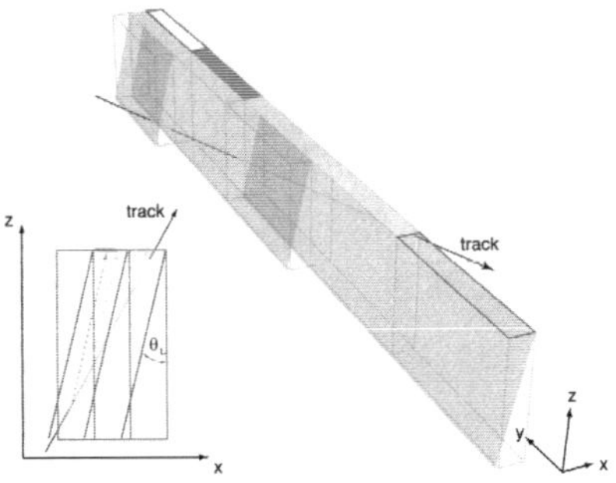

Figure 8. In a fast digitization method for silicon detectors the particle path length within the silicon material is projected onto the readout surfaces. This projection is corrected by the Lorentz angle drift and random smearing is applied to achieve a readout signal.

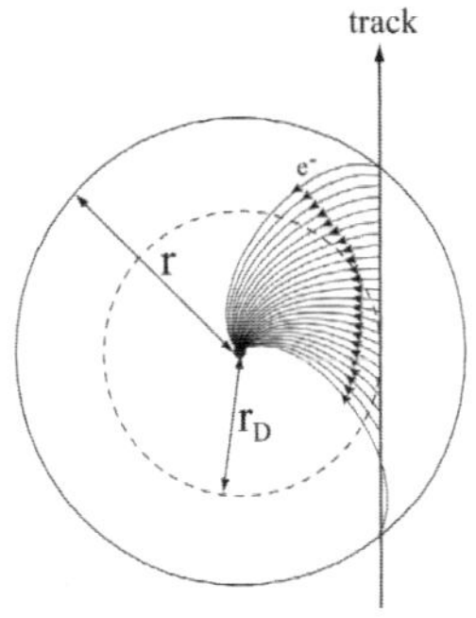

Figure 9. The fast digitization method for the transition radiation tracking detector computes the closest approach radius to the center wire and the measurement uncertainty directly from the detector simulation output. The closest approach radius is an estimate for the drift radius r_D which is forwarded to the reconstruction algorithms.

production chain. Generally speaking, fast digitization methods try to emulate the output of the full digitization by the use of simplifications and smearing approaches. The fast digitization approaches can be enabled optionally inside the same framework as the classical full digitization methods. Thus, the fast methods are not intended to replace the full digitization algorithms. For the ATLAS inner detector technologies, two different fast digitization approaches are currently being investigated:

Fast digitization in silicon tracking detectors converts the simulated path length of a particle track inside the silicon directly into a readout signal. To do so, the actual simulated energy deposits will be ignored, but the simulated particle position and path length inside the silicon is projected onto the readout surface. A Lorentz angle drift correction [14] is applied and the resulting projected path length will be smeared randomly. The resulting readout signal is made proportional to the smeared value. The schematic in figure 8 explains this approach for the ATLAS pixel detector.

Fast digitization in transition radiation tracking (TRT) detectors converts the simulated closest approach radius inside a TRT straw into the reconstruction-level equivalent drift radius (figure 9). This approach skips the computation of the time-over-threshold information which is generated by the classical TRT digitization. The fast digitization is therefore capable of generating reconstruction-level information directly from the simulation output, skipping the step where the detector response is simulated.

5. Fast Reconstruction

The reconstruction step takes a significant fraction of the overall computing time in the MC production chain (see figure 2). The dominating reconstruction algorithm is the inner detector track reconstruction which identifies track candidates from tracking detector hits. The required time to solve this combinatorial problem grows faster than linear with the increase of in-time pileup. Thus the reconstruction will take an even bigger fraction of the MC production time for the next MC campaigns. Once fast simulation and fast digitization techniques that were

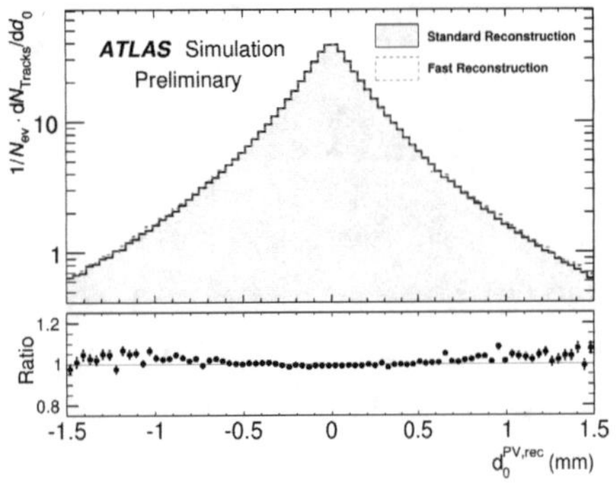

Figure 10. The truth-seeded track reconstruction uses detector simulation information to associate individual tracking detector hits with truth-level particle tracks. This fast reconstruction approach shows a very good agreement with the standard reconstruction output. The figure shows the reconstructed track impact parameter d_0 with respect to the primary vertex using simulated minimum bias samples at $\sqrt{s} = 8$ TeV.

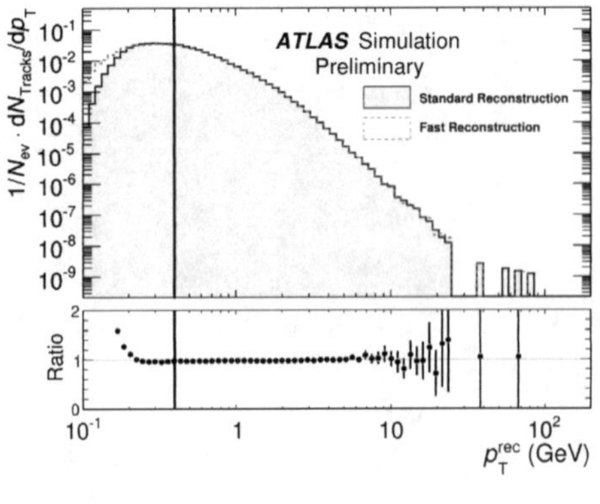

Figure 11. The transverse momentum p_T for reconstructed tracks shows very good agreement between standard reconstruction and the fast reconstruction. The truth-seeded tracking is too efficient for identifying low-p_T tracks. Physics objects are only formed from tracks with $p_T > 400$MeV/c, hence this deviation does not affect physics studies.

discussed in the previous sections are used in production, the reconstruction time will even dominate the MC production chain time consumption.

All this, similar to the fast digitization efforts, motivates the development and use of fast reconstruction algorithms. A truth-seeded tracking algorithm [15] is currently being validated for large scale Grid production and physics analysis use. This approach removes the very time consuming combinatorial task of assigning tracking detector hits to individual particle tracks. The fast reconstruction approach uses the link between the sensitive detector hits and the MC truth representation from the detector simulation step, to find all tracking detector hits caused by each individual particle. Doing so, bypasses the much slower standard track seeding which is done for track reconstruction in recorded data where there is no such information available. Since the truth-seeded tracking algorithm is not capable of producing fake particle tracks, it is only applied to the pileup part of the event, whereas the hard scattering event will be reconstructed by the standard algorithm. This minimizes the potential impact on physics studies and still provides most of the possible speedup.

A comparison of $\sqrt{s} = 8$ TeV simulated minimum bias samples shows an impressive agreement for fundamental track parameters between the standard reconstruction and the fast reconstruction algorithms (see figures 10 and 11). For particles with low transverse momenta (p_T) the truth-seeded tracking algorithm causes the identification efficiency to be higher than in the standard reconstruction algorithm. However, this is of no concern for the use in ATLAS since only particles with $p_T > 400$ MeV/c are considered for building physics objects used for any subsequent analysis.

6. Fast Monte Carlo Production Chain

With the significant speed improvements that were presented in the previous sections, the design of the current Monte Carlo production chain has to be adapted accordingly. Considering that the detector simulation setups have the potential to be up to a factor of 3000 faster than

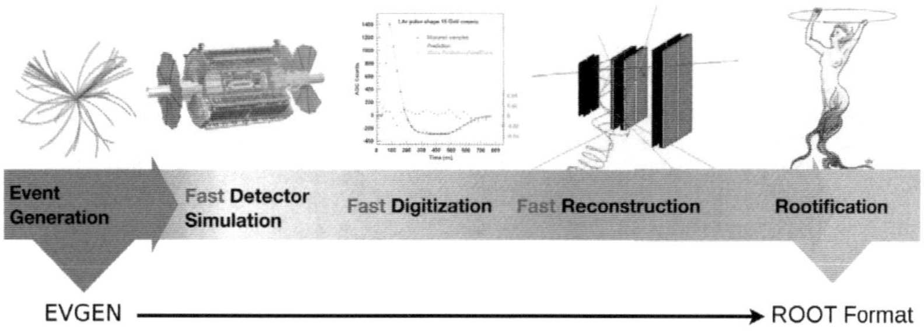

Figure 12. In order to fully utilize the speedup achieved by fast simulation, digitization and reconstruction techniques, the fast Monte Carlo chain will run all the different steps from EVGEN files to ROOT n-tuples in one job. Intermediate data formats do not get stored persistently.

full simulation (see figure 7), it will become inefficient to store the simulation output in the intermediate HITS file format. Eventually, the same holds for digitization (RDO) and the current reconstruction (AOD/ESD) output files. Thus the design of a very fast simulation chain implies that no intermediate file formats will be persistently stored. The fast Monte Carlo production chain will use event generator (EVGEN) input files to run detector simulation, digitization, reconstruction and AOD conversion in one job. The only output generated for persistent storage will be ROOT-readable n-tuples at the end of the chain which will be of the same format as the future standard ATLAS reconstruction output.

Current estimates suggest a processing time of a few seconds per event for the complete chain is feasible with this approach.

7. Conclusions

With the Monte Carlo simulation being the dominant consumer of computing resources in the past years, a paradigm shift towards fast simulation methods is imperative to guarantee the success of future research with the ATLAS experiment. This argument is even more enforced by the ever growing amount of recorded data and the increase of pileup over the next operational period of the LHC accelerator. This document describes approaches taken by the ATLAS experiment to speed up detector simulation, digitization and reconstruction algorithms.

Using parameterized detector responses and simplified detector geometries give FastCaloSim and Fatras a much improved ATLAS detector simulation speed over conventional Geant4. The Integrated Simulation Framework offers the flexibility to combine the available ATLAS detector simulators within one event to guarantee high accuracy where needed and high simulation speed where approximations are sufficient. The partial simulation of events within the ISF has shown to give a great improvement in simulation speed.

Fast digitization techniques will use simplified detector response models to convert simulation output into reconstruction-level information.

A truth-seeded track reconstruction removes the most time consuming part in the event reconstruction by using detector hit associations with the simulated particle causing the hit.

A re-designed fast Monte Carlo production chain will remove all intermediate file formats and generate the reconstruction output (ROOT n-tuples) directly from the input event generator (EVGEN) files. With an estimated processing time of a few seconds per event for the complete fast MC chain, both, adequate statistics and accuracy of MC samples for data analysis will become feasible.

References

[1] The ATLAS Collaboration 2008 The ATLAS Experiment at the CERN Large Hadron Collider *JINST* **3** S08003, doi:10.1088/1748-0221/3/08/S08003
[2] The ATLAS Collaboration 2010 The ATLAS Simulation Infrastructure *Eur. Phys. J. C* **70** 823-874 53 p, CERN-PH-EP-2010-044
[3] S. Agostinelli A *et al* 2003 Geant4 a simulation toolkit *Nucl. Instrum. Methods Phys. Res. A* **506** 250303
[4] Allison J *et al* 2006 Geant4 developments and applications *IEEE Trans. Nucl. Sci.* **53** 270278
[5] Antcheva I *et al* 2011 ROOT A C++ framework for petabyte data storage, statistical analysis and visualization *Comput. Phys. Commun.* **182** 1384-1385
[6] Barberio E *et al* 2007 *The Geant4-Based ATLAS Fast Electromagnetic Shower Simulation* (Geneva: CERN) ATL-SOFT-CONF-2007-002
[7] Edmonds K *et al* 2008 *The Fast ATLAS Track Simulation (FATRAS)* (Geneva: CERN) ATL-SOFT-PUB-2008-001
[8] Salzburger A 2007 *The ATLAS Track Extrapolation Package* (Geneva: CERN) ATL-SOFT-PUB-2007-005
[9] Salzburger A *et al* 2007 *The ATLAS Tracking Geometry Description* (Geneva: CERN) ATL-SOFT-PUB-2007-004
[10] Beckingham M *et al* 2010 *The simulation principle and performance of the ATLAS fast calorimeter simulation FastCaloSim* (Geneva: CERN) ATL-PHYS-PUB-2010-013
[11] Froidevaux D *et al* 1996 *ATLFAST 1.0 A package for particle-level analysis* (Geneva: CERN) ATL-PHYS-96-079
[12] Froidevaux D *et al* 1998 *ATLFAST 2.0 a fast simulation package for ATLAS* (Geneva: CERN) ATL-PHYS-98-131
[13] The ATLAS Collaboration 2005 ATLAS Computing Technical Design Report ATLAS-TDR-017, CERNLHCC-2005-022
[14] The ATLAS Collaboration 2010 The ATLAS Inner Detector commissioning and calibration *Eur. Phys. J. C* **70** 787-821. 34 p, CERN-PH-EP-2010-043
[15] Jansky R (to be published) *Truth seeded reconstruction for fast simulation in the ATLAS experiment*

Optimizing ATLAS code with different profilers

S Kama[1], R Seuster[2], G A Stewart[3] and R A Vitillo[4]

[1] Southern Methodist University, Dallas, TX, USA
[2] TRIUMF, Vancouver BC V6T 2A3, Canada
[3] CERN and School of Physics and Astronomy, University of Glasgow, Glasgow G12 8QQ
[4] Lawrence Berkeley National Laboratory, USA

E-mail: sami.kama@cern.ch

Abstract. After the current maintenance period, the LHC will provide higher energy collisions with increased luminosity. In order to keep up with these higher rates, ATLAS software needs to speed up substantially. However, ATLAS code is composed of approximately 6M lines, written by many different programmers with different backgrounds, which makes code optimisation a challenge. To help with this effort different profiling tools and techniques are being used. These include well known tools, such as the Valgrind suite and Intel Amplifier; less common tools like Pin, PAPI, and GOoDA; as well as techniques such as library interposing. In this paper we will mainly focus on Pin tools and GOoDA. Pin is a dynamic binary instrumentation tool which can obtain statistics such as call counts, instruction counts and interrogate functions' arguments. It has been used to obtain CLHEP Matrix profiles, operations and vector sizes for linear algebra calculations which has provided the insight necessary to achieve significant performance improvements. Complimenting this, GOoDA, an in-house performance tool built in collaboration with Google, which is based on hardware performance monitoring unit events, is used to identify hot-spots in the code for different types of hardware limitations, such as CPU resources, caches, or memory bandwidth. GOoDA has been used in improvement of the performance of new magnetic field code and identification of potential vectorization targets in several places, such as Runge-Kutta propagation code.

1. Introduction

The Large Hadron Collider (LHC) [1] is a proton collider built about 100m underground near Geneva, Switzerland. It has a 27 km circumference and it is designed to collide protons every 25 ns at a center-of-mass energy of 14 TeV with a luminosity of 10^{34} cm^{-2}s^{-1}. It started operation in 2010 and gradually increased the collision energy and luminosity. In 2011 LHC delivered a peak luminosity of 3.42×10^{33} cm^{-2}s^{-1} in 31 weeks of proton-proton collision runs at $\sqrt{s}$ = 7 TeV with 50 ns bunch crossing interval. In 2012, the center-of-mass energy was increased to 8 TeV and the peak instantaneous luminosity exceeded 7.7×10^{33} cm^{-2}s^{-1}. From March 2013, it has been shutdown for two years for maintenance and upgrades. It will operate at a higher beam energy and higher luminosity after the shutdown.

There are 4 detectors located at the LHC. ATLAS [2] is one of the two large general purpose detectors. It is composed of different co-centric cylindrical detectors of about 100 M readout channels in total. The ATLAS detector has a three-level trigger system to do event selection and reduce the enormous amounts of data produced to manageable sizes. Level 1 of the trigger is based on hardware and located on the detector. It has a design input rate of 40 MHz and an output rate of 75 kHz. Level 2 and Level 3 are based on software running on a PC farm of about 16k cores. They reduce final event rate to 600 Hz at ~1.6 MB per event. Selected events are stored and processed offline in more detail. Both online and offline selection is done using the same software framework called Athena, with different configurations. Up to the shutdown in 2013 ATLAS had stored and processed ~22 PB of raw data.

2. ATLAS Software

The ATLAS software framework is based on the Gaudi framework [3] and comprises more than 6 million lines of C++ and Python, with a small amount of FORTRAN code. It is spread over about 2000 packages, producing more than 4000 libraries of various sizes. It has been evolving for more than 10 years and was written by people with varied programming backgrounds. Some parts of the code were written by expert level programmers, while some were written by people with minimal programming knowledge. Throughout its development, detailed knowledge of packages has been frequently lost due to authors and maintainers changing topics, institutes or leaving the field. Athena configuration is done in Python and reconstruction of a 2012 high pile-up sample($\langle \mu \rangle \sim 35$) with 64-bit application consumes about ~4 GB of virtual memory and has approximately 2.7 GB resident set size(RSS).

With the increase in LHC energy, collision rate, event complexity and trigger output, ATLAS software needs to speed up considerably. Due to sheer size of the codebase and the domain expertise required it is not practical to go through the codebase to fix even the simplest perfomance issues. So instead, efforts should be focused on so called hot-spots, where a small section of source code contributes significantly to overall application run time. In order to monitor the performance of Athena, ATLAS uses various tools. These range from commonly available tools such as Valgrind suite, Google perf tools, oprofile, Intel VTUNE, igprof, PAPI and Pin [4] to in house developed tools PerfMon and Generic Optimization Data Analyzer (GOoDA) [5], which is developed in collaboration with Google. However, the enormous size of Athena creates significant challanges for most of these tools. In this report we will be describing our experience of improving the performance of Athena using GOoDA and Pin.

3. GOoDA

The Generic Optimization Data Analyzer is an open source project developed by a collaboration between ATLAS and Google. It uses Linux `perf` tool to configure and collect detailed performance monitoring unit(PMU) information from hardware monitoring units built in CPUs. Then it analyses the collected monitoring information, grouping it into a few issue classes. These issue classes provide a more general understanding of the code's performance than CPU specific PMU values. Results of the analysis are written in report spreadsheets which can we viewed through a web browser either locally or remotely. An example of such a report is shown in figure 1.

GOoDA reports contain hot-spots for each type of performance issue down to instruction level. It can also display source line if sources are available. Figure 1 shows a profile of Athena offline reconstruction job with high number of proton collisions (pile-up) per input event. Investigation of the profile shows several groups of code with rather specific issues. The most frequently encountered code domain is the tracking code. Being composed of a high number of arithmetic calculations, due to vector and matrix operations, this code frequently has instruction related issues, such as instruction starvation. It is possible to overcome such issues by using *single instruction multiple data* (SIMD) operations. Due to the intrinsic geometric nature of the tracking problem, representing tracking operations in terms of primitives of a SIMD vector math library is the most feasible approach, both from maintainability and development time points of view. However, such vector math libraries are not equally efficient for all types of operations or for all sizes of vectors and matrices. How the decision on which vector math library to use was taken is explained in the next section.

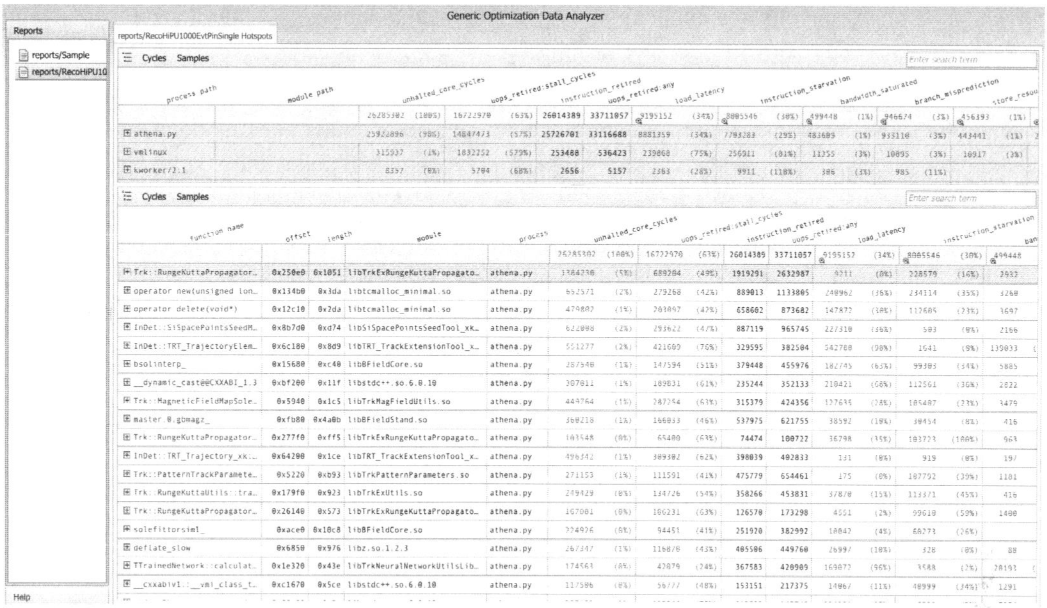

Figure 1. Main page of GOoDA profile. The left frame displays available reports. The top right frame displays processes in the report and their respective PMU event counts. The bottom right frame shows list of hot-spot functions in the selected process and their respective event counts.

The second group is the frequent use of new/delete operators, which is due to the design of the event data model. Frequent memory allocation and deallocation was already identified previously and some actions, such as replacing standard malloc with thread caching malloc (tcmalloc) from Google, were used to mitigate the effect of this. However, it still remains a significant problem and the profile helps to quantify it. In order to further reduce these frequent allocation/deallocation effects, there have been some changes in the event data model. There are also some studies, such as using memory arenas and different memory allocators, ongoing to reduce the impact on the performance.

The third group is the magnetic field code. This was written in FORTRAN and most of the time was actually spent on interfacing the old FORTRAN code with the C++ framework. So, rather than optimizing FORTRAN, the code was re-written in C++ from scratch. The new code was already 2x faster than the FORTRAN implementation due to a better design and native interfacing. After profiling a simple test job, which is querying the magnetic field at random points in the detector to stress it, further hot-spots were identified (see figure 2). The most notable point in this table is that about 70% of stall cycles are due to instruction latency coming from division operations. Following up the report points to the source lines 101-108 in figure 3. However, even though replacing division operators with inverse multiplications is straight forward and common performance practice, it doesn't maximize the performance. Further profiles have shown that there is instruction starvation around the lines shown in figure 3. After a detailed investigation of the source code, it can be seen that the complex calculations in lines 84-95 can be written as a dot product of two vectors with properly arranged coefficients. Profiling the version with the vector products and inverse multiplications is shown in the bottom table in figure 2. It is clearly visible here, that total unhalted core cycles count is reduced by 42%

with respect to initial profile. The final version is about 40% faster than initial code and overall effect of the new code with respect to the FORTRAN implementation is measured between 5% to 20% improvement in the runtime of simulation jobs.

function name	unhalted_core_cycles	uops_retired:stall_cycles	instruction_retired	uops_retired:any	load_latency	instruction_starvation	bandwidth_saturated	branch_misprediction	store_resources_saturated	instruction_latency	arith:cycles_div_busy	arith:div	
	189750443 (99%)	135036818 (71%)	135006354	171163932	1685841 (0%)	525883 (0%)	2942 (0%)	1173179 (0%)	460754 (0%)	87043710 (45%)	87043918 (45%)	5670196	48857
BFieldCache::getB(double ...	110657873 (100%)	77277673 (69%)	89363774	101452126	115206 (0%)	39329 (0%)	547 (0%)	15908 (0%)	116371 (0%)	53349337 (48%)	53349461 (48%)	3238233	162941
MagField::IMagFieldSvc::g...	26676464 (100%)	20986468 (78%)	14775629	19667221	66770 (0%)	24991 (0%)		328974 (1%)	22589 (0%)	17073200 (64%)	17073241 (64%)	999788	20858
MagField::AtlasFieldSvc::...	6399933 (100%)	4263998 (66%)	5628152	6579907	911735 (14%)	14433 (0%)	1094 (0%)	214006 (3%)	281013 (4%)	2667794 (41%)	2667810 (41%)	164328	5217

function name	unhalted_core_cycles	uops_retired:stall_cycles	instruction_retired	uops_retired:any	load_latency	instruction_starvation	bandwidth_saturated	branch_misprediction	store_resources_saturated	instruction_latency	arith:cycles_div_busy	arith:div	
	94257696 (100%)	61237244 (64%)	91887940	105711527	956148 (1%)	282528 (0%)	6704 (0%)	435646 (0%)	303020 (0%)	24884101 (26%)	24884136 (26%)	1367684	96165
BFieldCache::getB(double ...	64322753 (100%)	37891791 (58%)	70511475	78526221	59705 (0%)	25697 (0%)		9183 (0%)	93522 (0%)	10632460 (16%)	10632493 (16%)	580526	40112
MagField::MagFieldTestbed...	20231953 (100%)	15839738 (78%)	12068630	14851856	34362 (0%)	14225 (0%)	191 (0%)	1295438 (6...	34717 (0%)	10925070 (53%)	10925109 (53%)	571357	11936
MagField::AtlasFieldSvc::...	4984516 (100%)	3256834 (65%)	4553073	5494224	636914 (12%)	9538 (0%)	409 (0%)	156742 (3%)	152489 (3%)	1746017 (35%)	1746039 (35%)	98775	2289

Figure 2. Profile report of new magnetic field test job. The top profile is before optimization. The first hot-spot shows high percentage of stall cycles mostly originating from instruction latency coming from division operations. The bottom profile is after optimizing divisions and multiplications.

```
        float dBdz[3], dBdr[3], dBdphi[3];
        for ( int j = 0; j < 3; j++ ) { // Bz, Br, Bphi components
            dBdz[j]   = sz*( gr*( gphi*(m_field[4][j]-m_field[0][j]) +
                                  fphi*(m_field[5][j]-m_field[1][j]) ) +
                             fr*( gphi*(m_field[6][j]-m_field[2][j]) +
                                  fphi*(m_field[7][j]-m_field[3][j]) ) );
            dBdr[j]   = sr*( gz*( gphi*(m_field[2][j]-m_field[0][j]) +
                                  fphi*(m_field[3][j]-m_field[1][j]) ) +
                             fz*( gphi*(m_field[6][j]-m_field[4][j]) +
                                  fphi*(m_field[7][j]-m_field[5][j]) ) );
            dBdphi[j] = sphi*( gz*( gr*(m_field[1][j]-m_field[0][j]) +
                                    fr*(m_field[3][j]-m_field[2][j]) ) +
                               fz*( gr*(m_field[5][j]-m_field[4][j]) +
                                    fr*(m_field[7][j]-m_field[6][j]) ) );
        }
        // convert to cartesian coordinates
        float cc = c*c;
        float cs = c*s;
        float ss = s*s;
        deriv[0] = cc*dBdr[1] - cs*dBdr[2] - cs*dBdphi[1]/r + ss*dBdphi[2]/r + s*B[1]/r;
        deriv[1] = cs*dBdr[1] - ss*dBdr[2] + cc*dBdphi[1]/r - cs*dBdphi[2]/r - c*B[1]/r;
        deriv[2] = c*dBdz[1] - s*dBdz[2];
        deriv[3] = cs*dBdr[1] + cc*dBdr[2] - ss*dBdphi[1]/r - cs*dBdphi[2]/r - s*B[0]/r;
        deriv[4] = ss*dBdr[1] + cs*dBdr[2] + cs*dBdphi[1]/r + cc*dBdphi[2]/r + c*B[0]/r;
        deriv[5] = s*dBdz[1] + c*dBdz[2];
        deriv[6] = c*dBdr[0] - s*dBdphi[0]/r;
        deriv[7] = s*dBdr[1] + c*dBdphi[0]/r;
        deriv[8] = dBdz[0];
    }
}
```

Figure 3. Source code of hot-spots in the magnetic field code. Division operations are the cause of the instruction latency and complex calculations are the cause of instruction starvation. These complex calculations can be re-written as a dot product of two vectors.

4. Pin Tools

Pin is a dynamic binary instrumentation framework from Intel. It instruments binaries at run-time, eliminating need to modify or recompile the code. It can instrument the binary from instruction level to function level and supports dynamically generated code. Pin can access function parameters and register contents and it can work with threaded programs. If available, it provides limited access to symbol and debug information in the original binary. Instrumentation is done on an in-memory copy of the binary. Pin inspects the instructions in original binary and inserts calls to analysis functions. Due to its powerful features it has been used in computer architecture, security, emulation and parallel program analysis tools, such as Intel's Parallel Inspector, Parallel Amplifier, Trace Analyzer and Collector as well as CMP$im and many others. It has detailed documentation and an active user support community called PinHeads.

As mentioned in previous section, tracking code is largely composed of vector and martix operations. In Athena, CLHEP library [6] is used for most of the vector and matrix representation and their operations. However, the CLHEP library is not performance optimized and does not vectorize well. There are other SIMD vector and matrix libraries available, each having different performance for different primitive sizes and operations. In order to quantify most common vector dimensions and operation we instrumented CLHEP vector classes in Athena with Pin.

Table 1. Top 10 most frequently called CLHEP functions in Athena for 20 events with total instructions and average instructions per call, sorted by total instruction count.

Calls	Instr	<instr>/call	Function
1778523	6392431813	3594.24	operator*(HepMatrix const&, HepSymMatrix const&)
671676353	5988139520	8.92	Hep2Vector::operator()(int) const
232093102	5956556656	25.66	Transform3D::operator()(int, int) const
285282108	3709057782	13.00	Hep3Vector::operator()(int)
15815930	3179001930	201.00	HepRotation::rotateAxes(Hep3Vector const&, Hep3Vector const&, Hep3Vector const&)
20529818	2422518524	118.00	Transform3D::inverse() const
31612743	2212258670	69.98	HepSymMatrix::HepSymMatrix(HepSymMatrix const&)
28914115	1929106393	66.72	HepVector::HepVector(int, int)
51974716	1819115060	35.00	operator*(Transform3D const&, Point3D<double>const&)
27652274	1506352669	54.47	HepVector::HepVector(HepVector const&)

Table 1 shows the most frequently called CLHEP functions, together with their instruction costs and averge number of instructions per event, sorted by their total instruction counts. The multiplication operator for multiplication of a matrix with a symmetric matrix is the most expensive operation. These matrices usually have dimensions 5×5, 5×3, 3×3 and 3×5. Accessor operators for vectors and 3D transformation matrices also contribute significantly to instruction counts. The number of instructions can be significantly reduced by inlining.

Using information gained from Pin instrumentation, we devised a benchmark for various SIMD vector math libraries. The benchmark composes several operations such as $A_{5\times3} \times B_{3\times5}$, $A_{5\times3} \times B_{3\times5} + \alpha C_{5\times5}$ and $A_{4\times4} \times B_{4\times4}$. The 4×4 matrices are benchmarked in place of 3×3 matrices since a translation and rotation operation can be combined in this way. We implemented these operations using various vector math libraries such as CLHEP, the Intel Math Kernel Library(MKL) [7], the S-Matrix library [8] and the Eigen [9] library. The speedup ratio with respect to CLHEP for each library is shown in figure 4. Intel MKL was between 5-20% slower than CLHEP for small matrix operations. This is probably due to its tuning for rather large sized matrix operations. Nevertheless, for $A_{5\times3} \times B_{3\times5} + \alpha C_{5\times5}$ it was about 2 times faster than CLHEP. The S-Matrix library, which is included in ROOT framework and based on C++ expression templates, provided a speed up between 3.2 to 4.4 times compared to CLHEP. The best speedup is achieved by Eigen library which achieved 6 to 12 times speed up with respect to CLHEP depending on the operation. Eigen is similar to S-Matrix in that it uses expression templates, however, on top of that, it also contains explicit vectorization with SIMD instructions,

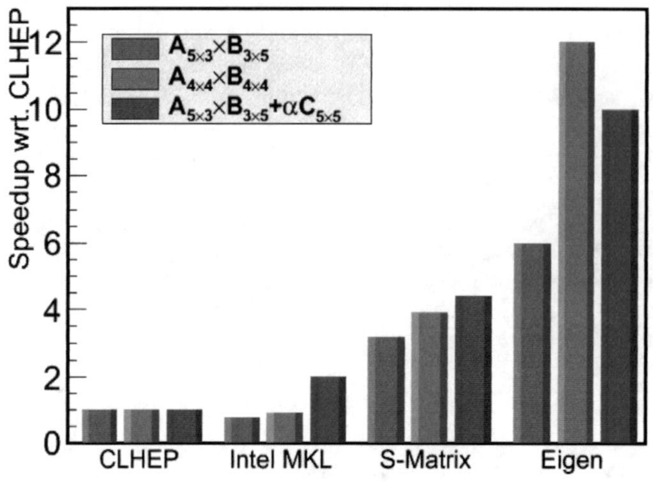

Figure 4. Speedup ratios of various libraries for different operations with respect to CLHEP library. The matrix sizes were found by instrumenting Athena and correspond to most commonly used matrix sizes in ATLAS reconstruction. The Eigen library performs significantly better than the other candidates.

avoids dynamic memory allocations and unrolls loops whenever it makes sense for fixed size matrices. Thanks to its significant speedup with respect to CLHEP, Eigen was choosen as a replacement for CLHEP. Migration from CLHEP to Eigen library in Athena framework is in progress.

5. Conclusions

The optimization of many-developer large code bases such as Athena is a challenging task. Using GOoDA we identified several performance problems and focused our efforts on these points. Replacement of magnetic field code alone provided up to 20% improvement in overall runtime while maintaining same physics performance and being used as the default field implementation in recent Athena releases. Using Pin we analyzed the most frequently used vector and matrix operations in Athena and better constrained the requirements of the vector math library. Using this information we picked the Eigen library as a replacement for CLHEP.

Both GOoDA and Pin are proven to be valuable tools for analyzing and improving Athena framework. Much information can be gained with little effort. Both tools can be used by average programmer at a basic level after some study, however it requires a certain level of expertise to fully exploit either tool.

References

[1] Evans L and Bryant P 2008 *JINST* **3** S08001
[2] Aad G *et al.* (ATLAS Collaboration) 2008 *JINST* **3** S08003
[3] Clemencic M, Degaudenzi H, Mato P, Binet S, Lavrijsen W *et al.* 2010 *J.Phys.Conf.Ser.* **219** 042006
[4] Luk C, Cohn R, Muth R, Patil H, Klauser A, Lowney G, Wallace S, Reddi V J and Hazelwood K 2005 *Programming Language Design and Implementation (PLDI)* 190–200
[5] Calafiura P, Eranian S, Levinthal D, Kama S and Vitillo R 2012 *J.Phys.Conf.Ser.* **396** 052072
[6] *CLHEP Web Page* URL `http://wwwasd.web.cern.ch/wwwasd/lhc++/clhep/index.html`
[7] *Intel Math Kernel Library* URL `http://software.intel.com/en-us/intel-mkl`
[8] *ROOT Manual* URL `http://root.cern.ch/root/html/MATH_SMATRIX_Index.html`

[9] *Eigen Web Page* URL http://eigen.tuxfamily.org/index.php

ACAT2013 IOP Publishing
Journal of Physics: Conference Series **523** (2014) 012037 doi:10.1088/1742-6596/523/1/012037

Modern Summation Methods for Loop Integrals in Quantum Field Theory: The Packages Sigma, EvaluateMultiSums and SumProduction

C Schneider
Research Institute for Symbolic Computation (RISC)
Johannes Kepler University Linz
Altenbergerstr. 69, 4040 Linz, Austria

E-mail: `Carsten.Schneider@risc.jku.at`

Abstract. A large class of Feynman integrals, like e.g., two-point parameter integrals with at most one mass and containing local operator insertions, can be transformed to multi-sums over hypergeometric expressions. In this survey article we present a difference field approach for symbolic summation that enables one to simplify such definite nested sums to indefinite nested sums. In particular, the simplification is given –if possible– in terms of harmonic sums, generalized harmonic sums, cyclotomic harmonic sums or binomial sums. Special emphasis is put on the developed Mathematica packages `Sigma`, `EvaluateMultiSums` and `SumProduction` that assist in the task to perform these simplifications completely automatically for huge input expressions.

1. Introduction

This survey article aims at giving an overview of the available summation tools and packages in the setting of difference fields and aims at providing the basic insight how they work and how they can be applied to Feynman integrals.

The symbolic summation approach under consideration started with Karr's telescoping algorithm in $\Pi\Sigma^*$-fields [1]. There one can treat indefinite nested product-sum expressions. More precisely, one can search for optimal sum representations for these expressions such that the arising sums have minimal nesting depth [2–4], the polynomial expressions in the sums have minimal degrees [5, 6] and such that the sums used are algebraically independent [7, 8]. In particular, one can discover and prove definite sums over such expressions by exploiting the summation paradigms of parameterized telescoping and recurrence finding [9, 10]. In a nutshell, this toolkit implemented in the summation package `Sigma` [11] can be considered as a generalization of the well known hypergeometric summation machinery [12–14]. Meanwhile these difference field tools have been applied successfully, e.g., in combinatorics, number theory or numerics; for instance, we refer to [15–18].

In the last 7 years I have pushed forward these summation technologies and have developed new packages [8, 19–21] to carry out challenging calculations in particle physics. In cooperation with J. Blümlein (DESY, Zeuthen), e.g., Feynman integrals with at most one mass have been considered. They are defined in D-dimensional Minkowski space with one time- and $(D-1)$ Euclidean space dimensions with $\varepsilon = D-4 \in \mathbb{R}$ where $|\varepsilon| \ll 1$. The integrals depend on a discrete

Mellin parameter n which comes from local operator insertions. As worked out in [22, 23] these integrals can be transformed to integrals of the form

$$\mathcal{I}(\varepsilon, n) = C(\varepsilon, n, M) \int_0^1 dx_1 \ldots \int_0^1 dx_m \frac{\sum_{i=1}^k \prod_{l=1}^{r_i} [P_{i,l}(x_1, \ldots, x_m)]^{\alpha_{i,l}(\varepsilon,n)}}{[Q(x_1, \ldots, x_m)]^{\beta(\varepsilon)}}, \tag{1}$$

with $k \in \mathbb{N} = \{0, 1, 2, \ldots\}$, $r_1, \ldots, r_k \in \mathbb{N}$ and the following ingredients. $C(\varepsilon, n, M)$ is a factor, which depends on the dimensional parameter ε, the integer parameter n and the mass M (being possible 0). The $P_i(x_1, \ldots, x_m)$ and $Q(x_1, \ldots, x_m)$ are polynomials in the integration variables x_i. Further, the exponent $\beta(\varepsilon)$ is a rational function in ε, i.e., $\beta(\varepsilon) \in \mathbb{Q}(\varepsilon)$, and similarly $\alpha_{i,l}(\varepsilon, n) = n_{i,l}n + \overline{\alpha}_{i,l}$ with $n_{i,l} \in \{0, 1\}$ and $\overline{\alpha}_{i,l} \in \mathbb{Q}(\varepsilon)$. For integrals without local operator insertions see also [24, 25].
Then given such an integral we seek for a Laurent series expansion for some $t \in \mathbb{Z}$:

$$\mathcal{I}(\varepsilon, n) = I_t(n)\varepsilon^t + I_{t+1}(n)\varepsilon^{t+1} + I_{t+2}(n)\varepsilon^{t+2} + \ldots \tag{2}$$

More precisely, the crucial task is to determine the first coefficients $I_i(n)$ in closed form.

Remark. There are computer algebra tools [19] available to solve this problem for integrals of the form (1). Namely, we can calculate a recurrence in n for (1) using Ablinger's `MultiIntegrate` package [26], an optimized and refined version of the Almkvist–Zeilberger algorithm [27]. Given such a recurrence, we can calculate the desired ε-expansion by algorithms [23] given in `Sigma`. We remark that the method of hyperlogarithms [28] can be also used to evaluate integrals of the form (1) for specific values $n \in \mathbb{N}$ if one can set $\varepsilon = 0$. An adaption of this method for symbolic n has been described and applied to massive 3-loop ladder graphs in [29].

For some types of integrals these approaches work nicely. However, for many cases it was more suitable to proceed as follows. Using the method in [23] the given integral (1) can be transformed to proper hypergeometric multi-sums of the format[1]

$$\mathcal{S}(\varepsilon, n) = \sum_{n_1=1}^{\infty} \ldots \sum_{n_r=1}^{\infty} \sum_{k_1=1}^{L_1(n)} \ldots \sum_{k_v=1}^{L_v(n,k_1,\ldots,k_{v-1})} \sum_{i=1}^{l} C_i(\varepsilon, n, M) \frac{\Gamma(t_{1,i}) \ldots \Gamma(t_{v',i})}{\Gamma(t_{v'+1,i}) \ldots \Gamma(t_{w',i})}. \tag{3}$$

Here the upper bounds $L_1(n), \ldots, L_v(n, k_1, \ldots, k_{v-1})$ are ∞ or integer linear expressions in the dependent parameters (i.e., linear combinations of the variables over the integers), and $t_{l,i}$ are linear combinations of the $n_1, \ldots, n_r, k_1, \ldots, k_v, \varepsilon$ over $\mathbb{Q}$.

Remark. At this level, one can apply similar techniques as for the integral representation (1). Namely, as carried out in [23] we can calculate recurrence relations for such multi-sums (3) by using either the WZ-approach [30] and efficient and refined algorithms developed in [31]. Similarly, we can use a common framework [32] within the summation package `Sigma` that combines difference field [33] and holonomic summation techniques [34] to compute recurrences for the given sum (3). Then given such a recurrence, one can use again `Sigma`'s recurrence solver to calculate the coefficients of the Laurent-series expansion (2).

Finding recurrence relations for the integrals of the form (1) and sums of the form (3) involving the ε-parameter is a rather tough problem. However, we can get rid of the parameter ε, if the sums (3) are uniformal convergent: First one expands the summand of (3), say

$$F(n, n_1, \ldots, n_r, k_1, \ldots, k_v) = F_t(n, n_1, \ldots, k_v)\varepsilon^t + F_{t+1}(n, n_1, \ldots, k_v)\varepsilon^{t+1} + \ldots \tag{4}$$

[1] For convenience, we assume that the summand is written in terms of the Gamma function $\Gamma(x)$. Later, also Pochhammer symbols or binomial coefficients are used which can be rewritten in terms of Gamma-functions.

with $t \in \mathbb{Z}$ by using formulas such as [23]

$$\Gamma(k+1+\bar{\varepsilon}) = \frac{\Gamma(k)\Gamma(1+\bar{\varepsilon})}{B(k,1+\bar{\varepsilon})} \text{ and } B(k,1+\bar{\varepsilon}) = \frac{1}{k}\exp\left(\sum_{i=1}^{\infty}\frac{(-\bar{\varepsilon})^i}{i}S_i(k)\right) = \frac{1}{k}\sum_{i=0}^{\infty}(-\bar{\varepsilon})^i S_{\underbrace{1,\dots,1}_{i}}(k)$$

with $\bar{\varepsilon} = r\varepsilon$ for some $r \in \mathbb{Q}$. Here $B(x,y) = \Gamma(x)\Gamma(y)/\Gamma(x+y)$ denotes the Beta-function and $S_{1,\dots,1}(n)$ is a special instance of the harmonic sums [35, 36] defined by

$$S_{c_1,\dots,c_r}(k) = \sum_{i_1=1}^{k}\frac{\text{sign}(c_1)^{i_1}}{i_1^{|c_1|}}\sum_{i_2=1}^{i_1}\frac{\text{sign}(c_2)^{i_2}}{i_2^{|c_2|}}\cdots\sum_{i_r=1}^{i_{r-1}}\frac{\text{sign}(c_r)^{i_r}}{i_r^{|c_r|}} \tag{5}$$

with $c_1, \dots, c_r$ being nonzero integers. If there are no poles[2] within the summation range, one can apply the summation signs to each of the coefficients in (4). Namely the ith coefficient of the ε-expansion (2) of (3) can be written in the form

$$I_i(n) = \sum_{n_1=1}^{\infty}\dots\sum_{n_r=1}^{\infty}\sum_{k_1=1}^{L_1(n)}\dots\sum_{k_v=1}^{L_v(n,k_1,\dots,k_{v-1})}\sum_{i=1}^{l}F_i(n,n_1,\dots,n_r,k_1,\dots,k_v) \tag{6}$$

where the summand is given by hypergeometric terms (e.g., in terms of the Gamma-function) and harmonic sums that arise in the numerators. Then the essential problem is the simplification of these sums to special functions, like harmonic sums (5), generalized harmonic sums [37, 38]

$$S_{c_1,\dots,c_r}(x_1,\dots,x_r;k) = \sum_{i_1=1}^{k}\frac{x_1^{i_1}}{i_1^{c_1}}\sum_{i_2=1}^{i_1}\frac{x_2^{i_2}}{i_2^{c_2}}\cdots\sum_{i_r=1}^{i_{r-1}}\frac{x_r^{i_r}}{i_r^{c_r}} \tag{7}$$

with $c_i \in \mathbb{N}\setminus\{0\}$ and $x_i \in \mathbb{R}\setminus\{0\}$, or cyclotomic harmonic sums [39]

$$S_{(a_1,b_1,c_1),\dots,(a_r,b_r,c_r)}(x_1,\dots,x_r;k) = \sum_{i_1=1}^{k}\frac{x_1^{i_1}}{(a_1i_1+b_1)^{c_1}}\sum_{i_2=1}^{i_1}\frac{x_2^{i_2}}{(a_2i_2+b_2)^{c_2}}\cdots\sum_{i_r=1}^{i_{r-1}}\frac{x_r^{i_r}}{(a_ri_r+b_r)^{c_r}} \tag{8}$$

where the a_i, c_i are positive integers, the b_i are non-negative integers with $a_i > b_i$, and $x_i \in \mathbb{R}\setminus\{0\}$. More generally, also binomial sums might arise [40]; for a detailed survey see [41]. For special cases (i.e., if the Gamma-functions in (3) arise in certain form), this simplification can be carried out with efficient methods; see, e.g., [35, 37, 42]. For more general classes the algorithms and packages presented in this article have been heavily used [29, 40, 43–52].

The backbone of all these calculations are the summation tools of `Sigma` based on difference fields. In Section 2 we will get a glimpse of how the difference field machinery works and elaborate the crucial summation techniques based on this approach.
In order to apply these tools systematically, the package `EvaluateMultiSums` [8, 21] has been developed. In this way, one obtains a completely automatic method to simplify multi-sums of the form (6) in terms of indefinite nested product-sums expressions covering as special cases the sums (5), (7) and (8). The underlying ideas of this package are presented in Section 3.
During our calculations in [40, 49, 52] we were faced with expressions with several thousand multi-sums. In order to assist this mass-production the package `SumProduction` [20] has been developed. In particular, it contains a multi-sum simplifier (again based on the package Sigma) that crunches such large expressions to few master sums and the `EvaluateMultiSums` package is only applied to those sums. In addition, a useful framework is provided to apply all the calculations in parallel on distributed machines. This latter package is presented in Section 4. A conclusion of the symbolic summation approach in difference fields and the produced packages is given in Section 5.

[2] If there are poles, these extra evaluations are treated separately by first plugging the values into f and by expanding this expression afterwards.

2. Symbolic summation in difference fields: the Sigma package

An important milestone of symbolic summation is Gosper's telescoping algorithm [53]: given a hypergeometric expression[3] $f(k)$, it finds –in case of existence– a hypergeometric expression $g(k)$ such that the telescoping equation

$$f(k) = g(k+1) - g(k) \tag{9}$$

holds. Then given $g(k)$, one can sum (9) over k and obtains, e.g., the identity

$$\sum_{k=1}^{a} f(k) = g(a+1) - g(1). \tag{10}$$

Moreover, the breakthrough concerning applications was lead by Zeilberger's extension of Gosper's algorithm to creative telescoping [12] in the framework of his holonomic system approach [54]: it enables one to derive recurrence relations for definite hypergeometric sums. In particular, solving such recurrences in terms of hypergeometric expressions [13] gave rise to the following toolbox: given a definite proper hypergeometric sum, one can decide algorithmically if it can be simplified in terms of a linear combination of hypergeometric expressions. For details on this machinery we refer to the pioneering book [14]; for a most recent point of view see [55]. In the last decades many further improvements and generalizations have been accomplished, like for holonomic sequences [34, 56], for non-holonomic sequences like the Stirling numbers [57] or for expressions represented in terms of difference fields. The latter approach started with Karr's telescoping algorithm in $\Pi\Sigma^*$-fields [1]. There one can treat indefinite nested product-sums in a very elegant way; for details see [8].

Definition. Let $f(k)$ be an expression that evaluates at non-negative integers (from a certain point on) to elements of a field $\mathbb{K}$ containing as subfield the rational numbers $\mathbb{Q}$. Then $f(k)$ is called indefinite nested product-sum expression w.r.t. k (over $\mathbb{K}$) if it is composed by elements from the rational function field $\mathbb{K}(k)$, the four operations $(+, -, \cdot, /)$, and indefinite sums and products of the type $\sum_{i=l}^{k} h(i)$ or $\prod_{i=l}^{k} h(i)$ where $l \in \mathbb{N}$ and where $h(i)$ is an indefinite nested product-sum expression w.r.t. i over $\mathbb{K}$ which is free of k.

Typical examples are the sums given in (5), (7) and (8).

2.1. The basic mechanism in difference fields

The basic ideas in the setting of difference fields are presented. This part can be omitted by those readers who are mainly interested in the application of the summation tools.

Consider the following indefinite summation problem: simplify $\sum_{k=1}^{n} f(k)$ with $f(k) = k\, S_1(k)$ where $S_1(k) = \sum_{i=1}^{k} \frac{1}{i}$ denotes the k-th harmonic numbers. To accomplish this task, we hunt for a solution $g(k)$ in terms of $S_1(k)$ such that (9) holds. In terms of the shift operator $\mathscr{S}_k$ w.r.t. k equation (9) reads as follows:

$$f(k) = \mathscr{S}_k g(k) - g(k). \tag{11}$$

First, the summation objects will be represented step by step in a field $\mathbb{F}$, and along with that the shift operator $\mathscr{S}_k$ is rephrased by a field automorphism $\sigma : \mathbb{F} \to \mathbb{F}$.

i) We start with the rational numbers $\mathbb{Q}$ and define the (only possible) field automorphism $\sigma : \mathbb{Q} \to \mathbb{Q}$ with $\sigma(q) = q$ for all $q \in \mathbb{Q}$.

[3] A sequence $f(k)$ (resp. an expression that evaluates to a sequence) is called hypergeometric if there is a rational function $r(x)$ and a $\lambda \in \mathbb{N}$ such that $r(k) = \frac{f(k+1)}{f(k)}$ for all $k \in \mathbb{N}$ with $k \geq \lambda$.

ii) Next, we need to model k with the shift behavior $\mathscr{S}_k k = k + 1$: Since all elements in $\mathbb{Q}$ are constant (i.e., $\sigma(q) = q$ for all $q \in \mathbb{Q}$), we adjoin a variable t_1 to $\mathbb{Q}$ and extend the automorphism to $\sigma : \mathbb{Q}(t_1) \to \mathbb{Q}(t_1)$ with $\sigma(t_1) = t_1 + 1$. In other words, for $f \in \mathbb{Q}(t_1)$ the element $\sigma(f)$ is obtained by replacing any occurrence of t_1 by $t_1 + 1$.

iii) Finally, we represent $S_1(k)$ with the shift behavior $\mathscr{S}_k S_1(k) = S_1(k) + \frac{1}{k+1}$: First, we try to find such an element in $\mathbb{Q}(t_1)$. Namely we look for a $\gamma \in \mathbb{Q}(t_1)$ such that $\sigma(\gamma) = \gamma + \frac{1}{t_1+1}$ or equivalently $\sigma(\gamma) - \gamma = \frac{1}{t_1+1}$ holds. Using our telescoping algorithms from [58] (or, e.g., Gosper's algorithm) proves that such an element γ does not exist. Therefore we adjoin the variable t_2 to $\mathbb{Q}(t_1)$ and extend the automorphism $\sigma : \mathbb{Q}(t_1)(t_2) \to \mathbb{Q}(t_1)(t_2)$ subject to $\sigma(t_2) = t_2 + \frac{1}{t_1+1}$. In other words, for $f \in \mathbb{Q}(t_1)(t_2)$ the element $\sigma(f)$ can be calculated by replacing any occurrences of t_1 by $t_1 + 1$ and of t_2 by $t_2 + \frac{1}{t_1+1}$.

In short, we have constructed a difference field $(\mathbb{F}, \sigma)$ with the rational function field $\mathbb{F} = \mathbb{Q}(t_1)(t_2)$ together with a field automorphism σ. There k and $S_1(k)$ are represented by t_1 and t_2, respectively, and the shift operator $\mathscr{S}_k$ with $\mathscr{S}_k k = k + 1$ and $\mathscr{S}_k S_1(k) = S_1(k) + \frac{1}{k+1}$ is reflected by σ. In this setting, $f(k)$ is given by $\phi = t_1^2 t_2$ and one seeks $\gamma \in \mathbb{Q}(t_1)(t_2)$ such that $\phi = \sigma(\gamma) - \gamma$. With our algorithm we calculate $\gamma = \frac{1}{4}(t_1 - 1)t_1\big(2t_2 - 1\big)$ which delivers the solution $g(k) = \frac{1}{4}(k-1)k\big(2S_1(k) - 1\big)$ for (11) and thus for (9). As a consequence we get (10) which produces the simplification

$$\sum_{k=1}^{n} kS_1(k) = \frac{1}{4}\big(2n(n+1)S_1(n) - (n-1)n\big). \tag{12}$$

Summarizing, we applied the following strategy; for more details we refer to [8].

(1) Represent the involved indefinite nested product-sum expressions, whose evaluation leads to elements from a field $\mathbb{K}$, in a difference field $(\mathbb{F}, \sigma)$. Here $\mathbb{F} = \mathbb{K}(t_1)\dots(t_e)$ is a rational function field where the generators t_i represent the sums and products. Moreover, the shift behavior of the objects is described by a field automorphism $\sigma : \mathbb{F} \to \mathbb{F}$ where for $1 \le i \le e$ either the sum relation $\sigma(t_i) = t_i + a_i$ or the product relation $\sigma(t_i) = a_i\, t_i$ with $0 \ne a_i \in \mathbb{F}_{i-1} := \mathbb{K}(t_1)\dots(t_{i-1})$ hold. As indicated in the example above, it is crucial that a new variable t_i with $\sigma(t_i) = t_i + a_i$ (similar for products) is only adjoined to $\mathbb{F}_{i-1}$ if there is no $\gamma \in \mathbb{F}_{i-1}$ with $\sigma(\gamma) = \gamma + a_i$. Exactly this problem is solvable [1]; for improved algorithms see [3, 58] and references therein.

(2) Solve the underlying summation problem (e.g., again telescoping, but also parameterized telescoping and recurrence solving given below) in this setting.

(3) Reformulate the solution to an expression in terms of indefinite nested product-sum expressions that yields a solution of the given summation problem.

Exactly this construction produces difference fields $(\mathbb{K}(t_1)\dots(t_e), \sigma)$ whose constants remain unchanged; see [1, 9, 59]. Difference fields with this property are also called $\Pi\Sigma^*$-fields. Formally, they can be defined as follows.

Definition. Let $\mathbb{F}$ be a field with characteristic 0 (i.e., the rational numbers are contained as sub-field) and let σ be a field automorphism of $\mathbb{F}$. The constant field of $\mathbb{F}$ is defined by $\mathbb{K} = \{f \in \mathbb{F} | \sigma(f) = f\}$. A difference field $(\mathbb{F}, \sigma)$ with constant field $\mathbb{K}$ is called a $\Pi\Sigma^*$-field if $\mathbb{F} = \mathbb{K}(t_1)\dots(t_e)$ where for all $1 \le i \le e$ each $\mathbb{F}_i = \mathbb{K}(t_1)\dots(t_i)$ is a transcendental field extension of $\mathbb{F}_{i-1} = \mathbb{K}(t_1)\dots(t_{i-1})$ (we set $\mathbb{F}_0 = \mathbb{K}$) and σ has the property that $\sigma(t_i) = a\, t_i$ or $\sigma(t_i) = t_i + a$ for some $a \in \mathbb{F}_{i-1}$.

In conclusion, all the summation paradigms presented in the next subsections rely on this mechanism: Reformulate the given problem in a $\Pi\Sigma^*$-field (or ring), solve it there and formulate the result back such that it is a solution of the input problem. In the following we will give an overview of `Sigma`'s symbolic summation toolbox that is based on the difference field (resp. ring) approach introduced above.

2.2. Simplification of indefinite nested product-sum expressions

Whenever `Sigma` deals with indefinite nested product-sum expressions, in particular when it outputs such expressions, the following problem is solved implicitly.

Problem EAR: Elimination of algebraic relations. *Given* an indefinite nested product-sum expression $f(k)$. *Find* an indefinite nested product-sum expression $F(k)$ and $\lambda \in \mathbb{N}$ such that $f(k) = F(k)$ for all $k \geq \lambda$ and such that the occurring sums are algebraically independent.

As worked out above, the following mechanism is applied: $f(k)$ is rephrased in a suitable $\Pi\Sigma^*$-field $(\mathbb{K}(t_1)\dots(t_e),\sigma)$ (or ring); this is accomplished by solving iteratively the telescoping problem. Here the sums (similarly the products) are represented by the variables t_i. Finally, reinterpreting the t_i as sums produces an alternative expression $F(k)$ of $f(k)$ where the occurring sums are algebraically independent; the solution to this problem relies on results of [7] and is worked out in detail in [8]. We remark that that the found relations for harmonic sums, cyclotomic sums and generalized harmonic sums coincide with the derived relations [38, 39, 60, 61] that are obtained by using the underlying quasi-shuffle algebras [35, 37, 62].

After loading the `Sigma` package into Mathematica:

In[1]:= **<< Sigma.m**

`Sigma − A summation package by Carsten Schneider© RISC`

the simplification in (12) can be accomplished as follows[4]:

In[2]:= $\mathbf{SigmaReduce}[\sum_{k=1}^{n} k\,S[1,k], n]$

Out[2]= $\frac{1}{4}(2n(n+1)S[1,n] - (n-1)n)$

Here the sum $\sum_{k=1}^{n} k\,S_1(k)$ has been reduced in terms of the objects k and $S_1(k)$. In particular, by difference field theory it follows that the sequence given by $S_1(k)$ is transcendental (resp. algebraically independent) over the rational sequences, more precisely, the sequences that one obtains by evaluating the elements of $\mathbb{Q}(k)$.

In particular, using improved telescoping algorithms, the underlying $\Pi\Sigma^*$-field can be constructed in such a way that the sums are given with certain optimality criteria.

Refined telescoping. *Given* an indefinite nested product-sum expression $f(k)$. *Find* an indefinite nested product-sum expression $g(k)$ such that (9) holds and such that $g(k)$ is as "simple" as possible.

Then summing the found equation (9) over k yields, e.g., the identity (10) where also the right hand side is as simple as possible. Subsequently, we present the main features of `Sigma`; some of these simplifications are carried out by default, some must be activated explicitly (see below).

2.2.1. Sum representations with optimal nesting depth The found expression $F(k)$ can be given with minimal nesting depth and for any occurring sum in $F(k)$ there is no other indefinite nested sum representation with lower nesting depth; for details see [2–4]. A typical example is as follows.

In[3]:= $\mathbf{SigmaReduce}[2^n \sum_{i=1}^{n} \frac{1}{i} \sum_{j=1}^{i} \frac{1}{2^j} \sum_{k=1}^{j} \frac{2^k}{k} \sum_{l=1}^{k} \frac{1}{2^l}, n]$

Out[3]= $2^n\Big(\big(\sum_{i=1}^{n}\frac{1}{i}\big)^2 + \big(\sum_{i=1}^{n}\frac{1}{2^i i}\big)\sum_{i=1}^{n}\frac{1}{i} + \sum_{i=1}^{n}\frac{1}{i^2} + \sum_{i=1}^{n}\frac{1}{2^i i^2} - 3\sum_{i=1}^{n}\frac{\sum_{j=1}^{i}\frac{1}{2^j j}}{i} - \sum_{i=1}^{n}\frac{\frac{1}{2^i}\sum_{j=1}^{i}\frac{2^j}{j}}{i}\Big)$

Here the product $2^n = \prod_{i=1}^{n} 2$ and the arising sums are algebraically independent over the rational sequences, and the sums have minimal nesting depth. In particular, the expression

[4] Here $S[c_1,\dots,c_r,k]$ denotes the harmonic sum (5).

in Out[3] cannot be written in terms of indefinite nested sums with lower nesting depth. Similarly, the following sum expression can be reduced to its optimal nesting depth.

In[4]:= $\textbf{SigmaReduce}[\sum_{k=0}^{a}\Big(\sum_{j=0}^{k}\frac{x^j}{\binom{n}{j}}\Big)^2]$

Out[4]= $\frac{-2x^{a+1}(a+1)}{(x+1)\binom{n}{a}}\sum_{i_1=0}^{a}\frac{x^{i_1}}{\binom{n}{i_1}}+\frac{(n+x+1)}{x+1}\sum_{i_1=0}^{a}\frac{x^{2i_1}}{\binom{n}{i_1}^2}+\frac{(x-1)}{x+1}\sum_{i_1=0}^{a}\frac{x^{2i_1}i_1}{\binom{n}{i_1}^2}+\frac{(ax+a-n+2x)}{x+1}\Big(\sum_{i_1=0}^{a}\frac{x^{i_1}}{\binom{n}{i_1}}\Big)^2$

We emphasize that this depth-optimal representation leads to efficient algorithms for telescoping, creative telescoping and recurrence solving; for details we refer to [3, 63]. Thus by efficiency reasons this feature of `Sigma` is activated by default.

2.2.2. Reducing the number of objects and the degrees in the summand The depth-optimal representation can be refined further as follows: *find* an alternative sum representation such that for the outermost summands the number of occurring objects is as small as possible. This problem was originally solved in [5]. For a more efficient and simplified algorithm see [58].
In the following example we can eliminate $S_1(k)$ from the summand:

In[5]:= $\textbf{SigmaReduce}[\sum_{k=0}^{a}(-1)^k S[1,k]^2\binom{n}{k}, a, \textbf{SimplifyByExt} \to \textbf{DepthNumber}]$

Out[5]= $-(a-n)\big(n^2 S[1,a]^2+2n\,S[1,a]+2\big)\frac{(-1)^a\binom{n}{a}}{n^3}-\frac{2}{n^2}-\frac{1}{n}\sum_{i_1=1}^{a}\frac{(-1)^{i_1}}{i_1}\binom{n}{i_1}$

Furthermore, one can calculate representations such that the degrees (w.r.t. the top extension of a $\Pi\Sigma^*$-field) in the numerators and denominators of the summands are minimal [6]:

In[6]:= $\textbf{SigmaReduce}[\sum_{k=0}^{a}(-1)^k S[1,k]^3\binom{n}{k}]$

Out[6]= $-\frac{3}{n^2}\sum_{i_1=1}^{a}\frac{(-1)^{i_1}\binom{n}{i_1}}{i_1}-\frac{3}{n}\sum_{i_1=1}^{a}\frac{(-1)^{i_1}\binom{n}{i_1}S[1,i_1]}{i_1}+\frac{1}{n}\sum_{i_1=1}^{a}\frac{(-1)^{i_1}\binom{n}{i_1}}{i_1^2}+(-1)^a\binom{n}{a}\Big(\frac{6(n-a)}{n^4}+\frac{6(n-a)S[1,a]}{n^3}-\frac{3(a-n)S[1,a]^2}{n^2}+\frac{(n-a)S[1,a]^3}{n}\Big)-\frac{6}{n^3}$

For algorithms dealing with the product case we point to [64, 65].

2.2.3. Minimal degrees w.r.t. the summation index By default `Sigma` outputs sums such that the denominators have minimal degrees w.r.t. the summation index (i.e., if possible, the denominator w.r.t. the summation index is linear). This feature is of particular importance to rewrite sums in terms of harmonic sums and their generalized versions. A typical example is

In[7]:= $\textbf{SigmaReduce}[\sum_{k=1}^{a}\Big(\frac{-2+k}{10(1+k^2)}+\frac{(1-4k-2k^2)S[1,k]}{10(1+k^2)(2+2k+k^2)}+\frac{(1-4k-2k^2)S[3,k]}{5(1+k^2)(2+2k+k^2)}\Big), a]$

Out[7]= $\frac{a^2+4a+5}{10(a^2+2a+2)}S[1,a]-\frac{(a-1)(a+1)}{5(a^2+2a+2)}S[3,a]-\frac{2}{5}\sum_{k=1}^{a}\frac{1}{k^2}$

2.3. The summation paradigms for definite summation
Definite sums over indefinite nested product sums[5], like

$$A(n)=\sum_{k=0}^{n}\binom{n}{k}S_1(k)^2=\sum_{k=0}^{n}\Big(\prod_{i=1}^{k}\frac{n+1-i}{i}\Big)\Big(\sum_{i=1}^{k}\frac{1}{i}\Big)^2, \tag{13}$$

can be handled by the following summation paradigms (see Fig. 1).

2.3.1. Finding recurrences by parameterized (creative) telescoping First, there is the following tool in the setting of $\Pi\Sigma^*$-fields to obtain recurrences; for the most recent summary see [58].

[5] Definite means that the upper bound is ∞ or consists of parameters that occur also inside of the sum.

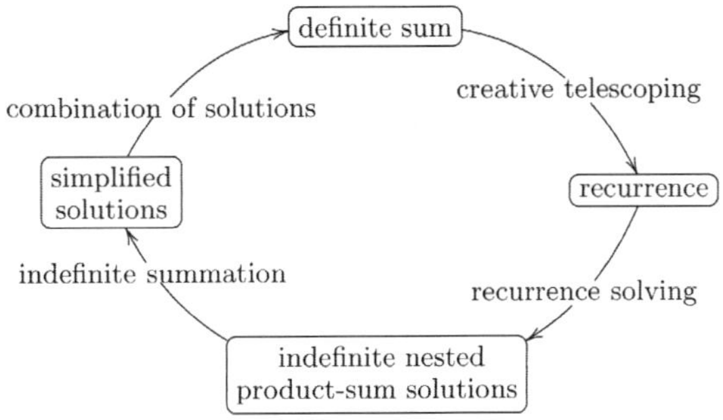

Figure 1. Sigma's summation spiral; see [10].

Problem PT: Parameterized Telescoping. *Given* indefinite nested product-sum expressions $f_0(k), \dots, f_\delta(k)$. *Find* constants $a_0, \dots, a_\delta$, not all 0 and all free of k, and *find* an indefinite nested product-sum expression $g(k)$ being not more complicated than the $f_i(k)$ such that

$$a_0\, f_0(k) + a_1\, f_1(k) + \dots + a_\delta\, f_\delta(k) = g(k+1) - g(k). \tag{14}$$

For simplicity suppose that the found relation (14) holds for $0 \le k \le a$. Then by telescoping one gets, e.g., the sum relation

$$a_0 \sum_{k=0}^{a} f_0(k) + a_1 \sum_{k=0}^{a} f_1(k) + \dots + a_\delta \sum_{k=0}^{a} f_\delta(k) = g(a+1) - g(0) \tag{15}$$

where the right hand side is simpler than the sums of the left hand side.
Specializing to $f_i(k) := f(n+i, k)$ for a bivariate expression yields the creative telescoping paradigm. Here one loops over[6] $\delta = 0, 1, 2, \dots$ and tries to solve the corresponding parameterized telescoping problem. If the method stops, then we can deduce (15), and we get the recurrence

$$a_0(n)\, A'(n) + a_1(n)\, A'(n+1) + \dots + a_\delta(n)\, A'(n+\delta) = g(a+1) - g(0)$$

for the sum $A'(n) = \sum_{k=0}^{a} f(n,k)$. To this end, specializing $a = n$ and taking care of extra terms yield a recurrence for the sum $A(n) = \sum_{k=0}^{n} f(n,k)$ of the form

$$a_0(n)A(n) + a_1(n)A(n+1) + \dots + a_\delta(n)A(n+\delta) = h(n). \tag{16}$$

For instance, take $f_i(k) = \binom{n+i}{k} S_1(k)^2 = \prod_{j=1}^{i} \frac{n+j}{n-k+j} \binom{n}{k} S_1(k)^2$. Then we find a parameterized telescoping solution (14) for $\delta = 4$. Performing the steps above, we finally get a recurrence of order 4 for our sum (13). All these steps can be carried out for $A(n) =$ `SUM[n]` as follows:

In[8]:= $\textbf{mySum} = \sum_{k=1}^{n} \binom{n}{k} \mathbf{S[1,k]^2};$

In[9]:= $\textbf{rec = GenerateRecurrence[mySum, n][[1]]}$

Out[9]= $8(n+1)(n+3)\mathrm{SUM}[n] - 4(5n^2+25n+29)\mathrm{SUM}[n+1] + 2(3n+8)(3n+10)\mathrm{SUM}[n+2] - (7n^2+49n+86)\mathrm{SUM}[n+3] + (n+4)^2\mathrm{SUM}[n+4] = 1$

We remark that the refined telescoping algorithms from Subsection 2.2 provide also refined tools for parameterized/creative telescoping [3, 58]: For our sum (13) we can calculate a recurrence of order 2 instead of order 4 by using extra sum extensions:

[6] Note that the special case $\delta = 0$ is telescoping.

In[10]:= **GenerateRecurrence[mySum, n, SimlifyByExt → DepthNumber][[1]]**

Out[10]= $4(n+1)\mathrm{SUM}[n] - 2(2n+3)\mathrm{SUM}[n+1] + (n*2)\mathrm{SUM}[n+2] == \frac{(3n+4)}{n+2}\sum_{i_1=0}^{n}\frac{\binom{n}{i_1}}{1+n-i_1} + \sum_{i_1=1}^{n}\frac{\binom{n}{i_1}}{i_1} + \frac{1}{n+2}$

Note that the found sums on the right hand side are definite. Simplifying these sums (by using just the methods that we describe here), we end up at the recurrence

$$4(n+1)A(n) - 2(2n+3)A(n+1) + (n+2)A(n+2) = S_1(2;n) - S_1(n) + \tfrac{2^{n+1}(3n+4)-(2n+3)}{(n+1)(n+2)}.$$

Usually such shorter recurrences are easier to solve. In order to demonstrate the summation tools below, we will continue with the recurrence given in Out[9].

2.3.2. Solving recurrences Next, we can apply the following recurrence solver [66] which is based on [33, 67] and generalizes ideas of [13, 68].

> **Problem RS: Recurrence Solving.**
> *Given* a recurrence of the form (16) where the coefficients $a_i(n)$ and $h(n)$ are given in terms of indefinite nested product-sum expressions. *Find* all solutions that are expressible in terms of indefinite nested product sum expressions.

This solver finds, if possible, a particular solution of (16) in terms of indefinite nested product-sum expressions; and it finds a linear independent set of expressions in terms of indefinite nested product-sums with the following property: their linear combinations produce all solutions of the homogeneous version of (16) that can be expressed in terms of indefinite nested product-sum expressions. The sequences generated by these solutions are called d'Alembertian solutions, a subclass of Liouvillian solutions [69]. For details dealing with the rational case see [70].

Our recurrence rec=Out[9] can be solved with the following command of Sigma:

In[11]:= **recSol = SolveRecurrence[rec, SUM[n], IndefiniteSummation → False]**

Out[11]= $\{\{0, 2^n\}, \{0, 2^n\sum_{i=1}^{n}\frac{1}{i}\}, \{0, 2^n\sum_{i=1}^{n}\frac{\sum_{j=1}^{i}\frac{1}{2^j}}{i}\}, \{0, 2^n\sum_{i=1}^{n}\frac{\sum_{j=1}^{i}\frac{1}{2^j}\sum_{k=1}^{j}\frac{2^k}{k}}{i}\}, \{1, 2^n\sum_{i=1}^{n}\frac{\sum_{j=1}^{i}\frac{1}{2^j}\sum_{k=1}^{j}\frac{2^k\sum_{l=1}^{k}\frac{1}{2^l}}{k}}{i}\}\}$

More precisely, we obtain the general solution

$$2^n\left[c_1 + c_2\sum_{i=1}^{n}\frac{1}{i} + c_3\sum_{i=1}^{n}\frac{1}{i}\sum_{j=1}^{i}\frac{1}{2^j} + c_4\sum_{i=1}^{n}\frac{1}{i}\sum_{j=1}^{i}\frac{1}{2^j}\sum_{k=1}^{j}\frac{2^k}{k} + \sum_{i=1}^{n}\frac{1}{i}\sum_{j=1}^{i}\frac{1}{2^j}\sum_{k=1}^{j}\frac{2^k}{k}\sum_{l=1}^{k}\frac{1}{2^l}\right], \quad c_1, c_2, c_3, c_4 \in \mathbb{Q}.$$

By construction the solutions are highly nested: in particular the depth of the particular solution equals usually the recurrence order plus the nesting depth of the inhomogeneous part of the recurrence. It is therefore a crucial (and often the most challenging) task to simplify these solutions further by solving Problem EAR and applying the refined telescoping algorithms from Subsection 2.2. For the simplification of the particular solution see Out[3] from above. With Sigma the recurrence rec given in Out[9] is solved (see Problem RS) and the found solutions are simplified (with the default options) by the following function call:

In[12]:= **recSol = SolveRecurrence[rec, SUM[n]]**

Out[12]= $\{\{0, 2^n\}, \{0, 2^n\sum_{i=1}^{n}\frac{1}{i}\}, \{0, 2^n(2\sum_{i=1}^{n}\frac{1}{i} - 2\sum_{i=1}^{n}\frac{1}{2^i i})\}, \{0, 2^n((\sum_{i=1}^{n}\frac{1}{i})^2 + \sum_{i=1}^{n}\frac{1}{i^2} - \sum_{i=1}^{n}\frac{1}{2^i i}\sum_{j=1}^{i}\frac{2^j}{j})\},$
$\{1, 2^n((\sum_{i=1}^{n}\frac{1}{i})^2 + \sum_{i=1}^{n}\frac{1}{i^2} + (\sum_{i=1}^{n}\frac{1}{i})\sum_{i=1}^{n}\frac{1}{2^i i} + \sum_{i=1}^{n}\frac{1}{2^i i^2} - \sum_{i=1}^{n}\frac{1}{2^i i}\sum_{j=1}^{i}\frac{2^j}{j} - 3\sum_{i=1}^{n}\frac{1}{i}\sum_{j=1}^{i}\frac{1}{2^j j})\}\}$

Example. In [45, 46] we computed a large amount of initial values (up to 3500) of the massless Wilson coefficients to 3-loop order for individual color coefficients by taking the result given in [71]. Next, we guessed recurrences of minimal order for these coefficients by using Kauer's package `Guess` [72]. Then, e.g., the largest recurrence of order 35 could be solved completely in about 1 day. This yielded 35 linearly independent solutions in terms of sums up to nesting depth 34. To this end, their simplifications in terms of harmonic sums took about 5 days.

2.3.3. Combining the solutions Finally, we take the linear combination of the homogeneous solutions (the entries with a 0) plus the particular solution (the entry with a 1) such that it agrees with $A(n)$ for $n = 1, 2, 3, 4$. This combination can be calculated by calling the function

In[13]:= **sol = FindLinearCombination[recSol, mySum, n, 4]**

Out[13]= $2^n\left(\left(\sum_{i=1}^{n}\frac{1}{i}\right)^2+\sum_{i=1}^{n}\frac{1}{i^2}+\left(\sum_{i=1}^{n}\frac{1}{i}\right)\sum_{i=1}^{n}\frac{1}{2^i i}+\sum_{i=1}^{n}\frac{1}{2^i i^2}-\sum_{i=1}^{n}\frac{1}{2^i i}\sum_{j=1}^{i}\frac{2^j}{j}-3\sum_{i=1}^{n}\frac{1}{i}\sum_{j=1}^{i}\frac{1}{2^j j}\right)$

Since the sum (13) and the derived expression in Out[13] agree for $n = 1, 2, 3, 4$ and since both are solutions of the recurrence Out[9], they evaluate to the same sequence for all $n \in \mathbb{N}$.

In order to rewrite the found expression Out[13] in terms of harmonic sums and their generalized versions, we load in the `HarmonicSums` package and execute the following function

In[14]:= **<< HarmonicSums.m**

`HarmonicSums by Jakob Ablinger — © RISC`

In[15]:= **TransformToSSums[sol]**

Out[15]= $2^n\left(\mathtt{S}[1,n]^2+\mathtt{S}[2,n]+\mathtt{S}[1,n]\mathtt{S}[1,\{\tfrac{1}{2}\},n]\right)+\mathtt{S}[2,\{\tfrac{1}{2}\},n]-\mathtt{S}[1,1,\{\tfrac{1}{2},2\},n]-3\mathtt{S}[1,1,\{1,\tfrac{1}{2}\},n]\big)$

To sum up, using the summation paradigms given in Fig. 1, we computed for the definite sum (13) the closed form in terms of generalized harmonic sums:

$$A(n) = 2^n\left[S_1(n)^2 + S_2(n) + S_1(n)S_1\left(\tfrac{1}{2};n\right) + S_2\left(\tfrac{1}{2};n\right) - S_{1,1}\left(\tfrac{1}{2},2;n\right) - 3S_{1,1}\left(1,\tfrac{1}{2};n\right)\right)\right]. \quad (17)$$

Remark. We highlight that all the calculation steps can be verified independently of the way how the (complicated) algorithms work. In this way, we obtain rigorous computer proofs.

3. Automatic simplification of multiple sums: the `EvaluateMultiSums` package

Using `Sigma`'s summation tools (see Fig. 1) the derivation of the right hand side of (17) can be done completely automatically with the package

In[16]:= **<< EvaluateMultiSums.m**

`EvaluateMultiSums by Carsten Schneider — © RISC`

Namely, our sum (13) can be simplified to indefinite nested sums by executing

In[17]:= **EvaluateMultiSum[$\sum_{k=1}^{n}$ S[1, k]$^2\binom{n}{k}$, {{k, 0, n}}, {n}, {0}, {∞}]**

Out[17]= $2^n\left(\mathtt{S}[1,n]^2+\mathtt{S}[2,n]+\mathtt{S}[1,n]\mathtt{S}[1,\{\tfrac{1}{2}\},n]\right)+\mathtt{S}[2,\{\tfrac{1}{2}\},n]-\mathtt{S}[1,1,\{\tfrac{1}{2},2\},n]-3\mathtt{S}[1,1,\{1,\tfrac{1}{2}\},n]\big)$

The underlying method applies systematically the tools of `Sigma` (taking care of all the different options to find the optimal treatment) until the simplification is accomplished.

The key point is that this method can be applied iteratively to multiple sums. E.g., in recent calculations we succeeded in deriving the massive 3-loop OMEs $A_{qq,Q}^{(3),\mathrm{NS}}$ and $A_{qq,Q}^{(3),\mathrm{NS,TR}}$ for general values of N, in particular we obtained the Wilson coefficient $L_{qq,Q}^{(3),\mathrm{NS}}$; for further comments we refer to [40]. Here one (of many multi-sums) is the hypergeometric quadruple sum

$$\mathcal{I}(\varepsilon,n) = \sum_{m=0}^{n}\sum_{j=0}^{m}\sum_{k=0}^{\infty}\sum_{i=0}^{k} f(\varepsilon,n,m,j,k,i) \quad (18)$$

where the summand is given in terms of Gamma functions and binomial coefficients:

In[18]:= $$\mathbf{f} = \frac{(-1)^{i+m+1} e^{-\frac{3\varepsilon\gamma}{2}} \binom{k}{i}\binom{m}{j}\binom{N}{m}\Gamma(1-\frac{\varepsilon}{2})\Gamma(\frac{\varepsilon}{2})\Gamma(i-\frac{\varepsilon}{2})}{k!\Gamma(-\frac{\varepsilon}{2}+i+1)\Gamma(-\frac{3\varepsilon}{2}+i+j+2)} \frac{\Gamma(-\varepsilon+i+j+1)\Gamma(k-\frac{3\varepsilon}{2})\Gamma(\frac{\varepsilon}{2}+j+k+1)\Gamma(m+2)\Gamma(-\varepsilon-j+m+1)}{\Gamma(\frac{\varepsilon}{2}+m+2)\Gamma(-\frac{\varepsilon}{2}+k+m+2)}$$

In this instance, the first 5 coefficients $I_{-3}(n), \ldots, I_1(n)$ of its Laurent series expansion (2) with $t = -3$ were needed. For this task (see Section 1) we first expand the summand

$$f(\varepsilon, n, m, j, k, i) = F_{-3}(n, m, j, k, i)\varepsilon^{-3} + \cdots + F_1(n, m, j, k, i)\varepsilon^1 + O(\varepsilon^2). \qquad (19)$$

Then the coefficients $I_i(n)$ are given by applying the sums to the $F_i(n, m, j, k, i)$. E.g., we compute $F_{-1}(n, m, j, k, i)$ by using the following function of the package `EvaluateMultiSums`:

In[19]:= $\mathbf{F_{-1} = SeriesForProduct[f, \{\varepsilon, -1, -1\}]}$

Out[19]= $$\frac{(-1)^{i+m+1} 2\binom{k}{i}\binom{m}{j}\binom{n}{m}(i+j)!(k-1)!(j+k)!(m-j)!}{i(i+j+1)!k!(k+m+1)!}$$

In other words, the main task is to simplify the definite multi-sum

$$\overbrace{\sum_{m=0}^{n}}^{(d)} \overbrace{\sum_{j=0}^{m}}^{(c)} \overbrace{\sum_{k=0}^{\infty}}^{(b)} \overbrace{\sum_{i=1}^{k}}^{(a)} \frac{(-1)^{i+m+1} 2\binom{k}{i}\binom{m}{j}\binom{n}{m}(i+j)!(k-1)!(j+k)!(m-j)!}{i(i+j+1)!k!(k+m+1)!}; \qquad (20)$$

note that the inner sum (a) starts with $i = 1$ since there is a pole at $i = 0$. Exactly here (as for all the other multi-sums in this context) the presented toolbox can be exploited. The sum (a) is a definite sum over an indefinite nested product-sum. Therefore we can activate our definite summation technology (Fig. 1): try to transform it to an indefinite nested expression w.r.t. k (which is the summation index of the next sum (b)). Namely, we execute the function[7]:

In[20]:= $\mathbf{sumA = EvaluateMultiSum[F_{-1}, \{\{i, 1, k\}\}, \{k, j, m, n\}, \{0, 0, 0, 2\}, \{\infty, m, n, \infty\}]}$

Out[20]= $$-2\frac{(-1)^m m!}{(k+m+1)!}\left(\left(\frac{1}{(j+1)^2 k} - \frac{S[1,k]}{(j+1)k}\right)\frac{\binom{n}{m}(j+k)!}{j!} - \frac{k!\binom{n}{m}}{(j+1)k(j+k+1)}\right)$$

Given this form, the next sum (b) fits again to our summation paradigm: it is a definite sum over an indefinite nested product sum-expression w.r.t. k. Hence with the function call

In[21]:= $\mathbf{sumB = EvaluateMultiSum[sumA, \{\{k, 0, \infty\}\}, \{j, m, n\}, \{0, 0, 2\}, \{m, n, \infty\}]}$

Out[21]= $$\frac{(-1)^m\binom{n}{m}}{(j+1)(m+1)}\Big(\sum_{i_1=1}^{j}\frac{1}{1+m-i_1}\Big)^2 - 2\frac{(-1)^m\binom{n}{m}}{(j+1)^2(m+1)}\sum_{i_1=1}^{j}\frac{1}{1+m-i_1} + \frac{2(-1)^m\binom{n}{m}}{(j+1)^3(m+1)}$$
$$+ (-1)^m\Big(2\frac{\binom{n}{m}(-m)_j}{(j+1)^2 j!} - \frac{2\binom{n}{m}}{(j+1)(m+1)}\Big)S[2,m] + \frac{(-1)^m\binom{n}{m}}{(j+1)(m+1)}\sum_{i_1=1}^{j}\frac{1}{(1+m-i_1)^2} - 2\frac{(-1)^m\binom{n}{m}(-m)_j}{(j+1)^2 j!}\sum_{i_1=1}^{j}\frac{i_1!}{(-m)_{i_1} i_1^2}$$
$$+ 2\frac{(-1)^m\binom{n}{m}}{(j+1)(m+1)}\sum_{i_1=1}^{j}\frac{\sum_{i_2=1}^{i_1}\frac{1}{1+m-i_2}}{i_1} + (-1)^m\Big(\frac{2\binom{n}{m}}{(j+1)(m+1)} - 2\frac{\binom{n}{m}(-m)_j}{(j+1)^2 j!}\Big)z_2$$

we obtain an indefinite nested product-sum expression w.r.t. j (if sums are free of j, they are indefinite nested w.r.t. m, etc.). Again we are ready to apply our summation toolkit:

In[22]:= $\mathbf{sumC = EvaluateMultiSum[sumB, \{\{j, 0, m\}\}, \{m, n\}, \{0, 2\}, \{n, \infty\}]}$

Out[22]= $$(-1)^m\Big(2\frac{(-1)^m(-n)_m}{(m+1)^4 m!} + 2\frac{(-1)^m(-n)_m}{(m+1)^3 m!}S[1,m] - 2 - \frac{(-1)^m(-n)_m}{(m+1)m!}S[2,1,m]\Big)$$

This yields an indefinite nested product sum expression w.r.t. m. To this end, the outermost sum (d) is transformed to an indefinite nested product-sum expression

In[23]:= $\mathbf{sumD = EvaluateMultiSum[sumC, \{\{m, 0, n\}\}, \{n\}, \{2\}, \{\infty\}]}$

Out[23]= $$\frac{2S[2,n]}{(n+1)^2} + \frac{2}{(n+1)^4}$$

which is nothing else than the simplification of (20). The full power of this machinery comes into action if the function is applied in one stroke by the following function call:

[7] The free integer parameters k, j, m, n are given explicitly where their lower values are $0, 0, 0, 2$ and their upper values are ∞, m, n, ∞, respectively. I.e., $0 \le k \le \infty$, $0 \le j \le m$, $0 \le m \le \infty$ and $2 \le n \le \infty$. In particular, the given order k, j, m, n specifies how the sums should be transformed: if a sum depends on k, it should be transformed to indefinite sums w.r.t. k; if it is free of k, but depends on j, it should be transformed w.r.t. j, etc.

In[24]:= **EvaluateMultiSum[F$_{-1}$, {{i, 1, k}, {k, 0, ∞}, {j, 0, m}, {m, 0, N}}, {n}, {2}, {∞}]**

Out[24]= $\frac{2\mathtt{S}[2,\mathtt{n}]}{(\mathtt{n}+1)^2}+\frac{2}{(\mathtt{n}+1)^4}$

More generally, we can deal with the following problem.

Problem EMS: EvaluateMultiSum. *Given*

$$F(\vec{m}) = \sum_{k_1=l_1}^{L_1(\vec{m})} \cdots \sum_{k_v=l_v}^{L_v(\vec{m},k_1,\ldots,k_{v-1})} f(\vec{m}, k_1, \ldots, k_v) \qquad (21)$$

with an indefinite nested product-sum expression f w.r.t. k_v, integer parameters $\vec{m} = (m_1 \ldots, m_r)$; $l_i \in \mathbb{N}$ and $L_i(\ldots)$ stands for ∞ or a linear combination of the involved parameters with integer coefficients. *Find* an indefinite nested product-sum expression[7] $\bar{F}(\vec{m})$ which evaluates to the same expression as $F(\vec{m})$.

[7] If a sum depends on m_r, it should occur only in the outermost bound. If it is free of m_r, but depends on m_{r-1}, the parameter m_{r-1} should only occur in the outermost bound, etc.

Moreover, if one uses for instance the option `ExpandIn`→ $\{\varepsilon, -3, 1\}$ also the expansion feature is applied. More precisely, the summand (19) is expanded and afterwards the summation machinery is applied. More precisely, with the following function call we arrive at the coefficients $I_{-3}(n), \ldots, I_1(n)$ in terms of harmonic sums and the Riemann Zeta values $z_i = \zeta(i) = \sum_{k=1}^{\infty} 1/k^i$:

In[25]:=**EvaluateMultiSum[f, {{i, 0, k}, {k, 0, ∞}, {j, 0, m}, {m, 0, N}}, {n}, {2}, {∞}, ExpandIn→{ε, −3, 1}]**

Out[25]=
$$\Big\{-\frac{8}{3(\mathtt{n}+1)^2}, -\frac{8\mathtt{S}[1,\mathtt{n}]}{3(\mathtt{n}+1)^2}, -\frac{2\mathtt{S}\mathtt{S}[1,\mathtt{n}]^2}{3(\mathtt{n}+1)^2}-\frac{2\mathtt{S}[2,\mathtt{n}]}{3(\mathtt{n}+1)^2}-\frac{\mathtt{z}_2}{(\mathtt{n}+1)^2}-\frac{8}{3(\mathtt{n}+1)^4},$$
$$-\frac{\mathtt{S}[1,\mathtt{n}]^3}{9(\mathtt{n}+1)^2}+\Big(-\frac{\mathtt{S}[2,\mathtt{n}]}{(\mathtt{n}+1)^2}-\frac{8}{3(\mathtt{n}+1)^4}\Big)\mathtt{S}[1,\mathtt{n}]-\frac{\mathtt{z}_2\mathtt{S}[1,\mathtt{n}]}{(\mathtt{n}+1)^2}-\frac{8\mathtt{S}[3,\mathtt{n}]}{9(\mathtt{n}+1)^2}-\frac{2\mathtt{S}[2,1,\mathtt{n}]}{3(\mathtt{n}+1)^2}-\frac{5\mathtt{z}_3}{3(\mathtt{n}+1)^2},$$
$$-\frac{\mathtt{S}[1,\mathtt{n}]^4}{72(\mathtt{n}+1)^2}+\Big(-\frac{\mathtt{S}[2,\mathtt{n}]}{4(\mathtt{n}+1)^2}-\frac{2}{3(\mathtt{n}+1)^4}\Big)\mathtt{S}[1,\mathtt{n}]^2+\Big(\frac{\mathtt{S}[2,\mathtt{n}]}{(\mathtt{n}+1)^3}+\frac{2\mathtt{S}[3,\mathtt{n}]}{9(\mathtt{n}+1)^2}-\frac{\mathtt{S}[2,1,\mathtt{n}]}{(\mathtt{n}+1)^2}\Big)\mathtt{S}[1,\mathtt{n}]-\frac{3\mathtt{S}[2,\mathtt{n}]^2}{8(\mathtt{n}+1)^2}-\frac{23\mathtt{z}_2^2}{16(\mathtt{n}+1)^2}-$$
$$\frac{2\mathtt{S}[2,\mathtt{n}]}{3(\mathtt{n}+1)^4}+\frac{\mathtt{S}[3,\mathtt{n}]}{(\mathtt{n}+1)^3}+\frac{\mathtt{S}[3,\mathtt{n}]}{12(\mathtt{n}+1)^2}-\frac{\mathtt{S}[2,1,\mathtt{n}]}{(\mathtt{n}+1)^3}-\frac{5\mathtt{S}[3,1,\mathtt{n}]}{3(\mathtt{n}+1)^2}+\frac{5\mathtt{S}[2,1,1,\mathtt{n}]}{3(\mathtt{n}+1)^2}+\Big(-\frac{\mathtt{S}[1,\mathtt{n}]^2}{4(\mathtt{n}+1)^2}-\frac{\mathtt{S}[2,\mathtt{n}]}{4(\mathtt{n}+1)^2}-\frac{1}{(\mathtt{n}+1)^4}\Big)\mathtt{z}_2+$$
$$\Big(\frac{2}{(\mathtt{n}+1)^3}-\frac{5\mathtt{S}[1,\mathtt{n}]}{3(\mathtt{n}+1)^2}\Big)\mathtt{z}_3-\frac{8}{3(\mathtt{n}+1)^6}\Big\}$$

We remark that during these calculations exceptional points at the summation borders are carefully treated (like the point $i = 0$ that does not hold for the sum representation (20)).

As demonstrated above, the definite summation spiral in Fig. 1 (finding a recurrence, solving the recurrence, combining the recurrence using initial values[8]) is applied iteratively from inside to outside, and the corresponding sums are transformed stepwise to indefinite nested product-sum expressions. For a description of the full method we refer to [20, 58]. Here we want to stress the following aspect: For an arbitrary input, the method might fail. First, there might not exist a recurrence for a certain subproblem. However, for Feynman integrals as given in Section 1 the arising multi-sums have the appropriate shape to guarantee that the recurrence finder is always successful; this follows by ideas from [14, 27, 30, 31]. Here only time and space resources might be the bottleneck. Second, our recurrence solver might fail to find sufficiently many solutions in terms of indefinite nested product-sum expressions. And exactly here the miracle happens –at least for the classes of Feynman integrals that we considered so far: Problem RS produces usually the full solution space for the occurring recurrences. Consequently, the solutions can be always combined to an alternative representation for the input sum. In summary, the presented method works very well for big classes of Feynman integrals.

[8] The initial values (e.g., $n = 1, 2, 3$) can be obtained by the same method (with one parameter less); see [73]. As a consequence, we find indefinite nested product-sum expressions which usually simplify to multiple zeta values, infinite versions of S-sums or cyclotomic sums. Then techniques from [35, 38, 39, 74] are heavily needed to rewrite the constants in this special form and to express them in terms of constants such that no further algebraic relations are known. In order to deal with such problems, Ablinger's `HarmonicSums` package is used.

4. Crunching sums and mass production: the package `SumProduction`

So far we presented symbolic summation technologies and the related packages that enables one to simplify multi-sums to expressions in terms of indefinite nested product-sum expressions. For various situations, in particular for our 2-loop calculations [44] (based on a careful preparation of my cooperation partners) and case studies of massive 3-loop scalar ladder integrals [29], this toolbox was sufficient to perform the necessary calculations.

However, in 3-loop calculations for complete physical problems in QCD the number of the occurring sums grows substantially. Here several thousands, even up to several hundred thousands of sums have to be simplified. Typical examples are, e.g., the calculations of the first two complete Wilson coefficients $L_{qq,Q}^{PS}$ and L_g^S for general values of the Mellin variable n; see [49]. Further examples are the current calculations of 3-loop graphs with two fermionic lines of equal mass and diagrams with two massive lines of different mass (for charm and bottom quarks); for examples see [40].

For all these problems the additional package `SumProduction` was heavily used to perform these large scale problems. Subsequently, the feature and usage of the package will be illustrated by the 3-loop corrections of $O(n_f T_F^2 C_{A,F})$ to the massive OMEs with local operator insertions on the gluonic lines, $A_{gq,Q}$ and $A_{gg,Q}$ at general values of the Mellin variable n [52]. One of the larger expressions (actually, it is one of the smallest examples in comparison to the other examples mentioned above) was produced with the help of FORM [75, 76]. I.e., it is a 2 GB expression of 2419 multi–sums. Each of them can be treated by EvaluateMultiSum, like e.g.,

In[26]:= $$\mathbf{EvaluateMultiSum}\Big[\frac{\pi 2^{\varepsilon+3}e^{-\frac{3\gamma\varepsilon}{2}}(-1)^{j_1}(j_2+1)\Gamma(2-\varepsilon)\Gamma(\frac{\varepsilon}{2}+2)\Gamma(-\frac{3\varepsilon}{2})\Gamma(-\frac{\varepsilon}{2}+j_1+4)\Gamma(-j_1+n-2)\Gamma(\varepsilon-j_1-j_2+n-5)}{(\varepsilon-10)(\varepsilon-8)(\varepsilon-2)\varepsilon\Gamma(\frac{5}{2}-\varepsilon)\Gamma(\frac{\varepsilon+5}{2})\Gamma(\frac{\varepsilon}{2}+n+1)\Gamma(-j_1-j_2+n-4)},$$
$$\{\{j_2,0,n-j_1-6\},\{j_1,0,n-5\}\},\{n\},\{5\},\mathbf{ExpandIn}\to\{\varepsilon,-3,-1\}\Big]$$

Out[26]= $$\Big\{0,\frac{16(-1)^n(3n^2+12n+11)}{135(n+1)(n+2)^2(n+3)^2}-\frac{16(n^8+6n^7-6n^6-80n^5-81n^4+178n^3+274n^2-4n-96)}{45(n-2)(n-1)^2n^2(n+1)(n+2)^2(n+3)^2}$$
$$\frac{16(n^2-n-8)}{45(n-1)n(n+2)(n+3)}S[1,n],-\frac{8(n^2-n-8)}{45(n-1)n(n+2)(n+3)}S[2,n]+\frac{2(-1)^n(187n+127)(3n^2+12n+11)c}{2025(n+1)^2(n+2)^2(n+3)^2}$$
$$+\Big(\frac{2(17n^6-231n^5+121n^4+2063n^3-1458n^2-2432n+960)}{675(n-2)(n-1)^2n^2(n+1)(n+2)(n+3)}-\frac{16(-1)^n(3n^2+12n+11)}{135(n+1)(n+2)^2(n+3)^2}\Big)S[1,n]$$
$$+\frac{2(43n^{12}+112n^{11}+263n^{10}-216n^9-11309n^8-16476n^7+55837n^6+78164n^5-95178n^4-116688n^3+51784n^2+30624n-23040)}{675(n-2)^2(n-1)^3n^3(n+1)^2(n+2)^2(n+3)^2}\Big\}$$

For details on the calculation steps for this particular sum we refer to [20]. Similarly, all the other sums could be treated step by step with a lot of computer resources. However, as worked out in [20] we can do it much better by using the toolbox of the package

In[27]:= **<< SumProduction.m**

`SumProduction − AsummationpackagebyCarstenSchneider© RISC − Linz`

4.1. Reduction to master sums.

First, we reduce the 2 GByte expression (stored in `expr` and being valid for $n \geq 6$) to master sums (resp. key sums) with the function call

In[28]:= **compactExpr = ReduceMultiSums[expr, {n}, {6}, {∞}];**

The reduced expression `compactExpr` is only 7.6 MByte large and it required 6 hours and 53 minutes to obtain this reduction.

Problem RMS: ReduceMultiSums. *Given* a linear combination of definite multi-sums[8] over indefinite nested product-sum expressions as given in (21). *Compactify* the expression, i.e., express the summands with objects such that no algebraic relations remain. In particular, synchronize the summation bounds and merge the sums to master sums.

[8] The sums can be of the form (3) like in our concrete example In[26], or the summands might also involve, e.g., harmonic sums in the numerators; cf. (6). In addition, further regulators (like ε) might be involved.

In this routine the merging can be done in different ways. By default, the summands are tried to be given in the form $r\, t_1^{m_1} \dots t_r^{m_r}$ where the t_i are indefinite nested sums or products, $m_i \in \mathbb{Z}$ and r is a rational function where the numerator and denominator are co-prime.
In short, we reduce the sum expressions to a minimal set of key sums/master sums where no algebraic relations occur within the summand objects. Given this reduced expression, only these master sums have to be treated for further simplification.

In our concrete example the 2419 sums are synchronized w.r.t. the occurring summation ranges (taking for each class the maximum of the lower bounds and the minimum of the upper bounds). As result, we obtained only 4 sums with synchronized ranges

$$\sum_{i_2=5}^{n-5}\sum_{i_1=0}^{i_2} h_1(\varepsilon, n, i_2, i_1), \quad \sum_{i_2=0}^{n-5}\sum_{i_1=0}^{n-i_2-5} h_2(\varepsilon, n, i_2, i_1), \quad \sum_{i_1=5}^{n-5} h_3(\varepsilon, n, i_1), \quad \sum_{i_1=0}^{\infty} h_4(\varepsilon, n, i_1)$$

plus a large term free of summation quantifiers. Next, all the occurring Pochhammer symbols, factorials/Γ-functions, and binomials are written in a basis of algebraically independent objects plus the extra object $(-1)^n$ (if necessary); for details see Problem EAR in Subsection 2.2. Finally, the expressions are split further to get the form $\sum h(n,(i_2,)i_1,\varepsilon)*r(n,(i_2,)i_1,\varepsilon)$ or $h(n,\varepsilon)*r(n,\varepsilon)$ where h stands for a (proper) hypergeometric term in n (and i_1, i_2). More precisely, it is a product of binomials/factorials/Pochhammers in the numerator and denominator, and $r(n,(i_2,i_1),\varepsilon)$ is a rational function in n, ε (and i_1, i_2); note that r might fill several pages. As final result we obtain an expression with only 29 sums and 15 terms being free of sums.

4.2. Automatic computation of the ε-expansions (in parallel)

Finally, the sums are simplified with `EvaluateMultiSums`. In particular, the coefficients of the required Laurent series expansion can be derived. In order to perform this calculation automatically, the following function call can be applied:

In[29]:= $\mathbf{ProcessEachSum[compactExpr, \{n\}, \{6\}, \{\infty\}, ExpandIn \to \{\varepsilon, -3, 0\}]}$

It sequentially applies `EvaluateMultiSum` with the corresponding input parameters to the occurring multi-sums in `compactExpr`. In our concrete example this step took in total 2 hours and 35 minutes.

Often the evaluation of one sum (in particular, for triple and quadruple sums) takes several hours. In order to utilize the benefit of the available computers, we emphasize that this function can be executed simultaneously with different Mathematica kernels, in particular, on different machines within a network. Here the following mechanism is applied. Internally, the function takes the first multi-sum and generates a file with the name SUM1. If the result is computed, the file is updated with the result. Then the routine continues with the second sum provided the file SUM2 is not existent on the hard disk. In this way, ProcessEachSum can be executed in parallel for mass productions.

4.3. Combination to the final result

Finally, the result of the sums (or the expansion of the sums) are read from the hard disk and are summed up to the final result. In particular, the expressions are reduced further by applying Problem EAR given in Subsection 2.2 to the occurring sums and products. As shortcut also the available algebraic relations for harmonic sums and their generalized versions [38, 39, 60, 61] are utilized. This last step can be carried out with the following function call

In[30]:= $\mathbf{result = CombineExpression[compactExpr, \{n\}, \{6\}, \{\infty\}];}$

This calculation took about 21 seconds. The final result can be expressed in terms of ζ_2, ζ_3, $(-1)^n$, $S_1(n)$, $S_2(n)$, $S_3(n)$, $S_{2,1}(n)$, $S_{3,1}(n)$, $S_{2,1,1}(n)$ and requires about 100 KByte memory. In summary, the total calculation took around 9 hours and 30 minutes.

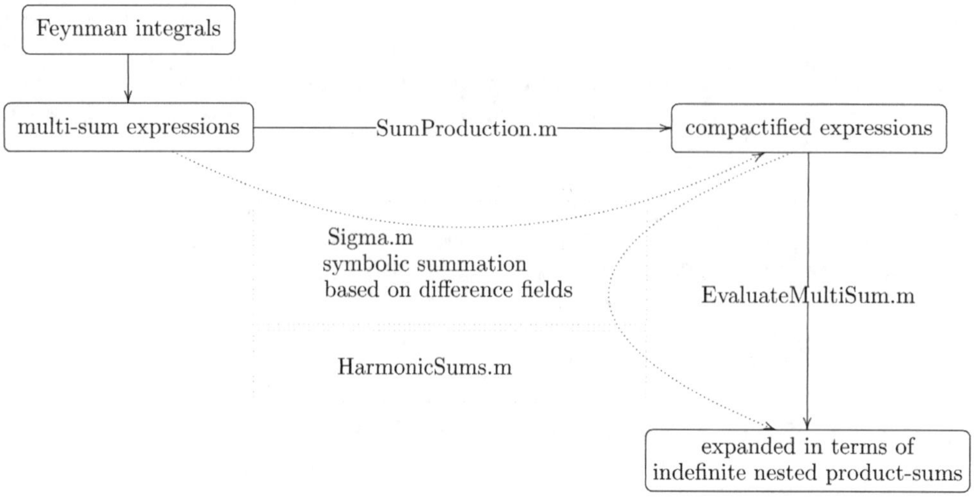

Figure 2. The packages in interaction

5. Conclusion

Summarizing, the whole interaction of the presented packages can be visualized in Fig. 2. Here the Feynman integrals are transformed to huge multi-sum expressions using mostly the computer algebra system FORM. These expressions are then loaded into the computer algebra system Mathematica and our machinery is activated. Using the package `SumProduction` the multi-sum expressions are crunched to expressions in terms of master sums/key sums. In addition, the package supports the user to perform the simplification completely automatically on distributed systems. In order to perform these simplifications, in particular to calculate the coefficients of the Laurent series expansion, the package `EvaluateMultiSums` is called accordingly. This summation technology relies heavily on the summation toolkit of the `Sigma` package that is based on difference field theory. In addition, special function algorithms are needed if infinite summations arise in the given expressions. Here we rely on Ablinger's HarmonicSums package [26, 38, 39, 61] that utilizes and generalizes ideas of [22, 35, 74, 77].

All the packages and underlying algorithms are steadily extended, improved and optimized to deal with more and more complicated Feynman integrals.

Acknowledgments

This work is supported by the Austrian Science Fund (FWF) grants P20347-N18 and SFB F50 (F5009-N15) and by the EU Network LHCPhenoNet PITN-GA-2010-264564.

References

[1] Karr M 1981 *J. ACM* **28** 305–350
[2] Schneider C 2005 *Proc. ISSAC'05* ed Kauers M (ACM) pp 285–292
[3] Schneider C 2008 *J. Symbolic Comput.* **43** 611–644 [arXiv:0808.2543v1]
[4] Schneider C 2010 *Motives, Quantum Field Theory, and Pseudodifferential Operators* (*Clay Mathematics Proceedings* vol 12) ed Carey A, Ellwood D, Paycha S and Rosenberg S (Amer. Math. Soc) pp 285–308 arXiv:0808.2543
[5] Schneider C 2004 *Proc. ISSAC'04* ed Gutierrez J (ACM Press) pp 282–289
[6] Schneider C 2007 *J. Algebra Appl.* **6** 415–441
[7] Schneider C 2010 *Ann. Comb.* **14** 533–552 [arXiv:0808.2596]
[8] Schneider C 2013 *Computer Algebra in Quantum Field Theory: Integration, Summation and Special Functions* Texts and Monographs in Symbolic Computation ed Schneider C and Blümlein J (Springer) pp 325–360 arXiv:1304.4134 [cs.SC]

[9] Schneider C 2011 *Symbolic Summation in Difference Fields* Ph.D. thesis RISC, Johannes Kepler University, Linz technical report 01-17
[10] Schneider C 2004 *Discrete Math. Theor. Comput. Sci.* **6** 365–386
[11] Schneider C 2007 *Sém. Lothar. Combin.* **56** 1–36 article B56b
[12] Zeilberger D 1991 *J. Symbolic Comput.* **11** 195–204
[13] Petkovšek M 1992 *J. Symbolic Comput.* **14** 243–264
[14] Petkovšek M, Wilf H S and Zeilberger D 1996 $A = B$ (Wellesley, MA: A. K. Peters)
[15] Paule P and Schneider C 2003 *Adv. in Appl. Math.* **31** 359–378
[16] Andrews G, Paule P and Schneider C 2005 *Advances in Applied Math.* **34** 709–739
[17] Driver K, Prodinger H, Schneider C and Weideman J A C 2006 *Ramanujan J.* **12** 299–314
[18] Osburn R and Schneider C 2009 *Math. Comp.* **78** 275–292 arXiv:math/0610281 [math.NT]
[19] Ablinger J, Blümlein J, Round M and Schneider C 2012 *Loops and Legs in Quantum Field Theory 2012* PoS(2012)50 pp 1–14 arXiv:1210.1685 [cs.SC]
[20] Blümlein J, Hasselhuhn A and Schneider C 2012 *Proceedings of RADCOR 2011* vol PoS(RADCOR2011)32 pp 1–9 arXiv:1202.4303 [math-ph]
[21] Ablinger J, Blümlein J, Klein S and Schneider C 2010 *Nucl. Phys. B (Proc. Suppl.)* **205-206** 110–115 arXiv::1006.4797 [math-ph]
[22] Blümlein J 2009 *Comput. Phys. Commun.* **180** [arXiv:0901.3106 [hep-ph]]
[23] Blümlein J, Klein S, Schneider C and Stan F 2012 *J. Symbolic Comput.* **47** 1267–1289 arXiv:1011.2656 [cs.SC]
[24] Bogner C and Weinzierl S 2010 *Int. J. Mod. Phys. A* **25** 2585–2618 [arXiv:1002.3458 [hep-ph]]
[25] Weinzierl S 2013 *Computer Algebra in Quantum Field Theory: Integration, Summation and Special Functions* Texts and Monographs in Symbolic Computation ed Schneider C and Blümlein J (Springer) pp 381–406
[26] Ablinger J 2012 *Computer Algebra Algorithms for Special Functions in Particle Physics* Ph.D. thesis J. Kepler University Linz
[27] Apagodu M and Zeilberger D 2006 *Adv. Appl. Math.* **37** 139–152
[28] Brown F 2009 *Math. Phys.* **287** 925–958
[29] Ablinger J, Blümlein J, Hasselhuhn A, Klein S, Schneider C and Wissbrock F 2012 *Nuclear Physics B* **864** 52–84 arXiv:1206.2252v1 [hep-ph]
[30] Wilf H and Zeilberger D 1992 *Invent. Math.* **108** 575–633
[31] Wegschaider K 1997 *Computer generated proofs of binomial multi-sum identities* Master's thesis RISC, J. Kepler University
[32] Schneider C 2005 *Adv. in Appl. Math.* **34** 740–767
[33] Schneider C 2005 *J. Differ. Equations Appl.* **11** 799–821
[34] Chyzak F 2000 *Discrete Math.* **217** 115–134
[35] Vermaseren J A M 1999 *Int. J. Mod. Phys.* **A14** 2037–2976 arXiv:hep-ph/9806280
[36] Blümlein J and Kurth S 1999 *Phys. Rev.* **D60** arXiv:hep-ph/9810241
[37] Moch S O, Uwer P and Weinzierl S 2002 *J. Math. Phys.* **43** 3363–3386
[38] Ablinger J, Blümlein J and Schneider C 2013 *J. Math. Phys.* **54** 1–74 arXiv:1302.0378 [math-ph]
[39] Ablinger J, Blümlein J and Schneider C 2011 *J. Math. Phys.* **52** 1–52 [arXiv:1007.0375 [hep-ph]]
[40] Ablinger J, Blümlein J, Freitas A D, Hasselhuhn A, von Manteuffel A, Raab C, Round M, Schneider C and Wissbrock F 2013 *XXI International Workshop on Deep-Inelastic Scattering and Related Subjects - DIS2013* arXiv:1307.7548 [hep-ph]
[41] Ablinger J and Blümlein J 2013 *Computer Algebra in Quantum Field Theory: Integration, Summation and Special Functions* Texts and Monographs in Symbolic Computation ed Schneider C and Blümlein J (Springer) pp 1–32
[42] Weinzierl S 2004 *J. Math. Phys.* **45** 2656–2673 arXiv:hep-ph/0402131
[43] Bierenbaum I, Blümlein J, Klein S and Schneider C 2007 *Proc. ACAT 2007* vol PoS(ACAT)082 pp 1–15 arXiv:0707.4659 [math-ph]
[44] Bierenbaum I, Blümlein J, Klein S and Schneider C 2008 *Nucl.Phys. B* **803** 1–41 [arXiv:hep-ph/0803.0273]
[45] Blümlein J, Kauers M, Klein S and Schneider C 2008 *Proc. of ACAT 2008* vol PoS(ACAT08)106 pp 1–7 arXiv:0902.4095 [hep-ph]
[46] Blümlein J, Kauers M, Klein S and Schneider C 2009 *Comput. Phys. Commun.* **180** 2143–2165 arXiv:0902.4091 [hep-ph]
[47] Ablinger J, Bierenbaum I, Blümlein J, Hasselhuhn A, Klein S, Schneider C and Wissbrock F 2010 *Nucl. Phys. B (Proc. Suppl.)* **205-206** 242–249 arXiv::1007.0375 [hep-ph]
[48] Ablinger J, Blümlein J, Klein S, Schneider C and Wissbrock F 2011 *19th International Workshop On Deep-Inelastic Scattering And Related Subjects (DIS 2011)* (American Institute of Physics (AIP)) arXiv:1106.5937 [hep-ph]
[49] Ablinger J, Blümlein J, Klein S, Schneider C and Wissbrock F 2011 *Nucl. Phys.* **B 844** 26–54 arXiv:1008.3347

[hep-ph]
[50] Ablinger J, Blümlein J, Freitas A D, Hasselhuhn A, Klein S, Raab C, Round M, Schneider C and Wissbrock F 2012 *Proc. Loops and Legs in Quantum Field Theory 2012* PoS(LL2012)033 pp 1–12 arXiv:1212.6823 [hep-ph]
[51] Ablinger J, Blümlein J, Freitas A D, Hasselhuhn A, Klein S, Schneider C and Wissbrock F 2012 *Proceedings of the 36th International Conference on High Energy Physics* vol PoS(ICHEP2012)270 pp 1–9 arXiv:1212.5950 [hep-ph]
[52] Blümlein J, Hasselhuhn A, Klein S and Schneider C 2013 *Nuclear Physics B* **866** 196–211 arXiv:1205.4184 [hep-ph]
[53] Gosper R W 1978 *Proc. Nat. Acad. Sci. U.S.A.* **75** 40–42
[54] Zeilberger D 1990 *J. Comput. Appl. Math.* **32** 321–368
[55] Kauers M and Paule P 2011 *The concrete tetrahedron* Texts and Monographs in Symbolic Computation (Springer)
[56] Koutschan C 2013 *Computer Algebra in Quantum Field Theory: Integration, Summation and Special Functions* Texts and Monographs in Symbolic Computation ed Schneider C and Blümlein J (Springer) pp 171–194 arXiv:1307.4554 [cs.SC]
[57] Kauers M 2007 *J. of Symbolic Comput.* **42** 948–970
[58] Schneider C 2013 *arXiv:1307.7887 [cs.SC]*
[59] Karr M 1985 *J. Symbolic Comput.* **1** 303–315
[60] Blümlein J 2004 *Comput. Phys. Commun.* **159** 19–54 [arXiv:hep-ph/0311046]
[61] Ablinger J, Blümlein J and Schneider C 2013 *In preparation: Structural Relations of Harmonic Sums*
[62] Hoffman M 2000 *J. Algebraic Combin.* **11** 49–68
[63] Schneider C 2010 *Appl. Algebra Engrg. Comm. Comput.* **21** 1–32
[64] Schneider C 2005 *Ann. Comb.* **9** 75–99
[65] Abramov S A and Petkovšek M 2010 *J. Symbolic Comput.* **45** 684–708
[66] Abramov S A, Bronstein M, Petkovšek M and Schneider C 2013 *In preparation*
[67] Bronstein M 2000 *J. Symbolic Comput.* **29** 841–877
[68] Abramov S A and Petkovšek M 1994 *Proc. ISSAC'94* ed von zur Gathen J (ACM Press) pp 169–174
[69] Hendriks P A and Singer M F 1999 *J. Symbolic Comput.* **27** 239–259
[70] Petkovšek M and Zakrajšek H 2013 *Computer Algebra in Quantum Field Theory: Integration, Summation and Special Functions* Texts and Monographs in Symbolic Computation ed Schneider C and Blümlein J (Springer) pp 259–284
[71] Moch S O, Vermaseren J A M and Vogt A 2004 *Nucl. Phys. B* **688** 101 arXiv:hep-ph/0403192v1
[72] Kauers M 2009 Guessing Handbook Tech. Rep. 09-07 RISC Report Series, University of Linz, Austria
[73] Pemantle R and Schneider C 2007 *Amer. Math. Monthly* **114** 344–350
[74] Blümlein J, Broadhurst D J and Vermaseren J A M 2010 *Comput. Phys. Commun.* **181** 582–625 [arXiv:0907.2557 [math-ph]]
[75] Vermaseren J A M 2000 *arXiv:math-ph/0010025*
[76] Vermaseren J A M 2013 *Computer Algebra in Quantum Field Theory: Integration, Summation and Special Functions* Texts and Monographs in Symbolic Computation ed Schneider C and Blümlein J (Springer) pp 361–379
[77] Remiddi E and Vermaseren J A M 2000 *Int. J. Mod. Phys.* **A 15** 725 arXiv:hep-ph/9905237v1

Analytical calculation of heavy quarkonia production processes in computer

V.V.Braguta, A.K.Likhoded, A.V.Luchinsky, S.V.Poslavsky
IHEP, Pobedy str. 1, Protvino, Moscow region, Russia, 142281

Abstract.
This report is devoted to the analytical calculation of heavy quarkonia production processes in modern experiments such as LHC, B-factories and superB-factories in computer. Theoretical description of heavy quarkonia is based on the factorization theorem. This theorem leads to special structure of the production amplitudes which can be used to develop computer algorithm which calculates these amplitudes automatically. This report is devoted to the description of this algorithm. As an example of its application we present the results of the calculation of double charmonia production in bottomonia decays and inclusive the χ_{cJ} mesons production in pp-collisions.

1. Introduction

It is well known that the study of heavy quarkonia lead to a great boost in the understanding of QCD [1]. This is connected to the fact that heavy quarkonia contains all manifestations of QCD. At the same time heavy quarkonia are nonrelativistic systems what makes the physics of heavy quarkonia much simpler that that for other mesons. For this reason so far the study of heavy quarkonia in different experiments is very important and intensive.

The study of heavy quarkonia production processes is based on the fact that the mass of heavy quark inside quarkonia is much larger than characteristic QCD energy scale ($m_Q \gg \Lambda_{QCD}$). This fact leads to the appearance of small parameter – the relative velocity of quark-antiquark pair inside quarkonia $v \ll 1$. For the charmonia mesons this parameter is $v \sim 0.3$, for bottomonia mesons $v \sim 0.1$. The appearance of the small parameter separates different energy scales involved in quarkonia physics and allows to build a rigorous formalism for studying quarkonia properties which is called nonrelativistic QCD (NRQCD) [2].

1.1. NRQCD formalism

NRQCD is based on the separation of different energy scales through the factorization theorem

$$T(e^+e^- \to \eta_c\gamma) = \sum_n C_n \times \langle \eta_c|\hat{O}_n|0\rangle, \tag{1}$$

where T is the amplitude of the process $e^+e^- \to \eta_c\gamma$[1], C_n are the Wilson coefficients which parametrize the physics of small distances, $\langle \eta_c|\hat{O}_n|0\rangle$ are different NRQCD operators which contribute to the process. The sum in (1) is infinite and it runs over all possible operators

[1] For definiteness below we consider the process $e^+e^- \to \eta_c\gamma$

with correct quantum numbers. In particular, for the production of the η_c meson the following operators contribute

$$\hat{O} \sim \chi^+\psi, \chi^+\mathbf{D^2}\psi, \chi^+\vec{H}\vec{\sigma}\psi, ...$$

However, if we restrict the study of the process under consideration by some power of relative velocity $O(v^n)$ then only finite number of the NRQCD operators contibute to (1). In particular, at the leading order approximation in relative velocity only the $\langle\eta_c|\chi^+\psi|0\rangle$ operator contributes.

If we omit the operators which assumes that quarkonia consist of more than quark-antiquark pair (for instance, $\chi^+\vec{H}\vec{\sigma}\psi$) then factorization theorem (1) can be rewritten in the following form

$$T(e^+e^- \to \eta_c\gamma) = \langle\eta_c|\chi^+\psi|0\rangle\Big(C_0 + C_2\langle q^2\rangle + C_4\langle q^4\rangle + ...\Big), \tag{2}$$

where the matrix elements $\langle q^n\rangle$ which control relativistic corrections can be written as follows

$$\langle q^n\rangle = m_c^n\langle v^n\rangle = \frac{\langle\eta_c|\chi^+(-\frac{i}{2}\mathbf{D})^{\mathbf{n}}\psi|0\rangle}{\langle\eta_c|\chi^+\psi|0\rangle} \tag{3}$$

The first few matrix elements $\langle v^n\rangle$ for the $1S, 2S, 1P$ charmonia mesons where calculated within QCD sum rules in papers [3, 4, 5, 6].

The Wilson coefficients C_n which parametrize short distance contribution to the amplitude can be expanded in the strong coupling constants α_s

$$C_n = c_n^{(0)} + c_n^{(1)}\alpha_s + c_n^{(2)}\alpha_s^2 + ... \tag{4}$$

The first aim of the present report is the description of the algorithm of analytical calculation of the coefficient $c_n^{(0)}$ for any process.

1.2. Light cone expansion formalism

If a heavy quarkonium production process contains energy scale E_h which is much larger than the masses of final mesons, one can apply light cone expansion formalism (LCF) to study such process [7, 8]. For instance, LCF can be applied if the energy of e^+e^- beam is much larger than the masses of charmonia mesons, what takes place at B-factories. The presence of high energy scale E_h, which is of order of characteristic energy of the hard exclusive process, allows one to apply factorization theorem for the amplitude of the process (1). As in the case of NRQCD, the coefficients C_n describe partons production at small distances, the matrix elements $\langle\eta_c|O_n|0\rangle$ describe hadronization of the partons which takes place at large distances. The sum is taken over all possible operators O_n. For instance, the operators

$$\hat{O}_n \sim \bar{Q}\gamma_\mu\gamma_5 Q, \bar{Q}\gamma_\mu\gamma_5 D_{\mu_1}D_{\mu_2}Q, \bar{Q}\sigma_{\mu\nu}\gamma_5 G_{\alpha\beta}Q \tag{5}$$

are few examples of the operator O_n. Actually, there are infinite number of the operators O_n that contribute to the pseudoscalar meson production.

The cross section of hard exclusive process can be expanded in the inverse powers of the high energy scale E_h

$$\sigma = \frac{\sigma_0}{E_h^n} + \frac{\sigma_1}{E_h^{n+1}} + ... \tag{6}$$

To determine if some operator contributes to a given term in $1/E_h$ expansion one uses the concept of the twist of this operator [9]. Thus only the leading twist – the twist-2 operators $\bar{Q}\hat{z}\gamma_5 Q, \bar{Q}\hat{z}\gamma_5(zD)Q, \bar{Q}\hat{z}\gamma_5(zD)^2Q, ...$[2] contribute to the leading term in expansion (6). From this

[2] z here is lightlike fourvector $z^2 = 0$.

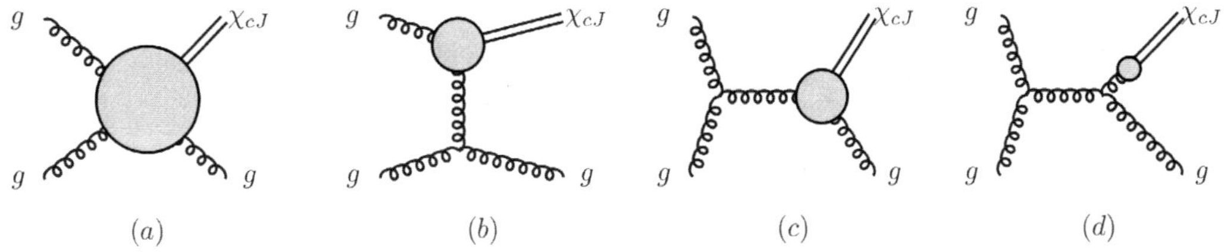

Figure 1. Feynman diagrams for $gg \to \chi_{cJ} g$ subprocess

one sees that already at the leading order approximation infinite number of operators contribute to the cross section. Nevertheless, it is possible to cope with infinite number of contributions if one parametrizes all the twist-2 operators by the moments of some function $\phi(x)$

$$\langle \eta_c(q)|\bar{Q}\hat{z}\gamma_5(-iD_{\mu_1})...(-iD_{\mu_n})Q|0\rangle \times z^{\mu_1}...z^{\mu_n} = if_\eta(qz)^{n+1}\int_0^1 dx\phi(x)(2x-1)^n, \quad (7)$$

where q is the momentum of the pseudoscalar meson η_c, f_η is the constant which is defined as $\langle \eta_c(q)|\bar{Q}\gamma_\mu\gamma_5 Q|0\rangle = if_\eta q_\mu$, x is the fraction of momentum of the whole meson η_c carried by quark. The function $\phi(x)$ is called the leading twist distribution amplitude (DA). One can think of it as about the Fourier transform of the wave function of the meson η_c with lightlike distance between quarks. Using the definition of this function, factorization theorem (1) can be rewritten as

$$T = \int_0^1 dx H(x)\phi(x), \quad (8)$$

where $H(x)$ is the hard part of the amplitude, which describes small distance effects. This part of the amplitude can be calculated within perturbative QCD. As it was noted, the leading twist DA parametrizes infinite set of the twist-2 operators. This part of the amplitude describes hadronization of quark-antiquark pair at large distances and parametrizes nonperturbative effects in the amplitude. DA of charmonia mesons were studied in papers [3, 4, 5, 6].

It should be noted that formula (8) resums the contributions of all twist-2 operators. If the meson η_c is a nonrelativistic meson, formula (8) resums relativistic corrections to the amplitude T. It should be also noted that if one takes into account the renormalization group running of the DA $\phi(x)$, formula (8) resums all leading logarithmic corrections to the amplitude of the process under considerations.

The second aim of the present report is the description of the algorithm of analytical calculation of the hard part of the amplitude $H(\xi)$ for any process.

1.3. Hadronic production of P-wave charmonium states

Another example is hadronic production of P-wave charmonium states at high energies. In NRQCD formalism χ_{cJ} mesons are considered as nonrelativistic quark-antiquark pairs in color singlet (CS) or color octet (CO) states

$$|\chi_{cJ}\rangle \sim |R'(0)|^2 \left|c\bar{c}\left[{}^3P_J^{[1]}\right]\right\rangle + \langle \mathcal{O}_S\rangle \left|c\bar{c}\left[{}^3S_1^{[8]}\right]g\right\rangle + \langle \mathcal{O}_P\rangle \left|c\bar{c}\left[{}^3S_0^{[8]}\right]g\right\rangle + \ldots, \quad (9)$$

The hadronization probability of each component are described by long distance matrix elements $|R'(0)|^2$, $\langle \mathcal{O}_S\rangle$, $\langle \mathcal{O}_P\rangle$, that are usually determined from fit of experimental data.

In high energy limit gluon-gluon interactions give the main contributions to hadronic production of χ_c-mesons (see diagrams shown in Fig.1). To obtain the analytic expressions for the amplitudes of these processes one can write down the amplitude for $c\bar{c}$ pair production in the reaction $gg \to c\bar{c}g$ and project it on the corresponding state. For example, in the case of color-singlet χ_{c0} meson one have the expression

$$\mathcal{A}(gg \to \chi_{c0} g) \quad \sim \quad \frac{\partial}{\partial q_\mu} \frac{\partial}{\partial \epsilon_{S\nu}} \mathcal{A}(gg \to c\bar{c}g) \left(\frac{P_\mu P_\nu}{P^2} - g_{\mu\nu} \right), \tag{10}$$

where P_μ, q_μ, and $\epsilon_{S\nu}$ are the momentum of final charmonium meson, relative momentum of quarks and spin polarization vector.

2. Analytical calculation

The main idea of the algorithm for analytical calculation within NRQCD and LCF is based on the factorization theorem (1). This theorem allows us to make factorization in the calculation of the amplitude which consists in the following. The amplitude is devided into two part. The first part parametrizes the small distance contribution. The calculation of this part is reduced to the calculation of some set of Feynman diagrams. The second part of the amplitude describes hadronization of quarks. In the calculation it is reduced to the some projection operators. In order to get total amplitude these two parts are convoluted by Dirac and Color indices. To carry out the calculation we use the Mathematica package FeynCalc [10] for analytical calculation in high energy physics.

As an example let us consider analytical calculation of $\langle H(p)\gamma(k)|J^\mu_{em}|0\rangle$ matrix elements, which can be used to build the whole amplitude of the processes $e^+e^- \to H(p)\gamma(k), H = \eta_c, \chi_{c0}, \chi_{c1}, \chi_{c2}$. The calculation will be carried out within LCF. Analogously one can repeat the algorithm for NRQCD.

The program which realizes the algorithm starts from the kinematics

- There is no transverse motion in the meson H ($p^\mu = \frac{\sqrt{s}}{2}(1, 0, 0, 1)$)
- The c-quark momentum is $p_{Q1} = x_1 p$, the $\bar{c}$-antiquark momentum $p_{Q2} = x_2 p$ $(x_1 + x_2 = 1)$
- The relative momentum of quark-antiquark pair is $q = \xi p, \quad \xi = x_1 - x_2$

and the definitions of momenta involved in the process

```
p[a_] = FourVector[p, a]; (* H - momentum *)
p1[a_] = x1*FourVector[p, a]; (* p1=x1*p quark momentum *)
p2[a_] = x2*FourVector[p, a]; (* p2=x2*p antiquark momentum *)
k[a_] = FourVector[k, a]; (* photon momentum *)
A[a_] = FourVector[A, a]; (* photon polarization *)

ScalarProduct[k, k] = 0;
ScalarProduct[p, p] = 0;
ScalarProduct[p1, p1] = 0;
ScalarProduct[p2, p2] = 0;
ScalarProduct[p, k] = s/2;
ScalarProduct[p1, k] = x1*s/2;
ScalarProduct[p2, k] = x2*s/2;
ScalarProduct[A, k] = 0;

(* Dirac matrixes *)
hp = GS[p];
hp1 = x1*GS[p];
hp2 = x2*GS[p];
```

Table 1. The results of the calculation of the bottomonia decays to double charmonia.

	$Br \cdot 10^{-5}$ NRQCD [12]	$Br \cdot 10^{-5}$ NRQCD [13]	$Br \cdot 10^{-5}$ [11]	$Br \cdot 10^{-5}$ Exp. [14]
$\chi_{b0} \to 2J/\psi$	0.5	1.9	1.9 ± 0.5	< 7.1
$\chi_{b2} \to 2J/\psi$	3.4	17.5	4.8 ± 1.2	< 4.5
$\chi_{b0} \to J/\psi\ \psi(2S)$	—	—	2.5 ± 0.7	< 12
$\chi_{b2} \to J/\psi\ \psi(2S)$	—	—	5.7 ± 1.8	< 4.9
$\chi_{b0} \to 2\psi(2S)$	—	—	0.8 ± 0.4	< 3.1
$\chi_{b2} \to 2\psi(2S)$	—	—	1.7 ± 0.8	< 1.6

```
hk = GS[k];
hA = GS[A];
```

After the definitions one can calculate the first part of the amplitude – Feynman diagrams of the processes involved

```
T = qc^2*e^2*ExpandScalarProduct[SpinorUBar[p1].GA[mu].(-hp2 - hk).
hA.SpinorV[p2]/(ScalarProduct[p2 + k, p2 + k]) +
SpinorUBar[p1].hA.(hp1 + hk).GA[mu].SpinorV[p2]
/(ScalarProduct[p1 + k, p1 + k])]* (SUNTrace[1]/SUNTrace[1]) /. {SUNN -> 3}
```

Note that thus calculated diagrams are very simple. If the corresponding diagrams are complicated and the number of diagrams is large then it is better to use package FeynArt (which is a part of FeynCalc [10]), where it is possible to calculate the diagrams automatically.

The second part of the amplitude which parametrizes hardonization of quarks is represented by projection operators

- the η_c meson: $P_{\beta\alpha} = (\hat{p}\gamma_5)_{\beta\alpha}\frac{f_P}{4}$
- the χ_{c0} meson: $P_{\beta\alpha} = (\hat{p})_{\beta\alpha}\frac{f_{\chi 0}}{4}$
- the χ_{c1} meson: $P_{\beta\alpha} = (\hat{p}\gamma_5)_{\beta\alpha}\frac{f_{\chi 1}}{4}$
- the χ_{c2} meson: $P_{\beta\alpha} = (\hat{p})_{\beta\alpha}\frac{f_{\chi 2}}{4}$

It is comfortable to write these projection operators in terms of the array **Project**

```
Project = {fP/4*hp.GA[5], fc0/4*hp, fc1/4*hp.GA[5], fc2/4*hp}
```

Now two parts of the amplitude can be convoluted by Dirac and color indices

```
Factor[Tr[T.Project[[1]]]] /. {x1 + x2 -> 1}
Factor[Tr[T.Project[[2]]]] /. {x1 + x2 -> 1}
Factor[Tr[T.Project[[3]]]] /. {x1 + x2 -> 1}
Factor[Tr[T.Project[[4]]]] /. {x1 + x2 -> 1}
```

At the end of the program one gets correct amplitudes of the processes involved.

Table 2. CS and CO model parameters for different values of the scale μ^2.

μ^2	$\|R'(0)\|^2$, GeV^5	$\langle \mathcal{O}_S \rangle$, GeV^3	$\langle \mathcal{O}_P \rangle$, GeV^5
M^2	0.22	7.6×10^{-5}	0.031
$m_T^2/2$	0.34	2.2×10^{-4}	0.043
m_T^2	0.41	2.7×10^{-4}	0.053
$2m_T^2$	0.50	3.3×10^{-4}	0.065

3. Applications

3.1. Double charmonia production in exclusive bottomonia decays

Using the algorithm described above in [11] all leading twist double charmonia production in exclusive bottomonia decays were calculated. The calculation was done both for NRQCD and LCF. It should be stressed that one program calculated the amplitudes and branching ratios for approximately 30 different processes of bottomonia decays. We are not going to present all the results obtained in paper [11] since it will take a lot of space. The only result that is presented in this report (see Table 1) is the branching ratio for the processes measured at Belle collaboration [14]. It is seen from Table 1 that our results (fourth column) are in agreement with the results obtained at Belle experiment (fifth column). It will be highly desirable to trace the origin of the disagreement in the results of papers [11, 12, 13]

3.2. Heavy quarkonia production in hadronic collisions

Using the formalism described above we have calculated also the cross sections of partonic reactions $gg \to \chi_{cJ} g$ with the help of FeynCalc and Redberry [15] packages. On the basis of these expressions one can calculate the cross sections of hadronic reactions

$$d\sigma(pp \to \chi_{cJ} + X) = \int_0^1 dx_1 dx_2 f_g(x_1) f_g(x_2) d\hat{\sigma}(gg \to \chi_{cJ} g), \tag{11}$$

where $x_{1,2}$ are momentum fractions of initial gluons and $f_g(x)$ is the hadronic structure function. Using the obtained in this way theoretical predictions for the transverse momentum distributions of the color singlet and octet cross sections one can determine from fit to existing experimental data the values of long distance matrix elements $|R'(0)|^2$, $\langle \mathcal{O}_S \rangle$, $\langle \mathcal{O}_P \rangle$. Such analysis was performed by our group in papers [16, 17]. Presented in these papers numbers are given in Table 2 and the comparison with experiment is shown in Fig.2 and Fig.3. One can see, that good agreement is achieved.

4. Conclusion

This report is devoted to the analytical calculation of heavy quarkonia production processes in modern experiments such as LHC, B-factories and superB-factories in computer. Theoretical description of heavy quarkonia is based on the factorization theorem. This theorem leads to special structure of the production amplitudes which can be used to develop computer algorithm which calculates these amplitudes automatically. In this report we described this algorithm. As an example of its application we presented the results of the calculation of double charmonia production in bottomonia decays and inclusive the χ_{cJ} mesons production in pp-collisions.

The authors would like to thank the ACAT organization committee for invitation to the conference and hospitality. The work was financially supported by Russian Foundation for Basic Research (grant #10-00061a), the grant of the president of Russian Federation (#MK-3513.2012.2), and FRRC grants.

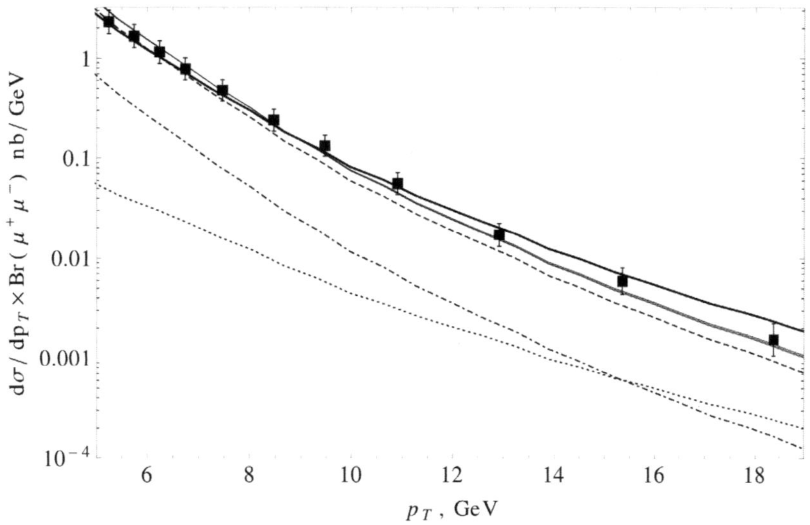

Figure 2. Transverse momentum distribution of J/ψ production in radiative χ_{cJ} decays at CDF in comparison with experimental data [18]. Solid lines correspond to the total $pp \to \chi_{cJ}+X \to J/\psi+X$ cross section in different choices of μ^2 (upper corresponds to $\mu^2 = m_T^2$, and lower three lines to other schemes). Dashed, dotted and dot-dashed lines correspond to CS, S-wave octet and P-wave octet contributions respectively (all at $\mu^2 = m_T^2$).

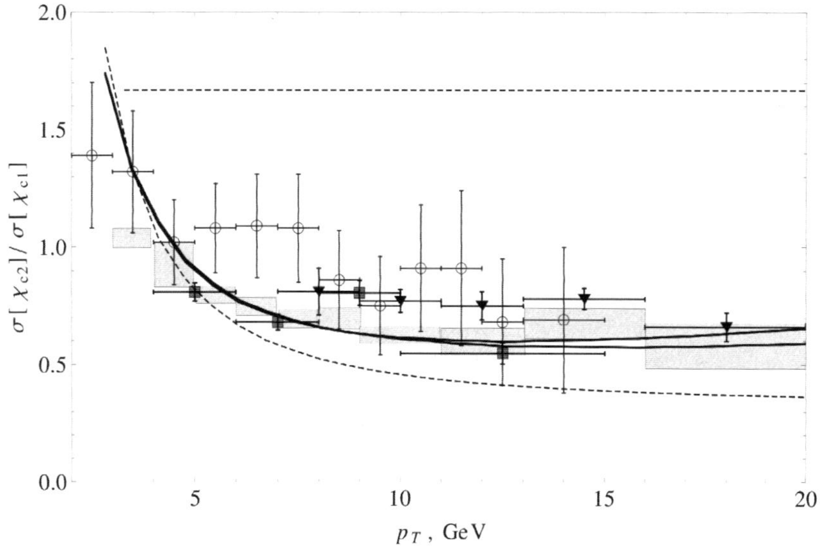

Figure 3. Theoretical results of the ratio χ_{c2}/χ_{c1} in comparison with experimental data from [19] (■), [20] (○), [21] (▽) and [22] (filled rectangles). Solid lines correspond to parameter values presented in Tab. 2, while dashed lines show predictions with only CS (lower curve) or CO (upper curve) contributions taken into account.

References

[1] Novikov V.A., Okun L.B., Okun M.A., Okun A.I., Voloshin M.B. and Zakharov V.I. 1978 Phys. Rept. **41** 1
[2] Bodwin G.T., Braaten E. and Lepage G.P. 1995 Phys. Rev. D **51** 1125 [1997 Erratum-ibid. D **55** 5853]
[3] Braguta V.V., Likhoded A.K. and Luchinsky A.V. 2007 Phys. Lett. B **646** 80
[4] Braguta V.V. 2007 Phys. Rev. D **75** 094016
[5] Braguta V.V. 2008 Phys. Rev. D **77** 034026
[6] Braguta V.V., Likhoded A.K. and Luchinsky A.V. 2009 Phys. Rev. D **79** 074004
[7] Lepage G.P. and Brodsky S.J. 1980 Phys. Rev. D **22** 2157
[8] Chernyak V.L. and Zhitnitsky A.R. 1984 Phys. Rept. **112** 173
[9] Braun V.M., Korchemsky G.P. and Mueller D. 2003 Prog. Part. Nucl. Phys. **51** 311
[10] Mertig R., Bohm M. and Denner A. 1991 Comp.Phys.Comm. 64 345 http://www.feyncalc.org/
[11] Braguta V.V., Likhoded A.K. and Luchinsky A.V. 2009 Phys. Rev. D **80**, 094008 [2012 Erratum-ibid. D **85** 119901]
[12] Zhang J., Dong H., Feng F. 2011 Phys.Rev. D84 094031
[13] Sang W., Rashidin R., Kim U., Lee J. 2011 Phys.Rev. D84 074026
[14] Shen C.P. *et al.* [Belle Collaboration] 2012 Phys. Rev. D **85** 071102
[15] Bolotin D.A. and Poslavsky S.V. 2013 [arXiv:1302.1219 [cs.SC]]
[16] Likhoded A.K., Luchinsky A.V. and Poslavsky S.V. 2012 Phys. Rev. D **86** 074027
[17] Likhoded A.K., Luchinsky A.V. and Poslavsky S.V. 2013 arXiv:1305.2389 [hep-ph]
[18] Abe F.*et al.* [CDF Collaboration] 1997 Phys. Rev. Lett. **79** 578
[19] Abulencia A. *et al.* [CDF Collaboration] 2007 Phys. Rev. Lett. **98** 232001
[20] Aaij R.*et al.* [LHCb Collaboration] 2012 Phys. Lett. B **714** 215
[21] Chatrchyan S. *et al.* [CMS Collaboration] 2012 Eur. Phys. J. C **72** 2251
[22] Aaij R. *et al.* [LHCb Collaboration] 2013 arXiv:1307.4285 [hep-ex]

ACAT2013 IOP Publishing
Journal of Physics: Conference Series **523** (2014) 012039 doi:10.1088/1742-6596/523/1/012039

Automatic calculation in quarkonium physics

Bin Gong, Lu-Ping Wan, Jian-Xiong Wang and Hong-Fei Zhang

Institute of High Energy Physics, Chinese Academy of Sciences, P.O. Box 918(4), Beijing, 100049, China.

E-mail: twain@ihep.ac.cn, wanlp@ihep.ac.cn, hfzhang@ihep.ac.cn and jxwang@ihep.ac

Abstract. In this report, an automatic calculating package based on REDUCE and RLISP, FDC, is introduced, especially its one-loop calculation part and its special treatment for quarkonium physics. With FDC, many works have been completed, most of them are very important in solve/clarify current puzzles in quarkonium physics.

1. FDC package and its progress

FDC, which stands for Feynman Diagram Calculation, is a package developed to do automatic calculation for physics processes, and it contained four parts: 1) prepare Lagrangian and deduce Feynman rules for first principle model, 2) generate all possible Feynman diagrams for the given process in a given model; 3) manipulate the amplitude of the process analytically; 4) generate FORTRAN code for final numerical results. It was started about twenty year ago [1] and written in REDUCE and RLISP to generate FORTRAN Code. First version of FDC has been presented at AIHENP93 [2]. FDC-MSSM part [3] was developed to automatic construct Lagrangian and deduce Feynman rules for supersymmetry model around 2000. FDC-PWA, a Partial Wave Analysis application for experiments, had been developed until 2002. A brief introduction on these early tree-level achievement could be found in Ref. [4]. In the following, loop-level achievement will be introduced since 2002.

1.1. One loop calculation in FDC

With the development of experiment in high energy physics, accuracy of the measurement has been gradually improved. This makes it more and more important to include higher order corrections in theoretical predictions. As an automatic calculating system, FDC has to be capable of one-loop calculation. The development of one-loop calculation part in FDC started from 2002, and finished in 2007. Thereafter, a few improvements have been introduced.

1.1.1. Real corrections A two-cutoff phase space slicing method [5] is realized in FDC to deal with IR divergence in real corrections. Two cutoffs, δ_s and δ_c are introduced to separate the phase space into three parts: soft, hard-collinear, hard-noncollinear. The hard-noncollinear part is finite, and can be calculated numerically with traditional Monte-Carlo method. Both the soft and hard-collinear part are factorized in soft/collinear limit, and added to corresponding virtual correction processes. All the divergence (including those in virtual corrections) are separated analytically, and then sum up to check if they are really cancelled with each other.

1.1.2. Virtual corrections First, with the input of renormalization constant, counter term diagrams will be generated automatically in FDC. There are two ways in FDC to generate square of amplitude: square the amplitude analytically; generate numerical result (in FORTRAN Code) of amplitude, then square it. These two ways will lead to different tensor reduction and then bring a cross-check inside FDC. Besides, all the divergence (both UV and IR) are separated during the calculation of amplitude analytically. Loop integrals are treated via following steps:

- For one and two point integrals, Feynman parametrization is used directly to derive the analytical results.
- Besides above, Passarino-Veltman reduction method is used to reduce tensors into scalars.
- Some other relations are used in the reduction of $N \geq 5$ point integrals.
- A special $i\epsilon$-regularization scheme is used to derive scalars under dimensional regularization.

1.1.3. $i\epsilon$-regularization scheme An N-point scalar integral under D-dimension regularization can be defined as

$$T_0^{(N)}(p_1,\ldots,p_{N-1},m_0,m_1,\ldots,m_{N-1}) = \mu^{4-D}\int \frac{\mathrm{d}^D q}{(2\pi)^D}\frac{1}{N_0\ldots N_{N-1}}, \tag{1}$$

where $N_n = (q+p_n)^2 - m_n^2 + i\epsilon, \quad n = 0,\ldots,N-1$ are the propagators.

According to Ref. [6], the IR singularities part of the scalar can be expressed as sum of a few 3-point scalars with IR singularities:

$$T_0|_{\text{sing}}^{(N)} = \sum_{n=0}^{N-1}\sum_{\substack{k=0\\ k\neq n,n+1}}^{N-1} A_{nk}\, C_0(p_0,\ldots,p_k,m_n,m_{n+1},m_k). \tag{2}$$

Thus, we can evaluate the scalar integral via

$$T_0^D = T_0^\epsilon - T_0|_{\text{sing}}^\epsilon + T_0|_{\text{sing}}^D, \tag{3}$$

where T_0^ϵ, $T_0|_{\text{sing}}^\epsilon$ are under $i\epsilon$-regularization, and $T_0|_{\text{sing}}^D$ is easily obtained in D-dimension regularization.

Here is the procedure to deal with the integral under $i\epsilon$ regularization:

- keep $i\epsilon$ in the propagators to make the scalar integral well defined.
- let the dimension back to 4.
- do the integration following the standard way described in Ref. [7].
- expand the results in $i\epsilon$.
- This way is suitable to program and is realized in FDC package.

1.1.4. Improvement in loop integrals In 2011, the reduction method for loop integrals proposed in Ref. [8] is implemented in FDC. It has many advantages:

- It can reduce integrals with abnormal dimension and denominators, which may appear e.g. in the calculation of processes involving P-wave states.
- It can reduce tensors with high (4^+) rank and more (5^+) external momenta at same time, where Passarino-Veltman reduction method may fail.
- It can further reduce some integrals, e.g. scalar integrals containing Coulomb singularities in the calculation of quarkonium physics.
- It is easy to obtain coefficients of certain Lorentz term with this method.

1.2. Special treatment for NRQCD

Quarkonium physics has been a hot topic since first quarkonium was found in 1974. As a factorization formalism, nonrelativistic QCD (NRQCD) [9] is now the most common framework to study quarkonium productions and decays. It allows consistent theoretical predictions to be made and improved perturbatively in the QCD coupling constant and the heavy-quark relative velocity. According to NRQCD factorization formalism, production (or decay) of a quarkonium H can be expressed as

$$d\sigma[H] = \sum_n f_n \langle \mathcal{O}_n^H \rangle, \tag{4}$$

where n represents possible $Q\bar{Q}$ intermediate states. f_n is the short-distance coefficient which can be calculated perturbatively, while $\langle \mathcal{O}_n^H \rangle$ is is the production (decay) matrix elements of H, which are fully governed by non-perturbative QCD effects. In order to calculate the short-distance coefficient, two spin projection operators [10]

$$\begin{aligned} \Pi_0(P,p) &= \frac{1}{2\sqrt{2}E(E+m)}\left(\frac{1}{2}\not{P}+m+\not{p}\right)\frac{\not{P}+2E}{4E}\gamma_5\left(\frac{1}{2}\not{P}-m-\not{p}\right), \\ \Pi_1(P,p,\epsilon) &= \frac{-1}{2\sqrt{2}E(E+m)}\left(\frac{1}{2}\not{P}+m+\not{p}\right)\frac{\not{P}+2E}{4E}\not{\epsilon}\left(\frac{1}{2}\not{P}-m-\not{p}\right), \end{aligned} \tag{5}$$

are implemented in FDC for spin singlet and triplet intermediate states, where m is the mass of heavy quark, P is the momentum of quarkonium, $p = (p_Q - p_{\bar{Q}})/2$ is the "relative momentum" between heavy quark pair, and $E \equiv \sqrt{m^2 - p^2}$ can be regarded as half of the "mass" of quarkonium.

2. Applications

Based on above progress in FDC, many important work are completed in quarkonium physics.

2.1. Exclusive J/ψ production at the B factories

2.1.1. $e^+e^- \to J/\psi + \eta_c$ Experiment data for this process measured at $\sqrt{s} = 10.6$ GeV by Belle [11] is

$$\sigma[J/\psi + \eta_c] \times \mathcal{B}^{\eta_c}[\geq 2] = (25.6 \pm 2.8 \pm 3.4) \text{ fb} \tag{6}$$

and by BABAR [12] is

$$\sigma[J/\psi + \eta_c] \times B^{\eta_c}[\geq 2] = (17.6 \pm 2.8^{+1.5}_{-2.1}) \text{ fb}, \tag{7}$$

where $B^{\eta_c}[\geq 2]$ denotes the branching fraction for the η_c decaying into at least two charged tracks. On the other hand, LO theoretical predictions in NRQCD are about $2.3 \sim 5.5$ fb [13, 14, 15], which is an order of magnitude smaller than the experimental results.

Such a large discrepancy between experimental results and theoretical predictions brings a challenge to the current understanding of charmonium production based on NRQCD. Many studies have been performed in order to resolve the problem. Braaten and Lee [13] have shown that the relativistic corrections would increase the cross section by a factor of about 2, which boost the cross section to 7.4 fb. And the NLO QCD correction of the process has been studied by Zhang *et al.* [16], which can enhance the cross section with a K factor (the ratio of NLO to LO) of about 2 and reduce the large discrepancy. Later the relativistic corrections have been studied again [17, 18], which are significant, and when combined with the NLO QCD corrections, may resolve the large discrepancy.

Due to the complexity and importance of NLO calculation for this process, we performed another independent calculation using FDC and obtained an analytic result [19]. The numerical

Table 1. Cross sections for $e^+e^- \to J/\psi + \eta_c$ with different charm quark mass m_c and renormalization scale μ. $\sqrt{s} = 10.6$ GeV is the center-of-mass energy.

m_c(GeV)	μ	$\alpha_s(\mu)$	σ_{LO}(fb)	σ_{NLO}(fb)	σ_{NLO}/σ_{LO}
1.5	m_c	0.369	16.09	27.51	1.710
1.5	$2m_c$	0.259	7.94	15.68	1.975
1.5	$\sqrt{s}/2$	0.211	5.27	11.14	2.114
1.4	m_c	0.386	19.28	34.92	1.811
1.4	$2m_c$	0.267	9.19	18.84	2.050
1.4	$\sqrt{s}/2$	0.211	5.76	12.61	2.189

results are shown in Table 1, which is consistent with Ref. [16]. Our work confirmed former calculation analytically, and is also another self check of FDC. Recently, this process is revisited by Dong *et al.* by including $\mathcal{O}(\alpha_s v^2)$ corrections [20]. The corrections mildly enhance the NRQCD predictions.

It should also be mentioned that same process has also been studied in another framework. Unlike NRQCD, Braguta *et al.* find that the cross section at LO can be 26.7 fb in the framework of the light cone (please check details in Ref. [21], and also Braguta's talk in this conference [22]).

2.1.2. $e^+e^- \to J/\psi + J/\psi$ LO NRQCD predictions for this channel was given by Bodwin *et al.* [23] in 2003. They found that the cross section of this channel may be larger than that of $J/\psi + \eta_c$ by a factor of 1.8, in spite of a suppression factor α^2/α_s^2. They suggested that a significant part of the discrepancy of $J/\psi + \eta_c$ production may be explained by this process. But new analysis performed by Belle [11] based on a 3 times larger data set in 2004 found no evidence for this channel. This becomes a new puzzle in double charmonia production at the B factories.

NLO QCD corrections and relativistic corrections for the dominant photon-fragmentation contribution diagrams are studied by Bodwin *et al.* [24], and the results show that the cross section is decreased by K factors of 0.39 and 0.78 for the NLO QCD and relativistic corrections respectively. In 2006, a more reliable estimate, 1.69 ± 0.35 fb, was given by Bodwin et al. [25]

Table 2. Cross sections for $e^+e^- \to J/\psi + J/\psi$ with different charm quark mass m_c and renormalization scale μ, and $\sqrt{s} = 10.6$ GeV.

m_c(GeV)	μ	$\alpha_s(\mu)$	σ_{LO}(fb)	σ_{NLO}(fb)	σ_{NLO}/σ_{LO}
1.5	m_c	0.369	7.409	−2.327	−0.314
1.5	$2m_c$	0.259	7.409	0.570	0.077
1.5	$\sqrt{s}/2$	0.211	7.409	1.836	0.248
1.4	m_c	0.386	9.137	−3.350	−0.367
1.4	$2m_c$	0.267	9.137	0.517	0.057
1.4	$\sqrt{s}/2$	0.211	9.137	2.312	0.253

With the help of FDC, we perform a full NLO QCD calculation to this process [26]. As shown in Table 2, the cross section would be much smaller than the rough estimate in Ref. [24].

Therefore it is easy to understand why there was no evidence for the process $e^+e^- \to J/\psi + J/\psi$ at B factories.

2.2. *Inclusive J/ψ production at the B factories*

The cross section for inclusive J/ψ production in e^+e^- annihilation was measured by BABAR [27, 12] as $2.54 \pm 0.21 \pm 0.21$ pb and Belle [28, 29] as $1.45 \pm 0.10 \pm 0.13$ pb. These measurements include both $J/\psi + c\bar{c} + X$ and $J/\psi + X_{\text{non}-c\bar{c}}$ parts in the final states. Many theoretical studies [30, 31, 32, 33, 34, 14, 35, 36, 37] have been performed on this production at LO in NRQCD and the results for inclusive J/ψ production cover the range $0.6 \sim 1.7$ pb depending on parameter choices. A further analysis by Belle [29] gives

$$\sigma(e^+e^- \to J/\psi c\bar{c} + X) = 0.87^{+0.21}_{-0.19} \pm 0.17 \text{ pb}. \tag{8}$$

It is about 5 times larger than the LO NRQCD prediction [14]. However, this large discrepancy was partially resolved by considering both NLO correction and feed-down from higher excited states [38].

2.2.1. *$J/\psi + X_{\text{non}-c\bar{c}}$*

The above measurements infer that $\sigma[e^+e^- \to J/\psi + X_{\text{non}-c\bar{c}}] = 0.6$ pb. For this part, the contributions from the color-singlet (CS) and color-octet (CO) channels are about 0.2 pb and 0.27 pb, respectively, at the LO in NRQCD [37]. However, the signal of the CO contribution was not found in the experiment [27, 28]. Therefore, the experimental measurement by Belle is about 3 times larger than the theoretical prediction from the CS channel at LO, and can be much more than 3 times by BABAR. To achieve a reasonable theoretical prediction for this channel, we perform a NLO QCD calculation in Ref. [39].

Table 3. Cross sections for $J/\psi + X_{\text{non}-c\bar{c}}$ part with different charm quark mass m_c where the renormalization scale $\mu = 2m_c$ and $\sqrt{s} = 10.6$ GeV.

m_c(GeV)	$\alpha_s(\mu)$	$\sigma^{(0)}$(pb)	$a(\hat{s})$	$\sigma^{(1)}$(pb)	$\sigma^{(1)}/\sigma^{(0)}$
1.4	0.267	0.341	2.35	0.409	1.20
1.5	0.259	0.308	2.57	0.373	1.21
1.6	0.252	0.279	2.89	0.344	1.23

The cross section at NLO can be expressed as

$$\sigma^{(1)} = \sigma^{(0)} \left\{ 1 + \frac{\alpha_s(\mu)}{\pi} \left[a(\hat{s}) + \beta_0 \ln\left(\frac{\mu}{2m_c}\right) \right] \right\}, \tag{9}$$

where β_0 is the one-loop coefficient of the QCD beta function, and $\hat{s} = s/(2m_c)^2$ is a dimensionless kinematic variable. The numerical results can be found in Table 3. And the scale dependence of the cross section is shown in Fig. 1 and it is improved significantly at NLO. Our final numerical result can be expressed as

$$\sigma^{(1)} = 0.373^{+0.036}_{-0.079} \text{ pb} \tag{10}$$

where the theoretical uncertainty is from the choices of m_c and μ, with $m_c = 1.4$ GeV and $\mu = 2m_c$ for the upper boundary and $m_c = 1.6$ GeV and $\mu = \sqrt{s}/2$ for the lower boundary.

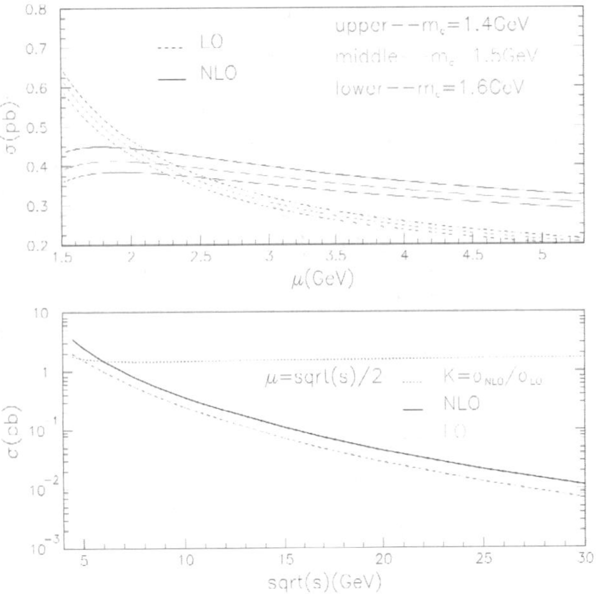

Figure 1. Cross sections for $J/\psi + X_{\text{non}-c\bar{c}}$ as function of the renormalization scale μ and the center-of-mass energy $\sqrt{s}$.

It should be mentioned that same work is also finished by Ma *et al.* [40] almost at the same time. Our results are in agreement with theirs. Also, most recent experimental measurement for this part is given by Belle[41] as

$$\sigma[e^+e^- \to J/\psi + X_{\text{non}-c\bar{c}}] = 0.43 \pm 0.09 \pm 0.09 \text{ pb}. \tag{11}$$

It seems that after including the NLO QCD corrections and feed-down contribution from higher charmonium states, the data can be almost saturated by CS channel, and leaves little room for CO contribution.

2.2.2. $J/\psi + c\bar{c}$ As mentioned above, the discrepancy between experimental data and theoretical predictions of this channel was partially resolved by considering both NLO correction and feed-down from higher excited states [38]. And more recent experimental measurement [41] gives

$$\sigma[e^+e^- \to J/\psi + c\bar{c}] = 0.74 \pm 0.08^{+0.09}_{-0.08} \text{ pb}, \tag{12}$$

which is even closer to the theoretical predictions. Since the calculation of the NLO QCD correction to this process is quite complicated and is very important in explaining the experimental data, it is desirable to have an independent calculation. A more important point is that there are already the momentum and angular distributions for J/ψ production in the new measurement. Therefore, we perform another study on the NLO QCD correction to $e^+e^- \to J/\psi + c\bar{c}$ with our FDC package [42].

Our results for the total cross section are shown in Table. 4, they are in agreement with the previous result in Ref. [38] when their renormalization scheme and input parameters are used. The scale dependence of this channel should be considered as scale dependence is still significant in this part. One possible way, although debatable, to choose the renormalization scale is the BLM scale [43]. A unique scale choice μ^* is obtained and the cross section at NLO is expressed as

$$\sigma^{(1)} = \sigma^{(0)}(\mu^*)[1 + \frac{\alpha_s(\mu^*)}{\pi} b(\hat{s})]. \tag{13}$$

Table 4. Cross sections for $J/\psi + c\bar{c}$ part with different charm quark mass m_c with the renormalization scale $\mu = 2m_c$ and $\sqrt{s} = 10.6$ GeV.

m_c(GeV)	$\alpha_s(\mu)$	$\sigma^{(0)}$(pb)	$a(\hat{s})$	$\sigma^{(1)}$(pb)	$\sigma^{(1)}/\sigma^{(0)}$
1.4	0.267	0.224	8.19	0.380	1.70
1.5	0.259	0.171	8.94	0.298	1.74
1.6	0.252	0.129	9.74	0.230	1.78

Table 5. Cross sections for $J/\psi + c\bar{c}$ part with different charm quark mass m_c. The renormalization scale $\mu = \mu^*$ is obtained by using BLM scale setting [43].

m_c(GeV)	$\alpha_s(\mu^*)$	$\sigma^{(0)}$(pb)	$b(\hat{s})$	$\sigma^{(1)}$(pb)	$\sigma^{(1)}/\sigma^{(0)}$	μ^*(GeV)
1.4	0.348	0.381	3.77	0.540	1.42	1.65
1.5	0.339	0.293	4.31	0.429	1.47	1.72
1.6	0.332	0.222	4.90	0.337	1.52	1.79

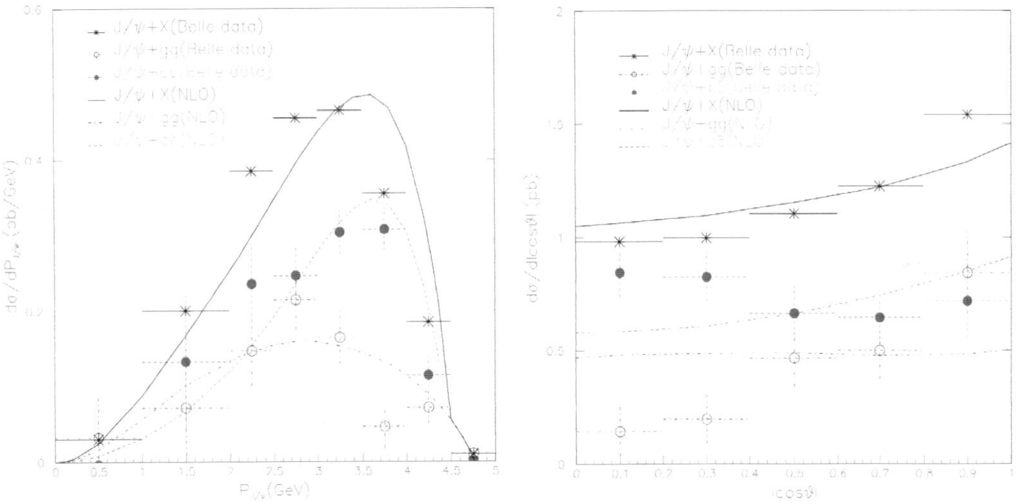

Figure 2. Momentum and angular distributions of inclusive J/ψ production at the B factories.

Corresponding total cross sections are shown in Table 5. From the relevant results listed, we can see that the K factors become smaller and the convergence for QCD perturbative expansions is improved.

Beside the total cross section, momentum and angular distributions of inclusive J/ψ production at the B factories are shown in Fig. 2, with $\mu = \mu^*$ and $m_c = 1.4$ GeV taken for the $J/\psi c\bar{c}$ part. It can be clearly seen that the momentum distributions roughly fits the experimental data, while for angular distributions, the predication for the total angular distribution agrees rather well with experimental measurement, but neither the $J/\psi + c\bar{c}$ nor the $J/\psi + X_{\text{non}-c\bar{c}}$ part can fit the experimental measurements.

2.3. Quarkonium production in hadron colliders
Based on NRQCD, the LO calculation predicts a sizable transverse polarization for J/ψ at high transverse momentum (p_t) region [44], while the CDF measurement [45] gives almost unpolarized result. Same problem was also found in bottomonium system. It is another challenging puzzle to our understanding on quarkonium production. Higher order corrections are highly expected to further clarify the situation. For J/ψ hadroproduction, up to LO in v^2, the intermediate state $(Q\bar{Q})_n$ can be ${}^3S_1^{[1]}, {}^3S_1^{[8]}, {}^1S_0^{[8]}$ and ${}^3P_J^{[8]}$. In Table 6, involving subprocess and number of corresponding Feynman diagrams are shown. The complexity and difficulty of corresponding calculation makes the work move slowly.

Table 6. Involving subprocesses and number of corresponding Feynman diagrams of quarkonium hadroproduction.

	${}^3S_1^{[1]}$	${}^3S_1^{[8]}$	${}^1S_0^{[8]}$	${}^3P_J^{[8]}$
$gg \to (Q\bar{Q})_n + g$	6/129	16/413	12/267	12/267
$gq \to (Q\bar{Q})_n + q$	—	5/111	2/49	2/49
$q\bar{q} \to (Q\bar{Q})_n + g$	—	5/111	2/49	2/49
$gg \to (Q\bar{Q})_n + gg$	60	123	98	98
$gg \to (Q\bar{Q})_n + q\bar{q}$	6	36	20	20
$gq \to (Q\bar{Q})_n + gq$	6	36	20	20
$q\bar{q} \to (Q\bar{Q})_n + gg$	6	36	20	20
$q\bar{q} \to (Q\bar{Q})_n + q\bar{q}$	—	14	4	4
$q\bar{q} \to (Q\bar{Q})_n + q'\bar{q}'$	—	7	2	2
$qq \to (Q\bar{Q})_n + qq$	—	14	4	4
$qq' \to (Q\bar{Q})_n + qq'$	—	7	2	2

2.3.1. CS channel In 2007, The NLO QCD corrections of CS channel are investigated by Campbell *et al.* [46]. The results show that the total cross section is boosted by a factor of about 2 and the p_t distribution is enhanced more and more as p_t becomes larger. Unfortunately, they do not have results for the polarization status. Soon after that, one-loop part of FDC package is completed. We perform a further study on this channel with FDC [47, 48]. Our predictions on the total and differential cross section are consistent with former work, and we also give predictions for the polarization for the first time. As shown in Fig. 3, the polarization is extremely changed from more transverse at LO into more longitudinal at NLO. Although it gives more longitudinal polarization than the recent experimental result [45] on the J/ψ polarization at Tevatron and the contribution of this channel is still too small to affect the total result, it sheds light on the solution to the large discrepancy.

2.3.2. S-wave CO channels Next step is on the S-wave CO channels [49, 50], which is thought to be dominant in total hadroproduction. Unfortunately, it is found that NLO QCD corrections in this channel are small. Neither the differential cross section nor the polarization is changed very much. Obvious gap is shown between theoretical predictions and experimental data. In the well established theoretical framework of NRQCD there still remains a narrow window which might make up the gap, one needs to investigate the NLO corrections to the P-wave CO channel ${}^3P_J^{[8]}$ and feed-down contribution from higher excited states.

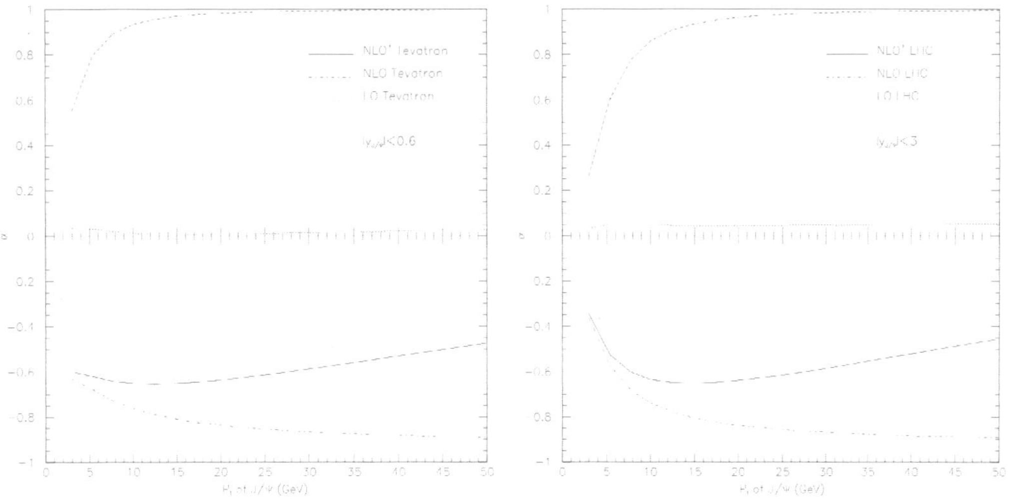

Figure 3. Transverse momentum distribution of polarization α at the Tevatron and LHC.

2.3.3. P-wave CO channel and complete NLO result But calculation on processes involving P-wave states is much more complicated than usual. It will cause abnormal denominator in the loop integral. FDC was unable to deal with such integral at that time. The study came to an impasse.

In 2010, complete NLO QCD corrections to J/ψ photoproduction are calculated by Butenschoen *et al.* [51]. This is for the first time such a complicated one-loop calculation involving P-wave state is completed. Later, NLO corrections for χ_{cJ} hadroproduction are studied by Ma *et al.* [52].

Then the complete NLO calculation for J/ψ hadroproduction is completed by Ma *et al.* [53, 54] and Butenschoen *et al.* [55] almost at the same time. Experiment measurement on the differential cross section of J/ψ hadroproduction can be described well, while NLO predictions for the polarization are still missing.

In 2011, a complete NLO result for polarization of J/ψ photoproduction is presented by Butenschoen *et al.* [56]. An overall agreement is found to recent measurement from H1 [57] and ZEUS [58]. The result inspires all the people concentrating on the J/ψ polarization puzzle.

In 2012, full NLO calculation on the polarization of J/ψ hadroproduction is completed by Butenschoen *et al.* [59]. Using the same LDMEs as those in above photoproduction case, their results disagree with CDF data [60, 45] at the Tevatron, but can account for ALICE data [61] at the LHC. Same work is also finished by Ma *et al.* [62] at the same time. Using a combined fit of J/ψ yield and polarization data from CDF, they find a possible solution to the polarization puzzle. The difference of the two groups comes from the fact that different NRQCD LDMEs are used in their works.

However, only direct J/ψ production is studied in both works, while there only exist polarization measurements for prompt (or even inclusive) J/ψ production [60, 45]. Moreover, it is known that among all the feeddown contributions to prompt J/ψ production, χ_{cJ} contributes more than $20 \sim 30\%$ [63, 64], and $\psi(2S)$ also contributes a small fraction, while others are negligible. The feeddown contribution is so large that it can drastically change the polarization results and must be considered.

Therefore, to test NRQCD factorization and solve (or clarify) the long-standing J/ψ polarization puzzle, it is a very important step to achieve the polarization predictions for prompt

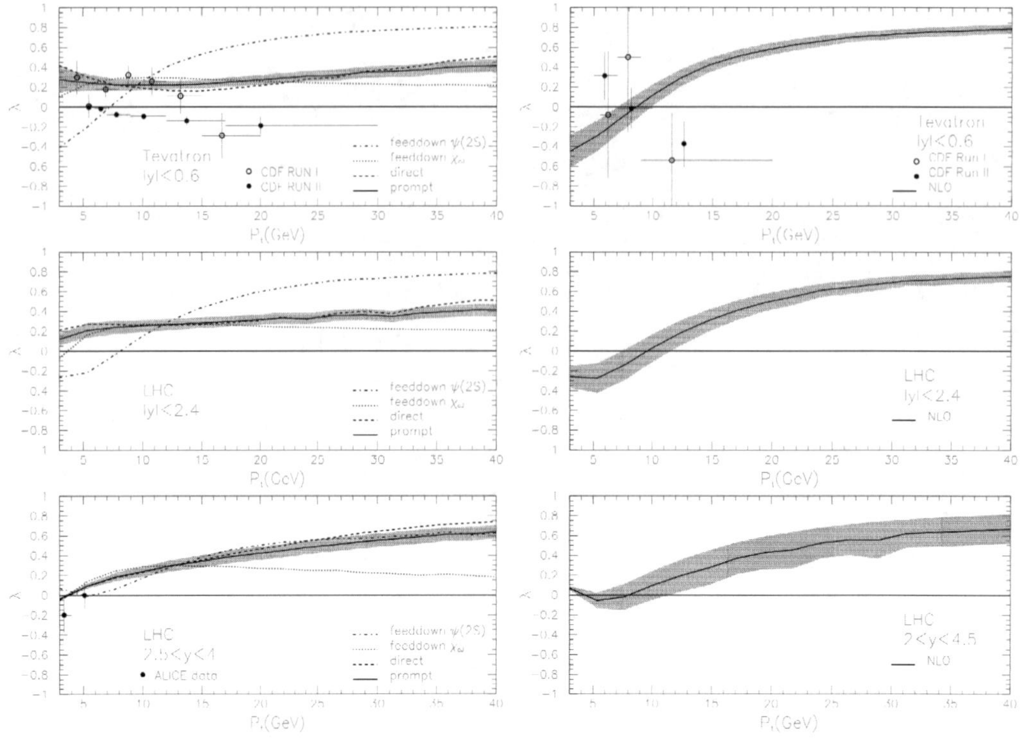

Figure 4. Polarization of J/ψ and $\psi(2S)$ hadroproduction at the Tevatron and LHC.

J/ψ hadroproduction. In 2011, FDC finished its upgrade on tensor reduction method. With this, we finished the study on polarization of prompt J/ψ hadroproduction [65]. In our result, not only all the direct channels, but also feed-down contribution from χ_{cJ} and $\psi(2S)$ are included. It is the first prediction for "prompt" J/ψ hadroproduction that can really compare with experimental data. As shown in Fig. 4, our predictions on the polarization of prompt J/ψ hadroproduction are in agreement with the CDF Run-I data, but in conflict with the CDF Run-II data. Up to now, it seems that in the framework of NRQCD, J/ψ polarization at the Tevatron and J/ψ cross section at the Tevatron/LHC can not be explained at same time.

For Υ hadroproduction, the CS channel [46, 47, 48] and S-wave CO channels have been investigated [50]. And NLO QCD correction to the yield for $\Upsilon(1S)$ via complete CO states (include ${}^3P_J^{[8]}$) is presented in Ref. [66]. The complete NLO QCD study on polarization of Υ hadroproduction has not yet been performed since there are more complicated feeddown and more difficulties to do experimental measurement than charmonium case at the Tevatron. However, the advantage for study on Υ is also obvious. Since bottom is almost three times as heavy as charm, both QCD coupling constant and v^2 are smaller in the bottomonium case. Then perturbative calculation on bottomonium should have better convergence in the double expansion of α_s and v^2 than that on charmonium. It is expected that the theoretical predictions on the polarization and yield of Υ at QCD NLO should be in better agreement with experimental measurement. Also it is found at the LHC that measurement on Υ is easier than that on J/ψ. As there are already polarization measurement on $\Upsilon(1S, 2S, 3S)$ by CMS [67], we perform a complete NLO study on the polarization and yield on Υ hadroproduction [68].

In Fig. 5, polarizations of $\Upsilon(1S, 2S, 3S)$ hadroproduction at the Tevatron and LHC are shown.

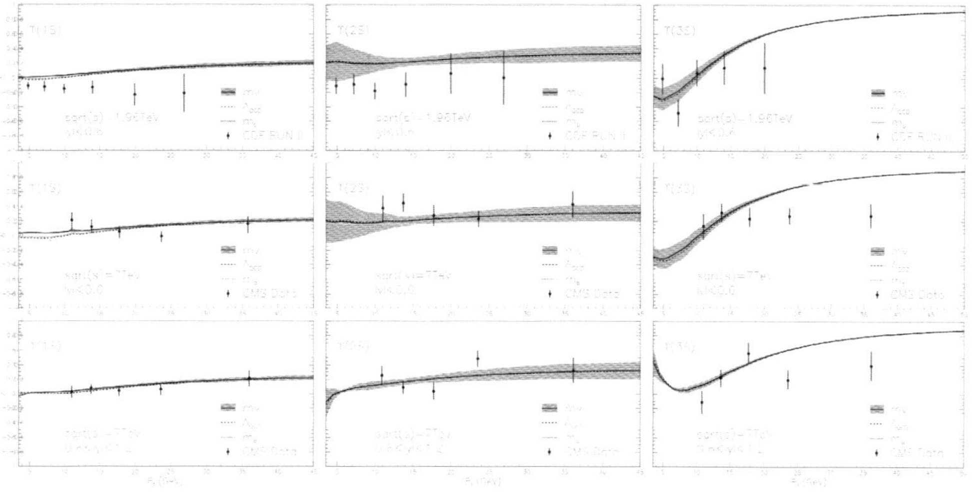

Figure 5. Polarization of $\Upsilon(1S, 2S, 3S)$ hadroproduction at the Tevatron and LHC.

For the polarization of $\Upsilon(3S)$, the theoretical prediction gives transverse polarization and goes far and far away from the data as p_t increases, from which we can conclude that the polarization can not be interpreted at LO in v^2 and NLO in α_s if unknown feeddown contribution from higher excited bottomonia is negligible. For $\Upsilon(1S, 2S)$, the predictions for polarization can well explain the CMS data, but still have some distance from the CDF data. The uncertainty from the badly known fraction of χ_{bJ} feeddown in the fits for $\Upsilon(1S, 2S)$ could be large which is not presented in the plots. Another uncertainty is from unknown feeddown contribution of higher excited states such as $\chi_{bJ}(3P)$. Therefore a further precise measurement on the fraction of χ_{bJ} feeddown or on direct Υ production will be very helpful to fix the polarization puzzle.

3. Summary

An automatic calculating system, FDC, is introduced. With its one-loop calculation part, and its upgrade on tensor reduction method, many important works in quarkonium physics have been studied. Most of them are very important in solve/clarify current puzzles in quarkonium physics.

4. Acknowledgments

We would like to thank the ACAT2013 organization committee for kindly invitation. The work is supported mainly by the National Natural Science Foundation of China (No. 10935012, and No. 11005137), DFG and NSFC (CRC110), and by CAS under Project No. INFO-115-B01.

References

[1] Wang J X 1993 *Comput.Phys.Commun.* **77** 263–285
[2] 3rd International Workshop on Software Engineering, Artificial Intelligence and Expert systems for High-energy and Nuclear Physics (AIHENP93), 4-8 Oct 1993, Oberammergau, Germany
[3] Wang J X 2001 *Proceedings of 6th Accelerator and Particle Physics Institute (APPI 2001), 20-23 Feb 2001, Japan* 108–121
[4] Wang J X 2004 *Nucl. Instrum. Meth.* **A534** 241–245
[5] Harris B W and Owens J F 2002 *Phys. Rev.* **D65** 094032
[6] Dittmaier S 2003 *Nucl. Phys.* **B675** 447–466
[7] 't Hooft G and Veltman M J G 1979 *Nucl. Phys.* **B153** 365–401

[8] Duplancic G and Nizic B 2004 *Eur.Phys.J.* **C35** 105–118 (*Preprint* hep-ph/0303184)
[9] Bodwin G T, Braaten E and Lepage G P 1995 *Phys. Rev.* **D51** 1125–1171
[10] Bodwin G T and Petrelli A 2002 *Phys.Rev.* **D66** 094011 (*Preprint* hep-ph/0205210)
[11] Abe K *et al.* (Belle Collaboration) 2004 *Phys.Rev.* **D70** 071102 (*Preprint* hep-ex/0407009)
[12] Aubert B *et al.* (BABAR) 2005 *Phys. Rev.* **D72** 031101
[13] Braaten E and Lee J 2003 *Phys. Rev.* **D67** 054007
[14] Liu K Y, He Z G and Chao K T 2003 *Phys. Lett.* **B557** 45–54
[15] Hagiwara K, Kou E and Qiao C F 2003 *Phys. Lett.* **B570** 39–45
[16] Zhang Y J, Gao Y j and Chao K T 2006 *Phys. Rev. Lett.* **96** 092001
[17] Bodwin G T, Kang D, Kim T, Lee J and Yu C 2007 *AIP Conf. Proc.* **892** 315–317 (*Preprint* hep-ph/0611002)
[18] He Z G, Fan Y and Chao K T 2007 *Phys. Rev.* **D75** 074011
[19] Gong B and Wang J X 2008 *Phys. Rev.* **D77** 054028 (*Preprint* 0712.4220)
[20] Dong H R, Feng F and Jia Y 2012 *Phys.Rev.* **D85** 114018 (*Preprint* 1204.4128)
[21] Braguta V, Likhoded A and Luchinsky A 2005 *Phys.Rev.* **D72** 074019 (*Preprint* hep-ph/0507275)
[22] Braguta V, Likhoded A, Luchinsky A and Poslavsky S 2013 (*Preprint* 1309.3093)
[23] Bodwin G T, Lee J and Braaten E 2003 *Phys. Rev. Lett.* **90** 162001
[24] Bodwin G T, Lee J and Braaten E 2003 *Phys. Rev.* **D67** 054023
[25] Bodwin G T, Braaten E, Lee J and Yu C 2006 *Phys. Rev.* **D74** 074014 (*Preprint* hep-ph/0608200)
[26] Gong B and Wang J X 2008 *Phys. Rev. Lett.* **100** 181803 (*Preprint* 0801.0648)
[27] Aubert B *et al.* (BABAR) 2001 *Phys. Rev. Lett.* **87** 162002 (*Preprint* hep-ex/0106044)
[28] Abe K *et al.* (BELLE) 2002 *Phys. Rev. Lett.* **88** 052001 (*Preprint* hep-ex/0110012)
[29] Abe K *et al.* (Belle) 2002 *Phys. Rev. Lett.* **89** 142001
[30] Driesen V M, Kuhn J H and Mirkes E 1994 *Phys. Rev.* **D49** 3197–3208
[31] Yuan F, Qiao C F and Chao K T 1997 *Phys. Rev.* **D56** 321–328 (*Preprint* hep-ph/9703438)
[32] Cho P L and Leibovich A K 1996 *Phys. Rev.* **D54** 6690–6695 (*Preprint* hep-ph/9606229)
[33] Schuler G A 1999 *Eur. Phys. J.* **C8** 273–281 (*Preprint* hep-ph/9804349)
[34] Baek S, Ko P, Lee J and Song H S 1998 *J. Korean Phys. Soc.* **33** 97–101 (*Preprint* hep-ph/9804455)
[35] Hagiwara K, Qi W, Qiao C F and Wang J X 2007 (*Preprint* arXiv:0705.0803 [hep-ph])
[36] Braaten E and Chen Y Q 1996 *Phys. Rev. Lett.* **76** 730–733 (*Preprint* hep-ph/9508373)
[37] Wang J X 2003 (*Preprint* hep-ph/0311292)
[38] Zhang Y J and Chao K T 2007 *Phys. Rev. Lett.* **98** 092003
[39] Gong B and Wang J X 2009 *Phys. Rev. Lett.* **102** 162003 (*Preprint* 0901.0117)
[40] Ma Y Q, Zhang Y J and Chao K T 2009 *Phys. Rev. Lett.* **102** 162002 (*Preprint* 0812.5106)
[41] Pakhlov P *et al.* (Belle Collaboration) 2009 *Phys.Rev.* **D79** 071101 (*Preprint* 0901.2775)
[42] Gong B and Wang J X 2009 *Phys. Rev.* **D80** 054015 (*Preprint* 0904.1103)
[43] Brodsky S J, Lepage G P and Mackenzie P B 1983 *Phys. Rev. D* **28** 228–235
[44] Beneke M and Rothstein I Z 1996 *Phys. Lett.* **B372** 157–164 (*Preprint* hep-ph/9509375)
[45] Abulencia A *et al.* (CDF) 2007 *Phys. Rev. Lett.* **99** 132001
[46] Campbell J, Maltoni F and Tramontano F 2007 *Phys. Rev. Lett.* **98** 252002
[47] Gong B and Wang J X 2008 *Phys. Rev. Lett.* **100** 232001 (*Preprint* 0802.3727)
[48] Gong B and Wang J X 2008 *Phys. Rev.* **D78** 074011 (*Preprint* 0805.2469)
[49] Gong B, Li X Q and Wang J X 2009 *Phys. Lett.* **B673** 197–200 (*Preprint* 0805.4751)
[50] Gong B, Wang J X and Zhang H F 2011 *Phys.Rev.* **D83** 114021 (*Preprint* 1009.3839)
[51] Butenschoen M and Kniehl B A 2010 *Phys.Rev.Lett.* **104** 072001 (*Preprint* 0909.2798)
[52] Ma Y Q, Wang K and Chao K T 2011 *Phys. Rev.* **D83** 111503 (*Preprint* 1002.3987)
[53] Ma Y Q, Wang K and Chao K T 2011 *Phys. Rev. Lett.* **106** 042002 (*Preprint* 1009.3655)
[54] Ma Y Q, Wang K and Chao K T 2011 *Phys. Rev.* **D84** 114001 (*Preprint* 1012.1030)
[55] Butenschoen M and Kniehl B A 2011 *Phys. Rev. Lett.* **106** 022003 (*Preprint* 1009.5662)
[56] Butenschoen M and Kniehl B A 2011 *Phys. Rev. Lett.* **107** 232001 (*Preprint* 1109.1476)
[57] Aaron F *et al.* (H1 Collaboration) 2010 *Eur.Phys.J.* **C68** 401–420 (*Preprint* 1002.0234)
[58] Chekanov S *et al.* (ZEUS Collaboration) 2009 *JHEP* **0912** 007 (*Preprint* 0906.1424)
[59] Butenschoen M and Kniehl B A 2012 *Phys.Rev.Lett.* **108** 172002 (*Preprint* 1201.1872)
[60] Affolder T *et al.* (CDF Collaboration) 2000 *Phys.Rev.Lett.* **85** 2886–2891 (*Preprint* hep-ex/0004027)
[61] Abelev B *et al.* (ALICE Collaboration) 2012 *Phys.Rev.Lett.* **108** 082001 (*Preprint* 1111.1630)
[62] Chao K T, Ma Y Q, Shao H S, Wang K and Zhang Y J 2012 *Phys.Rev.Lett.* **108** 242004 (*Preprint* 1201.2675)
[63] Abe F *et al.* (CDF Collaboration) 1997 *Phys.Rev.Lett.* **79** 578–583
[64] Aaij R *et al.* (LHCb Collaboration) 2012 *Phys.Lett.* **B718** 431–440 (*Preprint* 1204.1462)
[65] Gong B, Wan L P, Wang J X and Zhang H F 2013 *Phys. Rev. Lett 110,* **042002** (*Preprint* 1205.6682)
[66] Wang K, Ma Y Q and Chao K T 2012 *Phys.Rev.* **D85** 114003 (*Preprint* 1202.6012)

[67] Chatrchyan S *et al.* (CMS Collaboration) 2013 *Phys.Rev.Lett.* **110** 081802 (*Preprint* 1209.2922)
[68] Gong B, Wan L P, Wang J X and Zhang H F 2013 (*Preprint* 1305.0748)

ACAT2013 IOP Publishing
Journal of Physics: Conference Series **523** (2014) 012040 doi:10.1088/1742-6596/523/1/012040

Automated computation of scattering amplitudes

Giovanni Ossola

Physics Department, New York City College of Technology, The City University of New York, 300 Jay Street Brooklyn, NY 11201, USA

The Graduate School and University Center, The City University of New York, 365 Fifth Avenue, New York, NY 10016, USA

E-mail: gossola@citytech.cuny.edu

Abstract. We review some of the recent advances in the computation of one-loop scattering amplitudes which led to the construction of efficient and automated computational tools for NLO predictions. Particular attention is devoted to unitarity-based methods and integrand-level reduction techniques. Extensions of one-loop integrand-level techniques to higher orders are also briefly illustrated.

1. Introduction

The evaluation of scattering amplitudes allows us to test the theoretical models and compare their phenomenological prediction with the results of the experiments at particle colliders. The understanding of the structure of scattering amplitudes provides the theoretical framework to develop new techniques for their evaluation, and ultimately to design more efficient computational algorithms for the production of physical cross sections and differential distributions.

Theory predictions play a fundamental role in the particle physics experiments at current hadron colliders. The high luminosity accumulated by the experimental collaborations at the Large Hadron Collider (LHC), allowed for a very detailed investigation of the Standard Model of particle physics. In these analyses, for example to study the properties of the recently discovered Higgs boson [1, 2], theoretical predictions are not only needed for the signal, but also for the modeling of the relevant background processes, which share similar experimental signatures [3–5]. Further, precise theory predictions are important in order to constrain model parameters in the event that a signal of New Physics is detected.

The scope of this review talk is to summarize the recent progress in the evaluation of scattering amplitudes, which led to the development of powerful automated computational tools for Next-to-Leading Order (NLO) calculations. After a general overview of the many different strategies which are currently available for the evaluation of one-loop scattering amplitudes, in particular for the calculation of the virtual part, we will focus on the description of integrand-level techniques. The extensions of the integrand-level approach to higher orders in perturbation theory will also be briefly illustrated.

The study of scattering amplitudes has a long tradition and an extensive literature, which would be difficult to condense in one presentation. We refer the interested reader to detailed and comprehensive reviews [6, 7] for a more complete picture of the field.

2. Scattering Amplitudes at Next-to-Leading Order

In order to properly describe the data collected by the experimental collaborations at the LHC, theory predictions are not reliable without accounting for higher orders. Leading-order (LO) results, which for most processes can be obtained with a tree-level calculation, are affected by large theoretical errors.

There are two main sources of theoretical uncertainties: the *truncation error*, which is related to the size of the unknown missing terms due to the truncation of the perturbative expansion, and the *parametric error*, that depends on the uncertainty on the various input parameters which enter in the theoretical predictions. The most common strategy employed to assess the theoretical error is to study the dependence of the predictions upon variation of the unphysical factorization and renormalisation scales.

A further complication is represented by the fact that, in order to obtain results for hadronic collisions, the partonic matrix element (hard scattering process) should be convoluted with the parton distribution functions (PDF), which represent the low momentum scale dynamics of the partons inside the hadron. Such convolution is performed over the fractional momentum carried by the partons. While the hard scattering cross section can be obtained in perturbative QCD, the PDFs are obtained by fitting the data coming from a variety of experiments (deep-inelastic scattering, DrellYan, jet production) and represent a very sizable source of theoretical uncertainty in the final results for hadronic cross sections and distributions [8].

Tree-level cross sections suffer from a large scale dependence, which is contained in the running of the couplings. The scale dependence is further enhanced by the presence of multi-leg processes, i.e. additional jets in the final state, because of the insertion of additional powers of the coupling constants. The situation is substantially improved with the inclusion of NLO effects.

Aside from providing a reduction in the theoretical uncertainties, NLO corrections are often sizable, in particular in QCD. Moreover, for some processes, NLO contributions have a big effect on the shape of the the differential distributions.

2.1. The structure of NLO calculations

The computation of NLO matrix elements requires, in addition to the tree-level LO result, the evaluation of one-loop virtual corrections (virtual part), obtained adding a virtual particle to the LO diagrams, and contributions from real emission (real part), obtained by adding one additional particle in the final states. Both terms are separately divergent: the virtual part is both ultraviolet (UV) and infrared (IR) divergent, and the UV divergence is removed by the renormalization procedure; the real part is also IR divergent due to the presence of soft and collinear singularities. Only their combination leads to a finite physical result, since the IR divergences in the virtual part are cancelled by the ones which appear in the real part.

The cancellation among the IR singularities in the real and virtual parts involves integrals over different phase-space configurations. This problem can be solved by employing a subtraction method [9–14], namely by including a subtraction term which has the same singularity structure as the real-emission corrections and cancels the infrared divergences in the phase space integral. The same contribution is then added to the virtual term, after integrating out the additional particle. After carrying out this procedure, the phase-space integrals are infrared-safe and their evaluation can be performed numerically in four dimensions. Several automated programs, based on different subtraction methods, are currently available to perform this task [15–22].

While the LO matrix elements and the NLO real parts have been available for a long time, thanks to powerful tools for tree-level computations [23–34], until recently the evaluation of the virtual part of one-loop contributions represented the bottleneck towards the automation of NLO computation.

3. Calculation of NLO virtual corrections

The standard method for the computation of NLO virtual corrections relies on the evaluation of all the Feynman diagrams associated with the process. The general task of the calculation is to compute, for each diagram contributing to the amplitude and for each phase space point, the following integral:

$$\mathcal{M} = \int d^n\bar{q}\, A(\bar{q}) = \int d^n\bar{q} \frac{N(\bar{q})}{\bar{D}_0\bar{D}_1 \ldots \bar{D}_{m-1}} . \tag{1}$$

It is well known that the evaluation of the one-loop diagrams can be performed by decomposing each integral $\mathcal{M}$ in terms of a finite set of scalar master integrals (MI) [35, 36]. The traditional one-loop "master" formula

$$\begin{aligned}\mathcal{M} &= \sum_{i_0<i_1<i_2<i_3}^{m-1} \mathbf{d}(\mathbf{i_0 i_1 i_2 i_3}) \int d^n\bar{q} \frac{1}{\bar{D}_{i_0}\bar{D}_{i_1}\bar{D}_{i_2}\bar{D}_{i_3}} + \sum_{i_0<i_1<i_2}^{m-1} \mathbf{c}(\mathbf{i_0 i_1 i_2}) \int d^n\bar{q} \frac{1}{\bar{D}_{i_0}\bar{D}_{i_1}\bar{D}_{i_2}} \\ &+ \sum_{i_0<i_1}^{m-1} \mathbf{b}(\mathbf{i_0 i_1}) \int d^n\bar{q} \frac{1}{\bar{D}_{i_0}\bar{D}_{i_1}} + \sum_{i_0}^{m-1} \mathbf{a}(\mathbf{i_0}) \int d^n\bar{q} \frac{1}{\bar{D}_{i_0}} + \mathcal{R}\end{aligned} \tag{2}$$

allows one to express any one-loop integral in terms of 4-, 3-, 2- and 1-point scalar integrals, plus an additional rational function, known in the literature as *rational part* $\mathcal{R}$, which is a function of the masses and momenta appearing in the original amplitude, and does not contain any logarithm or poly-logarithm. The expressions for the finite parts and the UV and IR poles of all scalar integrals are well known, and have been codified in publicly available libraries for their numerical evaluation [37–41].

In general, each one-loop calculation consists of three main phases: i) the *generation* of the amplitudes, namely the evaluation of the unintegrated amplitudes $\mathcal{M}$, in particular their numerator functions $N(q)$ and the set of corresponding denominators; ii) the *reduction* of the amplitude to scalar MIs, whose task is to determine the all coefficients appearing in front of the scalar integral in the master formula of Eq. (2) and the rational term $\mathcal{R}$; iii) the *evaluation* of the MIs, which multiplied by the coefficients provide the final result.

The analytic expressions of the coefficients in Eq. (2), as well as the rational term $\mathcal{R}$, can be extracted in a fully algebraic way, separately for each process. This procedure, known as algebraic tensor reduction, works well for processes with a small number of particles, but was limited for multi-leg applications by its algebraic complexity and the appearance of spurious singularities. Nevertheless, improved tensorial reduction methods [42–45] led to the development of tools that are able to deal efficiently and precisely with processes of high complexity such as, for example, $pp \to t\bar{t}b\bar{b}$ [46, 47].

The knowledge of the general and process-independent form of any one-loop calculation, contained in Eq. (2), allowed for the development of new numerical and semi-numerical approaches that aim at the direct evaluation of all the coefficients without performing any algebraic reduction. A new powerful framework for one-loop calculation was developed by merging the idea of employing four-dimensional unitarity-cuts, which allow to explore the (poly)logarithmic structure of the amplitudes, contained in the cut-constructible term [48, 49], with the knowledge of the universal four-dimensional decomposition for the numerator of the integrand for any one-loop scattering amplitudes [50] contained in the so-called OPP method [51–53].

The only part of the one-loop decomposition in Eq. (2) that cannot be computed using four dimensional techniques is the rational part R. As is well known, even starting from a perfectly finite tensor integral, the tensor reduction may lead to integrals that need to be regularized. In dimensional regularization, this is achieved by upgrading the integration momentum to dimension $d = 4 - 2\epsilon$, both in the numerator function and in the set of denominators. Such procedure is responsible for the appearance of the rational part $\mathcal{R}$. The reconstruction of this term can be achieved by using *ad hoc* tree-level Feynman rules [53–60], by direct computation via tensorial reduction techniques [61,62], or by the bootstrapping method [63,64]. Other techniques employ d-dimensional cuts [65, 66], and evaluate the rational term either by performing a mass continuation generated by the additional components of the loop momentum [67–70], or by comparing the results obtained with cuts in two different integer dimensions [71].

Very significant improvements were achieved with the d-dimensional extension of unitarity methods [71–73]. The integrand-level decomposition performed in d dimensions, rather than in four, exposes a richer polynomial structure of the integrand (see Section 4) and allows for the combined determination of both cut-constructible and rational terms at once [74, 75].

Aside from a deeper understanding of the structure of scattering amplitudes, unitarity-based methods and integrand-level reduction techniques provided the theoretical framework for development of efficient computational algorithms for NLO calculations in perturbation theory, which have been implemented in various automated codes, such as ROCKET [74], BLACKHAT [76], FORMCALC [77, 78], HELAC-NLO [79, 80], MADLOOP [81], GOSAM [82], OPENLOOPS [83], RECOLA [84], NJET [85, 86]. The technical features of some of these tools, together with recent advances and calculations, have been described in dedicated talks [87–91] during this conference.

Driven by the progress in the technical treatment of both tensorial and unitarity-based algorithms for one-loop corrections and motivated by requirements of the LHC analyses, an impressive list of calculations have been preformed [92–116] in the past few years. Among the most challenging recent results at NLO QCD accuracy, let us mention the first computations of cross sections for processes with seven and eight external particles, namely the production of an electroweak gauge boson in association with four [117, 118] and five jets [119] jets, provided by the BLACKHAT collaboration, and the production, in the infinite top-mass approximation, of a Higgs boson in association with three jets in gluon-fusion [120], which contains over ten thousand diagrams including hexagons of rank seven, and was performed with GOSAM. Other important calculations were recently completed: the NLO QCD corrections to Higgs boson production in association with a top quark pair and a jet [121], which involve internal massive top quarks; the NLO QCD predictions for the production of a photon pair in association with two jets [122], which is an important background for Higgs boson production; the production of four [123] and five jets [124] in hadronic collisions at NLO in massless QCD; the complete results at NLO QCD accuracy for electroweak Higgs production, including vector boson fusion and Higgs-strahlung type contributions, in association with three jets [125].

The automated computation of physical observables at NLO accuracy, such as cross sections and differential distribution, requires to incorporate the one-loop results for the virtual amplitudes within a Monte Carlo framework (MC), such as SHERPA [31], POWHEG [126–128], MADGRAPH-MADEVENT [27, 129], HERWIG [130], or aMC@NLO [131], that can take care of the phase-space integration, and of the combination of the different pieces of the calculation. In several recent applications [132–137], the MC framework also provides the possibility of merging multiple NLO parton-level matrix elements with parton showers [138–142].

In order to facilitate the communication between the programs computing virtual one-loop amplitudes and the MC frameworks, a standard interface has been worked out during the workshop "Physics at TeV Colliders" at Les Houches in June 2009, called the *Binoth Les Houches Accord* (BLHA) [143], recently updated in Ref. [144].

Within the BLHA, the interaction between the One-loop Program (OLP) and the Monte Carlo framework (MC) proceeds in two phases. During the first phase, called pre-runtime phase, the two programs agree on the process and settings for the computation and make all necessary preparation work. In the pre-runtime phase, the MC creates an order file, which contains information about the setup and the subprocesses it will need from the OLP in order to perform the computation. The OLP reads the order file and checks availability for each item. Then it returns a contract file telling the MC what it can provide. In the second stage, after the contract has been "signed", the MC requires from the OLP the value of the virtual one-loop amplitude at specific phase-space points.

4. The Integrand-level approach

The reduction at the integrand level is based on the idea of expressing the numerator function of the amplitude in terms of the propagators that depend on the integration momentum, in order to identify before integration the structures that will generate the scalar integrals and their coefficients and those that will vanish upon integration of the loop momentum.

In this approach, the coefficients in Eq. (2) are numerically determined by solving a system of algebraic equations that are obtained by: i) the numerical evaluation of the numerator of the integrand at explicit values of the loop-variable; ii) and the knowledge of the most general polynomial structure of the integrand itself.

The solution of this system of equations becomes particularly simply if we evaluate the expressions for the numerator functions at the set of complex values of the integration momentum for which a given set of inverse propagators vanish, namely the integration momenta corresponding to the so-called quadruple, triple, double, and single cuts. This feature establish a strong connection between the integrand-level techniques and generalized unitarity methods, where the on-shell conditions are imposed at the integral level.

The strength of the method lies in the fact that the only information required in order to extract the coefficient of the MIs is the knowledge of the numerical value of the numerator function for a finite set of values of the integration momentum, that correspond to complex poles of the denominators.

4.0.1. Integrand-level Reduction in four dimensions The integral-level reduction algorithm for one-loop scattering amplitudes, also known as OPP method, was originally developed in four dimensions [51–53]. According to this approach, the numerator function $N(q)$ which appear in the integrand for any one-loop scattering amplitudes has a universal mathematical structure, independent from the particular process at hand.

Any four-dimensional numerator function $N(q)$ can be written, in terms of 4-dimensional denominators $D_i = (q+p_i)^2 - m_i^2$, as:

$$
\begin{aligned}
N(q) \;=\; & \sum_{i_0<i_1<i_2<i_3}^{m-1} \left[\mathbf{d}(\mathbf{i_0 i_1 i_2 i_3}) + \tilde{d}(q; i_0 i_1 i_2 i_3)\right] \prod_{i\neq i_0,i_1,i_2,i_3}^{m-1} D_i + \sum_{i_0<i_1<i_2}^{m-1} \left[\mathbf{c}(\mathbf{i_0 i_1 i_2}) + \tilde{c}(q; i_0 i_1 i_2)\right] \prod_{i\neq i_0,i_1,i_2}^{m-1} D_i \\
& + \sum_{i_0<i_1}^{m-1} \left[\mathbf{b}(\mathbf{i_0 i_1}) + \tilde{b}(q; i_0 i_1)\right] \prod_{i\neq i_0,i_1}^{m-1} D_i + \sum_{i_0}^{m-1} \left[\mathbf{a}(\mathbf{i_0}) + \tilde{a}(q; i_0)\right] \prod_{i\neq i_0}^{m-1} D_i \, . \qquad (3)
\end{aligned}
$$

The quantities $\tilde{d}$, $\tilde{c}$, $\tilde{b}$, $\tilde{a}$, that still depend on the integration momentum q, are called "spurious terms" because they vanish upon integration and do not contribute to the final result for the scattering amplitude. Their functional form, namely their expression in terms of the integration momentum q, is process-independent and it is provided in Ref. [51]. Inserted back in Eq. (1), this expression simply states the multi-pole nature of any m-point one-loop amplitude. The quantities d, c, b, and a, indicated in boldface characters in Eq. (3), do not depend on q and are exactly the set of coefficients which appear in front of the 4-, 3-, 2-, and 1-point one-loop scalar functions of Eq. (2).

Once Eq. (3) is established, the task of computing the one-loop amplitude is reduced to the algebraic problem of extracting all the coefficients by evaluating the function $N(q)$ a sufficient number of times at different values of q. This is achieved very efficiently if we employ values of q

such that a subset of denominators D_i vanish: such values correspond to the so-called quadruple, triple, double, and single cuts also used in the unitarity-cut method. Operating in this manner, the system of equations becomes triangular. First one determines all the coefficients of the 4-point functions, then moves on to the 3-point coefficients and so on.

The technique described above allows for the determination of the cut-constructible part, which can be fully achieved in four dimensions. The calculation of rational term $\mathcal{R}$ can be split in two separate parts, which have different origins. A first set $\mathcal{R}_1$ appears from the mismatch between the d-dimensional denominators of the master scalar integrals and the 4-dimensional denominators of Eq. (3). The term $\mathcal{R}_1$ can be recovered automatically by evaluating the amplitudes for a shifted value of the mass [51]. A second set $\mathcal{R}_2$ comes from the d-dimensionality of the numerator function, and can be recovered by means of *ad hoc* tree-level-like Feynman rules, that are provided in Refs. [53–60] for different models.

The four-dimensional integrand-level reduction algorithm has been implemented in the code CutTools [145], that is publicly available. The method itself does not provide specific recipe for the generation of the numerator function. Some of the early calculations based on CutTools that appeared in the literature [146,147] employ traditional Feynman diagrams for the generation of the amplitudes. More recently, CutTools has been incorporated within automated tools for the computation of NLO correction, such as FormCalc, which generates the unintegrated amplitudes algebraically [148], or Helac-NLO and MadLoop, which construct numerically the one-loop n-particle amplitudes starting from tree-order amplitudes with $n+2$ particles [79].

4.0.2. D-dimensional Integrand-level Reduction In the context of four-dimensional techniques, the evaluation of the cut-constructible term and the rational term, that escapes the four-dimensional detection, are necessarily performed separately. As discussed above, the reconstruction of the latter usually requires information from an extra source.

The idea of performing unitarity-cuts in d-dimension was the basis for the development of a new algorithm, called samurai [75], which relies on the extension of the OPP polynomial structures to include an explicit dependence on the extra-dimensional parameter μ needed for the automated computation of the full rational term according to the d-dimensional approach, the parametrization of the residue of the quintuple-cut in terms of the extra-dimension scale [149] and the numerical sampling of the multiple-cut solutions via Discrete Fourier Transform [150].

The d-dimensional numerator $N(\bar{q})$ can be expressed in terms of d-dimensional denominators $\bar{D}_i$, as follows

$$\begin{aligned} N(\bar{q}) &= \sum_{i<<m}^{n-1} \Delta_{ijk\ell m}(\bar{q}) \prod_{h\neq i,j,k,\ell,m}^{n-1} \bar{D}_h + \sum_{i<<\ell}^{n-1} \Delta_{ijk\ell}(\bar{q}) \prod_{h\neq i,j,k,\ell}^{n-1} \bar{D}_h + \\ &+ \sum_{i<<k}^{n-1} \Delta_{ijk}(\bar{q}) \prod_{h\neq i,j,k}^{n-1} \bar{D}_h + \sum_{i<j}^{n-1} \Delta_{ij}(\bar{q}) \prod_{h\neq i,j}^{n-1} \bar{D}_h + \sum_{i}^{n-1} \Delta_{i}(\bar{q}) \prod_{h\neq i}^{n-1} \bar{D}_h \,, \end{aligned} \tag{4}$$

where $i << m$ stands for a lexicographic ordering $i < j < k < \ell < m$ and a bar denotes objects in $d = 4 - 2\epsilon$ dimensions, following the prescription $\bar{\not{q}} = \not{q} + \not{\mu}$ with $\bar{q}^2 = q^2 - \mu^2$.

Eq. (4) is the d-dimensional counterpart of Eq. (3), where the functions $\Delta(\bar{q}) = \Delta(q, \mu^2)$ are polynomials in the components of q and in μ^2. Their detailed expression is provided in Ref. [75]. By substituting the decomposition of Eq. (4) in Eq. (1), the multi-pole nature of the integrand of any one-loop n-point amplitude becomes transparent:

$$A(\bar{q}) = \sum_{i<<m}^{n-1} \frac{\Delta_{ijk\ell m}(\bar{q})}{\bar{D}_i\bar{D}_j\bar{D}_k\bar{D}_\ell\bar{D}_m} + \sum_{i<<\ell}^{n-1} \frac{\Delta_{ijk\ell}(\bar{q})}{\bar{D}_i\bar{D}_j\bar{D}_k\bar{D}_\ell} + \sum_{i<<k}^{n-1} \frac{\Delta_{ijk}(\bar{q})}{\bar{D}_i\bar{D}_j\bar{D}_k} + \sum_{i<j}^{n-1} \frac{\Delta_{ij}(\bar{q})}{\bar{D}_i\bar{D}_j} + \sum_{i}^{n-1} \frac{\Delta_{i}(\bar{q})}{\bar{D}_i} \,. \tag{5}$$

Once the d-dimensional identity of Eq. (4) has been established, the calculation of a generic scattering amplitude amounts to the problem of extracting the coefficients of multivariate polynomials, generated at every step of the multiple-cut analysis.

After all coefficients contained in $\Delta_{ijk\ell m}$, $\Delta_{ijk\ell}$, Δ_{ijk}, Δ_{ij}, and Δ_i have been extracted, they are multiplied by the corresponding MIs. In addition to the standard scalar integrals already contained in the 4-dimensional decomposition, there are additional μ^2-dependent master integrals:

$$\int d^n\bar{q}\frac{\mu^2}{\bar{D}_i\bar{D}_j} \quad , \quad \int d^n\bar{q}\frac{\mu^2}{\bar{D}_i\bar{D}_j\bar{D}_k} \quad , \quad \int d^n\bar{q}\frac{\mu^4}{\bar{D}_i\bar{D}_j\bar{D}_k\bar{D}_l} \, ,$$

whose expressions are also well-known [67, 151]. The presence of these new contributions, together with the d-dimensional decomposition, account for the complete evaluation of the full rational term.

The SAMURAI code [75] is publicly available. The method itself does not provide specific recipe for the generation of the numerator function: SAMURAI can reduce integrands defined either as *numerator functions* sitting on products of denominators, to be used with calculations based on Feynman diagrams, or as *products of tree-level amplitudes* sewn along cut-lines, to be employed for the reduction of amplitudes generated with unitarity-based techniques. The reduction provided by SAMURAI has been has been employed within the GoSAM framework, as well as interfaced [152] with FormCalc.

The integrand decomposition was originally developed for renormalizable gauge theories, where, at one-loop, the rank of the numerator cannot be greater than the number of external legs. To deal with the presence of effective-gluon vertices generated by the large top-mass limit, required by the evaluation of $pp \to H + 2, 3$ jets in gluon fusion [116, 120], the reduction code within SAMURAI was upgraded to accommodate an extension of the polynomial residues and of the corresponding sampling required to fit all coefficients [153, 154].

4.0.3. Integrand Reduction via Laurent Expansion Elaborating on the the techniques proposed in [70, 155], a different approach to the integrand-reduction method for one-loop amplitudes was recently presented [154], which allows to extract the coefficients of the integrand decomposition by performing a Laurent expansion, whenever the analytic form of the numerator function is known.

In general, when the multiple-cut conditions do not fully constrain the loop momentum, the on-shell solutions are still functions of some free parameters, possibly the components of the momentum which are not frozen by the cut conditions. The integrand-reduction algorithm as implemented in the numerical codes of Refs. [75, 145] requires, in order to extract the value of the unknown coefficients, to solve a system of equations obtained by sampling the numerator on a finite set of values of such free parameters after subtracting all the non-vanishing contributions coming from higher-point residues.

The reduction algorithm can be simplified by exploiting the knowledge of the analytic expression of the integrand. Indeed, by performing a *Laurent expansion* with respect to one of the free parameters which appear in the solutions of the cut, both the integrand and the subtraction terms exhibit the same polynomial behavior of the residue. Moreover, the contributions coming from the subtracted terms can be implemented as *corrections at the coefficient level*, hence replacing the subtractions at the integrand level of the original algorithm. The parametric form of this corrections can be computed once and for all, in terms of a subset of the higher-point coefficients. With this method the number of coefficients entering in each subtracted term is significantly reduced. For instance, box and pentagons do not affect at all the computation of lower-points coefficients.

If either the analytic expression of the integrand or the tensor structure of the numerator is known, this procedure can also be implemented in a semi-numerical algorithm. Indeed, the coefficients of the Laurent expansion of a rational function can be computed, either analytically or numerically, by performing a polynomial division between the numerator and the denominator. This method has been implemented, within the GoSam framework, in the C++ library Ninja, showing an improvement in the computational performance, both in terms of speed and precision, with respect to the standard algorithms. The new library has been recently employed in the evaluation of NLO QCD corrections to $pp \to t\bar{t}Hj$ [121].

5. Beyond NLO

The Next-to-Next-to-Leading-Order (NNLO) computations are quite far from automation and only a few computations are available for processes at hadron colliders [156, 157].

At one-loop, the advantage of knowing that one complete basis of MIs is formed by scalar one-loop functions [35] and the availability of their analytic expression allowed the community to focus on the development of efficient algorithms for the extraction of the coefficients multiplying each MI. At higher-loop, a general basis of MIs is not known and they are only identified at the end of the reduction procedure. Moreover, many MIs do not have a known analytic expression and they should be evaluated numerically. The multi-loop reduction technique which is most often employed is the well-known Laporta algorithm [158], based on the solution of algebraic systems of equations obtained through integration-by-parts identities [159].

5.1. Integrand-Level Techniques Beyond One-Loop

Extensions of the integrand reduction method beyond one-loop, first proposed in Refs. [160, 161], have recently become the topic of several studies [162–168], thus providing a new direction in the study of multi-loop amplitudes (see also the presentation of S. Badger at this conference [169]). An alternative approach based on maximal unitarity has been developed in Refs. [170–174].

Higher-loop extension of the integrand-level techniques require a proper parametrization of the residues at the multi-particle poles [160]. The parametric form of the polynomial residues is *process-independent* and can be determined in a general way from the corresponding multiple cut: only the values of the coefficients which appear in the residues is *process-dependent*. Each residue can be written as a multivariate polynomial in the *irreducible scalar products* (ISP), namely the set of scalar products, involving one or more the loop momenta, which cannot be reconstructed in terms of denominators. The ISPs either yield *spurious contributions*, which vanish upon integration, or generate the integrals which form the basis of amplitude decomposition [160, 161].

In Refs. [162, 163], the general determination of the residues at the multiple cuts has been systematized within the mathematical framework of algebraic geometry as a problem of multivariate polynomial division. The use of these techniques proved that the integrand decomposition, originally formulated for one-loop amplitudes, is applicable at any order in perturbation theory, irrespective of the complexity of the topology of the diagrams involved, massless or massive, planar or non planar. The shape of the residues is uniquely determined by the on-shell conditions, without any additional constraint. An iterative integrand-recursion formula, based on successive divisions of the numerators modulo the Gröbner basis of the ideals generated by the cut denominators, can provide the required multi-particle pole decomposition for arbitrary amplitudes, independently of the number of loops. The shape of the residues is uniquely determined by the on-shell conditions, without any additional constraint.

The algorithm presented in Ref. [163] relies on general properties of the loop integrands, which for simplicity we write as:

$$\mathcal{I}_{i_1\cdots i_n} = \frac{\mathcal{N}_{i_1\cdots i_n}}{\bar{D}_{i_1}\cdots\bar{D}_{i_n}} \,. \tag{6}$$

i) When the number n of denominators $\bar{D}_i$ is larger than the total number of the components of the loop momenta, namely when the on-shell conditions $\bar{D}_1 = \bar{D}_2 = \ldots = \bar{D}_n$ have no solutions, the integrand $\mathcal{I}_{i_1\cdots i_n}$ is *reducible*: it can be written in terms of integrands with $(n-1)$ denominators (lower point functions). This is the case, for example, of six-point functions in the one-loop decomposition.
ii) When n is equal or less than the total number of components of the loop momenta, the on-shell conditions $\bar{D}_1 = \bar{D}_2 = \ldots = \bar{D}_n$ have solutions. To determine the corresponding residue, we divide the numerator $\mathcal{N}_{i_1\cdots i_n}$ modulo the Gröbner basis of the n-ple cut, namely modulo a set of polynomials that vanish on the same on-shell solutions as the cut denominators. The *remainder* of the division is the *residue* $\Delta_{i_1\cdots i_n}$ of the n-ple cut. The *quotients* generate integrands with $(n-1)$ denominators which should undergo the same decomposition. This allows us to cast the each numerator $\mathcal{N}_{i_1\cdots i_n}$, sitting on a set of denominators $\bar{D}_i$, in the form

$$\mathcal{N}_{i_1\cdots i_n} = \sum_{\kappa=1}^{n} \mathcal{N}_{i_1\cdots i_{\kappa-1}i_{\kappa+1}\cdots i_n} \bar{D}_{i_\kappa} + \Delta_{i_1\cdots i_n} , \tag{7}$$

which inserted in Eq. (6), provides the recurrence relation

$$\mathcal{I}_{i_1\cdots i_n} = \sum_{\kappa=1}^{n} \mathcal{I}_{i_1\cdots i_{\kappa-1}i_{\kappa+1}i_n} + \frac{\Delta_{i_1\cdots i_n}}{\bar{D}_{i_1}\cdots \bar{D}_{i_n}} . \tag{8}$$

iii) A specific set of on-shell cut conditions, labeled *maximum-cuts*, are defined, for each number of loops, by the maximum number of on-shell conditions which can be simultaneously satisfied by the loop momenta. The *Maximum Cut Theorem* [163] ensures that the residue at the maximum-cuts is parametrized by n_s coefficients, where n_s is the number of solutions of the multiple cut-conditions, and therefore can *always be reconstructed by evaluating the numerator at the solutions of the cut.* This theorem extends at all orders the features of the one-loop quadruple-cut [49, 51], where the only two complex solutions of the cut determine the two coefficients needed to parametrize the residue.

The integrand recurrence relation of Eq. (8) may be applied in two ways. When the parametric form of all residues and the solutions of all possible multiple cuts are known, all the coefficients which appear in the residues can be determined by evaluating the numerator at the solutions of the multiple cuts, as many times as the number of the unknown coefficients. This approach has been employed at one loop in the original integrand reduction [51], and the language of multivariate polynomial division provides its natural generalization at all loops.

As a different strategy [168], the decomposition of the amplitude can be obtained analytically by successive polynomial divisions, which at each step generate the actual residues. In this approach, the reduction algorithm described above is applied directly to the actual numerator functions, without requiring the knowledge of the parametric form of all residues or the solutions of the multiple cuts. This new technique can naturally be applied to integrands with denominators appearing with arbitrary powers, thus solving a long-standing problem within unitarity-based methods, and represents a viable starting point towards the construction of automated computational tools at higher loops.

6. Conclusions

There are several different approaches available for one-loop calculations. The *generation* of the amplitudes can be performed starting with traditional Feynman diagrams, by means of recursive relations, or by gluing tree-level sub-amplitudes, as in unitarity-based methods. Aside from traditional tensorial approaches, the coefficient can be also extracted by performing an integrand-level *reduction* in 4-dimensions, as in the original OPP method, or by employing a

d-dimensional basis, which accounts automatically for the rational terms. Finally, the integral basis of Eq. (2) can be upgraded to include tensorial terms, to improve stability or performance.

The combined developments of all these different techniques, together with an increase in the performance and availability of computational facilities, triggered the "NLO revolution" [175, 176]. The full automation of NLO calculations has been successfully achieved by means of different tools that, just like their tree-level predecessors, allow the user to compute full NLO virtual corrections at the simple effort of providing the code with the list of particles and some input parameters. It is indeed fascinating to witness the number and quality of advanced automated NLO calculations that have been performed with very different techniques in the past few years [177–179].

Automated codes for the evaluation of NLO QCD and EW corrections have been also successfully interfaced within MC tools to produce results that will be of great importance for the experimental analyses, starting with the studies of the properties of the Higgs boson and the searches for New Physics at the LHC.

Acknowledgments

I would like to thank Gionata Luisoni and Pierpaolo Mastrolia for their valuable comments and suggestions. This work is supported in part by the National Science Foundation under Grant No. PHY-1068550 and PSC-CUNY Award No. 65188-00 43.

References

[1] Aad G *et al.* (ATLAS Collaboration) 2012 *Phys.Lett.* **B716** 1–29 (*Preprint* 1207.7214)
[2] Chatrchyan S *et al.* (CMS Collaboration) 2012 *Phys.Lett.* **B716** 30–61 (*Preprint* 1207.7235)
[3] Dittmaier S *et al.* (LHC Higgs Cross Section Working Group) 2011 (*Preprint* 1101.0593)
[4] Dittmaier S, Dittmaier S, Mariotti C, Passarino G, Tanaka R *et al.* 2012 (*Preprint* 1201.3084)
[5] Heinemeyer S *et al.* (The LHC Higgs Cross Section Working Group) 2013 (*Preprint* 1307.1347)
[6] Bern Z, Dixon L J and Kosower D A 2007 *Annals Phys.* **322** 1587–1634 (*Preprint* 0704.2798)
[7] Ellis R K, Kunszt Z, Melnikov K and Zanderighi G 2012 *Phys.Rept.* **518** 141–250 (*Preprint* 1105.4319)
[8] Alekhin S, Alioli S, Ball R D, Bertone V, Blumlein J *et al.* 2011 (*Preprint* 1101.0536)
[9] Frixione S, Kunszt Z and Signer A 1996 *Nucl.Phys.* **B467** 399–442 (*Preprint* hep-ph/9512328)
[10] Catani S and Seymour M 1997 *Nucl.Phys.* **B485** 291–419 (*Preprint* hep-ph/9605323)
[11] Kosower D A 1998 *Phys.Rev.* **D57** 5410–5416 (*Preprint* hep-ph/9710213)
[12] Catani S, Dittmaier S, Seymour M H and Trocsanyi Z 2002 *Nucl.Phys.* **B627** 189–265 (*Preprint* hep-ph/0201036)
[13] Nagy Z and Soper D E 2007 *JHEP* **0709** 114 (*Preprint* 0706.0017)
[14] Chung C, Kramer M and Robens T 2011 *JHEP* **1106** 144 (*Preprint* 1012.4948)
[15] Gleisberg T and Krauss F 2008 *Eur.Phys.J.* **C53** 501–523 (*Preprint* 0709.2881)
[16] Seymour M H and Tevlin C 2008 (*Preprint* 0803.2231)
[17] Frederix R, Gehrmann T and Greiner N 2008 *JHEP* **0809** 122 (*Preprint* 0808.2128)
[18] Czakon M, Papadopoulos C and Worek M 2009 *JHEP* **0908** 085 (*Preprint* 0905.0883)
[19] Frederix R, Frixione S, Maltoni F and Stelzer T 2009 *JHEP* **0910** 003 (*Preprint* 0908.4272)
[20] Hasegawa K, Moch S and Uwer P 2010 *Comput.Phys.Commun.* **181** 1802–1817 (*Preprint* 0911.4371)
[21] Frederix R, Gehrmann T and Greiner N 2010 *JHEP* **1006** 086 (*Preprint* 1004.2905)
[22] Bevilacqua G, Czakon M, Kubocz M and Worek M 2013 (*Preprint* 1308.5605)
[23] Stelzer T and Long W 1994 *Comput.Phys.Commun.* **81** 357–371 (*Preprint* hep-ph/9401258)
[24] Kanaki A and Papadopoulos C G 2000 *Comput.Phys.Commun.* **132** 306–315 (*Preprint* hep-ph/0002082)
[25] Mangano M L, Moretti M, Piccinini F, Pittau R and Polosa A D 2003 *JHEP* **0307** 001 (*Preprint* hep-ph/0206293)
[26] Corcella G, Knowles I, Marchesini G, Moretti S, Odagiri K *et al.* 2002 (*Preprint* hep-ph/0210213)
[27] Maltoni F and Stelzer T 2003 *JHEP* **0302** 027 (*Preprint* hep-ph/0208156)
[28] Gleisberg T, Hoeche S, Krauss F, Schalicke A, Schumann S *et al.* 2004 *JHEP* **0402** 056 (*Preprint* hep-ph/0311263)
[29] Boos E *et al.* (CompHEP Collaboration) 2004 *Nucl.Instrum.Meth.* **A534** 250–259 (*Preprint* hep-ph/0403113)
[30] Pukhov A 2004 (*Preprint* hep-ph/0412191)

[31] Gleisberg T, Hoeche S, Krauss F, Schonherr M, Schumann S *et al.* 2009 *JHEP* **0902** 007 (*Preprint* 0811.4622)
[32] Alwall J, Demin P, de Visscher S, Frederix R, Herquet M *et al.* 2007 *JHEP* **0709** 028 (*Preprint* 0706.2334)
[33] Kilian W, Ohl T and Reuter J 2011 *Eur.Phys.J.* **C71** 1742 (*Preprint* 0708.4233)
[34] Cafarella A, Papadopoulos C G and Worek M 2009 *Comput.Phys.Commun.* **180** 1941–1955 (*Preprint* 0710.2427)
[35] Passarino G and Veltman M J G 1979 *Nucl. Phys.* **B160** 151
[36] 't Hooft G and Veltman M 1979 *Nucl.Phys.* **B153** 365–401
[37] van Oldenborgh G 1991 *Comput.Phys.Commun.* **66** 1–15
[38] Hahn T and Perez-Victoria M 1999 *Comput.Phys.Commun.* **118** 153–165 (*Preprint* hep-ph/9807565)
[39] Ellis R K and Zanderighi G 2008 *JHEP* **02** 002 (*Preprint* 0712.1851)
[40] van Hameren A 2011 *Comput.Phys.Commun.* **182** 2427–2438 (*Preprint* 1007.4716)
[41] Cullen G, Guillet J, Heinrich G, Kleinschmidt T, Pilon E *et al.* 2011 *Comput.Phys.Commun.* **182** 2276–2284 (*Preprint* 1101.5595)
[42] Denner A and Dittmaier S 2006 *Nucl.Phys.* **B734** 62–115 (*Preprint* hep-ph/0509141)
[43] Binoth T, Guillet J P, Heinrich G, Pilon E and Reiter T 2009 *Comput.Phys.Commun.* **180** 2317–2330 (*Preprint* 0810.0992)
[44] Heinrich G, Ossola G, Reiter T and Tramontano F 2010 *JHEP* **1010** 105 (*Preprint* 1008.2441)
[45] Fleischer J and Riemann T 2011 *Phys.Rev.* **D83** 073004 (*Preprint* 1009.4436)
[46] Bredenstein A, Denner A, Dittmaier S and Pozzorini S 2009 *Phys.Rev.Lett.* **103** 012002 (*Preprint* 0905.0110)
[47] Bredenstein A, Denner A, Dittmaier S and Pozzorini S 2010 *JHEP* **1003** 021 (*Preprint* 1001.4006)
[48] Bern Z, Dixon L J, Dunbar D C and Kosower D A 1994 *Nucl. Phys.* **B425** 217–260 (*Preprint* hep-ph/9403226)
[49] Britto R, Cachazo F and Feng B 2005 *Nucl. Phys.* **B725** 275–305 (*Preprint* hep-th/0412103)
[50] del Aguila F and Pittau R 2004 *JHEP* **0407** 017 (*Preprint* hep-ph/0404120)
[51] Ossola G, Papadopoulos C G and Pittau R 2007 *Nucl.Phys.* **B763** 147–169 (*Preprint* hep-ph/0609007)
[52] Ossola G, Papadopoulos C G and Pittau R 2007 *JHEP* **0707** 085 (*Preprint* 0704.1271)
[53] Ossola G, Papadopoulos C G and Pittau R 2008 *JHEP* **0805** 004 (*Preprint* 0802.1876)
[54] Draggiotis P, Garzelli M, Papadopoulos C and Pittau R 2009 *JHEP* **0904** 072 (*Preprint* 0903.0356)
[55] Garzelli M, Malamos I and Pittau R 2010 *JHEP* **1001** 040 (*Preprint* 0910.3130)
[56] Garzelli M, Malamos I and Pittau R 2011 *JHEP* **1101** 029 (*Preprint* 1009.4302)
[57] Garzelli M and Malamos I 2011 *Eur.Phys.J.* **C71** 1605 (*Preprint* 1010.1248)
[58] Shao H S, Zhang Y J and Chao K T 2011 *JHEP* **1109** 048 (*Preprint* 1106.5030)
[59] Shao H S and Zhang Y J 2012 *JHEP* **1206** 112 (*Preprint* 1205.1273)
[60] Page B and Pittau R 2013 (*Preprint* 1307.6142)
[61] Xiao Z, Yang G and Zhu C J 2006 *Nucl.Phys.* **B758** 1–34 (*Preprint* hep-ph/0607015)
[62] Binoth T, Guillet J and Heinrich G 2007 *JHEP* **0702** 013 (*Preprint* hep-ph/0609054)
[63] Bern Z, Dixon L J and Kosower D A 2006 *Phys.Rev.* **D73** 065013 (*Preprint* hep-ph/0507005)
[64] Badger S, Glover E N and Risager K 2007 *JHEP* **0707** 066 (*Preprint* 0704.3914)
[65] Britto R, Feng B and Yang G 2008 *JHEP* **0809** 089 (*Preprint* 0803.3147)
[66] Britto R, Feng B and Mastrolia P 2008 *Phys. Rev.* **D78** 025031 (*Preprint* 0803.1989)
[67] Bern Z and Morgan A G 1996 *Nucl. Phys.* **B467** 479–509 (*Preprint* hep-ph/9511336)
[68] Anastasiou C, Britto R, Feng B, Kunszt Z and Mastrolia P 2007 *Phys.Lett.* **B645** 213–216 (*Preprint* hep-ph/0609191)
[69] Anastasiou C, Britto R, Feng B, Kunszt Z and Mastrolia P 2007 *JHEP* **0703** 111 (*Preprint* hep-ph/0612277)
[70] Badger S D 2009 *JHEP* **01** 049 (*Preprint* 0806.4600)
[71] Giele W T, Kunszt Z and Melnikov K 2008 *JHEP* **0804** 049 (*Preprint* 0801.2237)
[72] Ellis R K, Giele W T and Kunszt Z 2008 *JHEP* **03** 003 (*Preprint* 0708.2398)
[73] Ellis R, Giele W T, Kunszt Z and Melnikov K 2009 *Nucl.Phys.* **B822** 270–282 (*Preprint* 0806.3467)
[74] Giele W and Zanderighi G 2008 *JHEP* **0806** 038 (*Preprint* 0805.2152)
[75] Mastrolia P, Ossola G, Reiter T and Tramontano F 2010 *JHEP* **1008** 080 (*Preprint* 1006.0710)
[76] Berger C, Bern Z, Dixon L, Febres Cordero F, Forde D *et al.* 2008 *Phys.Rev.* **D78** 036003 (*Preprint* 0803.4180)
[77] Hahn T 2001 *Comput.Phys.Commun.* **140** 418–431 (*Preprint* hep-ph/0012260)
[78] Agrawal S, Hahn T and Mirabella E 2012 *PoS* **LL2012** 046 (*Preprint* 1210.2628)
[79] van Hameren A, Papadopoulos C and Pittau R 2009 *JHEP* **0909** 106 (*Preprint* 0903.4665)
[80] Bevilacqua G, Czakon M, Garzelli M, van Hameren A, Kardos A *et al.* 2013 *Comput.Phys.Commun.* **184** 986–997 (*Preprint* 1110.1499)

[81] Hirschi V, Frederix R, Frixione S, Garzelli M V, Maltoni F *et al.* 2011 *JHEP* **1105** 044 (*Preprint* 1103.0621)
[82] Cullen G, Greiner N, Heinrich G, Luisoni G, Mastrolia P *et al.* 2012 *Eur.Phys.J.* **C72** 1889 (*Preprint* 1111.2034)
[83] Cascioli F, Maierhofer P and Pozzorini S 2012 *Phys.Rev.Lett.* **108** 111601 (*Preprint* 1111.5206)
[84] Actis S, Denner A, Hofer L, Scharf A and Uccirati S 2013 *JHEP* **1304** 037 (*Preprint* 1211.6316)
[85] Badger S, Biedermann B and Uwer P 2011 *Comput.Phys.Commun.* **182** 1674–1692 (*Preprint* 1011.2900)
[86] Badger S, Biedermann B, Uwer P and Yundin V 2013 *Comput.Phys.Commun.* **184** 1981–1998 (*Preprint* 1209.0100)
[87] Hahn T *Proceedings of ACAT2013*
[88] Cullen G, van Deurzen H, Greiner N, Heinrich G, Luisoni G *et al.* 2013 (*Preprint* 1309.3741)
[89] Kosower D *Proceedings of ACAT2013*
[90] Maierhofer P *Proceedings of ACAT2013*
[91] Yundin V *Proceedings of ACAT2013*
[92] Bevilacqua G, Czakon M, Papadopoulos C, Pittau R and Worek M 2009 *JHEP* **0909** 109 (*Preprint* 0907.4723)
[93] Lazopoulos A 2008 (*Preprint* 0812.2998)
[94] Giele W, Kunszt Z and Winter J 2010 *Nucl.Phys.* **B840** 214–270 (*Preprint* 0911.1962)
[95] van Hameren A 2009 *JHEP* **0907** 088 (*Preprint* 0905.1005)
[96] Berger C, Bern Z, Dixon L J, Febres Cordero F, Forde D *et al.* 2009 *Phys.Rev.Lett.* **102** 222001 (*Preprint* 0902.2760)
[97] Berger C, Bern Z, Dixon L J, Febres Cordero F, Forde D *et al.* 2009 *Phys.Rev.* **D80** 074036 (*Preprint* 0907.1984)
[98] Ellis R, Melnikov K and Zanderighi G 2009 *JHEP* **0904** 077 (*Preprint* 0901.4101)
[99] Ellis R, Melnikov K and Zanderighi G 2009 *Phys.Rev.* **D80** 094002 (*Preprint* 0906.1445)
[100] Berger C, Bern Z, Dixon L J, Cordero F, Forde D *et al.* 2010 *Phys.Rev.* **D82** 074002 (*Preprint* 1004.1659)
[101] Bevilacqua G, Czakon M, van Hameren A, Papadopoulos C G and Worek M 2011 *JHEP* **1102** 083 (*Preprint* 1012.4230)
[102] Denner A, Dittmaier S, Kallweit S and Pozzorini S 2011 *Phys.Rev.Lett.* **106** 052001 (*Preprint* 1012.3975)
[103] Melia T, Melnikov K, Rontsch R and Zanderighi G 2011 *Phys.Rev.* **D83** 114043 (*Preprint* 1104.2327)
[104] Bevilacqua G, Czakon M, Papadopoulos C and Worek M 2011 *Phys.Rev.* **D84** 114017 (*Preprint* 1108.2851)
[105] Greiner N, Guffanti A, Reiter T and Reuter J 2011 *Phys.Rev.Lett.* **107** 102002 (*Preprint* 1105.3624)
[106] Denner A, Dittmaier S, Kallweit S and Pozzorini S 2012 *JHEP* **1210** 110 (*Preprint* 1207.5018)
[107] Greiner N, Heinrich G, Mastrolia P, Ossola G, Reiter T *et al.* 2012 *Phys.Lett.* **B713** 277–283 (*Preprint* 1202.6004)
[108] Cullen G, Greiner N and Heinrich G 2013 *Eur.Phys.J.* **C73** 2388 (*Preprint* 1212.5154)
[109] Campanario F, Li Q, Rauch M and Spira M 2013 *JHEP* **1306** 069 (*Preprint* 1211.5429)
[110] Bevilacqua G and Worek M 2012 *JHEP* **1207** 111 (*Preprint* 1206.3064)
[111] Gehrmann T, Greiner N and Heinrich G 2013 *JHEP* **1306** 058 (*Preprint* 1303.0824)
[112] Campanario F, Kerner M, Ninh L D and Zeppenfeld D 2013 *Phys.Rev.Lett.* **111** 052003 (*Preprint* 1305.1623)
[113] Campanario F and Kubocz M 2013 (*Preprint* 1306.1830)
[114] Bevilacqua G, Czakon M, Krmer M, Kubocz M and Worek M 2013 *JHEP* **1307** 095 (*Preprint* 1304.6860)
[115] Greiner N, Heinrich G, Reichel J and von Soden-Fraunhofen J F 2013 (*Preprint* 1308.2194)
[116] van Deurzen H, Greiner N, Luisoni G, Mastrolia P, Mirabella E *et al.* 2013 *Phys.Lett.* **B721** 74–81 (*Preprint* 1301.0493)
[117] Berger C, Bern Z, Dixon L J, Cordero F, Forde D *et al.* 2011 *Phys.Rev.Lett.* **106** 092001 (*Preprint* 1009.2338)
[118] Ita H, Bern Z, Dixon L, Febres Cordero F, Kosower D *et al.* 2012 *Phys.Rev.* **D85** 031501 (*Preprint* 1108.2229)
[119] Bern Z, Dixon L, Febres Cordero F, Hoeche S, Ita H *et al.* 2013 *Phys.Rev.* **D88** 014025 (*Preprint* 1304.1253)
[120] Cullen G, van Deurzen H, Greiner N, Luisoni G, Mastrolia P *et al.* 2013 *Phys.Rev.Lett.* **111** 131801 (*Preprint* 1307.4737)
[121] van Deurzen H, Luisoni G, Mastrolia P, Mirabella E, Ossola G *et al.* 2013 (*Preprint* 1307.8437)
[122] Gehrmann T, Greiner N and Heinrich G 2013 (*Preprint* 1308.3660)
[123] Badger S, Biedermann B, Uwer P and Yundin V 2013 *Phys.Lett.* **B718** 965–978 (*Preprint* 1209.0098)
[124] Badger S, Biedermann B, Uwer P and Yundin V 2013 (*Preprint* 1309.6585)
[125] Campanario F, Figy T, Pltzer S and Sjdahl M 2013 (*Preprint* 1308.2932)
[126] Nason P 2004 *JHEP* **0411** 040 (*Preprint* hep-ph/0409146)
[127] Frixione S, Nason P and Oleari C 2007 *JHEP* **0711** 070 (*Preprint* 0709.2092)
[128] Alioli S, Nason P, Oleari C and Re E 2010 *JHEP* **1006** 043 (*Preprint* 1002.2581)
[129] Alwall J, Herquet M, Maltoni F, Mattelaer O and Stelzer T 2011 *JHEP* **1106** 128 (*Preprint* 1106.0522)

[130] Arnold K, d'Errico L, Gieseke S, Grellscheid D, Hamilton K *et al.* 2012 (*Preprint* 1205.4902)
[131] Pittau R 2012 (*Preprint* 1202.5781)
[132] Campbell J M, Ellis R K, Frederix R, Nason P, Oleari C *et al.* 2012 *JHEP* **1207** 092 (*Preprint* 1202.5475)
[133] Hamilton K, Nason P, Oleari C and Zanderighi G 2013 *JHEP* **1305** 082 (*Preprint* 1212.4504)
[134] Luisoni G, Nason P, Oleari C and Tramontano F 2013 (*Preprint* 1306.2542)
[135] Hoeche S, Huang J, Luisoni G, Schoenherr M and Winter J 2013 *Phys.Rev.* **D88** 014040 (*Preprint* 1306.2703)
[136] Cascioli F, Hoeche S, Krauss F, Maierhofer P, Pozzorini S *et al.* 2013 (*Preprint* 1309.0500)
[137] Cascioli F, Maierhoefer P, Moretti N, Pozzorini S and Siegert F 2013 (*Preprint* 1309.5912)
[138] Hamilton K, Nason P and Zanderighi G 2012 *JHEP* **1210** 155 (*Preprint* 1206.3572)
[139] Hoeche S, Krauss F, Schonherr M and Siegert F 2013 *JHEP* **1304** 027 (*Preprint* 1207.5030)
[140] Gehrmann T, Hoche S, Krauss F, Schonherr M and Siegert F 2013 *JHEP* **1301** 144 (*Preprint* 1207.5031)
[141] Frederix R and Frixione S 2012 *JHEP* **1212** 061 (*Preprint* 1209.6215)
[142] Lnnblad L and Prestel S 2013 *JHEP* **1303** 166 (*Preprint* 1211.7278)
[143] Binoth T, Boudjema F, Dissertori G, Lazopoulos A, Denner A *et al.* 2010 *Comput.Phys.Commun.* **181** 1612–1622 (*Preprint* 1001.1307)
[144] Alioli S, Badger S, Bellm J, Biedermann B, Boudjema F *et al.* 2013 (*Preprint* 1308.3462)
[145] Ossola G, Papadopoulos C G and Pittau R 2008 *JHEP* **03** 042 (*Preprint* 0711.3596)
[146] Binoth T, Ossola G, Papadopoulos C and Pittau R 2008 *JHEP* **0806** 082 (*Preprint* 0804.0350)
[147] Actis S, Mastrolia P and Ossola G 2010 *Phys.Lett.* **B682** 419–427 (*Preprint* 0909.1750)
[148] Hahn T 2008 *PoS* **ACAT08** 121 (*Preprint* 0901.1528)
[149] Melnikov K and Schulze M 2010 *Nucl.Phys.* **B840** 129–159 (*Preprint* 1004.3284)
[150] Mastrolia P, Ossola G, Papadopoulos C and Pittau R 2008 *JHEP* **0806** 030 (*Preprint* 0803.3964)
[151] Pittau R 1997 *Comput. Phys. Commun.* **104** 23–36 (*Preprint* hep-ph/9607309)
[152] Agrawal S, Hahn T and Mirabella E 2012 *J.Phys.Conf.Ser.* **368** 012054 (*Preprint* 1112.0124)
[153] Mastrolia P, Mirabella E, Ossola G, Peraro T and van Deurzen H 2012 *PoS* **LL2012** 028 (*Preprint* 1209.5678)
[154] Mastrolia P, Mirabella E and Peraro T 2012 *JHEP* **1206** 095 (*Preprint* 1203.0291)
[155] Forde D 2007 *Phys. Rev.* **D75** 125019 (*Preprint* 0704.1835)
[156] Czakon M, Fiedler P and Mitov A 2013 *Phys.Rev.Lett.* **110** 252004 (*Preprint* 1303.6254)
[157] Boughezal R, Caola F, Melnikov K, Petriello F and Schulze M 2013 *JHEP* **1306** 072 (*Preprint* 1302.6216)
[158] Laporta S 2000 *Int.J.Mod.Phys.* **A15** 5087–5159 (*Preprint* hep-ph/0102033)
[159] Tkachov F 1981 *Phys.Lett.* **B100** 65–68
[160] Mastrolia P and Ossola G 2011 *JHEP* **1111** 014 (*Preprint* 1107.6041)
[161] Badger S, Frellesvig H and Zhang Y 2012 *JHEP* **1204** 055 (*Preprint* 1202.2019)
[162] Zhang Y 2012 *JHEP* **1209** 042 (*Preprint* 1205.5707)
[163] Mastrolia P, Mirabella E, Ossola G and Peraro T 2012 *Phys.Lett.* **B718** 173–177 (*Preprint* 1205.7087)
[164] Kleiss R H, Malamos I, Papadopoulos C G and Verheyen R 2012 *JHEP* **1212** 038 (*Preprint* 1206.4180)
[165] Badger S, Frellesvig H and Zhang Y 2012 *JHEP* **1208** 065 (*Preprint* 1207.2976)
[166] Feng B and Huang R 2013 *JHEP* **1302** 117 (*Preprint* 1209.3747)
[167] Mastrolia P, Mirabella E, Ossola G and Peraro T 2013 *Phys.Rev.* **D87** 085026 (*Preprint* 1209.4319)
[168] Mastrolia P, Mirabella E, Ossola G and Peraro T 2013 (*Preprint* 1307.5832)
[169] Badger S *Proceedings of ACAT2013*
[170] Gluza J, Kajda K and Kosower D A 2011 *Phys.Rev.* **D83** 045012 (*Preprint* 1009.0472)
[171] Kosower D A and Larsen K J 2012 *Phys.Rev.* **D85** 045017 (*Preprint* 1108.1180)
[172] Larsen K J 2012 *Phys.Rev.* **D86** 085032 (*Preprint* 1205.0297)
[173] Caron-Huot S and Larsen K J 2012 *JHEP* **1210** 026 (*Preprint* 1205.0801)
[174] Johansson H, Kosower D A and Larsen K J 2013 (*Preprint* 1308.4632)
[175] Perret-Gallix D 2013 *J.Phys.Conf.Ser.* **454** 012051 (*Preprint* 1301.1211)
[176] Salam G P 2010 *PoS* **ICHEP2010** 556 (*Preprint* 1103.1318)
[177] Bern Z *et al.* (NLO Multileg Working Group) 2008 1–120 (*Preprint* 0803.0494)
[178] Andersen J *et al.* (SM and NLO Multileg Working Group) 2010 21–189 (*Preprint* 1003.1241)
[179] Alcaraz Maestre J *et al.* (SM AND NLO MULTILEG and SM MC Working Groups) 2012 (*Preprint* 1203.6803)

Automated One-loop Computation in Quarkonium Process within NRQCD Framework

Feng Feng

Center for High Energy Physics, Peking University, Beijing 100871, China

E-mail: F.Feng@outlook.com

Abstract. In last decades, it has been realized that the next-to-leading order corrections may become very important, and sometimes requisite, for some processes involving quarkoinum production or decay, e.g., $e^+e^- \to J/\psi + \eta_c$ and $J/\psi \to 3\gamma$. In this article, we review some basic steps to perform automated one-loop computations in quarkonium process within the Non-relativistic Quantum Chromodynamics (NRQCD) factorization framework[1], and we give an introduction to some related public tools or packages and their usages in each step. We start from generating Feynman diagrams and amplitudes with FeynArts for the quarkonium process, performing Dirac- and Color- algebras simplifications using FeynCalc and FeynCalcFormLink, and then to doing partial fractions on the linear-dependent propagators by APart, and finally to reducing the Tensor Integrals (TI) into Scalar Integrals (SI) or Master Integrals (MI) using Integration-By-Parts (IBP) method with the help of Fire. We will use a simple concrete example to demonstrate the basic usages of the corresponding packages or tools in each step.

1. Generating Feynman Diagram and Amplitude

Before starting to generate the corresponding Feynman diagram, we need to replace the incoming or outgoing hadronic states with corresponding partonic states, e.g., we need to replace J/ψ with a quark and anti-quark pair $c(p_1)\bar{c}(p_2)$ for color-singlet case, and the pair $c\bar{c}$ with an extra gluon $c(p_1)\bar{c}(p_2)g(k)$ for the color-octet case. We take the process $(e^+e^- \to J/\psi + \eta_c)$ as an example, for the color-singlet model, the Feynman diagrams we actually want to generate are such process:

$$e^+ + e^- \to \gamma^* \to c(\frac{p_3}{2} + q_3)\bar{c}(\frac{p_3}{2} - q_3) + c(\frac{p_4}{2} + q_4)\bar{c}(\frac{p_4}{2} - q_4)\,. \quad (1)$$

Currently, there are two tools or packages to automatically generate the Feynman diagrams:

- FeynArts[2] is a Mathematica package for the generation and visualization of Feynman diagrams and amplitudes, it can be downloaded from `http://www.feynarts.de/`.
- QGraf[3] is a computer program with the programming language: FORTRAN 77, that was written to assist in large perturbative calculations, in the context of Quantum Field Theory. It can generate Feynman diagrams and represent them by symbolic expressions, it can be downloaded from `http://cfif.ist.utl.pt/~paulo/qgraf.html`.

[1] The general desription of the applications to NRQCD to the quarkonium processes was presented in the Plenary talk of this Workshop [1].

Since we use MATHEMATICA as our computation environment, we give a demonstration of the usage of FEYNARTS to generate the Feynman diagrams and amplitudes for the process: $(e^+e^- \to \gamma^* \to J/\psi + \eta_c)$. The basic steps are as follows:

- The underlying partonic process is:

$$\gamma^* \to c(\frac{p_3}{2} + q_3)\bar{c}(\frac{p_3}{2} - q_3) + c(\frac{p_4}{2} + q_4)\bar{c}(\frac{p_4}{2} - q_4)$$

- Using `CreateTopologies` to generate the topologies for the case $1 \to 4$:

```
top=CreateTopologies[1,1->4,ExcludeTogologies->{WFCorrections, Tadpoles, V4onExt}];
```

  where we also exclude some topologies which don't contribute in our case.
- Using `InsertFields` to insert the fields in the corresponding model to the topologies we have just generated in the last step:

```
tmp=InsertFields[top, {V[1]}->{F[3],-F[3],F[3],-F[3]}, Model->"SMQCD",
    ExcludeParticles->{V[1|2|3|4],S[_],F[4]}, InsertionLevel->{Clases}];
```

- Selecting or removing other unwanted diagrams using `DiagramSelect` or `DiagramDelete`, e.g.,

```
all=DiagramDelete[tmp, 3...4, 13, 14, 25...25, 33...34, 42...43];
```

- Using `CreateFeynAmp` to generate Feynman amplitudes for each diagram:

```
amp=CreateFeynAmp[all, PreFactor->1]/.{FourMomentum[Incoming,1]->p3+p4,
    FourMomentum[Outgoing,1]->p3/2+q3, FourMomentum[Outgoing,2]->p3/2-q3,
    FourMomentum[Outgoing,3]->p4/2+q4, FourMomentum[Outgoing,4]->p4/2-q4,
    MQU[___]->mu, MQD[___]->md, EL->e, GS->GStrong};
```

- Exporting the amplitudes to output file for later processing in FEYNCALC[4].

2. Simplifying Dirac- and Color-Algebra

To perform the Dirac- and Color-algebra simplifications, we adopt the covariant spin projectors techniques[5, 6, 7] for the $q\bar{q}$ production:

$$v(\bar{p})\bar{u}(p) \quad \to \quad \frac{1}{4\sqrt{2}E(E+m_c)}(\not{\bar{p}} - m_c)\left[\gamma_5, \not{\epsilon}^*\right](\not{P} + 2E)(\not{p} + m_c) \tag{2}$$

where γ_5 and $\not{\epsilon}$ correspond to spin-singlet and spin-triplet respectively, and for $q\bar{q}$ decay, the projector reads:

$$u(p)\bar{v}(\bar{p}) \quad \to \quad \frac{1}{4\sqrt{2}E(E+m_c)}(\not{p} + m_c)(\not{P} + 2E)\left[\gamma_5, \not{\epsilon}\right](\not{\bar{p}} - m_c) \tag{3}$$

where E is defined by:

$$E = \sqrt{m_c^2 + \left(\frac{p-\bar{p}}{2}\right)^2} \tag{4}$$

Several packages or tools can be used to simplify the Dirac- and Color-algebra:

- FORM[8, 9, 10, 11, 12, 13] is a Symbolic Manipulation System written in C language, it can be downloaded from `http://www.nikhef.nl/~form/`.
- FEYNCALC[4] is a MATHEMATICA package for algebraic calculations in elementary particle physics, it can be downloaded from `http://www.feyncalc.org/`.
- FEYNCALC/FORMLINK[14] is developed to combine high-performance of FORM and user-friendliness of FEYNCALC, it can be downloaded from `http://www.feyncalc.org/formlink/`.

Note that there is also another package FormCalc[15] which uses Form from Mathematica. The difference between FeynCalc/FormLink and FormCalc is the way in which Mathematica and Form communicate with each other. FormCalc basically uses the method of input and output files, while FeynCalc/FormLink uses the piping method. The basic idea of FormLink is:

- FormLink creates two unnamed pipes: `r#` and `w#`.
- FormLink starts Form process with the command line: `form -pipe r#, w# init`, where `init` is just a initial Form file with extension `.frm`.
- Form sends its `Process ID(PID)` to FormLink in `w#`, and when FormLink receives the `PID`, it will responses two comma-separated `PID` to Form, the first one is the same as Form `PID`, and the last one corresponds to the `PID` of FormLink.
- Form start running the `init.frm` file with the following codes:

```
Off Statistics;
#ifndef 'PIPES_'
#message "No pipes found";
.end;
#endif
#if ('PIPES_' <= 0)
#message "No pipes found";
.end;
#endif
```

```
#procedure put(fmt, mexp)
#toexternal 'fmt', 'mexp'
#toexternal "#THE-END-MARK#"
#endprocedure
#setexternal 'PIPE1_';
#toexternal "OK"
#fromexternal
.end
```

The key statement is `#fromexternal`, when Form runs into this instruction, it will be blocked until the Form code has been sent from Mathematica through FormLink, and then Form will continue to execute the code which has been just sent.

Let us demonstrate the basic usage of FeynCalc/FormLink with a simple example, i.e., the trace of six Dirac gamma matrix, first we calculate the trace with FeynCalc:

```
<<HighEnergyPhysics'fc'
Tr[GS[p1, p2, p3, p4, p5, p6]]
```

It is also quite simple to perform the trace with FeynCalcFormLink, first we prepare the expression in FeynCalc syntax, i.e.,

```
exp = DiracTrace[GS[p1, p2, p3, p4, p5, p6]];
```

note that we use `DiracTrace` instead of `Tr` to prevent the evaluation of the trace, then we just use `FeynCalcFormLink` to calcuate the expression `exp`:

```
FeynCalcFormLink[exp]
```

`FeynCalcFormLink` first translate the `exp` in FeynCalc syntax to Form code, for this simple case, the translated Form code is as follows:

```
Vectors p1,p2,p3,p4,p5,p6;
Format Mathematica;
L resFL = (g_(1,p1)*g_(1,p2)*g_(1,p3)*g_(1,p4)*g_(1,p5)*g_(1,p6));
trace4,1;
contract 0;
.sort;
#call put("%E", resFL)
#fromexternal
```

Journal of Physics: Conference Series **523** (2014) 012041 doi:10.1088/1742-6596/523/1/012041

then the FORM code will be piped to FORM for execution, when FORM finishes running, it starts sending the result back to MATHEMATICA, and FEYNCALCFORMLINK will translate the result to FEYNCALC syntax.

`SUNSimplify` in FEYNCALC can be used to perform the simplification on the color-algebra, e.g., `SUNSimplify[SUNT[a,b,a,b]]` to get the result

$$T^a T^b T^a T^b = -\frac{1}{2} C_F (C_A - 2C_F). \tag{5}$$

Before we are going to do the loop momentum integrals, we can use another technique, the method of region expansion[16], to greatly simplify our calculations. Usually, we expand the relative momentum q between quark and anti-quark in quarkonium state after performing the loop integration, and then project the S-, P-, or D-waves. We can also expand the q before the loop integration, as long as only the hard region is concerned according to the method of region expansion[16]. So if the NRQCD factorization is valid, it will be safe to use the method of region expansion to compute the short-distance coefficients, which correspond to the hard region.

3. Passarino-Veltman Reduction for the Tensor Integrals

The generic one-loop integral looks like:

$$\mathcal{T}^{\mu_1 \cdots \mu_p} \equiv \frac{(2\pi\mu)^{4-d}}{i\pi^2} \int d^d k \frac{k^{\mu_1} \cdots k^{\mu_p}}{D_0 D_1 D_2 \cdots D_{n-1}} \tag{6}$$

where $D_i = (k + r_i)^2 - m_i^2 + i\varepsilon$, $r_i = \sum_{k=1}^{i} p_k$ $(i = 1, \cdots, n-1)$, $r_0 = 0$ and $r_{ij} = r_i - r_j$. These one-loop integrals can be characterized by the so-called n-point tensor integrals, e.g.,

$$\begin{aligned}
B^\mu(r_{10}^2, m_0^2, m_1^2 0 &= \frac{(2\pi\mu)^{4-d}}{i\pi^2} \int d^d k k^\mu \prod_{i=0}^{1} \frac{1}{(k+r_i)^2 - m_i^2} \qquad (7) \\
C^{\mu\nu}(r_{10}^2, r_{12}^2, r_{20}^2, m_0^2, m_1^2, m_2^2) &= \frac{(2\pi\mu)^{4-d}}{i\pi^2} \int d^d k k^\mu k^\nu \prod_{i=0}^{2} \frac{1}{(k+r_i)^2 - m_i^2} \\
D^{\mu\nu\rho\sigma}(r_{10}^2, r_{12}^2, r_{23}^2, r_{30}^2, r_{20}^2, r_{13}^2, m_0^2, m_1^2, m_2^2, m_3^2) &= \frac{(2\pi\mu)^{4-d}}{i\pi^2} \int d^d k k^\mu k^\nu k^\rho k^\sigma \prod_{i=0}^{3} \frac{1}{(k+r_i)^2 - m_i^2}
\end{aligned}$$

Generally, those n-point tensor integrals can be reduced to much simpler loop integrals, n-point scalar integrals:

$$\begin{aligned}
A_0(m_0^2) &= \frac{(2\pi\mu)^{4-d}}{i\pi^2} \int d^d k \frac{1}{k^2 - m_0^2} \qquad (8) \\
B_0(r_{10}^2, m_0^2, m_1^2) &= \frac{(2\pi\mu)^{4-d}}{i\pi^2} \int d^d k \prod_{i=0}^{1} \frac{1}{(k+r_i)^2 - m_i^2} \\
C_0(r_{10}^2, r_{12}^2, r_{20}^2, m_0^2, m_1^2, m_2^2) &= \frac{(2\pi\mu)^{4-d}}{i\pi^2} \int d^d k \prod_{i=0}^{2} \frac{1}{(k+r_i)^2 - m_i^2} \\
D_0(r_{10}^2, r_{12}^2, r_{23}^2, r_{30}^2, r_{20}^2, r_{13}^2, m_0^2, m_1^2, m_2^2, m_3^2) &= \frac{(2\pi\mu)^{4-d}}{i\pi^2} \int d^d k \prod_{i=0}^{3} \frac{1}{(k+r_i)^2 - m_i^2}
\end{aligned}$$

We take a rank 4 tensor integral $D^{\mu\nu\rho\sigma}$ as an example, the tensor integral $D^{\mu\nu\rho\sigma}$ can be expressed as follows according to the Lorentz invariance:

$$\begin{aligned} D^{\mu\nu\rho\sigma} &= (g^{\mu\nu}g^{\rho\sigma} + g^{\mu\rho}g^{\nu\sigma} + g^{\mu\sigma}g^{\nu\rho})D_{0000} + \sum_{i,j,k,l=1}^{3} r_i^\mu r_j^\nu r_k^\rho r_l^\sigma D_{ijkl} \\ &+ \sum_{i,j}^{3}(g^{\mu\nu}r_i^\rho r_j^\sigma + g^{\nu\rho}r_i^\mu r_j^\sigma + g^{\mu\rho}r_i^\nu r_j^\sigma + g^{\mu\sigma}r_i^\nu r_j^\rho + g^{\nu\sigma}r_i^\mu r_j^\rho + g^{\rho\sigma}r_i^\mu r_j^\nu)D_{00ij} \end{aligned} \tag{9}$$

where the D_{0000}, D_{00ij} and D_{ijkl} are some Lorentz scalar coefficients which can be expressed in terms of the n-point scalar integrals: A_0, B_0, C_0 and D_0. Such procedure is called Passarino-Veltman Reduction(PaVe-Reduction).

These coefficients can be achieved with the function PaVe in FEYNCALC, e.g., the

$$D_{0000}(r_{10}^2, r_{12}^2, r_{23}^2, r_{30}^2, r_{20}^2, r_{13}^2, m_0^2, m_1^2, m_2^2, m_3^2)$$

in FEYNCALC is expressed as:

$$\texttt{PaVe}[0,0,0,0,, \{r_{10}^2, r_{12}^2, r_{23}^2, r_{30}^2, r_{20}^2, r_{13}^2\}, \{m_0^2, m_1^2, m_2^2, m_3^2\}]$$

where the first part in the argument of PaVe function {0,0,0,0} is the subscript of the corresponding D_{0000} coefficient, and the remaining are the same as those in the argument of D_{0000} coefficient. It should be noted that the factor involving renormalization scale μ, i.e., $(2\pi\mu)^{4-d}$ has been dropped out in FEYNCALC.

Now to perform the PaVe-Reduction, we just use PaVeReduce,

```
PaVeReduce[PaVe[0,0,0,0, {1,2,3,4,5,6}, {1,1,1,1}]]
```

the output looks like

$$\begin{aligned} &-\frac{135\mathrm{C}_0(1,2,5,1,1,1)}{2401} - \frac{5751\mathrm{C}_0(1,4,6,1,1,1)}{192080} - \frac{88691\mathrm{C}_0(2,3,6,1,1,1)}{2650704} \\ &-\frac{5755\mathrm{C}_0(3,4,5,1,1,1)}{633864} + \frac{1587\mathrm{D}_0(1,2,3,4,5,6,1,1,1,1)}{38416} + \frac{51\mathrm{B}_0(1,1,1)}{1960} \\ &+\frac{907\mathrm{B}_0(2,1,1)}{27048} + \frac{1025\mathrm{B}_0(3,1,1)}{99176} + \frac{347\mathrm{B}_0(4,1,1)}{32340} - \frac{50\mathrm{B}_0(5,1,1)}{1617} - \frac{181\mathrm{B}_0(6,1,1)}{22540} + \frac{5}{72} \end{aligned} \tag{10}$$

where we take some special numerical values for the argument of D_{0000}. We can see that the coefficient D_{0000} is now expressed in terms of n-point scalar integrals: B_0, C_0 and D_0.

Since the PaVe-Reduction is based on solving linear equations, generally it will encounter some problems when the Gram determinant equals 0, which happens if we expand the relative momentum q before loop integration, due to taking the derivative over q. So we need some more general method called Integration-By-Parts (IBP) reduction to perform the reduction of tensor integrals.

4. Partial Fraction and IBP Reduction

Let us consider a general Feynman integral, here we adopt the notation as in [17],

$$F(a_1, \cdots, a_n) = \int \cdots \int \frac{d^d k_1 \cdots d^d k_h}{E_1^{a_1} \cdots E_n^{a_n}} \tag{11}$$

where k_i, $i = 1, \cdots, h$, are loop momenta and the denominators E_r, $r = 1, \cdots, n$, are either quadratic or linear with respect to the loop momenta k_i of the graph. Irreducible polynomials in the numerator can be represented as denominators raised to negative powers.

The basic idea of IBP reduction [18] is that, we know the integration of such derivative is 0, i.e.,

$$\int \cdots \int d^d k_1 d^d k_2 \cdots \frac{\partial}{\partial k_i} \left[\frac{p_j}{E_1^{a_1} \cdots E_n^{a_n}} \right] = 0 \tag{12}$$

where k_i are the loop momenta, and p_j are the momenta which can internal or external, so with different k_i and p_j, we can get a list of equations which can be expressed as follows:

$$\sum \alpha_i F(a_1 + b_{i,1}, \cdots, a_n + b_{i,n}) = 0 \tag{13}$$

By solving these equations, we can express the complicated loop integrals in terms of much simpler ones, which we call Master Integral (MI).

There are many packages or tools in the market which can be used to perform the IBP reduction, e.g.,

- AIR[19] is MAPLE package, which can be downloaded from `http://www.phys.ethz.ch/~pheno/air/`.
- FIRE[17] is MATHEMATICA package, which can be downloaded from `http://science.sander.su/FIRE.htm`.
- REDUZE[20] is written in `C`, which can be downloaded from `http://reduze.hepforge.org/`.
- LITERED[21] is another MATHEMATICA package, which can be downloaded from `http://www.inp.nsk.su/~lee/programs/LiteRed/`.
- Many other private codes.

There is a precondition to perform the IBP reduction, i.e., the propagators should be linear independent, so we need another MATHEMATICA package APART[22] to perform partial fraction on the propagators.

Let us take a simple physical loop integral to demonstrate the usage of APART and FIRE,

$$\texttt{exp} = \frac{(k \cdot p_1)\,(k \cdot p_2)}{k^2[(k+p_1)^2 - m^2][(k+p_2) - m^2]^2} \tag{14}$$

the linear independent variables involving loop momentum are k^2, $k \cdot p_1$, $k \cdot p_2$, which can be expressed in FEYNCALC:

```
xs = FCI/@{ SP[k], SP[k, p1], SP[k, p2] }
```

then to perform partial fraction on the loop integral is ready with APART:

```
$APart[exp, xs]
```

the result looks like:

$$\begin{aligned}
&\frac{1}{4}\left(m^2 - \mathrm{p1}^2\right) \left\| \frac{1}{\left(-k^2 + m^2 - \mathrm{p1}^2 - 2k\cdot\mathrm{p1}\right)\left(-k^2 + m^2 - \mathrm{p2}^2 - 2k\cdot\mathrm{p2}\right)^2} \right\| \\
&+\frac{1}{4}\left(m^2 - \mathrm{p1}^2\right) \left\| \frac{1}{k^2\left(k^2 - m^2 + \mathrm{p1}^2 + 2k\cdot\mathrm{p1}\right)\left(k^2 - m^2 + \mathrm{p2}^2 + 2k\cdot\mathrm{p2}\right)} \right\| \\
&+\frac{1}{2}\boxed{\left\| \frac{k\cdot\mathrm{p2}}{\left(-k^2 + m^2 - \mathrm{p1}^2 - 2k\cdot\mathrm{p1}\right)\left(k^2 - m^2 + \mathrm{p2}^2 + 2k\cdot\mathrm{p2}\right)^2} \right\|} \\
&+\frac{1}{4}\left(m^2 - \mathrm{p1}^2\right)\left(m^2 - \mathrm{p2}^2\right) \left\| \frac{1}{k^2\left(k^2 - m^2 + \mathrm{p1}^2 + 2k\cdot\mathrm{p1}\right)\left(k^2 - m^2 + \mathrm{p2}^2 + 2k\cdot\mathrm{p2}\right)^2} \right\| \\
&-\frac{1}{4}\left\| \frac{1}{\left(-k^2 + m^2 - \mathrm{p2}^2 - 2k\cdot\mathrm{p2}\right)^2} \right\| + \frac{1}{4}\left\| \frac{1}{k^2\left(k^2 - m^2 + \mathrm{p2}^2 + 2k\cdot\mathrm{p2}\right)} \right\| + \frac{1}{4}\left(m^2 - \mathrm{p2}^2\right)\left\| \frac{1}{k^2\left(k^2 - m^2 + \mathrm{p2}^2 + 2k\cdot\mathrm{p2}\right)^2} \right\|
\end{aligned} \tag{15}$$

we can see that there are at most three propagators in each term, and these propagators in each term are liner independent now.

Finally we can use FIRE to perform IBP reduction, we take the tensor integral framed with red box in Eq. (15) as an example, such integral can be expressed as `F[{-1,1,2}]` with `F` defined by:

$$\texttt{F[\{l, m, n\}]} = \int \frac{d^4k}{(2\pi)^4} \frac{(k \cdot p_2)^{-l}}{(m^2 - k^2 - 2k \cdot p_1 - p_1^2)^m(-m^2 + k^2 + 2k \cdot p_2 + p_2^2)^n} \tag{16}$$

The basic usage of FIRE is like this:

```
Replacement = {p1^2 -> m^2, p2^2 -> m^2, p1 p2 -> SP[p1, p2]};
Internal = {k};
External = {p1, p2};
Propagators = {k p2, -2 k p1 - k^2 + m^2 - p1^2, 2 k p2 + k^2 - m^2 + p2^2};
PrepareIBP[];
startinglist = {IBP[k, k], IBP[k, p1], IBP[k, p2]}/.Replacement;
Prepare[];
Burn[];
```

first we input the internal and external momenta, and provide the independent propagators in `Propagators`, then prepare the IBP equations with `startinglist`, finally `Burn` in FIRE, and now it is ready to get the result for `F[{-1, 1, 2}]`, just use the `F` function:

$$\texttt{F[\{-1, 1, 2\}]} = \frac{(d-2)G(\{0,0,1\})}{8\,(m^2 - \mathrm{p1}\cdot\mathrm{p2})} + \frac{(d-2)G(\{0,1,0\})}{8\,(m^2 - \mathrm{p1}\cdot\mathrm{p2})} + \frac{1}{4}(4-d)G(\{0,1,1\}) \tag{17}$$

where the definition G is the same as `F` in Eq. (16).

We can apply such procedure to each loop integral in Eq. (15) to get the finally IBP reduced result:

$$\begin{aligned}
&\left\| \frac{(k\cdot p_1)\,(k\cdot p_2)}{k^2[(k+p_1)^2 - m^2][(k+p_2) - m^2]^2} \right\| \\
\Rightarrow\ & \frac{(D-2)\left\|\frac{1}{-k^2-2k\cdot\mathrm{p1}}\right\|}{16\,(m^2 - \mathrm{p1}\cdot\mathrm{p2})} + \frac{(D-2)\left\|\frac{1}{k^2+2k\cdot\mathrm{p2}}\right\|}{16\,(m^2 - \mathrm{p1}\cdot\mathrm{p2})} \\
&+\frac{1}{8}(4-D)\left\|\frac{1}{(-k^2 - 2k\cdot\mathrm{p1})\,(k^2 + 2k\cdot\mathrm{p2})}\right\| + \frac{1}{4}\left\|\frac{1}{k^2\,(k^2 + 2k\cdot\mathrm{p2})}\right\|
\end{aligned} \tag{18}$$

where we set $p_1^2 = p_2^2 = m^2$ to simplify the result, and $\|\cdots\|$ is defined by

$$\|\mathrm{exp}\| = \int \frac{d^Dk}{(2\pi)^D}\ \texttt{exp} \tag{19}$$

It can be seen that the original tensor integral has been reduced to much simper scalar integrals or master integrals, and we can apply such procedure to each tensor integral in each Feynman diagram, and get the final expression expressed in terms of scalar integrals, which can be calculated, analytically or numerically, by any other means.

Acknowledgments

The author would like to thank Rolf Mertig for the collaboration in the FORMLINK and FORMLINKFEYNCALC project, and to Prof. Yu Jia, Postdoc Wen-Long Sang and Hai-Rong Dong for many useful discussions in the related works.

Finally, Feng Feng would like to commemorate his beloved mother and father.

References

[1] Braguta V V, Likhoded A K, Luchinsky A V and Poslavsky S V 2013 Analytical calculation of heavy quarkonia production processes in computer (*Preprint* arXiv:1309.3093 [hep-ph]).
[2] Hahn T 2001 Generating Feynman diagrams and amplitudes with FeynArts 3 *Comput. Phys. Commun.* **140** 418 (*Preprint* hep-ph/0012260).
[3] Nogueira P 1993 Automatic Feynman graph generation *J. Comput. Phys.* **105** 279.
[4] Mertig R, Bohm M and Denner A 1991 FEYN CALC: Computer algebraic calculation of Feynman amplitudes *Comput. Phys. Commun.* **64** 345.
[5] Bodwin G T and Petrelli A 2002 Order v^4 corrections to S wave quarkonium decay *Phys. Rev.* D **66** 094011 (*Preprint* hep-ph/0205210).
[6] Braaten E and Lee J 2003 Exclusive double charmonium production from e^+e^- annihilation into a virtual photon, *Phys. Rev.* D **67** 054007 [Erratum-ibid. D **72** 099901] (*Preprint* hep-ph/0211085).
[7] Bodwin G T, Garcia T i Tormo and Lee J 2010 Factorization in exclusive quarkonium production *Phys. Rev.* D **81** 114014 (*Preprint* arXiv:1003.0061 [hep-ph]).
[8] Kuipers J, Ueda T, Vermaseren J A M and Vollinga J 2012 FORM version 4.0 (*Preprint* arXiv:1203.6543 [cs.SC]).
[9] Vermaseren J A M 2000 New features of FORM (*Preprint* math-ph/0010025).
[10] Vermaseren J A M 2010 FORM development *PoS CPP* **2010** 012 (*Preprint* arXiv:1101.0511 [hep-ph]).
[11] Vermaseren J A M 2010 FORM facts *Nucl. Phys. Proc. Suppl.* **205-206** 104 (*Preprint* arXiv:1006.4512 [hep-ph]).
[12] Tentyukov M and Vermaseren J A M 2008 Current status of FORM parallelization *PoS ACAT* **08** 119.
[13] Vermaseren J A M 2008 The FORM project, *Nucl. Phys. Proc. Suppl.* **183** 19 (*Preprint* arXiv:0806.4080 [hep-ph]).
[14] Feng F and Mertig R 2012 FormLink/FeynCalcFormLink : Embedding FORM in Mathematica and FeynCalc (*Preprint* arXiv:1212.3522 [hep-ph]).
[15] Hahn T and Perez-Victoria M 1999 Automatized one loop calculations in four-dimensions and D-dimensions *Comput. Phys. Commun.* **118** 153 (*Preprint* hep-ph/9807565).
[16] Beneke M and Smirnov V A 1998 Asymptotic expansion of Feynman integrals near threshold *Nucl. Phys.* B **522** 321 (*Preprint* hep-ph/9711391).
[17] Smirnov A V 2008 Algorithm FIRE – Feynman Integral REduction, *JHEP* **0810** 107 (*Preprint* arXiv:0807.3243 [hep-ph]).
[18] Chetyrkin K G and Tkachov F V 1981 Integration by Parts: The Algorithm to Calculate beta Functions in 4 Loops *Nucl. Phys.* B **192** 159.
[19] Anastasiou C and Lazopoulos A 2004 Automatic integral reduction for higher order perturbative calculations *JHEP* **0407** 046 (*Preprint* hep-ph/0404258).
[20] Studerus C 2010 Reduze-Feynman Integral Reduction in C++ *Comput. Phys. Commun.* **181** 1293 (*Preprint* arXiv:0912.2546 [physics.comp-ph]).
[21] Lee R N 2012 Presenting LiteRed: a tool for the Loop InTEgrals REDuction (*Preprint* arXiv:1212.2685 [hep-ph]).
[22] Feng F 2012 $Apart: A Generalized Mathematica Apart Function *Comput. Phys. Commun.* **183**, 2158 (2012) (*Preprint* arXiv:1204.2314 [hep-ph]).

BCVEGPY and GENXICC for the hadronic production of the doubly heavy mesons and baryons

Xing-Gang Wu
Department of Physics, Chongqing University, Chongqing 401331, P.R. China
E-mail: wuxg@cqu.edu.cn

Abstract. Doubly heavy mesons and baryons provide a good platform for testing pQCD. Two high efficient generators BCVEGPY and GENXICC for simulating the hadronic production of the doubly heavy mesons and baryons have been developed in recent years. In this talk, we present their main idea and their recent progresses. The dominant gluon-gluon fusion mechanism programmed in those two generators are written based on the improved helicity amplitude approach, in which the hard scattering amplitude are dealt with directly at the amplitude level and the numerical efficiency are greatly improved by properly decomposing the Feynman diagrams and by fully applying the symmetries among them. Moreover, in comparison to the previous versions, we have updated the programs in order to generate the unweighted meson or baryon events much more effectively within various simulation environments. The generators can be conveniently imported into PYTHIA to do further hadronization and decay simulation, which have already been adopted by several collaborations as LHCb, CMS and etc. to do B_c and Ξ_{cc} simulations.

1. Introduction

Heavy quarkonium or baryon has attracted wide attention due to its special features, which provides a good platform for studying the perturbative QCD and the associated non-perturbative physics in the bound state system [1-4]. For example, the B_c meson and the Ξ_{cc} baryon can be a fruitful laboratory for testing various potential models and understanding the weak decay mechanism for heavy flavors. Systematic studies for the hadronic production of the doubly heavy quarkonium or baryon at the hadronic colliders as TEVATRON and LHC have been done in the literature [5-17]. In particular, two generators BCVEGPY and GENXICC have been completed and developed in recent years [18-24], which can be conveniently implemented into PYTHIA [25] for simulating the B_c meson and Ξ_{cc} baryon events with high efficiency. In the present paper, we present their main idea and their recent progresses. It is noted that in the case of productions of double heavy baryons the diquark is remaining in the coloured state. In this case the interactions with hadronic residue part may decrease the results of calculations due to disassociation effect [26]. Therefore, it is worth to mention, that the results of calculations of cross-sections of hadronic production of double heavy baryons are not absolutely identical to the case of production of B_c meson, and the obtained results for the cross-sections of the baryons production should be considered with care and may be treated as the estimate of upper bound of the values of the cross-sections.

There are gluon-gluon fusion mechanism, or the light quark-anti-quark mechanism, or the extrinsic or intrinsic heavy quark mechanism for the hadronic production of the doubly heavy meson or baryon respectively. All of which may provide sizable contribution in specific kinematic regions. It is noted that in most of the kinematic regions (e.g. large p_t region), the dominant production mechanism is the gluon-gluon fusion mechanism, e.g. for the production of the $(c\bar{b})$-quarkonium, it stands for the production via the process $gg \to (c\bar{b})[n] + b + \bar{c}$, where $[n]$ stands for the Fock-states $|(c\bar{b})_{\mathbf{1}}[^1S_0]\rangle$, $|(c\bar{b})_{\mathbf{8}}[^1S_0]g\rangle$, $|(c\bar{b})_{\mathbf{1}}[^3S_1]\rangle$, $|(c\bar{b})_{\mathbf{8}}[^3S_1]g\rangle$, $|(c\bar{b})_{\mathbf{1}}[^1P_1]\rangle$ and $|(c\bar{b})_{\mathbf{1}}[^3P_J]\rangle$ (with $J = (1, 2, 3)$) within the nonrelativistic QCD [27], respectively. The lowest Fock state $|(c\bar{b})_{\mathbf{1}}[^1S_0]\rangle$ corresponds to the usually called B_c meson. The cases for the hadronic production Ξ_{cc}, Ξ_{bc} and Ξ_{bb} baryons are similar, which can be obtained from the case of B_c meson production by properly transforming the anti-quark line to be the quark line and by properly dealing with the color flows [13]. So, in the following, we shall take the gluon-gluon fusion mechanism for the B_c meson production as an explicit example to explain our calculation technology and to show how to improve the events simulation efficiency.

2. Gluon-gluon fusion mechanism for the hadronic production

According to the pQCD factorization, the hadronic production cross section is formulated as

$$d\sigma = \int dx_1 \int dx_2\, F^g_{H_1}(x_1, \mu_F) \times F^g_{H_2}(x_2, \mu_F)\, d\hat{\sigma}_{gg\to(c\bar{b})[n]X}(x_1, x_2, \mu_F, \mu_R), \qquad (1)$$

where $F^g_H(x, \mu_F)$ is the gluon distribution function in hadron H, $d\hat{\sigma}_{gg\to B_cX}(x_1, x_2, \mu_F)$ is the cross section for the relevant inclusive production $(g + g \to (c\bar{b})[n] + b + \bar{c})$. μ_F and μ_R stand for the factorization scale and renormalization scale respectively. The generating of phase space and its integration can be done with the help of the routines RAMBOS [28] and VEGAS [29]. Then, the most important part left is to deal with the hard scattering amplitude of the process.

At the lowest α_s^4 order, there are totally 36 Feynman diagrams and hence 36 amplitudes for the gluon-gluon fusion $g + g \to (c\bar{b})[n] + b + \bar{c}$. According to the pQCD factorization, each amplitude can be factorized into two parts, that of perturbative $gg \to b + \bar{b} + c + \bar{c}$ (all the quarks are on shell) and that of the nonperturbative $c + \bar{b} \to (c\bar{b})[n]$ which can be represented by the universal matrix element for each Fock state. Using the conventional squared amplitude to deal with the amplitude is a tedious and time-consuming task. To save the time for events simulation, we can deal with the hard scattering amplitude for $gg \to b + \bar{b} + c + \bar{c}$ directly at the amplitude level. In the literature, two ways have been suggested, i.e. the improved trace technology [30-33] and the helicity amplitude approach [34,35]. In our generators, we adopt an improved version for the helicity amplitude approach as suggested in Ref.[18] to simplify the hard scattering amplitude, in which the massive quark lines are dealt with in a much more compact and effective way.

2.1. A short review of the improved helicity amplitude approach

Main idea of the improved helicity amplitude approach is shown in Fig.(1). According to the improved helicity amplitude approach [15], we first pick out all the basic spinor lines, which are constructed by the γ structure of Fermion line, from the whole amplitude, whose remaining parts include the color factors and the scalar part of the propagators that shall be dealt with separately. These basic spinor lines can be simplified by changing the involved space-like momenta to be light-like ones, and then use the relation $\not{k} = |k+\rangle\langle k + | + |k-\rangle\langle k - |$ (k is light-like) to transform and simplify all the spinor line into expressions of spinor products. This intermediate step can be done using the standard helicity amplitude approach, cf.Refs.[34,35]. Next, we construct the basic Feynman diagrams (nine for the gluon-gluon fusion) by proper combination of these basic spinor lines. Finally, we adopt the symmetries of the gluon-gluon exchanges, the quark line exchanges to construct all the amplitude of the process.

One subtle point. The authors of Ref.[34] group the Feynman diagrams of the process into gauge-invariant subsets. To achieve the goal, one have to introduce extra terms into the Feynman

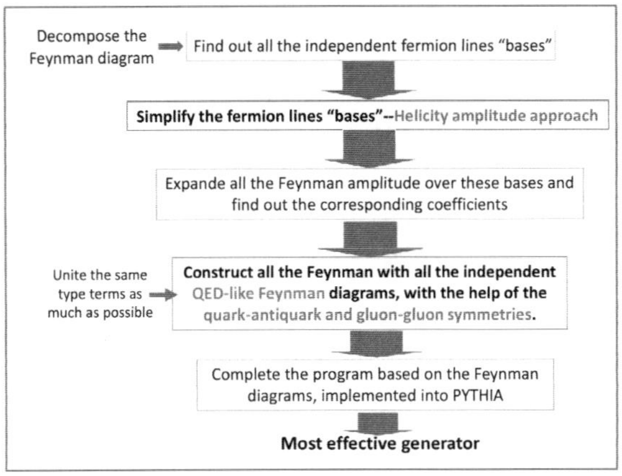

Figure 1. Main idea for the improved helicity amplitude approach [15].

rules for the three-gluon and four-gluon vertices, i.e.

$$\begin{aligned} gf_{abc}\tilde{T}^{\mu\nu\delta}(P,S,K) &= gf_{abc}T^{\mu\nu\delta}(P,S,K) + gf_{abc}G^{\mu\nu\delta}(P,S,K), \\ -ig\tilde{V}^{\lambda\mu\nu\delta}_{abcd}(P,Q,K_1,k_2) &= -igV^{\lambda\mu\nu\delta}_{abcd}(P,Q,K_1,k_2) - igG^{\lambda\mu\nu\delta}_{abcd}(P,Q,K_1,k_2), \end{aligned} \quad (2)$$

where P, S and etc. are input momenta for the gluons, $T^{\mu\nu\delta}(P,S,K)$ and $V^{\lambda\mu\nu\delta}_{abcd}(P,Q,K_1,k_2)$ are primary Feynman rules

$$\begin{aligned} T^{\mu\nu\delta}(P,S,K) &= (P-S)^{\delta}g^{\mu\nu} + (S-K)^{\mu}g^{\delta\nu} + (K-P)^{\nu}g^{\delta\mu} \\ V^{\lambda\mu\nu\delta}_{abcd}(P,Q,K_1,k_2) &= f_{abe}f_{cde}(g^{\lambda\nu}g^{\mu\delta} - g^{\lambda\delta}g^{\mu\nu}) + f_{ace}f_{dbe}(g^{\lambda\delta}g^{\mu\nu} - g^{\lambda\mu}g^{\nu\delta}) \\ &\quad + f_{ade}f_{bce}(g^{\lambda\mu}g^{\nu\delta} - g^{\lambda\nu}g^{\mu\delta}), \end{aligned} \quad (3)$$

and $G^{\mu\nu\delta}(P,S,K)$ and $G^{\lambda\mu\nu\delta}_{abcd}(P,Q,K_1,k_2)$ are modified parts

$$G^{\mu\nu\delta}(P,S,K) = (\pm)\left[P^{\mu}P^{\nu}P^{\delta}/(P\cdot K) + S^{\mu}S^{\nu}S^{\delta}/(S\cdot K)\right] \quad (4)$$

$$G^{\lambda\mu\nu\delta}_{abcd}(P,Q,K_1,k_2) = -f_{ace}f_{bde}\frac{S_1^{\lambda}S_1^{\mu}S_1^{\nu}S_1^{\delta}}{(S_1\cdot K_1)(S_1\cdot K_2)} - f_{ade}f_{bce}\frac{S_2^{\lambda}S_2^{\mu}S_2^{\nu}S_2^{\delta}}{(S_2\cdot K_1)(S_2\cdot K_2)}, \quad (5)$$

where $S_1 = P + K_1$ and $S_2 = S + K_2$. Because the modified part of the three gluon vertex is symmetric on the interchange of P and S, we must be careful to choose the sign of modified part in every associated diagrams so as to make all subsets be gauge invariant. When adding all extra terms together they will result in the required zero contribution. Then by choosing different gauge for each subset, one can obtain a compact result, when all terms of the amplitude are written according to the helicities of the fermions.

This method is most effective for massless cases. While, for the present massive case (the quark lines are massive), the question is much more involved. It is noted that the sum of all extra terms involving $G^{\mu\nu\delta}(P,S,K)$ and $G^{\lambda\mu\nu\delta}_{abcd}(P,Q,K_1,k_2)$ will have non-zero contributions, which complicates the final results. So in the improved helicity amplitude approach, we adopt a unique gauge for the whole amplitude. This gauge choice also avoids numerical cancelations between very large numbers and the numerical results are much more steady when doing the numerical calculations. This gauge independence can be adopted as one critical point on checking the rightness of the generator.

Another subtle point. We can further make the amplitude more symmetric by applying a proper decomposition of the amplitude. As shown by Fig.(2), this is achieved by decomposing

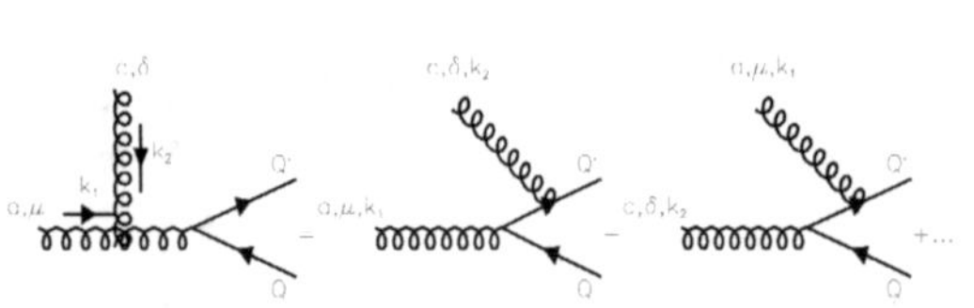

Figure 2. One typical three-gluon coupling vertex to be decomposed: the first two terms are the 'basic QED-like' terms and the 'remaining' terms are expressed by several extra basic functions from the diagrams involving the four-gluon vertex.

the terms involving three- and/or four-gluon vertices, into terms without self-interactions of gluons (i.e. into QED-like amplitudes). More over, it is noted that for the massive quark case, in addition to the QED-like Feynman diagrams, some more additional QED-like terms need to be introduced.

Fortunately, for the present subprocess $gg \to b+\bar{b}+c+\bar{c}$, these 'extra' terms are just parts of the diagrams with four-gluon vertices; i.e. they are one of the three terms within the amplitude involving the four-gluon vertex. This shows that the whole amplitude is just the recombination of all the basic functions already existed in the 36 Feynman amplitudes, which is like a magic. There are totally nine basic QED-like Feynman diagrams for the present process. Thus, only the analytic expressions for those 'basic Feynman diagrams' need to be put in the program precisely, while the non-basic ones can be generated by means of linear combination (the coefficients can be found in Ref.[18]) and/or by properly using the symmetry of all those basic ones.

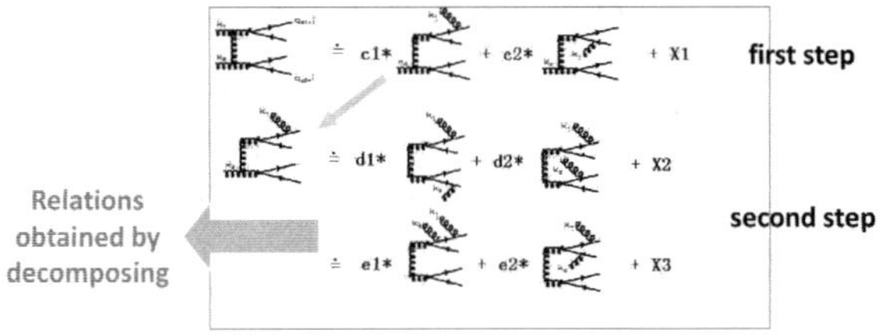

Figure 3. One example to decompose the Feynman diagram involving two three-gluon vertices to QED-like diagrams. Thus, we can obtain one relation among the QED-like diagrams.

As a final subtle point. As shown by Fig.(3), for the Feynman diagrams involving two three-gluon vertices, there are usually two ways to transform them into the QED-like diagrams; i.e. in the second step there are two ways to move the gluon either to the upper quark line or to the lower quark line respectively. Thus, two extra constraints can be found among the nine 'basic Feynman diagrams', the program can be further simplified and become more compact.

2.2. The structure and the efficiency of the BCVEGPY

The flowchart for BCVEGPY, which is implemented into PYTHIA, is shown in Fig.(4). The BCVEGPY (or GENXICC) is a Fortran programme written in a PYTHIA-compatible format and is written in a modularization structure, one may apply it to various situations or experimental environments conveniently. It is noted that the LHE format is a standard format that is proposed to store process and event information from the matrix-element-based generators. One can pass these parton-level information to the event generator as PYTHIA for further simulation. And BCVEGPY will generate a standard LHE data file [36] that contains useful information of the meson and its accompanying partons, which can be conveniently imported into PYTHIA to do further hadronization and decay simulation.

The schematic structure for the generator BCVEGPY is shown in Fig.(5). In general, the generator is systematically constructed in seven modules according to their purpose. Each

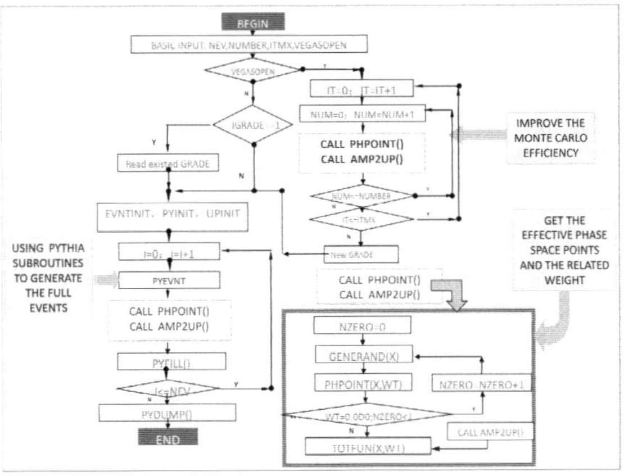

Figure 4. The flowchart for the generator BCVEGPY. The flowchart for the generator GENXICC is similar.

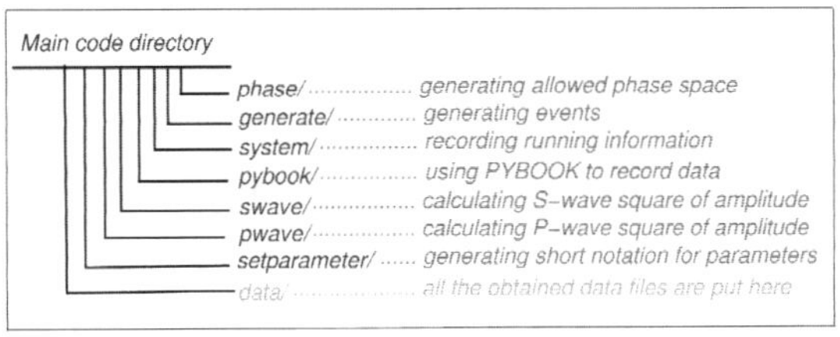

Figure 5. The schematic structure for the generator BCVEGPY. The schematic structure for the generator GENXICC is similar.

module contains necessary files to fulfill the specific tasks for generating events. The BCVEGPY is dominated by two blocks, i.e. the *vegas* block (in module **phase**) and the *event* block (in module **generate**). The *vegas* block is to generate the importance sampling function. The *event* block is to generate events by using PYTHIA, in a way that all the mentioned processes are implemented into PYTHIA as its external processes. This is achieved by properly setting the two PYTHIA subroutines UPINIT and UPEVNT.

Here we modify the *vegas* block not only to generate the sampling importance function but also to store an upper bound of the value of the cross sections in each cell [37-40]; i.e. an improved and more effective method to generate unweighted events has been programmed [21,24] based on the MINT package [40]. The importance sampling function is used to increase the simulation efficiency, while the upper bound value will be used to generate unweighted event if the user want to do the experimental analysis and further simulation. The upper bound value in each cell is an upper bound for the cross sections and also equals a multidimensional stepwise function, according to which it is easy to generate phase-space points. By using this new hit-and-miss technique, one can generate the points according to the original distribution. Moreover, in doing the initialization, we will call VEGAS twice to generate the upper bound grid XMAX and also a more precise importance sampling function. In VEGAS the integral together with its numerical error are related to the sampling numbers ncall and the iteration times itmx. So, to generate full events, we suggest the user to do a test running first so as to find an effective and time-saving parameters for VEGAS. Finally, one can generate events by calling the UPEVNT subroutine according to the probability proportional to the integrand. As we have shown that such treatment can largely improve the generation efficiency without affecting the total cross-section of the process [21,24].

3. Summary

The coming LHC experiment shall provide a better platform to check all the theoretical predications and to learn B_c, Ξ_{cc}, Ξ_{bc} and Ξ_{bb} properties in more detail. Recently some new measurements on B_c meson have been done by LHCb and CDF collaborations [41-45], in which BCVEGPY has been adopted for data analysis. The very recent measurement of Xi_{cc}^+ events by LHCb collaboration has adopted GENXICC to do analysis [46]. Due to their high running efficiency (because of the using of improved helicity amplitude approach), BCVEGPY and GENXICC are very useful for Monte Carlo simulation and also for theoretical studies. At the present, they have been adopted by most of the collaborations at LHC and TEVATRON, such as the LHCb, CMS, ATLAS and etc.. The suggesting future super Z factory, GIGAZ program, LEP3, and etc., shall also provide good platforms for doubly heavy meson and baryon productions, cf. Refs.[47-49].

4. Acknowledgments

This talk was based on the works done in collaborations with Profs.Chang CH, Wang JX, Qiao CF and Dr.Wang XY. This work was supported in part by the Fundamental Research Funds for the Central Universities under Grant No.CQDXWL-2012-Z002, by Natural Science Foundation of China under Grant No.11075225 and No.11275280, and by the Program for New Century Excellent Talents in University under Grant No.NCET-10-0882.

References

[1] Brambilla N *et al.* 2004 *Preprint* arXiv:0412158 [hep-ph]
[2] Bayatian G L *et al.* 2007 *J.Phys.G* **34** 995
[3] Brambilla N *et al.* 2011 *Eur.Phys.J.C* **71** 1534
[4] Aaij R *et al.* 2013 *Eur.Phys.J.C* **73** 2373
[5] Chang C H and Chen Y Q 1993 *Phys.Rev.D* **48** 4086
[6] Berezhnoy A V, Likhoded A K and Shevlyagin M V 1995 *Phys.Atom.Nucl.* **58** 672
[7] Kolodziej K, Leike A and Ruckl R 1995 *Phys.Lett.B* **355** 337
[8] Chang C H, Chen Y Q, Han G P and Jiang H T 1995 *Phys.Lett.B* **364** 78
[9] Chang C H, Chen Y Q and Oakes R J 1996 *Phys.Rev.D* **54** 4344
[10] Chang C H and Wu X G 2004 *Eur.Phys.J.C* **38** 267
[11] Chang C H, Wang J X and Wu X G 2004 *Phys.Rev.D* **70** 114013
[12] Chang C H, Qiao C F, Wang J X and Wu X G 2005 *Phys.Rev.D* **70** 074012
[13] Chang C H, Qiao C F, Wang J X and Wu X G 2005 *Phys.Rev.D* **72** 114009
[14] Chang C H, Qiao C F, Wang J X and Wu X G 2006 *Phys.Rev.D* **73** 094022
[15] Chang C H, Ma J P, Qiao C F and Wu X G 2007 *J.Phys.G* **34** 845
[16] Kiselev V V, Likhoded A K and Shevlyagin M V 1994 *Phys.Lett.B* **332** 411
[17] Zhang J W, Wu X G, Zhong T, Yu Y and Fang Z Y 2011 *Phys.Rev.D* **83** 034026
[18] Chang C H, Driouichi C, Eerola P and Wu X G 2004 *Comput.Phys.Commun.* **159** 192
[19] Chang C H, Wang J X and Wu X G 2006 *Comput.Phys.Commun.* **174** 241
[20] Chang C H, Wang J X and Wu X G 2006 *Comput.Phys.Commun.* **175** 624
[21] Wang X Y and Wu X G 2012 *Comput.Phys.Commun.* **183** 442
[22] Chang C H, Wang J X and Wu X G 2007 *Comput.Phys.Commun.* **177** 467
[23] Chang C H, Wang J X and Wu X G 2010 *Comput.Phys.Commun.* **181** 1144
[24] Wang X Y and Wu X G 2013 *Comput.Phys.Commun.* **184** 1070
[25] Sjostrand T, Mrenna S and Skands P 2006 *JHEP* **0605** 026
[26] Jiang J, Wu X G, Liao Q L, Zheng X C and Fang Z Y 2012 *Phys.Rev.D* **86** 054021
[27] Bodwin G T, Braaten E and Lepage G P 1995 *Phys.Rev.D* **51** 1125
[28] Kleiss R and Stirling W J 1986 *Comput.Phys.Commun* **40** 359
[29] Lepage G P 1978 *J.Comp.Phys.* **27** 192
[30] Kleiss R and Stirling W J 1985 *Nucl.Phys.B* **262** 235
[31] Chang C H and Chen Y Q 1992 *Phys.Rev.D* **46** 3845
[32] Chang C H, Wang J X and Wu X G 2008 **Phys.Rev.D** **77** 014022
[33] Deng L C, Wu X G, Yang Z, Fang Z Y and Liao Q L 2010 *Eur.Phys.J.C* **70** 113
[34] Xu Z, Zhang D H and Chang L 1987 *Nucl.Phys.B* **291** 392

[35] Qiao C F 2003 *Phys.Rev.D* **67** 097503
[36] Alwall J *et al.* 2007 *Comput.Phys.Commun.* **176** 300
[37] Setsuya K 1995 *Comput.Phys.Commun.* **88** 309
[38] Nason P 2004 *JHEP* **0411** 040
[39] Frixione S, Nason P and Oleari C 2007 *JHEP* **0711** 070
[40] Nason P 2007 *Preprint* arXiv:0709.2085 [hep-ph]
[41] Aaij R *et al.* 2012 *Phys.Rev.Lett.* **109** 232001
[42] Aaij R *et al.* 2012 *Phys.Rev.Lett.* **108** 251802
[43] Aaij R *et al.* 2013 *Phys.Rev.D* **87** 071103
[44] Aaltonen T *et al.* 2013 *Phys.Rev.D* **87** 011101
[45] Aaij R *et al.* 2013 *Preprint* arXiv:1306.6723 [hep-ex]
[46] Aaij R *et al.* 2013 *Preprint* arXiv:1310.2538 [hep-ex]
[47] Yang Z, Wu X G, Chen G, Liao Q L and Zhang J W 2012 *Phys.Rev.D* **85** 094015
[48] Jiang J, Wu X G, Wang S M, Zhang J W and Fang Z Y 2013 *Phys.Rev.D* **87** 054027
[49] Yang Z, Wu X G, Wang X Y 2013 *Preprint* arXiv:1305.4828 [hep-ph]

SANC system and its applications for LHC

R Sadykov, A Arbuzov, D Bardin, S Bondarenko, P Christova, L Kalinovskaya, V Kolesnikov, A Sapronov and E Uglov

Joint Institute for Nuclear Research, Joliot-Curie str. 6, Dubna, 141980, Russia

E-mail: renat.sadykov@cern.ch

Abstract. The SANC computer system is aimed at support of analytic and numeric calculations for experiments at colliders. The system is reviewed briefly. Recent results on high-precision description of the Drell-Yan processes at the LHC are presented. Special attention is paid to the evaluation of higher order final-state QED corrections to the single W and Z boson production processes. A new Monte Carlo integrator mcsanc suited for description of a series of high-energy physics processes at the one-loop precision level is presented.

1. Introduction

After the discovery at the LHC of a new particle which is very likely the long awaited Higgs boson, the precision tests of the Standard Model (SM) became of crucial importance. Moreover, the absence of clear signals of new physics at the LHC stimulates comparisons of high-precision predictions received within the SM with the experimental data. The rather accurate measurement of the Higgs boson mass performed at the LHC fixed the last free parameter of the SM. This potentially allows to compute theoretical predictions by means of perturbation theory to a very high precision level. The corresponding experimental accuracy is continuously growing up with accumulating of statistics and development of analysis techniques. To provide adequately accurate theoretical predictions, one has to take into account and combine different relevant effects.

2. SANC Project

SANC is a computer system for Support of Analytic and Numeric calculations for experiments at Colliders [1, 2]. It can be accessed through the Internet at http://sanc.jinr.ru/. The SANC system is suited for calculations of NLO QED, electroweak (EW), and QCD radiative corrections (RC) to various SM processes. Automatized analytic calculations in SANC provide FORM and FORTRAN modules [3]. Stand-alone FORTRAN modules for differential cross sections as well as Monte Carlo programs (integrators and generators) are available to download.

A schematic flowchart of the SANC system is presented in Fig. 1. In the core of the system, analytic calculations of amplitudes, cross sections, and other quantities for various processes are computed by means of the FORM language [4]. The results of analytic calculations are translated automatically into FORTRAN modules which can be used by external programs. The modules extensively use internal and external [5] FORTRAN libraries of Passarino-Veltman functions.

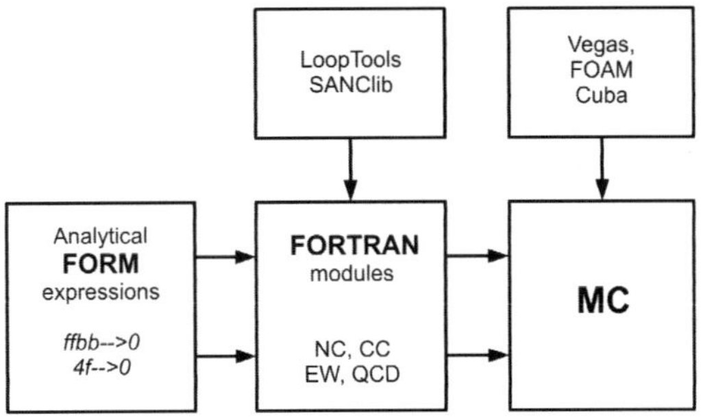

Figure 1. A schematic flowchart of the SANC system.

3. Description of Drell-Yan processes in SANC

High-precision theoretical description of Drell-Yan (DY) like processes (i.e. single Z and W boson production) at the LHC is of great importance. In fact, these processes have large cross sections, clear signatures in detectors, and they are rather well suited for tests of the SM and searches for new physics. So DY processes provide *standard candles* for detector calibration at the LHC and the Tevatron. Single Z and W boson production is also used at the LHC for extraction of parton distribution functions (PDF) in the kinematical region which has not been accessed by earlier experiments. It is planned to get the most precise experimental values for the mass and width of W boson using the future high statistics data on the charged current DY process. DY processes will also provide background to many other reactions being of interest at the LHC. Moreover, some new physics searches like the ones for contact four-fermion interactions will be performed in these channels. Therefore it is crucial to control the theoretical predictions for production cross sections and kinematic distributions of both the neutral current (NC) and charged current (CC) DY processes.

The experimental precision tag for inclusive observables in DY processes is about 1%. That means we need to provide the accuracy of theoretical predictions of about 0.3% in order not to spoil the results of the LHC data analysis. This is a challenge for the theory. Aiming at high precision of DY description we need to take into account the following effects:

- QCD contributions at LO, NLO and NNLO;
- parton showers and hadronization effects;
- EW RC at one-loop at least;
- most important higher order EW contributions (re-summed where possible);
- an interplay of QCD and EW corrections;
- a tuned input: coupling constants, the hadronic vacuum polarization, and PDF for the appropriate energy scales and x-values.

All relevant effects should be implemented in a Monte Carlo event generator which can be directly used in the experimental data analysis. Actually, treating all the listed effects in a single code is a very involved task. More realistic is the possibility to make a chain of event generators, which pass generated events one to another and *dress* them with additional effects.

With a tuned set-up we got an excellent agreement with several other groups in the values of EW RC to the CC [6] and NC [7] DY processes, see Table 1. Examples of differential distributions of the EW RC contributions are given in Fig. 2 and Fig. 3 for the CC channel and in Fig. 4 and

Table 1. Tuned comparisons between SANC and other codes in predictions on DY processes.

Processes	Codes	Ref.
CC DY (NLO EW)	SANC, HORACE, WGRAD2	[6]
NC DY (NLO EW)	SANC, HORACE, ZGRAD2	[7]
CC & NC DY (NLO QCD & EW)	SANC, HORACE, WZGRAD, RADY, FEWZ, DYNNLO, POWHEG-w, POWHEG-z	[8]

Fig. 5 for the NC one. The relative effect of EW RC is defined as

$$\Delta = \frac{\sigma(\mathcal{O}(\alpha)) - \sigma(\text{Born})}{\sigma(\text{Born})} \cdot 100\%. \tag{1}$$

A new series of tuned comparisons is going on within the W-mass workshop [8]. QCD corrections are also taken into consideration, and the set-up is adjusted to simulate the experimental conditions.

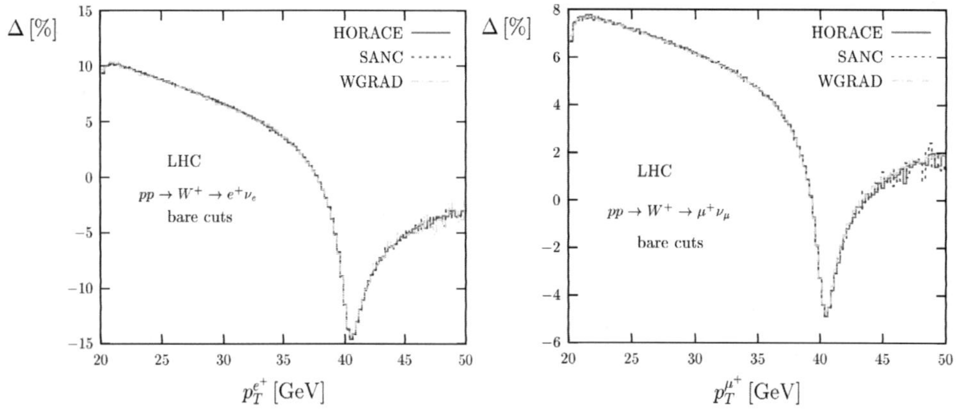

Figure 2. Tuned comparison of positron (left) and anti-muon (right) transverse momentum distribution in the CC DY process [6].

4. Final State Radiation in single W and Z production

Numerically, the effect of QED final-state radiation (FSR) off charged leptons at the LHC is very large. The corresponding correction can reach, *e.g.*, even about 200% in the distribution of e^+e^- invariant mass in NC DY. For this reason FSR deserves special attention. Recently a dedicated study [9] of the effect was performed by the joint effort of SANC and PHOTOS [10] groups.

Our goals were:

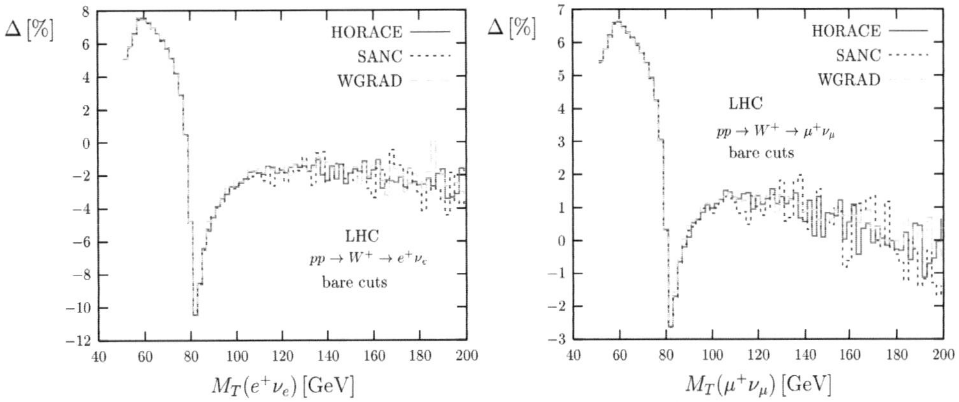

Figure 3. Tuned comparison of $e^+\nu_e$ (left) and $\mu^+\nu_\mu$ (right) transverse mass distribution in the CC DY process [6].

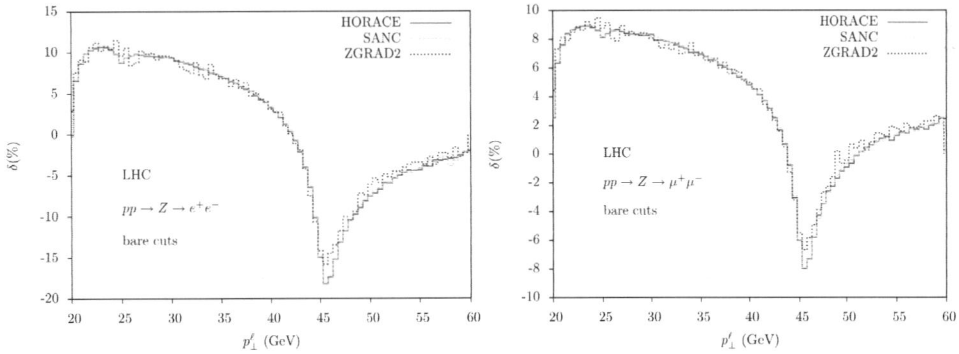

Figure 4. Tuned comparison of positron (left) and anti-muon (right) transverse momentum distribution in the NC DY process [7].

- to perform a comparison between SANC and PHOTOS in the *single* and *multiple* photon modes for FSR in DY processes both for neutral and charged currents;
- to check that QED FSR is properly installed in the programs;
- to tune separation of the FSR QED corrections from the complete EW NLO ones.

In PHOTOS the bremsstrahlung corrections to decays of W and Z bosons are treated in a special way in the next-to-leading approximation with exponentiation. PHOTOS is linked to the PYTHIA program in the standard Monte Carlo simulation through HepMC interface. In SANC the complete EW corrections at one-loop are calculated for single W and Z production. The FSR QED corrections can be separated from the rest of EW corrections by means of flags in SANC Monte Carlo event generators and integrators.

For a tuned comparison of FSR effects in SANC and PHOTOS we used the following set-up. Parton level cross-section was convoluted with CTEQ6L1 PDF set with running scale $Q^2 = s$, where s is the total energy in the center-of-mass system of partons. The C++ version of PHOTOS program was used together with Pythia8 which provided the Born-level events in HEPevt format that subsequently was passed to PHOTOS to simulate photon radiation off the final charged leptons.

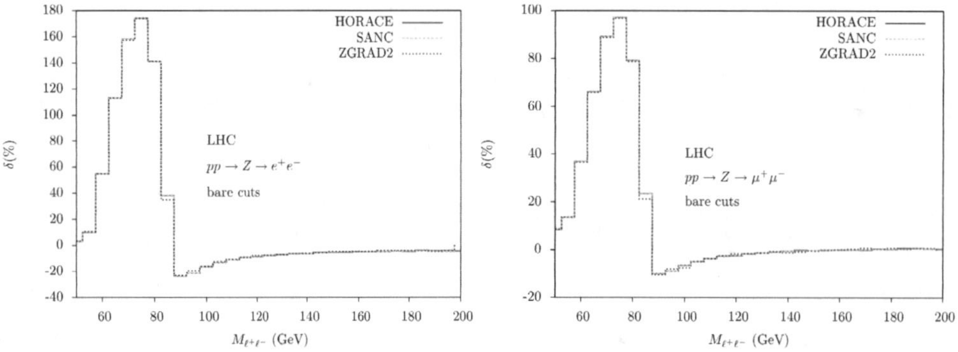

Figure 5. Tuned comparison of e^+e^- (left) and $\mu^+\mu^-$ (right) invariant mass distribution in the NC DY process [7].

PHOTOS was running in single and multiple photon modes with matrix-element corrections turned on. Here the relative corrections are defined as:

$$\delta = \frac{\sigma(\mathcal{O}(\alpha)FSR) - \sigma(\text{Born})}{\sigma(\text{Born})} \tag{2}$$

for single photon radiation, and

$$\delta_{h.o.} = \frac{\sigma(\text{h.o.FSR}) - \sigma(\mathcal{O}(\alpha)FSR)}{\sigma(\text{Born})} \tag{3}$$

for higher order FSR contributions.

Let us look at the comparison of a single photon, *i.e.* $\mathcal{O}(\alpha)$, FSR contribution to CC and NC DY processes presented in Fig. 6 and Fig. 7, respectively. The distributions in the lepton transverse momentum and transverse (invariant) mass agree pretty well. Nevertheless, the implementations of the FSR effect in SANC and PHOTOS are different: there is a small systematic shift, as can be seen from the corresponding distribution in the lepton pseudorapidity, see Fig. 8. This shift is due to the specific feature of the PHOTOS program: simulation of FSR by means of PHOTOS doesn't change the total cross section, while the true O(α) FSR makes a small change.

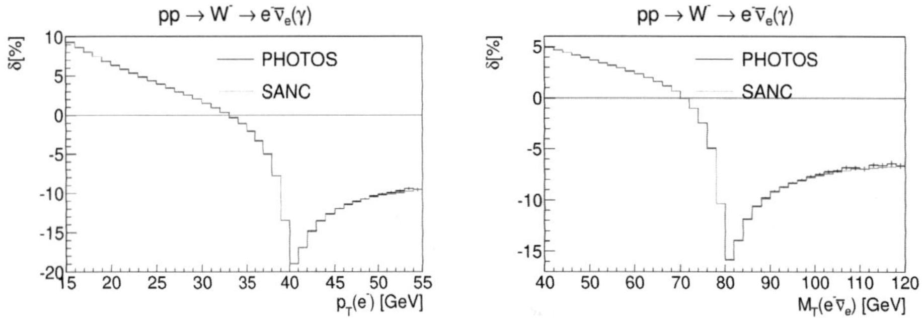

Figure 6. Tuned comparison of transverse momentum (left) and transverse mass (right) distributions in CC DY.

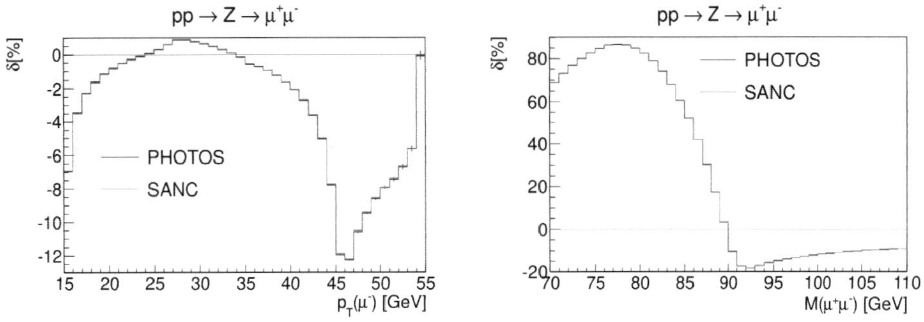

Figure 7. Tuned comparison of transverse momentum (left) and invariant mass (right) distributions in NC DY.

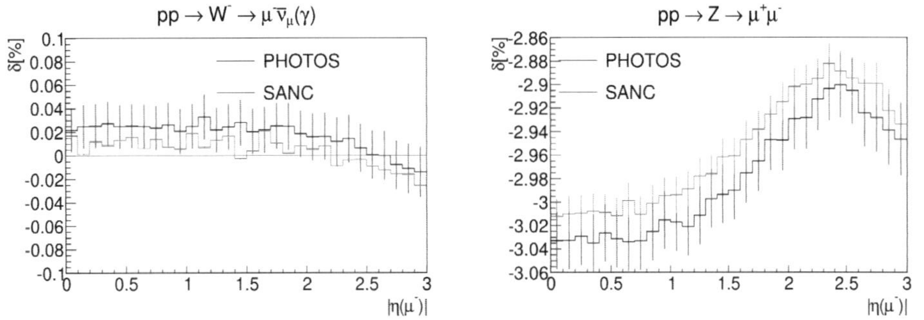

Figure 8. Tuned comparison of lepton pseudorapidity distribution in CC (left) and NC (right) DY processes.

Large corrections in the first order make it necessary to look at higher orders. Fig. 9 gives an example of comparison in higher orders. `SANC` computes the second and third order photonic FSR within the collinear leading logarithmic approximation, thus the results are sensitive to the choice of the QED factorization scale. `PHOTOS` performs exponentiation of the first order corrections. One should also keep in mind that in experimental analysis one typically applies the so-called calorimetric event selection (where charged leptons are re-combined with accompanying photons) which reduces the effect and uncertainties of the FSR considerably, as can be seen from Fig. 10 in comparison with Fig. 6.

5. Monte Carlo integrator `mcsanc`

Recently a Monte Carlo integrator (weighted events) based on `SANC` modules [3] was created [11, 12]. It features:

- calculations of fully differential cross sections for DY and inclusive cross sections of higgs-strahlung and single-top production;
- provides both NLO EW and QCD corrections;
- support of different EW schemes: $\alpha(0)$, $\alpha(M_Z)$, and G_μ;
- fixed and running factorization and renormalization scale options;
- kinematic cuts, recombination are foreseen;

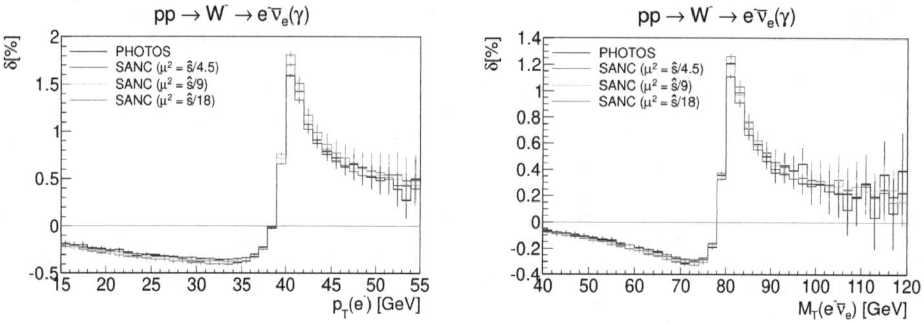

Figure 9. Comparison of higher order FSR contribution to transverse momentum (left) and transverse mass (right) distributions in CC DY.

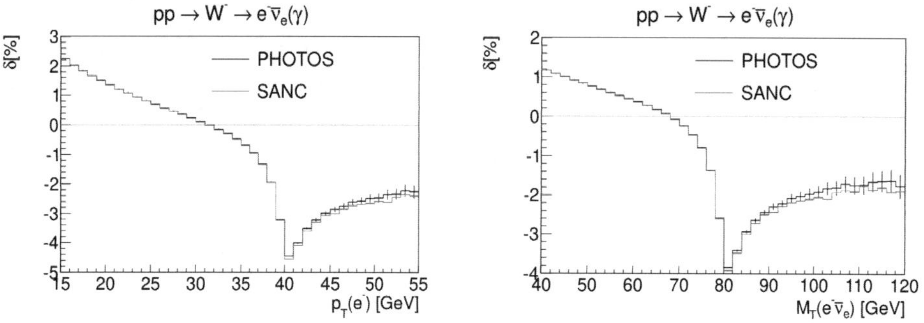

Figure 10. Tuned comparison of transverse momentum (left) and transverse mass (right) distributions in CC DY with calorimetric event selection.

- parallel calculation on multi-core machines thanks to `Cuba` library `http://www.feynarts.de/cuba/`;
- easy installation and configuration (GNU autotools, `LHAPDF`, input configurations for physics parameters, cuts, histogramming).

The code is suited to be run on multi-core computers. To get a high precision in differential distributions it takes many hours. That's why it is important to take care on stability and rescue. For this reason integration state files of the `Vegas` algorithm are saved while running `Cuba` after every iteration and upon run completion. The files can be used to increase statistics or restore from an interrupted run, *e.g.* when the batch time quota has been exceeded. When the code is run on a multi-core systems, the calculation is automatically split by the number of cores with an aid of `$CUBACORES` environment variable. The parallelization efficiency is limited due to inter-process communications: the optimal number of cores is 8, after which the run time doesn't reduce and efficiency (CPU load) remains below 50%.

The list of processes presently implemented in `mcsanc` is shown in Table 2. The process id notation is the following: first digit is the sign of EW current, and the last two digits specify the final particle choice:

$0xx$ means neutral current, $xx = 01(e), 02(\mu), 03(\tau), 04(HZ)$;

$\pm 1xx$ means charged current, $xx = 01(e)$, $02(\mu)$, $03(\tau)$, $04(HW)$, $05, 06$ (top quark production in s and t channels).

Table 2. List of processes implemented in `mcsanc`

pid	$f + f \to$	`SANC` ref.
001:003	$\ell^+ + \ell^-(\ell = e, \mu, \tau)$	[13, 14]
004	$Z^0 + H$	[2, 3]
± 101:103	$\ell^\pm + \nu_\ell$	[15]
± 104	$W^\pm + H$	-
105	$t + b$ (s-channel)	[11, 16]
106	$t + q$ (t-channel)	-//-
-105	$\bar{t} + b$ (s-channel)	-//-
-106	$\bar{t} + q$ (t-channel)	-//-

6. Conclusions

A brief overview of the `SANC` project is presented. The status of tuned comparisons between the results of `SANC` and other codes for EW corrections to Drell-Yan production is discussed.

The comparison of `SANC` and `PHOTOS` for single and multiple photon emission was performed for the neutral and charged current DY processes. The results agree within 0.1%.

New Monte Carlo integrator `mcsanc-v1.01` is available to download on `http://sanc.jinr.ru`. The tool is aimed for calculation of NLO EW and QCD corrections to Drell-Yan, higgs-strahlung, and single top quark production processes in pp collisions.

Acknowledgements. This work was supported by RFBR-CERN grant `12-02-91528-ЦЕРН_а` and by Heisenberg-Landau Program grant. AA and RS thank the ACAT organizing committee for partial covering of expenses during the conference. AA and AS are grateful for support to the Dynasty foundation.

References

[1] Andonov A et al 2006 *Comput. Phys. Commun.* **174** 481; 2007 *Erratum ibid.* **177** 623
[2] Bardin D et al 2007 *Comput. Phys. Commun.* **177** 738
[3] Andonov A, Arbuzov A, Bardin D, Bondarenko S, Christova P, Kalinovskaya L, Kolesnikov V and Sadykov R 2010 *Comput. Phys. Commun.* **181** 305
[4] Kuipers S, Ueda T, Vermaseren J A M and Vollinga J 2013 *Comput. Phys. Commun.* **184** 1453
[5] Hahn T and Rauch M 2006 *Nucl. Phys. Proc. Suppl.* **157** 236
[6] Gerber C E et al [TeV4LHC-Top and Electroweak Working Group Collaboration] arXiv:0705.3251 [hep-ph]
[7] Buttar C et al arXiv:0803.0678 [hep-ph]
[8] `http://lpcc.web.cern.ch/lpcc/index.php?page=electroweak_wg`
[9] Arbuzov A B, Sadykov R R and Was Z arXiv:1212.6783 [hep-ph] to appear in *Europ. Phys. J.* C
[10] Davidson N, Przedzinski T and Was Z arXiv:1011.0937 [hep-ph]
[11] Bardin D, Bondarenko S, Christova P, Kalinovskaya L, Rumyantsev L, Sapronov A, von Schlippe W 2012 *JETP Lett.* **96** 285
[12] Bondarenko S G and Sapronov A A 2013 *Comput. Phys. Commun.* **184** 2343
[13] Arbuzov A, Bardin D, Bondarenko S, Christova P, Kalinovskaya L, Nanava G and Sadykov R 2008 *Eur. Phys. J.* C **54** 451
[14] Andonov A, Arbuzov A, Bondarenko S, Christova P, Kolesnikov V, Nanava G and Sadykov R 2010 *Phys. Atom. Nucl.* **73** 1761
[15] Arbuzov A, Bardin D, S. Bondarenko, Christova P, Kalinovskaya L, Nanava G and Sadykov R 2006 *Eur. Phys. J.* C **46** 407; 2007 *Erratum-ibid.* C **50** 505
[16] Bardin D, Bondarenko S, Christova P, Kalinovskaya L, Kolesnikov V, von Schlippe V and Yordanova K 2012 *Phys. Part. Nucl. Lett.* **9** 472

New developments in FeynRules.

Adam Alloul[1,2], Neil D Christensen[3], Céline Degrande[4], Claude Duhr[5,6], Benjamin Fuks[1,7]

[1] Institut Pluridisciplinaire Hubert Curien/Département Recherches Subatomiques, Université de Strasbourg/CNRS-IN2P3, 23 Rue du Loess, F-67037 Strasbourg, France
[2] Groupe de Recherche de Physique des Hautes Énergies (GRPHE), Université de Haute-Alsace, IUT Colmar, 34 rue du Grillenbreit BP 50568, 68008 Colmar Cedex, France
[3] PITTsburgh Particle physics, Astrophysics and Cosmology Center (PITT PACC), University of Pittsburgh, Pittsburgh, PA 15260 USA
[4] Department of Physics, University of Illinois at Urbana-Champaign,1110 West green Street, Urbana, IL 61801, United States
[5] Institute for Theoretical Physics, ETH Zürich, 8093 Zürich, Switzerland
[6] Institute for Particle Physics Phenomenology, University of Durham, Durham, DH1 3LE, United Kingdom
[7] Theory Division, Physics Department, CERN, CH-1211 Geneva 23, Switzerland

E-mail: adam.alloul@iphc.cnrs.fr, neilc@pitt.edu, cdegrand@illinois.edu, duhrc@itp.phys.ethz.ch, benjamin.fuks@iphc.cnrs.fr, fuks@cern.ch

Abstract. The program FeynRules is a Mathematica package developed to facilitate the implementation of new physics theories into high-energy physics tools. Starting from a minimal set of information such as the model gauge symmetries, its particle content, parameters and Lagrangian, FeynRules provides all necessary routines to extract automatically from the Lagrangian (that can also be computed semi-automatically for supersymmetric theories) the associated Feynman rules. These can be further exported to several Monte Carlo event generators through dedicated interfaces, as well as translated into a Python library, under the so-called UFO model format, agnostic of the model complexity, especially in terms of Lorentz and/or color structures appearing in the vertices or of number of external legs. In this work, we briefly report on the most recent new features that have been added to FeynRules, including full support for spin-$\frac{3}{2}$ fermions, a new module allowing for the automated diagonalization of the particle spectrum and a new set of routines dedicated to decay width calculations.

1. Introduction

The high-energy particle physics community is living nowadays in one of the key moments of its history. The analysis of data acquired by the most powerful particle collider ever built, the Large Hadron Collider at CERN, is on-going and exciting results have already been established. For instance, the discovery of a scalar particle with a mass of about 125 GeV featuring the properties expected from the Standard Model (SM) Higgs boson was one of the major announcements in 2012 [1, 2]. In parallel, a major effort is put into the exploration of possible beyond the Standard Model (BSM) signatures hidden in data by comparing data to theoretical predictions.

In this context, Monte Carlo event generators play an important role as they allow to simulate the behavior of both the Standard Model background and the new physics signals. While these tools are ready to be used to simulate any SM process, the situation is however different for

BSMs where these models are not yet implemented in the Monte Carlo event generators. The FeynRules package [3, 4, 5, 6, 7, 8, 9] has been created with the aim of facilitating this task. Starting by providing basic model information such as its underlying gauge symmetries, field content, parameters and Lagrangian (that can also be semi-automatically derived for supersymmetric theories), FeynRules then extracts all the interaction vertices from the model Lagrangian and exports the associated Feynman rules, through dedicated interfaces, to various high-energy physics tools. Such interfaces exist for the CalcHep [10, 11], FeynArts [12], MadGraph [13, 14, 15, 16, 17, 18], Sherpa [19, 20] and Whizard [21, 22] programs. Moreover, the model can also be translated to a Python library under the Universal FeynRules Output (UFO) format [23]. The UFO has been developed to overcome all restrictions imposed by previously designed model formats for Monte Carlo programs. It indeed allows for arbitrary color and/or Lorentz structures in the interaction vertices and any number of external particles.

In this paper, we briefly review some of the new functionalities that will be included in the new version of the package, FeynRules 2.0, that can already be downloaded from the website `http://feynrules.irmp.ucl.ac.be`. We first describe the FeynRules program in section 2. Next, we detail the support for spin-$\frac{3}{2}$ fields in section 3 and a new module allowing for automated spectrum generation in section 4. Section 5 is dedicated to automated decay width calculations and we finally summarize our work in section 6.

2. The FeynRules package

The FeynRules program is a Mathematica package allowing for extracting Feynman rules from a quantum field theory Lagrangian. Additionaly to the improvements described in this paper, the latest version of FeynRules comes with a set of tools directly connecting the model definition in FeynRules to a series of Monte Carlo programs. Furthermore, it is also possible to store the model information in the form of a Python library, under the so-called Universal FeynRules output (UFO) format. This takes advantage of the object-oriented nature of the Python language and its ability to be interfaced with any other program.

Proper FeynRules model files follow a syntax inspired by the FeynArts model format and encompass information about the model gauge symmetries, field content and parameters. From the input of the Lagrangian describing the dynamics of the model particles, possibly given in a very compact form by means of quantities such as gauge covariant derivatives, (super)field strengh tensors or even user-defined functions, the FeynRules package is able to extract all the Feynman rules associated with the underlying interaction vertices. The latter can then be exported, together with all the relevant model information to several Monte Carlo and symbolic tools via dedicated interfaces. In addition, supersymmetric models can be naturally and easily implemented in a superspace module shipped with the code allowing to declare chiral and vector superfields as any other field and containing a set of methods dedicated to superspace computations.

All model publicly available can be found in the database located in the FeynRules website.

3. Spin-3/2 Rarita-Schwinger fields

It is of critical importance for phenomenological purposes to have tools able to handle any new physics models, no matter the structure of the interactions or the particle content. In particular, an automated treatment, starting at the level of FeynRules, for spin-$\frac{3}{2}$ particles was still missing until very recently. The version 2.0 of FeynRules fills this gap. The user can now use two new particle classes `RW` and `R` to declare two- and four-component spin-$\frac{3}{2}$ fields, respectively. In parallel, the UFO conventions have been updated so that they account for the arbitrariness associated with spin-$\frac{3}{2}$ propagators, and recent versions of both Monte Carlo event generators CalcHep and MadGraph 5 support spin-$\frac{3}{2}$ fields and their interactions [9].

The superspace module of FEYNRULES [6] has also been upgraded, and computations related to goldstino and gravitino fields are now simplified within the FEYNRULES framework. In the case of local supersymmetry, the interactions of the goldstino, a massless fermionic field arising from supersymmetry breaking and those of the gravitino, a spin-$\frac{3}{2}$ field predicted by supergravity theories, are derived from the knowledge of the supercurrent, a Majorana fermionic quantity $(J^\mu, \bar{J}^\mu)$ defined by

$$J^\mu_\alpha = \sqrt{2} D_\nu \phi_i^\dagger (\sigma^\nu \bar{\sigma}^\mu \psi^i)_\alpha + \sqrt{2} (\sigma^\mu \bar{\psi}_i)_\alpha W^{*i} + g\phi^\dagger T_a \phi (\sigma^\mu \bar{\lambda}^a)_\alpha - \frac{i}{2} (\sigma^\rho \bar{\sigma}^\nu \sigma^\nu \bar{\lambda}_a)_\alpha v^a_{\nu\rho} \ .$$

In this expression, a sum over the model particle content is understood, and $v_{\nu\rho}$ denotes the field strength tensor for the vector boson v_μ, D_ν stands for the gauge-covariant derivative, σ^μ is the four-vector built upon the Pauli matrices and T_a are generators of the model gauge group taken in the appropriate representation. Moreover, we generically denote scalar fields by ϕ, gauginos by λ, the Weyl components of the chiral supermultiplets by ψ and W^{*i} is the derivative of the hermitian conjugate of the superpotential W. The interactions of the gravitino $\tilde{G}$ and the goldstino χ fields are then given by

$$\begin{aligned} \mathcal{L}_\chi &= \frac{i}{2}\Big(\chi\sigma^\mu\partial_\mu\bar{\chi} - \partial_\mu\chi\sigma^\mu\bar{\chi}\Big) + \frac{1}{2v_1}\Big[\chi\cdot\partial_\mu\tilde{J}^\mu + \bar{\psi}_1\cdot\partial_\mu\bar{\tilde{J}}^\mu\Big] \ , \\ \mathcal{L}_G &= \frac{1}{2}\epsilon^{\mu\nu\rho\sigma}(\psi_{\tilde{G}\mu}\sigma_\rho\partial_\sigma\bar{\psi}_{\tilde{G}\nu} + \partial_\sigma\psi_{\tilde{G}\nu}\sigma_\rho\bar{\psi}_{\tilde{G}\mu}) + \mathcal{N}\Big[\psi_{\tilde{G}\mu}\cdot J^\mu + \bar{\psi}_{\tilde{G}\mu}\cdot\bar{J}^\mu\Big] \ , \end{aligned}$$

that can be implemented in FEYNRULES following standard techniques.

The supercurrent J^μ_α can be derived in FEYNRULES by calling, in a MATHEMATICA session, the function

```
SuperCurrent[lv,lc,lw,sp,mu]
```

which takes as arguments the Lagrangians `lv` associated with the gauge sector of the theory, `lc` associated with its chiral sector and the superpotential Lagrangian `lw`. Moreover, the supercurrent spin and Lorentz indices are given by `sp` and `mu`. As it was shown in Ref. [6], the different pieces of the supersymmetric Lagrangian can be straightforwardly obtained by means of standard methods of the superspace module of FEYNRULES.

4. Automated spectrum generation

When implementing in a given Monte Carlo tool a specific model, it is important to pay special attention to the numerical values of the various model parameters. This usually requires to compute and diagonalize mass matrices associated with the different gauge-eigenstates of the model, which often mix. In practice, such calculations cannot be solved analytically and it is necessary to rely on a numerical code. This issue has implied the development of a new module of FEYNRULES allowing, on the one hand, for the analytical extraction of all the model mass matrices from the Lagrangian, and, on the other hand, to generate a self-contained C++ package, dubbed ASPERGE[1], diagonalizing the mass spectrum and outputting both masses and mixing matrices under a Les Houches compliant format [8].

In order to employ these new functionalities, it is necessary to reorganize the information contained in the model file. As an example, let us consider the mixing of the neutral SM gauge bosons B_μ and W^3_μ to the photon A_μ and the Z-boson Z_μ states,

$$\begin{pmatrix} A_\mu \\ Z_\mu \end{pmatrix} = U_w \begin{pmatrix} B_\mu \\ W^3_\mu \end{pmatrix} \ ,$$

[1] The acronym ASPERGE stands for **A**utomated **Spe**ct**r**um **Ge**nerator.

which involves the mixing matrix U_w. Initially, these relations could have been implemented in the definition of the fields themselves, as in

```
V[11] == {
  ClassName -> B,
  Definitions -> { B[mu_] -> -sw Z[mu]+cw A[mu]}, ....};

V[12] == {
  ClassName -> Wi,
  Definitions -> { Wi[mu_,1] -> (Wbar[mu]+W[mu])/Sqrt[2],
                   Wi[mu_,2] -> (Wbar[mu]-W[mu])/(I*Sqrt[2]),
                   Wi[mu_,3] -> cw Z[mu] + sw A[mu]}, ...},
```

where `B` (`Wi`) stands for the gauge boson associated with the $U(1)$ ($SU(2)_L$) subgroup of the SM gauge group, `cw` and `sw` are the cosine and sine of the weak mixing angle, respectively, and the dots stand for other options irrelevant with respect to the mixings (more details can be found in Ref. [3]). The new functionalities of FeynRules 2.0 offer a second way to treat particle mixings. It relies on a variable `M$MixingsDescription` which consists of a list where each element is dedicated to a specific mixing relation. Turning to our example, this list is given by

```
M$MixingsDescription == {
  Mix["1u"] == {MassBasis -> {A, Z}, GaugeBasis -> {B, Wi[3]},
                MixingMatrix -> UG, BlockName -> WEAKMIX}}
```

We attribute a unique symbol (here `"1u"`) to each mixing relation and define its properties through the options `MassBasis` (refers to the mass eigenstates), `GaugeBasis` (refers to the mixing gauge eigenstates), `MixingMatrix` (defines the symbol to use for the mixing matrix which does not have to be declared as a model parameter) and `BlockName` (refers to the name of the Les Houches block containing the numerical values of the elements of the mixing matrix that will be calculated by ASperGe). Additional options are also available and special cases are possible, depending on the spin of the fields under consideration. We refer to Ref. [8] for their extensive description.

In addition to defining all mixing relations within the variable `M$MixingsDescription`, a correct extraction of the model mass matrices implies to provide some knowledge about the fields getting a vacuum expectation value (vev) at the ground state of the theory. This information is included in the variable `M$vevs`, a list of two-component elements. The first of these components is the name of a field and the second one is the symbol referring to its vev. The model mass matrices can then be obtained by issuing, in the Mathematica session,

```
ComputeMassMatrix[lagr]
```

where `lagr` stands for the symbol of the model Lagrangian and information associated with a specific mixing relation (here `"1u"`) can further be accessed by typing

```
MixingSummary["1u"]
```

The source code for the ASperGe package is generated via the command

```
WriteASperGe[lagr]
```

and consists in a set of model-dependent and model-independent files stored in a directory whose name is made up from the model name and the suffix `_MD`. The model-dependent files consist of C++ commands allowing to define the various parameters of the model and the mass matrices to be diagonalized. In contrast, the model-independent files contain all necessary routines allowing ASperGe to compute both the mass eigenvalues and the rotation matrices from the mass matrices on the basis of the QR reduction algorithm as implemented in GSL. In addition, these files also include methods to store these results in a Les Houches compliant format.

Finally, the numerical code can be compiled and executed by either typing `RunASperGe[]` in the MATHEMATICA session or executing in a shell, after moving to the dicrectory where it is stored,

```
make; ./ASperGe input/externals.dat output/output.dat
```

The first of the two files (`input/externals.dat`) contains, under the Les Houches format, the model free parameters together with their numerical values and the second one (`output/output.dat`) collects the results of the execution of ASPERGE.

5. Automatic calculation of decay widths

A careful design of a benchmark scenario for a given model requires, in addition to providing a numerical value for the model external parameters and the calculation of the mixing matrices necessary to diagonalize the mass spectrum, the computation of the particle total widths. Furthermore, the latter must in general be given together with the decay table of all the unstable particles. To this aim, the version 2.0 of FEYNRULES includes a series of functions dedicated to width calculations [24]. The advantages of using FEYNRULES for such a purpose are twofold. First, the new module calculates the analytical formulas associated to all tree-level $1 \to 2$ processes, so that the results are independent of any specific benchmark point. Second, the calculations can be performed once for a given model and reused every time the benchmark scenario is modified.

Assuming that the model Feynman rules have been stored and expanded over all flavor indices in a variable denoted by `verts`, the associated set of tree-level $1 \to 2$ decay widths are obtained by issuing

```
res = ComputeDecays[verts]
```

The results (stored both in a variable `res` and in the FEYNRULES variable `FR$PartialWidths`) consist of a list of two-component elements. Their first piece is another list of three fields {`F1,F2,F3`} indicating the decay process of interest (`F1` $\to$ `F2 F3`) while their second piece is the analytical formula of the corresponding partial width. Each possible decay of a field into two other fields is considered regardless of the numerical values of the particle masses. The results thus include all possible cyclic permutations of `F1`, `F2` and `F3`.

A specific partial width or branching ratio associated with the decay `F1` $\to$ `F2 F3` can then be retrieved analytically by respectively issuing, in the MATHEMATICA session, one of the two commands

```
PartialWidth[{F1, F2, F3}]
BranchingRatio[{F1, F2, F3}]
```

while the total width of a field `F1` can be computed by typing

```
TotWidth[F1]
```

The corresponding numerical values are obtained by making use of the standard FEYNRULES command `NumericalValue`. In order to update the numerical values of the width parameters from the content of the variable `FR$PartialWidths`, the user only has to call the function `UpdateWidths[ ]`.

In addition, the decay information is by default also included in the UFO library generated from FEYNRULES, this option being switched off by setting the option `AddDecays` of the function `WriteUFO` to `False`. The generated UFO files include a new file, `decays.py`, with all the calculated analytical formulas stored in the variable `FR$PartialWidths`.

6. Conclusion

The constant and fast evolution in new physics model building activities has triggered the development of programs such as the MATHEMATICA package FEYNRULES, which aims to facilitate the development of BSM models and their implementation within Monte Carlo tools. The soon-to-be-released 2.0 version of FEYNRULES contains several important new features with respect to the previous release, in addition to the already available superspace module, interfaces to the CALCHEP, FEYNARTS, MADGRAPH, SHERPA and WHIZARD programs and possibility to translate the full model under the form of a PYTHON library, the so-called UFO format. Among the major new functionalities, we have presented in this report a brief description of two modules, a first one dedicated to the automated computation of decay widths and branching ratios and a second one addressing the extraction of the model mass matrices and their diagonalization. In addition, we have also described some improvements related to spin-$\frac{3}{2}$ particles.

Acknowledgments

We are grateful to the organizers of the ACAT2013 conference for arranging such a nice event. This work has been partially supported by the Theory-LHC France initiative of the CNRS/IN2P3 and the French ANR 12 JS05 002 01 BATS@LHC. A.A. has been supported by a Ph.D. fellowship of the French Ministry for Education and Research. Ce. D. was supported in part by the U. S. Department of Energy under Contract No. DE-FG02-13ER42001 and by the NSF grant PHY-0757889; N.D.C. was supported in part by the LHC-TI under U.S. National Science Foundation, grant NSF-PHY-0705682, by PITT PACC, and by the U.S. Department of Energy under grant No. DE-FG02-95ER40896 and C.D. was supported by the ERC grant "IterQCD".

References

[1] Aad G et al. (ATLAS Collaboration) 2012 Phys.Lett. **B716** 1–29 (*Preprint* 1207.7214)
[2] Chatrchyan S et al. (CMS Collaboration) 2012 Phys.Lett. **B716** 30–61 (*Preprint* 1207.7235)
[3] Christensen N D and Duhr C 2009 Comput.Phys.Commun. **180** 1614–1641 (*Preprint* 0806.4194)
[4] Christensen N D, de Aquino P, Degrande C, Duhr C, Fuks B et al. 2011 Eur.Phys.J. **C71** 1541 (*Preprint* 0906.2474)
[5] Christensen N D, Duhr C, Fuks B, Reuter J and Speckner C 2012 Eur.Phys.J. **C72** 1990 (*Preprint* 1010.3251)
[6] Duhr C and Fuks B 2011 Comput.Phys.Commun. **182** 2404–2426 (*Preprint* 1102.4191)
[7] Fuks B 2012 Int.J.Mod.Phys. **A27** 1230007 (*Preprint* 1202.4769)
[8] Alloul A, D'Hondt J, De Causmaecker K, Fuks B and Rausch de Traubenberg M 2013 Eur.Phys.J. **C73** 2325 (*Preprint* 1301.5932)
[9] Christensen N D, de Aquino P, Deutschmann N, Duhr C, Fuks B et al. 2013 (*Preprint* 1308.1668)
[10] Pukhov A 2004 (*Preprint* hep-ph/0412191)
[11] Belyaev A, Christensen N D and Pukhov A 2012 (*Preprint* 1207.6082)
[12] Hahn T 2001 Comput.Phys.Commun. **140** 418–431 (*Preprint* hep-ph/0012260)
[13] Stelzer T and Long W 1994 Comput.Phys.Commun. **81** 357–371 (*Preprint* hep-ph/9401258)
[14] Maltoni F and Stelzer T 2003 JHEP **0302** 027 (*Preprint* hep-ph/0208156)
[15] Alwall J, Demin P, de Visscher S, Frederix R, Herquet M et al. 2007 JHEP **0709** 028 (*Preprint* 0706.2334)
[16] Alwall J, Artoisenet P, de Visscher S, Duhr C, Frederix R et al. 2009 AIP Conf.Proc. **1078** 84–89 (*Preprint* 0809.2410)
[17] Alwall J, Herquet M, Maltoni F, Mattelaer O and Stelzer T 2011 JHEP **1106** 128 (*Preprint* 1106.0522)
[18] de Aquino P, Link W, Maltoni F, Mattelaer O and Stelzer T 2012 Comput.Phys.Commun. **183** 2254–2263 (*Preprint* 1108.2041)
[19] Gleisberg T, Hoeche S, Krauss F, Schalicke A, Schumann S et al. 2004 JHEP **0402** 056 (*Preprint* hep-ph/0311263)
[20] Gleisberg T, Hoeche S, Krauss F, Schonherr M, Schumann S et al. 2009 JHEP **0902** 007 (*Preprint* 0811.4622)
[21] Moretti M, Ohl T and Reuter J 2001 (*Preprint* hep-ph/0102195)
[22] Kilian W, Ohl T and Reuter J 2011 Eur.Phys.J. **C71** 1742 (*Preprint* 0708.4233)
[23] Degrande C, Duhr C, Fuks B, Grellscheid D, Mattelaer O et al. 2012 Comput.Phys.Commun. **183** 1201–1214 (*Preprint* 1108.2040)
[24] Alwall J, Duhr C, Fuks B, Mattelaer O, Öztürk G and Shen C H In preparation

Three-loop beta-functions and anomalous dimensions in the Standard Model

A.V. Bednyakov[1], A.F. Pikelner[1], and V.N. Velizhanin[2]

[1] Joint Institute for Nuclear Research, 141980 Dubna, Russia
[2] Theoretical Physics Department, Petersburg Nuclear Physics Institute, Orlova Roscha, Gatchina, 188300 St. Petersburg, Russia

E-mail: bednya@theor.jinr.ru

Abstract. In this talk the methods and computer tools which were used in our recent calculation of the three-loop Standard Model renormalization group coefficients are discussed. A brief review of the techniques based on special features of dimensional regularization and minimal subtraction schemes is given. Our treatment of γ_5 is presented in some details. In addition, for a reasonable set of initial parameters the numerical estimates of the obtained three-loop contributions are presented.

The renormalization group (RG) proves to be a useful and powerful tool in studying high-energy behavior of the Standard Model. Before the discovery of the Higgs boson RG equations (RGE) were used, among other things, to bound the value of the Higgs self-coupling. However, the bounds significantly depend on the scale at which one expects the appearance of New Physics. The observation of the Higgs boson in 2012 [1, 2] in some sense finalizes the SM since the information about the values of all SM couplings become available from experiments.

Due to this fact, the interest to RG studies of the Standard Model arises again, but at a new level of precision. One- and two-loop results for SM beta-functions have been known for quite a long time [3] and are summarized in Ref. [4].

The first paper with full three-loop calculation of gauge coupling beta-functions within the SM was published in Ref. [5]. The next step was carried out by another group from Karlsruhe [6], which considered, in the specific limit, the three-loop beta-functions for the top quark Yukawa coupling, the Higgs self-coupling and mass parameter. At this state of things our group entered the game. We not only confirmed the results of Refs. [5, 6, 7] but also provided the three-loop expressions for beta-functions of all the Yukawa couplings [8] corresponding to the fermions of third generation. Contrary to K. Chetyrkin and M. Zoller [6], we include the dependence on electroweak couplings.

In the beginning of 2013 our group started the calculation of missing three-loop terms in beta-functions of the Higgs potential parameters. It turns out that the same problem was considered by the authors from Karlsruhe. Unfortunately to us, they managed to obtain and make their results [9] public a week earlier than our group [10]. Nevertheless, in such a complicated calculation it is important to have a confirmation from an independent source. In what follows, we are going to discuss the peculiarities of the procedure used to obtain the results published by our group in a series of papers [8, 10, 11].

First of all, let us mention that the calculation of the three-loop SM beta-functions requires evaluation of millions of Feynman diagrams. This task definitely requires automatization by means of a computer. Fortunately, all necessary tools were available on the market, so we only needed to combine them in a proper way.

It is due to nice features of dimensional regularization and minimal subtraction scheme one can significantly simplify the calculation. Since in the $\overline{\text{MS}}$-scheme all renormalization constants can be extracted from the ultraviolet (UV) divergent parts of the corresponding Green functions, one can modify infra-red (IR) structure of the model to simplify the calculation of counter-terms. This is the essence of the so-called infrared-rearrangement (IRR) trick, which was originally proposed by A.A Vladimirov[12]. This kind of modifications can lead to a spurious IR divergences which should be removed consistently by the so-called R^* operation [13]. However, in many practical cases one can avoid this kind of complications. In our series of paper we made use of two variants of (naive) IRR procedure.

For the calculation of gauge and Yukawa coupling beta-functions it is possible to convert all required two- and three- point Green functions to the massless propagator-type Feynman integrals. It is done via neglecting all internal masses and setting the Higgs boson external momenta entering Yukawa vertex to zero. The evaluation of massless three-loop propagators is performed via a FORM [14] package `MINCER` [15]. This kind of manipulations does not introduce spurious IR divergences. Moreover, since we are interesting only in UV counter-terms it is possible to work from the very beginning within the "unbroken phase" of the SM, in which all fields are massless and the Higgs doublet Φ does not have a vacuum expectation value.

The second approach to IRR, which was used in calculation of Higgs potential parameter beta-functions, is the introduction of an auxiliary mass parameter M in every propagator via iterative application of the following formula [16, 17]

$$\frac{1}{(q-p)^2} = \frac{1}{q^2 - M^2} + \frac{2qp - p^2 - M^2}{q^2 - M^2} \times \frac{1}{(q-p)^2} \tag{1}$$

where q and p are linear combinations of internal and external momenta, respectively. It is clear that if one applies this kind of decomposition a sufficient number of times the last term can be neglected in the calculation of UV divergences (after subtraction of subdivergences). It turns out that for the scalar four-point Green functions considered only the first term in Eq. (1) is necessary. Consequently, we are left with massive vacuum integrals, which can be calculated by either public `MATAD` package [18] or private `BAMBA` code written by V. Velizhanin. In such an approach no spurious IR divergences appear so it can be used in the situations when a naive application of the first variant of IRR fails. However, the price to pay for this advantage is the necessity of explicit calculation of diagrams with counter-term insertions. This is due to the fact that one needs to introduce counter-terms for divergences contributing to the auxiliary masses for vector and scalar bosons. For further details see Ref. [16].

It is worth mentioning that we can still exploit the symmetries of the unbroken SM. For example, all components of the Higgs boson doublet should have the same auxiliary mass counter-term. The same is true for the SU(2) gauge bosons.

Moreover, as it is stated in Ref. [17], the auxiliary mass appearing in the numerator in RHS of Eq. (1) can be safely neglected since it can only contribute to the unphysical mass counter-terms. In addition, one can also skip Feynman diagrams with vacuum subdiagrams. In spite of the fact that these subdiagrams are non-zero when the auxiliary mass is introduced, they still can be neglected due to the same reasons.

As it was noticed above there are a lot of diagrams which should be generated and evaluated in order to find three-loop contributions to the considered quantities. In our calculation we made use of two popular codes, `FeynArts`[19] and `DIANA` [20]/`QGRAF` [21], which generate necessary diagrams and produce corresponding analytic expressions. Both packages require a model file prepared in a special format to do their job. Since we wanted to simplify the calculation as much as possible we prepared a model file for the unbroken SM in the background field gauge (BFG) [22, 23]. This kind of gauge allows one to find gauge coupling beta-functions solely from UV-divergences of the corresponding gauge field propagators. We used a very fast `LanHEP` code [24] by A. Semenov to derive all SM vertices from the considered SM Lagrangian (see Ref. [11]) in `FeynArts` notation. It is worth mentioning that the Karlsruhe group made use of alternative package `FeynRules` [25] to solve similar problem. Latter on a simple script was written to convert the `FeynArts` model file to that of `DIANA`.

A typical problem which arises in this kind of calculation is internal momenta identification. In order to evaluate a Feynman diagram one needs to use the momenta notation of the chosen code (`MINCER/MATAD/BAMBA`). In the case of gauge and Yukawa couplings the problem was solved with the help of routine, `MapMincer`, which associate with every `FeynArts` topology the corresponding `MINCER` topology and distribute `MINCER` momenta accordingly. It is worth mentioning that `MapMincer` can deal with three-point vertices with one external leg carrying zero momentum (i.e., when internal lines have dots). The corresponding routine for `DIANA/QGRAF`, `MapDiana`, performs similar task, but maps every generated topology to fully massive vacuum integrals, which appear after the mentioned"exact" decomposition (1) of internal propagators[1].

Given a model file for `FeynArts/DIANA` together with the correct mapping of internal momenta to the notation of the utilized three-loop codes it is straightforward to generate and calculate one-, two-, and three-loop contributions to the 1PI Green-functions presented in Fig. 1. For the SU(3) color algebra the `COLOR` [26] package was used.

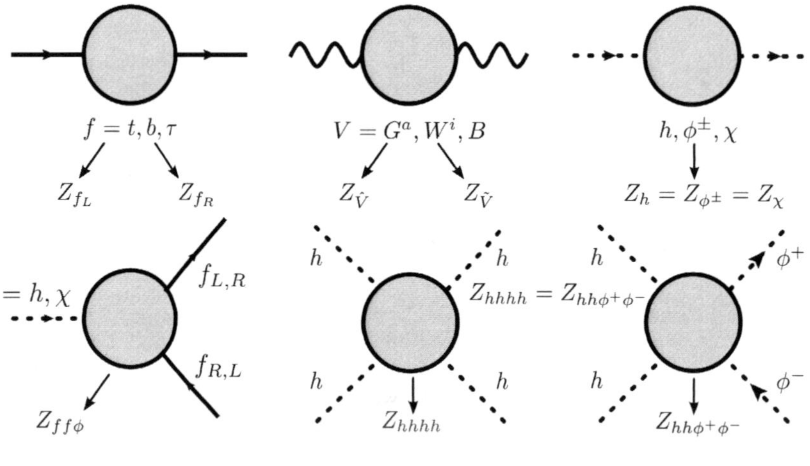

Figure 1: The Green functions, considered in the beta-function calculations, together with the corresponding renormalization constants. Left- and right-handed fermions, denoted by $f_{L,R}$, renormalise differently in the SM. The same is true for background $\hat{V}$ and quantum $\tilde{V}$ gauge fields in BFG employed. The SU(2) invariance implies the presented equalities, which serve as an additional cross-check.

From two-point Green functions we extract the corresponding wave function renormalization constants in the $\overline{\text{MS}}$-scheme. It is worth pointing that the self-energies of both background and quantum gauge fields are considered. The renormalization constants of the former, $Z_{\hat{V}}$, are directly connected to that of gauge couplings and the renormalization of the latter corresponds to the Z-factors of three gauge-fixing parameters, ξ_G, ξ_W, and ξ_B, which we keep non-zero during

[1] Both `MapMincer` and `MapDiana` are coded by A. Pikelner.

the whole stage of calculation. The independence of the final results on these parameters serves as an important cross-check of the obtained expressions.

The three-point 1PI functions for the Yukawa vertices with neutral Higgs bosons h and χ are also presented in Fig. 1. It is interesting to note that our calculation explicitly demonstrated that the semi-naive treatment of γ_5 discussed below is only applicable if one takes into account the gauge anomaly cancellation condition $N_C = 3$. Here N_C denotes the number of colors.

The beta-functions of the SM parameters are extracted from the corresponding renormalization constants. For the gauge and Yukawa couplings we used the following relations

$$Z_{g_1} = Z_{\hat{B}_1}^{-1/2}, \qquad Z_{g_2} = Z_{\hat{W}}^{-1/2}, \qquad Z_{g_s} = Z_{\hat{G}}^{-1/2}, \qquad Z_{y_f} = \frac{Z_{ff\phi}}{\sqrt{Z_{f_L} Z_{f_R} Z_\phi}}, \tag{2}$$

where g_1, g_2 are SU(2) and U(1) gauge couplings, respectively, g_s denotes the strong coupling, and y_f is the Yukawa coupling associated with the (right-handed) fermion $f = t, b, \tau$. Both neutral components, $\phi = h, \chi$, gave the same result, and, thus, provided us with a confirmation of the validity of the obtained expressions. For the Higgs self-coupling it is impossible to use `MINCER` naively, so Feynman diagrams for the four-point functions (see Fig. 1) converted to the fully massive vacuum integrals were calculated with the help of private code `BAMBA` (by V. Velizhanin, who considered the $hh\phi^+\phi^-$ vertex) and public package `MATAD` (by A. Bednyakov and A. Pikelner who considered the fully symmetric $hhhh$ vertex). These two independent evaluations lead to the same final expression for the vertex renormalization constants, i.e., confirming the SU(2) relation $Z_{hhhh} = Z_{hh\phi^+\phi^-}$.

A comment on the Higgs mass parameter m^2 is in order. It is possible to obtain the corresponding anomalous dimension by considering the renormalization of the $|\Phi|^2$ composite operator within the unbroken(=massless) SM (see, e.g., Ref. [6]). This kind of result can be found at almost no cost from the calculation of $hh\phi^+\phi^-$ vertex. It is sufficient to select the diagrams, which have ϕ^+ and ϕ^- external fields connected to the same four-point vertex (see Fig. 2), and weight different contributions with a correct combinatorial factor. This restricted set of diagrams gives rise to the $Z_{hh[\phi^+\phi^-]}$ renormalization constant.

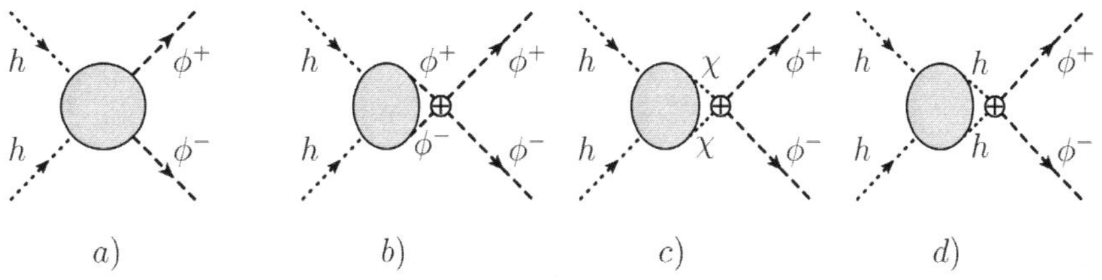

Figure 2: The restricted set of Feynman diagrams which is used to obtain the renormalization constant $Z_{hh[\phi^+\phi^-]}$. The diagrams of the first type have to be multiplied by 1/2.

At the end of the day, the renormalization constants for λ and m^2 are obtained with the help of the following relations

$$Z_\lambda = \frac{Z_{hhhh}}{Z_h^2} = \frac{Z_{hh\phi^+\phi^-}}{Z_h Z_{\phi^\pm}}, \qquad Z_{m^2} = \frac{Z_{hh[\phi^+\phi^-]}}{Z_h} \tag{3}$$

From renormalization constants Z_{a_i} for the dimensionless SM parameters,

$$a_i = \left(\frac{5}{3}\frac{g_1^2}{16\pi^2}, \frac{g_2^2}{16\pi^2}, \frac{g_s^2}{16\pi^2}, \frac{y_t^2}{16\pi^2}, \frac{y_b^2}{16\pi^2}, \frac{y_\tau^2}{16\pi^2}, \frac{\lambda}{16\pi^2} \right), \tag{4}$$

it is straightforward to obtain the corresponding beta-functions

$$\beta_i(a_k) = \left.\frac{da_i(\mu,\epsilon)}{d\ln\mu^2}\right|_{\epsilon=0}, \quad \beta_i = \sum_l a_l \frac{\partial c_i^{(1)}}{\partial a_l} - c_i^{(1)}, \quad a_{i,\text{Bare}}\mu^{-2\epsilon} = Z_{a_i} a_i = a_i + \sum_{n=1}^{\infty} \frac{c_i^{(n)}}{\epsilon^n} \tag{5}$$

Here μ is the $\overline{\text{MS}}$ renormalization scale, $\epsilon = (4-D)/2$ is the parameter of dimensional regularization, and c_i denotes the coefficient for the single pole in ϵ in the expression for Z_{a_i}, which enters the relation between the bare parameters $a_{i,\text{Bare}}$ and the renormalized ones.

For the anomalous dimension of the Higgs mass parameter m^2 one can use similar formulae

$$\gamma_{m^2}(a_k) = \left.\frac{d\ln m^2}{d\ln\mu^2}\right|_{\epsilon=0} = \sum_l a_l \frac{\partial c_{m^2}^{(1)}}{\partial a_l}, \qquad m^2_{\text{Bare}} = Z_{m^2} m^2 = m^2\left(1 + \sum_{n=1}^{\infty}\frac{c_{m^2}^{(n)}}{\epsilon^n}\right) \tag{6}$$

The full analytical results for the considered quantities can be found in ancillary files of the arXiv versions of our papers. The intermediate expressions, e.g., all renormalization constants, can also be obtained, if needed, from the authors. It is worth mentioning that the beta-functions of all fundamental SM parameters are free from gauge–parameter dependence, which is a crucial test for our calculation. In addition, the anomalous dimension of the Higgs doublet can be of some interest, so we also include the corresponding expression in the ancillary files of Ref. [10].

Before going to the results we would like to touch on the issue related to the definition of the γ_5 matrix within the dimensional regularization. It is known from the literature (see, e.g., Ref. [27] and recent explicit calculation [6]) that the traces with an odd number of γ_5 appearing for the first time in the three-loop diagrams require special treatment. We closely follow the semi-naive approach presented in Refs. [6, 7]. First of all, we anticommute γ_5 to the rightmost position in a fermion chain and use $\gamma_5^2 = 1$. In the "even" traces all γ_5 are contracted with each other, so the corresponding expressions can be treated naively in dimensional regularization. In "odd" traces we are left with only one γ_5. These traces are evaluated as in four dimensions and produce totally antisymmetric tensors via the relation

$$\text{tr}\left(\gamma^\mu\gamma^\nu\gamma^\rho\gamma^\sigma\gamma_5\right) = -4i\epsilon^{\mu\nu\rho\sigma} \tag{7}$$

with $\epsilon^{0123} = 1$.

Due to the fact that we use both the γ_5 anticommutativity and the four-dimensional relation (7), the cyclicity of the trace should be relinquished [28]. One has to choose a certain "reading prescription" for an "odd" Dirac trace, i.e., start reading a closed fermion chain from a proper place, in order to achieve the correct final result. However, in our calculations the problem of γ_5 positioning within the "odd" traces is solved implicitly since the diagram generation routines split the traces for us at certain points. It should be stressed that a non-trivial contribution to the considered quantities can only appear when there are two "odd" traces in a diagram, since two ϵ-tensors should be "contracted" with each other to produce a non-zero effect. We can distinguish two situations. Two "odd" traces in our three-loop calculations can appear either as two internal fermion loops in a diagram with external bosons or, if one considers Green functions with two external fermions, one Dirac trace from internal fermion loop can be combined with the trace appearing after contraction with an appropriate projector.

It is easy to convince oneself that in the case of two internal traces only triangle subloops with three external vector particles can potentially produce "eps"-tensors. However, it is known that in the SM these kind of traces cancel upon summation over all fermion species due to the absence of gauge anomalies [29, 30]. The same is true if one considers fermion self-energies up to three-loops.

It turns out that the non-trivial contribution due to the contraction of two ϵ-tensor appears from the Yukawa vertex (see Fig. 3). This kind of diagrams was also considered in Ref. [6].

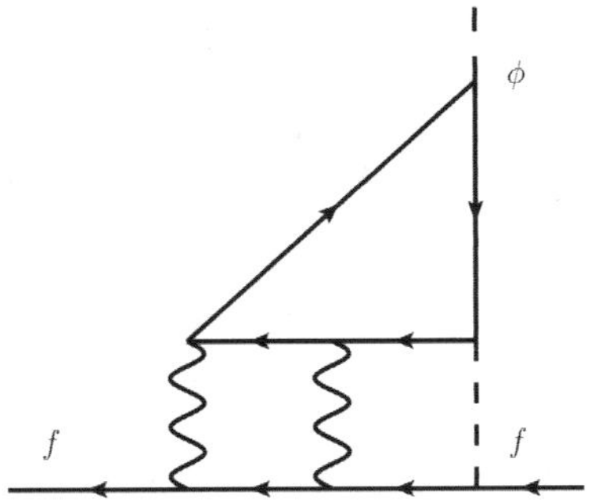

Figure 3: The diagrams for the fermion-fermion-Higgs vertex ($ff\phi$), which produce a non-zero contribution due to contraction of two ϵ-tensors appearing from two Dirac traces with γ_5. The ambiguity in positioning of γ_5 within the traces does not affect the UV-divergent part.

A final remark about γ_5 is again related to the fact that the ambiguity in the choice of "reading" point in the "odd" traces can only spoil our result for the three-loop UV-divergence in the case of the mentioned triangle subgraphs, for which one can set all "odd" traces to zero from the very beginning due to anomaly cancellations. For the only non-trivial case all "reading" points are equivalent since the difference gives the contribution $\mathcal{O}(\epsilon^0)$ which we neglect here.

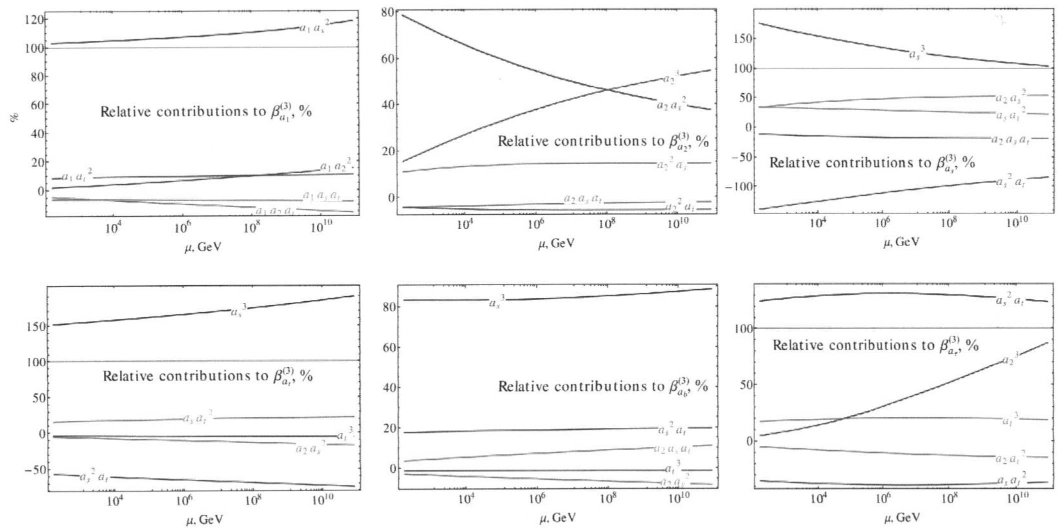

Figure 4: The scale dependence of the relative contributions of different terms in the three-loop corrections to the beta-functions of the (squared) gauge and Yukawa couplings. Only the most sizable corrections are shown. The boundary conditions at scale $\mu = 100$ GeV are given in (8).

It is obvious, that the resulting expression for the three-loop contributions to the considered renormalization group quantities are too lengthy to be presented here. For the demonstration purposes only, we would like to show how different relative contributions to the three-loop corrections to the SM coupling beta-functions evolve with renormalization scale μ (see Fig. 4). For a reasonable choice of the SM initial running parameters[2] at the scale $\mu = 100$ GeV

$$g_1 = 0.3576,\ g_2 = 0.6514,\ g_s = 1.2063,\ y_t = 0.9665,\ y_b = 0.016,\ y_\tau = 0.01,\ \lambda = 0.13 \qquad (8)$$

[2] obtained with the help of F. Bezrukov code http://www.inr.ac.ru/~fedor/SM

we solve the corresponding RGE numerically up $5\cdot 10^{10}$ GeV. With the help of these solutions one can evaluate the three-loop beta-functions at any scale and find how different terms contribute to the total value of $\beta_i^{(3)}$. From Fig. 4 one can see that the dominant contributions is due to the strong and top Yukawa couplings. However, with the increase of the renormalization scale the SU(2) coupling can also play a role.

It is fair to say that the most interesting SM beta-function is that of the Higgs self-coupling, since from the evolution of the latter one can deduce the so-called "vacuum stability bound" (see recent papers [31, 32, 33, 35] and references therein). In Fig. 5 one can find the same evolution of the relative contributions to $\beta_\lambda^{(3)}$. In addition, the slice (λ, β_λ) of the phase space (a_i, β_{a_i}) is presented together with trajectories, obtained with the help of one-, two-, and three-loop evolution from 100 GeV to $5 \cdot 10^{10}$ GeV. One can see that for the given set of initial conditions the running $\lambda(\mu)$ is driven to zero faster when one-loop RGEs are employed instead of two- or three-loop ones. The difference between two- and three-loop evolution is not sizable, but the fact that with three-loop corrections the scale at which λ hits zero slightly higher, favours the latter.

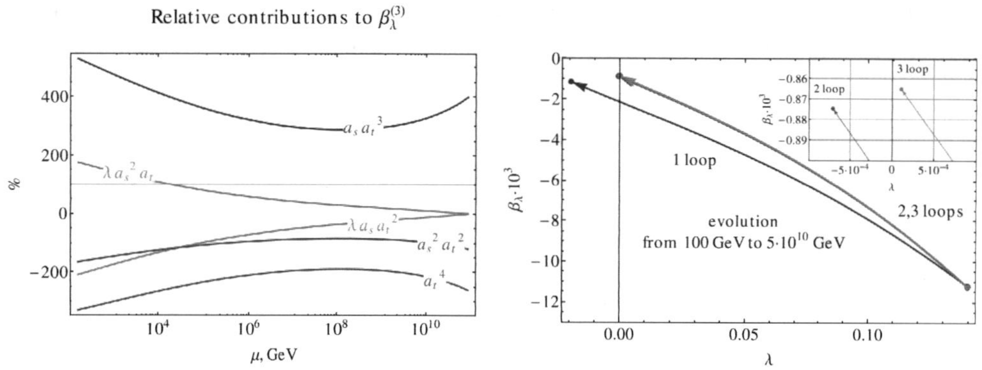

Figure 5: The scale dependence of the relative contributions to the three-loop corrections $\beta_\lambda^{(3)}$. Only the most sizable corrections are shown. The phase space trajectories from 1-,2-,and 3-loop evolution are provided in the plane (λ, β_λ). The corresponding boundary conditions are given in Eq. (8)

To conclude, we obtained the three-loop beta-functions for the SM parameters. The results for gauge and Higgs-potential couplings coincide with that obtained by two Karlsruhe groups. The beta-functions for Yukawa couplings were obtained for the first time. Moreover, we established a framework that allow us to carry out a similar calculation within an "arbitrary" QFT model. However, it should be stressed that in a self-consistent RG analysis of the chosen model the obtained RGEs should be accompanied by the so-called threshold (matching) corrections (see, e.g., Refs. [31, 34, 32, 35] for the recent SM results).

Acknowledgments

AVB is grateful to the organizers and conveniers of the ACAT2013 workshop for the invitation and for arranging such a nice event. In addition, we would like to thank M. Kalmykov for drawing our attention to the problem and stimulating discussions. The work is supported in part by RFBR grants 11-02-01177-a, 12-02-00412-a, and by JINR Grant No. 13-302-03.

References

[1] Aad G *et al.* (ATLAS Collaboration) 2012 *Phys.Lett.* **B716** 1–29 (*Preprint* 1207.7214)
[2] Chatrchyan S *et al.* (CMS Collaboration) 2012 *Phys.Lett.* **B716** 30–61 (*Preprint* 1207.7235)
[3] Gross D and Wilczek F 1973 *Phys.Rev.Lett.* **30** 1343–1346
Politzer H D 1973 *Phys.Rev.Lett.* **30** 1346–1349
Jones D 1974 *Nucl.Phys.* **B75** 531
Tarasov O and Vladimirov A 1977 *Sov.J.Nucl.Phys.* **25** 585
Caswell W E 1974 *Phys.Rev.Lett.* **33** 244
Egorian E and Tarasov O 1979 *Teor.Mat.Fiz.* **41** 26–32
Jones D 1982 *Phys.Rev.* **D25** 581
Fischler M S and Hill C T 1981 *Nucl.Phys.* **B193** 53
Machacek M E and Vaughn M T 1983 *Nucl.Phys.* **B222** 83
Machacek M E and Vaughn M T 1984 *Nucl.Phys.* **B236** 221
Luo M x, Wang H w and Xiao Y 2003 *Phys.Rev.* **D67** 065019 (*Preprint* hep-ph/0211440)
Jack I and Osborn H 1985 *Nucl.Phys.* **B249** 472
Gorishnii S, Kataev A and Larin S 1987 *Theor.Math.Phys.* **70** 262–270
Arason H, Castano D, Keszthelyi B, Mikaelian S, Piard E *et al.* 1992 *Phys.Rev.* **D46** 3945–3965
[4] Luo M x and Xiao Y 2003 *Phys.Rev.Lett.* **90** 011601 (*Preprint* hep-ph/0207271)
[5] Mihaila L N, Salomon J and Steinhauser M 2012 *Phys.Rev.Lett.* **108** 151602 (*Preprint* 1201.5868)
[6] Chetyrkin K and Zoller M 2012 *JHEP* **1206** 033 (*Preprint* 1205.2892)
[7] Mihaila L N, Salomon J and Steinhauser M 2012 *Phys.Rev.* **D86** 096008 (*Preprint* 1208.3357)
[8] Bednyakov A, Pikelner A and Velizhanin V 2013 *Phys.Lett.* **B722** 336–340 (*Preprint* 1212.6829)
[9] Chetyrkin K and Zoller M 2013 *JHEP* **1304** 091 (*Preprint* 1303.2890)
[10] Bednyakov A, Pikelner A and Velizhanin V 2013 (*Preprint* 1303.4364)
[11] Bednyakov A, Pikelner A and Velizhanin V 2013 *JHEP* **1301** 017 (*Preprint* 1210.6873)
[12] Vladimirov A 1980 *Theor.Math.Phys.* **43** 417
[13] Chetyrkin K and Smirnov V A 1984 *Phys.Lett.* **B144** 419–424
[14] Vermaseren J 2000 (*Preprint* math-ph/0010025)
[15] Gorishnii S, Larin S, Surguladze L and Tkachov F 1989 *Comput.Phys.Commun.* **55** 381–408
[16] Misiak M and Munz M 1995 *Phys.Lett.* **B344** 308–318 (*Preprint* hep-ph/9409454)
[17] Chetyrkin K G, Misiak M and Munz M 1998 *Nucl.Phys.* **B518** 473–494 (*Preprint* hep-ph/9711266)
[18] Steinhauser M 2001 *Comput.Phys.Commun.* **134** 335–364 (*Preprint* hep-ph/0009029)
[19] Hahn T 2001 *Comput.Phys.Commun.* **140** 418–431 (*Preprint* hep-ph/0012260)
[20] Tentyukov M and Fleischer J 2000 *Comput.Phys.Commun.* **132** 124–141 (*Preprint* hep-ph/9904258)
[21] Nogueira P 1993 *J.Comput.Phys.* **105** 279–289
[22] Abbott L 1981 *Nucl.Phys.* **B185** 189
[23] Denner A, Weiglein G and Dittmaier S 1995 *Nucl.Phys.* **B440** 95–128 (*Preprint* hep-ph/9410338)
[24] Semenov A 2010 (*Preprint* 1005.1909)
[25] Christensen N D and Duhr C 2009 *Comput.Phys.Commun.* **180** 1614–1641 (*Preprint* 0806.4194)
[26] van Ritbergen T, Schellekens A and Vermaseren J 1999 *Int.J.Mod.Phys.* **A14** 41–96 (*Preprint* hep-ph/9802376)
[27] Jegerlehner F 2001 *Eur.Phys.J.* **C18** 673–679 (*Preprint* hep-th/0005255)
[28] Korner J, Kreimer D and Schilcher K 1992 *Z.Phys.* **C54** 503–512
[29] Bouchiat C, Iliopoulos J and Meyer P 1972 *Phys.Lett.* **B38** 519–523
[30] Gross D J and Jackiw R 1972 *Phys.Rev.* **D6** 477–493
[31] Bezrukov F, Kalmykov M Y, Kniehl B A and Shaposhnikov M 2012 *JHEP* **1210** 140 (*Preprint* 1205.2893)
[32] Degrassi G, Di Vita S, Elias-Miro J, Espinosa J R, Giudice G F *et al.* 2012 *JHEP* **1208** 098 (*Preprint* 1205.6497)
[33] Alekhin S, Djouadi A and Moch S 2012 *Phys.Lett.* **B716** 214–219 (*Preprint* 1207.0980)
[34] Jegerlehner F, Kalmykov M Y and Kniehl B A 2013 *Phys.Lett.* **B722** 123–129 (*Preprint* 1212.4319)
[35] Buttazzo D, Degrassi G, Giardino P P, Giudice G F, Sala F *et al.* 2013 (*Preprint* 1307.3536)

Development of an object oriented lattice QCD code "Bridge++"

S Ueda[1], S Aoki[2], T Aoyama[3], K Kanaya[4], H Matsufuru[5], S Motoki[6], Y Namekawa[7], H Nemura[7], Y Taniguchi[4] and N Ukita[7]

[1] Theory Center, IPNS, High Energy Accelerator Research Organization (KEK), Tsukuba 305-0810, Japan
[2] Yukawa Institute for Theoretical Physics, Kyoto University, Kyoto 606-8502, Japan
[3] Kobayashi-Maskawa Institute for the Origin of Particles and the Universe (KMI), Nagoya University, Nagoya 464-8602, Japan
[4] Graduate School of Pure and Applied Sciences, University of Tsukuba, Tsukuba 305-8571, Japan
[5] Computing Research Center, High Energy Accelerator Research Organization (KEK), Tsukuba 305-0801, Japan
[6] University of Aizu, Aizu-Wakamatsu 965-8580, Japan
[7] Center for Computational Sciences, University of Tsukuba, Tsukuba 305-8577, Japan

E-mail: `sueda@post.kek.jp`

Abstract. We are developing a new lattice QCD code set "Bridge++" aiming at extensible, readable, and portable workbench for QCD simulations, while keeping a high performance at the same time. Bridge++ covers conventional lattice actions and numerical algorithms. The code set is constructed in C++ with an object oriented programming. In this paper we describe fundamental ingredients of the code and the current status of development.

1. Introduction

A perturbative approach to QCD succeeds in the high energy physics such as the deep inelastic scattering. On the other hand, it does not apply to the low energy phenomena such as the quark confinement and the light hadron spectrum. Some nonperturbative methods are necessary. Lattice QCD is a SU(3) gauge theory defined on 4-dimensional Euclidean lattice, which provides a nonperturbative framework of the calculation [1]. The path integral quantization enables Monte Carlo simulations.

Rapid increase of the processor performance and the integration technology of the large-scale cluster system enables us to apply the lattice simulation to wide range of research area: matrix elements to test the standard model in LHC and B factory experiments, the phase diagram and the nature of QCD at finite temperature and density, the force among nucleons, and so on. The lattice framework is also applied to gauge theories other than QCD, such as a technicolor theory to explain the mechanism of the Higgs sector of the standard model.

While numerical lattice simulations have become an indispensable tool for theoretical analysis, the programming technique has become more and more involved. The hybrid parallelization with MPI and OpenMP is inevitable for the current large-scale system such as K-Computer and Blue Gene/Q. GPGPUs require an additional development of the code with CUDA or OpenCL. The

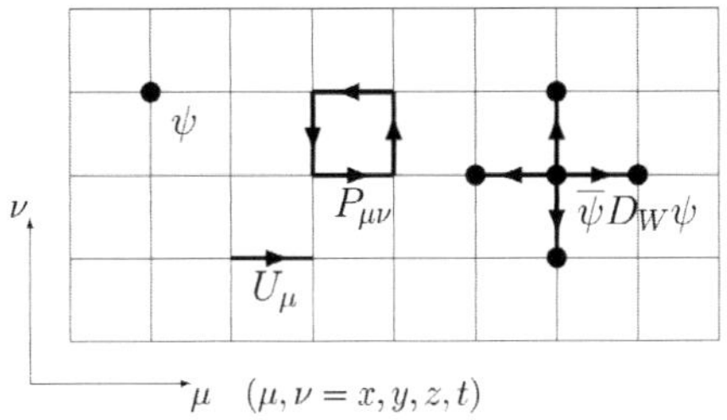

Figure 1. Gluon field U_μ and quark ψ on the lattice. The plaquette $P_{\mu\nu}$ as well as an interaction of fermion operator $\bar{\psi}D_W\psi$ are also given.

research groups have to develop new codes for updated machines. It makes these codes hard to be understood for beginners. In particular, it is a serious problem that new ideas or new physical quantities cannot be tested quickly.

In view of this programming situation, we started a project to develop a new common code set "Bridge++" for lattice QCD simulations. We adopt C++ language to make use of the object-oriented design. Bridge++ has a great deal of readability while keeping sufficient performance for frontier works. The machine dependent part of the code is hidden as much as possible. The first version of the code was released in July 2012 under the GNU General Public License. It is actively being developed to extend functions, brush up its design and the implementation, improve the performance, and provide more documents. The latest version is 1.1.1. The code is available from our website [2].

This paper is organized as follows. In the next section, the core constitution of the lattice QCD simulation is briefly summarized. In Sec. 3, the policy of development of Bridge++ project is described. Section 4 summarizes our current status of the project and future prospects.

2. Lattice QCD simulations

2.1. Principle of lattice QCD simulations

K.G. Wilson proposed to formulate QCD on a discretized Euclidean spacetime, namely on a lattice [3]. It enables a nonperturbative analysis of QCD by numerical simulations.

In lattice QCD, the gauge field A_μ is represented as 3×3 complex matrices on the link connecting the nearest neighbor sites such that $U_\mu(n) = \exp(igA_\mu(n))$, where g is the strong coupling constant and $n = (n_x, n_y, n_z, n_t)$ is a lattice site (Fig. 1). The number of lattice sites is finite in the numerical simulation by the lattice size L_μ, $n_\mu = 1, \ldots, L_\mu, \mu = x, y, z, t$. The quark field ψ is represented as a complex vector on a lattice site which carries 3 components of color and 4 components of spinor.

Applying the path integral quantization, an expectation value of observable $\mathcal{O}$ is given by a functional integral

$$\langle \mathcal{O} \rangle = \frac{1}{Z}\int \mathcal{D}\overline{\psi}\mathcal{D}\psi\mathcal{D}U_\mu \mathcal{O}(\overline{\psi}, \psi, U_\mu)e^{-S_{\text{lat}}} = \frac{1}{Z}\int \mathcal{D}U_\mu \mathcal{O}(U_\mu)\det(D)e^{-S_G}, \tag{1}$$

where Z is the partition function and $S_{\text{lat}} = S_G + S_F$ with the gauge part S_G and the fermion part $S_F = \bar{\psi}D\psi$ of the lattice QCD action, respectively. The quark field is integrated by hand and arrives at the second equality of Eq. (1). The quark propagator D^{-1} expresses a propagation from a site to another site. Hadron spectrum can be extracted from the hadron correlation functions that are composed of quark propagators.

The expectation value is calculated by generating ensemble of gauge fields with Monte Carlo method under the probability

$$P(U) \propto \det(D[U])e^{-S_G[U]}, \tag{2}$$

where U represents a gauge configuration $\{U_\mu(x)\}$. Once the ensemble of independent configurations $U^{(i)}$ $(i = 1, \ldots, N)$ are in hand, an expectation value of the physical observable is obtained by

$$\langle \mathcal{O} \rangle = \frac{1}{N} \sum_{i=1}^{N} \mathcal{O}[U^{(i)}]. \tag{3}$$

It is noted that the expectation value $\langle \mathcal{O} \rangle$ has not only the statistical error but also several systematic errors. Numerical simulations are inevitably performed at finite lattice spacing and finite volume. It is imperative to control these errors for a precise study.

2.2. Lattice QCD actions

We introduce the fundamental form of the gauge and fermion actions. An action on the lattice is constructed under the following principles.

- The lattice action must be invariant under the gauge transformation.
- The lattice action must coincide with the continuum theory in the continuum limit, i.e. the lattice spacing $a \to 0$.
- The lattice action should retain symmetries in the continuum theory as much as possible.

The lattice action is not unique, because there is freedom to add a term which vanishes in the continuum limit. Variety of lattice QCD simulations originate from the type of the lattice actions.

The standard gauge action was given by Wilson, and thus called Wilson (plaquette) action:

$$S_G = -\frac{N_c}{g^2} \sum_n \sum_{\mu \neq \nu} P_{\mu\nu}(n), \quad P_{\mu\nu}(n) = \frac{1}{N_c} \operatorname{tr} \left[U_\mu(n) U_\nu(n+\hat{\mu}) U_\mu^\dagger(n+\hat{\nu}) U_\nu^\dagger(n) \right], \tag{4}$$

where $N_c = 3$ is the number of color, n is a site, $\hat{\mu}$ is a unit vector in μ-direction. $P_{\mu\nu}(n)$ is the smallest closed loop made of link variables called plaquette. It corresponds to the field strength. This action agrees with the continuum gauge action up to $O(a^2)$ corrections. By adding closed loops, such as rectangular loops, improved gauge action is constructed so as to reduce the $O(a^2)$ effects.

The lattice fermion action is more complicated. A naive discretization of the continuum fermion action results in so-called doubling problem: the propagator possesses undesirable additional poles called doublers corresponding to particles at the edges of the Brillouin zone. There are other ways to circumvent the doubling problem. It provides a type of fermion actions. The simplest idea is to add a second derivative term to eliminate doublers. This leads to the Wilson fermion action constructed from the Wilson fermion operator D_W which connects a site m to n such that

$$D_W(m,n) = \delta_{m,n} - \kappa \sum_{\mu=1}^{4} \left[(1-\gamma_\mu) U_\mu(m) \delta_{m+\hat{\mu},n} + (1+\gamma_\mu) U_\mu^\dagger(m-\hat{\mu}) \delta_{m-\hat{\mu},n} \right]. \tag{5}$$

γ_μ is a 4×4 matrix acting on the spinor space, and κ is a parameter related to the quark mass m_q through $\kappa = 1/(2m_q + 8)$. It should be noticed that the second derivative term violates the chiral symmetry on the lattice, though it vanishes in the continuum limit. It has recently become popular to adopt other fermion operators which respects the chiral symmetry on the lattice at the high numerical cost, such as the domain-wall and overlap operators.

2.3. Generating gauge configurations

The simplest hybrid Monte Carlo (HMC) algorithm is explained. HMC is a standard algorithm to generate gauge field configuration from $U^{(i)}$ to $U^{(i+1)}$.

Since $\det D$ is hard to be evaluated, we represent it as an integration of boson fields, called the pseudo-fermion fields. For two flavors of degenerate quark masses, $\det(D^\dagger D)$ is to be evaluated as

$$Z = \int \mathcal{D}U \det(D^\dagger D) e^{-S_G} = \int \mathcal{D}U \mathcal{D}\phi^\dagger \mathcal{D}\phi e^{-S_G - \phi^\dagger (D^\dagger D)^{-1} \phi} \tag{6}$$

We need to integrate the link variable U and pseudo-fermion ϕ, whose action contains an inverse of the fermion operators, $(D^\dagger D)^{-1}$.

The new configuration is brought by evolutions of U and its conjugate momentum P. The Hamiltonian of P and U governs the evolution with respect to the fictitious time that is independent from the original temporal coordinate. During the evolution of P and U, ϕ is kept fixed as external field. P and ϕ are given randomly at the beginning of the evolution equation. In solving the evolution equations numerically, a finite step size provides a systematic error. This error can be compensated by a Metropolis test of the difference of Hamiltonian at the beginning and the end of the evolution.

To summarize, the gauge configuration is generated in the following steps.

- Generate the pseudo-fermion ϕ under the probability $P \propto \exp[-\phi^\dagger (D^\dagger D)^{-1} \phi]$, this can be realized by Gaussian random number ξ, under the probability $P = \exp(-\xi^\dagger \xi)$, and $\phi = D^{-1}\xi$.
- Set the conjugate momentum P with Gaussian probability distribution.
- Solve the evolution equation for U and P with respect to the time step. To evaluate the force of P, one needs to solve a linear equation for a large sparse matrix, $x = (D^\dagger D)^{-1}\phi$.
- After getting a candidate of new configuration $\{P', U'\}$, calculate the deference of Hamiltonian $\Delta H = H[U', P', \phi] - H[U, P, \phi]$ and accept U' as new configuration with the probability $\max\{1, e^{-\Delta H}\}$. If rejected, the old configuration is adopted as the new one.

Using the gauge configurations, expectation values of observables can be computed according to Eq. (3).

3. Lattice QCD code set Bridge++

3.1. Bridge++ project

Bridge++ project starts on the following background. There are many lattice QCD codes, such as MILC code (in C, USA) [4], CPS++ (C++, USA) [5], Chroma (C++, Europe) [6], and Lattice QCD Toolkit (Fortran, Japan) [7]. We have realized, however, it is problematic that there has been no genuine code set other than Lattice QCD Toolkit in Japan, although Japan is one of the centers of lattice QCD simulations. It is desirable to possess a code set under our management, because it can instantly reflect a feedback of modifications, machine specific tunings, and other improvements. Another reason is that it has no black box. We can confirm and understand the code genuinely. In addition, a new code does not have a phantasm from the development history, which sometimes discourages a beginner.

For these reasons, we have started a development of our new code set, named Bridge++, in 2009. We aimed that this general-purpose code set should have the followings features:

- Readability: the code structure is transparent so as to be understandable even for beginners.
- Extensibility: the code is easy to be modified for testing new ideas.
- Portability: the code runs not only on laptop PC but also on supercomputers.
- High-performance: the code has a high performance enough for productive research.

To achieve these goals, we adopt an object oriented design in C++ programming language. MPI is used for distributed memory machines. In Bridge++ code set, we use several simple idioms repeatedly. Such repetition helps beginners to understand the idioms and the code structure. The code is managed with a repository and bug tracking system.

In July 2012, the first version of Bridge++ was released. It contains fundamental lattice actions, linear algebraic algorithms, and various physical quantities. The documents are compiled on wiki as well as TeX based reports. We also provide a HTML document generated by doxygen [9] from the comments in the source file.

3.2. Object-oriented programing

Object-oriented programing (OOP) is employed in Bridge++. It is based on "Objects", which are sets of data fields and methods. A problem is solved using an interaction between objects. Data field is a characteristic variable of the object, called a member data or a member variable. Method is a function that defines a behavior of the object.

OOP is characterized by the following properties.

- Encapsulation: data is handled through the interface, that defines how to use the object.
- Inheritance: object is expandable by adding new functions.
- Polymorphism: objects with the same kind of behavior can be handled through the same interface.

These properties are mechanisms to maximize reusability of the code. Using OOP, an interface is separated from details of implementation, and localize the latter which is sometimes specific to the architecture due to optimization. The polymorphism allows us to implement an algorithm with a set of objects that have the common interface.

There is a compilation of wisdom to make use of these virtues of OOP. An example is so-called design patterns [10, 11]. The design pattern is a kind of programing idiom that frequently appears as a good solution to certain kind of problems. Their efficiency has been realized after the GoF's publication [10] that classified such idioms into 23 design patterns. The design pattern enables us to use the benefit of OOP easily, as well as to make our code transparent.

In the following, we introduce two design patterns and show how they are applied to our problems.

3.3. Bridge pattern and linear equation solver

Bridge pattern is a design pattern which decouples an interface from its implementation so that they can switch independently. Bridge++ employs this bridge pattern. It has advantages for readability and further developments.

A typical example of the bridge pattern is an implementation of a solver algorithm and a fermion operator. In lattice QCD simulation, the most time consuming part is solving a linear equation for the fermion operator. An example is the Wilson fermion operator in Eq. 5, which contains couplings up to the nearest neighbors, and thus is a large sparse complex matrix with a rank of $3 \times 4 \times N_x \times N_y \times N_z \times N_t$. The propagator is the inverse of the quark operator, $S_q(m,n) = D^{-1}(m,n)$. It is obtained by solving a linear equation

$$\sum_l D(m,l) S_q(l,n) = \delta_{m,n}, \tag{7}$$

where color and spin indexes are omitted. There are numbers of fermion operators and solver algorithms. The best solver algorithm depends on a type of the fermion operator. The solver algorithm should be implemented so as to be applied to any kind of fermion operators, and simultaneously solver algorithms must be changed easily.

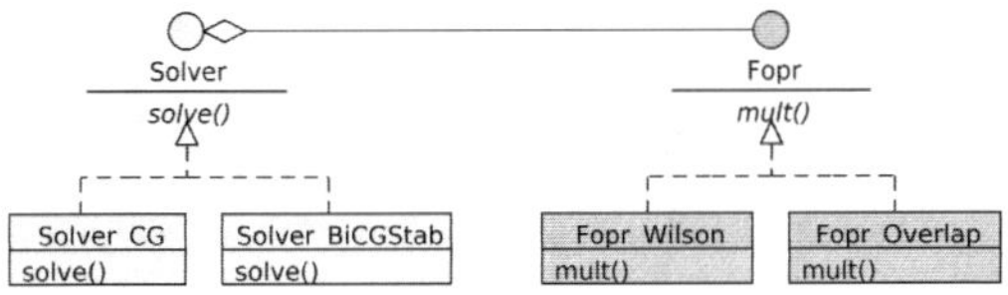

Figure 2. Class diagram of the Bridge pattern applied to a linear solvers.

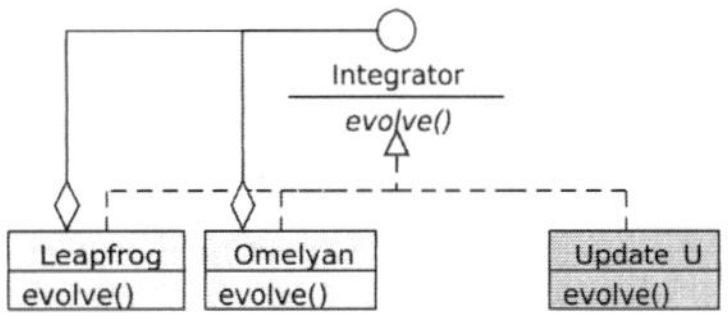

Figure 3. Class diagram of the Composite pattern applied to the HMC integrators.

Figure 2 shows the class diagram of the Bridge pattern based on UML (Unified Modeling Language). The interface of the fermion operator is `Fopr`. `mult()` is a function that applies the fermion operator to a given vector and returns the resultant vector. The practical implementation of `mult()` is given in a subclass of `Fopr`, such as `Fopr_Wilson` for the Wilson fermion and `Fopr_Overlap` for the overlap fermion. Similarly the base class `Solver` defines the interface of the solver algorithms. `solve()` function returns the solution of a linear equation for a given source vector. Again the practical implementation is given in subclasses, `Solver_CG` and `Solver_BiCGStab`. Bridge++ users can select any combination such as `Fopr_Wilson` with `Solver_CG`.

3.4. Composite pattern and HMC integrator

Composite pattern is a design pattern that treats a group of objects as a single instance of an object. It is convenient to implement a tree structure or nested objects. In Bridge++, it is used in the HMC algorithm.

HMC evolves $\{P, U\}$ according to the Hamiltonian equation discretized by a leapfrog integrator,

$$\begin{pmatrix} P(\tau) \\ U(\tau) \end{pmatrix} = V(\tau) \begin{pmatrix} P(0) \\ U(0) \end{pmatrix}, \qquad V(\tau) = \left[V_P\left(\frac{\Delta\tau}{2}\right) V_U(\Delta\tau) V_P\left(\frac{\Delta\tau}{2}\right) \right]^N, \tag{8}$$

where N is the total number of the leapfrog steps and $\Delta\tau = \tau/N$. V describes the evolution,

$$V_P(\Delta\tau)\cdot P(\tau) = P(\tau) + \Delta\tau F(\tau), \qquad V_U(\Delta\tau)\cdot U(\tau) = \exp(iP\Delta\tau)U(\tau). \tag{9}$$

$F(\tau) = \delta S_{\rm lat}/\delta U$ is the force of the conjugate momentum.

Instead of the simple integrator, multi-time step integrator can be adopted. It varies the number of time steps for the action terms, according to the contribution to the force. The integrator is

$$V = \left[V_P^{(F)}\left(\frac{\Delta\tau}{2}\right) V_1(\Delta\tau) V_P^{(F)}\left(\frac{\Delta\tau}{2}\right) \right]^N, \quad V_1 = \left[V_P^{(G)}\left(\frac{\Delta\tau}{2N_1}\right) V_U(\Delta\tau/N_1) V_P^{(G)}\left(\frac{\Delta\tau}{2N_1}\right) \right]^{N_1}, \tag{10}$$

where $V_P^{(G)}$ and $V_P^{(F)}$ are the integrator of P with the gauge and fermion forces, respectively. Adjusting N_1 by monitoring the size of the forces enables us to reduce the step size error, keeping the total numerical cost fixed. In addition, the leapfrog integrator can be replaced with an improved integrator, such as Omelyan integrator, which reduces the finite step size corrections.

Figure 3 shows the class diagram of the Composite pattern applied to the integrator. The base class of the integrator is defined as `Integrator`, which has a virtual method `evolve()`. Its subclass `Update_U` plays the role of V_U. The evolution operators V and V_1 are implemented by `Leapfrog` or `Omelyan`. An object of these classes may have other objects of subclasses of `Integrator`. This relation is represented in Figure 3 as lines with diamond. In analogy to the file system, V play a role of the directory, and V_P and V_U are files in the directory.

Table 1. Features and our development status of the fermion actions. In the last column, 'public' means that is included in the public version and 'testing' has already been implemented and being confirmed.

Action	chiral symmetry	numerical cost	Status
Wilson	violated	middle	public
Clover	violated	middle	public
Twisted mass	violated	middle	testing
Staggered	partially remains	low	testing
Domain-wall	almost exact	high	testing
Overlap	exact	very high	testing

3.5. Parallelization
For a current computational environment, parallelization is inevitable. Large-scale computers consist of nodes which have own memory devices and communicate with each other via high speed network. In addition, each node has many processor cores sharing memory.

Bridge++ works on one node as well as parallel nodes. In Bridge++, we adopt MPI (Message Passing Interface) for a distributed memory parallelization, i.e. for parallel nodes. An inter-node communication, such as broadcast and one-to-one data transfer, is performed through the functions of the Communicator class, which wraps API functions of MPI. If a machine-specific efficient library is available, the implementation of Communicator class is replaced with the one that uses the library. For an environment without MPI, so-called stub implementation of Communicator class is provided.

It is noted that a shared memory parallelization of Bridge++ is now under development. Hybrid parallelization with MPI and multi-threading has been considered. Two multi-thread libraries, Pthread and OpenMP, are compared based on a working implementation of the Wilson fermion operator. While Pthread can accomplish better performance, it is not easy to apply it to the entire code set. OpenMP has been selected as a primary method to multi-threading and performing performance tuning toward incorporating in the next release.

3.6. I/O format
In Bridge++, simulation parameters are given by ASCII files in YAML format. The parameters are held in `Parameter` object. It passes the parameters to each object. Extraction of values from files is handled by a parameter manager class. For the I/O of field objects, several formats are available including ILDG standard format for the gauge configuration. ILDG (International Lattice Data Grid) is an activity to promote sharing configuration data and standardizing data format and description of metadata [8]. The standard output is classified with a verbose level. At present 4 levels are prepared and the output level is selected for each object, such as a linear solver or an observable. There is a extra mode to generate a message with pragma for ILDG metadata generation.

4. Current status and future prospects
The current public version contains major algorithms and observables, though the fermion actions are limited to Wilson and clover fermions. Other fermion actions have been implemented and now being tested. Our status is summarized in Table 1, together with the features of each action. These fermion action can be improved by smearing link variable. The smearing has been implemented in Bridge++.

For linear algorithms, major iterative algorithms, such as CG, BiCGStab, GMRES, are available. Multi-shift solver with CG algorithm and eigensolver with Implicitly restarted Lanczos algorithm are also ready.

The status of performance is as follows. As a typical example, we quote the sustained performance for the HMC with clover fermions. On Hitachi SR16000, the rate to the peak performance is about 5% on 1 node (32 cores). On Blue Gene/Q, the public version runs with 2-3% on 32-nodes, while the latest code under development has improved the most time consuming solver part to more than 10%.

Toward the next public release, we are now incorporating the multi-threading to Bridge++. Performance tuning is now in progress. Active investigation to make use of GPGPU is also underway.

Acknowledgments
In addition to the authors of this paper, this project has been contributed by many of our colleagues, as listed in a document enclosed in the source code. This project is supported by H20 Grant-in-Aid for Scientific Research on Innovative Areas 'Research on the Emergence of Hierarchical Structure of Matter by Bridging Particle, Nuclear and Astrophysics in Computational Science', Joint Institute for Computational Fundamental Science and HPCI Strategic Program Field 5 'The origin of matter and the universe'. The code was developed and tested on Hitachi SR16000 and IBM System Blue Gene/Q at KEK under a support of its Large-scale Simulation Program (No.12/13-15), Hitachi SR16000 at YITP in Kyoto University, K-computer at RIKEN Advanced Institute for Computational Science, HA-PACS at University of Tsukuba under a support for its Interdisciplinary Computational Science Program (No.13a-25) and FX10 at University of Tokyo. This work is supported in part by the Grand-in-Aid for Scientific Research of the Japan (Nos.20105005, 24540250, 25400284).

References

[1] There are many textbooks of lattice QCD simulations, for example, Montvay I and Münster G 1994 *Quantum Fields on a Lattice* (Cambridge: Cambridge Univ. Press); DeGrand T and DeTar C 2006 *Lattice Methods for Quantum Chromodynamics* (Singapore: World Scientific Pub.).
[2] Bridge++ website, `http://bridge.kek.jp/Lattice-code/`.
[3] Wilson K G 1974 Confinement of Quarks *Phys. Rev.* D **10** 2445.
[4] `http://www.physics.utah.edu/~detar/milc/`
[5] `http://qcdoc.phys.columbia.edu/cps.html`
[6] `http://usqcd.jlab.org/usqcd-docs/chroma/`
[7] `http://nio-mon.riise.hiroshima-u.ac.jp/~LTK/LTKf90.html`
[8] `http://www.usqcd.org/ildg/`
[9] `http://www.stack.nl/~dimitri/doxygen/index.html`
[10] Gamma E, Helm R, Johnson R, and Vlissides J 1995 *Design Patterns: Elements of Reusable Object-Oriented Software.* (Boston: Addison-Wesley).
[11] Trott A and Shalloway J R 2004 *Design Patterns Explained: A New Perspective on Object-Oriented Design, 2nd ed.* (Boston: Addison-Wesley Professional)

Recent developments on FORM

Takahiro Ueda[1] and Jos Vermaseren[2]

[1] Institut für Theoretische Teilchenphysik, Karlsruhe Institute of Technology (KIT), D-76128 Karlsruhe, Germany

[2] Nikhef Theory Group, Science Park 105, 1098 XG Amsterdam, The Netherlands

E-mail: `takahiro.ueda@kit.edu`, `t68@nikhef.nl`

Abstract. We report on recent developments on the open source computer algebra system FORM. Especially we focus on the code optimization implemented after the release of FORM version 4.0 in March 2012.

1. Introduction

FORM [1] is one of widely used programs for symbolic manipulation of huge formulas with an arbitrary number of terms appearing in theoretical calculations based on perturbative quantum field theories. Since March 2012 FORM version 4.0 has been released. The new features introduced in this version include:

- Polynomial factorization. FORM has a functionality of factorization of multivariate polynomials at three levels: function arguments, expressions and $-variables.
- Rational function arithmetic. FORM has a mode in which rational functions are considered as coefficients of terms in sorting. There are also functions which provide basic operations on polynomials: polynomial division, polynomial GCD and so on.

There are parallel versions of FORM, which automatically distribute the computation over many CPUs. One is TFORM [2], which is based on POSIX threads and works with multi-core processors. The other is PARFORM [3, 4] based on MPI, which can run on computer clusters with fast network connections. The above newly introduced features in the version 4.0 should also work with the parallel versions.

FORM is continuously being developed even after the release of the version 4.0, for some bug fixes and adding more features[1]. One of the recently added features, which will be considered as a part of the next release FORM 4.1, is code optimization.

After performing symbolic manipulation on formulas in an analytic way, one occasionally needs to export the result as a program code, e.g., in Fortran or C, for further numerical computations such as Monte Carlo integrations. If a lengthy formula will be evaluated millions of times, it is worth compactifying the expression such that the number of operations needed to evaluate it becomes as small as possible for fast computation, even if the optimization needs a

[1] One can access to the Git repository from `https://github.com/vermaseren/form`. For example, a bug in the disk-to-disk sorting, which may give a wrong result, was fixed on 23 July 2013. On the other hand, a new feature introduced on 13 August 2013 changed the store file format and executables compiled from the sources after this date cannot read old store files.

non-negligible time. FORM 4.1 will be equipped with the code optimization to reduce the cost of evaluations of multivariate polynomials, which includes an implementation of Monte Carlo tree search (MCTS) for finding a suitable multivariate Horner scheme [5, 6].

2. Code optimization in FORM

As the first example, let us consider the optimization of a polynomial F of three variables x, y and z:

```
Symbols x,y,z;
ExtraSymbols vector,v;
Local F = 6*y*z^2 + 3*y^3 - 3*x*z^2 + 6*x*y*z - 3*x^2*z + 6*x^2*y;
Format Fortran;
Format O1,stats=on;
Print;
.end
```

The special command for the optimization here is `Format O1,stats=on`. This enables the code optimization in printing the expression with reporting statistics. FORM has several optimization options and parameters. FORM also defines optimization levels `O0`, `O1`, `O2` and `O3` for user convenience, which are actually certain combinations of the options and parameters. Increasing the optimization level makes FORM try to give the output with fewer operations, at the cost of more time. The code optimization is internally applied to polynomials containing only numbers and symbols. Other objects are converted into symbols temporarily, which are substituted back at the end. For practical use generating a complete code, `#Optimize` and `#Write` preprocessor instructions are provided in order to give more control to users, which we will not discuss here. The declaration `ExtraSymbols vector,v` specifies the name of the array that will be introduced for storing intermediate results during the evaluation.

The output of the above example is as follows:

```
      v(1)=y*z
      v(2)= - z + 2*y
      v(2)=x*v(2)
      v(3)=z**2
      v(1)=v(2) - v(3) + 2*v(1)
      v(1)=x*v(1)
      v(2)=y**2
      v(2)=2*v(3) + v(2)
      v(2)=y*v(2)
      v(1)=v(2) + v(1)
      F=3*v(1)
*** STATS: original  1P 16M 5A : 23
*** STATS: optimized 0P 10M 5A : 15
```

The numerical evaluation of F needs 18 multiplications and 5 additions, i.e., 23 operations in total, if it is done in the original expanded form. FORM tries to reduce the number of total operations assuming that the cost of addition and multiplications are equal. In the optimized form, the number of operations reduces to 15 operations in total, using the array v for temporary variables. By changing the optimization level to `O2`, one will obtain an output with 14 operations. `O3` gives an output with 12 operations.

3. Algorithms

The predefined optimization levels in FORM are summarized as follows:

- `O0`: No optimization.
- `O1`: The occurrence and reverse occurrence orders followed by common subexpression elimination are tried for a Honer scheme.
- `O2`: Besides `O1`, an extra "greedy" optimization is performed to find more common subexpressions at the end.
- `O3`: Monte Carlo tree search is used for finding a good Horner scheme with common subexpressions. At the end the greedy optimization is performed.

Here we will explain each algorithm.

3.1. Multivariate Horner schemes

For optimizing evaluations of univariate polynomials, it is well-known that a *Horner scheme* gives an efficient form for numerical computations:

$$a(x) = \sum_{i=0}^{n} a_i x^i = a_0 + x\Big(a_1 + x\big(a_2 + x(\cdots + x \cdot a_n)\big)\Big). \quad (1)$$

This form takes n multiplications and n additions to calculate a value for a dense polynomial of degree n.

Horner schemes could be extended for multivariate polynomials rather straightforwardly. One can choose one of variables of a polynomial and apply a Horner scheme to it, regarding other variables as constants. Then one can choose another variable from the remaining variables for the next Horner scheme. This is repeated until a Horner scheme is applied to all the variables. However, the number of operations that one can obtain after applying a multivariate Horner scheme depends on in which order the variables are processed.

3.1.1. Occurrence order Traditionally, the *occurrence order* has been often used for multivariate Horner schemes. All variables are ordered in how many times they appear in a polynomial. Applying a Horner scheme to the most-occurring variable gives the largest decrease of the number of operations at each step, because it factors out the most-occurring variable from the polynomial.

On the other hand, one may consider the *reverse occurrence order*. Applying a Horner scheme to the least-occurring variable at each step, the result may contain more common subexpressions within the inner parentheses, which can be detected by *common subexpression elimination* (CSE). Even though this order usually results more operations as a Horner scheme, this order could give fewer operations at the end if it is used in a combination with CSE.

3.1.2. Monte Carlo tree search Both occurrence order and reverse occurrence order may not give the optimal result. For a multivariate polynomial with N variables, the number of possible orders is $N!$. One could try all the orders to find the optimal one when N is small, but obviously it is not practical for large N. Because the orders of the variables can be represented by a *search tree*, one may expect a sophisticated tree search algorithm in computer science could be applied to this problem.

Recently, a new tree search technique, Monte Carlo tree search (MCTS) [7], which was developed in artificial intelligence and game theory, was applied to this problem and it turned out that MCTS actually can give a good order within a reasonable amount of time [5]. Roughly speaking, MCTS picks up some samples from all the orders, at first randomly, but it stochastically gives priority to the neighborhood of good solutions as more samples are collected. It works efficiently if one can expect that good solutions are clustered in branches of the search tree. Note that MCTS is a Monte Carlo algorithm, therefore it may give a different result at each run.

Table 1. FORM run time, gcc compilation time (compile) and the time to evaluate the compiled formula 10^5 times (run) for the resultant with $m = 7$ and $n = 6$. All times are in seconds on an Opteron 2.6GHz processor. The gcc version is 4.6.2.

	Format O0	Format O1	Format O2	Format O3
Operations	587880	71262	55685	36146
FORM time	0.12	1.66	65.43	2398
gcc -O0 compile	29.02	6.33	5.64	3.36
run	119.66	13.61	12.24	7.52
gcc -O1 compile	3018.6	295.96	199.47	80.82
run	24.30	6.88	6.12	3.58
gcc -O2 compile	3104.4	247.60	163.79	65.21
run	21.09	7.00	6.22	3.93
gcc -O3 compile	3125.4	276.77	179.24	71.02
run	21.02	6.95	6.19	3.93

3.2. Common subexpression elimination

In large expressions, the same subexpressions may appear many times. Most of them can be detected and eliminated by a method of *common subexpression elimination* (CSE), which takes time proportional to the input size [8]. Furthermore, for common subexpressions that cannot be detected by CSE, a "greedy" optimization is implemented. This is more or less similar to the algorithm in [9, 10] and much slower than CSE.

4. Benchmark

We performed two kind of benchmark tests of the implemented optimization. One is optimization of the resultant of two polynomials, coming from an example in [11]:

$$\sum_{i=0}^{n} a_i x^i, \qquad \sum_{j=0}^{m} b_j x^j, \tag{2}$$

which is the determinant of a $(m+n) \times (m+n)$ matrix. For this case, we measured the time used in FORM for the optimization, the time used in the compilation and the time to evaluate the compiled formula 10^5 times as well as the number of operations in the optimized code.

Table 1 shows a result for $m = 7$ and $n = 6$, which gives a 13×13 determinant with 43166 terms. One can see that `O2` gives a better optimization than `O1` with respect to the number of operations, and `O3` gives an even better optimization. On the other hand, increasing the optimization level takes more time for the optimization. One can also see that the compilation time as well as the time to evaluate the compiled formula decreases when increasing the FORM optimization level, in all the compiler optimization levels. This result suggests that the best choice of the FORM and compiler optimizations depends on the number of function evaluations needed in the specific problem, but that FORM is more efficient at it than the compiler.

The other benchmark comes from the GRACE system [12, 13], which can generate matrix elements containing one loop diagrams. Each formula corresponds to each diagram and is expressed in terms of Feynman parameters introduced for the loop momentum integration. Within the tensor reduction method used in GRACE, the coefficient of each combination of Feynman parameters must be evaluated separately. In this case the best way is to optimize these coefficients simultaneously, because they may share common subexpressions.

Table 2. Results for three formulas from the GRACE system. For the cases that the number of variables is presented in the form of a sum, the first number is the number of Feynman parameters and the coefficients of combinations of the Feynman parameters were simultaneously optimized. For the left column of HEP(σ), `Bracket` statement was not used. Therefore 35 coefficients with 11 parameters were merged into an expression with 15 parameters. This was done just for comparison, even though it does not make a sense for the GRACE system.

	HEP(σ)	HEP(σ)	F_{13}	F_{24}
Variables	15	4+11	5+24	5+31
Expressions	1	35	56	56
Terms	5717	5717	105114	836010
Format O0	47424	33798	812645	5753030
Format O1	6099	5615	71989	391663
Format O2	4979	4599	46483	233445
Format O3	3423	3380	41666	195691

For this purpose, one can combine two or more expressions into one expression and use the `Bracket` statement just before the optimization:

```
Symbol u;
Local F, G;
Local H = u * F + u^2 * G;
Bracket u;
```

where symbol `u` is used for a label to distinguish the original expressions in `F` and `G`. If `Bracket u` is used, FORM does not perform code optimization with respect to `u`. Therefore the optimized code for `H` still contains `u`. One can split `H` into `F` and `G` by using the exponent of `u`.

For the benchmark test, we picked up three formulas of different sizes. We will call them as HEP(σ), F_{13} and F_{24}. Table 2 shows the result of simultaneous optimizations for the coefficients of combinations of Feynman parameters in these formulas.

In some cases, like expressions appearing in physics, the user can do some work before using the optimization in FORM. One may have knowledge about the problem that can be used for simplification. In the above GRACE example, a set of shifts of variables such as

$$x_i \to x_i + ax_j, \qquad x_i \to x_i + a, \tag{3}$$

can give a significant improvement for the result.

5. Comparison with other programs

We have compared the code optimization implemented in FORM with those of other programs we had access to, with respect to the number of operations in the output. It is summarized in Table 3, which shows that FORM `O3` can give a better result than any other programs.

6. Conclusions

FORM is continuously being developed for bug fixes and adding more features. Recently code optimization was implemented as a new feature. It uses various simplification techniques such as a multivariate Horner scheme, CSE and so on, and is useful when one needs to numerically evaluate lengthy formulas repeatedly or one wants to shrink the sizes of executables. MCTS used in `O3` option gives a better optimization than anything we could find in the literature at some cost of time. This feature will become available as a part of the next version FORM 4.1.

Table 3. The numbers of operations after optimization by various programs. For m-n resultants and a formula appearing in high energy physics (the same one as the left column of HEP(σ) in Table 2). The number for the 7-5 resultant with 'Maple tryhard' is taken from [11]. For the 7-4 resultant they obtain 6707 operations, which must be due to a different way of counting. The same holds for the 7-6 resultant as [11] (HG + cse) starts with 601633 operations. For FORM O3, the parameter C_p was set as $C_p = 0.07$ instead of its default value, and 10×400 tree expansions were used.

	7-4 resultant	7-5 resultant	7-6 resultant	HEP(σ)
Original	29163	142711	587880	47424
FORM O1	4968	20210	71262	6099
FORM O2	3969	16398	55685	4979
FORM O3	3015	11171	36146	3524
Maple	8607	36464	-	17889
Maple tryhard	6451	$\mathcal{O}(27000)$	-	5836
Mathematica	19093	94287	-	38102
HG + cse	4905	19148	65770	-
Haggies [14]	7540	29125	-	13214

References

[1] Kuipers J, Ueda T, Vermaseren J A M and Vollinga J 2013 FORM version 4.0 *Comput. Phys. Commun.* **184** 1453 (*Preprint* 1203.6543 [cs.SC])

[2] Tentyukov M and Vermaseren J A M 2010 The Multithreaded version of FORM *Comput. Phys. Commun.* **181** 1419 (*Preprint* hep-ph/0702279)

[3] Tentyukov M, Fliegner D, Frank M, Onischenko A, Retey A, Staudenmaier H M and Vermaseren J A M ParFORM: Parallel Version of the Symbolic Manipulation Program FORM (*Preprint* cs/0407066)

[4] Tentyukov M, Staudenmaier H M and Vermaseren J A M, 2006 ParFORM: Recent development *Nucl. Instrum. Meth.* A **559** 224

[5] Kuipers J, Vermaseren J A M, Plaat A and van den Herik H J 2012 Improving multivariate Horner schemes with Monte Carlo tree search (*Preprint* 1207.7079 [cs.SC])

[6] Kuipers J and Vermaseren J A M (*in preparation*)

[7] Kocsis L and Szepesvári C 2006 Bandit based Monte-Carlo Planning *LNCS* **4212** 282

[8] Aho A V, Sethi R and Ullman J D 1986 *Compilers: Principles, Techniques, and Tools* (Addison-Wesley)

[9] Breuer M A 1969 Generation of optimal code for expressions via factorization *ACM Commum.* **12** 333

[10] van Hulzen J A 1983 Code optimization of multivariate polynomial schemes: A pragmatic approach *LNCS* **162** 286

[11] Leiserson C E, Li L, Maza M M and Xie Y 2010 Efficient Evaluation of Large Polynomials *LNCS* **6327** 342

[12] Yuasa F *et al* 2000 Automatic computation of cross-sections in HEP: Status of GRACE system *Prog. Theor. Phys. Suppl.* **138** 18 (*Preprint* hep-ph/0007053)

[13] Fujimoto J *et al* 2006 GRACE with FORM *Nucl. Phys. Proc. Suppl.* **160** 150

[14] Reiter T 2010 Optimising Code Generation with haggies *Comput. Phys. Commun.* **181** 1301 (*Preprint* 0907.3714 [hep-ph])

Numerical multi-loop calculations with SecDec

Sophia Borowka, Gudrun Heinrich[1]
Max Planck Institute for Physics, Föhringer Ring 6, 80805 Munich, Germany

Abstract. The new version 2.1 of the program SecDec is described, which can be used for the factorisation of poles and subsequent numerical evaluation of multi-loop integrals, in particular massive two-loop integrals. The program is not restricted to scalar master integrals; more general parametric integrals can also be treated in an automated way.

1. Introduction

In the absence of smoking gun signals of New Physics at the LHC so far, precision measurements play a crucial role in the search for less direct signs of New Physics and in exploring the Higgs sector. Certainly this is only possible if precise theory predictions are available. In most cases, this means that next-to-leading order (NLO) calculations are necessary, ideally matched to a parton shower. However, there is a number of examples where NNLO precision is required. While in the case of jet production the main challenge at NNLO consists in the treatment of the doubly unresolved real radiation part, processes involving massive particles – e.g. top quarks or W/Z bosons – contain another bottleneck, given by the two-loop integrals entering the virtual corrections. The recent result for $t\bar{t}$ production at NNLO [1] is based on a semi-numerical representation of the integrals [2], while fully analytical representations for most of the master integrals needed for this process have been worked out in [3, 4, 5, 6, 7]. Planar two-loop master integrals entering the production of two massive vector bosons at NNLO have been calculated in [8], and in the high energy limit in [9] for the case of W pair production.

However, non-planar massive two-loop integrals which cannot be expressed entirely in terms of generalized harmonic polylogarithms still pose a problem for a fully analytical evaluation. Numerical methods on the other hand do not have these difficulties. For the latter the problem rather lies in the isolation of the singularities, as well as in numerical efficiency and accuracy.

The program SecDec, presented in [10, 11], performs the task of isolating singularities which, if regulated by dimensional regularisation, appear as poles in $1/\epsilon$, in an automated way, based on the algorithm of sector decomposition [12, 13, 14]. Other implementations of sector decomposition into public programs are also available [15, 16, 17]. However, the latter are more or less restricted to the Euclidean region, while SecDec can deal with physical kinematics.

In this talk the new features of the program SecDec 2.1 will be presented, along with some results.

[1] Speaker; presented at the conference ACAT 2013, Beijing, China, May 2013.

2. Structure, usage and new features of SecDec version 2.1

2.1. Structure

As can be seen from Fig. 1, the program contains two sections, one for loop integrals, one for more general parametric functions (corresponding to the directories *loop* and *general*). The *loop* part has been extended in version 2.1 to be able to treat parametric integrals which are not in the canonical form as obtained directly from standard Feynman parametrisation. They can have a different format coming for example from variable transformations and/or analytical integrations over some Feynman parameters. Hence, as these functions differ from the standard representation which the program would derive from the propagators in an automated way, they have to be defined in an input file by the user. The detailed syntax to define these functions is given in subsection 2.4.1. Contour deformation is available for these functions, as they are assumed to originate from a Feynman integral structure where poles on the real axis are protected by the infinitesimal $i\delta$ prescription.

The part in the directory *general* can deal with even more general parametric functions, where the integrand can consist of a product of arbitrary length of polynomial functions to some power. However, these functions should have only endpoint singularities (i.e. dimensionally regulated singularities at the integration boundaries). Contour deformation is not available in this case because the correct sign of the imaginary part steering the deformation into the complex plane cannot be inferred if the assumption of an underlying loop integral structure is dropped.

The procedure of iterated sector decomposition and subsequent treatment is the same for all the different types of input functions, and is described in [10, 11].

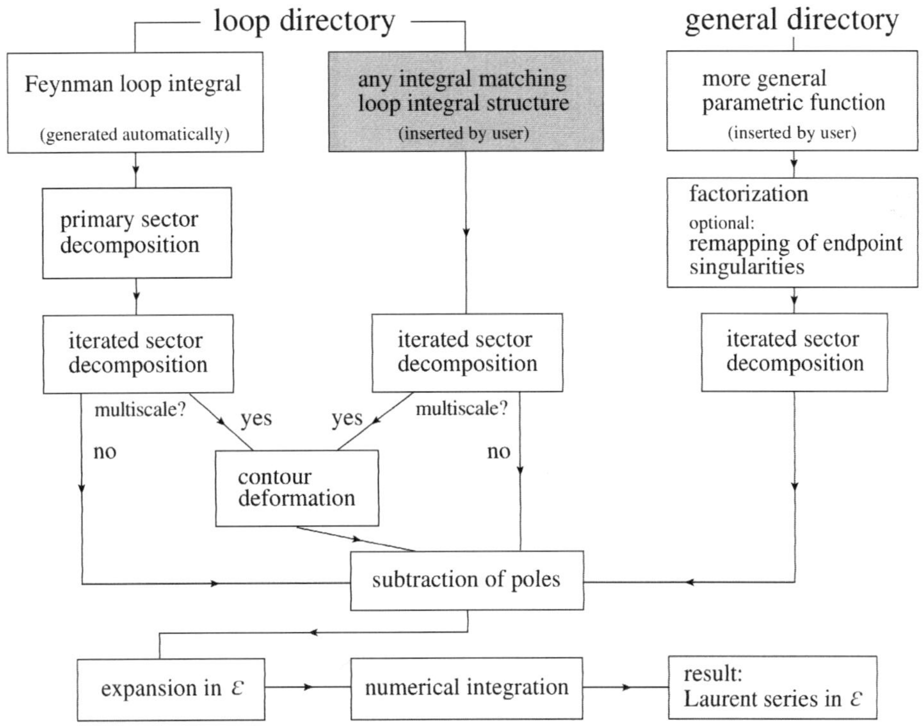

Figure 1. Flowchart showing the structure of the program SecDec.

2.2. Installation
The current version of the program can be downloaded from
`http://projects.hepforge.org/secdec`.

Unpacking the tar archive via *tar xzvf SecDec-2.1.2.tar.gz* will create a directory called `SecDec-2.1.2`. Changing to the `SecDec-2.1.2` directory and running *./install* will compile the CUBA library [18] needed for the numerical integration.

Prerequisites are Mathematica, version 6 or above, Perl (installed by default on most Unix/Linux systems) and a C++ compiler. If the Fortran option is used, a Fortran compiler is obviously also required.

2.3. Usage
There are two files to be edited: a text file `param*.input` where the parameters are set, and a Mathematica file `template*.m` where the funtion(s) to be integrated are defined. In more detail:

(i) In the `loop` directory, edit the files `paramloop.input` and `templateloop.m` if you want to compute a Feynman loop integral in a fully automated way. If you would like to define a set of own functions rather than a standard loop integral, take the files `paramuserdefined.input` and `templateuserdefined.m` as a starting point. In the `general` directory, the analogous files are `param.input` and `template.m`.
Let us call the files edited by the user `myparamfile.input` and `mytemplatefile.m`.

(ii) Execute the command *./launch -p myparamfile.input -t mytemplatefile.m* in the shell. If you add the option *-u* in the `loop` directory, user defined functions are computed.
If your files *myparamfile.input, mytemplatefile.m* are in a different directory, say, *myworkingdir*, use the option **-d myworkingdir**, i.e. the full command then looks like *./launch -d myworkingdir -p myparamfile.input -t mytemplatefile.m*, executed from the directory `SecDec/loop` or `SecDec/general`.

(iii) Collect the results. Depending on whether you have used a single machine or submitted the jobs to a cluster, the following actions will be performed:
- If the calculations are done sequentially on a single machine, the results will be collected automatically (via the corresponding `results*.pl` called by `launch`). The output file will be displayed with your specified text editor.
- If the jobs have been submitted to a cluster, when all jobs have finished, execute the command *./results.pl [-d myworkingdir -p myparamfile]* in the general, and *./resultsloop.pl [-d myworkingdir -p myparamfile]* or *./resultsuserdefined.pl [-d myworkingdir -p myparamfile]* in the loop directory, respectively. This will create the files containing the final results in the `graph` subdirectory specified in the input file.

(iv) After the calculation and the collection of the results is completed, you can use the shell command *./launchclean[graph]* to remove obsolete files.

More details about the usage can be found in Ref. [19], and also in the documentation coming with the program.

2.4. New Features
Version 2.1 of SECDEC contains the following new features, also partially illustrated by examples in Section 3.

2.4.1. Evaluation of user-defined functions in the loop part: If the user would like to calculate a "standard" loop integral, it is sufficient to specify the propagators, and the program will construct the integrand in terms of Feynman parameters automatically. However, analytical manipulations on the integrand before starting the iterated sctor decomposition can be helpful when dealing with complicated integrals. For example, integrating out one Feynman parameter analytically reduces the number of integration variables for the subsequent Monte Carlo integration and therefore can improve numerical efficiency. This implies that the constraint $\delta(1-\sum_i x_i)$ has been used already to achieve a convenient parametrisation, and therefore no primary sector decomposition to eliminate the δ-constraint is needed anymore. In such a case, the user can skip the primary sector decomposition step and insert the integrand functions directly into the Mathematica input file.

The detailed syntax is as follows. The user-defined functions should be polynomial in the Feynman parameters and can also involve kinematic invariants, i.e. should be structurally similar to the functions of type $\mathcal{U}$ and $\mathcal{F}$ (see e.g. eq. (4)). They can be raised to some power which does not have to be an integer. These functions in addition can be multiplied by an arbitrary "numerator" function. The only condition the latter must fulfill is that it should not contain any singularities or thresholds. In case the functions $\mathcal{U}$ and $\mathcal{F}$ contain thresholds, a deformation of the integration contour into the complex plane will be performed. The setup is such that the integration contour will be formed based on the function $\mathcal{F}$. The list of user-defined functions must be inserted into `mytemplatefile.m` using the following syntax
`functionlist=`{*function_1,function_2,...,function_i,...*};
with *function_i*={*# of function*,{*list of exponents*},{{*function* $\mathcal{U}$, *exponent of* $\mathcal{U}$, *decomposition flag*},{*function* $\mathcal{F}$,*exponent of* $\mathcal{F}$,*decomposition flag*}}, *numerator*}.
The *# of function* is an integer labelling the function, where the default is sequential numbering.

Each entry in the comma separated *list of exponents* corresponds to an exponent of a Feynman parameter occurring as a monomial in the Feynman integral. The *decomposition flag* should be `A` if no iterated sector decomposition is desired, and `B` if the function needs further decomposition. The *numerator* may contain several functions, with different exponents, as long as the functions are non-singular. An example including detailed comments can be found in `/loop/templateuserdefined.m` and `/loop/paramuserdefined.input`.

2.4.2. Tensor integrals: For the computation of tensor integrals, where the tensor is contracted with external momenta and/or loop momenta, the construction of the Feynman integral via topological cuts needs to be switched off, which corresponds to `cutconstruct=0`.
To define the numerator in `mytemplatefile.m`, each scalar product of loop momenta contracted with either external momenta or loop momenta should be given as an entry of a list:
`numerator=`{*prefactor*, *comma separated list of scalar products*}. For example, a numerator of the form $2\,(k_1\cdot p_1)(k_1\cdot k_2)$, where p_1 is an external momentum and the k_i are loop momenta, should be given as `numerator=`$\{2,\ k1*p1,\ k1*k2\ \}$.

2.4.3. Error estimates: When dealing with complicated integrands it can happen that the error given by the Monte Carlo integration program – which is based on the number of sampling points only – underestimates the true error. The numerical integrators contained in the CUBA library [18, 20] give an estimate how trustworthy the stated error is. The new SECDEC version 2.1 collects the maximal error probability for each computed order in the Laurent series in ϵ and writes it to the result files `*.res`. In the generic case, the information on the reliability of the stated error is given as a probability with values between 0 and 1. If the integrator returns a value larger than one, the integration has not come to successful completion, and a warning is written to the result files. This feature should enable the user to assess the reliability of the numerical result, respectively the given error estimate.

3. Examples and Results

3.1. Non-planar massive two-loop diagrams entering NNLO $t\bar{t}$ production

The sector decomposition algorithm is described in detail in [12, 21], while the general structure of the program SecDec is described in [10, 11]. Therefore we concentrate mainly on the new features and results here.

The most complicated master topologies occurring in the two-loop corrections to $t\bar{t}$ production in the gg channel are the non-planar seven-propagator integrals shown in Fig.2.

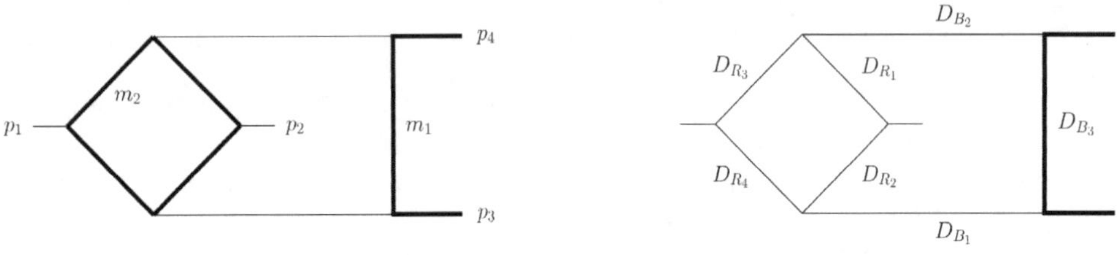

(a) ggtt1 (b) ggtt2

Figure 2. Massive non-planar two-loop box diagrams entering the heavy (a) and light (b) fermionic correction to the $gg \to t\bar{t}$ channel; the bold lines denote massive particles.

Analytic results for the integral containing a sub-diagram with a massive loop (called *ggtt1* here), are not available, while analytic results for the diagram corresponding to massless fermionic corrections in a sub-loop (called *ggtt2* here) have become available only very recently [3]. However, the numerical evaluation of *ggtt1* with SecDec is much easier than the one of *ggtt2*, due to its less complicated infrared singularity structure. While the leading poles of *ggtt2* are of order $1/\epsilon^4$, and intermediate expressions during sector decomposition contain (spurious) poles where the degree of divergence is higher than logarithmic, the integral *ggtt1* is finite and free from spurious poles. Therefore we can evaluate *ggtt1* with SecDec 2.1 using the fully automated setup. In contrast, for *ggtt2* it turned out to be advantageous to make some analytical manipulations beforehand. In particular, it was useful to perform one parameter integration analytically before feeding the integral into the decomposition and numerical integration algorithm. Further, we introduced special transformations to reduce the occurrence of spurious singularities, which are described in detail in [19]. These manipulations lead to functions which were not in the "standard form" of Feynman parameterised loop integrals anymore. Therefore this entailed the development of a setup for "non-standard" parametric functions, which has been made available for the user as described in section 2.4.1, as it can be beneficial in other contexts as well.

The expression for the scalar integral *ggtt2* in momentum space is given by

$$\mathcal{G}_{ggtt2} = \left(\frac{1}{\mathrm{i}\pi^{\frac{D}{2}}}\right)^2 \int \frac{\mathrm{d}^{\mathrm{D}}k_1\, \mathrm{d}^{\mathrm{D}}k_2}{D_{R_1}D_{R_2}D_{R_3}D_{R_4}D_{B_1}D_{B_2}D_{B_3}} \tag{1}$$

where $\mathrm{D} = 4 - 2\epsilon$. Integrating out the loop momentum k_2 (corresponding to massless sub-loop) first, we are left with an expression containing k_1 and external momenta only:

$$\mathcal{I}_R = \frac{1}{\mathrm{i}\pi^{\mathrm{D}/2}} \int \frac{\mathrm{d}^{\mathrm{D}}k_2}{D_{R_1}D_{R_2}D_{R_3}D_{R_4}} = \Gamma(2+\epsilon) \int \prod_{i=1}^{4} \mathrm{d}x_i \delta(1 - \sum_{j=1}^{4} x_j) \mathcal{F}(\vec{x}, k_1)^{-2-\epsilon} \;, \tag{2}$$

with

$$-\mathcal{F}(\vec{x}, k_1) = (k_1 - p_3)^2 x_1 x_2 + (k_1 + p_1 + p_4)^2 x_1 x_3 + (k_1 + p_2 + p_4)^2 x_2 x_4 + (k_1 + p_4)^2 x_3 x_4 \;.$$

We eliminate the δ-function in eq. (2) with the substitution

$$x_1 = t_2(1-t_3)\ , \qquad x_2 = t_1 t_3\ , \qquad x_3 = (1-t_1)t_3\ , \tag{3}$$

which allows us to integrate out the parameter t_3 analytically. Then we combine the expression for $\mathcal{I}_R$ with the remaining k_1-dependent propagators to obtain, after integrating out k_1,

$$\begin{aligned}\mathcal{G}_{NP} =& \frac{2\,\Gamma(3+2\epsilon)\Gamma^2(1-\epsilon)}{\epsilon\qquad \Gamma(1-2\epsilon)} \int_0^1 \mathrm{d}t_1 \int_0^1 \mathrm{d}t_2\ \times \\ & \prod_{i=1}^{4}\int_0^1 \mathrm{d}z_i\, z_4^{1+\epsilon}\,\delta(1-\sum_{j=1}^{4} z_j)\ \mathcal{F}_{\mathcal{NP}}(\vec{z},t_1,t_2)^{-3-2\epsilon}\ \mathcal{U}_{\mathcal{NP}}(\vec{z})^{1+3\epsilon}\ ,\end{aligned} \tag{4}$$

$$\begin{aligned}\mathcal{U}_{\mathcal{NP}}(\vec{z}) = \sum_{j=1}^{4} z_j\ ,\ \mathcal{F}_{\mathcal{NP}}(\vec{z},t_i) = -s_{12}z_2z_3 - Tz_1z_4 - S_1z_2z_4 - S_2z_3z_4 + m^2z_1(z_1+z_4Q)\ ,\\ T = s_{13}\bar{t}_1t_2 + s_{23}t_1\bar{t}_2\ ,\ S_1 = s_{12}t_1t_2\ ,\ S_2 = s_{12}\bar{t}_1\bar{t}_2\ ,\ Q = t_1\bar{t}_2 + \bar{t}_1t_2\ ,\ s_{ij} = (p_i+p_j)^2\ .\end{aligned} \tag{5}$$

We use the shorthand notation $\bar{t}_i = 1 - t_i$. Now we eliminate the δ-function in eq. (4) by performing a primary sector decomposition [12] in $z_1, \ldots, z_4$.

Note that, due to the substitutions made in (3), the integral $\mathcal{G}_{NP}$ can have singularities both at zero and one in t_1 and t_2. With the sector decomposition algorithm, only singularities at zero are factorised automatically. Consequently, we remap the singularities located at the upper integration limit to the origin of parameter space by splitting the integration region at $\frac{1}{2}$ and transforming the integration variables to remap the integration domain to the unit cube [21]. This procedure results in 12 integrals, some of which already being finite, such that no subsequent sector decomposition is required. Some of the integrals however lead to singularities of the type $\int_0^1 dx\, x^{-2-\epsilon}$ after sector decomposition, which we call *linear* divergences. These singularities are spurious, so it would be better to avoid this type of singularity from scratch. In the following section we sketch a method which has proven useful in this respect [19].

3.1.1. Removal of double linear divergences via backwards transformation: To explain the type of transformation advocated here, we use as an example the following function, which is part of eq. (4) after primary sector decomposition

$$\begin{aligned}\mathcal{F}(\vec{z}) &= -s_{12}z_3 - t\,z_1z_4 - s_1\,z_4 - s_2\,z_3z_4 + m^2z_1(z_1 + z_4(z_2\bar{z}_5 + z_5\bar{z}_2))\\ t &= s_{13}\bar{z}_5z_2 + s_{23}z_5\bar{z}_2\ ,\ s_1 = s_{12}z_5z_2\ ,\ s_2 = s_{12}\bar{z}_5\bar{z}_2\ .\end{aligned} \tag{6}$$

Concerning the Feynman parameters, we can identify the following structure in eq. (6):

$$\mathcal{F}(\vec{z}) = z_4\Big(P(\vec{z}_{14}) + z_1Q(\vec{z}_{14})\Big) + R(\vec{z}_{14})\ , \tag{7}$$

where P, Q and R are polynomials of arbitrary degree of Feynman parameters and kinematic invariants, with $\vec{z}_{14} = \{z_2, z_3, z_5\}$, i.e. P, Q and R do not depend on z_1 and z_4. In eq. (7), all terms multiplied by the Feynman parameter z_1 are also multiplied by the Feynman parameter z_4. Hence, the sector decomposition method can be applied "backwards". To explain this in more detail, consider the following function:

$$J = \int_0^1 \mathrm{d}z_1 \ldots \mathrm{d}z_5\, z_4^{-1-\epsilon}\,[z_4P + z_1Q + R]^{-3-2\epsilon}\,[\underbrace{\Theta(z_1-z_4)}_{(1)} + \underbrace{\Theta(z_4-z_1)}_{(2)}] \tag{8}$$

Now we substitute $z_4 = z_1 t_4$ in sector (1) and $z_1 = z_4 t_1$ in sector (2), to obtain, after renaming again t_i into z_i

$$J = \int_0^1 \mathrm{d}z_1 \ldots \mathrm{d}z_5 \, z_4^{-1-\epsilon} \left[z_4 P + z_1 Q + R \right]^{-3-2\epsilon} \tag{9}$$

$$= \int_0^1 \mathrm{d}z_1 \ldots \mathrm{d}z_5 \, z_4^{-1-\epsilon} z_1^{-\epsilon} \left[z_1 (z_4 P + Q) + R \right]^{-3-2\epsilon} \tag{10}$$

$$+ \int_0^1 \mathrm{d}z_1 \ldots \mathrm{d}z_5 \left[z_4 (P + z_1 Q) + R \right]^{-3-2\epsilon} . \tag{11}$$

We observe that the term in (11) is the same as in eq. (7). Therefore we can replace it by the expressions (9) minus (10). Doing so, the effect is twofold: The degree of the polynomial in $(z_1 z_4)$ is reduced in eq. (9), and in eq. (10) z_4 is traded for z_1 to multiply P and Q, which can be beneficial in view of further decompositions.

After all transformations of this type we arrive at a total of 15 functions partly needing an iterated sector decomposition. Making use of the new feature of *user-defined functions* described above, we were able to compute the *ggtt2* diagram much more efficiently than with the standard setup.

3.1.2. The ggtt2 diagram: Numerical results for the *ggtt2* diagram are shown in Fig. 3, where we used the numerical values $p_3^2 = p_4^2 = m^2 = 1, s_{23} = -1.25, s_{13} = 2m^2 - s_{12} - s_{23}$, and we extract an overall factor of $-16\,\Gamma(1+\epsilon)^2$. We only show results for the finite part here, as it is the most complicated one and therefore more interesting than the pole coefficients.

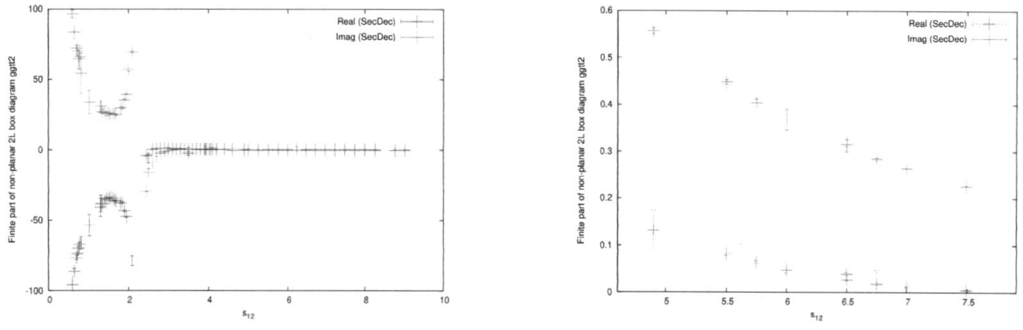

(a) finite part, including threshold (b) finite part, beyond threshold

Figure 3. Results for the finite part of the scalar integral *ggtt2*, (a) for a larger kinematic range, (b) zoom into a region further away from threshold. The vertical bars denote the numerical integration errors.

3.1.3. The ggtt1 diagram: Numerical results for the diagram *ggtt1* (see Fig. 2(a)) are shown in Fig. 4 for both the scalar integral and an irreducible rank two tensor integral. For the results shown in Fig. 4 we used the numerical values $m_1^2 = m_2^2 = m^2 = 1, s_{23} = -1.25, s_{13} = 2\,m^2 - s_{12} - s_{23}$. We set $m_1^2 = m_2^2$ for the results shown in Fig. 4 because this is the only case occurring in the process $gg \to t\bar{t}$ at two loops if the b-quarks are assumed to be massless.

The results shown in Fig. 4(b) correspond to the rank two tensor integral with a factor of $k_1 \cdot k_2$ in the numerator. The numerical integration errors are shown as horizontal markers on

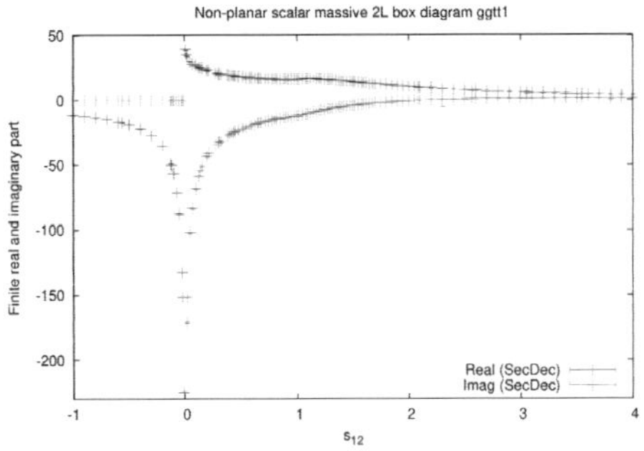

(a) scalar integral

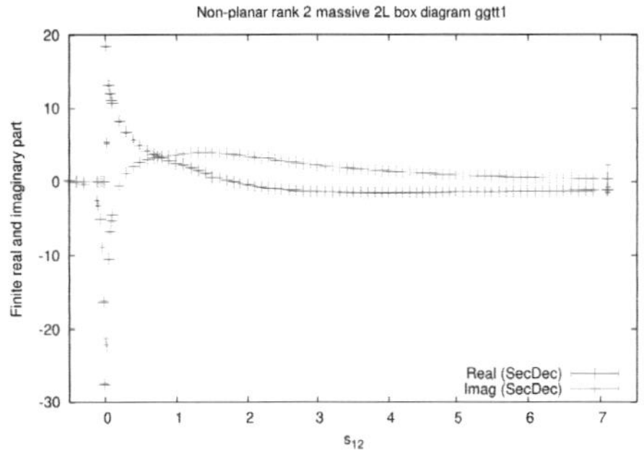

(b) rank 2 tensor integral

Figure 4. Results for the scalar integral *ggtt1* shown in Fig. 2(a), and the corresponding rank two tensor integral *ggtt1* with $k_1 \cdot k_2$ in the numerator. We vary s_{12} and fix $s_{23} = -1.25, m_2 = m_1, p_3^2 = p_4^2 = m_1^2 = 1$.

the vertical lines. The absence of such markers means that the numerical errors are smaller than visible in the plot.

The timings for one kinematic point for the scalar integral in Fig. 4(a) range from 11-60 secs for points far from threshold to 1.6×10^3 seconds for a point very close to threshold, with an average of about 500 secs for points in the vicinity of the threshold. A relative accuracy of 10^{-3} has been required for the numerical integration, while the absolute accuracy has been set to 10^{-5}. For the tensor integral, the timings are better than in the scalar case, as the numerator function present in this case smoothes out the singularity structure. The timings were obtained on a single machine using Intel i7 processors and 8 cores.

3.2. Massive tensor two-loop two-point functions

Here we show that the option to evaluate integrals with a non-trivial numerator can also be applied to calculate two-loop two-point functions involving different mass scales, without the need for a reduction to master integrals. This fact can be used for instance to calculate two-loop corrections to mass parameters in a straightforward way.

As an example we give results for a scalar integral and a rank three tensor integral, where the tensor integral is given by

$$\mathcal{G}_B = \left(\frac{1}{i\pi^{\frac{D}{2}}}\right)^2 \int \frac{d^Dk_1\, d^Dk_2\ (k_1 \cdot k_2)\,(k_1 \cdot p_1)}{D_1 \ldots D_5} \; ; \tag{12}$$
$$D_1 = k_1^2 - m_1^2, D_2 = (k_1 + p_1)^2 - m_1^2, D_3 = (k_1 - k_2)^2 - m_3^2,$$
$$D_4 = (k_2 + p_1)^2 - m_2^2, D_5 = k_2^2 - m_2^2 \,.$$

The fact that this tensor integral is reducible does not play a role here, because our purpose is to demonstrate that reduction may become obsolete considering the short integration times for the tensors. Results are shown in Fig. 5.

4. Conclusions

We have presented new features of the program SecDec, which can be used to calculate multi-loop integrals numerically in an automated way. Among the new features are the option to directly calculate tensor integrals, and to deal with more general types of integrands than the

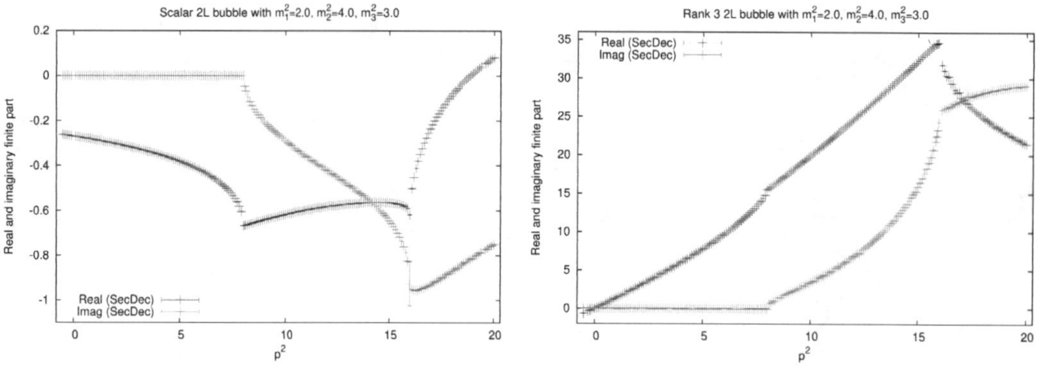

(a) scalar integral

(b) rank 3 tensor integral

Figure 5. Results for the rank 3 two-loop bubble diagram with three non-vanishing masses. (a) scalar case, (b) tensor case with numerator $(k_1 \cdot k_2)\,(k_1 \cdot p_1)$. The masses are $m_1^2 = 2, m_2^2 = 4, m_3^2 = 3$.

ones coming directly from standard Feynman integrals. These integrands can be defined by the user, while still profiting from automated processing for the factorisation of the poles and for the numerical integration.

The program is publicly available at `http://projects.hepforge.org/secdec`.

Acknowledgments
We would like to thank Andreas von Manteuffel for comparisons with analytic results, and the organizers of ACAT2013 for the nice conference.

References

[1] Czakon M and Mitov A 2013 (*Preprint* 1303.0693)
[2] Czakon M 2008 *Phys.Lett.* **B664** 307–314 (*Preprint* 0803.1400)
[3] von Manteuffel A and Studerus C 2013 (*Preprint* 1306.3504)
[4] von Manteuffel A and Studerus C 2012 Proceedings of the conference "Loops and Legs in Quantum Field Theory", 2012 (*Preprint* 1210.1436)
[5] Bonciani R, Ferroglia A, Gehrmann T, Manteuffel A and Studerus C 2011 *JHEP* **1101** 102 (*Preprint* 1011.6661)
[6] Bonciani R, Ferroglia A, Gehrmann T and Studerus C 2009 *JHEP* **0908** 067 (*Preprint* 0906.3671)
[7] Bonciani R, Ferroglia A, Gehrmann T, Maitre D and Studerus C 2008 *JHEP* **0807** 129 (*Preprint* 0806.2301)
[8] Gehrmann T, Tancredi L and Weihs E 2013 *JHEP* **1308** 070 (*Preprint* 1306.6344)
[9] Chachamis G, Czakon M and Eiras D 2008 *JHEP* **0812** 003 (*Preprint* 0802.4028)
[10] Carter J and Heinrich G 2011 *Comput.Phys.Commun.* **182** 1566–1581 (*Preprint* 1011.5493)
[11] Borowka S, Carter J and Heinrich G 2013 *Comput.Phys.Commun.* **184** 396–408 (*Preprint* 1204.4152)
[12] Binoth T and Heinrich G 2000 *Nucl. Phys.* **B585** 741–759 (*Preprint* hep-ph/0004013)
[13] Roth M and Denner A 1996 *Nucl. Phys.* **B479** 495–514 (*Preprint* hep-ph/9605420)
[14] Hepp K 1966 *Commun. Math. Phys.* **2** 301–326
[15] Smirnov A, Smirnov V and Tentyukov M 2011 *Comput.Phys.Commun.* **182** 790–803 (*Preprint* 0912.0158)
[16] Bogner C and Weinzierl S 2008 *Comput. Phys. Commun.* **178** 596–610 (*Preprint* 0709.4092)
[17] Gluza J, Kajda K, Riemann T and Yundin V 2011 *Eur.Phys.J.* **C71** 1516 (*Preprint* 1010.1667)
[18] Hahn T 2005 *Comput. Phys. Commun.* **168** 78–95 (*Preprint* hep-ph/0404043)
[19] Borowka S and Heinrich G 2013 *Comput.Phys.Commun.* **184** 2552–2561 (*Preprint* 1303.1157)
[20] Agrawal S, Hahn T and Mirabella E 2011 (*Preprint* 1112.0124)
[21] Heinrich G 2008 *Int. J. Mod. Phys.* **A23** 1457–1486 (*Preprint* 0803.4177)

Five-loop numerical evaluation of critical exponents of the φ^4 theory

L.Ts. Adzhemyan, M.V. Kompaniets

Department of Theoretical Physics, St. Petersburg State University, Uljanovskaja 1, St. Petersburg, Petrodvorez, 198504 Russia

E-mail: mkompan@gmail.com

Abstract. We present a new approach to calculation of anomalous dimensions in the framework of ϵ-expansion and renormalization group method. This approach allows one to skip the calculation of renormalization constants and express anomalous dimensions in terms of renormalized diagrams, which are presented in a form suitable for numerical calculations. This approach can be easily automated and extended to a wide range of models. The power of this approach is illustrated on 5 loop calculations of beta-function and anomalous dimensions in ϕ^4 model.

The renormalization group approach is one of the most effective methods of quantum field theory. The most studied in the framework of this approach is the model φ^4. On the one hand, this model describes the second order phase transition for different systems [1, 2, 3], on the other hand, it is usually used as a testing ground for approbation of new methods of Feynman diagrams calculations. In particular, in the framework of this model were tested the methods like Gegenbauer polynomial x-space technique (GPTX) [4, 5], integration-by-parts method [6] and infrared R^*-operation [7]. These methods allowed to reach maximal accuracy in critical exponents (5th order of ε-expansion) available now [5, 8, 9]. Critical exponent η was calculated with 5-loop accuracy in [5], then calculation of β-function in 5th order of perturbation theory was performed in [8]. Later on some inaccuracies were found in this calculation [9] and results for index η and β-function were corrected. In calculations [9] the same method, namely R^*-operation technique, as in [8] was used, that is why [9] cannot be treated as fully independent check of the preceding works [5, 8].

Nowadays methods [4, 6, 7] are successfully applied to different models of quantum field theory, however application of these methods to the models of critical dynamics [1, 10] (which are the main object of our interest) encounters significant problems due to some specific features of these models. In [11, 12] the new approach to calculating of critical exponents was proposed. This approach allows one to skip the calculation of renormalization constants and express anomalous dimensions in terms of renormalized diagrams, which are presented in a form suitable for numerical calculations. The main benefit of this approach is that it can be easily automated and extended to a wide range of models.

In order to test our approach we performed 5-loop calculation of critical exponents. On the one hand it allows us to examine the efficiency of our approach in high orders of perturbation theory, and on the other hand approach discussed in this paper deals nothing with the R^*-operation and can be used as a fully independent check of the previous results [5, 8, 9].

The results obtained for model φ^4 using our approach ($d = 4 - \varepsilon$)

$$\eta = 0.0185185185\ \epsilon^2 + 0.0186900(6)\epsilon^3 - 0.0083286(2)\epsilon^4 + 0.025656(2)\epsilon^5$$

$$\omega = \epsilon - 0.62962963(8)\epsilon^2 + 1.6182211(7)\epsilon^3 - 5.23513(2)\epsilon^4 + 20.7499(9)\epsilon^5$$

are in good agreement with the results obtained in [9]

$$\eta = 0.0185185185\ \epsilon^2 + 0.0186899862\ \epsilon^3 - 0.008328770\ \epsilon^4 + 0.025656451\ \epsilon^5$$

$$\omega = \epsilon - 0.629629629\ \epsilon^2 + 1.61822067\ \epsilon^3 - 5.2351359\ \epsilon^4 + 20.74984\ \epsilon^5$$

and confirm inaccuracies found in [5, 8].

The plan of the paper is as follows. In the next section we shortly describe our approach and renormalization scheme used. In the second section we present examples of diagram calculation in discussed approach. The last section is devoted to comparison of our results with results of [5, 8, 9] in more details.

1. Normalization point approach

Let us consider φ^4 theory in euclidean space with $d = 4 - \varepsilon$ dimensions. Basic action for this theory has the following form [1]

$$S_B = -\frac{1}{2}m^2\varphi^2 - \frac{1}{2}(\partial\varphi)^2 - \frac{1}{4!}g\mu^\varepsilon\varphi^4. \tag{1}$$

Diagrams of Green functions Γ_i calculated in the theory (1) have ultraviolet divergences, these divergences can be removed by adding counterterms to action (1). This leads to renormalized action of the following form

$$S = -\frac{1}{2}(m^2 Z_1 + \delta m^2)\varphi^2 - \frac{1}{2}Z_2(\partial\varphi)^2 - \frac{1}{4!}Z_3 g\mu^\varepsilon\varphi^4, \tag{2}$$

where renormalization constants Z_1, Z_2, Z_3 are expressed in terms of renormalization constants of mass Z_{m^2}, field Z_φ and coupling constant Z_g by the relations

$$Z_1 = Z_{m^2}Z_\varphi^2, \qquad Z_2 = Z_\varphi^2, \qquad Z_3 = Z_g Z_\varphi^4. \tag{3}$$

Particular choice of renormalization constants is determined by renormalization scheme. In this work we will use normalization point (NP) scheme defined as follows. The mass shift δm^2 is determined by the condition

$$\Gamma_2^R\,|_{p=0,m=0} = 0\,. \tag{4}$$

Renormalization constants Z_i are defined in such a way that at normalization point $p = 0$, $\mu = m$ the renormalized one-particle irreducible (1-PI) two-point function Γ_2^R and 1-PI four-point function Γ_4^R are equal to their loopless terms:

$$\Gamma_2^R\,|_{p=0,\mu=m} = -m^2\,, \quad \partial_{p^2}\Gamma_2^R\,|_{p=0,\mu=m} = -1\,, \quad \Gamma_4^R\,|_{p=0,\mu=m} = -gm^{2\varepsilon}\,. \tag{5}$$

Henceforth it is convenient to use normalized functions

$$\bar\Gamma_1 = -\left(\frac{\Gamma_2 - \Gamma_2|_{m=0}}{m^2}\right), \quad \bar\Gamma_2 = -\partial_{p^2}\Gamma_2\,, \quad \bar\Gamma_4 = \Gamma_4/(-g\mu^{2\varepsilon})\,, \tag{6}$$

which, according to (4), (5), satisfy the following relations

$$\bar\Gamma_2^R|_{p=0,m=0} = 0, \quad \bar\Gamma_1^R|_{p=0,\mu=m} = 1, \quad \bar\Gamma_2^R|_{p=0,\mu=m} = 1, \quad \bar\Gamma_4^R|_{p=0,\mu=m} = 1. \tag{7}$$

Renormalization constants defined by conditions (7) do not depend on m, like in minimal subtraction (MS) scheme. Renormalization group equations also coincide with that in MS scheme:

$$(\mathcal{D}_\mu + \beta\partial_g - \gamma_{m^2}\mathcal{D}_{m^2})\,\Gamma_i^R = n\gamma_\varphi\Gamma_i^R, \tag{8}$$

where $\mathcal{D}_{m^2} \equiv m^2\partial_{m^2}\mid_{\mu,g}, \mathcal{D}_\mu \equiv \mu\partial_\mu\mid_{m,g}, \gamma_i = \beta\partial_g \ln Z_i, \beta = -g(\varepsilon+\gamma_g)$. Using (8), one can obtain for normalized functions (6)

$$(\mathcal{D}_\mu + \beta\partial_g - \gamma_{m^2}\mathcal{D}_{m^2})\,\bar{\Gamma}_i^R = \gamma_i\bar{\Gamma}_i^R, \tag{9}$$

where, according to (3),

$$\gamma_1 = \gamma_{m^2} + 2\gamma_\varphi, \qquad \gamma_2 = 2\gamma_\varphi, \qquad \gamma_3 = \gamma_g + 4\gamma_\varphi. \tag{10}$$

Let us consider eqs. (9) at normalization point $p = 0, \mu = m$. Taking into account (7) we can express RG-functions γ_i in terms of renormalized functions $\bar{\Gamma}_i^R$ at normalization point [11, 12]:

$$\gamma_i = \frac{2F_i}{1+F_2-F_1}, \qquad F_i \equiv \left(-m^2\partial_{m^2}\bar{\Gamma}_i^R\right)\mid_{p=0,\mu=m}, \qquad i = 1,\ 2,\ 4. \tag{11}$$

For renormalized Green functions Γ_i^R it is possible to replace addition of counterterms (renormalization constants) by R-operation acting on Green functions Γ_i calculated in the theory (1):

$$\Gamma_i^R = R\Gamma_i. \tag{12}$$

Using R-operation let us define the following functions

$$f_i = R\left[\left(-m^2\partial_{m^2}\bar{\Gamma}_i\right)\right]\mid_{p=0,\mu=m}. \tag{13}$$

It is shown in [11, 12] that interrelations between these functions and functions F_i (11) are

$$f_i - F_i = f_iF_1, \qquad i = 2, 4. \tag{14}$$

This allows us to rewrite (11) in the following form

$$\gamma_i = \frac{2f_i}{1+f_2}, \qquad i = 2,\ 4. \tag{15}$$

Relations (13) and (15) will be used in further numerical calculations. The advantage of these relations with respect to (11) is that R-operation in (13) is taken at normalization point, and thus has more simple form. R-operation in (13) can be expressed in terms of product of $1 - K_i$ operations, which removes all divergences from diagrams [13]

$$R\Gamma = \prod_i (1 - K_i)\Gamma, \tag{16}$$

where product is taken over all relevant (UV-divergent) subgraphs of a particular diagram, including diagram as whole. Differentiation of diagram lines in (13) with respect to squared mass $(-\partial_{m^2})$ is equivalent to insertion of unit vertex into a line ("dot insertion"). That is why in such a graphs there are quadratically divergent 1-PI two-point subgraphs and logarithmically divergent 1-PI four-point subgraphs and two-point subgraphs with inset dot.

In renormalization scheme under consideration a subtraction operation for graphs taken at $\mu = m$ is equivalent to the subtraction of initial part of Taylor series over external momenta

$$(1 - K_i)F(k) = F(k) - \sum_{m=0}^{n} \frac{k^m}{m!} F^{(m)}|_{k=0}, \tag{17}$$

where $n = 0$ for logarithmically divergent subgraphs and $n = 2$ for quadratically divergent ones. It is possible to rewrite equation (17) using representation for reminder of Taylor series

$$(1 - K_i)F(k) = \frac{1}{n!} \int_0^1 da(1-a)^n \partial_a^{n+1} F(ak). \tag{18}$$

Combining (16) and (18) we can construct the following representation for R-operation [13]

$$R\chi = \prod_i \frac{1}{n_i!} \int_0^1 da_i (1-a_i)^{n_i} \partial_{a_i}^{n_i+1} \chi(\{a\}), \tag{19}$$

where product is taken over all 1-PI divergent subgraphs $\chi_{(i)}$ (including diagram χ as whole) with canonical dimension $n_i \geq 0$, and a_i – parameter that stretches moments flowing into i-th subgraph. The main advantage of such a representation is that the renormalized function is expressed by integrals finite at $\varepsilon = 0$, and there is no cancellation of large terms in integrand ("theory without divergences" [13]).

2. Example of a diagram calculation

Let us consider 2-loop contribution to f_4 from (13) determined by 2-loop four-point diagram of Γ_4

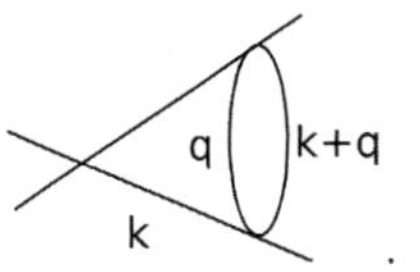

. (20)

The integral that corresponds to the diagram (20) is

$$J = \int \frac{d\mathbf{k}}{(2\pi)^d} \int \frac{d\mathbf{q}}{(2\pi)^d} I(\mathbf{k}, \mathbf{q}), \tag{21}$$

where

$$I(\mathbf{k}, \mathbf{q}) = \frac{1}{(k^2 + m^2)^2} \cdot \frac{1}{q^2 + m^2} \cdot \frac{1}{(\mathbf{k} + \mathbf{q})^2 + m^2}. \tag{22}$$

At $\varepsilon \to 0$ there is logarithmic superficial divergence in the diagram and logarithmic divergence in subgraph. According to (13) contribution of this diagram to function f_4 is given by

$$\delta f_4 = 3\, g^2 m^{2\varepsilon} R(-m^2 \partial_{m^2} J), \tag{23}$$

here multiplier 3 is a symmetry coefficient for diagram (20). Differentiation with respect to m^2 removes superficial divergence. Performing differentiation over m^2 and taking into account diagram symmetry we obtain two terms for this contribution

$$\delta f_4 = \delta f_4^{(1)} + \delta f_4^{(2)}$$

by the diagrams

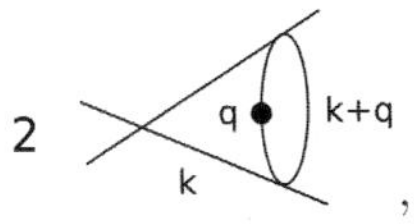

, (24)

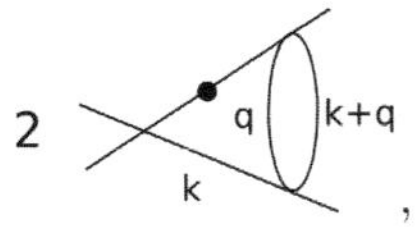

, (25)

with the following integrands in (21)

$$I^{(1)}(\mathbf{k},\mathbf{q}) = 2\frac{1}{(k^2+m^2)^2}\cdot\frac{1}{(q^2+m^2)^2}\cdot\frac{1}{(\mathbf{k}+\mathbf{q})^2+m^2}, \tag{26}$$

$$I^{(2)}(\mathbf{k},\mathbf{q}) = 2\frac{1}{(k^2+m^2)^3}\cdot\frac{1}{q^2+m^2}\cdot\frac{1}{(\mathbf{k}+\mathbf{q})^2+m^2}. \tag{27}$$

Dot insertion in the diagram (24) removes subgraph divergence. This results in the following: integral (26) is UV-finite and R-operation in (23) for this term is trivial ($R \equiv 1$). This allows us to calculate $\delta f_4^{(1)}$ numericaly as Taylor series over ε.

In diagram (25) divergence in subgraph is still present and R-operation is nontrivial. To construct R-operation of type (19) that will remove divergence in subgraph we must introduce streching parameter a for momenta flowing into this subgraph

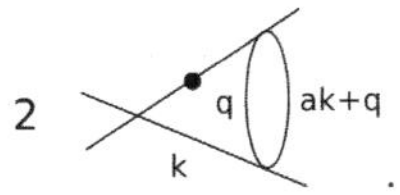

. (28)

Applying R-operation (19) to this diagram we obtain its contribution to function f_4 (13):

$$\delta f_4^{(2)} = 3\,g^2 m^{2\varepsilon}\int\frac{d\mathbf{k}}{(2\pi)^d}\int\frac{d\mathbf{q}}{(2\pi)^d}\int_0^1 da\,\partial_a\, I_2(\mathbf{k},\mathbf{q},a)\,, \tag{29}$$

where

$$I^{(2)}(\mathbf{k},\mathbf{q},a) = 2\frac{1}{(k^2+m^2)^3}\cdot\frac{1}{q^2+m^2}\cdot\frac{1}{(a\mathbf{k}+\mathbf{q})^2+m^2}. \tag{30}$$

Performing differentiation with respect to a and transition to dimensionless variables $\mathbf{k}\to m\mathbf{k}$, $\mathbf{q}\to m\mathbf{q}$ we obtain

$$\delta f_4^{(2)} = -3\,g^2\int\frac{d\mathbf{k}}{(2\pi)^d}\int\frac{d\mathbf{q}}{(2\pi)^d}\int_0^1 da\,\frac{1}{(k^2+1)^3}\cdot\frac{1}{q^2+1}\cdot\frac{4ak^2+4\mathbf{kq}}{[(a\mathbf{k}+\mathbf{q})^2+1]^2}. \tag{31}$$

This integral is finite at $\varepsilon = 0$ and can be evaluated numerically.

In practice for numerical calculations it is more convenient to use Feynman parameters u_i. It is well known that it is possible to write integrand in Feynman representation directly from diagram (without necessity of writing momentum representation), it turns out that it may be extended to integrands of type (30) with streching parameters.

Let us illustrate this on diagram (28)

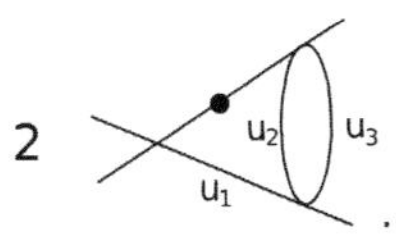

. (32)

Contribution $\delta f_4^{(2)}$ in Feynman representation will have the following form:

$$\delta f_4^{(2)} = 3\,g^2 \frac{\Gamma(1+\varepsilon)\Gamma^2(2-\varepsilon/2)}{4} \cdot J_4\,,$$
$$J_4 = 2\int_0^1 du_1 \int_0^1 du_2 \int_0^1 du_3 \int_0^1 da\,\partial_a \frac{u_1^2\,\delta(u_1+u_2+u_3-1)}{(u_1u_2+u_1u_3+au_2u_3)^{2-\varepsilon/2}}\,. \quad (33)$$

This integral is improper, because of denominators zeroes. In the absence of differentiation with respect to streching parameter denominators zero at $u_1 = 1, u_2 = u_3 = 0$ will produce a pole on ε. But after differentiation over a we got the following expression:

$$J_4 = (\varepsilon-4)\int_0^1 du_1 \int_0^1 du_2 \int_0^1 du_3 \int_0^1 da \frac{u_2u_3u_1^2\,\delta(u_1+u_2+u_3-1)}{(u_1u_2+u_1u_3+au_2u_3)^{3-\varepsilon/2}}\,. \quad (34)$$

One can see that this singularity becomes integrable and it is possible to calculate this integral as Taylor series on ε.

Presence of integrable singularities does not allow to reach desired accuracy in the numerical calculations. This problem is solved by applying Sector Decomposition technique [14]. In presence of streching parameters it is not possible to use existing sector decomposition strategies, but we've developed generalization of strategy S [15] that can be applied to this type of integrals. This allows us to calculate anomalous dimensions at 5th order of perturbation theory.

3. Diagram-by-diagram comparison with calculation in MS scheme. 5-loop results.

For more precise check of the results obtained in [9] we performed diagram-by-diagram comparison of our results and results of [9][1]. To perform diagram-by-diagram comparison from values calculated in NP scheme we need to construct counterterm contributions in MS scheme. Because of the fact that in NP scheme we calculate only finite parts of diagrams (13), first of all we need to construct counterterms in NP scheme, and then recalculate them from NP scheme to MS one.

The first problem (construction of counterterms in NP scheme) can be solved using representation for couterterms obtained in [11]

$$Z_i^{(n)} = \frac{2}{n\epsilon}\left(\mathcal{N}\bar{\Gamma}_i^{(n)} - \mathcal{J}\bar{\Gamma}_i^{(n)}\right), \quad (35)$$

where n – number of loops, $\mathcal{N}\bar{\Gamma}_i = f_i$, and the second term is defined by sum of diagrams from lower orders of perturbation theory (this reflects the fact that all high order poles can be expressed in terms of lower order diagrams).

The next step is recalculation of counterterms from NP scheme to MS one. It can be performed using so called R^{-1} operation [16]. This operation allows one to recursively recalculate counterterms from one renormalization scheme to another (without necessity of calculation of diagrams).

We've performed recalculation of our results to MS scheme and found that results are in good agreement with [3, 9]: all diagrams coincides with at least 10^{-5} accuracy. This confirms corrections made in [9].

In [9] it is stated that there are inaccuracies in 6 diagrams in [5, 8]. In tables 1, 2 one can find diagrams corrected in [9], values from previous papers [5, 8], and results of our numerical calculations with error estimations.

Nickel index mentioned in the tables 1,2 is the most efficient way to identify diagrams (to be more precise it is adjacency list ordered in special way), one can find its definition in [17]. The numbering of diagrams (first column) in table 1 is based on papers [5], in table 2 – on papers [8].

[1] values for particular diagrams are presented in [3]

Table 1. Diagrams from two-point Green function calculated in [5] and corrected in [9], compared with results obtained from normalization point scheme

N	Nickel index		original value	corrected value	current work
9	e112-23-34-44-e-	$1/\epsilon$	0.4	-0.93333	-0.93335(9)
		$1/\epsilon^2$	1.466666	1.466666	1.466669(9)
		$1/\epsilon^3$	-1.066666	-1.066666	-1.0666665(8)
		$1/\epsilon^4$	0.533333	0.533333	0.533333
10	e112-34-334-4-e-	$1/\epsilon$	0.3563478	0.0316508	0.03164(7)
		$1/\epsilon^2$	0.641666	0.641666	0.641670(8)
		$1/\epsilon^3$	-0.7666666	-0.7666666	-0.7666665(5)
		$1/\epsilon^4$	0.4	0.4	0.4

Table 2. Diagrams from four-point Green function calculated in [8] and corrected in [9], compared with results obtained from normalization point scheme ("s.c." = symmetry coefficient)

N	Nickel index		original value	corrected value	current work
2	e123-e23-45-45-e5-e-	s.c.	3	3/2	3/2
		$1/\epsilon$	20.807147	20.807147	20.807141(50)
24	e112-34-e35-45-e5-e-	$1/\epsilon$	14.246950	5.860538	5.860525(14)
		$1/\epsilon^2$	-4.14771102	-4.14771102	-4.1477109(10)
32	ee12-234-34-45-5-ee-	$1/\epsilon$	12.36505	1.995772	1.99578(20)
		$1/\epsilon^2$	-16.734897	-16.734897	-16.734892(6)
		$1/\epsilon^3$	11.539746	11.539746	11.539747(1)
32	ee12-ee3-445-455-5–	$1/\epsilon$	-1.70502	-1.12169	-1.12165(10)
		$1/\epsilon^2$	2.133333	2.133333	2.133338(20)
		$1/\epsilon^3$	-1.333333	-1.333333	-1.333333(1)
		$1/\epsilon^4$	0.533333	0.533333	0.533333

It can be stated that discussed approach shows its efficiency and can be used in wide range of models.

Acknowledgments

The authors are grateful to N.V. Antonov, K.G. Chetyrkin, M. Hnatic, A.L. Kataev and D.I. Kazakov for helpful discussions and advices, to ACAT'13 Organizing Committee, and in particular Jian-Xiong Wang (IHEP, Beijing) for support and hospitality, to the members of ACAT'13 Int. Advisory Committee, and in particular A.B. Arbuzov (JINR, Dubna) and A.L. Kataev (INR RAS) for the support to present this talk on ACAT'13. The work was supported in part by the Russian Foundation for Fundamental Research (project 12-02-00874-a). Research was carried out using computer resources provided by Resource Center "Computer Center of SPbU" (http://cc.spbu.ru)

References

[1] Vasilev A N1998 *Quantum Field Renormalization Group in Critical Behavior Theory and Stochastic Dynamics* [in Russian] (St.Petersburg: Petersburg Inst. Nucl. Phys.)
English transl.: 2004 *The Field Theoretic Renormalization Group in Critical Behavior Theory and Stochastic Dynamics*, (Boca Raton: Chapman & Hall/CRC)

[2] Zinn-Justin J 1989, 1993, 1996, 2002 *Quantum Field Theory andv Critical Phenomena.* (Oxford: Clarendon Press)
[3] Kleinert H, Schulte-Frohlinde V 2001 *Critical Properties of* ϕ^4*-Theories*, (Singapore: World Scientifc)
[4] Chetyrkin K G, Kataev A L, Tkachev F V 1980 *Nucl. Phys.* B **174** 345-377
[5] Chetyrkin K G, Kataev A L , Tkachev F V 1981 *Phys.Lett.* B **99** 147; B **101** 457(E)
[6] Tkachev F V 1981 *Phys. Lett.* B **100** 65
Chetyrkin K G, Tkachev F V 1981 *Nucl. Phys.* B **192** 159-204
[7] Chetyrkin K G and Tkachov F V 1982 *Phys. Lett.* B **114** 340.
Chetyrkin K G and Smirnov V A 1984 *Phys. Lett.* B **144** 419.
[8] Chetyrkin K G, Gorishny S G, Larin S A and Tkachov F V 1983 *Phys. Lett.* B **132** 351
Kazakov D I 1983 *Phys. Lett.* B **133** 406; 1984 *Theor.Math.Phys.* **58** 223-230; 1984 *Teor.Mat.Fiz.* **58** 343-353
Chetyrkin K G, Gorishny S G, Larin S A and Tkachov F V 1986 *Preprint* INR P-0453, Moscow.
[9] Kleinert H , Neu J, Shulte-Frohlinde V, Chetyrkin K G, Larin S A 1991 *Phys.Lett.* B **272** 39; Erratum 1993 B **319**, 545
[10] Folk R, Moser G 2006 *J.Phys.*, A **39**, R207-R313
[11] Adzhemyan L Ts, Kompaniets M V 2011 *Theor. Math. Phys.* **169**, 1450–1459
[12] Adzhemyan L Ts, Kompaniets M V, Novikov S V, Sazonov V K 2013 *Theor. Math. Phys.*, **175(3)** 719–728
[13] Zav'yalov O I 1979 *Renormalized Quantum Field Theory* [in Russian], (Moscow: Nauka); English transl.: 1990 (Dordrecht :Kluwer)
[14] Binoth T and Heinrich G 2000 *Nucl. Phys.* **B585** 741–759 (*Preprint* hep-ph/0004013).
[15] Smirnov A V, Smirnov V A 2009 *JHEP* 0905 p. 004, (*Preprint* hep-ph/0812.4700)
[16] Pivovarov G B, Tkachev F V 1988 *Theor. Mat. Fiz.* **77** 51
Chetyrkin K G 1990 *Il Nuovo Cimento* A **103** Issue 12 1653-1667
[17] Nickel B, Meiron D, Baker G *Preprint* University of Guelf Report,1977
`http://users.physik.fu-berlin.de/~kleinert/nickel/guelph.pdf`

ACAT2013 IOP Publishing
Journal of Physics: Conference Series **523** (2014) 012050 doi:10.1088/1742-6596/523/1/012050

FormCalc 8: Better Algebra and Vectorization

B. Chokoufe Nejad[1], T. Hahn[2], J.-N. Lang[1], E. Mirabella[2]

[1] Physikalisches Institut, Universität Würzburg, Emil-Hilb-Weg, D–97082 Würzburg, Germany
[2] Max-Planck-Institut für Physik, Föhringer Ring 6, D–80805 Munich, Germany

E-mail: {Bijan.Chokoufenejad,Jean-Nicolas.Lang}@physik.uni-wuerzburg.de,
{hahn,mirabell}@mpp.mpg.de

Abstract. We present Version 8 of the Feynman-diagram calculator FormCalc. New features include in particular significantly improved algebraic simplification as well as vectorization of the generated code. The Cuba Library, used in FormCalc, features checkpointing to disk for all integration algorithms. Report MPP-2013-273

1. Introduction

The Mathematica package FormCalc [1] simplifies Feynman diagrams generated with FeynArts [2] up to one-loop order. It provides the analytical results and can generate Fortran code for the numerical evaluation of the squared matrix element. Cuba is a library for multidimensional numerical integration which is included in FormCalc but can also be used independently.

This note presents the following features new in FormCalc 8 and Cuba 3.2:

- Significant improvement of the algebraic simplification with FORM 4 features.
- Vectorization of the helicity loop.
- Automated C-code generation.
- Optimizations for unitarity methods.
- Checkpointing for all Cuba algorithms.

2. Improvements in the Algebraic Simplification

The algebraic simplification of Feynman amplitudes is split between Mathematica and FORM. In a preprocessing stage, Mathematica translates the elements of a FeynArts amplitude into FORM syntax and writes them to an input file for FORM. (Note that none of the FeynArts symbols are directly redefined, such that processing does not start automatically.) FORM then does the major part of the symbolic simplification. In a postprocessing step, the FORM output is read and returned to Mathematica by FormCalc's ReadForm MathLink utility.

Many new and useful features were introduced in FORM 4 [3], most notably abbreviationing and factorization. The FORM part of FormCalc 8 has been rewritten to take advantage of these facilities, resulting in significantly improved algebraic simplification.

2.1. Abbreviationing

Once a partial expression is considered final at a particular point in the FORM program it is abbreviated, i.e. substituted by a symbol. This not only shortens the active expressions but makes the abbreviated parts inert, such that subsequent id-statements do not spend time on matching these, thus making the FORM code run faster.

A similar technique has been used since Version 6 [5], where the FORM expressions were sent on a round-trip to Mathematica halfway through the evaluation for introducing abbreviations. Since this involved quite some transmission overhead, it was performed only once during each FORM run. With abbreviationing built into FORM now, abbreviations are introduced whenever possible, thereby obviating the extra pass to Mathematica.

Abbreviationing also serves to prevent FORM's automatic expansion of expressions, i.e. it preserves a (pre)factorized structure, which is particularly useful in combination with the new factorization available in FORM (see Sect. 2.2 below).

What is more, since Mathematica receives an expression in many small pieces rather than one large chunk, more aggressive simplification functions can be applied upon return to Mathematica at reasonable efficiency. To this end, FormCalc wraps a zoo of simplification functions around various parts of the amplitude. All of these are 'transparent' in the sense that they can be replaced by `Identity` without affecting the numerical result. The three most important ones are listed below, a complete inventory is given in the FormCalc manual.

- `FormSub` is applied to subexpressions of an amplitude.
- `FormDot` is applied to combinations of dot products in an amplitude.
- `FormMat` is applied to the coefficients of matrix elements ('`Mat`') in an amplitude.

On the technical side, since the abbreviations are also transmitted to Mathematica as such (i.e. not back-substituted into the expressions), the volume of data transferred is significantly reduced and the final expression is stored efficiently as multiple instances of a subexpression are taken care of by reference count (same as with `Share[]`).

At the moment, FormCalc does not use FORM 4's "`format On`" output optimization, as it is not yet clear how to combine it with the postprocessing in ReadForm and Mathematica.

2.2. Factorization

FORM's new 'full' factorization (over the rationals) makes it possible to simplify expressions much better already inside of FORM. The old 'simple' factorization (pulling out common symbols from an expression) is still used in instances where full factorization is too expensive.

Potentially time-consuming instances of the `factarg` command in the FORM code can be suppressed by setting the `CalcFeynAmp` option `NoCostly` → `True`. This is occasionally necessary in models with more complex couplings such as the MSSM.

Plain factorization is not a cure-all for arbitrary expressions, however. For example, while the following expression is not factorizable as a whole,

```
-2*e2.k5*S35 + 2*e2.k5*T24 + 2*e2.k5*T14 - 2*e2.k5*MT2 +
2*e2.k5*S - 3*e2.k6*S35 - e2.k6*S45 - e2.k6*T25 - e2.k6*T15 +
4*e2.k6*T24 + 4*e2.k6*T14 - 4*e2.k6*MT2 + 4*e2.k6*S
```

it easily admits further compactification by collecting with respect to the dot products first:

```
-2*(MT2 - S + S35 - T14 - T24)*e2.k5 -
(4*MT2 - 4*S + 3*S35 + S45 - 4*T14 + T15 - 4*T24 + T25)*e2.k6
```

FormCalc takes typical objects such as dot products into account, of course. Still, for a general expression it is not straightforward to find a suitable simplification procedure, which is why it is useful to have functions like `FormDot` through which one can apply more sophisticated functions such as Mathematica's `Simplify`.

3. Vectorization of the Helicity Loop

The assembly of the squared matrix element in FormCalc can be sketched as in the following figure, where the helicity loop sits at the center of the calculation:

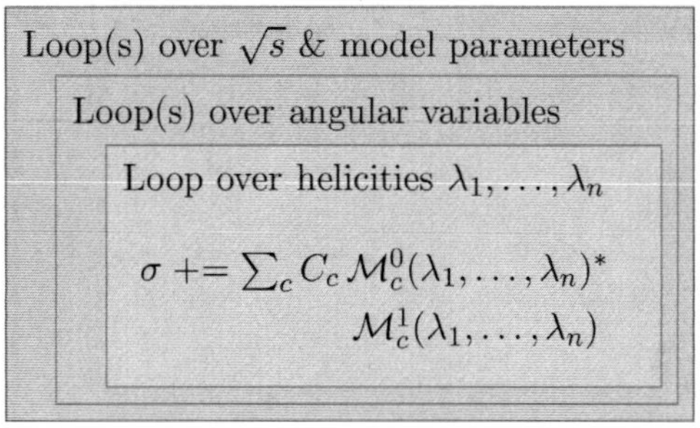

The helicity loop is not only strategically the most desirable but also the most obvious candidate for concurrent execution, as FormCalc does not insert explicit helicity states during the algebraic simplification [4]. That is, the amplitude is a numerical function of the helicities λ_i and not a bunch of (different) functions for each helicity combination,

$$\mathcal{M} = \mathcal{M}(\lambda_1, \lambda_2, \dots) \neq \{\mathcal{M}_{--\dots}, \mathcal{M}_{+-\dots}, \mathcal{M}_{-+\dots}, \mathcal{M}_{++\dots}\}.$$

Such a design is known as Single Instruction Multiple Data (SIMD) in computer science since a single code ($\mathcal{M}$) is independently run for multiple data (λ_i), and is conceptually easy to parallelize or vectorize.

Parallelization on the CPU's cores using `fork/wait` has been available from Version 7.5 on [7]. The drawback of this method is that it competes for compute cores in particular with Cuba. For better efficiency the cores should be assigned to Cuba since it computes entire phase-space points in parallel, not just the helicity loop.

GPU parallelization was attempted using OpenCL but we found that it was not too efficient. Since the transfer of data between the CPU and the GPU is relatively time-consuming, we believe that the distribution of the helicity-independent variables from the CPU to the GPU outweighed the parallelization gains.

Eventually the best speedup we could achieve was with vectorization. With Intel's x86 vector instructions, there is essentially no overhead.

3.1. Vectorization in C

Our implementation in C is based on the vector data type extensions offered by gcc and Intel's icc. Unfortunately, these are restricted to real algebra, even in C99.

Complex addition is obviously no problem and complex multiplication might have been solved through C++'s operator overloading. Since we wanted to stick to C to avoid linking hassles with Fortran object files, we adjusted the C-code generation in Mathematica to insert explicit macros for the multiplication of complex vectors: `SxH` stands for "scalar times helicity vector" and `HxH` for "helicity vector times helicity vector." Helicity vectors are declared with `HelType`. (To avoid confusion with Minkowski four-vectors, we use prefixes '`Hel`' or '`H`' to denote helicity vectors in the SIMD sense.)

Depending on the hardware features indicated by preprocessor flags these macros emit explicit SSE3 or AVX instructions. For SSE3 the maximum vector length is 1 (2 doubles per operation) which may at first not seem very useful, but besides performing addition twice as fast there exists an efficient complex multiplication routine with 2.5 instructions instead of 6. For AVX (requires i7 'Sandy Bridge' or higher) the maximum vector length is 2 (4 doubles per operation). Again the complex multiplication can be formulated fairly efficiently using Intel's vector instructions. Overall we found a speedup of 3.7 out of theoretical 4 with AVX for the helicity loop.

Currently the configure script does not automatically add flags to switch on SSE3 or AVX instructions, e.g. gcc needs the extra flag `-march=corei7-avx` to enable the latter. This may change in the future. A related question is which default to choose for executables that could potentially be run on a cluster of computers with differing SIMD capabilities.

3.2. Vectorization in Fortran

Vector data types are standard fare in Fortran 90 and so not only complex vectors are allowed but one can, in principle, choose arbitrary vector lengths. On the downside, the actual deployment of vector instructions is at the discretion of the compiler and may not be chosen for vector lengths incommensurable with the hardware.

Even though Fortran 90 is an effective requirement for vectorized computation, the code is still generated in fixed format and can be made compatible with Fortran 77 through preprocessor definitions, e.g. for inclusion in legacy packages.

4. Automated C-Code Generation

C-code generation has been available from FormCalc 7 on [6] but now its use is mostly automatic, i.e. also drivers and utility files are available in C. In fact, only the declarations needed to be translated as the initialization still takes place in Fortran and the C object files are simply linked in. For this to work, the layout of C's structs must match Fortran's common blocks, of course. Private declarations, e.g. for new models, are not automatically translated, but this is fairly straightforward as can be seen by comparing the C and Fortran versions of e.g. the Standard Model declarations.

To switch from Fortran to C output, the following statement needs to precede the output commands (e.g. `WriteSquaredME`, `WriteRenConst`):

```
SetLanguage["C"]
```

The output is by default in C99, because of complex numbers, but can easily be made to work with C++ by redefining the abstract data type `ComplexType`.

Even without SIMD vectorization, and perhaps remarkably so for Fortran aficionados, C and Fortran versions of the same amplitude show very similar performance figures, i.e. there is no penalty for using C.

5. Optimizations for Unitarity Methods

FormCalc can generate amplitudes for evaluation with the OPP (Ossola, Papadopoulos, Pittau [10]) unitarity method as implemented in the two libraries CutTools [11] and Samurai [12]. Instead of introducing tensor coefficients [9], the whole numerator is placed in a subroutine, as in:

$$\begin{aligned} \varepsilon_1^\mu \varepsilon_2^\nu B_{\mu\nu}(p, m_1^2, m_2^2) &= (\varepsilon_1 \cdot \varepsilon_2) B_{00} + (\varepsilon_1 \cdot p)(\varepsilon_2 \cdot p) B_{11} && \text{(tensor coeff.)} \\ &= B_{\text{cut}}(2, N, p, m_1^2, m_2^2)\,, && \text{(OPP)} \\ \text{where} \quad N(q_\mu) &= (\varepsilon_1 \cdot q)\,(\varepsilon_2 \cdot q)\,. \end{aligned}$$

The numerator subroutine N will be sampled by the OPP function (B_{cut} in this example). The first argument of B_{cut}, 2, refers to the maximum power of the integration momentum q in N.

The OPP procedure indeed generates significantly fewer terms than the traditional Passarino–Veltman decomposition, nevertheless a naive implementation runs quite a bit slower than its counterpart with tensor coefficients. This section describes our attempts to optimize the OPP performance. We were able to bring the slowdown from originally a factor 10 to about a factor 3 for multiplicities such as $2 \to 3$ and hope to improve matters further. To be fair, OPP was in the first place designed to increase the reach of one-loop calculations to higher-leg multiplicities and not so much to speed up the ones with not so many legs.

- The major part of the slowdown (at least half of that factor 10) comes from the fact that the OPP master integrals (the scalar integrals A_0, B_0, C_0, D_0) are naively computed over and over again. This is because the OPP functions must be evaluated inside the helicity loop since the numerator subroutine depends on the helicities. The scalar integrals contain only the denominators, however, and thus could simply be moved outside the helicity loop. In FormCalc, we generate code that foresees a split between the computation of the masters and their use in assembling the tensor integrals, for example:

```
ComplexType mas145(Mcc)
...
call Cmas(mas145, (C0 args))
...
call Ccut(mas145, num, (C0 args))
```

  The complex array `mas145` stores the master integrals computed by `Cmas` (outside the helicity loop) and used by `Ccut` (inside the helicity loop). Unfortunately, so far none of the available OPP libraries allows this decomposition even though the two tasks '`Cmas`' and '`Ccut`' must be completed internally in some way or another already now. LoopTools alleviates the situation by retrieving recurring masters from its cache, though even here the lookup time could be eliminated with the above construction.

- Some packages address this problem by moving the helicity sum into the numerator. This works if only the interference term is sought since then the amplitude contains at most one loop integration in each term:

$$\sum_\lambda 2\,\mathrm{Re}\,\mathcal{M}_0^* \underbrace{\int \mathrm{d}^4q \frac{N}{D\cdots}}_{\sim\mathcal{M}_1} = \int \mathrm{d}^4q \frac{\sum_\lambda 2\,\mathrm{Re}\,\mathcal{M}_0^* N}{D\cdots}$$

  In FormCalc we don't pursue this strategy, firstly because it is not applicable if the tree-level contribution is zero (or so small that including the loop-squared part becomes necessary) and secondly because it is not obvious how this evaluation fits into the present abbreviation concept.

- Subexpressions of the numerator function (coefficients, summands, etc.) independent of q are pulled out and computed once, ahead of invoking the OPP function, using FormCalc's abbreviationing machinery [13]. In particular in BSM theories, these coefficients can be lengthy such that pulling them out significantly increases performance.

- Our implementation admits mixing Passarino–Veltman decomposition with OPP in the sense that one chooses an integer n starting from which an n-point function is treated with OPP methods. For example, `OPP` $\to$ `4` means that A, B, C functions are treated with Passarino–Veltman and D and up with OPP.

- We optimize OPP calls to reduce sampling effort, e.g. by collecting denominators, as in:

$$\frac{N_4}{D_0 D_1 D_2 D_3} + \frac{N_3}{D_0 D_1 D_2} \to \frac{N_4 + D_3 N_3}{D_0 D_1 D_2 D_3}$$

  Depending on N and rank, joining integrals is not universally better. Rather, we tabulated the sampling behavior of Samurai and CutTools such that the algorithm can determine the optimal splitting.

- The Ninja library implements the D-dimensional integrand reduction via Laurent expansion [14], which constructs the tensor integral from fewer samples in a numerically more stable way. Ninja requires a slightly modified numerator subroutine which is currently being implemented in FormCalc.
- The profiler pointed us to a bottleneck in fermion chains: Before, we were using elementary operations to build up the fermion chain, which we have now merged into a single inlined function call:

$$\begin{aligned}\langle u|\,\sigma_\mu\overline{\sigma}_\nu\sigma_\rho\,|v\rangle\,k_1^\mu k_2^\nu k_3^\rho &= \langle u|\,k_1\overline{k}_2k_3\,|v\rangle\\ \text{(old)} &= \texttt{SxS(}u\texttt{, VxS(}k_1\texttt{, BxS(}k_2\texttt{, VxS(}k_3\texttt{, }v\texttt{))))}\\ \text{(new)} &= \texttt{ChainV3(}u, k_1, k_2, k_3, v\texttt{)}\;.\end{aligned}$$

  The elementary operations could not be inlined in Fortran because they returned a 2-component spinor, not a scalar value.
- The helicity information of an OPP integral's prefactor is taken into account in an extra argument. That is, if a term in the amplitude is known to become zero for a particular helicity combination due to its prefactor (because of massless external particles, say), the evaluation of the loop integral therein is cut short. For example, the following `Dcut` is actually computed only if `Hel1` $\neq 1$:

$$\texttt{Dcut}(3, N, 1-\texttt{Hel1}, \dots)$$

6. Checkpointing in Cuba

Cuba's current Version 3.2 allows checkpointing for all routines. Checkpointing means writing out the integrator's complete state to disk to be able to recover from the last state after a crash. In a long-running calculation this may mean losing one hour instead of one day.

Checkpointing is enabled by specifying a name for the state file. Note that, since only Vegas had this functionality in Version 3.0, the invocation of the other routines has changed to incorporate the extra state file argument. We always write the state to a new file and remove the former state file only when the new one is successfully stored. This makes checkpointing failsafe since even a crash during saving is recoverable. If the integration finishes successfully, the state file is removed.

The checkpoints have been implemented in the serial regions of the code which ensures reliable behavior regardless of parallelization.

Version 3.2 furthermore relaxes several restrictions on the compiler, it is now fully C99-compliant and uses no gcc extensions.

7. Summary

FormCalc 8 (`http://feynarts.de/formcalc`) has many new and improved features, most notably better algebra, a vectorized helicity loop, and OPP improvements. The Cuba library (`http://feynarts.de/cuba`), also included in FormCalc, adds checkpointing for all four integration algorithms, which is useful for resuming interrupted long-running integrations.

[1] Hahn T, Pérez-Victoria M, 1999, *Comput. Phys. Commun.* **118** 153 [hep-ph/9807565].
[2] Hahn T, 2001, *Comput. Phys. Commun.* **140** 418 [hep-ph/0012260].
[3] Kuipers J, Ueda T, Vermaseren JAM, Vollinga J, 2013, *Comput. Phys. Commun.* **184** 1453 [arXiv:1203.6543].
[4] Hahn T, 2003, *Nucl. Phys. Proc. Suppl.* **116** 363 [hep-ph/0210220]
[5] Hahn T, 2008, *PoS* **ACAT 2008** 121 [arXiv:0901.1528].
[6] Agrawal S, Hahn T, Mirabella E, 2012, *J. Phys. Conf. Ser.* **368** 012054 [arXiv:1112.0124].
[7] Agrawal S, Hahn T, Mirabella E, 2012, *PoS* **LL 2012** 046 [arXiv:1210.2628].

[8] Hahn T, 2005, *Comput. Phys. Commun.* **168** 78 [hep-ph/0404043].
[9] Passarino G, Veltman M, 1979, *Nucl. Phys. B* **160** 151.
[10] Ossola G, Papadopoulos C, Pittau R, 2007, *Nucl. Phys. B* **763** 147 [hep-ph/0609007].
[11] Ossola G, Papadopoulos C, Pittau R, 2008, *JHEP* **0803** 042 [arXiv:0711.3596].
[12] Mastrolia P, Ossola G, Reiter T, Tramontano F, 2010, *JHEP* **1008** 080 [arXiv:1006.0710].
[13] Hahn T, 2010, *PoS* **ACAT 2010** 078 [arXiv:1006.2231]
[14] Mastrolia P, Mirabella E, Peraro T, 2012, *JHEP* **1206** 095, Erratum: 2012, ibid. **1211**, 128 [arXiv:1203.0291].

UCLA/13/TEP/109 SB/F/423-13 SLAC-PUB-15760

IPPP-13-85 Saclay–IPhT–T13/229

The BlackHat Library for One-Loop Amplitudes

Z Bern[1], L J Dixon[2], F Febres Cordero[3], S Höche[4], H Ita[5], D A Kosower[6,*], D Maître[7], K J Ozeren[8]

[1,8]Department of Physics and Astronomy, UCLA, Los Angeles, CA 90095-1547, USA
[2,4]SLAC National Accelerator Laboratory, Stanford University, Stanford, CA 94309, USA
[3]Departamento de Física, Universidad Simón Bolívar, Caracas 1080A, Venezuela
[5]Physikalisches Institut, Albert-Ludwigs-Universität Freiburg, D–79104 Freiburg, Germany
[6]Institut de Physique Théorique, CEA–Saclay, F–91191 Gif-sur-Yvette cedex, France
[7]Institute for Particle Physics Phenomenology, University of Durham, Durham DH1 3LE, UK

Abstract. We present recent next-to-leading order (NLO) results in perturbative QCD obtained using the BLACKHAT software library. We discuss the use of n-tuples to separate the lengthy matrix-element computations from the analysis process. The use of n-tuples allows many analyses to be carried out on the same phase-space samples, and also allows experimenters to conduct their own analyses using the original NLO computation.

1. Introduction

Experimental studies of Standard-Model processes at the Large Hadron Collider (LHC), whether as standard candles, as backgrounds to new physics, or as part of a program of precision measurements, require a theoretical counterpart. The dominant contributions to theoretical predictions arise from perturbative quantum chromodynamics (QCD). In QCD, unlike the case for other components of the Standard Model, leading-order (LO) calculations, which rely on tree-level amplitudes in QCD, do not offer a quantitatively reliable prediction. This problem is a result of the uncompensated dependence of the prediction on the renormalization and factorization scales. At LO this dependence arises only via the running coupling and the evolving parton distribution functions. Next-to-leading order (NLO) calculations in QCD cure this problem, and are thus the the first order in perturbation theory to provide a quantitatively reliable estimate of backgrounds due to Standard-Model processes [1, 2].

Recent years have seen remarkable progress in NLO calculations, thanks to advances in computing the one-loop matrix elements which lie at their core [3, 4, 5, 6, 7, 8, 9, 10]. The BLACKHAT collaboration has developed its eponymous software [11] to implement computations of one-loop amplitudes in a numerical approach. For multiplicities beyond $W+2$-jet production, essentially only tree-level amplitudes and integrals are implemented in computer code directly from analytic formulas. The library implements so-called on-shell methods, which rely only

*Presenter at ACAT 2013, Beijing, China, May 16–21, 2013

on information present in amplitudes with on-shell external states in order to arrive at values for more complex amplitudes. This keeps the relative simplicity — compared to the factorial complexity one might have feared from a Feynman-diagrammatic expansion — manifest at every stage of a calculation. This simplification is particularly evident in high-multiplicity calculations, for which the frontier has moved from W + 2-jet production [4, 12, 13] a decade and a half ago to W + 5-jet production earlier this year [14].

2. N-Tuples

While numerical one-loop libraries have made previously-unattainable NLO calculations feasible, they are not the only ingredient needed. In addition to the tree-level matrix elements for the basic process, we also need tree-level matrix elements with an additional parton emission: a gluon emitted, or a gluon splitting to a quark–antiquark pair. We use the COMIX [15] library within SHERPA [16] to compute these matrix elements. Different contributions to an NLO calculation have infrared divergences: in the virtual corrections, they are present explicitly, in the form of poles in the dimensional regulator ϵ; in the real-emission corrections, they arise in phase-space integrals over squared matrix elements. These divergences cancel in physical quantities such as cross sections and distributions of infrared-safe observables. In order to make this cancellation manifest in an operationally-tractable manner, one subtracts local approximations to the squared real-emission matrix elements, and adds back the analytic integrals of these local approximations. These integrals have manifest singularities, again in the form of poles in ϵ, which cancel those in the virtual corrections. The local subtraction terms and the computation of the appropriate subtraction integrals (using analytic formulas) are also performed by the COMIX library. We thus have four types of contributions to integrate numerically: Born (B), virtual corrections (V), integrated subtraction (I), and subtracted real emission (R).

A typical calculation at high multiplicity will involve dozens of subprocesses (two-quark, four-quark, six-quark, etc. in all possible crossings). Each must be integrated over final-state phase space, subject to cuts simulating the experimental ones. We use SHERPA to manage the subprocesses and perform the required numerical integrations. We use the FastJet library [17] to apply a jet algorithm to phase-space configurations of partons. The different contributions span a large numerical range; to focus computer time on the largest contributions, it is helpful to further split up each of the four types (B,V,I,R) into parts, corresponding to leading and subleading color, or to different initial states, etc.

This subdivision does exact a penalty, in the form of the increased complexity of managing a calculation, tracking all the different parts, and ensuring that each part was run correctly. In a physics study conducted using stand-alone NLO codes such as `NLOJET++` [18], one would write code for a set of observables, and obtain distributions in those observables by running the complete code which computes both squared matrix elements and observables. Computing new observables as a physics study evolves; varying renormalization and factorization scales; computing uncertainties due to our imprecise knowledge of parton distribution functions (PDFs), would all require rerunning the code from scratch. This would force recomputation of the short-distance matrix elements, which are much more expensive to compute than observables. It would also force us to confront the subprocesses' and subparts' management complexity anew each time.

It would simplify the calculation to separate the computation of the squared matrix elements (along with the generation of phase-space configurations) from the computation of observables. We do this by saving generated phase-space configurations, along with their weights, in a set of Root [19] n-tuple files. The weights are obtained by combining the squared matrix elements with the phase-space measure, parton-distribution values, and appropriate Jacobian factors. In addition, we include a set of auxiliary coefficients which allow the recomputation of the weights using different renormalization and factorization scales, or different PDF sets. The files are quite voluminous, especially for higher multiplicities, so the compression offered by ROOT yields a significant reduction compared to a naive binary format both in disk-space usage and in transmission times. A lightweight analysis code scans the resulting files, and computes cross sections or histograms of observables. This code can be rerun as needed for new observables or for estimating scale-sensitivity of observables, at relatively low cost in computer time.

The use of n-tuple files has a bonus: we can hand over sets of n-tuple files to our experimenter colleagues, who can then run their own analyses on the events they contain. This avoids the experimenters having to learn how to run a somewhat complicated set-up, and avoids them having to adopt our code or framework for performing analyses. The analyses are of parton-level fixed-order events, so that only distributions [of infrared-safe observables] can be compared in a meaningful way to experimental data. (The n-tuples we have computed to date are not suitable as input to existing NLO-matched parton-shower codes.)

3. Vector Boson+Jets

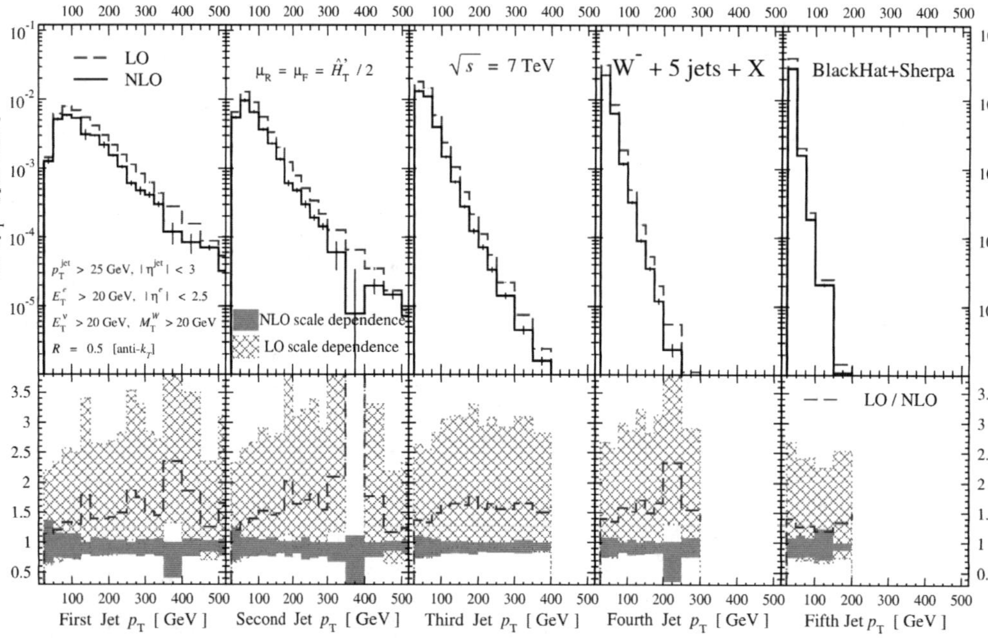

Figure 1. The p_T distributions of the leading five jets in W^- + 5-jet production, ordered in decreasing p_T from left to right. In the upper panels, the blue (dashed) line shows the LO prediction, and the black (solid) line, the NLO one. The lower panels show the ratio of the central LO prediction to the NLO one, along with the LO scale variation band in orange-brown (hatched) and the NLO scale variation band in gray (solid).

As an example of recent studies we have performed using the BLACKHAT library running under SHERPA's control, and using n-tuple files, we show the distribution of the leading five jets in an NLO computation [14] of W^- + 5-jet+X production in fig. 1. Successive jet distributions fall more steeply. This reflects the fact that moving out on the distribution of, say, the third jet allows the fourth and fifth jet to remain at low p_T, thus forcing up the center-of-mass energy of the short-distance partonic collision less than moving out on the distribution of the fifth jet. As the center-of-mass energy grows, the short-distance cross section decreases rapidly, both because of direct dependence in the squared matrix element and flux factors, and because the parton distributions decrease as well.

For the leading through next-to-softest jet (the fourth jet in this case), the NLO distributions fall somewhat more steeply, as reflected in the upward tilt of the LO/NLO ratio in the bottom panels of fig. 1. This continues a pattern seen at lower multiplicity, for example in the $W+$ 4-jet calculation [20].

The scale-variation bands shrink dramatically in the NLO calculation as compared to the LO one. This fulfills one of the primary motivations for pursuing this calculation: we have achieved a 10–15% level of reliability, within expected experimental uncertainties for this high-multiplicity process.

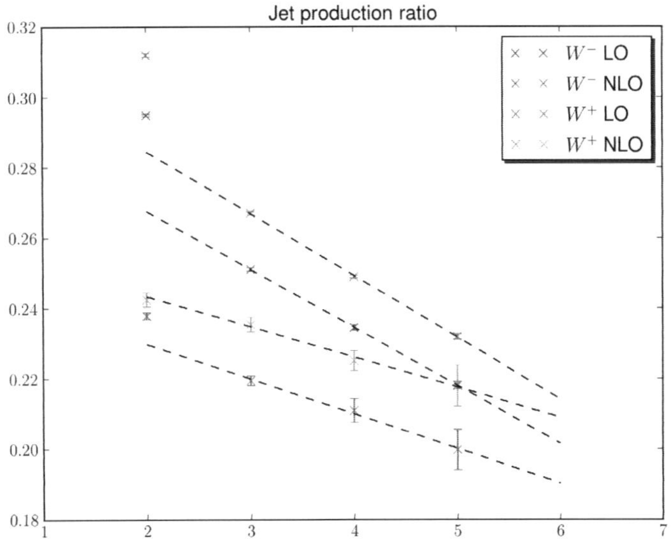

Figure 2. Ratios of $W+$ n- to $W+$ $(n-1)$-jet production at LO and NLO. The uppermost set of points shows the ratios for W^+ production at LO; the next set, for W^- production at LO; the third, for W^+ production at NLO; and the bottommost, W^- production at NLO.

There is an added bonus, if we consider ratios of cross sections for $W+$ n- to $W+$ $(n-1)$-jet production. This "jet-production" ratio [21] is sometimes called the "Berends" or "staircase" ratio. We should expect the $W+$ 1-jet cross section to behave quite differently from cross sections with higher multiplicity, because some subprocesses (for example, gg-initiated ones) are missing at LO, and also because of kinematic constraints on phase space both at LO and at NLO. This in turn means that the $W+$ 2- to $W+$ 1-jet ratio should be expected to behave quite differently than higher-multiplicity ratios. There are also kinematic constraints evident in $W+$ 2-jet production at LO, but these are largely relaxed at NLO; so we might expect to see signs of regularity or universality appearing with the $W+$ 3- to $W+$ 2-jet ratio at NLO. We consider the ratios of total cross sections with the cuts as shown in the leftmost panel of fig. 1. If we plot these ratios, as shown in fig. 2, regularity is exactly what we see: the ratios for $W+$ 3- to $W+$ 2-jet, $W+$ 4- to $W+$ 3-jet, and $W+$ 5- to $W+$ 4-jet production all lie on straight lines, both at LO and NLO. The former is a bit surprising, but it is the latter that is most significant for applications to LHC physics. It allows us to extrapolate linearly, and make a prediction for

W + 6-jet production. In this extrapolation, the W + 5-jet calculation plays a crucial role in establishing the reasonableness of a linear extrapolation. We find (see ref. [14] for details),

$$\begin{aligned} W^- + 6 \text{ jets}: \quad & 0.15 \pm 0.01 \text{ pb}, \\ W^+ + 6 \text{ jets}: \quad & 0.30 \pm 0.03 \text{ pb}, \end{aligned} \tag{1}$$

using the same cuts as shown in fig. 1. The uncertainties are estimated using a Monte-Carlo procedure with synthetic data sets.

Acknowledgments
This research was supported by the US Department of Energy under contracts DE–AC02–76SF00515 and DE–FG02–13ER42022. DAK's research is supported by the European Research Council under Advanced Investigator Grant ERC–AdG–228301. DM's work was supported by the Research Executive Agency (REA) of the European Union under the Grant Agreement number PITN–GA–2010–264564 (LHCPhenoNet). SH's work was partly supported by a grant from the US LHC Theory Initiative through NSF contract PHY–0705682. This research used resources of Academic Technology Services at UCLA, and of the National Energy Research Scientific Computing Center, which is supported by the Office of Science of the U.S. Department of Energy under Contract No. DE–AC02–05CH11231.

References
[1] Bern Z *et al.* [NLO Multileg Working Group Collaboration], arXiv:0803.0494 [hep-ph]
[2] Andersen J R *et al.* [SM and NLO Multileg Working Group Collaboration], arXiv:1003.1241 [hep-ph]
[3] Bern Z, Dixon L J, Dunbar D C and Kosower D A, 1994 *Nucl. Phys.* B **425** 217 [hep-ph/9403226]; 1995 *Nucl. Phys.* B **435** 59 [hep-ph/9409265]
[4] Bern Z, Dixon L J and Kosower D A, 1998 *Nucl. Phys.* B **513** 3 [hep-ph/9708239]
[5] Britto R, Cachazo F and Feng B, 2005 *Nucl. Phys.* B **725** 275 [hep-th/0412103]
[6] Berger C F, Bern Z, Dixon L J, Forde D and Kosower D A, 2006 *Phys. Rev.* D **74** 036009 [hep-ph/0604195]
[7] Bern Z and Morgan A G, 1996 *Nucl. Phys.* B **467** 479 [hep-ph/9511336]; Bern Z, Dixon L J, Dunbar D C and Kosower D A, 1997 *Phys. Lett.* B **394** 105 [hep-th/9611127]; Anastasiou C, Britto R, Feng B, Kunszt Z and Mastrolia P, 2007 *Phys. Lett.* B **645** 213 [hep-ph/0609191]; Britto R and Feng B, 2008 *JHEP* **0802** 095 [arXiv:0711.4284 [hep-ph]]; Giele W T, Kunszt Z and Melnikov K, 2008 *JHEP* **0804** 049 [arXiv:0801.2237 [hep-ph]]; Britto R, Feng B and Mastrolia P, 2008 *Phys. Rev.* D **78** 025031 [arXiv:0803.1989 [hep-ph]]; Ellis R K, Giele W T, Kunszt Z and Melnikov K, 2009 *Nucl. Phys.* B **822** 270 [arXiv:0806.3467 [hep-ph]]
[8] Forde D, 2007 *Phys. Rev.* D **75** 125019 [arXiv:0704.1835 [hep-ph]]
[9] Ossola G, Papadopoulos C G and Pittau R, 2007 *Nucl. Phys.* B **763** 147 [hep-ph/0609007]
[10] Badger S D, 2009 *JHEP* **0901** 049 [arXiv:0806.4600 [hep-ph]]
[11] Berger C F, Bern Z, Dixon L J, Febres Cordero F, Forde D, Ita H, Kosower D A and Maître D, 2008 *Phys. Rev.* D **78** 036003 [arXiv:0803.4180 [hep-ph]]
[12] Glover E W N and Miller D J, 1997 *Phys. Lett.* B **396** 257 [hep-ph/9609474]; Bern Z, Dixon L J, Kosower D A and Weinzierl S, 1997 *Nucl. Phys.* B **489** 3 [hep-ph/9610370]; Campbell J M, Glover E W N and Miller D J, 1997 *Phys. Lett.* B **409** 503 [hep-ph/9706297]
[13] Campbell J M and Ellis R K, 2002 *Phys. Rev.* D **65** 113007 [hep-ph/0202176]
[14] Bern Z, Dixon L J, Febres Cordero F, Höche S, Ita H, Kosower D A, Maître D and Ozeren K J, 2013 *Phys. Rev.* D **88** 014025 [arXiv:1304.1253 [hep-ph]]
[15] Gleisberg T and Hoeche S, 2008 *JHEP* **0812** 039 [arXiv:0808.3674 [hep-ph]]
[16] Gleisberg T, Höche S, Krauss F, Schälicke A, Schumann S and Winter J C, 2004 *JHEP* **0402** 056 [hep-ph/0311263]; Gleisberg T, Höche S, Krauss F, Schönherr M, Schumann S, Siegert F and Winter J, 2009 *JHEP* **0902** 007 [arXiv:0811.4622 [hep-ph]]
[17] Cacciari M, Salam G P and Soyez G, 2012 *Eur. Phys. J.* C **72** 1896 [arXiv:1111.6097 [hep-ph]]
[18] Nagy Z, 2002 *Phys. Rev. Lett.* **88** 122003 [hep-ph/0110315]; 2003 *Phys. Rev.* D **68** 094002 [hep-ph/0307268]
[19] Brun R and Rademakers F, 1997 *Nucl. Instrum. Meth.* A **389** 81 ; Antcheva I *et al.* 2011 *Comput. Phys. Commun.* **182** 1384
[20] Berger C F, Bern Z, Dixon L J, Febres Cordero F, Forde D, Gleisberg T, Ita H, Kosower D A and Maître D, 2011 *Phys. Rev. Lett.* **106** 092001 [arXiv:1009.2338 [hep-ph]]
[21] Ellis S D, Kleiss R and Stirling W J, 1985 *Phys. Lett.* B **154** 435 ; Berends F A, Giele W T, Kuijf H, Kleiss R and Stirling W J, 1989 *Phys. Lett.* B **224** 237 ; Berends F A, Kuijf H, Tausk B and Giele W T, 1991 *Nucl. Phys.* B **357** 32 ; Abouzaid E and Frisch H J, 2003 *Phys. Rev.* D **68** 033014 [hep-ph/0303088]; Gerwick E, Plehn T, Schumann S and Schichtel P, 2012 *JHEP* **1210** 162 [arXiv:1208.3676 [hep-ph]]

Distributed and multi-core computation of 2-loop integrals

E de Doncker[1] **and F Yuasa**[2]

[1] Department of Computer Science, Western Michigan University, Kalamazoo MI 49008, U. S. A.
[2] High Energy Accelerator Research Organization (KEK), Oho 1-1, Tsukuba, Ibaraki, 305-0801, Japan

E-mail: elise.dedoncker@wmich.edu, fukuko.yuasa@kek.jp

Abstract. For an automatic computation of Feynman loop integrals in the physical region we rely on an extrapolation technique where the integrals of the sequence are obtained with iterated/repeated adaptive methods from the QUADPACK 1D quadrature package. The integration rule evaluations in the outer level, corresponding to independent inner integral approximations, are assigned to threads dynamically via the OpenMP runtime in the parallel implementation. Furthermore, multi-level (nested) parallelism enables an efficient utilization of hyperthreading or larger numbers of cores. For a class of loop integrals in the unphysical region, which do not suffer from singularities in the interior of the integration domain, we find that the distributed adaptive integration methods in the multivariate PARINT package are highly efficient and accurate. We apply these techniques without resorting to integral transformations and report on the capabilities of the algorithms and the parallel performance for a test set including various types of two-loop integrals.

1. Introduction

The integral for an n-dimensional scalar Feynman diagram with L loops and N internal lines can be represented in Feynman parameter space as

$$\mathcal{I} = \frac{\Gamma\left(N - \frac{nL}{2}\right)}{(4\pi)^{nL/2}}(-1)^N \int_0^1 \prod_{j=1}^{N} dx_j \, \delta(1 - \sum x_i) \frac{C^{N-n(L+1)/2}}{(D - i\varepsilon C)^{N-nL/2}}. \quad (1)$$

Here the functions $\mathcal{C}$ and $\mathcal{D}$ are polynomials determined by the topology of the corresponding Feynman diagram [1] and we consider $n = 4$.

Section 2 of this paper outlines a multi-core algorithm for iterated/repeated integration and extrapolation, which allows for parallelization at multiple levels of the integration scheme. A parameter ε is introduced in the integrand denominator and decreased according to a geometric sequence. The corresponding integrals are evaluated by iterated integration, using a one-dimensional adaptive program from the QUADPACK package [2]. The ϵ-algorithm [3, 4] is applied for convergence acceleration of the integral sequence as ε tends to zero.

Section 3 gives a brief description of the distributed adaptive algorithm in the PARINT multivariate integration (cubature) package [5]. Written in C and layered over MPI [6], PARINT consists of parallel adaptive, Monte Carlo and quasi-Monte Carlo methods. Not unlike the QUADPACK routines, these are implemented as tools for *automatic* integration, where the user defines the integrand function and the domain, and specifies a relative and absolute error tolerance for the computation (t_r and t_a, respectively). Denoting the integral by $\mathcal{I}f$ over the domain $\mathcal{D}$,

$$\mathcal{I}f = \int_{\mathcal{D}} f(\vec{x}) d\vec{x}, \quad (2)$$

it is then the objective to return an approximation Qf and absolute error estimate E_af such that $|\, Qf - \mathcal{I}f \,| \leq \max\{\, t_a, t_r|\, \mathcal{I}f \,|\,\} \leq E_af$, or indicate with an error flag that the requested accuracy cannot be achieved. When a relative accuracy (only) needs to be satisfied we set $t_a = 0$ (and vice-versa).

For iterated integration over a d-dimensional product region we will express (2) as

$$\mathcal{I}f = \int_{\alpha_1}^{\beta_1} dx_1 \int_{\alpha_2}^{\beta_2} dx_2 \ldots \int_{\alpha_d}^{\beta_d} dx_d\, f(x_1, x_2, \ldots, x_d), \tag{3}$$

where the limits of integration are given by functions $\alpha_j = \alpha_j(x_1, x_2, \ldots, x_{j-1})$ and $\beta_j = \beta_j(x_1, x_2, \ldots, x_{j-1})$. In particular, the boundaries of the d-dimensional unit simplex $\mathcal{S}_d$ are $\alpha_j = 0$ and $\beta_j = 1 - \sum_{k=1}^{j-1} x_k$, i.e.,

$$\mathcal{S}_d = \{\, (x_1, x_2, \ldots, x_d) \in \mathbb{R}^d \mid \sum_{j=1}^{d} x_j \leq 1 \text{ and } x_j \geq 0 \,\}. \tag{4}$$

Note that, after removing the δ-function in (1) and eliminating one of the variables in view of $\sum_{j=1}^{N} x_j = 1$, the integration is over a d-dimensional simplex of the form (4) with $d = N - 1$.

The multivariate adaptive approach yields excellent results for a set of two-loop box integrals specified by Laporta [7]. On the other hand, the iterated scheme with extrapolation has been shown to succeed for various classes of integrals in the physical region [8, 9, 10, 11, 12], where multivariate adaptive integration fails.

Results for the Laporta diagrams are reported in Sections 4 and 5 of this paper. The treatment of the sample integrals does not rely on integral transformations as proposed in [12] to improve the integrand behavior. We obtained most of the numerical approximations and timings on the HPCS (High Performance Computational Science) cluster at WMU. For this purpose we used 16-core cluster nodes with Intel Xeon E5-2670, 2.6GHz processors and 128GB of memory, and the cluster's Infiniband interconnect for message passing via MPI.

Some sample sequential results were also collected from runs on a PC node at minamivt005.kek.jp with 6-core Intel Xeon X5680@3.33GHz CPU, and on a 2.6GHz Intel Core i7 Mac-Pro with 4 cores. For the inclusion of OpenMP [13] multi-threading compiler directives in the iterated integration code (based on the Fortran version of QUADPACK) we used the (GNU) *gfortran* compiler and the Intel Fortran compiler, with the flags *-fopenmp* and *-openmp*, respectively. PARINT and its integrand functions were compiled with gcc. Code sections of the sample integrands are supplied in Appendix A.

2. Parallel iterated integration and extrapolation

The numerical integration over the interval $[\alpha_j, \beta_j], 1 \leq j \leq d$ in (3) can be performed with the 1D adaptive integration code DQAGE from the QUADPACK package [2] in each coordinate direction. Via an input parameter to DQAGE we select the 15-point Gauss-Kronrod rule pair for the integral (and error) approximation on each subinterval. If an interval $[a, b]$ arises in the partitioning of $[\alpha_j, \beta_j]$, then the local integral approximation over $[a, b]$ is of the form

$$\int_a^b dx_j\, F(c_1, \ldots, c_{j-1}, x_j) \approx \sum_{k=1}^{K} w_k F(c_1, \ldots, c_{j-1}, x^{(k)}), \tag{5}$$

where the w_k and $x^{(k)}, 1 \leq k \leq K (= 15)$, are the weights and abscissae of the local rule scaled to the interval $[a, b]$ and applied in the x_j-direction. For $j = 1$ this is the outer integration direction. The function evaluation

$$F(c_1, \ldots, c_{j-1}, x^{(k)}) = \int_{\alpha_{j+1}}^{\beta_{j+1}} dx_{j+1} \ldots \int_{\alpha_d}^{\beta_d} dx_d\, f(c_1, \ldots, c_{j-1}, x^{(k)}, x_{j+1}, \ldots, x_d), \quad 1 \leq k \leq K, \tag{6}$$

is itself an integral in the $x_{j+1}, \ldots, x_d$-directions for $1 \leq j < d$, and is computed by the method(s) for the inner integrations. For $j = d$, (6) is the evaluation of the integrand function

$$F(c_1, \ldots, c_{d-1}, x^{(k)}) = f(c_1, \ldots, c_{d-1}, x^{(k)}).$$

Note that successive coordinate directions may be combined into layers in the iterated integration scheme. Furthermore, the error incurred in any inner integration will contribute to the integration error in all of its subsequent outer integrations [14, 15, 16].

Since the $F()$ evaluations on the right of (5) are independent of one another they can be evaluated in parallel. Important benefits of this approach include that

(i) the granularity of the parallel integration is large, especially when the inner integrals $F()$ are of dimension two or greater;

(ii) the points where the function F is evaluated in parallel are the same as those of the sequential evaluation; i.e., apart from the order of the summation in (5), the parallel calculation is essentially the same as the sequential one. This important property facilitates the debugging of parallel code. As another characteristic the parallelization does not increase the total amount of computational work.

In addition, the memory required for the procedure is determined by (the sum of) the amounts of memory needed for the data pertaining to the subintervals incurred in each coordinate direction (corresponding to the length of the recursion stack for a recursive implementation). Consequently the total memory increases linearly as a function of the dimension d.

For a numerical extrapolation with the ϵ-algorithm [3, 4] we generate a sequence of integral values $\mathcal{I}(\varepsilon_\ell)$, using a geometric progression of ε_ℓ which tends to zero with increasing ℓ (see also [8, 10]). The extrapolation results given here are obtained with a version of the ϵ-algorithm code (DQEXT) from QUADPACK. Note that the accuracy of the extrapolated result is generally limited by the accuracy of the input sequence, which in turn is determined in PARINT by the user-prescribed error tolerance for the integration.

The input parameters of DQAGE include the maximum number (*limit*) of subdivisions allowed in the partitioning of the given interval by the adaptive algorithm. A related output parameter is the error code *iflag*, which returns 1 to indicate that the number of subdivisions reached its maximum, *limit*, upon termination (and is 0 otherwise). In an iterated integration it is sometimes beneficial to keep track of the number of times *limit* is reached throughout the integration on each particular level, as an indication of the difficulty of the integrand in different coordinate directions.

3. PARINT parallel adaptive integration

Layered over MPI [6] for distributed computing, the PARINT adaptive integration executable launches a user-specified number of processes which assume the role of controller and workers. The integration domain is divided initially among the workers. Each on its own part of the domain, the workers engage in an adaptive partitioning strategy similar to that of DQAGE, DCUHRE [17] and HALF [18] by successive bisections, and thus each generate a local priority queue of subregions as a task pool. The priority queue is implemented as a max-heap keyed with the estimated integration errors over the subregions, so that the subregion with the largest estimated error is stored in the root of the heap. However, if the user specifies a maximum size for the heap structure on the worker, the task pool is stored as a *deap* or *double-ended heap* which allows deleting of the maximum as well as the minimum element efficiently (in order to maintain the size of the data structure once it reaches its maximum).

Each task consists of the subdivision (bisection) of the associated subregion (generating two children regions), integration over the children, deletion of the parent region (root of the heap) and insertion of the children back into the heap. The bisection of a region is performed perpendicularly to the coordinate direction in which the integrand is found to vary the most, according to fourth-order differences computed

Table 1. Times for $N = 6$ and $N = 7$ parallel iterated integration (HPCS cluster)

Diagram:	Fig 1(b)			Fig 1(c)	Fig 1(d)	Fig 1(e)
N =	6	6	6	7	7	7
limit =	10	15	15	10	14	14
t_r (1D) =	10^{-8}		10^{-11}	10^{-6}	10^{-8}	10^{-11}
ε_0 =	1.2×10^{-10}		1.2×10^{-20}	1.2×10^{-15}	1.2×10^{-14}	1.2×10^{-13}
$\|Error\|$ =	5.0×10^{-6}	5.0×10^{-6}	1.8×10^{-10}	1.2×10^{-5}	2.6×10^{-9}	4.6×10^{-9}
# THREADS			TIME (S)			
1	1155.6	1167.5	37381.0	4151.0	15316.8	
2	716.2	627.2	36808.4	3620.0	14706.1	
4	360.5	360.4	18248.5	1994.0	7443.2	
6	255.4	254.8	12068.2	1011.3	4793.6	
8	226.5	221.4	9045.1	800.7	3847.1	
10	206.8	213.7	7409.4	649.2	3482.7	
12	201.4	207.2	6978.2	587.2	3287.3	
15	193.9	198.8	5634.6	549.9	3191.1	
$(15, 15)$	109.0	115.4	3237.7	282.5	1216.8	5837.9

in each direction [17, 18]. The integration rule on a subregion is determined by the type of region and the user-selected polynomial degree. The available cubature rules in PARINT include a set of rules for the d-dimensional cube [19, 20, 21], the Quadpack rules for $d = 1$, and a set of rules for the d-dimensional simplex [22, 23]. The latter are the simplex rules by Grundmann and Möller [23] (which were also found by de Doncker [24] via extrapolation).

An important mechanism of the distributed integration algorithm is the load balancing strategy, which is receiver-initiated and geared to keeping the loads on the worker task pools balanced, particularly in cases of irregular integrand behavior. The message passing is performed in a non-blocking and asynchronous manner, and permits overlapping of computation and communication. The user has the option of turning load balancing on or off, as well as allowing or dis-allowing the controller to also act as a worker. Optionally the installation can be configured to use long doubles instead of doubles.

The multivariate partitioning (sequentially or in parallel) exhibits a dimensional effect (exponential as a function of d) with respect to the total number of regions generated and thus the required total memory. To some extent this is helped by the distribution of the subregions over the participating processors' local memories in the parallel system.

As a result of the asynchronous processing and message passing on MPI, PARINT executes on a hybrid platform (multi-core and distributed) by assigning multiple processes to each node. However, it is generally not possible to repeat an execution exactly. The parallelization also leads to *breaking loss* or extra work performed at the end of the computation, due to the time elapsed while workers continue to generate subregions after the termination message has been issued but before receiving it. This may increase the total amount of work (compared to the sequential work), particularly when the number of processes is large.

4. Results of multi-threaded iterated integration

We consider five two-loop box diagrams from Laporta [7], given in Fig 1. Table 1 gives an overview of the problem specifications, the execution times for a varying number of threads, and the absolute accuracy achieved (|Error|) with respect to the values C_0 pp. 46-47 in [7] for the diagrams in Fig 1(b-d). For Fig 1(e) we compare with the result from PARINT executed in long double precision (cf., Section 5 below). Detailed results for the $N = 5$ diagram of Fig 1(a) are listed in Table 2.

The multi-threading is invoked in the outer (x_1) integration in (3), except for the entry denoted $(15, 15)$ which represents nested parallelization applied in the outer two (x_1 and x_2) integrations. It emerges that the latter outperforms the single level parallelization, showing a more efficient use of the available cores. The integrations are performed over the simplex $\mathcal{S}_d$ for $d = N - 1$ (see (4)). Appendix A.1-5 provides code segments for the integrands corresponding to Fig 1(a-e). The integrand evaluation in

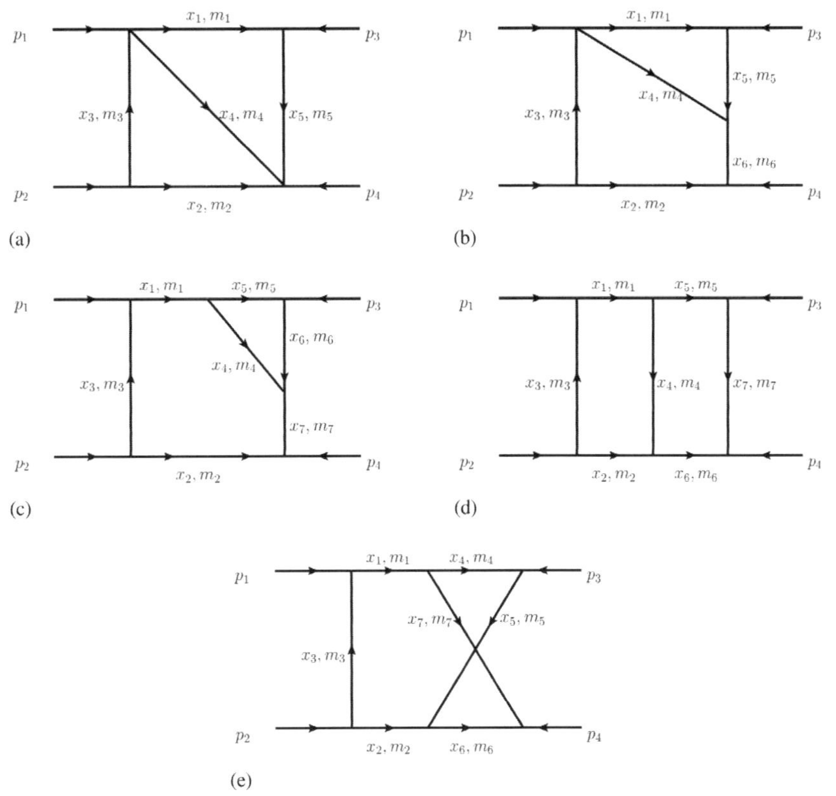

Figure 1. Diagrams (a) $N = 5$, (b) $N = 6$, (c) $N = 7$, (d) $N = 7$ ladder, (e) $N = 7$ crossed

the code corresponds to an expression of the real part integrand where the factor εC in the denominator of (1) is replaced by ε for the extrapolation as $\varepsilon \to 0$. Note also that the factor $1/(4\pi)^{nL/2}$ is omitted.

The problem specifications in Table 1 include: the number of internal lines N, the relative error tolerance t_r requested for the 1D integrations of the iterated scheme, the maximum value *limit* for the number of subdivisions of the adaptive integration procedure in each coordinate direction, and the starting value of $\varepsilon = \varepsilon_0$ for the extrapolation sequence. For the sequence we use $\varepsilon_\ell = \varepsilon_0 \times 1.2^{-\ell}$, $0 \le \ell < \ell_{\max}$, where the length of the sequence ($\ell_{\max}$) is around 15. In view of the trade-off between the amount of work done and the accuracy achieved, we choose smaller ε_0 and larger *limit* for higher accuracy work.

4.1. Diagram with five internal legs

For a sequential run on minamivt005, with target relative accuracy of 10^{-6} for the 1D integrations, and the maximum number of subdivisions *limit* = 10, 10, 20, 20 in the x_1, x_2, x_3 and x_4 coordinate directions, respectively, extrapolation with the ϵ-algorithm on a sequence of 15 integrals for $\varepsilon_\ell = 1.2^{-15-\ell}$, $0 \le \ell < 15$ converges with relative error 4×10^{-6} in a total elapsed time of 30.3 seconds.

A more accurate computation leads to the convergence results in Table 2, where in successive columns the following are given: ℓ, ε_ℓ, the integration time, the integral approximation corresponding to ε_ℓ and the extrapolated value for C_0. The integration time listed is that of a sequential run on a node of the HPCS cluster. The target 1D relative accuracy for the integrations was 10^{-10}, $\varepsilon_0 = 1.2^{-20}$ and the maximum number of subdivisions *limit* = 30, 30, 50, 50 in the successive coordinate directions.

Table 2. Extrapolation, $N = 5$, $\varepsilon_0 = 1.2^{-20}$, rel. tol. 10^{-10}, seq. times (HPCS cluster)

ℓ	ε_ℓ	TIME (s)	$\mathcal{I}(\varepsilon_\ell)$ Approx.	C_0 Approx.
0	0.260841e-01	15.0	0.879998554543172662	
1	0.217367e-01	15.2	0.892831432362168109	
2	0.181139e-01	16.7	0.903287762406520423	0.9492935019767
3	0.150949e-01	23.0	0.911812341919760572	0.9494302717234
4	0.125791e-01	24.1	0.918769150341187113	0.9490308675735
⋮	⋮	⋮	⋮	⋮
12	0.292550e-02	65.1	0.943930996955826496	0.9509233190442
13	0.243792e-02	82.8	0.945122576092676248	0.9509234745063
14	0.203160e-02	99.5	0.946108275775448759	0.9509235737400
15	0.169300e-02	105.4	0.946924415927191898	0.9509235597296
16	0.141083e-02	108.9	0.947600717832237871	0.9509235626613
	TOTAL TIME:	843.4	LAPORTA VALUE:	0.9509235623171

Table 3. Extrap. results $N = 6$, $\varepsilon_0 = 1.2^{-20}$, rel. tol. 10^{-11}, *limit* = 15, # threads (reg. nodes) = 15

ℓ	ε_ℓ	TIME (s)	$\mathcal{I}(\varepsilon_\ell)$ Approx.	C_0 Approx.	$\mid(C_0(\ell) - C_0(\ell-1))\mid$
0	0.260841e-01	65.4	0.221532578551221299		
1	0.217367e-01	99.9	0.230888677472657844		
2	0.181139e-01	109.1	0.238665585251643331	0.276963858955	
3	0.150949e-01	161.6	0.245113995206973190	0.276414002035	5.50e-04
4	0.125791e-01	185.5	0.250453078514603744	0.275897645038	5.16e-04
5	0.104826e-01	209.5	0.254870342845598397	0.275953613377	5.60e-05
⋮	⋮	⋮	⋮	⋮	⋮
10	0.421272e-02	687.1	0.267830329163125991	0.276209336599	1.08e-06
11	0.351060e-02	841.0	0.269250457421247535	0.276209925289	5.89e-07
12	0.292550e-02	721.5	0.270427723407943632	0.276209228098	6.97e-07
13	0.243792e-02	904.6	0.271404161999583338	0.276209225178	2.92e-09
TOTAL TIME (15t):		5634.6	LAPORTA VALUE:	0.276209225359	

The extrapolation converges within eight to nine digits which agree with Laporta's result. In view of the small computation times of the individual integrals, parallelization is not highly beneficial. Some speedup is observed for up to four threads with a parallel time of 445s, compared to 536s for two threads, and 843s for one thread (i.e., the sequential time of Table 2. Parallel speedup is defined as the ratio of sequential to parallel time.

4.2. *Diagram with six internal legs*

In test runs for a relative tolerance of 10^{-8}, and allowed maximum number of subdivisions *limit* = 10 in each coordinate direction, extrapolation on the sequence with $\varepsilon_\ell = 1.2^{-10-\ell}$, $0 \leq \ell < 15$ converges to Laporta's result within 4 to 5 digits, in about 1,156 seconds on a node of the HPCS cluster for the sequential computation (764 seconds on Mac Pro). Timings for *limit* = 10 and 15, and for a varying number of threads are given in columns 2 and 3 of Table 1.

We achieved a higher accuracy (differing 1.8×10^{-10} from Laporta's result, i.e., with a relative accuracy of 6.5×10^{-10}), as specified in column 4 of Table 1. Note how the time decreases overall from 37,381 seconds for the sequential run to 3,238 seconds with the parallel version, yielding a speedup of about 12.6. The convergence results in Table 3 and execution times per iteration, using 15 threads, were obtained on the HPCS cluster. Corresponding to $C_0 = C_0(\ell)$, the last column represents $|C_0(\ell) - C_0(\ell - 1)|$ which serves for the purpose of gauging the convergence. Since it is based on only two consecutive elements it may not be a good estimate of the error. The ϵ-algorithm routine DQEXT in QUADPACK produces a more sophisticated (and more conservative) error estimate.

Table 4. Extrap. results $N = 7$ crossed, $\varepsilon_0 = 1.2^{-13}$, rel. tol. 10^{-11}, *limit* = 14, # threads $(15, 15)$

ℓ	ε_ℓ	TIME (s)	CUMULATIVE TIME (s)	$\mathcal{I}(\varepsilon_\ell)$ Approx.	C_0 Approx.	$\mid (C_0(\ell) - C_0(\ell-1)) \mid$
0	0.934639e-01	56.4	56.4	0.162481351192060378e-01		
1	0.778866e-01	100.7	157.1	0.284733506638041936e-01		
2	0.649055e-01	109.2	266.3	0.398260398940242025e-01	0.1875391249e+00	
3	0.540879e-01	118.6	384.9	0.497257434763870229e-01	0.1171759149e+00	7.04e-02
4	0.450732e-01	147.0	531.9	0.579677963500485893e-01	0.9894834504e-01	1.82e-02
5	0.375610e-01	159.1	691.0	0.645980481345204693e-01	0.8382861180e-01	1.51e-02
6	0.313009e-01	268.8	959.8	0.697965854019316212e-01	0.8555394058e-01	1.73e-03
7	0.260841e-01	195.1	1154.8	0.737945973869281180e-01	0.8541352574e-01	1.40e-04
8	0.217367e-01	333.7	1488.5	0.768245432456762939e-01	0.8535482088e-01	5.87e-05
⋮	⋮	⋮	⋮	⋮	⋮	
14	0.727958e-02	831.2	5032.2	0.840912496690573441e-01	0.8535140207e-01	3.10e-08
15	0.606632e-02	805.6	5837.9	0.844433778969480436e-01	0.8535139544e-01	6.63e-09

4.3. Diagrams with seven internal legs

The integrands for the ladder diagram of Fig 1(d) and the crossed diagram of Fig 1(e) correspond to those for the planar and the non-planar diagram, respectively, before the transformations applied in [12].

4.3.1. Diagram of Fig 1(c). For the diagram Fig 1(c), a sequential run on minamivt005 generated a result with relative error of 1.2×10^{-5} in 14,719 seconds, using relative integration error tolerance 10^{-6}, maximum number of subdivisions *limit* = 10 in all coordinate directions, and extrapolation on an input sequence of 15 integrals starting at $\varepsilon_0 = 1.2^{-15}$.

By examining the *iflag* parameters, the output from the execution reveals that the maximum number of subdivisions is reached particularly toward the inner integrations. For this problem, reversing the order of the integration variables as shown in the function definition in Appendix A.3, improves the performance dramatically. Keeping all other inputs the same, the sequential execution time decreases to 4,151 seconds while the accuracy of the result is maintained. Parallel times obtained when varying the number of threads (on HPCS cluster nodes) are given in Table 1 (column of Fig 1(c)). A speedup of about 14.7 is achieved with nested parallelism.

4.3.2. Two-loop ladder box diagram Fig 1(d). For the diagram of Fig 1(d) a preliminary minamivt005 run with *limit* = 5 in all directions, relative integration tolerance 10^{-6} and 15 integrations starting at $\varepsilon = 1.2^{-10}$, gave the result 0.1036438 which, compared to Laporta's listed value of 0.10364072, has an absolute accuracy of about 3×10^{-6} (relative accuracy 3×10^{-5}), after an execution time of 7,848 seconds. The *iflag* parameters indicated difficult behavior in the inner integrations for this problem. By changing the order of the integration variables as listed in the code segment of Appendix A.4 for the integrand function, while keeping all other inputs unchanged, the program runs in 1,072 seconds on Mac Pro and delivers the extrapolation result 0.1036451 (with relative error 4.2×10^{-5}).

We produced a more accurate result and timed the executions with *limit* = 14 in all directions, relative 1D integration tolerance 10^{-8}, for 17 integrations starting at $\varepsilon_0 = 1.2^{-14}$. For the timing results see Table 1 (column of Fig 1(d)). The extrapolated integral, 0.1036407236, has a relative error of 2.5×10^{-8}. Note the overall gain in performance by the parallelization, from 15,317 seconds (sequential) to 1,217 seconds (parallel), or a speedup of about 12.6.

4.3.3. Two-loop crossed box diagram Fig 1(e). A sequential run on minamivt005 with *limit* = 5 in all directions, relative integration tolerance 10^{-6}, $\varepsilon_0 = 1.2^{-10}$ generated the result 0.85354533e-01 at $\ell = 15$, in 2,901 seconds. The integration variables were set as: $x1 = x(1)$, $x2 = 1 - x(1) - x(2) - x(3) - x(4) - x(5) - x(6)$, $x3 = x(2)$, $x4 = x(3)$, $x5 = x(4)$, $x6 = x(5)$, $x7 = x(6)$. With the order of the integration variables changed to that of the code segment in Appendix A.5 for the integrand, the program returned the result 0.85355139e-01 at $\ell = 14$, and 0.85349174e-01 at $\ell = 15$, in a total time of 801 seconds on Mac Pro.

Higher accuracy runs were performed with relative integration tolerance 10^{-11}, *limit* = 14 and $\varepsilon_0 = 1.2^{-13}$, resulting in 0.85351402e-01 at $\ell = 14$ in 5,032 seconds, and 0.85351395e-01 at $\ell = 15$ in 5,838 seconds (total). Table 4 gives an overview of the convergence and the times for the individual iterations using nested multi-threading $(15, 15)$. As another comparison, since the problem is actually unphysical for the specified kinematics, it is possible to set $\varepsilon = 0$ in the integrand. With 1D relative error tolerance 10^{-8} and *limit* = 15, this gives 0.853513981536e-01 (with absolute error estimate of the outer integration 2.2e-09) in 4,383 seconds, using $(15, 15)$ nested multi-threading. Laporta's method did not solve this problem at the time of the writing of his article.

5. PARINT results

PARINT contains a parallel adaptive method with load balancing for d-dimensional rectangular regions roughly based on the subdivision strategy of [19, 20, 21] and for simplex regions [22, 25], with results provided by d-dimensional integration rules of degrees 7 and 9 for the cube, and 3, 5, 7 and 9 for the simplex. For the problems at hand, some testing reveals that it is more efficient for PARINT to transform the integral from the (unit) simplex to the (unit) cube, followed by an integration over the cube. We use the transformation $(x(1), \ldots, x(d)) \rightarrow (x_1, (1-x_1)x(2), \ldots, (1-x_1-\ldots-x_{d-1})x(d))$ with Jacobian $(1-x_1)(1-x_1-x_2)\ldots(1-x_1-\ldots-x_{d-1})$.

Since the integrals with kinematics specified by Laporta [7] are unphysical we set $\varepsilon = 0$, or *eps* = 0 in the integrand codes of Appendix A.1-5. However, care needs to be taken to handle boundary singularities in case the integration rule employs function evaluations on the boundary. For example, if $x_1 = x_4 = x_5 = 0$ (and *eps* = 0) in the definition of *fc* in A.1, then *cc* = *dd* = 0. We set the integrand to 0 to avoid a divide-by-0 floating point exception.

Fig 2 presents timing results obtained with PARINT on the HPCS cluster, corresponding to the five diagrams in Fig 1. Up to 64 MPI processes were spawned on up to 4 cluster nodes, using 16 procs (slots) per node. When referring to numbers of integrand evaluations, *million* and *billion* are abbreviated by "M" and "B", respectively. Since the times vary when the same run is repeated, the plots show timings usually averaged over three or four runs for the same p. The times T_p are expressed in seconds, as a function of the number of MPI processes p, $1 \leq p \leq 64$.

Table 5 gives a brief overview of pertinent test specifications, T_1, T_p and the speedup $S_p = T_1/T_p$ for $p = 64$. For example, the times of the $N = 5$ diagram decrease from 32.6s at $p = 1$ to 0.74s at $p = 64$ for $t_r = 10^{-10}$ (reaching a speedup of 44); whereas the $N = 7$ crossed diagram times range between 27.6s and 0.49s for $t_r = 10^{-7}$ (with speedup exceeding 56).

For the two-loop crossed box problem we also ran PARINT in *long double* precision. The results for an increasing allowed maximum number of evaluations and increasingly strict (relative) error tolerance t_r (using 64 processes) are given in Table 6, as well as the corresponding double precision results. Listed are: the integral approximation, actual relative error estimate E_r, number of function evaluations reached

Table 5. Test specifications and range of times in Fig 2(a-d)

DIAGRAM	N	REL TOL t_r	MAX EVALS	$T_1[s]$	$T_{64}[s]$	SPEEDUP S_{64}
Fig 1(a) / 2(a)	5	10^{-10}	400M	32.6	0.74	44.1
Fig 1(e) / 2(a) crossed	7	10^{-7}	300M	27.6	0.49	56.3
Fig 1(b) / 2(b)	6	10^{-9}	3B	213.6	5.06	42.2
Fig 1(d) / 2(b) ladder	7	10^{-8}	2B	189.9	4.33	43.9
Fig 1(c) / 2(c)	7	10^{-8}	5B	507.9	8.83	57.5
Fig 1(e) / 2(d) crossed	7	10^{-9}	20B	1892.5	34.6	54.7

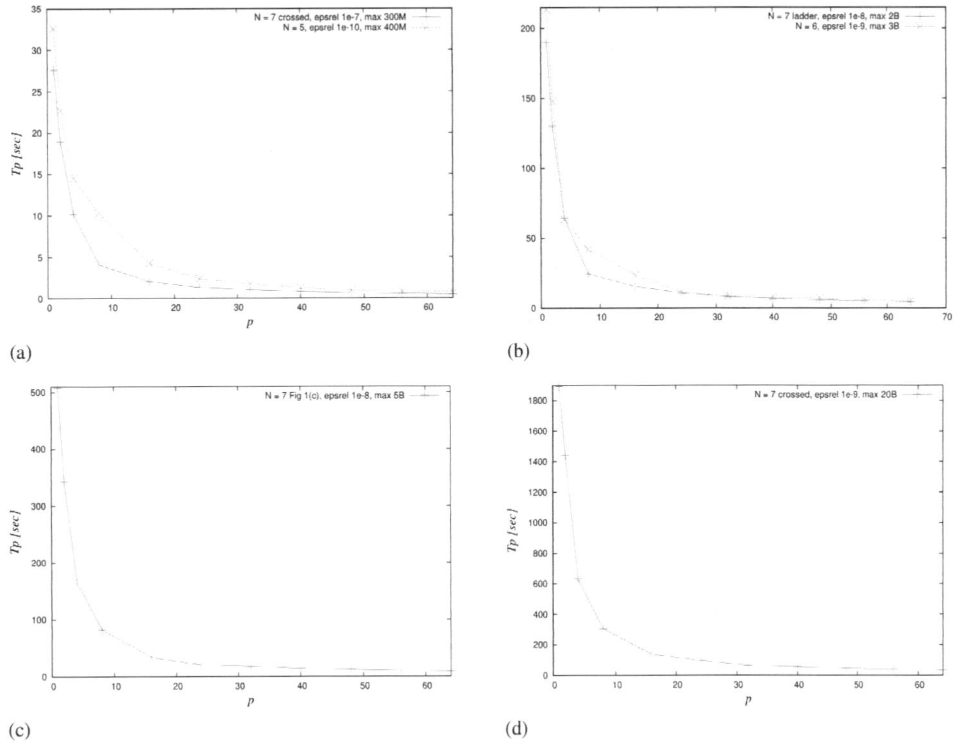

Figure 2. (a) $N = 5, t_r = 10^{-10}$ and $N = 7$ crossed, $t_r = 10^{-7}$; (b) $N = 6, t_r = 10^{-8}$ and $N = 7$ ladder, $t_r = 10^{-8}$; (c) $N = 7$ [Fig 1(c)], $t_r = 10^{-8}$; (d) $N = 7$ crossed, $t_r = 10^{-9}$

Table 6. PARINT double and long double results for crossed diagram Fig 1(e)

t_r	Max		*long double* precision					*double* precision		
	Evals	INTEGRAL APPROX	E_r	*iflag*	#EVALS	TIME	E_r	*iflag*	#EVALS	TIME
10^{-08}	600M	0.085351397048123	3.5e-07	1	600000793	1.7	3.4E-07	1	600001115	0.95
	1B	0.085351397753978	1.7e-07	1	1000000141	2.9	1.6e-07	1	1000000141	1.6
	2B	0.085351398064779	2.9e-08	1	2000000443	6.0	2.9e-08	1	2000000765	3.3
	6B	0.085351398130559	5.6e-09	0	4164032999	14.3	8.0e-09	0	4424455329	8.6
10^{-09}	10B	0.085351398143465	5.5e-09	1	10000001571	29.7	5.5e-09	1	10000000927	16.4
	50B	0.085351398152623	9.9e-10	0	35701321579	124.4	4.5e-10	0	9799638359	16.0
10^{-10}	80B	0.085351398153315	5.5e-10	1	80000001137	240.2	5.5e-10	1	80000000171	133.5
	100B	0.085351398153507	4.1e-10	1	100000000093	302.0	4.1e-10	1	100000000093	168.3
	300B	0.085351398153798	9.1e-11	0	238854968513	642.3	1.3e-10	1	300000000279	587.2

and time taken in *long double* precision, followed by the relative error estimate, number of function evaluations reached and running time in *double* precision. Allowing for 16 processes per node and up to 64 processes total, the MPI host file has four lines of the form *nx slots=16* where *nx* represents a selected node name. The running time is reported (in seconds) from the PARINT executable, and comprises all computation not inclusive the process spawning and PARINT initialization. For a comparable number of function evaluations, the time using long doubles is slightly less than twice the time taken using doubles. The *iflag* parameter returns 0 when the requested accuracy is assumed to be achieved, and 1 otherwise. Reaching the maximum number of evaluations results in abnormal termination with *iflag* = 1.

The integral approximation for the *double* computation is not listed in Table 6; it is consistent with the *long double* result within the estimated error (which appears to be over-estimated). Using doubles the program terminates abnormally for the requested relative accuracy of $t_r = 10^{-10}$. Normal termination is achieved in this case around 239B evaluations with long doubles.

6. Concluding remarks

In this paper we explored *parallel* integration methods for the *automatic* computation of a class of two-loop box integrals [7]: (i) multi-threaded iterated/repeated integration with extrapolation, and (ii) distributed adaptive multivariate integration. With regard to (i) we demonstrated the capabilities of the extrapolation for convergence acceleration, and of the parallelization on multiple levels of the iterated integration scheme to speed up the execution.

For physical kinematics we rely on calculating a sequence of integrals for extrapolation as the parameter ε decreases [8, 9, 10, 11, 12], and it has generally not been possible to calculate the integrals for small enough ε using adaptive partitioning of the multivariate domain. However, the test cases specified in [7] allow setting $\varepsilon = 0$ in the integrand, and it emerges that the distributed adaptive approach of PARINT solves these problems very efficiently (although iterated integration has an advantage with respect to low memory usage). PARINT executes on multiple nodes, and multiple cores per node. Future work in this area will address, for example, the distributed (non-adaptive) quasi-Monte Carlo and Monte Carlo methods in the current PARINT package, parallel random number generators and utilizing GPUs and other accelerators for the computations.

Future work on the iterated strategy will include improved error handling across the levels of the iterated scheme. Furthermore, the parallelization of the iterated strategy can be extended to handle infrared divergence as presented in [27]. Higher-dimensional problems can be attempted with iterated integration by combining some successive 1D into multivariate layers as was done in [8]. Evaluating the points in the 2D (or 3D) outer layer on GPUs may yield an alternative in some cases, if the structure of the computation is sufficiently similar for all evaluation points [26]. In view of the computationally intensive nature of the underlying problems, roundoff error guards and suitable summation techniques need to be considered. Finally, the multi-threaded iterated integration can be incorporated within the distributed setting of PARINT on MPI [6].

Acknowledgments

We acknowledge the support from the National Science Foundation under Award Number 1126438. This work is further supported by Grant-in-Aid for Scientific Research (24540292) of JSPS, and the Large Scale Simulation Program No.12-07 of KEK. We also thank Dr. L. Cucos and O. Olagbemi for help with PARINT.

Appendix A. Code of two-loop box integrands

Code segments for the integrands corresponding to the two-loop box diagrams of Fig 1(a-e) are given below. Note that '&' indicates continuation, the *sqeps* variable represents ε^2 and *sqrt3* represents $\sqrt{3}$.

Appendix A.1. Diagram Fig 1(a) (N = 5)

```
      x1 = x(1)
      x2 = 1-x1-x(2)-x(3)-x(4)
      x3 = x(2)
      x4 = x(3)
      x5 = x(4)
      cc = (x1+x5)*(x2+x3+x4)+x4*(x2+x3)
      dd =
     & -x1**2*x2-x1**2*x3-x1**2*x4-x1*x2**2-x1*x2*x3-2*x1*x2*x4-x1*x2*x5
     & -x1*x3**2-2*x1*x3*x4-x1*x3*x5-x1*x4**2-x1*x4*x5-x2**2*x4-x2**2*x5
```

```
&  -x2*x3*x4-x2*x3*x5-x2*x4**2-2*x2*x4*x5-x2*x5**2-x3**2*x4-x3**2*x5
&  -x3*x4**2-2*x3*x4*x5-x3*x5**2-x4**2*x5-x4*x5**2

   fc = -dd/cc/(dd**2+sqeps)
```

Appendix A.2. Diagram Fig 1(b) (N = 6)

```
   x6 = x(1)
   x5 = x(2)
   x4 = x(3)
   x3 = x(4)
   x2 = x(5)
   x1 = 1-x6-x2-x3-x4-x5

   dd = (x1**2+x5**2+x1*x5)*(x2+x3+x4+x6)
&       +(x2**2+x3**2+x6**2+x2*x3+x2*x6+x3*x6)*(x1+x4+x5)
&       +x4**2*(x1+x2+x3+x5+x6)+3*x4*x5*x6+
&       2*x1*x4*(x2+x3+x6)+2*x4*x5*(x2+x3)

   fd = (dd-eps)*(dd+eps)/(dd**2+sqeps)**2
```

Appendix A.3. Diagram of Fig 2(a) (N = 7)

```
   x7 = x(1)
   x6 = x(2)
   x5 = x(3)
   x4 = x(4)
   x3 = x(5)
   x2 = x(6)
   x1 = 1-x7-x2-x3-x4-x5-x6
   dd =
&  -(x1**2+x2**2+x3**2+x7**2+x1*x2+x1*x3+x1*x7+x2*x3+x2*x7+x3*x7)
&  *(x4+x5+x6)-x4**2*(1-x4)-(x5**2+x6**2+x5*x6)*(1-x5-x6)
&  -3*x4*(x1*x5+x6*x7)-2*((x1+x2+x3)*x4*x6+(x2+x3+x7)*x4*x5)

   cc = x1*x4+x1*x5+x1*x6+x2*x4+x2*x5+x2*x6+x3*x4+x3*x5+x3*x6+x4*x5
&     +x4*x6+x4*x7+x5*x7+x6*x7

   fg = -2*cc*dd*(dd-sqrt3*eps)*(dd+sqrt3*eps)/(dd**2+eps**2)**3
```

Appendix A.4. Ladder box diagram Fig 1(d) (N = 7)

```
   x7 = x(1)
   x6 = x(2)
   x5 = x(3)
   x4 = x(4)
   x3 = x(5)
   x2 = x(6)
   x1 = 1-x2-x3-x4-x5-x6-x7

   x1234 = x1+x2+x3+x4
   x45 = x4+x5
   x67 = x6+x7
   x4567 = x45 + x67

   cc = x1234*x4567 - x4**2
   dd = cc*(x1+x2+x3+x4+x5+x6+x7) - (x1*x2*x4567
&       + x5*x6*x1234 + x1*x4*x6 + x2*x4*x5 + x3*x4*x7
&       + x3*(x1*x4 + x1*x5 + x1*x6 + x1*x7 + x4*x5)
&       + x3*(x2*x4 + x2*x5 + x2*x6 + x2*x7 + x4*x6)
&       + x7*(x1*x4 + x1*x5 + x2*x5 + x3*x5 + x4*x5)
&       + x7*(x1*x6 + x2*x4 + x2*x6 + x3*x6 + x4*x6))

   fh = 2*cc*dd*(dd-sqrt3*eps)*(dd+sqrt3*eps)/(dd**2+eps**2)**3
```

Appendix A.5. Crossed box diagram Fig 1(e) (N = 7)

```
      x7 = x(1)
      x6 = x(2)
      x5 = x(3)
      x4 = x(4)
      x3 = x(5)
      x2 = x(6)
      x1 = 1-x2-x3-x4-x5-x6-x7

      cc = (x1+x2+x3+x4+x5)*(x1+x2+x3+x6+x7) - (x1+x2+x3)**2

      dd = cc - (x1*x2*x4 + x1*x2*x5 + x1*x2*x6 + x1*x2*x7
     &     + x1*x5*x6 + x2*x4*x7 - x3*x4*x6
     &     + x3*(-x4*x6 + x5*x7)
     &     + x3*(x1*x4 + x1*x5 + x1*x6 + x1*x7 + x4*x6 + x4*x7)
     &     + x3*(x2*x4 + x2*x5 + x2*x6 + x2*x7 + x4*x6 + x5*x6)
     &     + x1*x4*x5 + x1*x5*x7 + x2*x4*x5 + x2*x4*x6
     &     + x3*x4*x5 + x3*x4*x6 + x4*x5*x6 + x4*x5*x7
     &     + x1*x4*x6 + x1*x6*x7 + x2*x5*x7 + x2*x6*x7
     &     + x3*x4*x6 + x3*x6*x7 + x4*x6*x7 + x5*x6*x7)

      fi = 2*cc*dd*(dd-sqrt3*eps)*(dd+sqrt3*eps)/(dd**2+eps**2)**3
```

References

[1] Nakanishi N 1971 *Graph Theory and Feynman Integrals* (Gordon and Breach, New York)
[2] Piessens R, de Doncker E, Überhuber C W and Kahaner D K 1983 *QUADPACK, A Subroutine Package for Automatic Integration (Springer Series in Computational Mathematics* vol 1) (Springer-Verlag)
[3] Shanks D 1955 *J. Math. and Phys.* **34** 1–42
[4] Wynn P 1956 *Mathematical Tables and Aids to Computing* **10** 91–96
[5] de Doncker E, Kaugars K, Cucos L and Zanny R 2001 *Proc. of Computational Particle Physics Symposium (CPP 2001)* pp 110–119
[6] Open-MPI http://www.open-mpi.org
[7] Laporta S 2000 *Int. J. Mod. Phys. A* **15** 5087–5159 arXiv:hep-ph/0102033v1
[8] de Doncker E, Shimizu Y, Fujimoto J and Yuasa F 2004 *Computer Physics Communications* **159** 145–156
[9] de Doncker E, Fujimoto J, Kurihara Y, Hamaguchi N, Ishikawa T, Shimizu Y and Yuasa F 2010 *XIV Adv. Comp. and Anal. Tech. in Phys. Res.* PoS (ACAT10) 073
[10] de Doncker E, Fujimoto J, Hamaguchi N, Ishikawa T, Kurihara Y, Shimizu Y and Yuasa F 2011 *Journal of Computational Science (JoCS)* **3** 102–112
[11] de Doncker E, Yuasa F and Assaf R 2013 *Journal of Physics: Conf. Ser.* **454**
[12] Yuasa F, de Doncker E, Hamaguchi N, Ishikawa T, Kato K, Kurihara Y and Shimizu Y 2012 *Journal Computer Physics Communications* **183** 2136–2144
[13] OpenMP website, http://www.openmp.org
[14] Fritsch F N, Kahaner D K and Lyness J N 1981 *ACM TOMS* **7** 46–75
[15] Kahaner D, Moler C and Nash S 1988 *Numerical Methods and Software* (Prentice Hall)
[16] de Doncker E and Kaugars K 2010 *Procedia Computer Science* **1** 117–124
[17] Berntsen J, Espelid T O and Genz A 1991 *ACM Trans. Math. Softw.* **17** 452–456
[18] De Ridder L and Van Dooren P 1976 *Journal of Computational and Applied Mathematics* **2** 207–210
[19] Genz A and Malik A 1980 *Journal of Computational and Applied Mathematics* **6** 295–302
[20] Genz A and Malik A 1983 *SIAM J. Numer. Anal.* **20** 580–588
[21] Berntsen J, Espelid T O and Genz A 1991 *ACM Trans. Math. Softw.* **17** 437–451
[22] Genz A 1990 *Lecture Notes in Computer Science* vol 507 ed Sherwani N A, de Doncker E and Kapenga J A pp 279–285
[23] Grundmann A and Möller H 1978 *SIAM J. Numer. Anal.* **15** 282–290
[24] de Doncker E 1979 *Math. Comp.* **33** 1003–1018
[25] Genz A and Cools R 2003 *ACM Trans. Math. Soft.* **29** 297–308
[26] Yuasa F, Ishikawa T, Hamaguchi N, Koike T and Nakasato N 2013 *Journal of Physics: Conf. Ser.* **454**
[27] de Doncker E, Yuasa F and Kurihara Y 2012 *Journal of Physics: Conf. Ser.* **368**

Electroweak and QCD Radiative Corrections to Drell–Yan Process for Experiments at the Large Hadron Collider

Vladimir A Zykunov

Belarussian State University of Transport, 246653 Gomel, Belarus

E-mail: vladimir.zykunov@cern.ch, zykunov@rambler.ru

Abstract. Next-to-leading order electroweak and QCD radiative corrections to the Drell–Yan process with high dimuon masses for experiments CMS LHC at CERN have been studied in fully differential form. The FORTRAN code READY for numerical analysis of Drell–Yan observables has been presented. The radiative corrections are found to become significant for CMS LHC experiment setup.

1. Introduction

For more than twenty years the Standard Model (SM) has had the status of a consistent and experimentally confirmed theory since the experimental data of past and present accelerators (LEP, SLC, Tevatron) has shown no significant deviation from SM predictions up to the energy scale of a few hundred GeV and, finally, LHC has discovered Higgs boson [1]. However, various New Physics (NP) models such as production of high-mass dilepton resonances [2], extra spatial dimensions [3] etc. suggest deviations beyond SM predictions and testing them at the new energy scale (the few thousand GeV region) is one of the main tasks of modern physics. The forthcoming experiments at the LHC with maximal energy would either provide the first data on NP or strengthen the current status of the SM.

The experimental investigation of the continuum for the Drell-Yan production of dileptons, i.e. data on the cross section and the forward-backward asymmetry of the reaction

$$pp \to (\gamma, Z) \to l^+l^-X \tag{1}$$

at large invariant mass of a dilepton pair (see [4] and references therein) is considered to be one of the most powerful tools in the experiments at the LHC from a NP exploration standpoint.

The studies of the NP effects are impossible without exact knowledge of the SM predictions including higher-order electroweak (EW) and QCD radiative corrections. Many programs have been developed for this: DYNNLO, FEWZ, HORACE, MC@NLO, POWHEG, RADY, READY, SANC, ZGRAD/ZGRAD2 et al. A large list of references quoted, for example, in recent papers [5, 6] dedicated to description of FEWZ and POWHEG, correspondingly. These codes were used for taking into account the uncertainty due to the EW and QCD corrections at recent measurements of the differential $d\sigma/dM$ (M is dilepton invariant mass) and double-differential $d^2\sigma/(dMdy)$ (y is dilepton rapidity) Drell–Yan cross sections at LHC energy $\sqrt{S} = 7$ TeV,

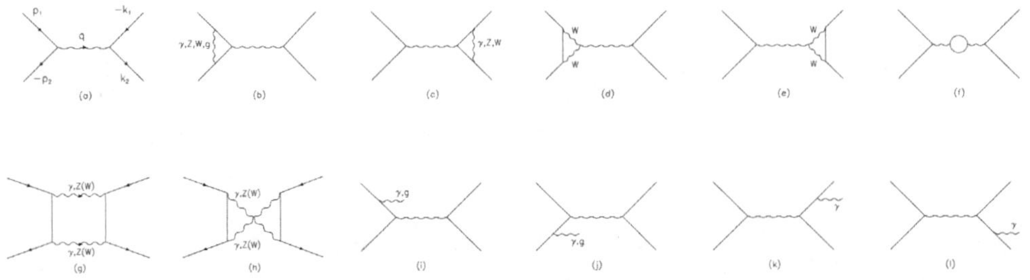

Figure 1. Feynman graphs for the Born (a), one-loop virtual diagrams (b-h), and bremsstrahlung diagrams for the Initial (ISR) (i, j) and Final State Radiation (FSR) (k, l). The unlabeled wavy lines stand for the virtual γ or Z boson.

$M \le 1.5$ TeV and integrated luminosity 4.5 fb^{-1} [7]. Measurements are in agreement with the SM predictions: all of them with next-to-next-to-leading order (NNLO) of FEWZ using MSTW2008 parton density functions (PDF) and double-differential observables with NLO of POWHEG using CT10 PDF.

At the edges of kinematical region (especially at extra large M) the important task is to make the correction procedure of background both accurate and fast. For the latter it is desirable to obtain the set of as much compact as possible formulas both for EW and QCD corrections. To get leading effect of weak corrections in the region of large invariant dilepton mass we actively used the so-called Sudakov logarithms (SL) [8] which grow with the energy scale and thus give one of the main effects in the region of large invariant dilepton mass. In addition, the collinear logarithms (CL) of the QED and QCD radiative corrections can compete with double SL in the investigated region. Such formulas have been obtained in previous papers [9]–[14] using the asymptotic approach for the most complicated weak components of EW corrections, and using the leading CL extraction [12, 13, 14] for the QED and QCD component. This paper is devoted to the analysis of the interplay of these effects for observable quantities of CMS LHC in general fully differential form.

2. Notations and cross sections with the Born kinematics

At LO the Drell–Yan process $p(P_A)+p(P_B)\to l^+(k_1)+l^-(k_2)+X$ in quark-parton model is described by Fig.1,a. Our notations are the following: $p_1(p_2)$ is the 4-momentum of the quark or antiquark with flavor q and mass m_q from the incoming proton with 4-momentum P_A or P_B; $k_1(k_2)$ is the 4-momentum of the final lepton $l^+(l^-)$ with mass m; $q=k_1+k_2$ is the 4-momentum of the i-boson with mass m_i ($i=\gamma,Z$; $m_\gamma=0$). We use the standard set of Mandelstam invariants for the partonic elastic scattering:

$$s=(p_1+p_2)^2,\ t=(p_1-k_1)^2,\ u=(k_1-p_2)^2, \tag{2}$$

and $S=(P_A+P_B)^2$ for hadron scattering. The invariant mass of the dilepton is $M=\sqrt{q^2}$.

Let us start by presenting the convolution formula for the total cross section with non-radiative kinematics:

$$\sigma_N=\frac{1}{3}\int\limits_0^1 dx_1\int\limits_0^1 dx_2\int\limits_{-S}^0 dt\sum_{q=u,d,\dots}\Big[f_q^A(x_1,Q^2)f_{\bar q}^B(x_2,Q^2)\sigma_N^{q\bar q}(t)+$$
$$+f_{\bar q}^A(x_1,Q^2)f_q^B(x_2,Q^2)\sigma_N^{\bar q q}(t)\Big]\theta(s+t)\theta_M\theta_D. \tag{3}$$

Here, $f_q^H(x, Q^2)dx$ is the probability of finding, in hadron H, a quark q at energy scale Q^2 carrying a momentum fraction between x and $x + dx$, and $\sigma^{q\bar{q}}$ and $\sigma^{\bar{q}q}$ are the cross sections at the quark-parton level. According to the quark-parton model rules, we take $s = x_1 x_2 S$. The function $\theta_M = \theta(s - M_1^2)\theta(M_2^2 - s)$ under the integral sign is determined by the kinematics of the parton reaction and provides the integration in the interval of invariant mass $M_1 \le M \le M_2$. The factor

$$\theta_D = \theta(\zeta^* - \zeta)\theta(\zeta^* + \zeta)\theta(\zeta^* - \psi)\theta(\zeta^* + \psi)\theta(p_T(l^+) - p_T^{\min})\theta(p_T(l^-) - p_T^{\min}) \qquad (4)$$

cuts the region of integration according to detector geometry. Here, $\zeta = \cos\theta_1$, $\psi = \cos\theta_2$, where $\theta_{1(2)}$ is the scattering angle of the lepton with 4-momenta $k_{1(2)}$ in the hadron center-of-mass frame. For the CMS detector the parameter $\zeta^* \approx 0.986614$ corresponds to the lepton rapidity limitation $y(l)^* = 2.5$. For the transverse components of lepton momenta we have the relations $p_T(l^+) = k_{10}\sin\theta_1$ and $p_T(l^-) = k_{20}\sin\theta_2$, and, for the CMS detector, $p_T^{\min} = 20$ GeV.

We use the common index for contributions with non-radiative kinematics $N = \{0, V, \text{soft}\}$, where 0 stands for the Born contribution, the special indices for contributions with at least one additional virtual particle $V = \{\text{BSE}, \text{HV}, b\}$ and separately for the contributions of boxes $b = \{\gamma\gamma, \gamma Z, ZZ, WW\}$. Abbreviations mean: BSE for boson self energies, HV for Heavy Vertices induced by at least one massive boson, "$\gamma\gamma$" for the infrared(IR)-finite part of $\gamma\gamma$-boxes, "γZ" for the IR-finite part of γZ-boxes and "ZZ" for the ZZ-boxes, "WW" for the WW-boxes, "soft" for the sum of Light Vertices (LV) induced by one massless photon or gluon, the IR-divergent parts of the $\gamma\gamma$-boxes, γZ-boxes, and of the (soft) bremsstrahlung cross section. The "soft"-part is IR-finite in sum and described by Born kinematics.

Now, let us present explicit formulae for $q\bar{q}$ cross sections given in (3), employing the notation $\sigma(t) \equiv d\sigma/dt$. To find the cross section for the $\bar{q}q$-case, we can use crossing rules. The Born cross section has the form

$$\sigma_0^{q\bar{q}}(t) = \frac{2\pi\alpha^2}{s^2} \sum_{i,j=\gamma,Z} D^i D^{j*} \sum_{\chi=+,-} \lambda_{q\chi}^{i,j} \lambda_{l\chi}^{i,j} B_\chi, \qquad (5)$$

where the boson propagators look like $D^j = (s - m_j^2 + i m_j \Gamma_j)^{-1}$, Γ_j is the j-boson width, $B_\chi = t^2 + \chi u^2$. The combinations of coupling constants for a f-fermion with an i- (or j-) boson have the form

$$\lambda_{f+}^{i,j} = v_f^i v_f^j + a_f^i a_f^j, \; \lambda_{f-}^{i,j} = v_f^i a_f^j + a_f^i v_f^j, \qquad (6)$$

where

$$v_f^\gamma = -Q_f, \; a_f^\gamma = 0, \; v_f^Z = \frac{I_f^3 - 2s_W^2 Q_f}{2s_W c_W}, \; a_f^Z = \frac{I_f^3}{2s_W c_W}, \qquad (7)$$

Q_f is the electric charge of fermion f in proton charge units e, ($e = \sqrt{4\pi\alpha}$), I_f^3 is the third component of the weak isospin of fermion f, and s_W (c_W) is the sine(cosine) of the weak mixing angle.

The BSE-part is

$$\sigma_{\text{BSE}}^{q\bar{q}}(t) = -\frac{4\alpha^2\pi}{s^2}\Big[\sum_{i,j=\gamma,Z} \Pi_S^i D^i D^{j*} \sum_{\chi=+,-} \lambda_{q\chi}^{i,j}\lambda_{l\chi}^{i,j} B_\chi + \\ +\Pi_S^{\gamma Z} D^Z \sum_{i=\gamma,Z} D^{j*} \sum_{\chi=+,-} (\lambda_{q\chi}^{\gamma,j}\lambda_{l\chi}^{Z,j} + \lambda_{q\chi}^{Z,j}\lambda_{l\chi}^{\gamma,j}) B_\chi\Big]. \qquad (8)$$

Here $\Pi_S^{\gamma, Z, \gamma Z}$ are connected with the renormalized photon–, Z– and γZ–self energies [15, 16] as

$$\Pi_S^\gamma = \frac{\hat{\Sigma}^\gamma}{s}, \; \Pi_S^Z = \frac{\hat{\Sigma}^Z}{s - m_Z^2}, \; \Pi_S^{\gamma Z} = \frac{\hat{\Sigma}^{\gamma Z}}{s}.$$

The HV-part has the following form:

$$\sigma^{q\bar{q}}_{\rm HV}(t) = \frac{4\pi\alpha^2}{s^2}{\rm Re}\sum_{i,j=\gamma,Z} D^i D^{j*} \sum_{\chi=+,-} (\lambda^{{\rm F}i,j}_{q\,\chi}\lambda^{i,j}_{l\chi} + \lambda^{i,j}_{q\chi}\lambda^{{\rm F}i,j}_{l\,\chi})B_\chi, \tag{9}$$

where the form factors $\lambda^{{\rm F}i,j}_{f\,\pm}$ are given in [9]. The boxes can be presented as

$$\sigma^{q\bar{q}}_b(t) = \frac{2\alpha^3}{s^2}\sum_{k=\gamma,Z} D^{k*}[\delta^{b,k}(t,u,b_+,b_-) - \delta^{b,k}(u,t,b_-,b_+)], \tag{10}$$

where the functions $\delta^{b,k}(t,u,b_+,b_-)$, b_χ and all prescriptions for them can be found in [9, 10].

The QED "soft"-part (the result of infrared singularity cancellation of $\gamma\gamma$, γZ, photon LV and soft photon bremsstrahlung) is proportional to Born cross section:

$$\sigma^{q\bar{q}}_{\rm soft}(t) = \frac{\alpha}{\pi}\delta^{q\bar{q}}_{\rm soft}\sigma^{q\bar{q}}_0(t) \tag{11}$$

with corresponfing factor

$$\delta^{q\bar{q}}_{\rm soft} = 2\ln\frac{2\omega}{\sqrt{s}}\Big[Q_q^2(\ln\frac{s}{m_q^2}-1) - 2Q_qQ_l\ln\frac{t}{u} + Q_l^2(\ln\frac{s}{m^2}-1)\Big] + Q_q^2\Big(\frac{3}{2}\ln\frac{s}{m_q^2} - 2 + \frac{\pi^2}{3}\Big) -$$

$$-Q_qQ_l\Big(\ln\frac{s^2}{tu}\ln\frac{t}{u} + \frac{\pi^2}{3} + \ln^2\frac{t}{u} + 4\,{\rm Li}_2\frac{-t}{u}\Big) + Q_l^2\Big(\frac{3}{2}\ln\frac{s}{m^2} - 2 + \frac{\pi^2}{3}\Big), \tag{12}$$

where ω is a parameter that determines the "softness" of a photon – the maximal energy of a soft photon, and ${\rm Li}_2$ denotes the Spence dilogarithm. The QCD "soft"-part can be found from (12) by neglecting the FSR and interference parts and after substitution:

$$C_{\rm QED} = Q_q^2\frac{\alpha}{\pi} \to \sum_{a=1}^{N^2-1} t^a t^a \frac{\alpha_s}{\pi} = \frac{N^2-1}{2N} I\frac{\alpha_s}{\pi} \to \frac{4}{3}\frac{\alpha_s}{\pi} = C_{\rm QCD}, \tag{13}$$

where $N=3$, and $2t^a$ are Gell-Mann matrices.

3. Hard photons and gluons. Inverse gluon emission

Let us present the Drell–Yan cross section contribution induced by bremsstrahlung (Fig.1(i-l)). We introduce the total phase space of the reaction $p(P_A) + p(P_B) \to l^+(k_1) + l^-(k_2) + b(k) + X$, ($b=\gamma$ or g) as

$$I_\Omega[A] = \int_0^1 dx_1 \int_0^1 dx_2 \iiiint_\Omega dt dv dz du_1 \frac{1}{\pi\sqrt{R_{u_1}}}\theta(R_{u_1})\theta^R_M\theta^R_D\, A, \tag{14}$$

where $z=2k_1k$, $v=2k_2k$, $z_1=2p_1k$, $u_1=2p_2k$ (for radiative kinematics $v=s+t+u$ and $z+v=z_1+u_1$) and k is the 4-momentum of a real bremsstrahlung photon (gluon).

The factor θ^R_M for the radiative case has the form

$$\theta^R_M = \theta(s-z-v-M_1^2)\theta(M_2^2-s+z+v). \tag{15}$$

For θ_D^R, we use the "non-radiative" expression θ_D (4) with the angles and energies depending on additional "radiative" invariants:

$$\zeta = \frac{x_1 u - x_2 t}{x_1 u + x_2 t}, \ \psi = \frac{x_1(s+u-u_1) + x_2(u+z_1-v)}{x_1(s+u-u_1) - x_2(u+z_1-v)}, \tag{16}$$

$$k_{10} = -\frac{1}{2\sqrt{S}}\Big(\frac{t}{x_1} + \frac{u}{x_2}\Big), \ k_{20} = \frac{1}{2\sqrt{S}}\Big(\frac{s+t-z_1}{x_1} + \frac{s+u-u_1}{x_2}\Big). \tag{17}$$

The physical region Ω is determined by $\theta(R_{u_1})$, where $-R_{u_1}$ is the Gram determinant, which has the form

$$R_{u_1} = -A_{u_1}u_1^2 - 2B_{u_1}u_1 - C_{u_1}, \ \ A_{u_1} = -4m^2 s + (s-v)^2,$$

$$B_{u_1} = v[m^2(3s-v) + (s-v)(m_q^2 - s - t + v)] + z[m^2(s-v) - m_q^2(s+v) + st + v(s+t-v)],$$

$$C_{u_1} = z^2[(m^4 + m_q^4 - 2m^2(m_q^2 + t - v) - 2m_q^2(t+v) + (t-v)^2] +$$

$$+ 2zv[m^4 + m_q^4 + m_q^2(s-2t) - m^2(2m_q^2 + s + 2t - 2v) + (t-v)(s+t-v)] +$$

$$+ v^2[m^4 - 2m^2(m_q^2 + s + t - v) + (m_q^2 - s - t + v)^2]. \tag{18}$$

Then the total bremsstrahlung cross section has the form

$$\sigma_R = \frac{\alpha^3}{3} I_\Omega[T], \ \ T = \frac{1}{s^2} \sum_{\chi=+,-} \sum_{q=u,d,\ldots} \sum_{i,j=\gamma,Z} \lambda_{q\chi}^{i,j} \lambda_{l\chi}^{i,j} \times$$

$$\Big([f_q^A(x_1,Q^2) f_{\bar q}^B(x_2,Q^2) + \chi f_{\bar q}^A(x_1,Q^2) f_q^B(x_2,Q^2)][Q_q^2 R_{qk}{}_\chi^{q\bar q} \Pi^i \Pi^{j*} + Q_l^2 R_l{}_\chi^{q\bar q} D^i D^{j*}]$$

$$+ [f_q^A(x_1,Q^2) f_{\bar q}^B(x_2,Q^2) - \chi f_{\bar q}^A(x_1,Q^2) f_q^B(x_2,Q^2)] Q_l Q_q R_{int}{}_\chi^{q\bar q} \frac{\Pi^i D^{j*} + D^i \Pi^{j*}}{2} \Big). \tag{19}$$

Subscripts at R (they can be found in Appendix A of [11]) indicate the origin of the emitted particle: qk – quark for ISR both for photon and gluon [taking into account (13)], l and int – lepton and interference term only for photon, respectively. The boson propagators corresponding to the radiative case look like

$$\Pi^j = \frac{1}{s - z - v - m_j^2 + i m_j \Gamma_j}. \tag{20}$$

We use the standard (noncovariant) method of IR singularity separation dissecting the region of integration with the help of the function $\theta_\omega = \theta(\frac{v+z}{2\sqrt{s}} - \omega)$ and dividing the cross section (19) into two parts: the first one corresponds to soft photons (gluons) with energy less then ω (it goes to IR singularity cancellation in formula (12)) and the second one corresponds to hard ones with energy larger than ω.

To finalize the calculation we have to take into consideration the inverse gluon emission (see, Fig. 2). Methods of calculation are similar to cited above nonsinglet-channel (according to the QCD terminology) $q\bar q$-contributions. All formulas for cross sections and kinematics can be found in [14].

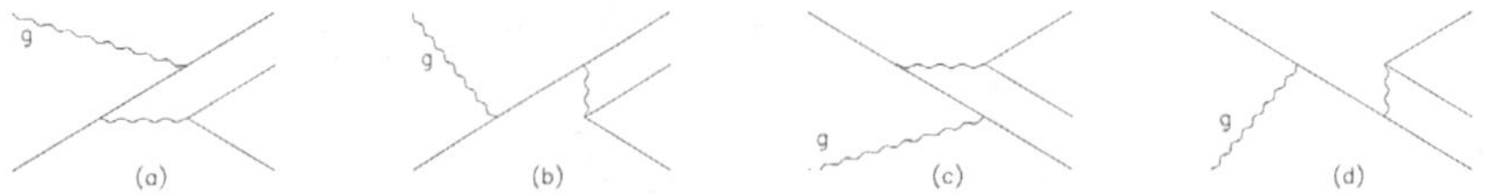

Figure 2. Feynman graphs of gq-type (a, b) and qg-type (c, d) for inverse gluon bremsstrahlung. The unlabeled wavy lines stand for the virtual γ or Z boson.

4. Fully differential cross section

Here we rebuild all of the cross sections to fully differential form

$$\sigma_C \to \sigma_C^{(3)} \equiv \frac{d^3\sigma_C}{dMdyd\psi}, \tag{21}$$

where y is the dilepton rapidity $y \equiv y(l^+l^-)$, $\sigma_C^{(3)}$ are the corresponding contributions (or sum of contributions) of the differential cross section, for example, if $C = 0$, $\sigma_0^{(3)}$ is the differential Born cross section and so on.

For the part of the cross section with non-radiated kinematics, the translation to differential form is easy to do using the Jacobian J_N:

$$dx_1dx_2dt = J_N dMdyd\psi, \quad J_N = \frac{4M^3e^{2y}}{S[1+\psi+(1-\psi)e^{2y}]^2}. \tag{22}$$

Then we get the following correlations:

$$t = -\frac{M^2(\psi+1)}{1+\psi+(1-\psi)e^{2y}}, \; x_1 = e^y\frac{M}{\sqrt{S}}, \; x_2 = e^{-y}\frac{M}{\sqrt{S}}, \tag{23}$$

and the differential cross section with non-radiative kinematics looks like

$$\sigma_N^{(3)} = \frac{1}{3}J_N \sum_{q=u,d,...} [f_q^A(x_1,Q^2)f_{\bar{q}}^B(x_2,Q^2)\sigma_N^{q\bar{q}}(t) + f_{\bar{q}}^A(x_1,Q^2)f_q^B(x_2,Q^2)\sigma_N^{\bar{q}q}(t)]\theta_D. \tag{24}$$

For the radiative case, we rebuild the cross section to fully differential form in a way analogical to the non-radiative case:

$$dx_1dx_2dt = J_R^{(3)}dMdyd\psi, \tag{25}$$

where we use the correlations

$$t = \frac{(u_1+M^2)(e^{2y}(\psi-1)(v-u_1)+(1+\psi)(z_1+M^2))}{e^{2y}(\psi-1)(u_1+M^2)-(1+\psi)(z_1+M^2)}, \tag{26}$$

$$x_1 = \frac{e^y\sqrt{(u_1+M^2)(v+z+M^2)}}{\sqrt{S}\sqrt{z_1+M^2}}, \quad x_2 = \frac{e^{-y}\sqrt{(z_1+M^2)(v+z+M^2)}}{\sqrt{S}\sqrt{u_1+M^2}}. \tag{27}$$

These are obtained from: 1) the formula for the radiative s: $s = M^2+v+z$, 2) the formula for the radiative ψ (16), 3) the definition of dilepton rapidity y: $e^{2y} = (x_1/x_2)(z_1+M^2)/(u_1+M^2)$. Then, the Jacobian in radiative case can be expressed as

$$J_R^{(3)} = \frac{4Me^{2y}}{S}\frac{(v+M^2)(z_1+M^2)(u_1+M^2)}{[(1+\psi)(z_1+M^2)+(1-\psi)e^{2y}(u_1+M^2)]^2}, \tag{28}$$

and satisfies $\lim_{k\to 0} J_R^{(3)} = J_N$. The differential cross section corresponding hard bremsstralung is now given by

$$\sigma_R^{(3)} = \frac{\alpha^3}{3} \iiint \frac{1}{\pi\sqrt{R_{u_1}}} \theta_D^R T \theta_\omega J_R^{(3)} dv dz du_1. \tag{29}$$

The remaining triple integral over the physical bremsstrahlung region $\iiint ...dvdzdu_1$ has to be computed numerically due to the complexity of the integration region and form of the integrand, and due to the presence in the integrand of the intricate PDF, which are z, u_1-dependent (see (27)). We can realize this numerical integration by Monte Carlo routine based on the VEGAS algorithm [18], or simplify the hard bremsstrahlung contribution extracting the leading logarithm part and integrating $\iint ...dzdu_1$ analytically. For further estimations we choose the last option. Exact formulas for QED CL parts can be found in [12], results for nonsinglet QCD and singlet IGE CL parts can be found in [13] and [14], respectively.

At last, starting with the fully differential cross sections, we can construct the distributions over y and (or) M:

$$\frac{d\sigma_C}{dMdy} = \int_{-\zeta^*}^{\zeta^*} d\psi \sigma_C^{(3)} \theta_D; \quad \frac{d\sigma_C}{dM} = \int_{-\zeta^*}^{\zeta^*} d\psi \int_{y_-}^{y_+} dy\ \sigma_C^{(3)} \theta_D,\ y_\pm = \pm \ln \frac{\sqrt{S}}{M}. \tag{30}$$

5. Independence from unphysical parameters

The proof of independence of the results from the parameter ω is rather simple and can be done numerically or analytically (see, for example, [11, 12]). For the soft-hard photon separator we use $\omega = 0.1$ GeV; however the results presented below do not depend on ω in a wide interval: 1 GeV $\leq \omega \leq$ 0.0001 GeV.

In order to solve the problem of quark mass singularity (QS), we used the $\overline{\text{MS}}$ scheme [19], as in paper [20]. After all of the prescribed manipulations, the part of the cross section that must be subtracted in order to avoid the dependence on the quark mass assumes the form

$$\sigma_{\text{QED(QCD)},\text{QS}} = \frac{1}{3} \int_0^1 dx_1 \int_0^1 dx_2 \int_{-S}^0 dt \int_0^{1-2\omega/M} d\eta \sum_{q=u,d,...} \Big[\Big(q(x_1)\Delta\bar{q}(x_2,\eta) +$$

$$+\Delta q(x_1,\eta)\bar{q}(x_2)\Big)\sigma_0^{q\bar{q}}(t) + (q \leftrightarrow \bar{q}) \Big]\theta(s+t)\theta_M\theta_D, \tag{31}$$

$$\Delta q(x,\eta) = \frac{1}{2} C_{\text{QED(QCD)}} \left[\frac{1}{\eta} q(\frac{x}{\eta}, M_{\text{sc}}^2)\theta(\eta - x) - q(x, M_{\text{sc}}^2) \right] \frac{1+\eta^2}{1-\eta} \left(\ln \frac{M_{\text{sc}}^2}{m_q^2} - 2\ln(1-\eta) - 1 \right), \tag{32}$$

where $q(x) \equiv f_q(x, Q^2)$, and M_{sc} is the factorization scale [19], which should be equal to Q [12]. For the quark masses we used $m_q = m_u$, although our numerical results practically do not depend on m_q within the interval $0.01m_u \leq m_q \leq 10m_u$. For IGE the result of QS-term substraction is trivial:

$$\sigma_{\text{IGE}}^{(3)} - \sigma_{\text{IGE,QS}}^{(3)} = \sigma_{\text{IGE}}^{(3)}(m_q \to M_{\text{sc}}).$$

6. Discussion of numerical results

We investigate the scale of EW and QCD corrections and their effect on the differential observables of the Drell-Yan processes for CMS experiment using the FORTRAN program READY (Radiative corrEctions to lArge invariant mass Drell-Yan process) with the following set of parameters and prescriptions:

- SM input electroweak parameters: $\alpha = 1/137.035999679$, $m_W = 80.398$ GeV, $m_Z = 91.1876$ GeV, $\Gamma_W = 2.141$ GeV, $\Gamma_Z = 2.4952$ GeV, $m_H = 125.7$ GeV;
- muon mass $m_\mu = 0.105658367$ GeV, masses of the other fermions for loop contributions to the BSE: $m_e = 0.51099891$ keV, $m_\tau = 1.77699$ GeV, $m_u = 0.06983$ GeV, $m_c = 1.2$ GeV, $m_t = 174$ GeV, $m_d = 0.06984$ GeV, $m_s = 0.15$ GeV, $m_b = 4.6$ GeV; (the light quark masses provide $\Delta\alpha^{(5)}_{had}(m_Z^2)$=0.0276);
- modern MSTW2008 set of PDF [21] with the choice $Q = M$;
- taking into account 5 flavors of valence and sea quarks in the proton (with the exception of the t flavor) and set their masses as regulators of the collinear singularity to $m_q = m_u$;
- using "bare" setup for leptons identification requirements (no smearing, no recombination of lepton and photon).

In Table 1 as example of READY output we show the relative corrections (RC) to Born differential cross section

$$\delta_{\rm C} = \sigma^{(3)}_{\rm C}/\sigma^{(3)}_0 \tag{33}$$

via different y, ψ and M with the muons in the final state ($l = \mu$), and the energy $\sqrt{S} = 14$ TeV planned at the LHC in 2015.

Table 1. Relative corrections $\delta_{\rm NLO}$ via different y, ψ and M.

y	ψ	$\delta_{\rm NLO}$ at M=1 TeV	$\delta_{\rm NLO}$ at M=3 TeV	$\delta_{\rm NLO}$ at M=5 TeV
0.0	−0.8	$-0.035+0.320-0.134$	$-0.191+0.442-0.068$	$-0.329+0.625-0.054$
0.0	−0.4	$-0.043+0.320-0.089$	$-0.171+0.442-0.061$	$-0.264+0.625-0.051$
0.0	0.0	$-0.036+0.320-0.073$	$-0.154+0.442-0.059$	$-0.231+0.625-0.051$
0.0	0.4	$-0.043+0.320-0.089$	$-0.171+0.442-0.061$	$-0.264+0.625-0.051$
0.0	0.8	$-0.035+0.320-0.134$	$-0.191+0.442-0.068$	$-0.329+0.625-0.054$
0.6	−0.8	$0.008+0.456-0.267$	$-0.146+0.582-0.145$	$-0.101+0.770-0.163$
0.6	−0.4	$-0.014+0.453-0.196$	$-0.138+0.577-0.114$	$-0.114+0.773-0.123$
0.6	0.0	$-0.024+0.453-0.140$	$-0.128+0.569-0.090$	$-0.142+0.767-0.090$
0.6	0.4	$-0.028+0.443-0.078$	$-0.145+0.565-0.064$	$-0.223+0.759-0.061$
0.6	0.8	$-0.052+0.429-0.059$	$-0.207+0.555-0.052$	$-0.335+0.754-0.049$
1.2	0.0	$0.003+0.595-0.295$	$-0.015+0.849-0.215$	
1.2	0.4	$0.009+0.596-0.205$	$-0.045+0.850-0.147$	
1.2	0.8	$-0.024+0.577-0.047$	$-0.183+0.850-0.056$	

Numbers presented in 3rd, 4th and 5th columns correspond to sum of all NLO contributions: NLO = EW+QCD($q\bar{q}$)+QCD(qg). Results for all RCs strongly depend on kinematical position, the necessary symmetry for different contributions is conserved. Using different PDFs (CTEQ6, MRST2004, MSTW2008) we did not mark any significant effect for RCs in the whole kinematical region of CMS.

Let us now compare, as example, our EW results with the numbers of several leading world groups HORACE, SANC and ZGRAD presented in [22]. All parameters and detector conditions here are taken to be the same as in [22]. Our results for the relative correction to $d\sigma/dM$ at the point $M = 1$ TeV ($l = \mu$, $\sqrt{S}$=14 TeV) is different by $\sim 1.5\%$ comparing with HORACE and SANC. At $M = 2$ TeV, this difference is $\sim 0.5\%$. The numbers of the ZGRAD group in the region 0.9 TeV $\leq M \leq$ 1.8 TeV are larger and are in better agreement with ours. We find such agreement to be satisfactory, because, for the weak component of corrections, we use the asymptotic approach [10], which greatly simplifies the formulas and accelerates the calculation, but only works well in the region $M > 0.5$ TeV, this explains why the agreement becomes better with increasing M.

7. Conclusions

The complete NLO EW and QCD radiative corrections to the Drell-Yan process at large invariant dilepton mass is studied in fully differential form. The results for weak, QED, QCD parts are the compact expressions, they expand in Sudakov and collinear logarithms. Using the FORTRAN code READY, the numerical analysis is performed in the high-energy region corresponding to the CMS experiment at the CERN LHC. Both EW and QCD RCs are found to become large at high dilepton mass M and to have the same order of magnitude as the systematic uncertainty expected on CMS [23]. Such large scale of RC does not allow neglecting the radiative correction procedure in the future experiments on the Drell-Yan process with high dimuon masses at CMS LHC. The exact NNLO QCD and $\mathcal{O}(\alpha\alpha_s)$ (see, for example [24], where one of first understanding of role of such effects at high energies had been achieved) corrections would be desirable for a better control of theory vs. experiment.

8. Acknowledgments

I thank local organizing committee of ACAT2013 and personally Prof. Jian-Xiong Wang and Bin Gong for help, finacial support and, as result, a happy possibility to take part in conference. I am grateful to Prof. A. L. Kataev for the interest to the work and support. I would like to thank A. Aleksejevs, A. Arbuzov, S. Barkanova, E. Dydyshko, E. Kuraev, A. Lanyov, S. Pozzorini, V. Mossolov, S. Shmatov and N. Shumeiko for the stimulating discussions. I am grateful to A. Arbuzov, S. Bondarenko and D. Wackeroth for a detailed comparison of part of the results. I thank CERN (CMS Group), where part of this work was carried out, for warm hospitality during my visits. Part of this work was supported by Belarus scientific program "Convergence".

References

[1] ATLAS Collaboration 2012 *Phys. Lett.* B **716** 1; CMS Collaboration 2012 *Phys. Lett.* B **716** 30
[2] Leike A 1999 *Phys. Rep.* **317** 143
[3] Arkani-Hamed N *et al* 1998 *Phys. Lett.* B **429** 263; Antoniadis I *et al* 1998 *Phys. Lett.* B **436** 257; Randall L and Sundrum R 1999 *Phys. Rev. Lett.* **83** 3370; Kokorelis C 2004 *Nucl. Phys.* B **677** 115
[4] Belotelov I *et al* CERN-CMS-NOTE-2006-123
[5] Li Y, Petriello F 2012 *Phys. Rev.* D **86** 094034
[6] Barze L *et al* CERN-PH-TH-2013-027
[7] CMS Collaboration CMS-PAS-EWK-11-007
[8] Sudakov V 1956 *Sov. Phys. JETP* **3** 65
[9] Zykunov V A 2006 *Yad. Fiz.* **69** 1557 (Engl. vers.: 2006 *Phys. of Atom. Nucl.* **69** 1522)
[10] Zykunov V A 2007 *Phys. Rev.* D **75** 073019
[11] Zykunov V A 2008 *Yad. Fiz.* **71** 757 (Engl. vers.: 2008 *Phys. of Atom. Nucl.* **71** 732)
[12] Zykunov V A 2010 *Yad. Fiz.* **73** 1617 (Engl. vers.: 2010 *Phys. of Atom. Nucl.* **73** 1571)
[13] Zykunov V A 2010 *Yad. Fiz.* **73** 1269 (Engl. vers.: 2010 *Phys. of Atom. Nucl.* **73** 1229)
[14] Zykunov V A 2011 *Yad. Fiz.* **74** 72 (Engl. vers.: 2011 *Phys. of Atom. Nucl.* **74** 72)
[15] Böhm M, Spiesberger H and Hollik W 1986 *Fortschr. Phys.* **34** 687
[16] Hollik W 1990 *Fortschr. Phys.* **38** 165
[17] Baur U *et al* 1998 *Phys. Rev.* D **57** 199
[18] Lepage G 1978 *J. Comput. Phys.* **27** 192
[19] Bardeen W A *et al* 1978 *Phys. Rev.* D **18** 3998
[20] Arbuzov A *et al* 2008 *Eur. Phys. J.* C **54** 451
[21] Martin A D *et al* 2009 *Eur. Phys. J.* C **63** 189
[22] Buttar C *et al* 2007 *Proc. of Les Houches 2007, Physics at TeV colliders* p 121
[23] CMS Physics TDR: V. II, Physics Performance, The CMS Collaboration 2007 *J. Phys. G.* **34** 995
[24] Kataev A L 1992 *Phys. Lett.* B **287** 209

Precise calculation for heavy gauge boson production in the LHT model

Guo Lei, Zhang Ren-You and Ma Wen-Gan

Department of Modern Physics, University of Science and Technology of China (USTC), Hefei, Anhui 230026, P.R.China

E-mail: guolei@mail.ustc.edu.cn, zhangry@ustc.edu.cn, mawg@ustc.edu.cn

Abstract. In the framework of the littlest Higgs model with T parity, we study the W_H/Z_H+q_- and W_H-pair productions at the CERN Large Hadron Collider up to the QCD next-to-leading order (NLO). The kinematic distributions of final decay products and the theoretical dependence of the cross section on the factorization/renormalization scale are analyzed. We adopt the PROSPINO scheme in the QCD NLO calculations to avoid double counting and keep the convergence of the perturbative QCD description. By using the subtraction scheme, the QCD NLO corrections enhance the leading order cross section with a K-factor in the range of $1.00 \sim 1.43$ for $W_H(Z_H)q_-$ production process, and in the range of $1.09 \sim 1.22$ for the W_H pair production process.

1. Introduction

Although the standard model (SM) [1, 2] provides a remarkably successful description of high energy physics phenomena at the energy scale up to 100 GeV, it leaves a number of theoretical problems unsolved. Many extended models are proposed to deal with these problems such as grand unified theories [3], supersymmetric models [4], extra dimensions models [5], left-right symmetric models [6], B-L (baryon number minus lepton number) extended SM models [7], little Higgs models [8] and many more. Each of these models has motivation to solve one or more of the problems that the SM encounters. Among them the little Higgs models deserve attention due to their elegant solution to hierarchy problem and are proposed as one kind of electroweak symmetry breaking (EWSB) models without fine-tuning in which the Higgs boson is naturally light as a result of nonlinearly realized symmetry [9]-[14]. The littlest Higgs (LH) model [15], an $SU(5)/SO(5)$ nonlinear sigma model [11], is the most simplest version of little Higgs models, in which a set of new heavy gauge bosons (A_H, W_H, Z_H) and a vector-like quark (T) are introduced to cancel the quadratic divergence contribution to Higgs boson mass from the SM gauge boson loops and the top quark loop respectively. However, this model predicts large corrections to electroweak precision observables and the scale of the global symmetry breaking f, is constrained by experimental data [12], which set severe constraints on the new heavy particle masses and the model parameters. For instance, recent experimental measurements on the decay processes of $W_H^{\mp} \to l^{\mp} \overset{(-)}{\nu}$ and $Z_H \to l^+l^-$ provide the constraints of $M_{W_H} > 2.18\ TeV$ and $M_{Z_H} > 1.83\ TeV$ [16, 17]. These constraints would enforce the symmetry breaking scale f, which characterizes the mass of new particles, to be larger than 2.5 TeV and 3 TeV respectively.

Consequently, the cutoff scale $\Lambda \sim 4\pi f$ becomes so large that calls for the fine-tuning between the electroweak scale and the cutoff scale again.

By introducing a discrete symmetry, the T parity, the littlest Higgs model with T parity (LHT) [18]-[22] offers a viable solution to the naturalness problem of the SM, and also predicts a set of new heavy fermions, gauge bosons as well as a candidate for dark matter. In the LHT, all the SM particles are T-even and almost all the new heavy particles are T-odd. Due to the different T parity quantum numbers, the SM gauge bosons cannot mix with the new gauge bosons in the LHT. This would alleviate the constraints from the electroweak precision tests and thus allows the scale f to be significantly lower than 1 TeV [21]. For instance, due to the T parity conservation, the processes $W_H^{\mp} \to l^{\mp} \overset{(-)}{\nu}$ and $Z_H \to l^+l^-$ are forbidden, and the only decay modes of these T-odd heavy gauge bosons are $W_H \to A_H W$ and $Z_H \to A_H H$. In this case, the leptons are produced from the decays of W and H, but not from the heavy gauge bosons directly. Therefore, these T-even gauge bosons escape from the experimental constraints shown in Refs.[16, 17]. Furthermore, as a lightest T-odd particle, the heavy photon A_H cannot further decay into other particles, and would be a good candidate for the dark matter [23]. Since the CERN Large Hadron Collider (LHC) has potential to detect the signals of new gauge bosons and fermions, the phenomenology of the LHT would be quite interesting and a number of phenomenological works has been presented [20, 24, 25, 26].

In this paper, we present the QCD NLO corrections to the processes $pp \to W_H(Z_H)q_- + X$ [27] and $pp \to W_H^+W_H^- + X$ [28].

2. The related LHT theory

Before our calculations, we will briefly recapitulate the LHT theory which is relevant to our work. The details of the LHT can be found in Refs.[18, 20, 21, 24].

At some high scale f the global symmetry $SU(5)$ is broken down to $SO(5)$, leading to 14 massless Nambu-Goldstone bosons. Four of them are manifested as the longitudinal modes of the heavy gauge bosons. The other 10 decompose into a T-even $SU(2)$ doublet h, identified as the SM Higgs field, and a complex T-odd $SU(2)$ triplet Φ, which obtains a mass of $m_\Phi = \sqrt{2}m_h f/v_{SM}$, with m_h and v_{SM} being SM Higgs mass and the electroweak symmetry break scale, respectively.

The additional discrete symmetry, T-parity, is in analogy to the R-parity in the minimal supersymmetric standard model (MSSM) [18, 20, 22]. The T-parity transformations for gauge sector are defined as the exchange between the gauge bosons of the two $SU(2) \times U(1)$ groups, i.e., $W_1^a \leftrightarrow W_2^a$ and $B_1 \leftrightarrow B_2$. Thus their T-odd and T-even combinations can be obtained as

$$
\begin{aligned}
W_H^a &= \frac{1}{\sqrt{2}}(W_1^a - W_2^a), \quad B_H = \tfrac{1}{\sqrt{2}}(B_1 - B_2), \quad (T - odd), \\
W_L^a &= \frac{1}{\sqrt{2}}(W_1^a + W_2^a), \quad B_L = \tfrac{1}{\sqrt{2}}(B_1 + B_2), \quad (T - even).
\end{aligned} \tag{1}
$$

The mass eigenstates of the gauge sector in the LHT are expressed as

$$
\begin{aligned}
W_H^{\pm} &= \frac{1}{\sqrt{2}}(W_H^1 \mp iW_H^2), \quad Z_H = s_H B_H + c_H W_H^3, \quad A_H = c_H B_H - s_H W_H^3, \\
W_L^{\pm} &= \frac{1}{\sqrt{2}}(W_L^1 \mp iW_L^2), \quad Z_L = -s_w B_L + c_w W_L^3, \quad A_L = c_w B_L + s_w W_L^3,
\end{aligned} \tag{2}
$$

where $s_w = \sin\theta_W$, $c_w = \cos\theta_W$, $s_H = \sin\theta_H$, $c_H = \cos\theta_H$, θ_W is the Weinberg angle, and the mixing angle θ_H at the $\mathcal{O}(v^2/f^2)$ is expressed as

$$
\sin\theta_H \simeq \left[\frac{5gg'}{4(5g^2 - g'^2)}\frac{v_{SM}^2}{f^2}\right]. \tag{3}
$$

Then the gauge sector consists of T-odd heavy new gauge bosons $W_H^{\pm}$, Z_H, A_H and T-even light gauge bosons identified as SM gauge bosons, $W^{\pm}$, Z^0 and one massless photon. The T parity partner of the photon, A_H, is the lightest T-odd particle, therefore, the candidate of dark matter in the LHT. The masses of the T parity partners of the photon, Z^0- and $W^{\pm}$-boson are expressed as [24]

$$m_{W_H} \simeq m_{Z_H} \simeq gf\left(1-\frac{1}{8}\frac{v_{SM}^2}{f^2}\right), \quad m_{A_H} \simeq \frac{1}{\sqrt{5}}g'f\left(1-\frac{5}{8}\frac{v_{SM}^2}{f^2}\right), \tag{4}$$

where $v_{SM} = 246\ GeV$. At the tree level the SM gauge boson masses can be expressed as $m_W = \frac{gv_{SM}}{2}$ and $m_Z = \frac{v_{SM}\sqrt{g^2+g'^2}}{2}$.

In the LHT, the fermion sector of the first two generations in the SM is remained unchanged and the third generation of quarks is modified. We introduce two fermion doublets q_1 and q_2 for each fermion generation. The T parity transformation to these fermion doublets is defined as $q_1 \leftrightarrow -q_2$. Therefore, the T-odd and T-even combinations can be constructed as $q_- = \frac{1}{\sqrt{2}}(q_1+q_2)$ and $q_+ = \frac{1}{\sqrt{2}}(q_1 - q_2)$, where q_+ is the doublet for the SM fermions and q_- for their T-odd partners. We take the Lagrangian suggested in Refs.[18, 20, 21] to generate the masses of the T-odd fermion doublets,

$$-\kappa f(\bar{\Psi}_2\xi\Psi_c + \bar{\Psi}_1\Sigma_0\Omega\xi^\dagger\Omega\Psi_c) + \text{h.c.}, \tag{5}$$

where $\Omega = diag(1,1,-1,1,1)$, $\Psi_c = (q_c, \chi_c, \tilde{q}_c)^T$, and the $SU(5)$ multiplets Ψ_1 and Ψ_2 are expressed as

$$\Psi_1 = \begin{pmatrix} q_1 \\ 0 \\ \mathbf{0}_2 \end{pmatrix}, \quad \Psi_2 = \begin{pmatrix} \mathbf{0}_2 \\ 0 \\ q_2 \end{pmatrix}. \tag{6}$$

The interaction Lagrangian in Eq.(5) can be proofed to be invariant under T-parity, and T-odd quark doublet q_- gets a Dirac mass with $\tilde{q}_c \equiv (id_{R_-}, -iu_{R_-})^{\mathrm{T}}$ from Eq.(5) expressed as [24]

$$m_{U_-} \simeq \sqrt{2}\kappa f\left(1-\frac{1}{8}\frac{v_{SM}^2}{f^2}\right), \qquad m_{D_-} = \sqrt{2}\kappa f, \tag{7}$$

where the lower indexes $U_- = u_-, c_-, t_-$ and $D_- = d_-, s_-, b_-$, which represent the T-odd heavy partners of the SM quarks, and κ is the mass coefficient in Lagrangian of the quark sector. As we know in the LHT $f > 500\ GeV$ [29], it is evident from Eq.(7) that the T-odd up- and down-type heavy partners have nearly equal masses.

In order to avoid the large radiative correction to Higgs boson mass induced by top-quark loop, the top sector must be additionally modified. We introduce the following two multiplets,

$$\mathcal{Q}_1 = \begin{pmatrix} q_1 \\ U_{L1} \\ \mathbf{0}_2 \end{pmatrix}, \quad \mathcal{Q}_2 = \begin{pmatrix} \mathbf{0}_2 \\ U_{L2} \\ q_2 \end{pmatrix}, \tag{8}$$

where U_{L1} and U_{L2} are the singlet fields and the q_1 and q_2 are the doublets. Under the $SU(5)$ and the T parity transformations, $\mathcal{Q}_1$ and $\mathcal{Q}_2$ behave themselves same as Ψ_1 and Ψ_2.

In addition to the T-even SM top quark right-handed $SU(2)$ singlet u_R, the LHT contains two $SU(2)$ singlet fermions U_{R1} and U_{R2} of hypercharge 2/3, which transform under T parity as

$$U_{R1} \leftrightarrow -U_{R2}. \tag{9}$$

The T parity invariant Yukawa Lagrangian of the top sector can be written as

$$\mathcal{L}_t^Y = \frac{\lambda_1 f}{2\sqrt{2}}\epsilon_{ijk}\epsilon_{xy}[(\bar{\mathcal{Q}}_1)_i\Sigma_{jx}\Sigma_{ky} - (\bar{\mathcal{Q}}_2\Sigma_0)_i\tilde{\Sigma}_{jx}\tilde{\Sigma}_{ky}]u_R$$
$$+\lambda_2 f(\bar{U}_{L1}U_{R1} + \bar{U}_{L2}U_{R2}) + \text{h.c.}\,. \quad (10)$$

where $\tilde{\Sigma} = \Sigma_0\Omega\Sigma^\dagger\Omega\Sigma_0$ is the image of the Σ field under T parity, and i, j and k run over $1-3$ and x and y over $4-5$. The T parity eigenstates are constructed as

$$q_\pm = \frac{1}{\sqrt{2}}(q_1 \mp q_2), \quad U_{L\pm} = \frac{1}{\sqrt{2}}(U_{L1} \mp U_{L2}), \quad U_{R\pm} = \frac{1}{\sqrt{2}}(U_{R1} \mp U_{R2}). \quad (11)$$

The T-odd states U_{L-} and U_{R-} combine to form a Dirac fermion T_-, and we obtain the mass of the T_- quark from the Lagrangian of Eq.(10) as

$$m_{T_-} = \lambda_2 f. \quad (12)$$

The left-handed (right-handed) top quark t is a linear combination of u_{L+} and U_{L+} (u_{R+} and U_{R+}), and another independent linear combination is a heavy T-even partner of the top quark T_+:

$$\begin{pmatrix} t_X \\ T_{+X} \end{pmatrix} = \begin{pmatrix} c_X & -s_X \\ s_X & c_X \end{pmatrix}\begin{pmatrix} u_{X_+} \\ U_{X_+} \end{pmatrix}, \quad (X = L, R), \quad (13)$$

where the mixing matrix elements are approximately expressed as

$$s_L = s_\alpha^2\frac{v_{SM}}{f} + \cdots, \quad s_R = s_\alpha\left[1 - \frac{c_\alpha^2(c_\alpha^2 - s_\alpha^2)}{2}\frac{v_{SM}^2}{f^2} + \cdots\right]. \quad (14)$$

There we define $s_\alpha = \lambda_1/\sqrt{\lambda_1^2 + \lambda_2^2}$ and $c_\alpha = \lambda_2/\sqrt{\lambda_1^2 + \lambda_2^2}$. The t is identified with the SM top and T_+ is its T-even heavy partner. Then the masses of the top quark and T-even heavy top quark can be obtained as

$$m_t \simeq \frac{\lambda_1\lambda_2 v_{SM}}{\sqrt{\lambda_1^2 + \lambda_2^2}}, \quad m_{T_+} \simeq f\sqrt{\lambda_1^2 + \lambda_2^2}. \quad (15)$$

The couplings of the T-odd $SU(2)$ doublet quarks and gauge bosons to the T-even SM particles used in our calculations are listed in Table 1 [20, 30], where $(V_{Hu})_{ij}$ and $(V_{Hd})_{ij}$ are the matrix elements of the CKM-like unitary mixing matrices V_{Hu} and V_{Hd}, respectively. The two mixing matrices satisfy $V_{Hu}^\dagger V_{Hd} = V_{CKM}$ [30], therefore, they cannot simultaneously be set to the identity. In the following calculations we take V_{Hu} to be a unit matrix, then we have $V_{Hd} = V_{CKM}$.

3. Renormalization and PROSPINO scheme

The strong coupling constant, the masses and wave functions of the relevant colored particles in the LHT are renormalized to remove the UV divergences of the virtual corrections. In our calculations, the following renormalization constants are introduced:

$$\psi_{q(q_-)}^{0,L,R} = \left(1 + \frac{1}{2}\delta Z_{q(q_-)}^{L,R}\right)\psi_{q(q_-)}^{L,R}, \quad m_{q_-}^0 = m_{q_-} + \delta m_{q_-},$$
$$G_\mu^0 = \left(1 + \frac{1}{2}\delta Z_g\right)G_\mu, \quad g_s^0 = g_s + \delta g_s, \quad (16)$$

Interaction	Feynman rule	Interaction	Feynman rule
$W_H^{+\mu}\bar{u}^i_- d^j$	$i\frac{g}{\sqrt{2}}(V_{Hd})_{ij}\gamma^\mu P_L$	$W_H^{-\mu}\bar{d}^i_- u^j$	$i\frac{g}{\sqrt{2}}(V_{Hu})_{ij}\gamma^\mu P_L$
$Z_H^\mu \bar{u}^i_- u^j$	$i(\frac{gC_H}{2} - \frac{g'S_H}{10})(V_{Hu})_{ij}\gamma^\mu P_L$	$Z_H^\mu \bar{d}^i_- d^j$	$i(-\frac{gC_H}{2} - \frac{g'S_H}{10})(V_{Hd})_{ij}\gamma^\mu P_L$
$\bar{q}^\alpha_- q^\beta_- G^a_\mu$	$ig_s(T^a)_{\alpha\beta}\gamma^\mu$		

Table 1. The related LHT Feynman rules used in our calculations, where $q_- = u_-, d_-, c_-, s_-, t_-, b_-$, i and j are the generation indices and $C_H^2 = 1 - S_H^2$.

where g_s denotes the strong coupling constant, m_{q_-} is the T-odd quark mass, $\psi^{L,R}_{q(q_-)}$ and G_μ denote the fields of the SM quark, T-odd heavy quark and gluon, respectively. The masses and wave functions of the colored fields are renormalized by adopting the on-shell scheme, then the relevant renormalization constants are expressed as

$$\delta Z_q^{L,R} \equiv \delta Z_q = -\frac{\alpha_s(\mu_r)}{3\pi}\Big[\Delta_{UV} - \Delta_{IR}\Big], \tag{17}$$

$$\delta Z_{q_-}^{L,R} \equiv \delta Z_{q_-} = -\frac{\alpha_s(\mu_r)}{3\pi}\left[\Delta_{UV} + 2\Delta_{IR} + 4 + 3\ln\left(\frac{\mu_r^2}{m_{q_-}^2}\right)\right], \tag{18}$$

$$\frac{\delta m_{q_-}}{m_{q_-}} = -\frac{\alpha_s(\mu_r)}{3\pi}\left\{3\left[\Delta_{UV} + \ln\left(\frac{\mu_r^2}{m_{q_-}^2}\right)\right] + 4\right\}, \tag{19}$$

$$\delta Z_g = -\frac{\alpha_s(\mu_r)}{2\pi}\left\{\frac{3}{2}\Delta_{UV} + \frac{5}{6}\Delta_{IR} + \frac{1}{3}\ln\left(\frac{\mu_r^2}{m_t^2}\right) + \frac{1}{3}\sum_{T=T_+}^{T_-}\ln\left(\frac{\mu_r^2}{m_T^2}\right) + \frac{1}{3}\sum_{q_-}\ln\frac{\mu_r^2}{m_{q_-}^2}\right\},$$
$$(q_- = u_-, d_-, c_-, s_-, t_-, b_-), \tag{20}$$

where $\Delta_{UV} = 1/\epsilon_{UV} - \gamma_E + \ln(4\pi)$ and $\Delta_{IR} = 1/\epsilon_{IR} - \gamma_E + \ln(4\pi)$.

For the renormalization of the strong coupling constant g_s, we adopt the $\overline{MS}$ scheme at the renormalization scale μ_r, except that the divergences associated with the massive top-quark, T-odd $SU(2)$ doublet quarks ($u_-, d_-, c_-, s_-, t_-, b_-$) and $T_\pm$ loops are subtracted at zero momentum [31]. Then the renormalization constant of the strong coupling constant can be obtained as

$$\frac{\delta g_s}{g_s} = -\frac{\alpha_s(\mu_r)}{4\pi}\left[\frac{3}{2}\Delta_{UV} + \frac{1}{3}\ln\frac{m_t^2}{\mu_r^2} + \frac{1}{3}\sum_{T=T_+}^{T_-}\ln\frac{m_T^2}{\mu_r^2} + \frac{1}{3}\sum_{q_-}\ln\frac{m_{q_-}^2}{\mu_r^2}\right],$$
$$(q_- = u_-, d_-, c_-, s_-, t_-, b_-). \tag{21}$$

In the calculation, we can find there are additional on-shell T-odd quark resonance in the real light-quark emissions. We adopt the PROSPINO scheme [32, 33] to remove them. The PROSPINO scheme is defined as a replacement of the Breit-Wigner propagator [33]

$$\frac{|\mathcal{M}|^2(s_{V_Hq})}{(s_{V_Hq} - m_{q_-}^2)^2 + m_{q_-}^2\Gamma_{q_-}^2} \to \frac{|\mathcal{M}|^2(s_{V_Hq})}{(s_{V_Hq} - m_{q_-}^2)^2 + m_{q_-}^2\Gamma_{q_-}^2} - \frac{|\mathcal{M}|^2(m_{q_-}^2)}{(s_{V_Hq} - m_{q_-}^2)^2 + m_{q_-}^2\Gamma_{q_-}^2}\Theta(\hat{s} - 4m_{q_-}^2)\Theta(m_{q_-} - m_{V_H}), \tag{22}$$

where s_{V_Hq} is the squared momentum flowing through the intermediate q_- propagator.

4. Numerical results of $pp \to W_H(Z_H)q_- + X$ process

Due to the additional T-odd quark resonance in light-quark emission subprocesses, we apply three schemes in considering the QCD NLO corrections in this work. In scheme (I) (denoted as "QCD NLO I") we include all light-quark emission contributions in the QCD NLO corrections. In scheme (II) (denoted as "QCD NLO II") we exclude the contributions of the partonic processes of light-quark emission. The PROSPINO scheme for light-quark emission is used in scheme (III).

In the study of the dependence of the QCD NLO corrected cross section on the factorization and renormalization scales, we set the two unphysical scales equal to a common value ($\mu_f = \mu_r = \mu$) and do not vary them in an independent way for simplicity.

We take one-loop and two-loop running α_s in the LO and QCD NLO calculations, respectively [34]. The central value of the factorization/renormalization scale μ is chosen as $\mu_0 = (m_{W_H} + m_{d_-})/2$. We adopt the CTEQ6L1 and CTEQ6M parton densities with five flavors in the LO and NLO calculations, respectively [35]. The strong coupling constant $\alpha_s(\mu)$ is determined by the QCD parameter $\Lambda_5^{LO} = 165\ MeV$ for the CTEQ6L1 at the LO and $\Lambda_5^{\overline{MS}} = 226\ MeV$ for the CTEQ6M at the NLO [34]. We ignore the masses of u-, d-, c-, s-, b-quarks, and take $\alpha_{ew}(m_Z^2)^{-1}|_{\overline{MS}} = 127.925$, $m_W = 80.399\ GeV$, $m_Z = 91.1876\ GeV$, $m_t = 171.2\ GeV$ and $\sin^2\theta_W = 1 - \left(\frac{m_W}{m_Z}\right)^2 = 0.222646$.

The colliding energy in the proton-proton center-of-mass system is taken as $\sqrt{s} = 7\ TeV$ for the early LHC and $\sqrt{s} = 14\ TeV$ for the later running at the LHC. The Cabibbo-Kobayashi-Maskawa (CKM) matrix elements are taken as

$$V_{CKM} = \begin{pmatrix} V_{ud} & V_{us} & V_{ub} \\ V_{cd} & V_{cs} & V_{cb} \\ V_{td} & V_{ts} & V_{tb} \end{pmatrix} = \begin{pmatrix} 0.97418 & 0.22577 & 0 \\ -0.22577 & 0.97418 & 0 \\ 0 & 0 & 1 \end{pmatrix}. \tag{23}$$

4.1. Dependence on factorization/renormalization scale

In Figs.1(a,b,c) and Figs.2(a,b,c) we present the LO, QCD NLO corrected cross sections and the corresponding K-factors for the $pp \to W_H q_- + X$ and $pp \to Z_H q_- + X$ processes as the functions of the factorization/renormalization scale at the LHC with $\sqrt{s} = 7\ TeV$ and $14\ TeV$, respectively. In Figs.1(a,b) and Figs.2(a,b) the LHT input parameters are taken as $f = 500\ GeV$ and $\kappa = 1$, while in Fig.1(c) and Fig.2(c) we take $f = 1\ TeV$ and $\kappa = 1$. In these figures the curves labeled by "NLO I", "NLO II" and "NLO III" are for the QCD NLO corrected cross sections using the (I), (II) and (III) schemes, respectively. The figures show that by using the (II) and (III) subtraction schemes we can get almost the same and moderate QCD NLO corrections to the production rate with a strongly reduced factorization/renormalization scale uncertainty in the plotted range of μ, while the QCD NLO corrections using the scheme (I) do not obviously improve the scale dependence of the LO cross section and destroy the perturbative convergence in some range of μ. In the following analysis we set the factorization/renormalization scale μ as its central value $\mu_0 = (m_{W_H} + m_{d_-})/2$.

4.2. Dependence on LHT parameters

We depict the LO, QCD NLO corrected cross sections and the corresponding K-factors for the $pp \to W_H q_- + X$ and $pp \to Z_H q_- + X$ processes as the functions of f, the $SU(5)$ global symmetry breaking scale of the LHT, at the LHC with $\sqrt{s} = 7\ TeV$ and $14\ TeV$ in Figs.3(a,b,c) and Figs.4(a,b,c), respectively. In Figs.3(a,b) and Figs.4(a,b) the parameter κ is set to be 1, while in Fig.3(c) and Fig.4(c) we take $\kappa = 3$. The curves labeled by "NLO I", "NLO II" and "NLO III" are for the QCD NLO corrected cross sections using the (I), (II) and (III) schemes, respectively. One can conclude from these figures that the cross section for the $pp \to W_H(Z_H)q_- + X$ process decreases quickly with the increment of f, because the two final T-odd particles become heavier

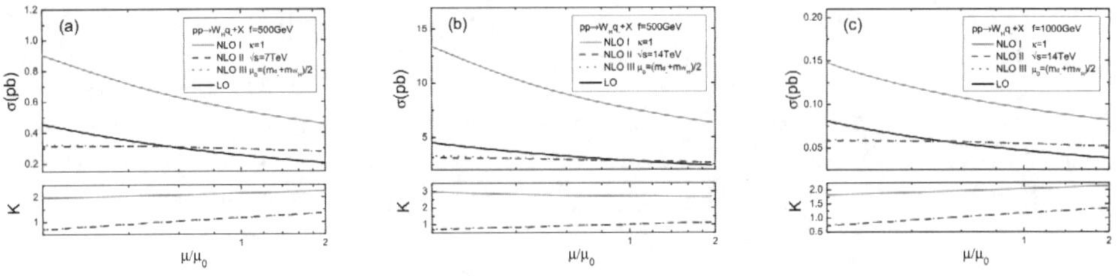

Figure 1. The dependence of the cross sections and the corresponding K-factors for the $pp \to W_H q_- + X$ process on the factorization/renormalization scale μ at the LHC. (a) $f = 500\ GeV$, $\kappa = 1$ and $\sqrt{s} = 7\ TeV$. (b) $f = 500\ GeV$, $\kappa = 1$ and $\sqrt{s} = 14\ TeV$. (c) $f = 1\ TeV$, $\kappa = 1$ and $\sqrt{s} = 14\ TeV$.

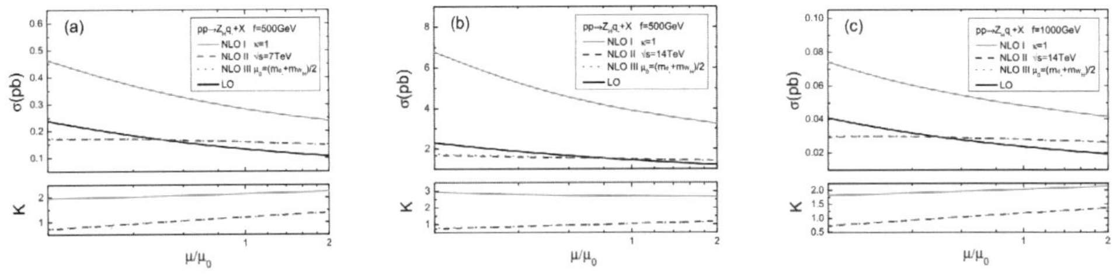

Figure 2. The dependence of the cross sections and the corresponding K-factors for the $pp \to Z_H q_- + X$ process on the factorization/renormalization scale μ at the LHC. (a) $f = 500\ GeV$, $\kappa = 1$ and $\sqrt{s} = 7\ TeV$. (b) $f = 500\ GeV$, $\kappa = 1$ and $\sqrt{s} = 14\ TeV$. (c) $f = 1\ TeV$, $\kappa = 1$ and $\sqrt{s} = 14\ TeV$.

with the increment of f. However, in the plotted range of f we could have observable production rates for the $pp \to W_H q_- + X$ and $pp \to Z_H q_- + X$ processes, especially when $\kappa = 1$.

5. Numerical results of $pp \to W_H^+ W_H^- + X$ process

We take $\alpha_{ew}(m_Z^2)^{-1} = 127.916$, $m_W = 80.399\ GeV$, $m_Z = 91.1876\ GeV$, $\sin^2\theta_W = 1 - \left(\frac{m_W}{m_Z}\right)^2 = 0.2226$ and $m_t = 171.2\ GeV$ [34]. The masses of all the SM leptons and quarks except top quark are neglected. The center-of-mass energies $\sqrt{s}$ of proton-proton collision are taken to be 14 TeV and 8 TeV for the future and early LHC, separately. We set the factorization and renormalization scale to be equal ($\mu_r = \mu_f$) and define $\mu_0 = m_{W_H}$. We employ CTEQ6L1 and CTEQ6M in the the LO and NLO calculations respectively [35], and fix the LHT parameters $\kappa = 1$ and $s_\alpha = c_\alpha = \frac{\sqrt{2}}{2}$. In this work we only use the PROSPINO scheme to deal with the real-quark emission subprocesses.

5.1. Dependence on factorization/renormalization scale

In Figs.5(a,b) we present the dependence of the LO, QCD NLO corrected integrated cross sections and the corresponding K-factor ($K \equiv \sigma_{NLO}/\sigma_{LO}$) on the factorization/renormalization

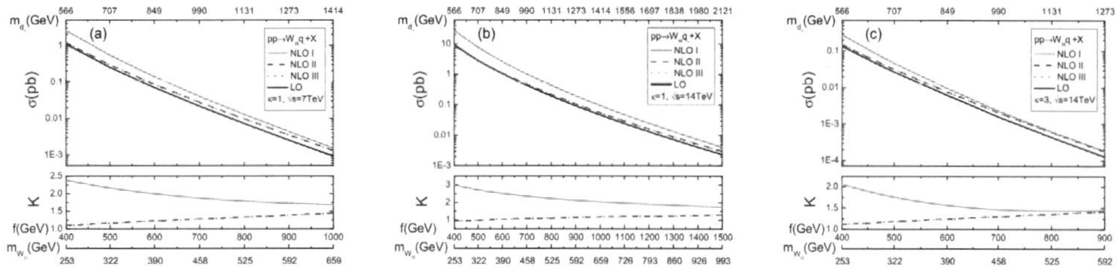

Figure 3. The cross sections and the corresponding K-factors for the $pp \to W_H q_- + X$ process as the functions of the LHT parameter f at the LHC. The corresponding m_{W_H} and m_{d_-} values are also scaled on the x-axis. (a) $\kappa = 1$ and $\sqrt{s} = 7\ TeV$. (b) $\kappa = 1$ and $\sqrt{s} = 14\ TeV$. (c) $\kappa = 3$ and $\sqrt{s} = 14\ TeV$.

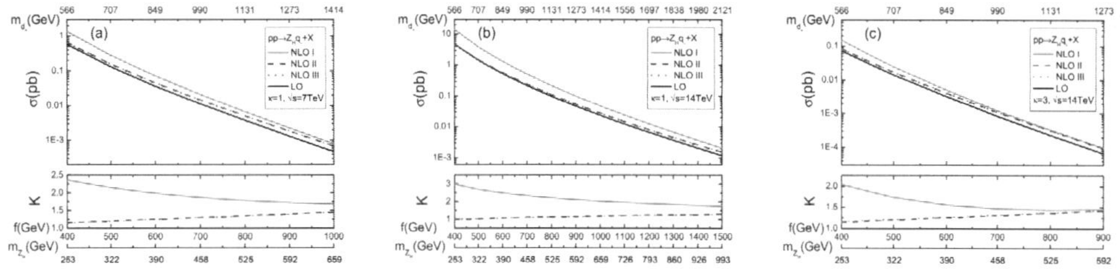

Figure 4. The cross sections and the corresponding K-factors for the $pp \to Z_H q_- + X$ process as the functions of the LHT parameter f at the LHC. The corresponding m_{W_H} and m_{d_-} values are also scaled on the x-axis. (a) $\kappa = 1$ and $\sqrt{s} = 7\ TeV$. (b) $\kappa = 1$ and $\sqrt{s} = 14\ TeV$. (c) $\kappa = 3$ and $\sqrt{s} = 14\ TeV$.

scale μ for the process $pp \to W_H^+ W_H^- + X$ at the $\sqrt{s} = 14\ TeV$ and the $\sqrt{s} = 8\ TeV$ LHC separately, where we take the LHT parameters $f = 800\ GeV$, $\kappa = 1$ and $s_\alpha = c_\alpha = \frac{\sqrt{2}}{2}$. From the curves in Figs.5(a,b), we find that QCD NLO corrections to the $pp \to W_H^+ W_H^- + X$ process significantly reduce the scale uncertainty. We can read out from the figures that the LO and QCD NLO corrected cross sections at $\mu_0 = m_{W_H}$ are $\sigma_{LO}(\sqrt{s} = 14TeV) = 32.63^{+9.56}_{-6.38}\ fb$, $\sigma_{NLO}(\sqrt{s} = 14TeV) = 37.43^{+2.19}_{-2.83}\ fb$ and $\sigma_{LO}(\sqrt{s} = 8TeV) = 5.54^{+2.71}_{-1.51}\ fb$, $\sigma_{NLO}(\sqrt{s} = 8TeV) = 6.14^{+0.26}_{-0.70}\ fb$, where the uncertainties describe the missing higher-order corrections estimated via scale variations in the range of $0.1\mu_0 < \mu < 10\mu_0$. The K-factor varies from 0.94 (0.77) to 1.32 (1.35) at the $\sqrt{s} = 14\ TeV$ ($8\ TeV$) LHC, when μ/μ_0 goes from 0.1 to 10. With the definition of scale uncertainty as $\eta = \frac{|\sigma(0.1\mu_0) - \sigma(10\mu_0)|}{\sigma(\mu_0)}$, we obtain that the scale uncertainties are reduced from 48.88% (LO) to 13.40% (NLO) at the $\sqrt{s} = 14\ TeV$ and from 76.23% (LO) to 14.54% (NLO) at the $\sqrt{s} = 8\ TeV$ LHC, respectively. In Table 2 we list some numerical results of the cross sections and K-factors for some typical values of μ/μ_0, which are read out from Figs.5(a,b). In order to investigate the contribution from the $pp \to gg \to W_H^+ W_H^- + X$ process, which is considered as a component of the QCD NLO corrections to the parent process $pp \to W_H^+ W_H^- + X$, we also present the cross sections for the $pp \to gg \to W_H^+ W_H^- + X$ process ($\sigma(gg)$) in this table. We can obtain from the data that the QCD NLO correction part from

$\sqrt{s}$ (TeV)	μ/μ_0	σ_{LO} (fb)	σ_{NLO} (fb)	$\sigma(gg)$ (fb)	K
	0.1	42.190(1)	39.62(2)	0.993(1)	0.939
	0.5	35.091(1)	38.17(2)	0.3946(6)	1.09
14	1	32.626(1)	37.43(2)	0.2810(5)	1.15
	2	30.444(1)	37.53(2)	0.2056(4)	1.20
	10	26.242(1)	34.60(2)	0.1081(2)	1.32
	0.1	8.2548(4)	6.333(3)	0.0805(1)	0.767
	0.5	6.1850(3)	6.300(3)	0.02842(3)	1.019
8	1	5.5417(2)	6.143(3)	0.01943(3)	1.11
	2	5.0006(2)	5.947(3)	0.01371(2)	1.21
	10	4.0304(2)	5.440(2)	0.00671(1)	1.40

Table 2. The numerical results of σ_{LO}, σ_{NLO} and the corresponding K-factors at the 14 TeV and the 8 TeV LHC by taking $f = 800\ GeV$, $\kappa = 1$, $s_\alpha = c_\alpha = \frac{\sqrt{2}}{2}$ and some typical values of factorization/renormalization scale μ. $\sigma(gg)$ is the cross section for the $pp \to gg \to W_H^+ W_H^- + X$ process, which is considered as a component of the QCD NLO correction to the parent process $pp \to W_H^+ W_H^- + X$.

the $pp \to gg \to W_H^+ W_H^- + X$ process at $\mu = \mu_0$ is about 5.85% (3.23%) of the total QCD NLO correction ($\Delta\sigma_{NLO}$) at the 14 TeV (8 TeV) LHC. We can see also that the NLO theoretical uncertainty due the choice of μ mainly comes from the genuine QCD NLO corrected cross section for the $pp \to q\bar{q} \to W_H^+ W_H^- + X$ process, while the contribution from the $pp \to gg \to W_H^+ W_H^- + X$ process is relatively small. In further numerical calculations we fix the renormalization and factorization scales being equal to their central value, i.e., $\mu = \mu_r = \mu_f = \mu_0 = m_{W_H}$.

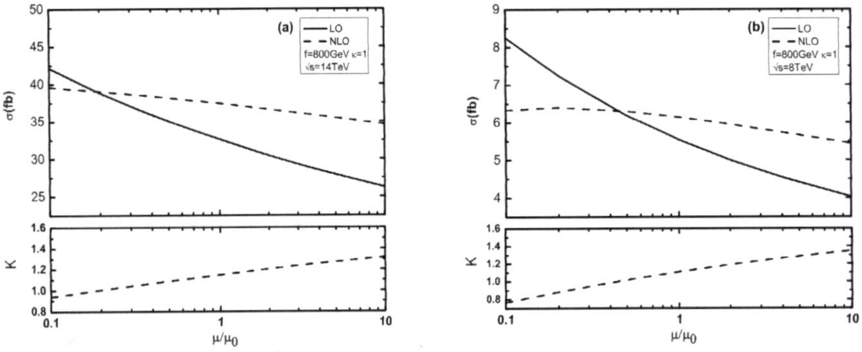

Figure 5. The dependence of the LO, QCD NLO corrected integrated cross sections and the corresponding K-factors for the $pp \to W_H^+ W_H^- + X$ process on the factorization/renormalization scale μ at the LHC with $f = 800\ GeV$, $\kappa = 1$ and $s_\alpha = c_\alpha = \frac{\sqrt{2}}{2}$. (a) $\sqrt{s} = 14\ TeV$. (b) $\sqrt{s} = 8\ TeV$.

5.2. Dependence on global symmetry breaking scale f

The LO and QCD NLO corrected integrated cross sections together with the corresponding K-factor as functions of the scale f at the $\sqrt{s} = 14\ TeV$ and the $\sqrt{s} = 8\ TeV$ LHC are depicted in Figs.6(a) and (b), respectively. We can see from Fig.6 that the LO and NLO total cross sections for the $pp \to W_H^+ W_H^- + X$ process decrease drastically when f goes up. This is because the mass of final W_H becomes heavier as the increment of f, therefore the phase-space becomes smaller. We can read out from the figures that the corresponding K-factor varies from 1.22 to 1.10 at the $\sqrt{s} = 14\ TeV$ LHC and from 1.17 to 1.10 at the $\sqrt{s} = 8\ TeV$ LHC in the plotted f range. In Table 3, we list some numerical results of the LO, NLO cross sections and the corresponding K-factors for some typical values of f which are shown in Figs.6(a,b).

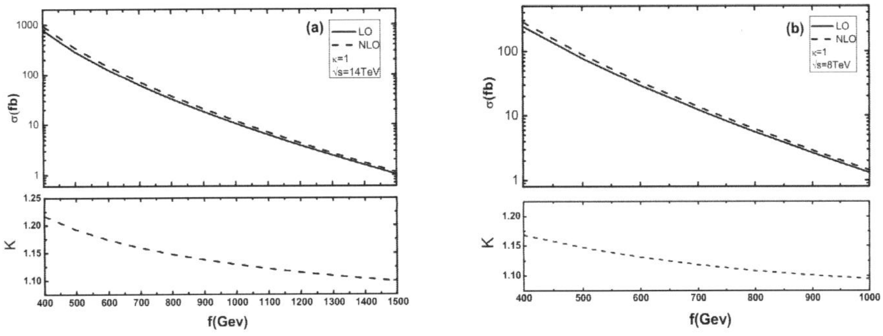

Figure 6. The LO, QCD NLO corrected integrated cross sections and the corresponding K-factors for the $pp \to W_H^+ W_H^- + X$ process as the functions of the global symmetry breaking scale f at the LHC with $\kappa = 1$ and $s_\alpha = c_\alpha = \frac{\sqrt{2}}{2}$. (a) $\sqrt{s} = 14\ TeV$. (b) $\sqrt{s} = 8\ TeV$.

6. Summary

In this paper, we present the calculation of $W_H(Z_H)q_-$ and W_H-pair productions at the $\sqrt{s} = 14\ TeV$ and the $\sqrt{s} = 8\ TeV$ LHC in QCD NLO. The dependence of the total cross section on the renormalization/factorization scale shows that the QCD NLO corrections can reduce significantly the uncertainty of the LO theoretical predictions. We also display the dependence of cross sections on global symmetry breaking scale f. The more calculation detail and numerical results were presented in Refs.[27, 28].

Acknowledgments: This work was supported in part by the National Natural Science Foundation of China (Contract No.11075150, No.11005101), and the Specialized Research Fund for the Doctoral Program of Higher Education (Contract No.20093402110030).

[1] S. L. Glashow, Nucl. Phys. **22** (1961) 579; S. Weinberg, Phys. Rev. Lett. **19** (1967) 1264; A. Salam, Proc. 8th Nobel Symposium Stockholm 1968,ed. N. Svartholm (Almquist and Wiksells, Stockholm 1968) p.367; H. D. Politzer, Phys. Rept. **14** (1974) 129.

[2] P. W. Higgs, Phys. Lett. **12** (1964) 132, Phys. Rev. Lett. **13** (1964) 508, Phys. Rev. **145** (1966) 1156; F. Englert and R. Brout, Phys. Rev. Lett. **13** (1964) 321; G. S. Guralnik, C. R. Hagen and T. W. B. Kibble, Phys. Rev. Lett. **13** (1964) 585; T. W. B. Kibble, Phys. Rev. **155** (1967) 1554.

$\sqrt{s}$ (TeV)	f (GeV)	σ_{LO} (fb)	σ_{NLO} (fb)	K
14	500	289.93(1)	345.6(2)	1.19
	700	62.053(3)	71.93(4)	1.16
	800	32.626(1)	37.43(2)	1.15
	900	18.1252(8)	20.61(1)	1.14
	1100	6.2964(3)	7.059(3)	1.12
	1300	2.44312(9)	2.709(1)	1.11
	1500	1.02314(4)	1.1246(5)	1.10
8	500	77.693(3)	89.12(5)	1.15
	700	12.2863(5)	13.741(7)	1.12
	800	5.5417(2)	6.143(3)	1.11
	900	2.6357(1)	2.901(1)	1.10

Table 3. The numerical results of σ_{LO}, σ_{NLO} and the corresponding K-factors at the $\sqrt{s} = 14\ TeV$ and the $\sqrt{s} = 8\ TeV$ LHC by taking $\kappa = 1$, $s_\alpha = c_\alpha = \frac{\sqrt{2}}{2}$, $\mu = \mu_0$ and some typical values of f.

[3] G.G. Ross, Grand Unied Theories (Addison-Wesley Publishing Company, Reading, MA, (1984); P. Langacker, Phys. Rep. 72 (1981) 185; H. Georgi, S.L Glashow, Phys. Rev. Lett. 32 (1974) 438; A.J. Buras, J. Ellis, M.K. Gaillard, D.V. Nanopoulos, Nucl. Phys. **B135**(1978) 66.

[4] S. P. Martin, arXiv:hep-ph/9709356; M. E. Peskin, arXiv:0801.1928 [hep-ph]; K. A. Olive, arXiv:hep-ph/9911307; M. Drees, arXiv:hep-ph/9611409; H. E. Haber and G. L. Kane, Phys. Rept. 117 (1985) 75; H. P. Nilles, Phys. Rept. 110, 1 (1984); A. Signer, J.Phys.**G36** (2009) 073002, arXiv:0905.4630 [hep-ph].

[5] N. Arkani-Hamed, S. Dimopoulos, G. Dvali, Phys. Lett. B429 (34): 263(1998), arXiv:hep-ph/9803315; N. Arkani-Hamed, S. Dimopoulos, G. Dvali, Phys. Rev. D59 086004(1999), arXiv:hep-ph/9807344; I. Antoniadis, N. Arkani-Hamed, S. Dimopoulos, G. Dvali, Phys. Lett. **B436** (34): 257(1998), arXiv:hep-ph/9804398; M. Shifman, Int. J. Mod. Phys. **A25** 199-225,2010, arXiv:0907.3074v2 [hep-ph].

[6] R. N. Mohapatra and J. C. Pati, Phys. Rev. **D11** (1975), 566571; G. Senjanovic and R. N. Mohapatra; Phys. Rev. **D12** (1975), 1502; A. Adulpravitchai, M. Lindner, A. Merle, and R. N. Mohapatra, Phys. Lett. **B680** (2009) 476, arXiv/hep-ph:0908.0470.

[7] L.Basso, S. Moretti, G.M. Pruna, Phys. Rev.**D83**(2011) 055014, arXiv:1011.2612v4 [hep-ph].

[8] N. Arkani-Hamed, A. G. Cohen and H. Georgi, Phys. Lett. **B513** (2001) 232; M. Schmaltz and D. Tucker-Smith, Ann. Rev. Nucl. Part. Sci. **55** (2005) 229; M. Perelstein, Prog. Part. Nucl. Phys. **58** (2007) 247; and references therein.

[9] Arkani-Hamed, A. G. Cohen, E. Katz, A. E. Nelson, T. Gregoire and J. G. Wacker, JHEP **08** (2002) 021.

[10] N. Arkani-Hamed, A. G. Cohen, T. Gregoire, J. G. Wacker and A. G. Cohen, JHEP **08** (2002) 020.

[11] I. Low, W. Skiba and D.Smith, Phys. Rev. **D66**, (2002) 072001.

[12] C. Csaki, J. Hubisz, G. D. Kribs, P. Meade and J. Terning, Phys. Rev. **D67** (2003) 115002, [arXiv:hep-ph/0211124].

[13] M. Schmaltz, Nucl. Phys. Proc. Suppl. **117** (2003) 40.

[14] T. Gregoire and J. G. Wacker, JHEP **08** (2002) 019.

[15] N. Arkani-Hamed, A. G. Cohen, E. Katz and A. E. Nelson, JHEP **07** (2002) 034.

[16] ATLAS Collaboration, Phys. Lett. **B705** (2011) 18-46.

[17] D. Olivito, for the ATLAS collaboration, Conference proceedings for the Meeting of the Division of Particles and Fields of the American Physical Society (DPF), 2011, arXiv:1109.0934.

[18] I. Low, JHEP **0410** (2004) 067.

[19] R. Barbieri and A. Strumia, IFUP-TH/2000-22 and SNS-PH/00-12, [arXiv:hep-ph/0007265].

[20] J. Hubisz and P. Meade, Phys. Rev. **D71** (2005) 035016.

[21] J. Hubisz, P. Meade, A. Noble and M. Perelstein, JHEP **01** (2006) 135.

[22] H.-C. Cheng and I. Low, **JHEP** 09 (2003) 051; **JHEP** 08 (2004) 061.

[23] A. Birkedal, A. Noble, M. Perelstein and A. Spray, Phys. Rev. **D74** (2006) 035002; M. Asano, S. Matsumoto, N. Okada, and Y. Okada, Phys. Rev. **D75** (2007) 063506.

[24] A. Belyaev, C. -R. Chen, K. Tobe and C. -P. Yuan, Phys. Rev. **D74**, (2006) 115020.
[25] A. Belyaev, C. -R. Chen, K. Tobe and C. -P. Yuan, in Proceedings of Monte Carlo Tools for Beyond the Standard Model Physics, Fermilab, 2006, given by A. Belyaev, http://theory.fnal.gov/mc4bsm/agenda.html; in Proceedings of Osaka University, 2006, Osaka, given by C. -P. Yuan, http://www-het.phys.sci.osaka-u.ac.jp/seminar/seminar/seminar.html; in Proceedings of the Summer Institute on Collider Phenomenology, National Tsing Hua University, Taiwan, 2006, given by K. Tobe, http://charm.phys.nthu.edu.tw/hep/summer2006/; in Proceedings of ICHEP'06, Moscow, 2006, given by A. Belyaev, http://ichep06.jinr.ru/reports/116_11s1_10p20_belyaev.pdf.
[26] C.-S. Chen, K. Cheung and T. -C. Yuan, Phys. Lett. **B644** (2007) 158.
[27] R.-Y. Zhang, H. Yan, W.-G. Ma, S.-M. Wang, L. Guo and L. Han, Phys. Rev. **D85** (2012)015017.
[28] S.-M. Du, L. Guo, W. Liu, W.-G. Ma and R.-Y. Zhang, Phys. Rev. **D86** (2013) 054027.
[29] J. Hubisz, P. Meade, A. Noble and M. Perelstein, JHEP **0601**, 135 (2006), arXiv:hep-ph/0506042.
[30] M. Blanke, A. J. Buras, A. Poschenrieder, S. Recksiegel, C. Tarantino, S. Uhlig and A. Weiler JHEP **01** (2007) 066.
[31] J. Collins, F. Wilczek, and A. Zee, Phys. Rev. **D18** (1978) 242; W. J. Marciano, Phys. Rev. **D29** (1984) 580; P. Nason, S. Dawson and R.K. Ellis, Nucl. Phys. **B327** (1989) 49, Nucl. Phys. **B335** (1990) 260(E).
[32] W. Beenakker, R. Höpker, M. Spira and P. M. Zerwas, Nucl. Phys. **B492** (1997) 51; W. Beenakker, M. Klasen, M. Krämer, T. Plehn, M. Spira and P. M. Zerwas, Phys. Rev. Lett. **83** (1999) 3780; http://www.thphys.uni-heidelberg.de/~ plehn/index.php?show=prospino.
[33] T. Plehn and C. Weydert, PoS **CHARGED2010** (2010) 026, [arXiv:1012.3761]; T. Binoth, D. Goncalves-Netto, D. Lopez-Val, K. Mawatari, T. Plehn and I. Wigmore, Phys. Rev. D 84 (2011) 075005, [arXiv:1108.1250].
[34] K. Nakamura, *et al.*, J. Phys. **G37** (2010) 075021.
[35] J. Pumplin, D. R. Stump, J. Huston, H. -L. Lai, P. Nadolsky and W. -K. Tung, JHEP **07** (2002) 012; D. Stump, J. Huston, J. Pumplin, W. -K. Tung, H. -L. Lai, S. Kuhlmann and J. F. Owens, JHEP **10** (2003) 046.

On cross-section computation in the brane-world models

Kirpichnikov Dmitry

Institute for Nuclear Research of the Russian Academy of Sciences, Prospect of the 60th Anniversary of October 7a, Moscow, Russia, 117312

E-mail: kirpich@ms2.inr.ac.ru

Abstract. We present Mathematica7 numerical simulation of the process $pp \to \text{jet} + \not{E}_T$ in the framework of modified Randall-Sundrum brane-world model with one infinite and n compact extra dimension. We compare the energy missing signature with the standard model background $pp \to \text{jet} + \nu\bar{\nu}$, which was simulated at CompHep. We show that the models with numbers of compact extra dimensions greater than 4 can be probed at the protons center-of-mass energy equal 14 TeV. We also find that testing the brane-world models at 7 TeV on the LHC appears to hopeless.

1. Introduction

There are a lot of softwares for simulating a "New physics" processes at the accelerating experiments. Such programs as **CompHep** [1] and **PYTHIA** [2] are among them. Nevertheless, one can consider an infinite extra spatial dimension in the brane-world models. **CompHep** and **PYTHIA** are not adopted for such class of models. In this paper we present the numerical simulations of the processes $pp \to \text{jet} + E_T^{miss}$ on **Mathematica7** in the backgound of modified Randall-Sundrum brane model with one infinite and n compact extra dimensions (RSII-n model). In this brane world model neutral particle such as Z boson and photon can leave our brane, escaping in to the extra dimension of infinite size. In our sumulation we use Gluck Reya Vogt leading order parton distribution functions. In Sec. 2 we discuss a motivation of using LO Gluck Reya Vogt PDF [3], by comparing the latter with CTEQ [4], MRST [5] and Alekhin's [6] LO PDFs. This PDFs coincide at large QCD scale parameter squared Q^2. In Sec. 3 we compare **CompHep** and **Mathematica7** numerical simulations of the process $pp \to \text{jet} + Z^0$ in the framework of standard model. We discuss RSII-n set up in Sec. 4. In Sec. 5 we present numerical simulations of the process $pp \to \text{jet} + E_T^{miss}$ in the framework of RSII-n model.

2. Comparison of PDFs at large Q^2.

In this section we compare LO Gluck Reya Vogt parton distribution functions with CTEQ, MRST and Alekhin's PDFs. Corresponding distributions for u and $\bar{u}$ quarks are presented on Fig. 1. We suppose that QCD scale parameter is fixed at $Q = 100$ GeV. These data are taken from an open High energy physics database. We can see from Fig. 1 that GRV LO PDFs coincide with CTEQ LO, MRST LO and Alekhin's LO PDFs at large Q^2. One can obtain an analogous distribution for d, $\bar{d}$ quarks and gluon. This means that in the numerical analysis we can use GRV LO PDFs as well as CTEQ, MRST and Alekhins PDFs if Q^2 is fixed at large values.

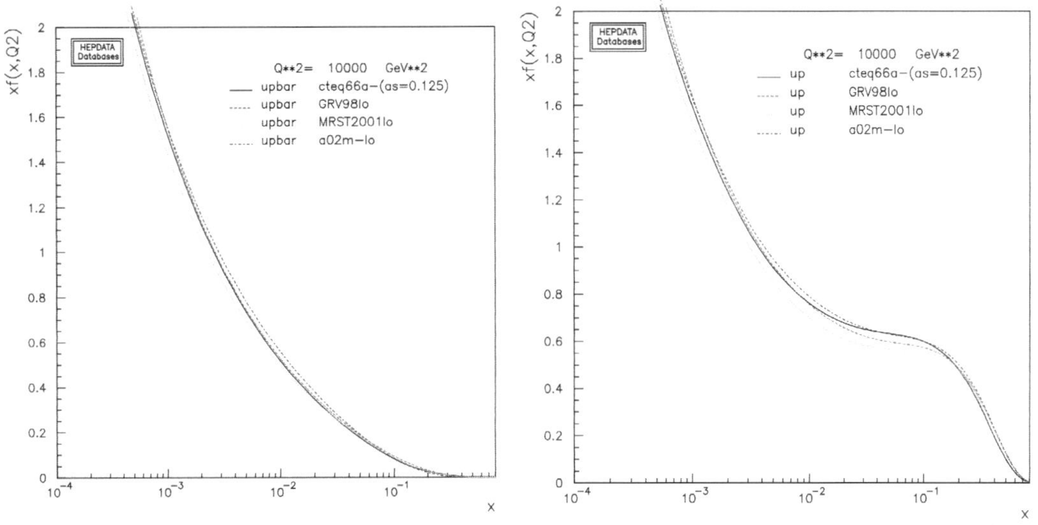

Figure 1. LO PDFs $xu(x,Q^2)$ and $x\bar{u}(x,Q^2)$ for GRV, CTEQ, MRST and Alekhin's collaborations.

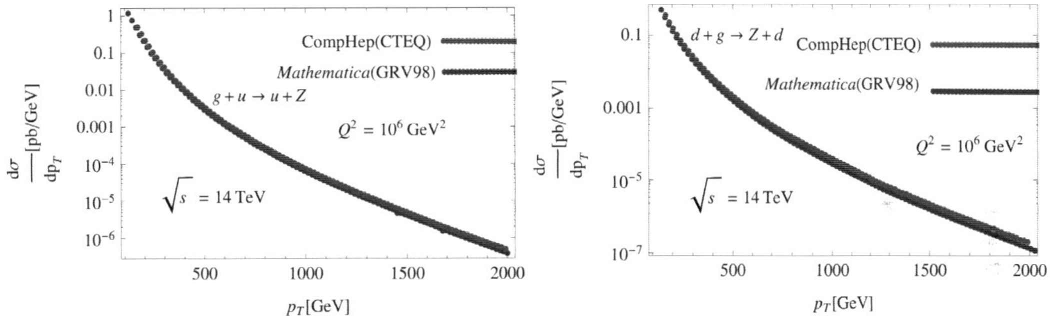

Figure 2. Comparison of differential distribution for the processes $dg \to Z^0 d$ and $ug \to Z^0 u$ simualated at **CompHep** and **Mathematica7** in the framework of standard model.

3. CompHep vs Mathematica7 in the SM framework.

In this section we compare **CompHep** and **Mathematica7** numerical simulations for the parton cross-sections in the framework of standard model. In Fig. 2 we show the differential cross-section of the processes $dg \to Z^0 d$ and $ug \to Z^0 u$ versus transverse momentum of Z^0 boson. The QCD scale is fixed at $Q = 1000$ GeV. The diagrams for GRV and CTEQ LO PDFs are coincided. But there is a small discreapancy in $xu(x,Q^2)$ distribution at high p_T values. The analogous distribution can be obtained for the other SM processes such as $\bar{d}g \to Z^0\bar{d}$, $\bar{u}g \to Z^0\bar{u}$, $\bar{d}d \to Z^0 g$ and $\bar{u}d \to Z^0 g$. Nevertheless, the main contribution to the process $pp \to \text{jet} + Z^0$ comes from the gluon cross-sections which are shown on Fig. 2.

Table 1. The lower bounds on the AdS curvature k for various n.

n	1	2	3	4	5	6
k, GeV	5.5×10^6	20×10^3	2.5×10^3	900	400	300

4. Modified Randall-Sundrum model

In this section we discuss a peculiar features of the modified RSII-n model. Let us consider a 3-brane with n compact dimensions embeeded in a $(5+n)$ space-time AdS metric

$$ds^2 = a(z)^2(\eta_{\mu\nu}dx^\mu dx^\nu - \delta_{ij}d\theta^i d\theta^j) - dz^2. \tag{1}$$

This metric was suggested by T. Gherghetta and M. Shaposhnikov [8]. Here z is the infinite extra dimension, θ_i are the compact extra-dimensions $\theta_i \in [0, 2\pi R_i]$, $i = \overline{1,n}$, R_i are the sizes of compact extra dimensions, n - is the number of compact extra dimensions, $a(z) = e^{-k|z|}$ is a warp factor from Randall-Sundrum model and k is a AdS curvature.

We put entire $SU(2) \times U(1)$ gauge sector, as well as the Higgs sector into the bulk space, but the fermions of the standard modell are supposed to be localized on the brane. The action of the model is

$$S = \int d^4x\, dz \prod_{i=1}^{n} \frac{d\theta_i}{2\pi R_i} \sqrt{g}\Big[-\frac{1}{2}|\, W_{MN}\,|^2 + m_W^2 |\, W_M\,|^2 - \frac{1}{4}Z_{MN}^2 + \frac{1}{2}m_Z^2 Z_M^2$$
$$-\frac{1}{4}F_{MN}^2 + \frac{1}{2}(\partial_M \chi)^2 - \frac{1}{2}m_\chi^2\chi^2 + \delta(z)\mathcal{L}_{ferm}\Big], \tag{2}$$

where $m_W^2 = \frac{1}{4}\tilde{g}_2^2 v^2$, $m_Z^2 = \frac{1}{4}(\tilde{g}_2^2 + \tilde{g}_1^2)v^2$ and $m_\chi^2 = \lambda v^2$ are the bulk masses of the gauge fields and Higgs respectively. We also suppose the size of the compact extra dimension to be $1/R_i \gg \sqrt{s}$. This means that corresponding KK excitations become infinitely heavy and dissappear from the spectrum.

Since the Z-boson is not exactly localized on the brane, one can obtain the bounds on number of compact extra dimension n and the AdS curvature k. We require that the invisible width decay of Z-boson in RSII-n model is bounded by the experimental uncertainty of the total Z-boson width decay [9]:

$$\Gamma_{RS}(M_Z) \leq \Delta\Gamma_{tot}^Z = 1.5\,\text{MeV}.$$

where

$$\Gamma_{RS}^Z = \frac{2\pi}{n\Gamma^2\left(\frac{n}{2}\right)} M_Z \left(\frac{M_Z}{2k}\right)^n, \tag{3}$$

is the invisible decay rate of Z_{bulk}^0 boson [10]. These bounds are shown in Tab. 1. We use these values of k when presenting the numerical simulations in Sec. 5

5. Numerical simulation of the process $pp \to$ jet $+ E_T^{miss}$.

In this section we discuss numerical simulation of the process $pp \to$ jet $+ Z_{bulk}(\gamma_{bulk})$ at the collider experiment in the frame work of RSII-n model. Here the jet originates from gluon or quark, and Z_{bulk} and γ_{bulk} are the particles which escape the brane. The differential cross-section

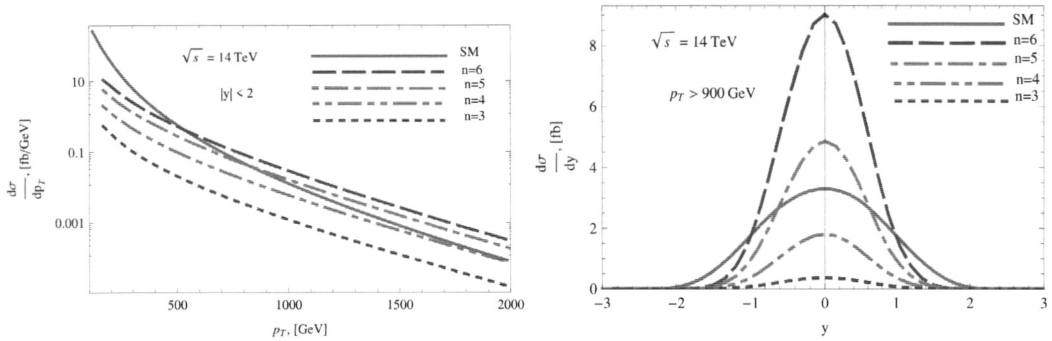

Figure 3. Differential cross section of the process $pp \to \text{jet} + Z_{bulk}(\gamma_{bulk})$ for $n = 3, 4, 5, 6$ at the center-of-mass energy of protons equal $\sqrt{s} = 14\text{TeV}$. The standard model background is $pp \to \text{jet} + \nu\bar{\nu}$.

Table 2. Integrated luminosity $\mathcal{L}$ and numbers of signal events N_S for various n, at the LHC center-of-mass energy $\sqrt{s} = 14$ TeV.

n	3	4	5	6
$\mathcal{L}$, fb^{-1}	7.1×10^2	3.7×10^1	7.5	3.1
N_S	3.6×10^2	10^2	5×10^1	3.8×10^1

of this process is

$$\frac{d^2\sigma}{dy\, p_T dp_T} = \int dx_1 dx_2 \sum_{i,j=q,\bar{q},g} \frac{f_i(x_1, Q^2)}{x_1} \frac{f_j(x_2, Q^2)}{x_2} \frac{1}{4\pi s\, m} \sum_k \overline{|\mathcal{M}_{ij\to k(bulk)}|^2} \tag{4}$$

where m is a bulk invariant mass of Z_{bulk} and γ_{bulk}:

$$m^2 = s x_1 x_2 - x_1 p_T \sqrt{s} e^{-y} - x_2 p_T \sqrt{s} e^{+y},$$

and $f_i(x_1, Q^2)$ are GRV LO PDFs [3]. The QCD scale parameter is fixed at $Q = 1$ TeV. We compare the distribution (4) with the standard model background $pp \to \text{jet} + \nu\bar{\nu}$. This background was computed by **CompHep** program [1].

Let us consider the case of the proton center-of-mass energy equal 14 TeV. In Fig. 3 we show p_T and jet rapidity distributions of the process $pp \to \text{jet} + Z_{bulk}(\gamma_{bulk})$. One can see from Fig. 3 that if $n = 6$ and $n = 5$, then the signal dominates over the background for $p_T > 500$ GeV and $p_T > 750$ GeV, correspondingly. And the background dominates over the signal for $n = 3, 4$. Once the larger values of k correspond to the smaller n (see Tab. 1), so that the signal cross section grows with the increase of n. It is clear from Fig. 3 that the jet rapidity distributions are correlated with the jet transverse momentum distributions. In Tab. 2 we show the integrated luminosity and number of signal events needed for 5σ discovery at the LHC center-of-mass energy equal 14 TeV. Where the cuts in jet rapidity and transverse momentum are $|y| < 2$, $p_T > 900$ GeV.

Now let us consider the case $\sqrt{s} = 7$ TeV. In Fig. 4 we show the jet transverse momentum and rapidity distributions. Only the case with $n \geq 6$ can be probed, since the background dominates over the signal for $n = 3, 4, 5$. The integrated luminosity required for 5σ discovery at the LHC is $\mathcal{L} = 200\,\text{fb}^{-1}$ even for $n = 6$.

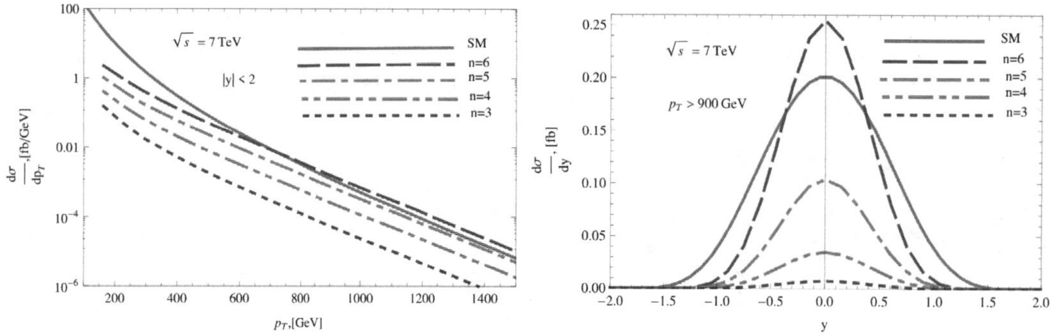

Figure 4. Same as in Fig. 3, but for the energy of protons equal 7 TeV.

6. Summary

In the framework of RSII-n model the distributions of the process $pp \to \text{jet} + Z^0_{bulk}(\gamma_{bulk})$ were simulated on **Mathematica7** with GRV LO PDFs implemented. The detection of the extra spatial dimension at 7 TeV appears to be hopeless. At the LHC energy equal 14 TeV the luminosity needed for 5σ discovery is in the range $(10 - 100)\text{fb}^{-1}$.

7. Acknowledgements

We are indebted to A. B. Arbuzov and A. L. Kataev, for helpful discussions and advices, to ACAT 2013 Organizing Comittee and in particular Bin Gong for support and hospitality. This work was supported in part by grants of Russian Ministry of Education and Science NS-5590.2012.2 and GK-8412, grants of the President of Russian Federation MK-2757.2012.2, and grants of RFBR 12-02-31595 MOL A and RFBR 13-02-01127 A.

References

[1] Boos E Bunichev E Dubinin M Dudko L Edneral V Ilyin V Kryukov A and Savrin V 2008 *PoS ACAT* **08** 008
[2] Sjostrand T Eden P Friberg C Lonnblad L Miu G Mrenna S and Norrbin E 2001 *Comput. Phys. Commun.* **135** 238
[3] Gluck M Reya E and Vogt A 1998 *Eur. Phys. J.* C **5** 461
[4] Pumplin J Stump D Huston J Lai H Nadolsky P and Tung W 2002 *JHEP* **0207** 012
[5] Martin A Roberts R Stirling W and Thorne R 2002 *Phys. Lett.* B **531** 216
[6] Alekhin S 2003 *Phys. Rev.* D **68** 014002
[7] Gherghetta T and Shaposhnikov M 2000 *Phys. Rev. Lett.* **85** 240
[8] Amsler C et al. 2008 *Phys. Lett.* **667** 1
[9] Kirpichnikov D 2012 *Phys. Rev.* D **85** 115008

Journal of Physics: Conference Series **523** (2014) 012056 doi:10.1088/1742-6596/523/1/012056

GoSam applications for automated NLO calculations

G Cullen[1], H van Deurzen[2], N Greiner[2], G Heinrich[2], G Luisoni[2], P Mastrolia[2,3], E Mirabella[2], G Ossola[4], T Peraro[2], J Reichel[2], J Schlenk[2], J F von Soden-Fraunhofen[2], F Tramontano[5]

[1] Deutsches Elektronen-Synchrotron DESY, Platanenallee 6, 15738 Zeuthen, Germany
[2] Max Planck Institute for Physics, Föhringer Ring 6, 80805 Munich, Germany
[3] Dipartimento di Fisica e Astronomia, Università di Padova, and INFN Sezione di Padova, via Marzolo 8, 35131 Padova, Italy
[4] Physics Department, New York City College of Technology, The City University of New York, 300 Jay Street Brooklyn, NY 11201, USA and
The Graduate School and University Center, The City University of New York, 365 Fifth Avenue, New York, NY 10016, USA
[5] Dipartimento di Scienze Fisiche, Università degli studi di Napoli and INFN, Sezione di Napoli, 80125 Napoli, Italy

Abstract. We present applications of the program GoSam for the automated calculation of one-loop amplitudes. Results for NLO QCD corrections to beyond the Standard Model processes as well as Higgs plus up to three-jet production in gluon fusion are shown. We also discuss some new features of the program.

1. Introduction

With the LHC data confirming the Standard Model to an almost incredible extent, precision measurements will be of enormous importance in order to scrutinize the Higgs properties and be sensitive to deviations from the Standard Model, and such measurements should come in hand with precision predictions. Therefore, it is of primary interest to provide tools which allow one to perform the comparison of LHC data to theory at NLO accuracy.

The automation of NLO calculations for multi-particle final states has seen an enormous level of progress in the past years, as can be seen from examples for recent achievements in terms of codes [1–10] providing one-loop amplitudes.

In this talk, we explain the usage and present applications of the program GoSam [4], which can generate and evaluate one-loop amplitudes for multi-particle processes in a fully automated way. GoSam also offers the possibility to be interfaced – via the Binoth Les Houches interface [11, 12] – to different Monte Carlo programs providing the real radiation and the infrared subtraction terms. Further, an interface to model files in Universal FeynRules Output (`UFO`) [13] or `LanHEP` [14] format allows to extend the application range of the code to Beyond the Standard Model physics.

2. The GoSam program

2.1. Generation of the virtual amplitudes

Amplitudes are expressed in terms of Feynman diagrams, where the integrand is generated analytically using **`QGRAF`** [15], **`FORM`** [16], **`spinney`** [17] and **`haggies`** [18].

The information about the model is either read from the built-in Standard Model file or is generated from a Universal FeynRules Output (`UFO`) [13, 19] or `LanHEP` [14] file. Precompiled `MSSM_UFO` and `MSSM_LHEP` files and examples for their import can also be found in the subdirectory `examples/model`.

The program offers the option to use different reduction techniques: either the unitarity-based integrand reduction [20, 21] as implemented in SAMURAI [22] or traditional tensor reduction as implemented in `golem95C` [23, 24] interfaced through tensorial reconstruction at the integrand level [25], or a combination of both.

2.2. Installation and Usage

The user can download the code either as a tar-archive or from the subversion repository at

`http://projects.hepforge.org/gosam/`

To install GOSAM, the user needs to run

```
python setup.py install --prefix MYPATH
```

If `MYPATH` is not among the system default paths, the environment variables `PATH`, `LD_LIBRARY_PATH` and `PYTHONPATH` might have to be set accordingly. For more details we direct the user to [4] and the reference manual coming with the code.

Prerequisites are a Linux/Unix environment, `Python` ($\geq$ 2.6), JAVA ($\geq$ 1.5), `Make`, and a `Fortran95` compiler. On top of a standard Linux environment, the programs `FORM` [16, 26] version $\geq$ 3.3, and `QGRAF` [15] need to be installed on the system. Further, at least one of the libraries SAMURAI [22] or `golem95C` [24] needs to be present at compile time of the generated code. For the user's convenience we have prepared a package `gosam-contrib-1.0.tar.gz` containing SAMURAI and `golem95C` together with the integral libraries `OneLOop` [27], `QCDLoop` [28] and `FF` [29]. The package is available from `http://projects.hepforge.org/gosam/`.

In order to generate the code for a process, the user needs to prepare an input file which we will call *process.in*, containing

- process specific information, such as a list of initial and final state particles, their helicities (optional), the order of the coupling constants, and the underlying model;
- scheme specific information, such as the regularisation and renormalisation schemes;
- system specific information, such as paths to programs and libraries or compiler options;
- optional information for optimisations which control the code generation.

The code can also generate a template input file. In order to import settings with system specific information in an automated way, the user can prepare a file *gosam.rc* which will be imported into *process.in* by `gosam.py -m gosam.rc -t process.in`. The virtual amplitude can then be generated and compiled by invoking

```
gosam.py process.in
make compile
```

2.3. Interfacing with Monte Carlo programs for the real radiation

The so-called "Binoth Les Houches Accord" (BLHA) [11, 12] defines an interface for a standardised communication between one-loop programs (OLP) and Monte Carlo (MC) tools, where the latter provide the Born amplitude, as well as the matrix elements for the NLO real radiation and the infrared subtraction terms. GOSAM can act as an OLP in the framework of the BLHA, such that the calculation of complete cross sections is straightforward.

In the BLHA setup, the MC writes an order file containing the process specifications, called for example `olp_order.lh`, which can be used by `gosam.py` to generate the virtual amplitude as follows:

Table 1. NLO calculations done by interfacing GoSam with different Monte Carlo programs.

GoSam + MadDipole/MadGraph4/MadEvent [30–35]
$pp \to b\bar{b}b\bar{b}$ [36, 37]
$pp \to W^+W^-$+2 jets [38]
$pp \to \gamma\gamma$+1,2 jets [39, 40]
SUSY QCD corrections to $pp \to \tilde{\chi}_1^0\tilde{\chi}_1^0$+jet [41]
QCD corrections to $pp \to$ graviton($\to \gamma\gamma$)+jet [42]
GoSam + Sherpa [43, 44]
$pp \to W^+W^+$+2 jets
$pp \to W^\pm b\bar{b}$
$pp \to W^+W^- b\bar{b}$ [45]
$pp \to W^+W^-$
$pp \to W^\pm$+0,1,2,3 jets
$pp \to Z/\gamma$+0,1,2 jets
$pp \to t\bar{t}$+0,1 jets [46]
$pp \to H$+2 jets (gluon fusion) [47]
$pp \to t\bar{t}H$+0,1 jet [48]
GoSam + MadDipole/MadGraph4/MadEvent+Sherpa
$pp \to H$+3 jets (gluon fusion) [49]
GoSam +Powheg [50, 51]
$pp \to HW/HZ$+0 and 1 jet [52]

```
gosam.py --olp --mc=MCname --config=YourPathTo/gosam.rc olp_order.lh
```

The sequence `--mc=MCname` is optional, but can facilitate to adapt to MC specific settings. Supported names at the moment are `sherpa` and `powhegbox`. If `gosam.rc` is in the current working directory or in the GoSam installation directory, its specification in the command line can be omitted. The following sequence of commands will generate and compile the files for the virtual matrix element:

```
sh ./autogen.sh --prefix = 'pwd'
make install
```

For more detailed information we refer to the `BLHA HowTo` on the GoSam webpage. Examples for full NLO calculations with GoSam interfaced to different MC programs are shown in Table 1. Further, pre-generated code to be used within Sherpa for a large number of processes can be downloaded from `http://projects.hepforge.org/gosam/proc`. To produce results for these processes, no GoSam installation is needed. Automated scripts coming with the process packages will ensure smooth running within the Sherpa framework.

3. Phenomenological results

In the following we show a selection of results obtained with GoSam interfaced to different Monte Carlo programs.

3.1. $W^+W^- b\bar{b}$

We calculated the NLO QCD corrections to the process $pp(p\bar{p}) \to W^+W^-b\bar{b} + X \to (e^+\nu_e)(\mu^-\bar{\nu}_\mu)b\bar{b} + X$, leading to a final state which is a signature of the decay of a $t\bar{t}$ pair with leptonic W boson decays, including singly-resonant and non-resonant contributions. Results are shown for the LHC at 7 TeV. All final state partons are clustered into jets with a separation $R > 0.5$ using the anti-k_T jet algorithm [53, 54] implemented in `Fastjet` [55]. Each event has to contain at least two b-jets with $p_{T,b} > 30\,\mathrm{GeV}$ and $\eta_b < 2.5$. Further cuts are $p_{T,l} > 20\,\mathrm{GeV}, \eta_l < 2.5, p_{T,\mathrm{miss}} > 20\,\mathrm{GeV}$.

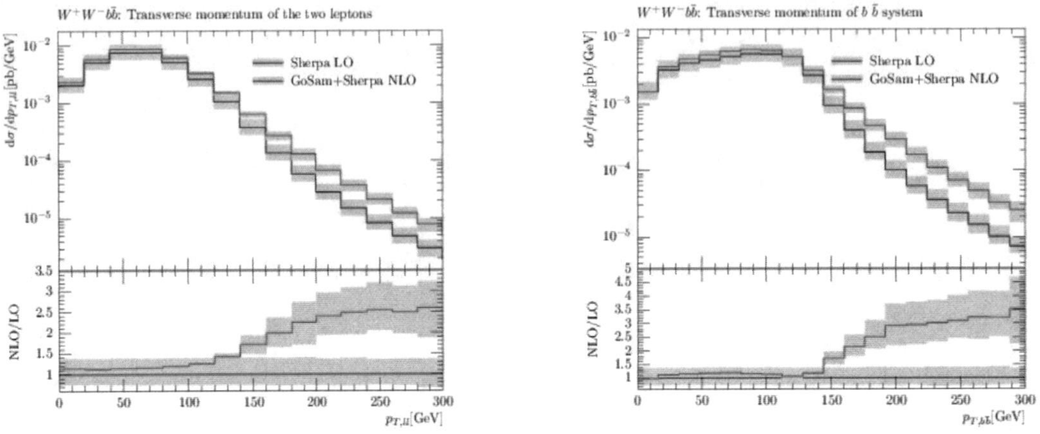

(a) (b)

Figure 1. Results for (a) the transverse momentum distribution of the lepton pair stemming from W^+W^- boson decay and (b) transverse momentum distribution of the sum of the two tagged b-jet momenta.

Figure 1 shows the distributions for the sum of the transverse momenta of the two leptons and the two b-jets, respectively. These observables receive large NLO corrections because most of the particles inherit their transverse momentum from a top quark pair. At LO the $t\bar{t}$ pair has zero transverse momentum, while it can obtain transverse momentum at NLO by recoiling against the real radiation. Note that the LO scale variations cannot account for this effect and therefore the uncertainty band on the LO distribution obtained by scale variations does not include the NLO result in the tail of these distributions.

3.2. Diphotons+jets

GoSam in combination with MadDipole/MadGraph4/MadEvent [30–35] has also been used to calculate the NLO QCD corrections to $pp \to \gamma\gamma + 1\,\mathrm{jet}$ [39] and $pp \to \gamma\gamma + 2\,\mathrm{jets}$ [40], where the former also includes the fragmentation component. The numerical results for $pp \to \gamma\gamma + 2\,\mathrm{jets}$ which are shown in Fig. 2 have been calculated at a center-of-mass energy $\sqrt{s} = 8\,\mathrm{TeV}$. The following kinematic cuts have been applied:

$$p_T^{\mathrm{jet}} > 30\ \mathrm{GeV}, \quad p_T^{\gamma,1} > 40\ \mathrm{GeV}, \quad p_T^{\gamma,2} > 25\ \mathrm{GeV},$$
$$|\eta^\gamma| \leq 2.5, \quad |\eta^j| \leq 4.7, \quad R_{\gamma,j} > 0.5, \quad R_{\gamma,\gamma} > 0.45.$$

For the photon isolation, we use the smooth cone isolation criterion [56] with $R = 0.4, n = 1$ and $\epsilon = 0.05$. Renormalization and factorization scales μ and μ_F have been set equal, with the central scale $\mu_0^2 = \frac{1}{4}\,(m_{\gamma\gamma}^2 + \sum_j p_{T,j}^2)$. Again, we observe a significant change of the shape at NLO, due to the extra QCD radiation in regions which are kinematically suppressed at LO.

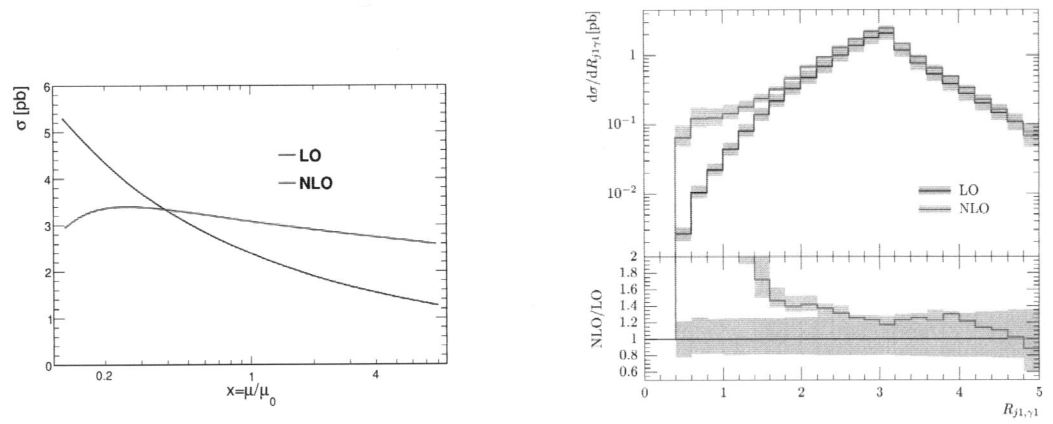

(a) (b)

Figure 2. (a) scale variations for diphoton+2 jets, (b) R-separation between the hardest jet and the hardest photon.

3.3. Associated Higgs production

After the recent discovery of a Higgs boson at the LHC, being able to disentangle signal processes from background ones and also the different production channels of the SM Higgs boson became of central relevance. The developments in GoSam allowed recently to compute the NLO QCD corrections to the production of $H + 2$ jets [47] and $H + 3$ jets [49] (in gluon-gluon fusion in the $m_{top} \to \infty$ limit) and also of $Ht\bar{t}$ and $Ht\bar{t}$+jet [48] for the LHC at 8 TeV. $H + 2$ jets and the processes involving the top quark were computed using a fully automated interface to the Sherpa MC event generator, based on the BLHA, whereas for $H + 3$ jets a hybrid setup combining the virtual part generated by GoSam with MadDipole/Madgraph4/MadEvent and Sherpa was used. The virtual corrections for $Ht\bar{t}(j)$ where computed using a new reduction algorithm based on an integrand decomposition via Laurent expansion [57], which was implemented in the library Ninja.

In the calculation of $H + 3$ jets the cteq6L1 and cteq6mE parton-distribution functions were used for LO and NLO respectively, and a minimal set of cuts based on the anti-k_T jet algorithm with $R = 0.5$, $p_{T,min} > 20$ GeV and $|\eta| < 4.0$ was applied. For $Ht\bar{t}(j)$ we used CT10 at NLO and a minumum transverse momeutm cut of $p_{T,min} > 15$ GeV.

Figure 3(a) shows the transverse momentum distributions at LO and NLO for the three leading jets in $H + 3$ jets. We observe that the NLO corrections drag the p_T-spectra towards smaller p_T values, as expected as an effect of additional QCD radiation. Figure 3(b) displays the transverse momentum distribution of the Higgs boson in $Httj$ at LO and NLO. In this case the NLO corrections become larger with increasing p_T. Furthermore the scale uncertainty is reduced by 60-70% in going from LO to NLO.

3.4. SUSY-QCD corrections to neutralino pair plus jet production

GoSam also has been used to calculate the NLO Susy-QCD corrections to the production of a pair of the lightest neutralinos plus one jet at the LHC at 8 TeV, appearing as a monojet signature in combination with missing energy. We fully included all non-resonant diagrams, i.e. we did not use the simplifying assumption that production and decay factorise. Examples of pentagon diagrams occurring in the virtual corrections, as well as the missing transverse energy distribution, are shown in Fig. 4. We observe that the NLO corrections are large, mainly due

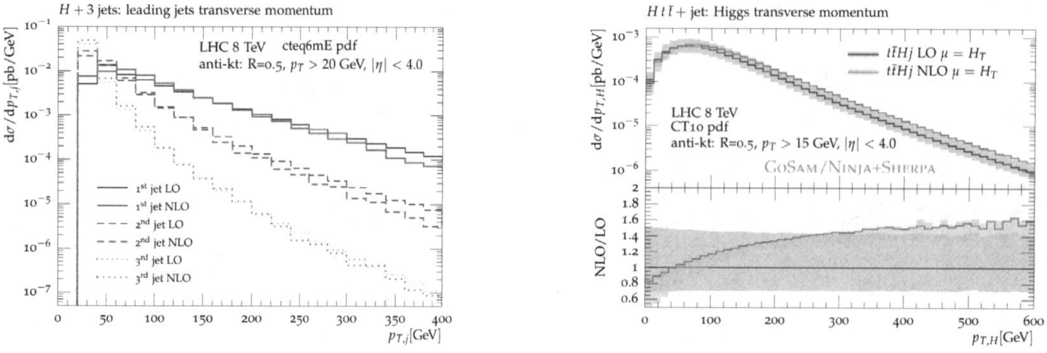

(a) (b)

Figure 3. (a) transverse momentum distributions of the leading jets in $H + 3$ jets production, (b) transverse momentum distribution of the Higgs boson in $Ht\bar{t}$+jet production.

to additional channels opening up at NLO. The detailed setup can be found in [41].

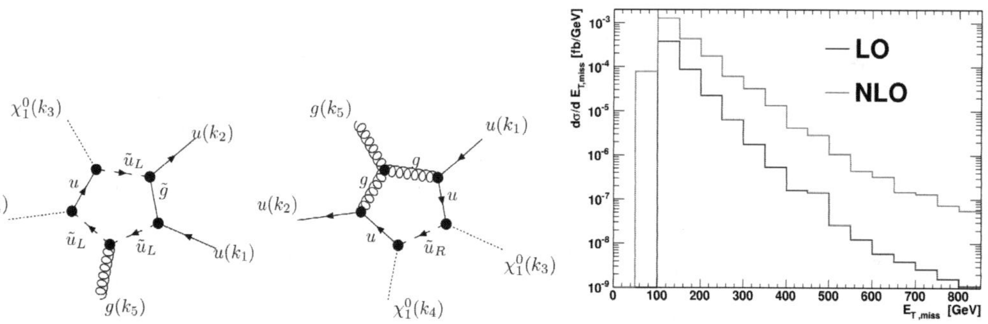

(a) (b)

Figure 4. (a) Two illustrative pentagon diagrams, (b) missing transverse energy $E_T^{\rm miss}$ for the process $pp \to \tilde{\chi}_1^0\tilde{\chi}_1^0$+jet at $\sqrt{s} = 8\,$TeV.

3.5. GoSam *and extra dimensions*

Another computationally intense calculation based on GoSam+ MadDipole/MadGraph4/MadEvent are the NLO QCD corrections to the production of a graviton in association with one jet [42], where the graviton decays into a photon pair, within ADD models of large extra dimensions [58, 59]. The calculation is quite involved due to the complicated tensor structure introduced by spin-2 particles, and the non-standard propagator of the graviton, coming from the summation over very densely distributed Kaluza-Klein modes. As can be seen from Fig. 5, the K-factors turn out *not* to be uniform over the range of the diphoton invariant mass distribution. As the latter in general is used to derive exclusion limits, the differential NLO corrections should be taken into account. For details we refer to [42].

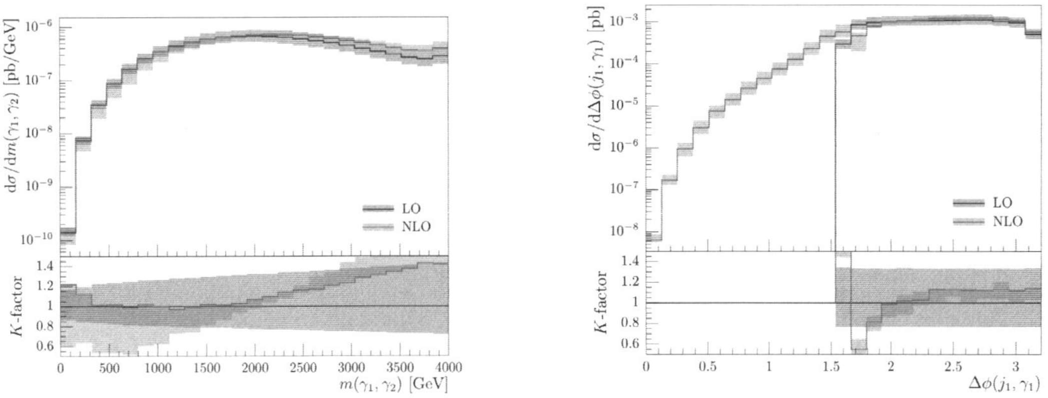

(a) (b)

Figure 5. (a) NLO QCD corrections to the invariant mass distribution of the photon pair stemming from graviton decay within the ADD model for $\delta = 4$ large extra dimensions. The bands show the scale variations by a factor of two around the central scale $\mu_0^2 = \mu_F^2 = \frac{1}{4}\left(m_{\gamma\gamma}^2 + p_{T,jet}^2\right)$. (b) Azimuthal angle distribution between the leading photon and the leading jet. The region $\Delta\Phi < \pi/2$ is kinematically inaccessible at LO.

4. Code development

We are working on a number of new features concerning code generation as well as integrand reduction. For instance, we implemented a new strategy to produce optimized `fortran95` code based on `FORM` version 4 [26] rather than `haggies` [18], leading to faster code generation and more compact code. Further, the possibility of parallelisation at diagram level, the option to have numerical polarisation vectors and the option to sum diagrams sharing the same propagators algebraically at `FORM` level lead to an enormous gain in code generation time and reduction of code size. Concerning the amplitude reduction, we implemented integrals where the rank exceeds number of propagators. An alternative reduction based on the Laurent expansion method developed in [57] also has been implemented and used successfully in [48].

A version GoSam 2.0 where all these new features are publicly available is in preparation.

5. Conclusions

We have presented applications of the program GoSam which can generate and evaluate one-loop matrix elements for multi-particle processes in an automated way. The program is publicly available at `http://projects.hepforge.org/gosam/` and can be used to produce NLO corrections within QCD, electroweak theory, or other models which can be imported via an interface to FeynRules. Monte Carlo programs for the real radiation are linked via the BLHA (Binoth Les Houches Accord) interface. This way GoSam is a very flexible and widely applicable tool for the automated calculation of multi-particle observables at next-to-leading order.

Acknowledgments

The work of G.C. was supported by the DFG SFB-TR-9 and the EU TMR Network LHCPhenoNet. P.M. is supported by the Alexander von Humboldt Foundation, in the framework of the Sofja Kovaleskaja Award Project "Advanced Mathematical Methods for Particle Physics", endowed by the German Federal Ministry of Education and Research. The work of G.O. was supported in part by the National Science Foundation under Grant PHY-1068550.

References

[1] Berger C, Bern Z, Dixon L, Febres Cordero F, Forde D *et al.* 2008 *Phys.Rev.* **D78** 036003 (*Preprint* 0803.4180)
[2] Bevilacqua G, Czakon M, Garzelli M, van Hameren A, Kardos A *et al.* 2013 *Comput.Phys.Commun.* **184** 986–997 (*Preprint* 1110.1499)
[3] Hirschi V, Frederix R, Frixione S, Garzelli M V, Maltoni F *et al.* 2011 *JHEP* **1105** 044 (*Preprint* 1103.0621)
[4] Cullen G, Greiner N, Heinrich G, Luisoni G, Mastrolia P *et al.* 2012 *Eur.Phys.J.* **C72** 1889 (*Preprint* 1111.2034)
[5] Cascioli F, Maierhofer P and Pozzorini S 2012 *Phys.Rev.Lett.* **108** 111601 (*Preprint* 1111.5206)
[6] Agrawal S, Hahn T and Mirabella E 2012 *PoS* **LL2012** 046 (*Preprint* 1210.2628)
[7] Badger S, Biedermann B, Uwer P and Yundin V 2013 *Comput.Phys.Commun.* **184** 1981–1998 (*Preprint* 1209.0100)
[8] Actis S, Denner A, Hofer L, Scharf A and Uccirati S 2013 *JHEP* **1304** 037 (*Preprint* 1211.6316)
[9] Campbell J M, Ellis R K, Frederix R, Nason P, Oleari C *et al.* 2012 *JHEP* **1207** 092 (*Preprint* 1202.5475)
[10] Arnold K, Bellm J, Bozzi G, Campanario F, Englert C *et al.* 2012 (*Preprint* 1207.4975)
[11] Binoth T, Boudjema F, Dissertori G, Lazopoulos A, Denner A *et al.* 2010 *Comput.Phys.Commun.* **181** 1612–1622 (*Preprint* 1001.1307)
[12] Alioli S, Badger S, Bellm J, Biedermann B, Boudjema F *et al.* 2013 (*Preprint* 1308.3462)
[13] Degrande C, Duhr C, Fuks B, Grellscheid D, Mattelaer O *et al.* 2012 *Comput.Phys.Commun.* **183** 1201–1214 (*Preprint* 1108.2040)
[14] Semenov A 2010 (*Preprint* 1005.1909)
[15] Nogueira P 1993 *J.Comput.Phys.* **105** 279–289
[16] Vermaseren J 2000 (*Preprint* math-ph/0010025)
[17] Cullen G, Koch-Janusz M and Reiter T 2011 *Comput.Phys.Commun.* **182** 2368–2387 (*Preprint* 1008.0803)
[18] Reiter T 2010 *Comput.Phys.Commun.* **181** 1301–1331 (*Preprint* 0907.3714)
[19] Christensen N D and Duhr C 2009 *Comput.Phys.Commun.* **180** 1614–1641 (*Preprint* 0806.4194)
[20] Ossola G, Papadopoulos C G and Pittau R 2007 *Nucl.Phys.* **B763** 147–169 (*Preprint* hep-ph/0609007)
[21] Ellis R, Giele W T, Kunszt Z and Melnikov K 2009 *Nucl.Phys.* **B822** 270–282 (*Preprint* 0806.3467)
[22] Mastrolia P, Ossola G, Reiter T and Tramontano F 2010 *JHEP* **1008** 080 (*Preprint* 1006.0710)
[23] Binoth T, Guillet J P, Heinrich G, Pilon E and Reiter T 2009 *Comput.Phys.Commun.* **180** 2317–2330 (*Preprint* 0810.0992)
[24] Cullen G, Guillet J, Heinrich G, Kleinschmidt T, Pilon E *et al.* 2011 *Comput.Phys.Commun.* **182** 2276–2284 (*Preprint* 1101.5595)
[25] Heinrich G, Ossola G, Reiter T and Tramontano F 2010 *JHEP* **1010** 105 (*Preprint* 1008.2441)
[26] Kuipers J, Ueda T, Vermaseren J and Vollinga J 2013 *Comput.Phys.Commun.* **184** 1453–1467 (*Preprint* 1203.6543)
[27] van Hameren A 2011 *Comput.Phys.Commun.* **182** 2427–2438 (*Preprint* 1007.4716)
[28] Ellis R K and Zanderighi G 2008 *JHEP* **02** 002 (*Preprint* 0712.1851)
[29] van Oldenborgh G and Vermaseren J 1990 *Z.Phys.* **C46** 425–438
[30] Frederix R, Gehrmann T and Greiner N 2008 *JHEP* **0809** 122 (*Preprint* 0808.2128)
[31] Frederix R, Gehrmann T and Greiner N 2010 *JHEP* **1006** 086 (*Preprint* 1004.2905)
[32] Gehrmann T and Greiner N 2010 *JHEP* **1012** 050 (*Preprint* 1011.0321)
[33] Stelzer T and Long W 1994 *Comput.Phys.Commun.* **81** 357–371 (*Preprint* hep-ph/9401258)
[34] Maltoni F and Stelzer T 2003 *JHEP* **0302** 027 (*Preprint* hep-ph/0208156)
[35] Alwall J, Demin P, de Visscher S, Frederix R, Herquet M *et al.* 2007 *JHEP* **0709** 028 (*Preprint* 0706.2334)
[36] Binoth T, Greiner N, Guffanti A, Reuter J, Guillet J P *et al.* 2010 *Phys.Lett.* **B685** 293–296 (*Preprint* 0910.4379)
[37] Greiner N, Guffanti A, Reiter T and Reuter J 2011 *Phys.Rev.Lett.* **107** 102002 (*Preprint* 1105.3624)
[38] Greiner N, Heinrich G, Mastrolia P, Ossola G, Reiter T *et al.* 2012 *Phys.Lett.* **B713** 277–283 (*Preprint* 1202.6004)
[39] Gehrmann T, Greiner N and Heinrich G 2013 *JHEP* **1306** 058 (*Preprint* 1303.0824)
[40] Gehrmann T, Greiner N and Heinrich G 2013 (*Preprint* 1308.3660)
[41] Cullen G, Greiner N and Heinrich G 2013 *Eur.Phys.J.* **C73** 2388 (*Preprint* 1212.5154)
[42] Greiner N, Heinrich G, Reichel J and von Soden-Fraunhofen J F 2013 (*Preprint* 1308.2194)
[43] Gleisberg T and Krauss F 2008 *Eur.Phys.J.* **C53** 501–523 (*Preprint* 0709.2881)

[44] Gleisberg T, Hoeche S, Krauss F, Schonherr M, Schumann S *et al.* 2009 *JHEP* **0902** 007 (*Preprint* 0811.4622)
[45] Heinrich G, Schlenk J and Winter J 2013 (*Preprint* in preparation)
[46] Hoeche S, Huang J, Luisoni G, Schoenherr M and Winter J 2013 *Phys.Rev.* **D88** 014040 (*Preprint* 1306.2703)
[47] van Deurzen H, Greiner N, Luisoni G, Mastrolia P, Mirabella E *et al.* 2013 *Phys.Lett.* **B721** 74–81 (*Preprint* 1301.0493)
[48] van Deurzen H, Luisoni G, Mastrolia P, Mirabella E, Ossola G *et al.* 2013 (*Preprint* 1307.8437)
[49] Cullen G, van Deurzen H, Greiner N, Luisoni G, Mastrolia P *et al.* 2013 (*Preprint* 1307.4737)
[50] Frixione S, Nason P and Oleari C 2007 *JHEP* **0711** 070 (*Preprint* 0709.2092)
[51] Alioli S, Nason P, Oleari C and Re E 2010 *JHEP* **1006** 043 (*Preprint* 1002.2581)
[52] Luisoni G, Nason P, Oleari C and Tramontano F 2013 (*Preprint* 1306.2542)
[53] Cacciari M and Salam G P 2006 *Phys.Lett.* **B641** 57–61 (*Preprint* hep-ph/0512210)
[54] Cacciari M, Salam G P and Soyez G 2008 *JHEP* **0804** 063 (*Preprint* 0802.1189)
[55] Cacciari M, Salam G P and Soyez G 2012 *Eur.Phys.J.* **C72** 1896 (*Preprint* 1111.6097)
[56] Frixione S 1998 *Phys.Lett.* **B429** 369–374 (*Preprint* arXiv hep-ph/9801442)
[57] Mastrolia P, Mirabella E and Peraro T 2012 *JHEP* **1206** 095 (*Preprint* 1203.0291)
[58] Arkani-Hamed N, Dimopoulos S and Dvali G 1998 *Phys.Lett.* **B429** 263–272 (*Preprint* hep-ph/9803315)
[59] Antoniadis I, Arkani-Hamed N, Dimopoulos S and Dvali G 1998 *Phys.Lett.* **B436** 257–263 (*Preprint* hep-ph/9804398)

Journal of Physics: Conference Series **523** (2014) 012057 doi:10.1088/1742-6596/523/1/012057

Computation of multi-leg amplitudes with NJet

S Badger[1], B Biedermann[2], P Uwer[3] and V Yundin[4,5]

[1] Theory Division, Physics Department, CERN, CH-1211 Geneva 23, Switzerland
[2] Universität Würzburg, Institut für Theoretische Physik und Astrophysik, Emil-Hilb-Weg 22, 97074 Würzburg, Germany
[3] Humboldt-Universität zu Berlin, Institut für Physik, Newtonstraße 15, 12489 Berlin, Germany
[4] Max-Planck-Institut für Physik, Föhringer Ring 6, 80805 Munich, Germany

E-mail: yundin@mpp.mpg.de

Abstract. In these proceedings we report our progress in the development of the publicly available C++ library NJet for accurate calculations of high-multiplicity one-loop amplitudes. As a phenomenological application we present the first complete next-to-leading order (NLO) calculation of five jet cross section at hadron colliders.

1. Introduction

NLO predictions of multi-jet production at hadron colliders have a long history. They are important processes for the LHC both as precision tests of QCD and direct probes of the strong coupling and also as background in many new physics searches. The LHC experiments have been able to measure jet rates for up to 6 hard jets which are now being used in new physics searches [1–3]. This presents a serious challenge for precise theoretical predictions since high multiplicity computations in QCD are notoriously difficult. Di-jet production has been known at NLO for more than 20 years [4] and has recently seen improvements via NLO plus parton shower (NLO+PS) description [5, 6] and steady progress towards NNLO QCD results [7]. The full three-jet computation was completed and implemented in a public code NLOJET++ 10 years ago [8]. Recently predictions for four-jet production have been presented by two independent groups [9, 10].

The advances in methods of evaluation of multi-leg virtual amplitudes [11–25] have inspired many efforts to automate NLO computations [26–31]. Processes with four final states, previously out of reach, can now be routinely used for phenomenological predictions [32–38]. We refer the reader to other contributions to these proceedings for further details on the current state-of-the art [39–41].

Five partons in the final state still constitute a considerable challenge, though steady progress in that direction gives hope for the same level of automation in the near future. Recent state-of-the-art calculations with five QCD partons in the final state include the NLO QCD corrections to $pp \to W+5j$ [34] by the BlackHat collaboration and NLO QCD corrections to $pp \to 5j$ [42].

[5] Speaker; talk given at the Workshop on Advanced Computing and Analysis Techniques in Physics (ACAT), Beijing, China, May 2013.

2. 5-Jet production at the LHC at 7 and 8 TeV

The different parts of the calculation, which contribute to NLO cross section can be schematically written as

$$\delta\sigma^{\mathrm{NLO}} = \int\limits_{n} \left(d\sigma_n^{\mathrm{V}} + \int\limits_{1} d\sigma_{n+1}^{\mathrm{S}}\right) + \int\limits_{n} d\sigma_n^{\mathrm{Fac}} + \int\limits_{n+1} \left(d\sigma_{n+1}^{\mathrm{R}} - d\sigma_{n+1}^{\mathrm{S}}\right). \tag{1}$$

We used the Sherpa Monte-Carlo event generator [43] to handle phase-space integration and generation of tree-level amplitudes and Catani-Seymour dipole subtraction terms as implemented in Comix [44, 45].

The one-loop matrix elements for the virtual corrections $d\sigma_n^{\mathrm{V}}$ are evaluated with the publicly available NJet[6] package [31] interfaced to Sherpa via the Binoth Les Houches Accord [46, 47]. NJet is based on the NGluon library [26] and uses an on-shell generalized unitarity framework [17–22] to compute multi-parton one-loop primitive amplitudes from tree-level building blocks [48]. The scalar loop integrals are obtained via the QCDLoop/FF package [49, 50].

NJet implements full-colour expressions for up-to five outgoing QCD partons. The complexity of high-multiplicity virtual corrections motivates us to explore ways to speed up the computation. One of the optimizations implemented in NJet is the usage of de-symmetrized colour sums for multi-gluon final states, which allows us to get full colour result at a small fraction of computational cost by exploiting the Bose symmetry of the phase space [31, 51]. Another possibility is to separate leading and sub-leading contributions, which enables Monte-Carlo integrator to sample the dominant however simpler terms more often and get the same statistical error with fewer evaluations of the expensive sub-leading part.

In our leading terms we include all multi-quark processes in the large N_c limit and processes with two or more gluons in the final state using the de-symmetrized colour sums. In Figure 1 we compare leading and full virtual contributions to the hadrest jet transverse momentum in $pp \to 5j$. The correction from the sub-leading part is around 10% at low p_T and shows a tendency to grow with increasing hardness of the jet. Considering that $d\sigma_n^{\mathrm{V}}$ contribute $\sim 50\%$ of the total NLO cross section for this process this translates to $5 - 10$ percent effect depending on the kinematic region.

The calculation is done in QCD with five massless quark flavours including the bottom-quark in the initial state. We neglect contributions from top quark loops. We set the renormalization scale equal to the factorization scale ($\mu_r = \mu_f = \mu$) and use a dynamical scale based on the total transverse momentum $\widehat{H}_T$ of the final state partons:

$$\widehat{H}_T = \sum_{i=1}^{N_{\mathrm{parton}}} p_{T,i}^{\mathrm{parton}}. \tag{2}$$

For the definition of physical observables we use the anti-kt jet clustering algorithm as implemented in FastJet [52, 53]. We apply asymmetric cuts on the jets ordered in transverse momenta, p_T, to match the ATLAS multi-jet measurements [1]:

$$p_T^{j_1} > 80 \text{ GeV} \qquad p_T^{j_{\geq 2}} > 60 \text{ GeV} \qquad R = 0.4 \tag{3}$$

The PDFs are obtained through the LHAPDF interface [54] with all central values using NNPDF2.1 [55] for LO ($\alpha_s(M_Z) = 0.119$) and NNPDF2.3 [56] for NLO ($\alpha_s(M_Z) = 0.118$) if not mentioned otherwise.

[6] To download NJet visit the project home page at `https://bitbucket.org/njet/njet/`.

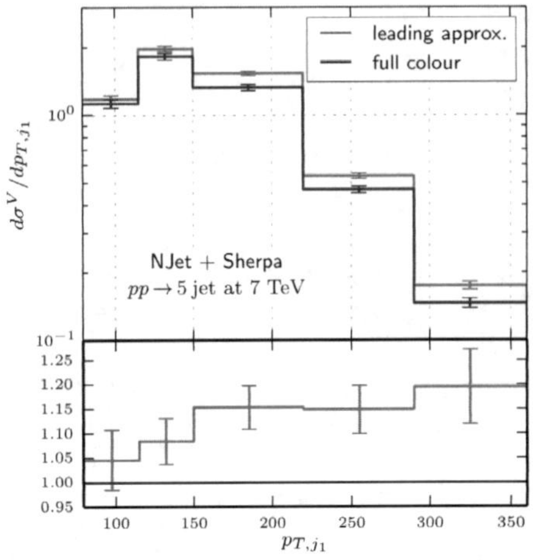

Figure 1: Full colour and leading approximation (as explained in the text) for the virtual corrections to the transverse momentum of the 1st jet in $pp \to 5j$.

Generated events are stored in ROOT Ntuple format [57] which allows for flexible analysis. Renormalization and factorization scales can be changed at the analysis level as well as the PDF set. This technique makes it possible to do extended analysis of PDF uncertainties and scale dependence, which would otherwise be prohibitively expensive for such high multiplicity processes.

2.1. Numerical results

Using the above setup we obtain for the 5-jet cross section at 7 TeV

$$\sigma_5^{\text{7TeV-LO}}(\mu = \widehat{H}_T/2) = 0.699(0.004)^{+0.530}_{-0.280}\ \text{nb}, \tag{4}$$

$$\sigma_5^{\text{7TeV-NLO}}(\mu = \widehat{H}_T/2) = 0.544(0.016)^{+0.0}_{-0.177}\ \text{nb}. \tag{5}$$

In parentheses we quote the uncertainty due to the numerical integration. The theoretical uncertainty has been estimated from scale variations over the range $\mu \in [\widehat{H}_T/4, \widehat{H}_T]$ and is indicated by the sub- and superscripts. As seen in Fig. 2 the total cross section at the scale $\mu = \widehat{H}_T$ is lower than the central value which is the origin of the zero value of the upper error bound. The total cross section at this scale is $\sigma_5^{\text{7TeV-NLO}}(\mu = \widehat{H}_T) = 0.544(0.016)$ nb. For a centre-of-mass energy of 8 TeV the results read:

$$\sigma_5^{\text{8TeV-LO}}(\mu = \widehat{H}_T/2) = 1.044(0.006)^{+0.770}_{-0.413}\ \text{nb}, \tag{6}$$

$$\sigma_5^{\text{8TeV-NLO}}(\mu = \widehat{H}_T/2) = 0.790(0.021)^{+0.0}_{-0.313}\ \text{nb}, \tag{7}$$

where we have found $\sigma_5^{\text{8TeV-NLO}}(\mu = \widehat{H}_T) = 0.723(0.011)$ nb.

As usual for a next-to-leading order correction a significant reduction of the scale uncertainty can be observed. In Fig. 2 the scale dependence of the LO and NLO cross section is illustrated. The dashed black line indicates the central scale $\mu = \widehat{H}_T/2$. The horizontal bands show the cross section uncertainty estimated by a scale variation within $\mu \in [\widehat{H}_T/4, \widehat{H}_T]$.

By comparing Figs. 2a and 2b we observe that a significant part of the NLO corrections comes from using NLO PDFs with the corresponding α_s. Similar to what has been found in Ref. [10] we conclude that using the NLO PDFs in the LO predictions gives a better approximation to the full result compared to using LO PDFs.

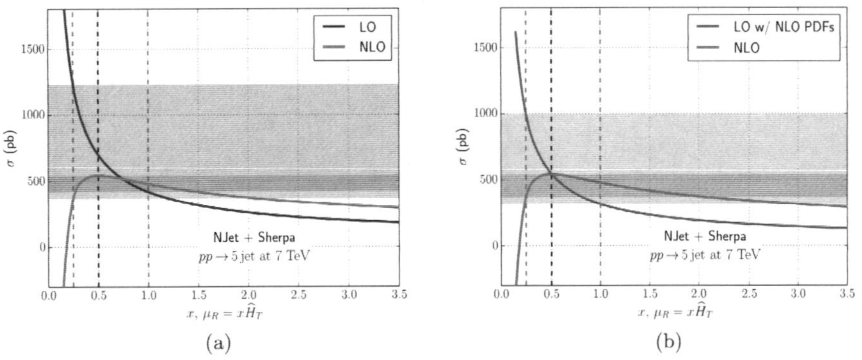

Figure 2: Residual scale dependence of the 5-jet cross section in leading and next-to-leading order using LO (a) and NLO (b) PDFs for LO prediction.

In Tab. 1 we show for completeness the cross sections for two, three and four-jet production as calculated with NJET using the same setup as in the five jet case.

Table 1: Cross sections for 2, 3 and 4 jets at 7 TeV.

n	$\sigma_n^{\text{7TeV-LO}}$ [nb]	$\sigma_n^{\text{7TeV-NLO}}$ [nb]
2	$768.0(0.9)^{+203.0}_{-151.3}$	$1175(3)^{+120}_{-129}$
3	$71.1(0.1)^{+31.5}_{-20.0}$	$52.5(0.3)^{+1.9}_{-19.3}$
4	$7.23(0.02)^{+4.37}_{-2.50}$	$5.65(0.07)^{+0}_{-1.93}$

The jet rates have been measured recently by ATLAS using the 7 TeV data set [1]. In Fig. 3a we show the data together with the theoretical predictions in leading and next-to-leading order. In case of the six jet rate only LO results are shown. In the lower plot the ratio of theoretical predictions with respect to data is given. With exception of the two-jet cross section the inclusion of the NLO results improves significantly the agreement with data.

In addition to inclusive cross sections it is useful to consider their ratios since many theoretical and experimental uncertainties may cancel between numerator and denominator. In particular we consider

$$\mathcal{R}_n = \frac{\sigma_{(n+1)\text{-jet}}}{\sigma_{n\text{-jet}}}. \tag{8}$$

This quantity is in leading order proportional to the QCD coupling α_s and can be used to determine the value of α_s from jet rates. In Fig. 3b we show QCD predictions in NLO using different PDF sets together with the results from ATLAS. The results obtained from NNPDF2.3 are also collected in Tab. 2 where, in addition, the ratios at leading order (using the LO setup with NNPDF2.1) are shown. In case of $\mathcal{R}_3$ and $\mathcal{R}_4$ perturbation theory seems to provide stable results. The leading order and next-to-leading order values differ by less than 10%. In addition NNPDF [56], CT10 [58] and MSTW08 [59] give compatible predictions. ABM11 [60] gives slightly smaller results for $\mathcal{R}_3$ and $\mathcal{R}_4$. Within uncertainties the predictions also agree with the ATLAS measurements. The poor description of $\mathcal{R}_2$ can be attributed to the inclusive two-jet cross section which seems to be inadequately described by a fixed order NLO calculation. As a

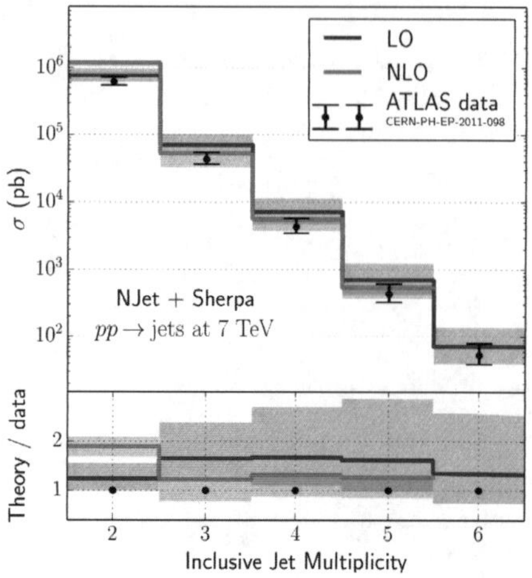

(a)

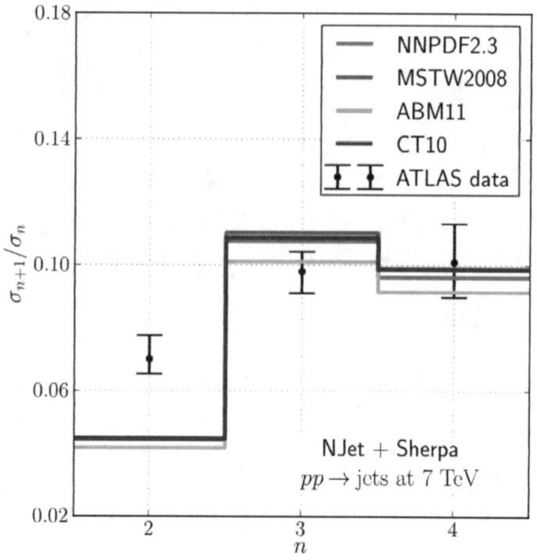

(b)

Figure 3: (a) LO and NLO cross sections for jet production calculated with NJET as well as results from ATLAS measurements [1]. (b) NLO NJET predictions with different PDF sets for the jet ratios $\mathcal{R}_n$ compared with recent ATLAS measurements [1].

Table 2: Results for the jet ratios $\mathcal{R}_n$ for the central scale of $\widehat{H}_T/2$ and NNPDF2.3 PDF set.

$\mathcal{R}_n$	ATLAS [1]	LO	NLO
2	$0.070^{+0.007}_{-0.005}$	0.0925(0.0002)	0.0447(0.0003)
3	$0.098^{+0.006}_{-0.007}$	0.102(0.000)	0.108(0.002)
4	$0.101^{+0.012}_{-0.011}$	0.097(0.001)	0.096(0.003)
5	$0.123^{+0.028}_{-0.027}$	0.102(0.001)	--

function of the leading jet p_T, all PDF sets agree well with the 3/2 ratio ATLAS data at large p_T as shown in Fig. 4a. In Fig. 4b we compare LO and NLO predictions for $\mathcal{R}_n$ as function of the leading jet p_T. While for $\mathcal{R}_3$ and $\mathcal{R}_4$ the corrections are moderate for all values of p_T we observe large negative corrections independent from p_T in case of $\mathcal{R}_2$.

In Fig. 5 we show the transverse momentum and rapidity distributions of the leading jet for five-jet production. Similarly to total cross section we observe significant reduction of the scale uncertainty when going from LO to NLO. Using again the NLO setup to calculate the LO predictions, the NLO calculation gives very small corrections. Over a wide range the LO predictions are modified by less than 10%. A remarkable feature observed already in the 4-jet calculation [9, 10] is the almost constant K-factor.

3. Conclusions

In this contribution we have presented first results for five-jet production at NLO accuracy in QCD. We find moderate corrections of the order of 10% at NLO with respect to a leading order computation using NLO PDFs. We have compared theoretical predictions for inclusive jet cross sections and jet rates with data from ATLAS. With the exception of quantities affected by the two-jet rate we find good agreement between theory and data.

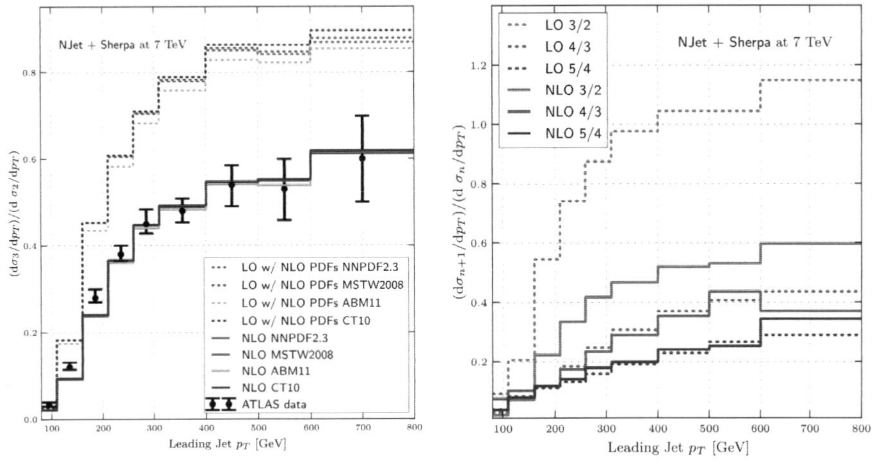

(a) (b)

Figure 4: (a) The 3/2 jet ratio as a function of the p_T of the leading jet compared with ATLAS data [1] ($R = 0.6$). (b) The $\mathcal{R}_n$ ratios as functions of the p_T of the leading jet ($R = 0.4$).

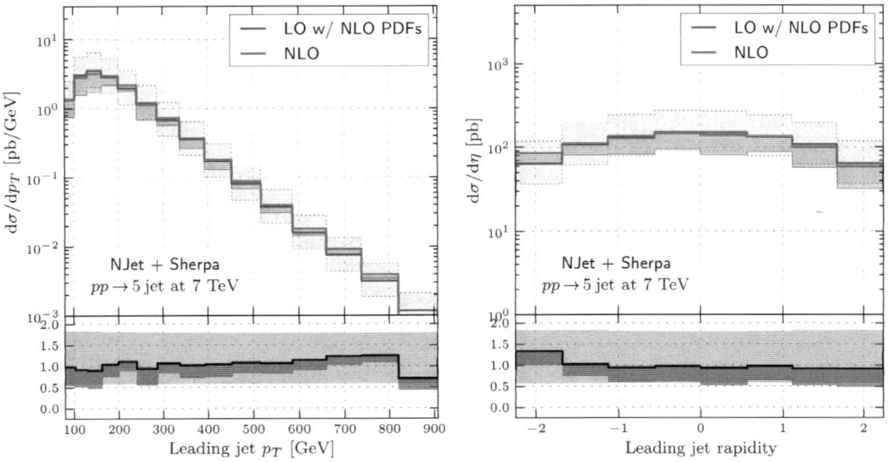

(a) (b)

Figure 5: The p_T and rapidity distributions of the leading jet. Both LO and NLO use the NNPDF2.3 PDF set with $\alpha_s(M_Z) = 0.118$

Acknowledgements
This work has been supported by the Helmholtz Gemeinschaft under contract HA-101 (Alliance Physics at the Terascale), by the German Research Foundation (DFG) through the transregional collaborative research centre "Computational Particle Physics" (SFB-TR9), by the European Commission through contract PITN-GA-2010-264564 (LHCPhenoNet) and by the Alexander von Humboldt Foundation, in the framework of the Sofja Kovaleskaja Award 2010, endowed by the German Federal Ministry of Education and Research.

References

[1] Aad G *et al.* (ATLAS Collaboration) 2011 *Eur.Phys.J.* **C71** 1763 (*Preprint* arXiv:1107.2092)

[2] Chatrchyan S *et al.* (CMS Collaboration) 2013 (*Preprint* arXiv:1311.1799)

[3] Chatrchyan S *et al.* (CMS Collaboration) 2013 (*Preprint* arXiv:1311.4937)

[4] Giele W, Glover E N and Kosower D A 1993 *Nucl.Phys.* **B403** 633–670 (*Preprint* hep-ph/9302225)

[5] Alioli S, Hamilton K, Nason P, Oleari C and Re E 2011 *JHEP* **1104** 081 (*Preprint* arXiv:1012.3380)

[6] Hoeche S and Schonherr M 2012 *Phys.Rev.* **D86** 094042 (*Preprint* arXiv:1208.2815)

[7] Gehrmann-De Ridder A, Gehrmann T, Glover E and Pires J 2013 *Phys.Rev.Lett.* **110** 162003 (*Preprint* arXiv:1301.7310)

[8] Nagy Z 2002 *Phys.Rev.Lett.* **88** 122003 (*Preprint* hep-ph/0110315)

[9] Bern Z, Diana G, Dixon L, Febres Cordero F, Hoeche S *et al.* 2012 *Phys.Rev.Lett.* **109** 042001 (*Preprint* arXiv:1112.3940)

[10] Badger S, Biedermann B, Uwer P and Yundin V 2013 *Phys.Lett.* **B718** 965–978 (*Preprint* arXiv:1209.0098)

[11] Bern Z, Dixon L J, Dunbar D C and Kosower D A 1994 *Nucl.Phys.* **B425** 217–260 (*Preprint* hep-ph/9403226)

[12] Bern Z, Dixon L J, Dunbar D C and Kosower D A 1995 *Nucl.Phys.* **B435** 59–101 (*Preprint* hep-ph/9409265)

[13] Cascioli F, Maierhofer P and Pozzorini S 2012 *Phys.Rev.Lett.* **108** 111601 (*Preprint* arXiv:1111.5206)

[14] Becker S, Goetz D, Reuschle C, Schwan C and Weinzierl S 2012 *Phys.Rev.Lett.* **108** 032005 (*Preprint* arXiv:1111.1733)

[15] Actis S, Denner A, Hofer L, Scharf A and Uccirati S 2013 *JHEP* **1304** 037 (*Preprint* arXiv:1211.6316)

[16] Mastrolia P, Ossola G, Reiter T and Tramontano F 2010 *JHEP* **1008** 080 (*Preprint* arXiv:1006.0710)

[17] Britto R, Cachazo F and Feng B 2005 *Nucl.Phys.* **B725** 275–305 (*Preprint* hep-th/0412103)

[18] Ellis R K, Giele W and Kunszt Z 2008 *JHEP* **0803** 003 (*Preprint* arXiv:0708.2398)

[19] Forde D 2007 *Phys.Rev.* **D75** 125019 (*Preprint* arXiv:0704.1835)

[20] Giele W T, Kunszt Z and Melnikov K 2008 *JHEP* **0804** 049 (*Preprint* arXiv:0801.2237)

[21] Badger S 2009 *JHEP* **0901** 049 (*Preprint* arXiv:0806.4600)

[22] Ossola G, Papadopoulos C G and Pittau R 2007 *Nucl.Phys.* **B763** 147–169 (*Preprint* hep-ph/0609007)

[23] Mastrolia P, Mirabella E, Ossola G and Peraro T 2012 *Phys.Lett.* **B718** 173–177 (*Preprint* arXiv:1205.7087)

[24] Mastrolia P, Mirabella E and Peraro T 2012 *JHEP* **1206** 095 (*Preprint* arXiv:1203.0291)

[25] van Hameren A 2009 *JHEP* **0907** 088 (*Preprint* arXiv:0905.1005)

[26] Badger S, Biedermann B and Uwer P 2011 *Comput.Phys.Commun.* **182** 1674–1692 (*Preprint* arXiv:1011.2900)

[27] Hirschi V, Frederix R, Frixione S, Garzelli M V, Maltoni F *et al.* 2011 *JHEP* **1105** 044 (*Preprint* arXiv:1103.0621)

[28] Berger C, Bern Z, Dixon L, Febres Cordero F, Forde D *et al.* 2008 *Phys.Rev.* **D78** 036003 (*Preprint* arXiv:0803.4180)

[29] Bevilacqua G, Czakon M, Garzelli M, van Hameren A, Kardos A *et al.* 2013 *Comput.Phys.Commun.* **184** 986–997 (*Preprint* arXiv:1110.1499)

[30] Cullen G, Greiner N, Heinrich G, Luisoni G, Mastrolia P *et al.* 2012 *Eur.Phys.J.* **C72** 1889 (*Preprint* arXiv:1111.2034)

[31] Badger S, Biedermann B, Uwer P and Yundin V 2013 *Comput.Phys.Commun.* **184** 1981–1998 (*Preprint* arXiv:1209.0100)

[32] Bevilacqua G and Worek M 2012 *JHEP* **1207** 111 (*Preprint* arXiv:1206.3064)

[33] Greiner N, Heinrich G, Mastrolia P, Ossola G, Reiter T *et al.* 2012 *Phys.Lett.* **B713** 277–283 (*Preprint* arXiv:1202.6004)

[34] Bern Z, Dixon L, Febres Cordero F, Hoeche S, Ita H *et al.* 2013 *Phys.Rev.* **D88** 014025 (*Preprint* arXiv:1304.1253)

[35] Cullen G, van Deurzen H, Greiner N, Luisoni G, Mastrolia P *et al.* 2013 *Phys.Rev.Lett.* **111** 131801 (*Preprint* arXiv:1307.4737)

[36] van Deurzen H, Luisoni G, Mastrolia P, Mirabella E, Ossola G *et al.* 2013 *Phys.Rev.Lett.* **111** 171801 (*Preprint* arXiv:1307.8437)

[37] Gehrmann T, Greiner N and Heinrich G 2013 (*Preprint* arXiv:1308.3660)

[38] Campanario F, Figy T, Pltzer S and Sjdahl M 2013 *Phys.Rev.Lett.* **111** 211802 (*Preprint* arXiv:1308.2932)

[39] Ossola G 2013 (*Preprint* arXiv:1310.3214)

[40] Cullen G, van Deurzen H, Greiner N, Heinrich G, Luisoni G *et al.* 2013 (*Preprint* arXiv:1309.3741)

[41] Bern Z, Dixon L, Cordero F F, Hoeche S, Ita H *et al.* 2013 (*Preprint* arXiv:1310.2808)

[42] Badger S, Biedermann B, Uwer P and Yundin V 2013 (*Preprint* arXiv:1309.6585)

[43] Gleisberg T, Hoeche S, Krauss F, Schonherr M, Schumann S *et al.* 2009 *JHEP* **0902** 007 (*Preprint* arXiv:0811.4622)

[44] Gleisberg T and Hoeche S 2008 *JHEP* **0812** 039 (*Preprint* arXiv:0808.3674)

[45] Gleisberg T and Krauss F 2008 *Eur.Phys.J.* **C53** 501–523 (*Preprint* arXiv:0709.2881)

[46] Binoth T, Boudjema F, Dissertori G, Lazopoulos A, Denner A *et al.* 2010 *Comput.Phys.Commun.* **181** 1612–1622 (*Preprint* arXiv:1001.1307)

[47] Alioli S, Badger S, Bellm J, Biedermann B, Boudjema F *et al.* 2013 (*Preprint* arXiv:1308.3462)

[48] Berends F A and Giele W 1988 *Nucl.Phys.* **B306** 759

[49] van Oldenborgh G 1991 *Comput.Phys.Commun.* **66** 1–15

[50] Ellis R K and Zanderighi G 2008 *JHEP* **0802** 002 (*Preprint* arXiv:0712.1851)

[51] Ellis R K, Melnikov K and Zanderighi G 2009 *JHEP* **0904** 077 (*Preprint* arXiv:0901.4101)

[52] Cacciari M, Salam G P and Soyez G 2012 *Eur.Phys.J.* **C72** 1896 (*Preprint* arXiv:1111.6097)

[53] Cacciari M, Salam G P and Soyez G 2008 *JHEP* **0804** 063 (*Preprint* arXiv:0802.1189)

[54] Whalley M, Bourilkov D and Group R 2005 (*Preprint* hep-ph/0508110)

[55] Ball R D *et al.* (NNPDF Collaboration) 2012 *Nucl.Phys.* **B855** 153–221 (*Preprint* arXiv:1107.2652)

[56] Ball R D, Bertone V, Carrazza S, Deans C S, Del Debbio L *et al.* 2013 *Nucl.Phys.* **B867** 244–289 (*Preprint* arXiv:1207.1303)

[57] Andersen J *et al.* (SM and NLO Multileg Working Group) 2010 21–189 (*Preprint* arXiv:1003.1241)

[58] Lai H L, Guzzi M, Huston J, Li Z, Nadolsky P M *et al.* 2010 *Phys.Rev.* **D82** 074024 (*Preprint* arXiv:1007.2241)

[59] Martin A, Stirling W, Thorne R and Watt G 2009 *Eur.Phys.J.* **C63** 189–285 (*Preprint* arXiv:0901.0002)

[60] Alekhin S, Blumlein J and Moch S 2012 *Phys.Rev.* **D86** 054009 (*Preprint* arXiv:1202.2281)

Automatic one-loop calculations with Sherpa+OpenLoops

F Cascioli[1], S Höche[2], F Krauss[3], P Maierhöfer[1,†], S Pozzorini[1] and F Siegert[4]

[1] Institut für Theoretische Physik, Universität Zürich, 8057 Zürich, Switzerland
[2] SLAC National Accelerator Laboratory, Menlo Park, CA 94025, USA
[3] Institute for Particle Physics Phenomenology, Durham University, Durham DH1 3LE, UK
[4] Physikalisches Institut, Albert-Ludwigs-Universität Freiburg, D-79104 Freiburg, Germany

E-mail: `cascioli@physik.uzh.ch`, `shoeche@slac.stanford.edu`, `frank.krauss@durham.ac.uk`, `philipp@physik.uzh.ch`, `pozzorin@physik.uzh.ch`, `frank.siegert@cern.ch`

Abstract. We report on the `OpenLoops` generator for one-loop matrix elements and its application to four-lepton production in association with up to one jet. The open loops algorithm uses a numerical recursion to construct the numerator of one-loop Feynman diagrams as functions of the loop momentum. In combination with tensor integrals this results in a highly efficient and numerically stable matrix element generator. In order to obtain a fully automated setup for the simulation of next-to-leading order scattering processes we interfaced `OpenLoops` to the `Sherpa` Monte Carlo event generator.

1. Introduction

The simulation of multi-particle scattering amplitudes with next-to-leading order (NLO) accuracy is a key requirement for the analysis of the data taken at the Large Hadron Collider. The necessity to manage the large number of processes to be considered at the experiments demands for integrated frameworks which automate the full tool-chain from the matrix element generation to hadronic final states via Monte Carlo event generators.

Regarding the NLO matrix elements, in the last few years the approach based on tensor integral reduction and algebraic methods was pushed to processes which involve up to 6 external particles [1, 2]. While this method can lead to efficient code its applicability is limited by expensive algebraic simplifications and the size of the process specific code. On the other hand the application of on-shell reduction techniques e. g. in combination with tree-level recursions lead to a high degree of automation of one-loop generators [3–8].

The open loops algorithm [9] exhibits a new way to calculate loop amplitudes using a tree-like recursion for loop momentum polynomials [10] and tensor integrals. The algorithm can be fully automated and achieves high efficiency and numerical stability. A generator based on a similar approach with a Dyson-Schwinger recursion and tensor integrals was presented in [11].

[†]Speaker; talk given at the International Workshop on Advanced Computing and Analysis Techniques in Physics Research (ACAT), Beijing, China, May 2013.

We interfaced our `OpenLoops` implementation [9] to `Sherpa` [12] which provides us with Monte Carlo integration, MC@NLO matching [13–15] to the `Sherpa` parton shower and MEPS@NLO multi-jet merging [16, 17]. Within this framework we calculated QCD corrections to four lepton production as a background to $\mathrm{H} \to \mathrm{WW}^* \to \ell\nu\ell\nu$ [18]. After the discovery of the Higgs boson [19, 20] this channel continues to play an important role in the investigation of its properties. To render the signal visible it is necessary to apply jet vetoes and perform the analysis in exclusive jet bins, in particular to suppress the background due to $\mathrm{t\bar{t}}$ production. The MEPS@NLO method allows us to retain NLO accuracy in the individual jet bins and resum potentially large logarithms.

The production of two W-bosons including leptonic decays was studied extensively in the literature [21–26] and also matched to parton showers with the MC@NLO [13] and with the POWHEG method [27, 28]. W-pair production in association with a jet [29–31] has been studied as well, however previously to our simulation not matched to a parton shower.

In section 2 we recapitulate the open loops algorithm and in section 3 we introduce the `Sherpa+OpenLoops` framework. Results for the four lepton production are shown in section 4.

2. The open loops algorithm and its implementation in `OpenLoops`

Tree-level amplitudes $\mathcal{M}$ and one-loop corrections $\delta\mathcal{M}$ are handled as sums of tree and one-loop Feynman diagrams,

$$\mathcal{M} = \sum_d \mathcal{M}^{(d)}, \qquad \delta\mathcal{M} = \sum_d \delta\mathcal{M}^{(d)}, \tag{1}$$

allowing for the factorisation of colour factors from the Lorentz structure of the diagrams. This way the colour reduction can be done once per process, resulting in negligible CPU cost for the colour summation [32].

Colour stripped tree diagrams are calculated by recursively connecting "sub-trees" starting from external wave functions. A sub-tree which is obtained from a tree diagram by cutting a line is represented as a complex n-tuple $w^\beta(i)$, where β is the spinor or Lorentz index of the cut propagator and the index i represents the topology, (off-shell) momentum, and particle content,

$w^\beta(i) =$ 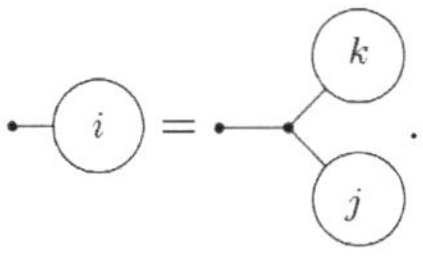 . (2)

Cut lines are marked by dots, and external lines are not depicted. For brevity, quartic vertices are not shown explicitly, but their inclusion is straightforward. In terms of n-tuples, the recursion step reads

$$w^\beta(i) = \frac{X^\beta_{\gamma\delta}(i,j,k)\, w^\gamma(j)\, w^\delta(k)}{p_i^2 - m_i^2 + \mathrm{i}\varepsilon}, \tag{3}$$

where $X^\beta_{\gamma\delta}/(p_i^2 - m_i^2 + \mathrm{i}\varepsilon)$ describes a vertex connecting i, j, k, and a propagator attached to i. The recursion terminates when the sub-trees which are needed to build all tree diagrams have been generated. Sub-trees which appear in more than one diagram are calculated once and used in all occurrences. The numerical implementation of the algorithm is based on universal routines which implement all wave functions, propagators and vertices corresponding to the Feynman rules of the theoretical model.

A colour stripped n-point one-loop diagram $\delta\mathcal{A}^{(d)}$ is regarded as an ordered set of n sub-trees, $\mathcal{I}_n = \{i_1, \ldots, i_n\}$, connected by loop propagators,

$$\delta\mathcal{A}^{(d)} = \int \frac{\mathrm{d}^D q\, \mathcal{N}(\mathcal{I}_n; q)}{D_0 D_1 \ldots D_{n-1}} = \qquad (4)$$

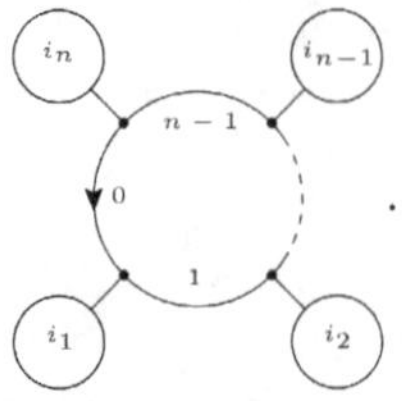

The denominators $D_i = (q+p_i)^2 - m_i^2 + \mathrm{i}\varepsilon$ depend on the loop momentum q, external momenta p_i, and internal masses m_i. All other contributions from loop propagators, vertices, and external sub-trees are summarised in the numerator, which is a polynomial of degree $R \leq n$ in the loop momentum,

$$\mathcal{N}(\mathcal{I}_n; q) = \sum_{r=0}^{R} \mathcal{N}_{\mu_1 \ldots \mu_r}(\mathcal{I}_n)\, q^{\mu_1} \ldots q^{\mu_r}. \qquad (5)$$

Momentum-shift ambiguities are eliminated by setting $p_0 = 0$, singling out the D_0 propagator. The loop momentum q flowing through this propagator is marked by an arrow in (4). After cutting the D_0 propagator the numerator function can be constructed by recursively attaching sub-trees to the loop

$$\mathcal{N}_\alpha^\beta(\mathcal{I}_n; q) = \qquad (6)$$

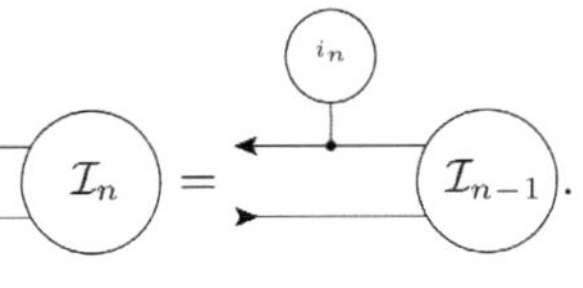

The indices α, β stand for the spinor rsp. Lorentz indices of the cut propagator. Analogously to eq. (3) the numerator $\mathcal{N}_\alpha^\beta$ is built by recursively attaching sub-trees,

$$\mathcal{N}_\alpha^\beta(\mathcal{I}_n; q) = X_{\gamma\delta}^\beta(\mathcal{I}_n, i_n, \mathcal{I}_{n-1})\, \mathcal{N}_\alpha^\gamma(\mathcal{I}_{n-1}; q)\, w^\delta(i_n), \qquad (7)$$

where $X_{\gamma\delta}^\beta$ are the same vertices as in the tree recursion. Separating the loop momentum from its coefficients

$$\mathcal{N}_\alpha^\beta(\mathcal{I}_n; q) = \sum_{r=0}^{R} \mathcal{N}_{\mu_1 \ldots \mu_r; \alpha}^\beta(\mathcal{I}_n)\, q^{\mu_1} \ldots q^{\mu_r}, \quad X_{\gamma\delta}^\beta = Y_{\gamma\delta}^\beta + Z_{\mu;\gamma\delta}^\beta q^\mu \qquad (8)$$

leads to a recursion relation for the so-called open loops which encode the functional dependence of the numerator on the loop momentum

$$\mathcal{N}_{\mu_1 \ldots \mu_r; \alpha}^\beta(\mathcal{I}_n) = \left[Y_{\gamma\delta}^\beta\, \mathcal{N}_{\mu_1 \ldots \mu_r; \alpha}^\gamma(\mathcal{I}_{n-1}) + Z_{\mu_1;\gamma\delta}^\beta\, \mathcal{N}_{\mu_2 \ldots \mu_r; \alpha}^\gamma(\mathcal{I}_{n-1}) \right] w^\delta(i_n). \qquad (9)$$

The recursion terminates with the contraction of the α, β indices $\mathcal{N}_{\mu_1 \ldots \mu_r} = \mathcal{N}_{\mu_1 \ldots \mu_r; \alpha}^\alpha$, resulting in the coefficients of the tensor integral representation of the diagram

$$\delta\mathcal{A}^{(d)} = \sum_{r=0}^{R} \mathcal{N}_{\mu_1 \ldots \mu_r}(\mathcal{I}_n)\, T_{n,r}^{\mu_1 \ldots \mu_r} \quad \text{with} \quad T_{n,r}^{\mu_1 \ldots \mu_r} = \int \frac{\mathrm{d}^D q\, q^{\mu_1} \ldots q^{\mu_r}}{D_0 D_1 \ldots D_{n-1}}. \qquad (10)$$

Table 1. Processes which are available to the ATLAS and CMS Monte Carlo working groups. Vector boson production (V = Z/W$^{\pm}$) includes leptonic decays except for VVV. Lower jet multiplicities are implicitly understood. Brackets denote that the process will be available with the next update.

W/Z	γ	jets	HQ pairs	single-top	Higgs
$V+3j$	$\gamma+3j$	$3(4)j$	$t\bar{t}+1j$	$tb+1j$	$(H+2j)$
$VV+1(2)j$	$\gamma\gamma+1(2)j$		$t\bar{t}V+0(1)j$	$t+1(2)j$	$VH+1j$
$gg\to VV+1j$	$V\gamma+1(2)j$		$b\bar{b}V+0(1)j$	$tW+0(1)j$	$t\bar{t}H$
$VVV+0(1)j$					$qq\to Hqq+0(1)j$

The tensor integrals $T^{\mu_1\ldots\mu_r}_{n,r}$ are subsequently reduced to m-point scalar integrals $T_{m,0}$ with $m = 1,2,3,4$. Alternatively, the OPP method [33] avoids tensor integrals through a direct connection between the numerator $\mathcal{N}(\mathcal{I}_n;q)$ and the scalar-integral representation of the amplitude. The coefficients of the scalar integrals are determined by multiple evaluations of $\mathcal{N}(\mathcal{I}_n;q)$ for loop momenta q which satisfy multiple-cut conditions of the form $D_i = D_j = \cdots = 0$.

We implemented the described algorithm in the program `OpenLoops`, a fully automatic generator for QCD corrections to Standard Model processes. Feynman diagrams are generated by `FeynArts` [34] and `Mathematica` organises the open loops recursion and generates `Fortran 90` code. For the reduction of the tensor integrals we use the `Collier` [35] library which implements the Denner-Dittmaier reduction procedure [36, 37] and the scalar integrals of ref. [38]. The `Collier` library cures numerical instabilities which arise due to vanishing Gram determinants and other kinematic quantities by applying expansions in these quantities, thus allowing for the numerically stable evaluation of tensor integrals in double precision.

Rational terms of type R_2 are reconstructed by counterterm-like Feynman rules [39]. In order to asses the performance and the numerical stability we considered the $2\to 2,3,4$ reactions $\mathrm{u\bar{u}}\to\mathrm{W^+W^-}+n\mathrm{g}$, $\mathrm{u\bar{d}}\to\mathrm{W^+g}+n\mathrm{g}$, $\mathrm{u\bar{u}}\to\mathrm{t\bar{t}}+n\mathrm{g}$, and $\mathrm{gg}\to\mathrm{t\bar{t}}+n\mathrm{g}$, with $n=0,1,2$ gluons [9]. For the most complicated $2\to 4$ processes the runtime per phase space point is below 1 second on an i5-750 CPU (single core) and the size of a compiled process library is of the order of at most 1 MB. The average number of correct digits ranges from 11 to 15 for the 12 processes and the probability to encounter numerical precision below 10^{-5} and 10^{-3} is less than 2 and 0.1, respectively.

3. NLO simulations with `Sherpa+OpenLoops`

For realistic simulations of NLO processes the matrix elements must be combined with parton showers and hadronisation. Especially for the description of exclusive observables the resummation of large logarithms as provided by a parton shower is imperative.

We wrote an interface to the `Sherpa` Monte Carlo event generator which provides us with Monte Carlo integration, infra-red subtraction, real corrections, MC@NLO matching to its parton shower and MEPS@NLO merging of different jet multiplicities, providing NLO plus parton shower accuracy in the individual jet bins. The interface works in a fully automatic way, loading process libraries on request at runtime. The matrix element generation is steered by standard `Sherpa` runcards.

The `Sherpa+OpenLoops` framework is available to the Monte Carlo working groups of the ATLAS and CMS collaborations, including the set of processes shown in table 1. All provided processes were thoroughly validated against an independent in-house matrix element generator.

Apart from the 4 lepton study the framework was applied to $\mathrm{t\bar{t}b\bar{b}}$ production with massive

Table 2. Exclusive 0- and 1-jet bin $\mu^+\nu_\mu e^-\bar{\nu}_e$+jets cross sections in the signal (S) and control (C) regions of the ATLAS analysis at 8 TeV. Fixed-order NLO results are compared to MC@NLO and MEPS@NLO predictions. Scale uncertainties are shown as $\sigma \pm \delta_{\rm QCD} \pm \delta_{\rm res}$, where $\delta_{\rm QCD}$ and $\delta_{\rm res}$ correspond to variations of the QCD ($\mu_{\rm R}, \mu_{\rm F}$) and resummation (μ_Q) scales, respectively. Statistical errors are given in parenthesis.

0-jet bin	NLO $4\ell(+1j)$	MC@NLO $4\ell(+1j)$	MEPS@NLO $4\ell(+1j)$
σ_S [fb]	$34.28(9)\,^{+2.1\%}_{-1.6\%}$	$32.52(8)\,^{+2.1\%}_{-0.8\%}\,^{+1.2\%}_{-0.7\%}$	$33.81(12)\,^{+1.4\%}_{-2.2\%}\,^{+2.0\%}_{-0.4\%}$
σ_C [fb]	$55.76(9)\,^{+2.0\%}_{-1.7\%}$	$52.28(9)\,^{+1.4\%}_{-0.7\%}\,^{+1.4\%}_{-1.1\%}$	$54.18(15)\,^{+1.4\%}_{-1.9\%}\,^{+2.5\%}_{-0.4\%}$
1-jet bin	NLO $4\ell(+1j)$	MC@NLO $4\ell(+1j)$	MEPS@NLO $4\ell(+1j)$
σ_S [fb]	$8.99(4)\,^{+4.9\%}_{-9.5\%}$	$8.02(4)\,^{+8.5\%}_{-6.4\%}\,^{+0\%}_{-3.1\%}$	$9.37(9)\,^{+2.6\%}_{-2.7\%}\,^{+2.5\%}_{-0.0\%}$
σ_C [fb]	$26.50(8)\,^{+6.4\%}_{-12.5\%}$	$24.58(8)\,^{+6.1\%}_{-6.5\%}\,^{+1.2\%}_{-3.0\%}$	$28.32(13)\,^{+3.1\%}_{-4.7\%}\,^{+4.1\%}_{-0.0\%}$

b quarks, matched to the parton shower [40].

4. Irreducible background to H → WW*

We used Sherpa+OpenLoops§ for the simulation of $\mu^+\nu_\mu e^-\bar{\nu}_e(+j)$ production (in the following referred to as 4 leptons or $4\ell(+j)$) as irreducible background to $\rm H \to WW^*$ at a centre-of-mass energy of 8 TeV, including off-shell and non-resonant contributions and all respective interferences [18]. To assess the effects of the parton shower and the merging we calculated the processes in three different approximations, fixed order NLO, MC@NLO, and MEPS@NLO.

As input parameters we use

$$M_{\rm W} = 80.399\,{\rm GeV}, \quad M_{\rm Z} = 91.1876\,{\rm GeV}, \quad \Gamma_{\rm W} = 2.0997\,{\rm GeV}, \quad \Gamma_{\rm Z} = 2.5097\,{\rm GeV} \tag{11}$$

for the vector boson masses and widths. The electroweak mixing angle is obtained from the complex W- and Z-boson masses [41] as

$$\cos^2\theta_{\rm w} = \frac{M_{\rm W}^2 - {\rm i}\Gamma_{\rm W} M_{\rm W}}{M_{\rm Z}^2 - {\rm i}\Gamma_{\rm Z} M_{\rm Z}}, \tag{12}$$

and the electromagnetic fine-structure constant is derived from the Fermi constant $G_\mu = 1.16637 \cdot 10^{-5}\,{\rm GeV}^{-2}$ in the G_μ-scheme as

$$\alpha^{-1} = \frac{\pi}{\sqrt{2}\,G_\mu M_{\rm W}^2}\left(1 - \frac{M_{\rm W}^2}{M_{\rm Z}^2}\right)^{-1} = 132.348905\,. \tag{13}$$

As for the parton distribution functions we chose five-flavour CT10 NLO [42] with the respective running strong coupling α_S. To avoid any overlap with $\rm t\bar{t}$ production, only partonic channels without external b quarks are considered. This requires a prescription to separate $\rm W^+W^- + j$ from top pair and single top production which treats infrared singularities and large logarithms arising from $\rm g \to b\bar{b}$ splittings in a well defined way [18]. Such a prescription is not unique and reflects in an ambiguity of the order of 1% which disappears when $\rm W^+W^- + j$ and $\rm W^+W^-b\bar{b}$ are consistently merged.

§Results were obtained with a pre-release version of Sherpa 2.0 corresponding to SVN revision 21825.

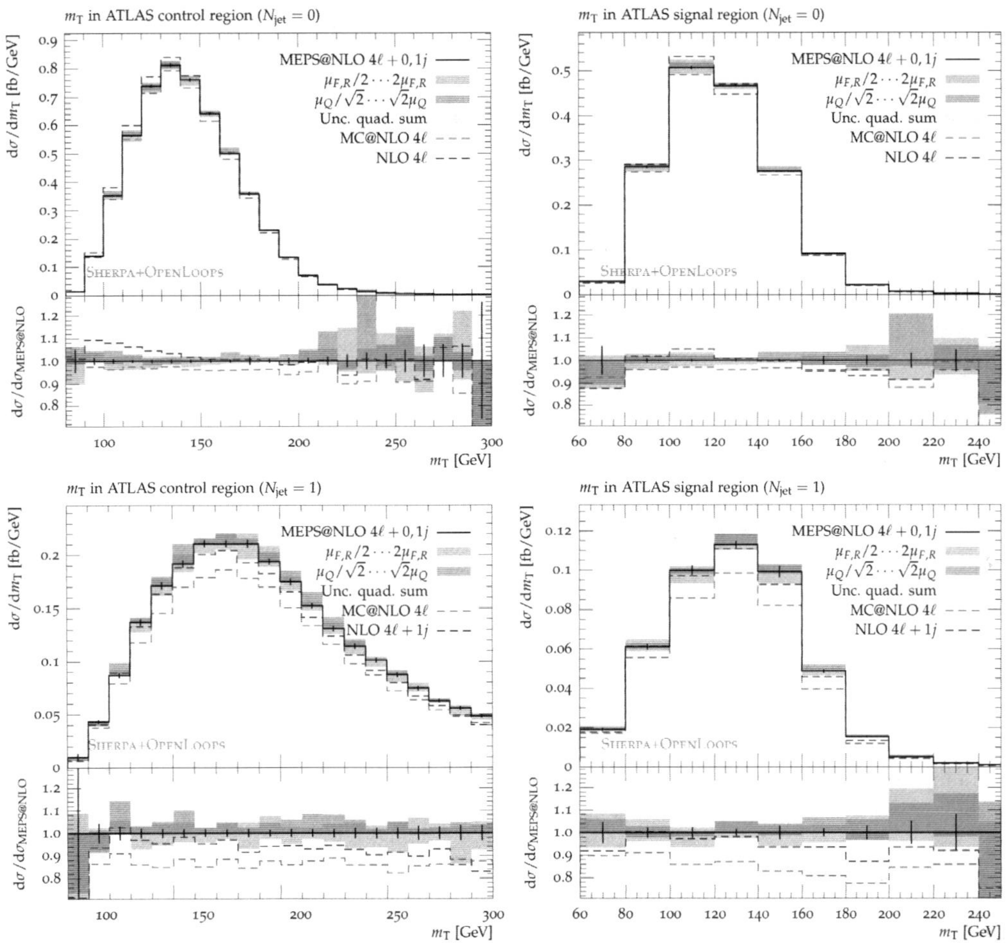

Figure 1. Transverse mass distributions for the ATLAS analysis at $8\,TeV$ after control (left) and signal (right) cuts in the 0-jet (top) and 1-jet (bottom) bins. The lines show MEPS@NLO results (black solid), MC@NLO (red dashed) and NLO (blue dashed) with error bands for μ_R and μ_F factor 2 variations (red) and μ_Q factor $\sqrt{2}$ variations (blue), and quadratically added errors (yellow). Error bands are colour additive (yellow+blue=green, yellow+red=orange, yellow+red+blue=brown).

Table 2 shows the cross sections for the three different simulations in the signal and control regions of the ATLAS analysis. Corresponding results for the CMS analysis can be found in [18]. The default renormalisation (μ_R), factorisation (μ_F) and resummation (μ_Q) scale is chosen as the average transverse energy of the W bosons,

$$\mu_0 = \frac{1}{2}(E_{T,\mathrm{W}^+} + E_{T,\mathrm{W}^-}), \quad \text{where} \quad E_{T,\mathrm{W}}^2 = M_{\mathrm{W}}^2 + (\vec{p}_{T,\ell} + \vec{p}_{T,\nu})^2. \tag{14}$$

QCD scale uncertainties are estimated by factor 2 variations of μ_R and μ_F, excluding opposite direction variations. The resummation scale μ_Q is varied by a factor $\sqrt{2}$. In the parton shower and for the jet emission in the $4\ell + j$ matrix elements for the MEPS@NLO simulation the scale choice is based on the CKKW technique which adapts the α_S scale to the transverse momentum of the jet. The merging scale $Q_{\rm cut}$ is set to 20 GeV. Figure 1 shows the m_T distributions in the 0- and 1-jet bin in the signal and control regions as defined by ATLAS.

The NLO and MEPS@NLO results agree fairly well with sizeable deviations only in the large m_T region in the 1-jet bin. The good agreement suggests that resummation effects are rather small. The discrepancies between MC@NLO and MEPS@NLO however reach up to 20% in the 1-jet bin with moderate shape distortions. This is not surprising given that the jet emission in the MC@NLO simulation is only leading order accurate.

In the full analysis [18] we also study squared quark-loop contributions which form a finite and gauge invariant subset of NNLO corrections and can have sizeable impact due to opening gluon fusion channels, especially with signal cuts applied. Furthermore several other observables relevant for the experimental analysis are considered as well as the impact of the different approximations on the description of jet veto effects.

5. Conclusions

We implemented a fully automatic generator for NLO QCD corrections to Standard Model processes based on the open loops algorithm to construct loop momentum polynomials by a numerical recursion. With its high performance and numerical stability it is suited to address problems which involve a large number of multi-leg partonic processes. The matrix element generator was interfaced to `Sherpa` and a library for a large set of processes is available to the ATLAS and CMS Monte Carlo working groups.

Within the `Sherpa+OpenLoops` framework we performed detailed simulations of the production of 4 leptons with up to one jet as a background for the $H \to WW^*$ analysis of the ATLAS and CMS experiments and studied the impact of parton shower and merging effects. The MEPS@NLO simulation which provides NLO accuracy and resummation in both jet bins is our best prediction and exhibits scale uncertainties below 5%. This approach is particularly suited to study exclusive observables and provides more realistic error estimates.

Acknowledgments

The research of F. Cascioli, P. Maierhöfer and S. Pozzorini was supported by the SNSF. Stefan Höche was supported by the U.S. Department of Energy under Contract No. DE–AC02–76SF00515. Frank Siegert's work was supported by the German Research Foundation (DFG) via grant DI 784/2-1.

References

[1] Bredenstein A, Denner A, Dittmaier A and Pozzorini A 2009 *Phys. Rev. Lett.* **103** 012002 (*Preprint* arXiv:0905.0110 [hep-ph])
[2] Denner A, Dittmaier S, Kallweit S and Pozzorini S 2011 *Phys. Rev. Lett.* **106** 052001 (*Preprint* arXiv:1012.3975 [hep-ph])
[3] Berger C F, Bern Z, Dixon L J, Febres F Cordero, Forde D, Ita H, Kosower D A and Maitre D 2008 *Phys. Rev.* D **78** 036003 (*Preprint* arXiv:0803.4180 [hep-ph])
[4] Ellis R K, Giele W T, Kunszt Z and Melnikov K 2009 *Nucl. Phys.* B **822** 270 (*Preprint* arXiv:0806.3467 [hep-ph])
[5] Hirschi V, Frederix R, Frixione S, Garzelli M V, Maltoni F and Pittau R 2011 *JHEP* **1105** 044 (*Preprint* arXiv:1103.0621 [hep-ph])
[6] Cullen G, Greiner N, Heinrich G, Luisoni G, Mastrolia P, Ossola G, Reiter T and Tramontano F 2012 *Eur. Phys. J.* C **72** 1889 (*Preprint* arXiv:1111.2034 [hep-ph])
[7] Bevilacqua G, Czakon M, Garzelli M V, van Hameren A, Kardos A, Papadopoulos C G, Pittau R and Worek M 2013 *Comput. Phys. Commun.* **184** 986 (*Preprint* arXiv:1110.1499 [hep-ph])
[8] Badger S, Biedermann B, Uwer P and Yundin V 2013 *Comput. Phys. Commun.* **184** 1981 (*Preprint* arXiv:1209.0100 [hep-ph])
[9] Cascioli F, Maierhoefer P and Pozzorini S 2012 *Phys. Rev. Lett.* **108** 111601 (*Preprint* arXiv:1111.5206)

[10] van Hameren A 2009 *JHEP* **0907** 088 (*Preprint* arXiv:0905.1005)
[11] Actis S, Denner A, Hofer L, Scharf A and Uccirati S 2013 *JHEP* **1304** 037 (*Preprint* arXiv:1211.6316 [hep-ph])
[12] Gleisberg T, Hoeche S, Krauss F, Schonherr M, Schumann S, et al. 2009 *JHEP* **0902** 007 (*Preprint* arXiv:0811.4622)
[13] Frixione S and Webber B R 2002 *JHEP* **0206** 029 (*Preprint* hep-ph/0204244)
[14] Hoeche S, Krauss F, Schonherr M and Siegert F 2012 *JHEP* **1209** 049 (*Preprint* arXiv:1111.1220)
[15] Hoeche S, Krauss F, Schonherr M and Siegert F 2013 *Phys. Rev. Lett.* **110** 052001 (*Preprint* arXiv:1201.5882)
[16] Gehrmann T, Hoeche S, Krauss F, Schonherr M and Siegert F 2013 *JHEP* **1301** 144 (*Preprint* arXiv:1207.5031)
[17] Hoeche S, Krauss F, Schonherr M and Siegert F 2013 *JHEP* **1304** 027 (*Preprint* arXiv:1207.5030)
[18] Cascioli F, Hoeche S, Krauss F, Maierhofer P, Pozzorini S and Siegert F *Preprint* arXiv:1309.0500 [hep-ph]
[19] ATLAS Collaboration, Aad G et al. 2012 *Phys. Lett.* B **716** 1–29 (*Preprint* arXiv:1207.7214)
[20] CMS Collaboration, Chatrchyan S et al. 2012 *Phys. Lett.* B **716** 30–61 (*Preprint* arXiv:1207.7235)
[21] Ohnemus J 1991 *Phys. Rev.* D **44** 1403–1414
[22] Frixione S 1993 *Nucl. Phys.* B **410** 280–324
[23] Ohnemus J 1994 *Phys. Rev.* D **50** 1931–1945 (*Preprint* hep-ph/9403331)
[24] Dixon L J, Kunszt Z and Signer A 1999 *Phys. Rev.* D **60** 114037 (*Preprint* hep-ph/9907305)
[25] Campbell J M and Ellis R K 1999 *Phys. Rev.* D **60** 113006 (*Preprint* hep-ph/9905386)
[26] Campbell J M, Ellis R K and Williams C 2011 *JHEP* **1107** 018 (*Preprint* arXiv:1105.0020)
[27] Frixione S, Nason P and Oleari C 2007 *JHEP* **0711** 070 (*Preprint* arXiv:0709.2092)
[28] Melia T, Nason P, Rontsch R and Zanderighi G 2011 *JHEP* **1111** 078 (*Preprint* arXiv:1107.5051)
[29] Campbell J M, Ellis R K and Zanderighi G 2007 *JHEP* **0712** 056 (*Preprint* arXiv:0710.1832)
[30] Dittmaier S, Kallweit S and Uwer P 2008 *Phys. Rev. Lett.* **100** 062003 (*Preprint* arXiv:0710.1577)
[31] Dittmaier S, Kallweit S and Uwer P 2010 *Nucl. Phys.* B **826** 18–70 (*Preprint* arXiv:0908.4124)
[32] Bredenstein A, Denner A, Dittmaier S and Pozzorini S 2010 *JHEP* **1003** 021 (*Preprint* arXiv:1001.4006 [hep-ph])
[33] Ossola G, Papadopoulos C G and Pittau R 2007 *Nucl. Phys.* B **763** 147–169 (*Preprint* hep-ph/0609007)
[34] Hahn T 2001 *Comput. Phys. Commun.* **140** 418-431 (*Preprint* hep-ph/0012260)
[35] Denner A, Dittmaier S and Hofer L *in preparation*
[36] Denner A and Dittmaier S 2003 *Nucl. Phys.* B **658** 175–202 (*Preprint* hep-ph/0212259)
[37] Denner A and Dittmaier S 2006 *Nucl. Phys.* B **734** 62–115 (*Preprint* hep-ph/0509141)
[38] Denner A and Dittmaier S 2011 *Nucl. Phys.* B **844** 199–242 (*Preprint* arXiv:1005.2076)
[39] Draggiotis P, Garzelli M, Papadopoulos C and Pittau R 2009 *JHEP* **0904** 072 (*Preprint* arXiv:0903.0356)
[40] Cascioli F, Maierhoefer P, Moretti N, Pozzorini S and Siegert F *Preprint* arXiv:1309.5912 [hep-ph]
[41] Denner A, Dittmaier S, Roth M and Wieders L 2005 *Nucl. Phys.* B **724** 247–294 (*Preprint* hep-ph/0505042)
[42] Lai H L, Guzzi M, Huston J, Li Z, Nadolsky P M, et al. 2010 *Phys. Rev.* D **82** 074024 (*Preprint* arXiv:1007.2241)

`LiteRed` 1.4: a powerful tool for reduction of multiloop integrals

Roman N. Lee

Budker Institute of Nuclear Physics, 630090, Novosibirsk, Russia

E-mail: `R.N.Lee@inp.nsk.su`

Abstract. We review the *Mathematica* package `LiteRed`, version 1.4.

1. Introduction

Multiloop integrals are the building blocks of many calculations of radiative corrections in QFT. One of the key approaches to the calculation of the multiloop integrals is the IBP reduction, based on the integration-by-parts identities between integrals [1, 2]. The IBP reduction almost necessarily should be done with the aid of various computer programs. One of the most successful methods of the IBP reduction is the Laporta algorithm [3]. The algorithm is easy to implement and to use, and allows for a number of programming improvements. These advantages explain why many modern most powerful reduction programs heavily rely on this algorithm, in particular, `AIR` [4], `FIRE` [5, 6], `Reduze` [7, 8], and many private versions.

However, the Laporta algorithm has some weak points, which may put restrictions on its application. In particular, being intrinsically a brute-force search, this algorithm is both time- and memory-consuming. Another approach to the reduction is a derivation of the symbolic reduction rules. Its advantages are obvious: nothing is being solved in the process of reduction, therefore, the reduction is very fast. Symbolic rules are small in size, so, they can be easily saved for future calculations. The bottleneck of this approach is the search of these symbolic rules. Much effort has been devoted to the developement of the approach connected with the notion of the Groebner basis [9, 10, 11, 12], but, for now, this approach is far from being satisfactory. Probably, the only, partly successful, attempt to implement this approach has been made in `FIRE`, where the notion of s-bases [11, 12] has been used.

Recently, the new IBP reduction package `LiteRed` has been presented in Ref. [13]. This package uses completely different approach to the reduction. At first stage it tries to find symbolic reduction rules using heuristics. Then it applies the rules to the specific reduction. Found rules are very lightweight and can be easily stored for the reusage. In this contribution we describe the version 1.4 of the `LiteRed` package. The package can be downloaded from `http://www.inp.nsk.su/~lee/programs/LiteRed/`.

2. Multiloop integral

To fix notation, let us assume that we are interested in the calculation of the L-loop integral depending on the E external momenta $p_1, \ldots, p_E$. There are $N = L(L+1)/2 + LE$ scalar products depending on the loop momenta l_i:

$$s_{ij} = l_i \cdot q_j \,, \quad 1 \leqslant i \leqslant L, j \leqslant L+E, \tag{1}$$

where $q_{1,\dots,L} = l_{1,\dots,L}$, $q_{L+1,\dots,L+E} = p_{1,\dots,E}$.

The general form of the integral is the following

$$\begin{aligned} J\left(\boldsymbol{n}\right) = J(n_1, n_2, \dots, n_N) &= \int \frac{d^d l_1 \dots d^d l_L}{D_1^{n_1} D_2^{n_2} \dots D_N^{n_N}} \,, \\ D_\alpha = a_\alpha^{ij} l_i \cdot l_j + 2b_\alpha^{ik} l_i \cdot p_k + c_\alpha \,. \end{aligned} \tag{2}$$

Here a_α, b_α, and c_α are $L \times L$ matrices, $L \times E$ matrices, and numbers, respectively. As usual, we assume that $D_1, \dots, D_N$ form a complete basis in the sense that any s_{ik} can be uniquely expressed in terms of D_α. The multiindex $\boldsymbol{n} = (n_1, \dots, n_N)$ can be thought of as a point in $\mathbb{Z}^N$. Some of D_α correspond to the denominators of the propagators, others may correspond to the irreducible numerators. E.g., the K-legged L-loop diagram with generic external momenta corresponds to $E = K - 1$ and the maximal number of denominators is $M = E + 3L - 2$, so that the rest $N - M = (L-1)(L+2E-4)/2$ functions correspond to irreducible numerators.

2.1. Differential equations

The differential equations can be used for finding the master integrals. The simplest type of such equations is the differential equation with respect to the mass. Probably, the first example of their application is presented in Refs. [14, 15, 16]. The differential equations with respect to the invariant constructed of the external momenta have been introduced and applied in Refs. [17, 18, 19]. In general case, when there are $E > 2$ external vectors, we have the following formulas:

$$\begin{aligned} \frac{\partial}{\partial\left(p_1 \cdot p_2\right)} J\left(\mathbf{n}\right) &= \sum \left[\mathbb{G}^{-1}\right]_{i2} p_i \cdot \partial_{p_1} J\left(\mathbf{n}\right) = \sum \left[\mathbb{G}^{-1}\right]_{i1} p_i \cdot \partial_{p_2} J\left(\mathbf{n}\right) \,, \\ \frac{\partial}{\partial\left(p_1^2\right)} J\left(\mathbf{n}\right) &= \frac{1}{2} \sum \left[\mathbb{G}^{-1}\right]_{i1} p_i \cdot \partial_{p_1} J\left(\mathbf{n}\right) \,. \end{aligned} \tag{3}$$

where $\mathbb{G} = \mathbb{G}\left(p_1, \dots, p_E\right) = \begin{pmatrix} p_1^2 & \cdots & p_1 \cdot p_E \\ \vdots & \ddots & \vdots \\ p_1 \cdot p_E & \cdots & p_E^2 \end{pmatrix}$ is a Gram matrix.

Acting by the operator on the right-hand side on the integrand and performing the IBP reduction, one obtains the differential equation for $J\left(\mathbf{n}\right)$.

2.2. Dimensional recurrences

Probably, the first appearance of the dimension shifting relations is in Ref. [20], where they have been derived for certain three-loop integrals in the momentum representation. Later, in Ref. [21] Tarasov derived dimensional relations using the paramentric representation. It is interesting that the first approach led to the lowering recurrence, while the latter one led to the raising recurrence. The lowering (raising) recurrence relates one integral in $d+2$ $(d-2)$ dimensions to several integrals in d dimensions.

As it was shown in Ref. [22], the two recurrences can be represented as

$$J^{(d-2)}\left(\mathbf{n}\right) = \left(\mu/2\right)^L \det \left\{ 2^{\delta_{ij}} \frac{\partial D_k}{\partial s_{ij}} A_k|_{i,j=1\dots L} \right\} J^{(d)}\left(\mathbf{n}\right). \tag{4}$$

$$J^{(d+2)}(\mathbf{n}) = \frac{(2\mu)^L \left[V(p_1,\dots,p_E)\right]^{-1}}{(d-E-L+1)_L} P(B_1,\dots,B_N)\, J^{(d)}(\mathbf{n}), \tag{5}$$

where $\mu = \pm 1$ for Euclidean/Minkovskian metric, $\alpha_L = \alpha(\alpha+1)\dots(\alpha+L-1)$ is the Pochhammer symbol, $V(v_1,\dots v_k) = \det \mathbb{G}(v_1,\dots v_k)$ is the Gram determinant, and $P(D_1,\dots,D_N) = V(q_1,\dots q_{L+E})$.

The operators A_α and B_α are defined as follows

$$\begin{aligned}\left(A_i J^{(\mathcal{D})}\right)(n_1,\dots,n_N) &= n_i J^{(\mathcal{D})}(n_1,\dots,n_i+1,\dots,n_N),\\ \left(B_i J^{(\mathcal{D})}\right)(n_1,\dots,n_N) &= J^{(\mathcal{D})}(n_1,\dots,n_i-1,\dots,n_N).\end{aligned} \tag{6}$$

3. Parametric representation

The parametric representation (or Feynman parametrization), is, of no doubt, one of the most useful tools for the multiloop calculations. It is important for both analytical (in particular, with subsequent Mellin-Barnes representation) and numerical (in particular, together with sector decomposition) calculations of the multiloop integrals. Moreover, it may serve as a fundamental definition of the multiloop integrals for the case of non-integer dimensionality d.

But it is also important that the parametric representation is very useful for revealing relations between the integrals. One example is Tarasov's derivation of the raising dimensional recurrence relation [21]. In Ref. [23] the algorithm, based on the use of parametric representation, for the identification of the master integrals has been introduced. In recent paper [24] it was shown that parametric representation allows for a simple determination of the number of master integrals in the given sector.

`LiteRed` uses parametric representation for two purposes. First, it finds equivalent simple sectors by comparing their parametric representation. To account for the possible permutation of the parameters, it uses an approach combining the ideas from Ref. [23] and Ref. [25]. Second, it uses parametric representation to determine zero sectors. In this section we describe shortly this new algorithm.

The parametric representation of the integral $J(\mathbf{n})$ has the form

$$J(\mathbf{n}) = \frac{\Gamma(\Sigma n - Ld/2)}{\prod_\alpha \Gamma(n_\alpha)} \int \prod_\alpha dz_\alpha z_\alpha^{n_\alpha-1} \delta(1-\Sigma z) \frac{F^{Ld/2-\Sigma n}}{U^{(L+1)d/2-\Sigma n}}, \tag{7}$$

where $\Sigma n = \sum_\alpha n_\alpha$, $\Sigma z = \sum_\alpha z_\alpha$, U and F are the homogeneous polynomials of degrees L and $L+1$, respectively. These polynomials can be expressed in terms of quantities

$$a^{ij} = \sum_\alpha z_\alpha a_\alpha^{ij}, \quad b^i = \sum_\alpha z_\alpha b_\alpha^{ij} p_j, \quad c = \sum_\alpha z_\alpha c_\alpha \tag{8}$$

as follows

$$U = \det(a), \quad F = c\det(a) - \left(a^{\mathrm{Adj}}\right)^{ij} b^i \cdot b^j, \tag{9}$$

where $a^{\mathrm{Adj}} = \det(a)\, a^{-1}$ is the adjoint matrix. The representation (7) does not make sense when some of n_α are nonpositive integers. In this case one has to replace the corresponding integration with the derivative at zero point. The resulting formula can be written as

$$J(\mathbf{n}) = \Gamma(\Sigma n - Ld/2) \prod_\alpha \hat{n}_\alpha \delta(1-\Sigma_+ z) \frac{F^{Ld/2-\Sigma n}}{U^{(L+1)d/2-\Sigma n}}, \tag{10}$$

where the functional $\hat{n}_\alpha$ is determined as

$$\hat{n}_\alpha[\phi(z_\alpha)] = \begin{cases} \int_0^\infty \frac{dz_\alpha z_\alpha^{n_\alpha-1}}{\Gamma(n_\alpha)}\phi(z_\alpha) & n_\alpha > 0 \\ (-1)^{n_\alpha}\phi^{(-n_\alpha)}(0) & n_\alpha \leqslant 0 \end{cases} \tag{11}$$

and the sum $\Sigma_+ z = \sum_\alpha \theta_\alpha z_\alpha$ ($\theta_\alpha = \Theta(n_\alpha - 1/2)$) goes over the variables, corresponding to the denominators. Remarkably, it is possible to rewrite Eq. (10) in the form, which contains U and F only in the combination $F + U$. Similar to Ref. [24], we have

$$J(\mathbf{n}) = \frac{\Gamma\left[d/2\right]}{\Gamma\left[(L+1)\,d/2 - \Sigma n\right]} \prod_\alpha \hat{n}_\alpha G^{-d/2}, \quad G = F + U\,. \tag{12}$$

The scaleless integral can be defined as the one which gains additional non-unity factor under some linear transformation of the loop momenta. In dimensional regularization scaleless integrals are set to zero. If $j\,(\theta_1, \ldots, \theta_N)$ is scaleless, then all integrals of the sector $(\theta_1, \ldots, \theta_N)$ are zero. We will call such a sector a *zero sector*.

A simple criterion of zero sectors has been formulated in Ref. [26]. According to this criterion, the sector is zero if the solution of the IBP equations in the corner point $(\theta_1, \ldots, \theta_N)$ result in the identity $j\,(\theta_1, \ldots, \theta_N) = 0$. Note that this criterion may miss some scaleless sectors. Let us explain on a simple example why this happens. Consider the massless one-loop onshell propagator integral

$$J\,(n_1, n_2) = \int \frac{d^d l}{\left[l^2\right]^{n_1} \left[(l-k)^2\right]^{n_2}}, \quad k^2 = 0\,.$$

Obviously, this integral is zero for any n_1 and n_2. However, it can be explicitely checked that the solution of the IBP identities in the corner point of the sector $(1,1)$ does not result directly to $J\,(1,1) = 0$. In order to prove that the integral $J\,(1,1)$ is scaleless, let us consider instead the following operator

$$O = \partial_l \cdot \left(l + (l\cdot k)\,\tilde{k} - \left(l \cdot \tilde{k}\right) k\right),$$

where $\tilde{k}$ is an auxiliary vector chosen to satisfy the conditions $\tilde{k}^2 = 0$ and $\tilde{k} \cdot k = 1$. It is easy to check that $Oj\,(1,1) = (d-4)\,j\,(1,1)$. Since the operator O is a generator of the linear transformation $l \to l + \epsilon\left(l + (l\cdot k)\,\tilde{k} - \left(l\cdot\tilde{k}\right)k\right)$, the integral $j\,(1,1)$ is scaleless. The reason why the IBP identities failed to lead to the identity $J\,(1,1) = 0$ is that the construction of this identity required introduction of the auxiliary vector $\tilde{k}$.

In some cases the number of zero sectors overlooked by the criterion of Ref. [26] is rather big. So, we formulate below another criterion, based on parametric representation, which detects virtually all zero sectors.

As we said above, for the detection of zero sectors it is sufficient to consider only the integral in the corner point of the sector. In particular, we may set all z_β, corresponding to numerators, to zero. In what follows we assume this is done and z_α denotes a parameter, corresponding to the denominator of the sector. Consider an infinitesimal scaling of these parameters

$$z_\alpha \to \tilde{z}_\alpha = (1 + k_\alpha \omega)\, z_\alpha\,. \tag{13}$$

Here ω is the infinitesimal parameter, and k_α are some finite coefficients. Suppose that we are able to find such k_α that the function G scales as follows:

$$G\,(\tilde{z}) = (1+\omega)\,G\,(z)\,. \tag{14}$$

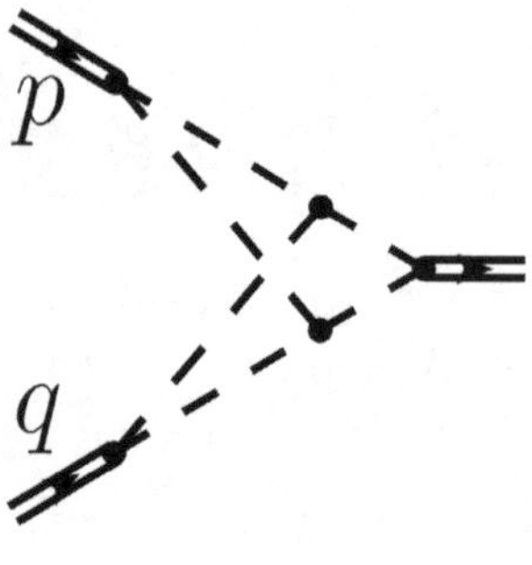

Figure 1. Two-loop massless onshell vertex.

Then, making the change (13) in Eq. (12), we get

$$J(\boldsymbol{\theta}) = \left[1 + \omega\left(\sum_{\alpha} k_{\alpha} - d/2\right)\right] J(\boldsymbol{\theta}). \tag{15}$$

The equation (14) does not depend on d, therefore, suitable k_α, if they exist at all, can be chosen also independent of d. Therefore, the coefficient in Eq. (15) in front of ω is not zero, and the integral $J(\boldsymbol{\theta})$ is scaleless.

Note that Eq. (14) can be cast as

$$\sum_{\alpha} k_\alpha z_\alpha \frac{\partial G(z)}{\partial z_\alpha} = G(z), \tag{16}$$

which should be understood as equality of the two polynomials of z_α. Collecting the coefficients in front of distinct monomials, we obtain a linear system of equation with respect to k_α. The existence of the solution of this system can be established by ordinary algebraic means.

Therefore, we get the following

Criterion of zero sector: *For a given sector, construct* $G = F + U$. *The sector is zero if Eq.*(16) *has a* z*-independent solution with respect to* k_α.

It is just this criterion which is implemented in `LiteRed1.4`.

4. How `LiteRed` finds reduction rules

When trying to find the complete set of the reduction rules, `LiteRed` performs, roughly speaking, the same steps as a person would do. Let us explain this on the example of the two-loop massless onshell vertex shown in Fig. 1. We choose the basis

$$\{D_1, \ldots, D_7\} = \left\{l^2, r^2, (p-l)^2, (q-r)^2, (-l+p+r)^2, (l+q-r)^2, (l-r)^2\right\}. \tag{17}$$

The function $D_7 = (l-r)^2$ corresponds to the irreducible numerator. The diagram in Fig. 1 corresponds to the sector $(1,1,1,1,1,1,0)$ (see the definition of sectors in Ref. [13]).

- First, `LiteRed` solves the IBP and LI identities in the general point $\mathbf{n} = (n_1, n_2, n_3, n_4, n_5, n_6, n_7)$ with respect to the most complex integrals. Then it shifts the indices in the rules found so that they all have the form $J(n_1, n_2, n_3, n_4, n_5, n_6, n_7) \to \ldots$
- After that it analyzes the right-hand side to determine when each rule is applicable. The inapplicability may come from zeros in the denominators, or from the positive shifts in the indices, corresponding to the numerator, e.g., if the right-hand side contains the

integral $J(n_1, n_2, n_3, n_4, n_5, n_6, n_7 + 1)$. Note that it often happens that the latter integral appears with the coefficient n_7 and then does not result in the applicability condition $n_7 \neq 0$. Acting in this way, `LiteRed` finds 9 rules with the following applicability conditions: $\{c_1, \ldots, c_7\} = \{n_6 \neq 1, n_5 \neq 1, \neg(n_3 = 1 \vee n_7 = 0), \neg(n_4 = 1 \vee n_7 = 0), n_6 \neq 1, \neg(n_1 = 1 \vee n_7 = 0), \neg(n_6 = 1 \vee n_7 = 0), \neg(n_2 = 1 \vee n_7 = 0), \neg(n_5 = 1 \vee n_7 = 0)\}$.

- None of the found rules is applicable when the condition $\neg(c_1 \vee \ldots \vee c_7)$ is fulfilled. `LiteRed` reduces this condition to disjunctive normal form: $(n_5 = 1 \wedge n_6 = 1 \wedge n_7 = 0) \vee (n_1 = 1 \wedge n_2 = 1 \wedge n_3 = 1 \wedge n_4 = 1 \wedge n_5 = 1 \wedge n_6 = 1)$.
- Then it takes the first alternative $(n_5 = 1 \wedge n_6 = 1 \wedge n_7 = 0)$ and tries to find the rules for the integral $J(n_1, n_2, n_3, n_4, 1, 1, 0)$. It starts from the IBP and LI identities at the point $(n_1, n_2, n_3, n_4, 1, 1, 0)$ and checks for the possibility to shift indices in order to reduce the found rules to the form $J(n_1, n_2, n_3, n_4, 1, 1, 0) \to \ldots$. In contrast to all-indeterminate case, this is not always possible, so if the appropriate rule is not found, `LiteRed` starts to generate and solve identities in the points neighboring $(n_1, n_2, n_3, n_4, 1, 1, 0)$. In fact, this search is the Laporta algorithm augmented by the procedure which checks for possible shifts.
- When the appropriate rule is found, the condition of its applicability is constructed. In our case, this condition looks like $n_5 = 1 \wedge n_6 = 1 \wedge n_7 = 0 \wedge n_4 \neq 1$.
- Performing the same steps, `LiteRed` finds 17 rules which reduce all integrals in the sector except $J(1, 1, 1, 1, 1, 1, 0)$ which it declares a master.

Note that `LiteRed` succesfully finds the reduction rules for much more complicated cases than the one presented above. In particular, it succeeds for the four-loop massless propagators and some other complicated cases.

5. Short reference guide for LiteRed package

The package is loaded by the command **<<LiteRed`**. Let us describe briefly the most important procedures of `LiteRed` .

Declare[*vars*, *type***]** — variable declaration. Here *vars* — either variable name or the list of variable names, *type* is either **Vector** or **Number**. Be sure to declare all variables that enter the basis definition (see below). Both vectors and numbers can be declared in one construct.
Example: **Declare[{l,q},Vector,mm,Number]**.

NewBasis[*name*, *{D1,D2,...}*, *loopmoms*, *options***]** — definition of the basis. Here *name* is the variable which will be the basis name. It will appear in many commands, associated with a given basis, like **GenerateIBP[***name***]**. Derived objects, like IBP identities, reduction rules, will be associated with this name. Make sure this name is not used anywhere else, but in the appropriate commands. *{D1,D2,...}* is a list of functions D_α. The scalar product is entered as **sp[p1,p2]**. Third argument *loopmoms* is a list of loop momenta. The following options can be appended:

- **Directory->"***dirname***"** — determine the directory, where all basis definitions will be saved.
- **Append->True** — if the set of the functions D_α, given in the first parameter, is not complete, append some automatically chosen numerators.
- **GenerateIBP->True** — generate IBP identities, see the corresponding procedure below.
- **AnalyzeSectors->True** — determine zero and simple sectors, see the corresponding procedure below.

- **`FindSymmetries->True`** — find equivalent sectors, see the corresponding procedure below.

In case of success, **`NewBasis`** creates objects **`Ds[`***name***`]`**, **`SPs[`***name***`]`**, **`LMs[`***name***`]`**, **`EMs[`***name***`]`**, and **`Toj[`***name***`]`**. Their meaning is explained in the output of the **`NewBasis`** procedure. The integrals of the basis are denoted as **`j[`***name*, *n1*, *n2*, ...**`]`**.
Example: **`NewBasis[b1,{sp[l,l]+mm,sp[l-q,l-q]},{l},Directory->"bdir"]`**
Convert the explicit expression to **`j`** form with **`Toj`**:
`Toj[b1,sp[l,q](sp[l,l]+mm)^-1(sp[l-q,l-q])^-2]`
gives **`j[b1,0,2]-j[b1,1,1]+(sp[q,q]-mm)j[b1,1,2]`**. The inverse transformation can be done with **`Fromj[`***expr***`]`**.

`GenerateIBP[`*name***`]`** generates IBP and LI identities for the basis. IBP identities in the point $(n_1, n2, \ldots)$ can be retrieved by **`IBP[`***name***`][`***n1*,*n2*,...**`]`**.

`AnalyzeSectors[`*name*, *pattern***`]`** finds zero sectors and some other objects as it reports in the output. **`ZeroSectors[`***name***`]`** is a list of zero sectors, each element have the form **`js[`***name*, θ_1, θ_2, ...**`]`**. The optional parameter *pattern* tells the procedure to analyze only sectors matching pattern. E.g. if the last two D_α correspond to irreducible numerators, use **`AnalyzeSectors[`***name*, {__, 0, 0}**`]`**. This procedure should be called before the call of **`FindSymmetries`**.

`FindSymmetries[`*name***`]`** finds equivalent sectors and forms. It generates several objects as it reports, including the list of unique sectors **`UniqueSectors[`***name***`]`** and the list **`MappedSectors[`***name***`]`** of sectors which can be mapped onto unique ones. For each mapped sector **`js[`***name*, θ_1, θ_2, ...**`]`** the mapping rules can be retrieved as **`jRules[`***name*, θ_1, θ_2, ...**`]`**.

`SolvejSector[js[`*name*, θ_1, θ_2, ...**`]`**, *options***`]`** is a procedure which performs a heuristic search of the reduction rules for a given sector. If it succeeds, the list of found rules can be retrieved as **`jRules[`***name*, θ_1, θ_2, ...**`]`**. It returns the number of master integrals found. Typical usage is **`SolvejSector/@UniqueSectors[`***name***`]`**. The useful options include

- **`Depth ->`** n set heuristic search depth. Default is $n = 2$
- **`SR -> True`** use internal symmetries of the sector.
- **`TimeConstrained ->`** n set time constraint in seconds.

`DiskSave[`*name***`]`** save all definitions to disk (see option **`Directory`** in **`NewBasis`**).

`IBPReduce[`*expr***`]`** performs the IBP reduction of the expression *expr*.

Several additional tools are included in the package:

`Dinv[j[`*name*, *n1*, *n2*, ...**`],sp[p,q]]`** returns the derivative with respect to the invariant constructed of the external momenta.

`RaisingDRR[`*name*, *n1*, *n2*, ...**`]`** returns the right-hand side of the dimensional recurrence relation (4). Note that the factor $\mu^L = -1$ for Minkovskian metrics and odd number of loops should be taken into account manually.

`LoweringDRR[`*name*, *n1*, *n2*, ...**`]`** returns the right-hand side of the dimensional recurrence relation (5).

`FeynParUF[js[`*name*, θ_1, θ_2, ...**`]]`** returns the list $\{U, F,$ **`{x1,x2,...}`**$\}$, where U and F are the functions entering the parametric representation of the integrals in the given sector, and **`x1,x2,...`** are the parameters.

Learning more One is encouraged to examine the examples that are included in the distribution. Another good starting point to know more about the functions of the package is to submit a command **`?LiteRed`*`**.

6. Conclusion

In this contribution we have reviewed the `LiteRed` package performing the IBP reduction of the multiloop integrals. We have described a new algorithm of detecting the zero sectors implemented in `LiteRed` version 1.4.

Acknowledgments I am grateful to the organizers of ACAT-2013 for the support and hospitality. This work is supported by the Russian Foundation for Basic Research through grant 11-02-01196 and by the Ministry of Education and Science of the Russian Federation.

References

[1] Chetyrkin K G and Tkachov F V 1981 *Nucl. Phys. B* **192** 159
[2] Tkachov F V 1981 *Phys. Lett. B* **100** 65–68
[3] Laporta S 2000 *Int. J. Mod. Phys. A* **15** 5087
[4] Anastasiou C and Lazopoulos A 2004 *J. High Energy Phys.* **0407** 046 (*Preprint* hep-ph/0404258)
[5] Smirnov A V 2008 *J. High Energy Phys.* **10** 107 (*Preprint* 0807.3243)
[6] Smirnov A and Smirnov V 2013 (*Preprint* 1302.5885)
[7] Studerus C 2010 *Comput. Phys. Commun.* **181** 1293–1300 (*Preprint* 0912.2546)
[8] von Manteuffel A and Studerus C 2012 (*Preprint* 1201.4330)
[9] Tarasov O V 1998 *Acta Phys. Polon. B* **29** 2655
[10] Gerdt V P 2004 *Nucl. Phys. Proc. Suppl.* **135** 232–237 (*Preprint* hep-ph/0501053)
[11] Smirnov A V and Smirnov V A 2006 S-bases as a tool to solve reduction problems for feynman integrals (*Preprint* hep-ph/0606247)
[12] Smirnov A V 2006 *J. High Energy Phys.* **0604** 026 (*Preprint* hep-ph/0602078)
[13] Lee R N 2012 (*Preprint* 1212.2685)
[14] Kotikov A V 1991 *Phys. Lett. B* **254** 158–164
[15] Kotikov A V 1991 *Phys. Lett. B* **259** 314–322
[16] Kotikov A V 1991 *Phys. Lett. B* **267** 123–127
[17] Gehrmann T and Remiddi E 2000 *Nucl. Phys. B* **580** 485 (*Preprint* hep-ph/9912329)
[18] Gehrmann T and Remiddi E 2001 *Nucl. Phys. B* **601** 248–286 (*Preprint* hep-ph/0008287)
[19] Gehrmann T and Remiddi E 2001 *Nucl. Phys. B* **601** 287–317 (*Preprint* hep-ph/0101124)
[20] Derkachov S E, Honkonen J and Pis'mak Y M 1990 *J. Phys. A: Mathematical and General* **23** 5563
[21] Tarasov O V 1996 *Phys. Rev. D* **54** 6479 (*Preprint* hep-th/9606018)
[22] Lee R N 2010 *Nucl. Phys. Proc. Suppl.* **205-206** 135–140 (*Preprint* 1007.2256)
[23] Pak A 2012 *J. Phys.: Conf. Series* **368** 012049 (*Preprint* 1111.0868)
[24] Lee R N and Pomeransky A A 2013 (*Preprint* 1308.6676)
[25] Nickel B, Meiron D and Baker G 1977 Compilation of 2-pt. and 4-pt. graphs for continuous spin models Tech. rep. University of Guelph
[26] Lee R N 2008 *J. High Energy Phys.* **07** 031

DESY 13-189, DO-TH 13/28, SFB/CPP-13-78, LPN13-077, arXiv:1310.5645[math-ph]

Generalized Harmonic, Cyclotomic, and Binomial Sums, their Polylogarithms and Special Numbers

J Ablinger[a], J Blümlein[b1], and C Schneider[a]

[a] Research Institute for Symbolic Computation (RISC) Johannes Kepler University, Altenbergerstraße 69, A-4040 Linz, Austria

[b] Deutsches Elektronen-Synchrotron, DESY, Platanenallee 6, D-15738 Zeuthen, Germany

E-mail: Johannes.Bluemlein@desy.de

Abstract. A survey is given on mathematical structures which emerge in multi-loop Feynman diagrams. These are multiply nested sums, and, associated to them by an inverse Mellin transform, specific iterated integrals. Both classes lead to sets of special numbers. Starting with harmonic sums and polylogarithms we discuss recent extensions of these quantities as cyclotomic, generalized (cyclotomic), and binomially weighted sums, associated iterated integrals and special constants and their relations.

1. Introduction

During the late 1990ies several massless and massive two-loop calculations in Quantum Chromodynamics reached a complexity, see e.g. [1–5], which made it necessary to introduce new functions in a systematic manner to represent the analytic results in an adequate form. Dilogarithms, polylogarithms [6–16] and Nielsen integrals [17–19] with complicated arguments did not allow to perform further calculations. Due to this harmonic sums, resp. specific types of Mellin transforms, were independently introduced in [20] and [21] as the basic building blocks. Shortly after the harmonic polylogarithms over the alphabet $\{1/x, 1/(1-x), 1/(1+x)\}$ were found [22]. These are iterated integrals of the Volterra-type having been studied by Poincaré [23–26] more than 100 years before.

Physics expressions, such as massless and massive Wilson coefficients in the asymptotic region to 2-loops, can be expressed in terms of harmonic sums only [27–31]. This also applies to the 3-loop anomalous dimensions [32–35]. However, in the calculation of the massless 3-loop Wilson coefficients in deep-inelastic scattering [36], resp. the massive case [34,35,37], generalized harmonic sums, also called S-sums [38,39], emerge at least in intermediate results. For massive 3-loop graphs $4th$ and $6th$ root of unity weights contribute. At the side of the nested sums they belong to the cyclotomic harmonic sums [40]. Furthermore, root-valued letters occur in the alphabets of iterated integrals [41]. They correspond to binomially-weighted generalized cyclotomic sums. Finally, also elliptic integrals emerge in the calculation of massive Feynman diagrams [42–47]. Special numbers are associated to the above nested sums in the limit $N \to \infty$

[1] Speaker

and the iterated integrals for $x = 1$. In the simplest case these are the multiple zeta values (MZVs) [48]. It is obvious, that more structures are expected to contribute calculating Feynman diagrams at even higher loops and for more legs. In this article we give a brief survey on the structures having been found so far.[2] The *method* to unravel these structures consists in applying modern summation techniques to the multiply nested sums, through which the corresponding Feynman diagrams are represented, and to solve them in difference fields using the algorithms [50–58] being encoded in the `Mathematica`-package `Sigma` [59, 60].

2. Harmonic Sums, Polylogarithms and Multiple Zeta Values

Let us consider the 2-point functions in Quantum Chromodynamics with local operator insertions. Already in case of the quark-quark anomalous dimension the most simple harmonic sum

$$S_1(N) = \sum_{k=1}^{N} \frac{1}{k}, \qquad N \in \mathbb{N} \tag{1}$$

occurs, cf. e.g. [5]. At higher orders more general harmonic sums contribute. They are defined by [20, 21]

$$S_{b,\vec{a}}(N) = \sum_{k=1}^{N} \frac{(\mathrm{sign}(b))^k}{k^{|b|}} S_{\vec{a}}(k), \quad S_\emptyset = 1 \quad b, a_i \in \mathbb{Z}\backslash\{0\}\ . \tag{2}$$

The Mellin transformation [61, 62]

$$\mathbf{M}[f(x)](N) = \int_0^1 dx\, x^N\, f(x) \tag{3}$$

relates harmonic sums to harmonic polylogarithms

$$S_{\vec{a}}(N) = \sum_c r_c \mathbf{M}[\mathrm{H}_{\vec{b}}(x)](N) + \sum_d r_d \sigma_{\vec{c}_d}, \qquad r_c \in \mathbb{Q}\ , \tag{4}$$

with $\sigma_{\vec{c}_d}$ multiple zeta values. The harmonic polylogarithms [22] may be defined as iterated integrals

$$\mathrm{H}_{b,\vec{a}}(x) = \int_0^x \frac{dy}{y-b} \mathrm{H}_{\vec{a}}(y), \qquad b \in \{0,1,-1\}, \qquad \mathrm{H}_\emptyset(x) = 1, \tag{5}$$

$$\mathrm{H}_{\underbrace{0,\ldots,0}_{k}}(x) = \frac{1}{k!} \ln^k(x)\ . \tag{6}$$

An example for relation (4) is

$$\begin{aligned} S_{-2,1,1}(N) &= -(-1)^N \mathbf{M}\left[\frac{\mathrm{H}_{0,1,1}(x)}{x+1}\right](N) + (-1)^N \zeta_3 \mathbf{M}\left[\frac{1}{x+1}\right](N) \\ &\quad - \mathrm{Li}_4\left(\frac{1}{2}\right) - \frac{\ln^4(2)}{24} + \frac{\ln^2(2)\zeta_2}{4} - \frac{7\ln(2)\zeta_3}{8} + \frac{\zeta_2^2}{8}\ . \end{aligned} \tag{7}$$

[2] For a recent review see [49].

Here also special constants occur, either as infinite nested harmonic sums or as values of the harmonic polylogarithms at $x = 1$ as long as these are defined,

$$\sigma_{\vec{a}} = \lim_{N \to \infty} S_{\vec{a}}(N) . \tag{8}$$

They are called multiple zeta values [48]. Since for $N \to \infty$ the Mellin transforms in (7) vanish one obtains

$$\sigma_{-2,1,1} = -\mathrm{Li}_4\left(\frac{1}{2}\right) - \frac{\ln^4(2)}{24} + \frac{\ln^2(2)\zeta_2}{4} - \frac{7\ln(2)\zeta_3}{8} + \frac{\zeta_2^2}{8} . \tag{9}$$

Here

$$\zeta_k = \sum_{l=1}^{\infty} \frac{1}{l^k}, \quad k \geq 2, \quad k \in \mathbb{N}, \qquad \mathrm{Li}_k(x) = \sum_{l=1}^{\infty} \frac{x^l}{l^k} \tag{10}$$

are the values of Riemann's ζ-function and Li_k denotes the polylogarithm. It is useful also to associate the symbol

$$\sigma_0 \equiv \sum_{k=1}^{\infty} \frac{1}{k} \tag{11}$$

to the multiple zeta values. All divergencies of the multiple zeta values can be expressed polynomially by σ_0.[3]

Harmonic sums obey quasi-shuffle algebras [63] as harmonic polylogarithms obey shuffle algebras. These are implied by their multiplication relations at equal argument N resp. x. The shuffle product is given by the sum of all permutations of indices of the two sets, which preserve the original ordering. In case of the quasi-shuffle (stuffle [64]) algebras additional terms occur, cf. [65]. The product of two harmonic polylogarithms is thus given by

$$\mathrm{H}_a(x) \cdot \mathrm{H}_{bcd}(x) = \mathrm{H}_a(x) ⧢ \mathrm{H}_{bcd}(x) = \mathrm{H}_{abcd}(x) + \mathrm{H}_{bacd}(x) + \mathrm{H}_{bcad}(x) + \mathrm{H}_{bcda}(x) . \tag{12}$$

The stuffle and shuffle relations imply relations between the harmonic sums and harmonic polylogarithms, respectively, which do not depend on their arguments N and x and are called algebraic relations [65]. Both these algebras can also be applied to the multiple zeta values. Their action is not identical.

The algebraic relations are not the only relations of the harmonic sums or polylogarithms. Other relations concern the argument of these quantities and are sometimes only valid for sub-classes of indices. They are called structural relations. In case of harmonic sums they are implied by the duplication relation [48] and differentiation w.r.t. N referring to an analytic continuation [21, 66–68]. At a given weight $\mathsf{w} = \sum_i |a_i|$ there are

$$N_{\mathsf{all}}(\mathsf{w}) = 2 \cdot 3^{\mathsf{w}-1} \tag{13}$$

harmonic sums. The algebraic (A), differential (D) and duplication (H) relations lead to

$$N_A(\mathsf{w}) = \frac{1}{\mathsf{w}} \sum_{d|\mathsf{w}} \mu\left(\frac{\mathsf{w}}{d}\right) 3^d, \quad N_D(\mathsf{w}) = 4 \cdot 3^{\mathsf{w}-2}, \quad N_H(\mathsf{w}) = 2 \cdot 3^{\mathsf{w}-1} - 2^{\mathsf{w}-1}, \tag{14}$$

[3] This also holds for the special numbers occurring in case of the cyclotomic harmonic sums.

independent sums individually. Here, $\mu(\xi)$ denotes the Möbius function [69]. Applying these relations one obtains

$$N_{ADH}(\mathsf{w}) = \frac{1}{\mathsf{w}}\sum_{d|\mathsf{w}}\mu\left(\frac{\mathsf{w}}{d}\right)\left[3^d-2^d\right]-\frac{1}{\mathsf{w}-1}\sum_{d|\mathsf{w}-1}\mu\left(\frac{\mathsf{w}-1}{d}\right)\left[3^d-2^d\right] \qquad (15)$$

independent sums. For $\mathsf{w} = 8$ the 4374 harmonic sums can thus be expressed by 486 basic sums.

The harmonic polylogarithms obey a general argument relation under the transformation

$$x \to \frac{1-x}{1+x}. \qquad (16)$$

An example is

$$\begin{aligned} \mathrm{H}_{-1,0,1}\left[\frac{1-x}{1+x}\right] &= -\mathrm{H}_{-1,1}(x)\,(\mathrm{H}_0(x)+\ln(2)) + \mathrm{H}_{-1}(x)\,[\mathrm{H}_{-1,1}(x)+\mathrm{H}_{0,-1}(x) \\ &\quad + \mathrm{H}_{0,1}(x)-\zeta_2] - 2\,\mathrm{H}_{-1,-1,1}(x) - \mathrm{H}_{0,-1,-1}(x) - \mathrm{H}_{0,1,-1}(x) \\ &\quad -\frac{1}{2}\mathrm{H}^2_{-1}(x)\,[\mathrm{H}_0(x)+\ln(2)] + \frac{1}{6}\mathrm{H}^3_{-1}(x) + \ln(2)\zeta_2 - \frac{5}{8}\zeta_3. \end{aligned} \qquad (17)$$

There are other relations for $x \to \{1-x, 1/x, x^2\}$ which are valid for special index sets [22].

For the multiple zeta values over the alphabet $\{0,1,-1\}$ the shuffle and stuffle relations imply all relations up to $\mathsf{w} = 7$. From $\mathsf{w} = 8$ the duplication and from $\mathsf{w} = 12$ the generalized duplication relation [48] leads to new relations. The number of newly contributing basis elements for the lowest weights are, [48],

w	1	2	3	4	5	6	7	8	9	10	11	12
# basis	2	1	1	1	2	2	4	5	8	11	18	25

I.e. up to w= 12 80 basis elements span the multiple zeta values. Up to w= 7 one possible representation reads [20]

$$\left\{\sigma_0, \ln(2), \zeta_2, \zeta_3, \mathrm{Li}_4\left(\frac{1}{2}\right), \zeta_5, \mathrm{Li}_5\left(\frac{1}{2}\right), \mathrm{Li}_6\left(\frac{1}{2}\right), \sigma_{-5,-1}, \zeta_7, \mathrm{Li}_7\left(\frac{1}{2}\right), \sigma_{-5,1,1}, \sigma_{5,-1,-1}\right\}. \qquad (18)$$

It is not proven at present, whether these are all relations. For special sequences of harmonic sums there are further relations, see e.g. [48]. A global property of the MZVs over the alphabet $\{0,1\}$, stating that they can be expressed in terms of MZVs having only indices $a_i = 2,3$, has been conjectured in [70] and recently proven in [71].

3. Cyclotomic Harmonic Sums, Polylogarithms, and Numbers

The denominators $(x-1)$ and $(x+1)$ appearing in the harmonic polylogarithms form the first two cyclotomic polynomials [72]. One may extend the alphabet in allowing all cyclotomic polynomials [40]. They are given by

$$\Phi_n(x) = \frac{x^n-1}{\prod_{d|n,d<n}\Phi_d(x)}, \quad d,n \in \mathbb{N}\backslash\{0\}. \qquad (19)$$

We define the corresponding set of letters by

$$f_0^0(x) = \frac{1}{x}, \quad f_k^l(x) = \frac{x^l}{\Phi_k(x)}, k \in \mathbb{N}\backslash\{0\},\ l \in \mathbb{N},\ l < \varphi(k), \qquad (20)$$

with $\varphi(k)$ being Euler's totient function [72]. A few early examples of Mellin transforms of cyclotomic polylogarithms were given in [73].

Iterating these letters one forms the cyclotomic polylogarithms which obey a shuffle algebra. Applying the Mellin transform (3) one obtains combinations of the cyclotomic harmonic sums and the associated constants. As an extension of the normal harmonic sums, the single cyclotomic sums are given by

$$S_{l,m,n}(N) = \sum_{k=0}^{N} \frac{(\text{sign}(n))^k}{(lk+m)^{|n|}}\,, \tag{21}$$

i.e. harmonic sums with periodic gaps in the terms accounted. By iteration of this structure the general cyclotomic sums are obtained. They occur in the calculation of massive Feynman diagrams. The cyclotomic sums obey algebraic and differentiation relations as well as three multiple argument relations [40], for which counting relations are available.

The special constants being associated to the cyclotomic sums and polylogarithms extend the multiple zeta values. The single sums at $\mathsf{w}=1$ can be expressed by $\sigma_0, \ln(2)$ and π, which replaces $\zeta_2 = \pi^2/6$ as a more fundamental constant. At higher cyclotomy l also the logarithms $\ln(3), \ln(\sqrt{2}-1), \ln(\sqrt{3}-1), \ln(\sqrt{5}-1)$ and several algebraic numbers occur. For $l \leq 6$ and $\mathsf{w} \geq 2$ the basic constants $\zeta_{2k+1}, \psi^{(2k+1)}(1/3)$, $\text{Ti}_{2k}(1), \psi^{(k)}(1/5), \psi^{(2k+1)}(2/5), \psi^{(k)}(1/8), \psi^{(2k)}(1/12)$ contribute. Here ψ denotes the di-gamma function and $\text{Ti}_l(1) = \sum_{k=0}^{\infty}(-1)^k/(2k+1)^l$, with $\text{Ti}_2(1) = \mathbf{C}$ being Catalan's constant [74]. These are the real representations of these constants. Likewise one may consider the infinite generalized harmonic sums with weights at the roots of unity

$$\lim_{N\to\infty} S_{k_1,\ldots,k_m}(x_1,\ldots,x_m;N) \equiv \sigma_{k_1,\ldots,k_m}(x_1,\ldots,x_m), \quad x_j \in \mathcal{C}_n, \ n \geq 1, \ k_1 \neq 1 \text{ for } x_1 = 1, \tag{22}$$

with $\mathcal{C}_n \in \{e_n | e_n^n = 1, e_n \in \mathbb{C}\}$. The real representations being discussed above are related to these complex representations. For lower weights they have been studied for cyclotomy $l \leq 20$ in [40].

4. Generalized Harmonic Sums, Polylogarithms, and Numbers

Generalized harmonic sums are defined by [38, 39]

$$S_{b,\vec{a}}(\zeta,\vec{\xi};N) = \sum_{k=1}^{N} \frac{\zeta^k}{k^b} S_{\vec{a}}(\vec{\xi};k), \quad b, a_i \in \mathbb{N}\backslash\{0\}; \quad \zeta, \xi_i \in \mathbb{R}\backslash\{0\}. \tag{23}$$

The corresponding iterated integrals are built over the alphabet $\{0,\zeta,\vec{\xi}\}$. To also associate the constants, i.e. the sums in the limit $N \to \infty$, one has to restrict the range of weights ζ, ξ_i accordingly to obtain convergent expressions. In intermediate physics results, however, divergent sums for $|\xi_i| > 1$ do occur and have to be dealt with [41]. Known examples refer to alphabets $\xi_i \in \{1,-1,1/2,-1/2,2,-2,1/3,-1/3,3,-3,\ldots\}$. [4] In some applications the weights ξ_i are general real numbers. One may generalize the sums (23) introducing cyclotomic denominators [39]

$$S_{\{a_1,b_1,c_1\},\ldots,\{a_l,b_l,c_l\}}(s_1,\ldots,s_l;N) = \sum_{k_1=1}^{N} \frac{s_1^k}{(a_1k_1+b_1)^{c_1}} S_{\{a_2,b_2,c_2\};\ldots;\{a_l,b_l,c_l\}}(s_2,\ldots,s_l;k_1), \tag{24}$$

[4] We would like to thank W. Bernreuther and O. Dekkers for a remark.

with $S_\emptyset = 1,\ \ a_i, c_i \in \mathbb{N}\backslash\{0\},\ b_i \in \mathbb{N},\ \ s_i \in \mathbb{R}\backslash\{0\}, a_i > b_i$. Also these sums are related to the corresponding polylogarithms by the inverse Mellin transform. The elements of both spaces obey (quasi)shuffle relations and a series of structural relations which were worked out in Ref. [39]. An even wider class of special numbers is associated to the generalized (cyclotomic) harmonic sums and polylogarithms. A convenient way to work with these and the more special functions being listed above is provided by the `Mathematica`-package `HarmonicSums` [39, 75].

5. Nested Binomial Sums

In case of some Feynman diagrams [41] contributing to the massive Wilson coefficient for the structure function $F_2(x, Q^2)$ at higher scales of Q^2 at 3-loop order further extensions of weighted generalized cyclotomic sums occur :

$$\sum_{i=1}^{N} \binom{2i}{i} (-2)^i \sum_{j=1}^{i} \frac{1}{j\binom{2j}{j}} S_{1,2}\left(\tfrac{1}{2}, -1; j\right) \tag{25}$$
$$= \int_0^1 dx \frac{x^N - 1}{x-1} \sqrt{\frac{x}{8+x}} \left[\mathrm{H}^*_{w_{17},-1,0}(x) - 2\mathrm{H}^*_{w_{18},-1,0}(x)\right]$$
$$+ \frac{\zeta_2}{2} \int_0^1 dx \frac{(-x)^N - 1}{x+1} \sqrt{\frac{x}{8+x}} \left[\mathrm{H}^*_{12}(x) - 2\mathrm{H}^*_{13}(x)\right] + c_3 \int_0^1 dx \frac{(-8x)^N - 1}{x + \frac{1}{8}} \sqrt{\frac{x}{1-x}},$$

where

$$w_{12} = \frac{1}{\sqrt{x(8-x)}}, \qquad w_{13} = \frac{1}{(2-x)\sqrt{x(8-x)}},$$
$$w_{17} = \frac{1}{\sqrt{x(8+x)}}, \qquad w_{18} = \frac{1}{(2+x)\sqrt{x(8+x)}}. \tag{26}$$

The iterated integrals H^* are defined here on the interval $[x, 1]$. The new element consists in binomial $\binom{2i}{i}$ terms emerging both in the numerators and denominators of the finite nested sums.[5] About 100 independent nested sums of similar type contribute. The associated iterated integrals request square-root valued alphabets with about 30 new letters, extending those in case of the generalized harmonic polylogarithms. Examples are given in Eqs. (26).

6. Elliptic Integrals

Nested binomial sums, weighting generalized cyclotomic sums, lead to square-root values letters. One may now imagine that Mellin convolutions of these quantities do also emerge, as it happens already for ordinary harmonic polylogarithms. Let us consider a simple case of this kind :

$$T(x) = \frac{1}{\sqrt{1-x}} \otimes \frac{1}{\sqrt{1-x}} = \int_x^1 \frac{dy}{y} \frac{1}{\sqrt{1-y}} \frac{1}{\sqrt{1-\frac{x}{y}}}$$
$$= 2i \left[\mathbf{F}\left(\arcsin\left(\frac{1}{\sqrt{x}}\right), x\right) - \mathbf{K}(x)\right]. \tag{27}$$

[5] Infinite single sums with binomial weights in the numerator and denominator over (generalized) harmonic sums were studied in [76, 77].

It involves the elliptic integrals

$$F(x;k) = \int_0^x \frac{dt}{\sqrt{(1-t^2)(1-k^2t^2)}}, \tag{28}$$

$$K(k) = F(1,k) = \frac{\pi}{2} {}_2F_1\left(\frac{1}{2},\frac{1}{2};1;k^2\right) . \tag{29}$$

On the other hand, the Mellin transform of $T(x)$ yields the following simple expression.

$$\mathbf{M}[T(x)](N) = \int_0^1 dx x^N T(x) = \frac{4^{2N}}{\binom{2N}{N}^2 (N+\frac{1}{2})^2} . \tag{30}$$

Higher powers of the binomial $\binom{2N}{N}$ emerge in Mellin space, which are seemingly one source of elliptic integrals in x-space. One has to contest, that the N-space expression is more simple here.

7. Analytic Continuations the various Sums

In Mellin space one may thoroughly perform the solution of the QCD evolution equations analytically, cf. e.g. [78]. The convolution of the Wilson coefficients with the evolved parton distribution functions is given by a simple product. This representation is therefore preferred in fitting the non-perturbative parton distribution functions. On the numerical side only one final contour integral around the singularities of the problem has to be performed. This requests the analytic continuation of the variable N in the nested sums to $N \in \mathbb{C}$. Observing the crossing relations of the respective process the analytic continuation proceeds either from the even *or* the odd values of N. First the singularities in the complex plane have to be determined. For the harmonic sums and cyclotomic harmonic sums the singularities are located at the non-positive integers and one obtains meromorphic functions. This is not necessarily the case for generalized sums since they may diverge in some cases exponentially as $N \to \infty$. Whenever an asymptotic expansion exists, it can be calculated analytically [39, 68, 75][6] and thus be given at arbitrary precision in principle. Starting with this representation, the shift-property of the nested sums for $N \to N+1$ allows to arrive at any non-singular point in the complex plane using a thoroughly analytic representation to be evaluated numerically. Physical quantities like the massless and the known massive Wilson coefficients and massive operator matrix elements [32–35, 81] possess regular asymptotic representations. For these quantities a corresponding representation is therefore possible. This also applies for the Wilson coefficients of the Drell-Yan process, hadronic Higgs-boson production [27] and time-like quantities [28]. Furthermore, precise representations can be derived also in case cross sections are given numerically only, cf. [82].

8. Conclusions

The mathematical functions expressing Feynman diagrams in N-space form a hierarchy starting with rational functions, harmonic sums, followed by generalized harmonic sums, cyclotomic sums, their generalization, binomially weighted generalized cyclotomic sums, etc. Accordingly, the corresponding iterated integrals and special numbers are organized. The relations of the different quantities can be illustrated by Figure 1 [49].

[6] Precise numerical implementations for the analytic continuations of special Mellin-transforms up to those needed to express the 3-loop anomalous dimensions were given in [79, 80].

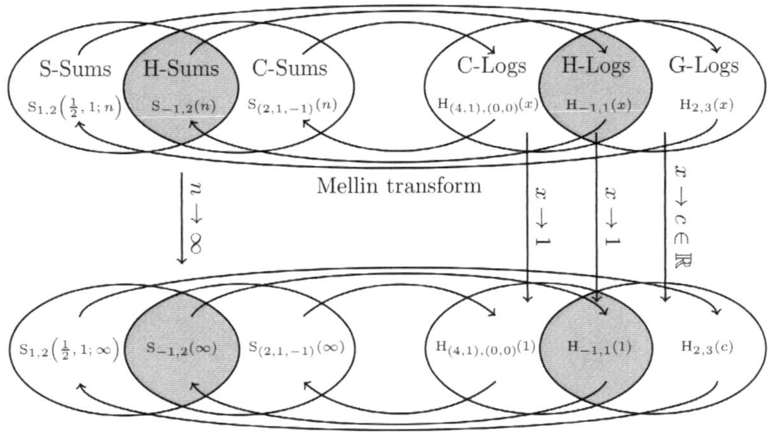

Figure 1. Connection between harmonic sums (H-Sums), S-sums (S-Sums) and cyclotomic harmonic sums (C-Sums), their values at infinity and harmonic polylogarithms (H-Logs), generalized harmonic polylogarithms (G-Logs) and cyclotomic harmonic polylogarithms (C-Logs) and their values at special constants.

The cyclotomic polynomials provide a natural extensions of the letters used with iterated integrals leading to harmonic polylogarithms. Corresponding terms occur in massive higher order calculations. The Mellin transform associates the nested sums and the iterated integrals. Both classes form quasi-shuffle resp. shuffle algebras and obey structural resp. argument-induced relations. Similar relations hold for the different sets of special numbers. One expects an even richer structure in case of multi-leg integrals at higher loop orders, a territory which is widely unexplored still. In this way Feynman diagrams generate a still growing number of new classes of mathematical structures. Knowing their relations greatly helps to simplify the theoretical calculations and also allows for better numerical representations.

Finally we would like to mention that the different functions being described in this paper are encoded in a series of codes which are in mutual use in multi-loop and multi-leg calculations of high-energy processes. A detailed survey on these codes, also of more general character, has been given revently in Ref. [83]. These are for harmonic sums `summer.h` [20] and for harmonic polylogarithms `harmpol.h` [22], `HPL` [84], and `CHAPLIN` [85]. For generalized harmonic sums codes are `nestedsums` [86] and `Xsummer` [87]. An extensive data base for multiple zeta values was given in [48]. A still growing tool allowing for extensive operations with harmonic sums, polylogarithms, generalized harmonic sums, their polylogarithms and numbers, (generalized) cyclotomic sums, polylogarithms and the associated constants, as well as the binomially weighted sums, iterated integrals and the corresponding numbers is `HarmonicSums` [39, 75].

Acknowledgment. We would like to thank A. De Freitas, C. Raab, and J. Vermaseren for discussions. This work has been supported in part by DFG Sonderforschungsbereich Transregio 9, Computergestützte Theoretische Teilchenphysik, by the Austrian Science Fund (FWF) grant P20347-N18, SFB F50 (F5009-N15) and by the EU Network LHCPHENOnet PITN-GA-2010-264564.

References

[1] Zijlstra E B and van Neerven, W L 1994 *Nucl. Phys.* B **417** 61 [Erratum-ibid. B **426** (1994) 245] [Erratum-ibid. B **773** (2007) 105].
[2] Zijlstra E B and van Neerven, W L 1991 *Phys. Lett.* B **273** 476.
[3] van Neerven W L and Zijlstra E B 1991 *Phys. Lett.* B **272** 127.
[4] Zijlstra E B and van Neerven W L 1992 *Nucl. Phys.* B **383** 525.
[5] See also the references given in Blümlein J 2013 *Prog. Part. Nucl. Phys.* **69** 28 [arXiv:1208.6087 [hep-ph]].
[6] Leibniz G W 1696 *Mathematische Schriften*, ed. C.J. Gerhardt, Vol. III, p. 351, (Verlag H.W.Schmidt, Halle, 1858).
[7] Euler L 1768 *Institutiones calculi integralis*, **Vol. I**, (Impensis Academiae Imperialis Scientiarum, Petropoli), 110; Mémoires de l'Académie de Sint-P'etersbourg, Vol. 3 (1809-1810) 26.
[8] Landen J 1760 *Phil. Trans. R. Soc. Lond.* **51** 553; *Mathematical Memoirs*, (Printed for the Author, J. Nourse, London, 1780), 112.
[9] Lewin L 1958 *Dilogarithms and associated functions*, (Macdonald, London).
[10] Kirillov A N 1995 *Prog. Theor. Phys. Suppl.* **118** 61[hep-th/9408113].
[11] Maximon L C 2003 *Proc. R. Soc.* **A459** 2807.
[12] Zagier D 1988 *Journal of Mathematical and Physical Sciences* **22** 131; in P. Cartier, B. Julia, B.; P. Moussa et al., Eds., *Frontiers in Number Theory, Physics, and Geometry II - On Conformal Field Theories, Discrete Groups and Renormalization*, (Springer, Berlin, 2007) 3.
[13] Spence W 1809 *An essay of the theory of the various orders of logarithmic transcendents* (J. Murray, London).
[14] Kummer E E 1840 *Journal für Math. (Crelle)* **21** 74.
[15] Lewin L 1981 *Polylogarithms and associated functions*, (North Holland, New York).
[16] Devoto A and Duke D W 1984 *Riv. Nuovo Cim.* **7N6** 1.
[17] Nielsen N 1909 *Nova Acta Leopold.* **XC** Nr. 3 125.
[18] Kölbig K S, Mignoco J A and Remiddi E 1970 *BIT* **10** 38.
[19] Kölbig K S 1986 *SIAM J. Math. Anal.* **17** 1232.
[20] Vermaseren J A M 1999 *Int. J. Mod. Phys.* A **14** 2037.[hep-ph/9806280].
[21] Blümlein J and Kurth S 1999 *Phys. Rev.* D **60** 014018 [hep-ph/9810241].
[22] Remiddi E and Vermaseren J A M 2000 *Int. J. Mod. Phys.* A **15** 725 [hep-ph/9905237].
[23] Poincaré H 1884 *Acta Math.* **4** 201.
[24] Lappo-Danilevsky J A 1953 *Mémoirs sur la Théorie des Systèmes Différentielles Linéaires*, (Chelsea Publ. Co, New York).
[25] Chen K T 1971 *Trans. A.M.S.* **156** (3) 359.
[26] Jonquière A 1889 *Bihang till Kongl. Svenska Vetenskaps-Akademiens Handlingar* **15** (1889) 1.
[27] Blümlein J and Ravindran V 2005 *Nucl. Phys.* B **716** 128 [hep-ph/0501178].
[28] Blümlein J and Ravindran V 2006 *Nucl. Phys.* B **749** 1 [hep-ph/0604019].
[29] Blümlein J and Klein S 2007 *PoS ACAT* 084 [arXiv:0706.2426 [hep-ph]].
[30] Bierenbaum I, Blümlein J, Klein S and Schneider C 2008 *Nucl. Phys.* B **803** 1 [arXiv:0803.0273 [hep-ph]].
[31] Czakon M, Gluza J and Riemann T 2005 *Phys. Rev.* D **71** 073009 [hep-ph/0412164].
[32] Moch S, Vermaseren J A M and Vogt A 2004 *Nucl. Phys.* B **688** 101 [hep-ph/0403192];
[33] Vogt A, Moch S and Vermaseren J A M 2004 *Nucl. Phys.* B **691** 129 [hep-ph/0404111].
[34] Ablinger J, Blümlein J, Klein S, Schneider C and Wißbrock F 2011 *Nucl. Phys.* B **844** 26 [arXiv:1008.3347 [hep-ph]].
[35] Blümlein J, Hasselhuhn A, Klein S, and Schneider C 2013 *Nucl. Phys.* B **866** 196 [arXiv:1205.4184 [hep-ph]].
[36] Vermaseren J A M, Vogt A and Moch S 2005 *Nucl. Phys.* B **724** 3 [arXiv:hep-ph/0504242] and refences therein.
[37] Ablinger J, Blümlein J, Hasselhuhn A, Klein S, Schneider C and Wißbrock F 2012 *Nucl. Phys.* B **864** 52 [arXiv:1206.2252 [hep-ph]].
[38] Moch S-O, Uwer P and Weinzierl S 2002 *J. Math. Phys.* **43** 3363 [hep-ph/0110083].
[39] Ablinger J, Blümlein J and Schneider C 2013 *J. Math. Phys.* **54** 082301 [arXiv:1302.0378 [math-ph]].
[40] Ablinger J, Blümlein J and Schneider C 2011 *J. Math. Phys.* **52** 102301 [arXiv:1105.6063 [math-ph]].
[41] Ablinger J, Blümlein J, Raab C, Schneider C and Wißbrock F, DESY 13–063.
[42] Laporta S 2002 *Phys. Lett.* B **549** 115 [hep-ph/0210336].
[43] Laporta S 2008 *Int. J. Mod. Phys.* A **23** 5007 [arXiv:0803.1007 [hep-ph]].
[44] Bailey D H, Borwein J M, Broadhurst D and Glasser M L 2008 arXiv:0801.0891 [hep-th].
[45] Müller-Stach S, Weinzierl S and Zayadeh R 2012 arXiv:1212.4389 [hep-ph].
[46] Adams L, Bogner C and Weinzierl S 2013 arXiv:1302.7004 [hep-ph].
[47] Czakon M and Mitov A 2010 *Nucl. Phys.* B **824** 111 [arXiv:0811.4119 [hep-ph]].

[48] Blümlein J, Broadhurst D J and Vermaseren J A M 2010 *Comput. Phys. Commun.* **181** 582 [arXiv:0907.2557 [math-ph]] and references therein.
[49] Ablinger J and Blümlein J 2013 arXiv:1304.7071 [math-ph], in: *Computer Algebra in Quantum Field Theory: Integration, Summation, and Special Functions*, Eds. C. Schneider and J. Blümlein, (Springer, Wien, 2013), p. 1.
[50] Karr M 1981 *J. ACM* **28** 305.
[51] Schneider C 2001 *Symbolic Summation in Difference Fields* Ph.D. Thesis RISC, Johannes Kepler University, Linz technical report 01-17.
[52] Schneider C 2005 *J. Differ. Equations Appl.* **11** 799.
[53] Schneider C 2007 *J. Algebra Appl.* **6** 415.
[54] Schneider C 2008 *J. Symbolic Comput.* **43** 611. [arXiv:0808.2543v1]
[55] Schneider C 2010 *Appl. Algebra Engrg. Comm. Comput.* **21** 1.
[56] Schneider C 2010 *Motives, Quantum Field Theory, and Pseudodifferential Operators (Clay Mathematics Proceedings* **12** ed Carey A, Ellwood D, Paycha S and Rosenberg S (Amer. Math. Soc) 285 arXiv:0808.2543
[57] Schneider C 2010 *Ann. Comb.* **14** (2010) 533.[arXiv:0808.2596]
[58] Schneider C 2013 arXiv:1307.7887 [cs.SC]; arXiv:1310.0160 [cs.SC].
[59] Schneider C 2007 *Sém. Lothar. Combin.* **56** 1,article B56b.
[60] Schneider C 2013 *Computer Algebra in Quantum Field Theory: Integration, Summation and Special Functions* Texts and Monographs in Symbolic Computation eds. C. Schneider and J. Blümlein (Springer, Wien) 325 arXiv:1304.4134 [cs.SC].
[61] Mellin H 1886 *Acta Soc. Fennicae* **21** (1886) 1.
[62] Mellin H 1902 *Acta Math.* **25** 139.
[63] Hoffman M E 2000 *J. Algebraic Combin.*, **11** 49 [arXiv:math/9907173 [math.QA]].
[64] Borwein J M, Bradley D M, Broadhurst D J and Lisonek P (2001) *Trans. Am. Math. Soc.* **353** (907 [math/9910045 [math-ca]].
[65] Blümlein J 2004 *Comput. Phys. Commun.* **159** 19 [hep-ph/0311046].
[66] Blümlein J 2009 *Comput. Phys. Commun.* **180** 2218 [arXiv:0901.3106 [hep-ph]].
[67] Blümlein J 2010 in : A. Carey, D. Ellwood, S. Paycha, and S. Rosenberg, editors, *Motives, Quantum Field Theory, and Pseudodifferential Operators*, **12** of *Clay Mathematics Proceedings, Amer. Math. Soc.*, pp. 167 arXiv:0901.0837 [math-ph].
[68] Ablinger J, Blümlein J and Schneider C, DESY 13–064.
[69] See e.g. Hardy G H and Wright E M 2002 *An Introduction to the Theory of Numbers*, (Calendron Press, Oxford).
[70] Hoffman M E 1997 *Journ. of Algebra* **194** 477.
[71] Brown F 2012 *Annals of Math.* **175** (1) 949.
[72] Lang S 2002 *Algebra*, (Springer, New York), 3rd edition.
[73] Nielsen N 1906 *Handbuch der Theorie der Gammafunktion*, (Teubner, Leipzig,); reprinted by (Chelsea Publishing Company, Bronx, New York, 1965).
[74] Catalan E 1883 *Mémoires de l'Academie imperiale des sciences de Saint-Pétersbourg*, Ser. **7** 1.
[75] Ablinger A 2013 arXiv:1305.0687 [math-ph].
[76] Davydychev I A and Kalmykov M Y 2004 *Nucl. Phys.* B **699** 3 [hep-th/0303162].
[77] Weinzierl S 2004 *J. Math. Phys.* **45** 2656 [hep-ph/0402131].
[78] Blümlein J and Vogt A 1998 *Phys. Rev.* D **58** 014020 [hep-ph/9712546].
[79] Blümlein J 2000 *Comput. Phys. Commun.* **133** 76 [hep-ph/0003100].
[80] Blümlein J and Moch S O 2005 *Phys. Lett.* B **614** 53 [hep-ph/0503188].
[81] Ablinger J, Blümlein J, De Freitas A, Hasselhuhn A, von Manteuffel A, Round M, Schneider C and Wißbrock F, DESY 13–185.
[82] Alekhin S I and Blümlein J 2004 *Phys. Lett.* B **594** 299 [hep-ph/0404034].
[83] Blümlein J 2011 *PoS RADCOR* **2011** 048 [arXiv:1205.4991 [hep-ph]].
[84] Maitre D 2006 *Comput. Phys. Commun.* **174** 222 [hep-ph/0507152].
[85] Buehler S and Duhr C 2011 arXiv:1106.5739 [hep-ph].
[86] Weinzierl S 2002 *Comput. Phys. Commun.* **145** 357 [math-ph/0201011].
[87] Moch S and Uwer P 2006 *Comput. Phys. Commun.* **174** 759 [math-ph/0508008].

Multi-loop Integrand Reduction with Computational Algebraic Geometry

Simon Badger[1], **Hjalte Frellesvig**[2,3], **Yang Zhang**[3]

[1] Theory Division, Physics Department, CERN, CH-1211 Geneva 23, Switzerland

[2] Instituto Nazionale di Fisica Nucleare, Sezione di Roma, P.le Aldo Moro 2, 00185 Roma, Italy

[3] Niels Bohr International Academy and Discovery Center, The Niels Bohr Institute, University of Copenhagen, Blegdamsvej 17, DK-2100 Copenhagen, Denmark

E-mail: simon.badger@cern.ch

Abstract. We discuss recent progress in multi-loop integrand reduction methods. Motivated by the possibility of an automated construction of multi-loop amplitudes via generalized unitarity cuts we describe a procedure to obtain a general parameterisation of any multi-loop integrand in a renormalizable gauge theory. The method relies on computational algebraic geometry techniques such as Gröbner bases and primary decomposition of ideals. We present some results for two and three loop amplitudes obtained with the help of the MACAULAY2 computer algebra system and the Mathematica package BASISDET.

1. Introduction

There has been huge progress recently in the programme to fully automate one-loop amplitude computations. Much of this success has been possible thanks to the development of new reductions methods such as unitarity [1, 2], generalized unitarity [3] and integrand reduction [4] which allow efficient generation of one-loop amplitudes from simpler tree-level building blocks. We refer to G. Ossola's contribution to these proceedings for more details of the status of this field [5].

There has been some interest in extending these techniques to multi-loop amplitudes in the hope of improving our predictions of multi-particle production beyond $2 \to 2$ scattering processes. Thanks to developments in infra-red subtraction techniques full NNLO QCD computations with coloured initial and final states have been now been achieved for $pp \to H + j$ [6], $pp \to t\bar{t}$ [7] and $pp \to 2j$ [8] which give hope that a similar level of precision prediction may also be possible for $2 \to 3$ processes in the near future. In order to combat the rapid growth of the Feynman diagrams in such computations, on-shell techniques only work with physical degrees of freedom and simplify intermediate steps.

The traditional approach to multi-loop Feynman diagram computations using integration-by-parts identities has been well studied since its proposal more than 30 years ago [9]. Several automated public codes have been written and new developments are constantly increasing the range of applicability, see for example the contribution of R. Lee to these proceedings [10, 11]. Nevertheless complete solutions for multi-scale problems are limited and are computationally intensive.

Since a complete integral basis for a general multi-loop amplitude is still unknown it has been difficult to proceed with the conventional unitarity construction of loop amplitudes. Nevertheless progress in understanding the planar basis by explicit construction of unitarity compatible IBP relations [12, 13]. This procedure has been a key ingredient in the maximal unitarity method proposed by Kosower and Larsen [14], where master integral coefficients are computed directly from on-shell cuts into tree-level amplitudes after finding a specific contour of a complex integral. The method has now been applied to a number of high-multiplicity cases of the two-loop planar double box [15, 16, 17] and also to non-planar topologies [18].

The extension of the OPP integrand reduction method was first proposed by Mastrolia and Ossola [19]. A key insight in how to construct a general procedure has come from algebraic geometry. As we will see, the problem reduces to operations on (quadratic) polynomial equations with many variables. Such operations are well known in mathematics and many techniques, and programs, now exist for dealing with computational algebraic geometry problems. The most well known of these is Buchberger's algorithm for the construction of Gröbner bases used in multivariate polynomial division. This direction has been explored both in the context of reducing Feynman diagrams [20, 21, 22] and using generalized unitarity [23, 24, 25]. More formal applications of computational algebraic geometry have also bee used to analyze the general structure of two and higher loop amplitudes [26, 27].

In these proceedings we present a brief overview of the methods presented in refs. [23, 24, 25]. We describe the integrand reduction procedure valid in D-dimensions for an arbitrary loop amplitude. We outline applications at both two and three loops and consider the outlook for the future.

2. Multi-loop integrand reduction with computational algebraic geometry

The multi-loop amplitudes in this paper will refer to colour ordered primitive amplitudes which have a fixed ordering for the external legs and a well defined set of internal propagators. An L-loop amplitude in D dimensions can be schematically represented as,

$$A_n^{(L),[D]} = \int \prod_{i=1}^{L} \frac{d^D k_i}{(2\pi)^D} \frac{N(\{k\},\{p\})}{\prod_{l=1}^{L(L+9)/2} D_l(\{k\},\{p\})}, \tag{1}$$

where D_l are the denominators appearing in the parent topology, $\{k\}$ are the loop integration variables and $\{p\}$ are the external momenta. The numerator function N is a function of Lorentz products of loop momenta and external momenta, external polarization vectors or external spinor wave-functions. All kinematic quantities can be represented in terms of invariants or scalar products of loop momenta with vectors in a basis $\{e_1, \ldots, e_4\}$ which can be constructed from the independent external momenta plus additional spurious directions as proposed by van Neerven and Vermaseren [28]. We will write all scalar products as $k_i \cdot e_j = x_{ij}$. We must also include extra-dimensional variables $\mu_{ij} = -k_i^{[-2\epsilon]} \cdot k_j^{[-2\epsilon]}$ where $k_i = \bar{k}_i + k_i^{[-2\epsilon]}$ and $\bar{k}_i$ is the four-dimensional components of k_i. The number of propagators in $D = 4 - 2\epsilon$ dimensions is $L(L+9)/2$ coming from the $4L$ components of the 4-dimensional loop momenta plus an additional $L(L+1)/2$ components needed to describe the extra dimensional terms.

The integrand reduction method begins by reducing the numerator into terms proportional to the denominators D_l plus a remainder which is integrand of the topology in question,

$$N(\{k\},\{p\}) = \Delta_M(\{k\},\{p\}) + \sum_{l=1}^{M} a_l D_l(\{k\},\{p\}) \tag{2}$$

where $M = \min(n + 3(L-1), L(L+9)/2)$ represents the maximum cut topology. The sum of terms proportional to denominators D_l will all vanish when all propagators are put on-shell.

To perform this operation we must first parametrise the integrand Δ_M in terms of rational coefficients multiplying monomials of scalar products,

$$\Delta_M(\{k\},\{p\}) = \sum_k c_k \times \prod_{i=1}^{L}\prod_{j=1}^{4} x_{ij}^{\alpha_{k;ij}} \prod_{l=1}^{L}\prod_{m=l}^{L}(\mu_{lm})^{\beta_{k,lm}}, \tag{3}$$

where c_i are rational functions of the external kinematics and the $\alpha_{k;ij}$ are integer powers of the Lorentz products x_{ij} and $\beta_{k,lm}$ are the powers of the extra dimensional parameters μ_{lm}. The notation used in eq. (3) includes an over complete set of scalar products (i.e. the matrices $\alpha_{k,ij}$ will contain many zeros). For this reason it is necessary to distinguish reducible scalar products (RSPs), which are linear in the propagators, from the irreducible scalar products (ISPs) which satisfy non-linear identities. The basis of ISPs can be constructed from the system of propagators as we will describe in the section 2.1.

Having obtained this representation we can then proceed to compute the coefficients c_k from products of tree-level amplitudes. This requires solving the system of on-shell constraints $D_l(\{k\},\{p\}) = 0$ which will in general have many possible families of solutions $\{k^{(s)}\}$. Each of these solutions can be parametrised in terms of a number of free variables, $\{\tau_1,\ldots,\tau_{L(L+9)/2-M}\}$ [1]. On each cut solution the integrand factorises into a product of tree-level amplitudes that can be written as a polynomial in τ_i,

$$\Delta_M(\{k^{(s)}\},\{p\}) = \prod_{V=1}^{M-1} A^{(0)}(V,\{\{k^{(s)}\}\}), \tag{4}$$

$$= \sum_i d_i \prod_{j=1}^{L(L+9)/2-M} \tau_j^{\gamma_{k;j}}, \tag{5}$$

where we have used a schematic notation for the on-shell tree-level amplitude appearing at each vertex of the cut-topology, $A^{(0)}(V,\{\{k^{(s)}\}\})$. $\gamma_{k;j}$ are integer powers that follow from $\alpha_{k;ij}$ and $\beta_{k;lm}$. Combining the information obtained in eqs. (3) and (5) allows to construct a linear system relating the rational coefficients of the integrand with the coefficients d_i that can be obtained from tree-level amplitudes,

$$B\cdot\vec{c} = \vec{d}, \tag{6}$$

where $\vec{d} = (d_1^{(1)},\ldots,d_{n(1)}^{(1)},d_1^{(2)},\ldots,d_{n(S)}^{(S)})$ and $n(s)$ is the dimension of the on-shell solution s of which there are total of S solutions. Solving this system then determines the coefficients c_i.

After the form of Δ_M has been determined we can proceed with lower propagator cuts by performing the OPP style subtraction of the singularities,

$$\Delta_m(\{k^{(s)}\},\{p\}) = \prod_{V=1}^{m-1} A^{(0)}(V,\{\{k^{(s)}\}\}) - \sum_{T=m+1}^{M}\sum_{t\in T}\frac{\Delta_{T;t}(\{k^{(s)}\},\{p\})}{\prod_{l\in P(t)/P(m)} D_l(\{k^{(s)}\},\{p\})} \tag{7}$$

where T is the number of propagators in the higher topologies. For each multiplicity we must sum over all the possible topologies t each of which will have a set of additional propagators in comparison with the topology m in question. The sum is represented as $P(t)/P(m)$.

If we restrict to the case of 4 dimensions $M = \min(n+3(L-1),4L)$ and all coefficients $\beta_{k,lm} = 0$. There are two important details that need to be clarified before we can show that such an approach will work:

[1] The existence of this parametrization is only guaranteed in $D = 4-2\epsilon$ dimensions. In cases where the cut equations have genus>1 this form will not exist (for example the 10-leg two-loop double box with 2 massless momenta at each of the 6 vertices).

(i) How do we define the integrand parameterisation? (i.e. compute $\alpha_{k,ij}$ and $\beta_{k,lm}$)

(ii) How can we determine the number of on-shell solutions S to define $\vec{d}$?

2.1. Integrand parameterisation

The solution to first question above has been presented using multi-variate polynomial division with Gröbner basis Zhang [24] and also Mastrolia, Mirabella, Ossola and Peraro [20].

In this section we will construct, using the language of algebraic geometry, an ideal from the set of propagators for a topology with m denominators,

$$P = \{D_1, \cdots, D_m\} \tag{8}$$

whose integrand we will denote as Δ_m. Using the van Neerven Vermaseren construction we are able to construct a basis where external momenta $\{p_1, \ldots, p_x\}$ span the physical space and the directions $\{\omega_1, \cdots, \omega_4\}$, orthogonal to the physical space, span the spurious, or trivial, space:

$$e = \{e_1, \cdots, e_4\}. \tag{9}$$

All of the Lorentz products appearing in the expansion of P can be written in terms of the basis products $k_i \cdot e_j = x_{ij}$ using the 4×4 Gram matrix $[G_4]_{ij} = e_i \cdot e_j$,

$$k_i \cdot v_j = (k_i \cdot e_1\, k_i \cdot e_2\, k_i \cdot e_3\, k_i \cdot e_4).G_4^{-1}. \begin{pmatrix} v_j \cdot e_1 \\ v_j \cdot e_2 \\ v_j \cdot e_3 \\ v_j \cdot e_4 \end{pmatrix} \tag{10}$$

and

$$k_i \cdot k_j = -\mu_{ij} + (k_i \cdot e_1\, k_i \cdot e_2\, k_i \cdot e_3\, k_i \cdot e_4).G_4^{-1}. \begin{pmatrix} k_j \cdot e_1 \\ k_j \cdot e_2 \\ k_j \cdot e_3 \\ k_j \cdot e_4 \end{pmatrix} \tag{11}$$

The above assumes that we have all external vectors in four dimensions and hence only obtain contributions from the $4D$ parts of the loop momenta $k_i \cdot e_j = \bar{k}_i \cdot e_j$. To find the constraints from the propagators on the integrand we will use eqs. (10) and (11) to translate eq. (8) into polynomials in x_{ij}. The RSPs could be identified from linear relations among the equations however an efficient method is to compute the Gröbner basis using a graded monomial ordering [24].

It is helpful to consider an example at this point so we take the case of the two-loop planar double box in four dimensions. The list of propagators is,

$$P = \{k_1^2, (k_1 - p_1)^2, (k_1 - p_1 - p_2)^2, k_2^2, (k_2 - p_4)^2, (k_2 - p_3 - p_4)^2, (k_1 + k_2)^2\} \tag{12}$$

and we will span k_1 and k_2 with the basis,

$$\vec{e} = (p_1, p_2, p_4, \omega) \tag{13}$$

where $\omega^\nu = 4i\varepsilon^{\nu 124}$. The Gram matrix is explicitly

$$G_4 = \frac{1}{2} \begin{pmatrix} 0 & s & t & 0 \\ s & 0 & u & 0 \\ 0 & u & t & 0 \\ 0 & 0 & 0 & stu \end{pmatrix} \tag{14}$$

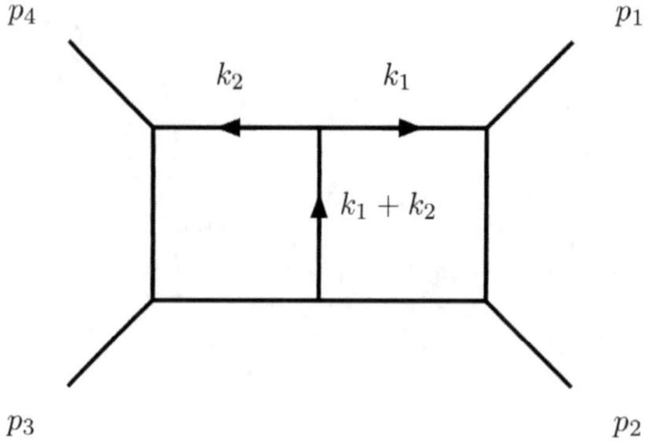

Figure 1. The two-loop planar double box configuration

where $u = -s - t$. It is then straightforward to apply eq. (10) and express P in terms of the 8 scalar products x_{ij}. We find four linear relations in the on-shell equations:

$$x_{11} = 0 \qquad 2x_{12} + s = 0 \qquad x_{23} = 0 \qquad 2x_{22} + 2x_{21} + s = 0 \tag{15}$$

which leaves us with a set of 4 ISPs $\{x_{24}, x_{14}, x_{21}, x_{13}\}$ and an ideal

$$I = \big\langle -4tx_{13} + 4x_{13}^2 - 4x_{14}^2 + t^2, -4tx_{21} + 4x_{21}^2 - 4x_{24}^2 + t^2,\\ st\,(2x_{13} + 2x_{21}) + s\,(4x_{13}x_{21} - 4x_{14}x_{24}) + 8tx_{13}x_{21} - st^2\big\rangle \tag{16}$$

which we should use to remove terms from the polynomial ansatz for Δ_7 restricted by imposing renormalizability constraints,

$$\Delta_7^{\text{ansatz}} = \sum_{a_1,\ldots,a_4} c_{a_1a_2a_3a_4}\, x_{14}^{a_1} x_{24}^{a_2} x_{13}^{a_3} x_{24}^{a_4} \tag{17}$$

where $a_1 + a_3 \leq 4, a_2 + a_4 \leq 4$ and $a_1 + a_2 + a_3 + a_4 \leq 6$ which is a sum of 160 terms. The polynomial division operation based on Gröbner bases is a standard algorithm in computer algebra packages and so it is now straightforward to obtain:

$$\Delta_7^{\text{ansatz}}/G(I) = \Delta_7 = \sum_{i=1} c_i\, x_{14}^{\alpha_{i1}} x_{24}^{\alpha_{i2}} x_{13}^{\alpha_{i3}} x_{21}^{\alpha_{i4}} \tag{18}$$

which contains 32 terms defined by:

$$\alpha = ((0,0,0,0),(0,0,0,1),(0,0,0,2),(0,0,0,3),(0,0,0,4),\\ (0,0,1,0),(0,0,1,1),(0,0,1,2),(0,0,1,3),(0,0,1,4),\\ (0,0,2,0),(0,0,2,1),(0,0,3,0),(0,0,3,1),(0,0,4,0),\\ (0,0,4,1),(0,1,0,0),(0,1,0,1),(0,1,0,2),(0,1,0,3),\\ (0,1,1,0),(0,1,1,1),(0,1,1,2),(0,1,1,3),(0,1,2,0),\\ (0,1,3,0),(0,1,4,0),(1,0,0,0),(1,0,0,1),(1,0,1,0),\\ (1,0,2,0),(1,0,3,0)) \tag{19}$$

The algorithm presented here is implemented in the public Mathematica package BasisDet by Yang Zhang. We note the the problem of determining the integrand parameterisation is rather efficient, especially when external kinematic quantities are computed numerically, and has been applied to many complicated multi-loop topologies in 4 and $4-2\epsilon$ dimensions. The integrand basis can be rather large for higher loop topologies. The three-loop triple box, shown in figure 2, considered in ref. [25] contains a total 398 monomials in terms of 7 ISPs. In this case IBP reduction was used subsequently to reduce the integrand onto a basis of three master integrals.

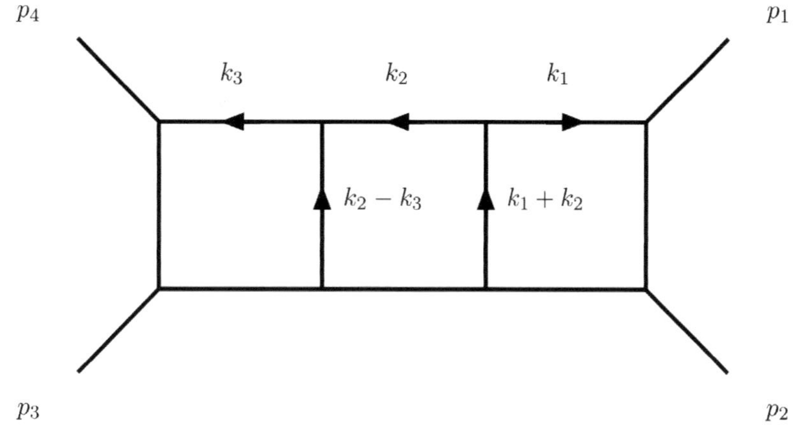

Figure 2. The three-loop triple box configuration

2.2. On-shell solutions

The solution to the second question has been formulated by Zhang in ref. [24]. In general the geometry of the solutions space can be very complicated but algebraic geometry again provides a technique to systematically analyze the zeros of the propagator ideal I, which will denote $\mathcal{Z}(I)$ or the zero locus of I. The famous Lasker-Noether theorem for primary decomposition of ideals proves that any ideal may be decomposed into a set of smaller primary ideals,

$$I = \bigcap_{s=1}^{S} I_s \tag{20}$$

where $\dim(I_s) = n(s)$. The zero locus of I follows from this since,

$$\mathcal{Z}(I) = \bigcup_{s=1}^{S} \mathcal{Z}(I_s). \tag{21}$$

Algorithms to compute the primary decomposition are available in the MACAULAY2 computer algebra package [29]. For the two loop double box considered in the previous section one can use this technique to reproduce the 6 independent solutions first obtained by Kosower and Larsen [14]. The three-loop triple box has a total of 14 solutions and explicit parameterisation have been presented in ref. [25].

2.3. D-dimensional cuts

One more aspect of the linear system in eq. (6) that remains is to prove that the matrix B does indeed have the property $\text{rank}(M) = \dim(\vec{c})$. This is not guaranteed in four dimensions. However, as shown recently in ref. [30], by going to D dimensions this problem is avoided at the cost of creating larger linear systems relating the integrand coefficients to the tree-level amplitudes. We have observed two cases where the procedure can break down:

(i) The propagator ideal is not radical, often written $I \neq \sqrt{I}$. In this case there will be terms in Δ that cannot be determined from the product of trees as they will vanish on each cut solution $\{k^{(s)}\}$.

(ii) The dimension of a branch of the cut solutions is not regular. For example, in four dimensions, we can find that $\dim(I_i) \neq 4L - m$ for all primary ideals in an m propagator cut. The two-loop pentagon-triangle is an example of this: though there are seven propagators we find two families of solutions with dimension 2 instead of dimension 1.

There is a possibility these problems could be avoided by considering different topologies simultaneously but it will complicate the integrand reduction procedure. In $4 - 2\epsilon$ dimensions however this problem is side stepped since one can prove that all ideals are radical ideals [30]. In this case there will always be exactly one solution to the on-shell constraints ($S = 1$).

In order to generate the extra-dimensional dependence in the tree amplitudes it is necessary at two-loops to use at least six dimensions. In general tree amplitudes will need to be computed in higher and higher dimensions as we increase the loop order.

3. Outlook

We have presented an overview of an integrand reduction procedure valid in D dimensions which allows the rational coefficients of the integrand monomials to be extracted from tree-level amplitudes. The method has been applied in the recent computation of the planar all-plus helicity five-gluon two-loop amplitude, the first example of a $2 \rightarrow 3$ helicity amplitude in QCD [30].

There are still many questions as to whether the method can be an efficient alternative to integration-by-parts identities in the future. Of course the integrand basis can always be further reduced by IBP to reduce the number of integrals that must be computed. There is also an open question concerning the treatment of doubled propagator diagrams that can appear in the input. In this case it has been shown that there is no obstacle in performing the polynomial division [22] yet the on-shell cuts into tree amplitudes will need regulating in order to follow the approach suggested here.

The use of algebraic geometry in multi-loop computations may have many applications in the future both for integrand computations and in the computations of the integrals themselves. Progress in the computation of the loop integrals themselves, for example the differential equation of Henn [31], give hope that precision predictions of multi-particle states at NNLO may be possible within the near future.

References

[1] Bern Z, Dixon L J, Dunbar D C and Kosower D A 1994 *Nucl.Phys.* **B425** 217–260 (*Preprint* hep-ph/9403226)
[2] Bern Z, Dixon L J, Dunbar D C and Kosower D A 1995 *Nucl.Phys.* **B435** 59–101 (*Preprint* hep-ph/9409265)
[3] Britto R, Cachazo F and Feng B 2005 *Nucl.Phys.* **B725** 275–305 (*Preprint* hep-th/0412103)
[4] Ossola G, Papadopoulos C G and Pittau R 2007 *Nucl.Phys.* **B763** 147–169 (*Preprint* hep-ph/0609007)
[5] Ossola G 2013 (*Preprint* 1310.3214)
[6] Boughezal R, Caola F, Melnikov K, Petriello F and Schulze M 2013 *JHEP* **1306** 072 (*Preprint* 1302.6216)
[7] Czakon M, Fiedler P and Mitov A 2013 *Phys.Rev.Lett.* **110** 252004 (*Preprint* 1303.6254)
[8] Ridder A G D, Gehrmann T, Glover E and Pires J 2013 *Phys.Rev.Lett.* **110** 162003 (*Preprint* 1301.7310)
[9] Chetyrkin K and Tkachov F 1981 *Nucl.Phys.* **B192** 159–204
[10] Lee R 2012 (*Preprint* 1212.2685)
[11] Lee R N 2013 (*Preprint* 1310.1145)
[12] Gluza J, Kajda K and Kosower D A 2011 *Phys.Rev.* **D83** 045012 (*Preprint* 1009.0472)
[13] Schabinger R M 2012 *JHEP* **1201** 077 (*Preprint* 1111.4220)
[14] Kosower D A and Larsen K J 2012 *Phys.Rev.* **D85** 045017 (*Preprint* 1108.1180)
[15] Caron-Huot S and Larsen K J 2012 *JHEP* **1210** 026 (*Preprint* 1205.0801)
[16] Johansson H, Kosower D A and Larsen K J 2013 *Phys.Rev.* **D87** 025030 (*Preprint* 1208.1754)
[17] Johansson H, Kosower D A and Larsen K J 2013 (*Preprint* 1308.4632)

[18] Søgaard, Mads 2013 *JHEP* **1309** 116 (*Preprint* 1306.1496)
[19] Mastrolia P and Ossola G 2011 *JHEP* **1111** 014 (*Preprint* 1107.6041)
[20] Mastrolia P, Mirabella E, Ossola G and Peraro T 2012 *Phys.Lett.* **B718** 173–177 (*Preprint* 1205.7087)
[21] Mastrolia P, Mirabella E, Ossola G and Peraro T 2013 *Phys.Rev.* **D87** 085026 (*Preprint* 1209.4319)
[22] Mastrolia P, Mirabella E, Ossola G and Peraro T 2013 (*Preprint* 1307.5832)
[23] Badger S, Frellesvig H and Zhang Y 2012 *JHEP* **1204** 055 (*Preprint* 1202.2019)
[24] Zhang Y 2012 *JHEP* **1209** 042 (*Preprint* 1205.5707)
[25] Badger S, Frellesvig H and Zhang Y 2012 *JHEP* **1208** 065 (*Preprint* 1207.2976)
[26] Feng B and Huang R 2013 *JHEP* **1302** 117 (*Preprint* 1209.3747)
[27] Huang R and Zhang Y 2013 *JHEP* **1304** 080 (*Preprint* 1302.1023)
[28] van Neerven W and Vermaseren J 1984 *Phys.Lett.* **B137** 241
[29] Grayson D R and Stillman M E Macaulay2, a software system for research in algebraic geometry Available at http://www.math.uiuc.edu/Macaulay2/
[30] Badger S, Frellesvig H and Zhang Y 2013 (*Preprint* 1310.1051)
[31] Henn J M 2013 *Phys.Rev.Lett.* **110** 251601 (*Preprint* 1304.1806)

On applications of Mathematica Package "FAPT" in QCD

Viacheslav Khandramai

Gomel State Technical University, Gomel 246746, Belarus

E-mail: v.khandramai@gmail.com

Abstract. We consider computational problems in the framework of nonpower Analityc Perturbation Theory and Fractional Analytic Perturbation Theory that are the generalization of the standard QCD perturbation theory. The singularity-free, finite couplings $\mathcal{A}_\nu(Q^2), \mathfrak{A}_\nu(s)$ appear in these approaches as analytic images of the standard QCD coupling powers $\alpha_s^\nu(Q^2)$ in the Euclidean and Minkowski domains, respectively. We provide a package "FAPT" based on the system Mathematica for QCD calculations of the images $\mathcal{A}_\nu(Q^2)$, $\mathfrak{A}_\nu(s)$ up to N^{3}LO of renormalization group evolution. Application of these approaches to Bjorken sum rule analysis and Q^2-evolution of higher twist μ_4^{p-n} is considered.

1. Introduction

The QCD perturbation theory (PT) in the region of space-like momentum transfer $Q^2 = -q^2 > 0$ is based on expansions in a series in powers of the running coupling $\alpha_s(\mu^2 = Q^2)$ which in the one-loop approximation is given by $\alpha_s^{(1)}(Q^2) = (4\pi/b_0)/L$ with b_0 being the first coefficient of the QCD beta function, $L = \ln(Q^2/\Lambda^2)$, and Λ is the QCD scale. The one-loop solution $\alpha_s^{(1)}(Q^2)$ has a pole singularity at $L = 0$ called the Landau pole. The ℓ-loop solution $\alpha_s^{(\ell)}(Q^2)$ of the renormalization group (RG) equation has an ℓ-root singularity of the type $L^{-1/\ell}$ at $L = 0$, which produces the pole as well in the ℓ-order term $d_\ell\, \alpha_s^\ell(Q^2)$. This prevents the application of perturbative QCD in the low-momentum space-like regime, $Q^2 \sim \Lambda^2$, with the effect that hadronic quantities, calculated at the partonic level in terms of a power-series expansion in $\alpha_s(Q^2)$, are not everywhere well defined.

In 1997, Shirkov and Solovtsov discovered couplings $\mathcal{A}_1(Q^2)$ free of unphysical singularities in the Euclidean region [1], and Milton and Solovtsov discovered couplings $\mathfrak{A}_1(s)$ in the Minkowski region [2]. Due to the absence of singularities of these couplings, it is suggested to use this systematic approach, called Analytic Perturbation Theory (APT), for all Q^2 and s. The APT yields a sensible description of hadronic quantities in QCD (see reviews [3–5]), though there are alternative approaches to the singularity of effective charge in QCD — in particular, with respect to the deep infrared region $Q^2 < \Lambda^2$. One of the main advantages of the APT analysis is much faster convergence of the APT nonpower series as compared with the standard PT power series (see [6]). Recently, the analytic and numerical methods, necessary to perform calculations in two- and three-loop approximations, were developed [7–9]. The APT approach was applied to calculate properties of a number of hadronic processes, including the width of the inclusive τ lepton decay to hadrons [10–14], the scheme and renormalization-scale dependencies in the Bjorken [15, 16] and Gross–Llewellyn Smith [17] sum rules, the width of Υ meson decay to hadrons [18], meson spectrum [19], etc.

The generalization of APT for the fractional powers of an effective charge was done in [20, 21] and called the Fractional Analytic Perturbation Theory (FAPT). The important advantage of FAPT in this case is that the perturbative results start to be less dependent on the factorization scale. This reminds the results obtained with the APT and applied to the analysis of the pion form factor in the $O(\alpha_s^2)$ approximation, where the results also almost cease to depend on the choice of the renormalization scheme and its scale (for a detailed review see [22] and references therein). The process of the Higgs boson decay into a $b\bar{b}$ pair of quarks was studied within a FAPT-type framework in the Minkowski region at the one-loop level in [23] and within the FAPT at the three-loop level in [21]. The results on the resummation of nonpower-series expansions of the Adler function of scalar D_S and a vector D_V correlators within the FAPT were presented in [24]. The interplay between higher orders of the perturbative QCD expansion and higher-twist contributions in the analysis of recent Jefferson Lab data on the lowest moment of the spin-dependent proton structure function, $^{p}_{1}(Q^2)$, was studied in [25] using both the standard PT and APT/FAPT. The FAPT technique was also applied to analyse the structure function $F_2(x)$ behavior at small values of x [26, 27] and calculate binding energies and masses of quarkonia [28]. All these successful applications of APT/FAPT necessitate to have a reliable mathematical tool for extending the scope of these approaches. In this paper, we present the theoretical background which is necessary for the running of $\mathcal{A}_\nu[L]$ and $\mathfrak{A}_\nu[L]$ in the framework of APT and its fractional generalization, FAPT, and which is collected in the easy-to-use Mathematica package "FAPT" [29]. This task has been partially realized for APT as the `Maple` package `QCDMAPT` in [30] and as the `Fortran` package `QCDMAPT_F` in [31]. We have organized "FAPT" in the same manner as the well-known package "RunDec" [32]. A few examples of APT and FAPT applications are given.

2. Theoretical framework

Let us start with the standard definitions used in "FAPT" for standard PT calculations. The QCD running coupling, $\alpha_s(\mu^2) = \alpha_s[L]$ with $L = \ln[\mu^2/\Lambda^2]$, is defined through

$$\frac{d\alpha_s[L]}{dL} = \beta\left(\alpha_s[L]; n_f\right) = -\alpha_s[L] \sum_{k\geq 0} b_k(n_f) \left(\frac{\alpha_s[L]}{4\pi}\right)^{k+1}, \tag{1}$$

where n_f is the number of active flavours. The β-function coefficients are given by (see [33])

$$\begin{aligned}
b_0(n_f) &= 11 - \frac{2}{3} n_f\,, \\
b_1(n_f) &= 102 - \frac{38}{3} n_f\,, \\
b_2(n_f) &= \frac{2857}{2} - \frac{5033}{18} n_f + \frac{325}{54} n_f^2\,, \\
b_3(n_f) &= \frac{149753}{6} + 3564\,\zeta_3 - \left[\frac{1078361}{162} + \frac{6508}{27}\zeta_3\right] n_f \\
&\quad + \left[\frac{50065}{162} + \frac{6472}{81}\zeta_3\right] n_f^2 + \frac{1093}{729} n_f^3\,.
\end{aligned} \tag{2}$$

ζ is Riemann's zeta function. We introduce the following notation:

$$\beta_f \equiv \frac{b_0(n_f)}{4\pi}\,, \quad a(\mu^2; n_f) \equiv \beta_f\, \alpha_s(\mu^2; n_f) \quad \text{and} \quad c_k(n_f) \equiv \frac{b_k(n_f)}{b_0(n_f)^{k+1}}\,. \tag{3}$$

Then Eq. (1) in the l-loop approximation can be rewritten as:

$$\frac{da_{(\ell)}[L;n_f]}{dL} \;=\; -\left(a_{(\ell)}[L;n_f]\right)^2 \left[1+\sum_{k\geq 1}^{\ell} c_k(n_f)\,\left(a_{(\ell)}[L;n_f]\right)^k\right]. \qquad (4)$$

In the one-loop ($\ell = 1$) approximation ($c_k(n_f) = b_k(n_f) = 0$ for all $k \geq 1$) we have the solution

$$a_{(1)}[L] \;=\; \frac{1}{L} \qquad (5)$$

with the Landau pole singularity at $L \to 0$. In the two-loop ($\ell = 2$) approximation ($c_k(n_f) = b_k(n_f) = 0$ for all $k \geq 2$) the exact solution of Eq. (1) is also known [34]

$$a_{(2)}[L;n_f] = \frac{-c_1^{-1}(n_f)}{1+W_{-1}\left(z_W[L]\right)} \quad \text{with} \quad z_W[L] = -c_1^{-1}(n_f)\,e^{-1-L/c_1(n_f)}\,, \qquad (6)$$

where $W_{-1}[z]$ is the appropriate branch of the Lambert function.

The three- ($c_k(n_f) = b_k(n_f) = 0$ for all $k \geq 3$) and higher-loop solutions $a_{(\ell)}[L;n_f]$ can be expanded in powers of the two-loop one, $a_{(2)}[L;n_f]$, as has been suggested and investigated in [8, 9, 14]:

$$a_{(\ell)}[L;n_f] = \sum_{n\geq 1} C_n^{(\ell)}\,\left(a_{(2)}[L;n_f]\right)^n. \qquad (7)$$

The coefficients $C_n^{(\ell)}$ can be evaluated recursively. As has been shown in [9], this expansion has a finite radius of convergence, which appears to be sufficiently large for all values of n_f of practical interest. Note here that this method of expressing the higher-ℓ-loop coupling in powers of the two-loop one is equivalent to the 't Hooft scheme, where one puts by hand all coefficients of the β-function, except b_0 and b_1, equal to zero and effectively takes into account all higher coefficients b_i by redefining perturbative coefficients d_i (see for more detail [35]).

The basic objects in the Analytic approach are the analytic couplings in the Euclidian $\mathcal{A}_\nu^{(\ell)}[L;n_f]$ and Minkowskian $\mathfrak{A}_\nu^{(\ell)}[L_s;n_f]$ domains calculated with the spectral densities $\rho_\nu^{(\ell)}(\sigma;n_f)$ which enter into the Kallen–Lehmann spectral representation:

$$\mathcal{A}_\nu^{(\ell)}[L;n_f] \;=\; \int_0^\infty \frac{\rho_\nu^{(\ell)}(\sigma;n_f)}{\sigma+Q^2}\,d\sigma = \int_{-\infty}^\infty \frac{\rho_\nu^{(\ell)}[L_\sigma;n_f]}{1+\exp(L-L_\sigma)}\,dL_\sigma\,, \qquad (8)$$

$$\mathfrak{A}_\nu^{(\ell)}[L_s;n_f] \;=\; \int_s^\infty \frac{\rho_\nu^{(\ell)}(\sigma;n_f)}{\sigma}\,d\sigma = \int_{L_s}^\infty \rho_\nu^{(\ell)}[L_\sigma;n_f]\,dL_\sigma\,. \qquad (9)$$

It is convenient to use the following representation for spectral functions:

$$\rho_\nu^{(\ell)}[L;n_f] \equiv \frac{1}{\pi}\,\mathbf{Im}\left(\alpha_s^{(\ell)}\,[L-i\pi;n_f]\right)^\nu = \frac{\sin[\nu\,\varphi_{(\ell)}[L;n_f]]}{\pi\,(\beta_f\,R_{(\ell)}[L;n_f])^\nu}\,. \qquad (10)$$

In the one-loop approximation the corresponding functions have the simplest form

$$\varphi_{(1)}[L] = \arccos\left(\frac{L}{\sqrt{L^2+\pi^2}}\right), \quad R_{(1)}[L] = \sqrt{L^2+\pi^2}\,, \qquad (11)$$

whereas at the two-loop order they have a more complicated form

$$R_{(2)}[L;n_f] \;=\; c_1(n_f)\left|1+W_1\left(z_W[L-i\pi;n_f]\right)\right|, \qquad (12)$$

$$\varphi_{(2)}[L;n_f] \;=\; \arccos\left[\mathbf{Re}\left(\frac{-R_{(2)}[L;n_f]}{1+W_1\left(z_W[L-i\pi;n_f]\right)}\right)\right] \qquad (13)$$

with $W_1[z]$ being the appropriate branch of Lambert function.

In the three- ($\ell = 3$) and four-loop ($\ell = 4$) approximation we use Eq. (7) and then obtain

$$R_{(\ell)}[L] = \left| \frac{e^{i\,\varphi_{(2)}[L]}}{R_{(2)}[L]} + \sum_{k\geq 3} C_k^{(\ell)} \frac{e^{i\,k\,\varphi_{(2)}[L]}}{R^k_{(2)}[L]} \right|^{-1}, \tag{14}$$

$$\varphi_{(\ell)}[L] = \arccos\left[\frac{R_{(\ell)}[L]\cos\left(\varphi_{(2)}[L]\right)}{R_{(2)}[L]} + \sum_{k\geq 3} C_k^{(\ell)} \frac{R_{(\ell)}[L]\cos\left(k\,\varphi_{(2)}[L]\right)}{R^k_{(2)}[L]} \right]. \tag{15}$$

Here we do not show explicitly the n_f dependence of the corresponding quantities — it goes inside through $R_{(2)}[L] = R_{(2)}[L;n_f]$, $\varphi_{(2)}[L] = \varphi_{(2)}[L;n_f]$, $C_k^{(3)} = C_k^{(3)}[n_f]$, $C_k^{(4)} = C_k^{(4)}[n_f]$, $c_k = c_k(n_f)$.

The package "FAPT" performs the calculations of the basic required objects: $\left(\alpha_s^{(\ell)}[L,n_f]\right)^\nu$ in Eqs. (5), (6) and (7), $\mathcal{A}_\nu^{(\ell)}[L,n_f]$ in Eq. (8) and $\mathfrak{A}_\nu^{(\ell)}[L,n_f]$ in Eq. (9) up to the N^{3}LO approximation ($\ell = 4$) with a fixed number of active flavours n_f and the global one with taking into account all heavy-quark thresholds (for more details and description of procedures see [29]). As an example, we present here the following Mathematica realizations for analytic coupling $\mathcal{A}_\nu^{(\ell)}[L,n_f]$ and $\mathfrak{A}_\nu^{(\ell)}[L,n_f]$:

- `AcalBar`ℓ`[L,Nf,Nu]` computes the ℓ-loop n_f-fixed analytic coupling $\mathcal{A}_\nu^{(\ell)}[L,n_f]$ in the Euclidean domain, where the logarithmic argument `L`$=\ln[Q^2/\Lambda^2]$, the number of active flavors `Nf`$=n_f$, and the power index `Nu`$=\nu$;
- `UcalBar`ℓ`[L,Nf,Nu]` computes the ℓ-loop n_f-fixed analytic coupling $\mathfrak{A}_\nu^{(\ell)}[L,n_f]$ in the Minkowski domain, where the logarithmic argument `L`$=\ln[s/\Lambda^2]$, the number of active flavors `Nf`$=n_f$, and the power index `Nu`$=\nu$.

3. APT and FAPT applications

As an example of the APT application, we present the Bjorken sum rule (BSR) analysis (see for more details [39]). The BSR claims that the difference between the proton and neutron structure functions integrated over all possible values

$$_1^{p-n}(Q^2) = \int_0^1 \left[g_1^p(x,Q^2) - g_1^n(x,Q^2)\right] dx\,, \tag{16}$$

of the Bjorken variable x in the limit of large momentum squared of the exchanged virtual photon at $Q^2 \to \infty$ is equal to $g_A/6$, where the nucleon axial charge $g_A = 1.2701 \pm 0.0025$ [33]. Commonly, one represents the Bjorken integral in Eq. (16) as a sum of perturbative and higher twist contributions

$$_1^{p-n}(Q^2) = \frac{g_A}{6}\left[1 - \Delta_{\rm Bj}(Q^2)\right] + \sum_{i=2}^{\infty} \frac{\mu_{2i}^{p-n}}{Q^{2i-2}}\,. \tag{17}$$

The perturbative QCD correction $\Delta_{\rm Bj}(Q^2)$ has a form of the power series in the QCD running coupling $\alpha_s(Q^2)$. At the up-to-date four-loop level in the massless case in the modified minimal subtraction ($\overline{\rm MS}$) scheme, for three active flavors, $n_f = 3$, it looks like [36]

$$\Delta_{\rm Bj}^{\rm PT}(Q^2) = 0.3183\,\alpha_s(Q^2) + 0.3631\,\alpha_s^2(Q^2) + 0.6520\,\alpha_s^3(Q^2) + 1.804\,\alpha_s^4(Q^2). \tag{18}$$

The perturbative representation (18) violates analytic properties due to the unphysical singularities of $\alpha_s(Q^2)$. To resolve the issue, we apply APT. In particular, the four-loop APT expansion for the perturbative part $\Delta_{\rm Bj}^{\rm PT}(Q^2)$ is given by the formal replacement

$$\Delta_{\rm Bj}^{\rm PT}(Q^2) = \sum_{k\le 4} c_k\, \alpha_s^k(Q^2) \quad \Rightarrow \quad \Delta_{\rm Bj}^{\rm APT}(Q^2) = \sum_{k\le 4} c_k\, \mathcal{A}_k(Q^2)\,. \tag{19}$$

Clearly, at low Q^2 a value of α_s is quite large, questioning the convergence of perturbative QCD series (18). The qualitative resemblance of the coefficients pattern to the factorial growth did not escape our attention although more definite statements, if possible, would require much more efforts. This observation allows one to estimate the value of $\alpha_s \sim 1/3$ providing a similar magnitude of three- and four- loop contributions to the BSR. To test that, we present in Figs. 1 and 2 the relative contributions of separate i-terms in the four-loop expansion in Eq. (18) for the PT case and in Eq. (19) for APT. As it is seen from Fig. 1, in the region $Q^2 < 1\,{\rm GeV}^2$ the

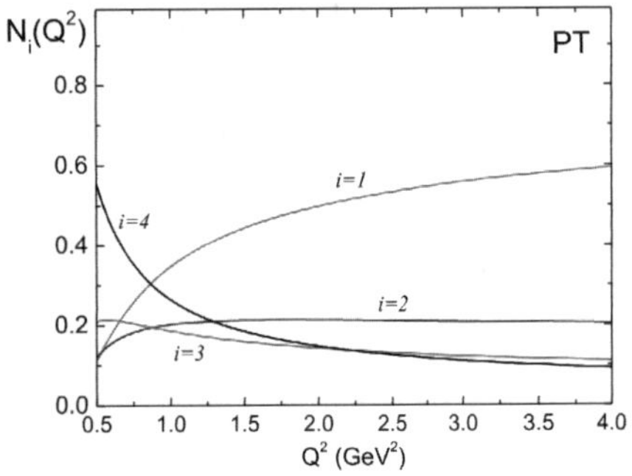

Figure 1. The Q^2-dependence of the relative contributions at the four-loop level in the PT approach.

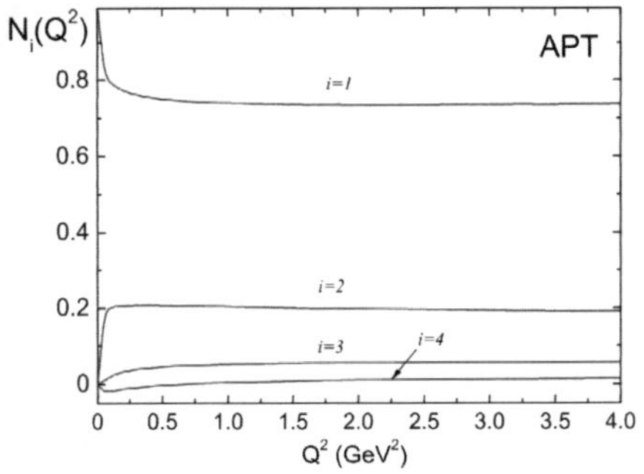

Figure 2. The Q^2-dependence of the relative contributions of the perturbative expansion terms in Eq. (19) in the APT approach.

dominant contribution to the pQCD correction $\Delta_{\rm Bj}(Q^2)$ comes from the four-loop term $\sim \alpha_s^4$. Moreover, its relative contribution increases with decreasing Q^2. In the region $Q^2 > 2\,{\rm GeV}^2$ the situation changes – the major contribution comes from one- and two-loop orders there. Analogous curves for the APT series given by Eq. (19) are presented in Fig. 2.

Figures 1 and 2 demonstrate the essential difference between the PT and APT cases, namely, the APT expansion obeys much better convergence than the PT one. In the APT case, the higher order contributions are stable at all Q^2 values, and the one-loop contribution gives about 70 %, two-loop – 20 %, three-loop – not exceeds 5%, and four-loop – up to 1 %.

One can see that the four-loop PT correction becomes equal to the three-loop one at $Q^2 = 2\,{\rm GeV}^2$ and noticeably overestimates it (note that the slopes of these contributions are quite close in the relatively wide Q^2 region) for $Q^2 \sim 1\,{\rm GeV}^2$ which may be considered as an extra argument supporting an asymptotic character of the PT series in this region. In the APT case, the contribution of the higher loop corrections is not so large as in the PT one. The four-loop order in APT can be important, in principle, if the theoretical accuracy to better than 1 % will be required.

Now we briefly discuss how the APT applications affect the values of the higher-twist coefficients μ_{2i}^{p-n} in Eq. (17) extracted from Jlab data. Previously, a detailed higher-twist analysis of the four-loop expansions in powers of α_s was performed in [39]. In Figs. 3 and 4 we present the results of 1- and 3-parametric fits in various orders of the PT and APT. The corresponding fit results for higher twist terms μ_{2i}^{p-n}, extracted in different orders of the PT and APT, are given in Table 1 (all numerical results are normalized to the corresponding powers of the nucleon mass M). From these figures and Table 1 one can see that APT allows one

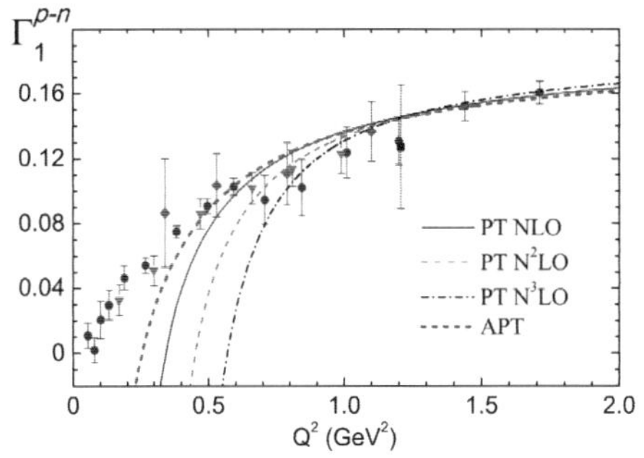

Figure 3. The one-parametric μ_4^{p-n}-fits of the BSR JLab data in various (NLO, N^{2}LO, N^{3}LO) orders of the PT and the all-order APT expansions.

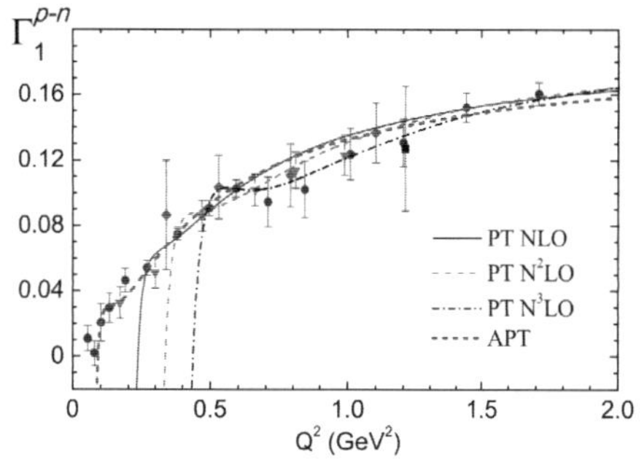

Figure 4. The three-parametric $\mu_{4,6,8}^{p-n}$-fits of the BSR JLab data in various (NLO, N^{2}LO, N^{3}LO) orders of the PT and the all-order APT expansions.

Table 1. Results of higher twist extraction from the JLab data on BSR in various (NLO, N^{2}LO, N^{3}LO) orders of the PT and all orders of APT.

Method	Q^2_{min},	μ_4^{p-n}/M^2	μ_6^{p-n}/M^4	μ_8^{p-n}/M^6
		The best μ_4^{p-n}-fit results		
PT NLO	0.5	−0.028(5)	–	–
PT N^{2}LO	0.66	−0.014(7)	–	–
PT N^{3}LO	0.71	0.006(9)	–	–
APT	0.47	−0.050(4)	–	–
		The best $\mu_{4,6,8}^{p-n}$-fit results		
PT NLO	0.27	−0.03(1)	−0.01(1)	0.008(4)
PT N^{2}LO	0.34	0.01(2)	−0.06(4)	0.04(2)
PT N^{3}LO	0.47	0.05(4)	−0.2(1)	0.12(6)
APT	0.08	−0.061(4)	0.009(1)	−0.0004(1)

to move down up to $Q^2 \sim 0.1\,\text{GeV}^2$ in description of the experimental data [39]. At the same time, in the framework of the standard PT the lower border shifts up to higher Q^2 scales when increasing the order of the PT expansion. This is caused by extra unphysical singularities in the higher-loop strong coupling. It should be noted that the magnitude of μ_4^{p-n}/M^2 decreases with an order of the PT and becomes compatible to zero at the four-loop level. It is interesting to mention that a similar decreasing effect has been found in the analysis of the experimental data for the neutrino-nucleon DIS structure function xF_3 [37] and for the charged lepton-nucleon DIS structure function F_2 [38].

Consider the application of the FAPT approach by the example of the RG-evolution of the non-singlet higher-twist $\mu_4^{p-n}(Q^2)$ in Eq. (17). The evolution of the higher-twist terms $\mu_{6,8,\ldots}^{p-n}$ is still unknown. The RG-evolution of $\mu_4^{p-n}(Q^2)$ in the standard PT reads

$$\mu_{4,PT}^{p-n}(Q^2) \; = \; \mu_{4,PT}^{p-n}(Q_0^2)\left[\frac{\alpha_s(Q^2)}{\alpha_s(Q_0^2)}\right]^{\nu}, \quad \nu = \gamma_0/(8\pi\beta_0)\,, \quad \gamma_0 = \frac{16}{3}C_F\,, \quad C_F = \frac{4}{3}. \tag{20}$$

In the framework of FAPT the corresponding expression reads as follows:

$$\mu_{4,APT}^{p-n}(Q^2) = \mu_{4,APT}^{p-n}(Q_0^2)\,\frac{\mathcal{A}_\nu^{(1)}(Q^2)}{\mathcal{A}_\nu^{(1)}(Q_0^2)}. \tag{21}$$

We present in Table 2 the best fits for $\mu_4^{p-n}(Q_0^2)$ taking into account the corresponding RG-evolution with $Q_0^2 = 1\,\mathrm{GeV}^2$ as a normalization point and without the RG-evolution. We do not

Table 2. Results of higher twist extraction from the JLab data on BSR with inclusion and without inclusion of the RG-evolution of $\mu_4^{p-n}(Q^2)$ normalized at $Q_0^2 = 1\,\mathrm{GeV}^2$.

Method	Q^2_{min}, GeV^2	μ_4^{p-n}/M^2	μ_6^{p-n}/M^4	μ_8^{p-n}/M^6
	0.47	−0.055(3)	0	0
NNLO APT	0.17	−0.062(4)	0.008(2)	0
no evolution	0.10	−0.068(4)	0.010(3)	−0.0007(3)
	0.47	−0.051(3)	0	0
NNLO APT	0.17	−0.056(4)	0.0087(4)	0
with evolution	0.10	−0.058(4)	0.0114(6)	−0.0005(8)

take into account the RG-evolution in μ_4^{p-n} for the standard PT calculations and compare with FAPT since the only effect of that would be the enhancement of the Landau singularities by extra divergencies at $Q^2 \sim \Lambda^2$, whereas at higher $Q^2 \sim 1\,\mathrm{GeV}^2$ the evolution is negligible with respect to other uncertainties. We see from Table 2 that the fit results become more stable with respect to Q_{min} variations, which reduces the theoretical uncertainty of the BSR analysis.

4. Summary

To summarize, APT and FAPT are the closed theoretical schemes without unphysical singularities and additional phenomenological parameters which allow one to combine RG-invariance, Q^2-analyticity, compatibility with linear integral transformations and essentially incorporate nonperturbative structures. The APT provides a natural way for the coupling constant and related quantities. These properties of the coupling constant are the universal loop-independent infrared limit and weak dependence on the number of loops. At the same time, FAPT provides an effective tool to apply the Analytic approach for RG improved perturbative amplitudes. This approaches are used in many applications. In particular, in this paper we consider the application of APT and FAPT to the RG-evolution of nonsinglet structure functions and Bjorken sum rule higher-twist analysis at the scale $Q^2 \sim \Lambda^2$ considered.

The singularity-free, finite couplings $\mathcal{A}_\nu(Q^2), \mathfrak{A}_\nu(s)$ appear in APT/FAPT as analytic images of the standard QCD coupling powers $\alpha_s^\nu(Q^2)$ in the Euclidean and Minkowski domains, respectively. In this paper, we presented the theoretical background, used in a package "FAPT" [29] based on the system Mathematica for QCD calculations in the framework of APT/FAPT, which are needed to compute these couplings up to N^3LO of the RG running. We hope that this will expand the use of these approaches.

Acknowledgments

I am grateful to the organizers and conveniers of the ACAT2013 workshop for the invitation and for arranging such a nice even. In addition, I would like to thank Alexander Bakulev, Sergei Mikhailov and Andrei Kataev for stimulating discussions and useful remarks. This work was supported in part by the Belarussian state fundamental research programm "Convergency", BelRFBR grants under Grant No. F12D-002 and No. F13M-143 and by RFBR Grant No. 11-01-00182.

References

[1] Shirkov D V and Solovtsov I L 1997 *Phys. Rev. Lett.* **79** 1209
[2] Milton K A and Solovtsov I L 1997 *Phys. Rev.* **D55** 5295
[3] Solovtsov I L and Shirkov D V 1999 *Theor. Math. Phys.* **120** 1220
[4] Shirkov D V 2001 *Theor. Math. Phys.* **127** 409
[5] Shirkov D V and Solovtsov I L 2007 *Theor. Math. Phys.* **150** 132
[6] Shirkov D V 2012 *Nucl. Phys. Proc. Suppl.* **225-227** 5
[7] Magradze B A 2000 *Int. J. Mod. Phys.* **A15** 2715
[8] Kourashev D S and Magradze B A 2003 *Theor. Math. Phys.* **135** 531
[9] Magradze B A 2006 *Few Body Syst.* **40** 71
[10] Milton K A, Solovtsov I L, Solovtsova O P and Yasnov V I 2000 *Eur. Phys. J.* **C14** 495
[11] Milton K A, Solovtsov I L and Solovtsova O P 2002 *Phys. Rev.* **D65** 076009
[12] Cvetic G and Valenzuela C 2006 *Phys. Rev.* **D74** 114030
[13] Cvetic G, Kogerler R and Valenzuela C 2010 *J. Phys.* **G37** 075001
[14] Magradze B A 2010 *Few Body Syst.* **48** 143
[15] Milton K A, Solovtsov I L and Solovtsova O P 1998 *Phys. Lett.* **B439** 421
[16] Pasechnik R S, Shirkov D V and Teryaev O V 2008 *Phys. Rev.* **D78** 071902
[17] Milton K A, Solovtsov I L and Solovtsova O P 1999 *Phys. Rev.* **D60** 016001
[18] Shirkov D V and Zayakin A V 2007 *Phys. Atom. Nucl.* **70** 775
[19] Baldicchi M, Nesterenko A V, Prosperi G M, Shirkov D V and Simolo C 2007 *Phys. Rev. Lett.* **99** 242001
[20] Bakulev A P, Mikhailov S V and Stefanis N G 2005 *Phys. Rev.* **D72** 074014, 2005 *Erratum: ibid.* **D72** 119908(E)
[21] Bakulev A P, Mikhailov S V and Stefanis N G 2007 *Phys. Rev.* **D75** 056005, 2008 *Erratum: ibid.* **D77** 079901(E)
[22] Bakulev A P 2009 *Phys. Part. Nucl.* **40** 715
[23] Broadhurst D J, Kataev A L and Maxwell C J 2001 *Nucl. Phys.* **B592** 247
[24] Bakulev A P, Mikhailov S V and Stefanis N G 2010 *JHEP* **1006** 085
[25] Pasechnik R S, Shirkov D V, Teryaev O V, Solovtsova O P and Khandramai V L 2010 *Phys. Rev.* **D81** 016010
[26] Cvetic G, Illarionov A Y, Kniehl B A and Kotikov A V 2009 *Phys. Lett.* **B679** 350
[27] Kotikov A V, Krivokhizhin V G and Shaikhatdenov B G 2012 *Phys. Atom. Nucl.* **75** 507
[28] Ayala C and Cvetic G 2013 *Phys. Rev.* **D87** 054008
[29] Bakulev A P and Khandramai V L 2013 *Comput. Phys. Commun.* **184** 183
[30] Nesterenko A V and Simolo C 2010 *Comput. Phys. Commun.* **181** 1769
[31] Nesterenko A V and Simolo C 2011 *Comput. Phys. Commun.* **182** 2303
[32] Chetyrkin K G, Kuhn J H and Steinhauser M 2000 *Comput. Phys. Commun.* **133** 43
[33] Beringer J *et al.* (Particle Data Group Collaboration) 2012 *Phys. Rev.* **D86** 010001
[34] Gardi E, Grunberg G and Karliner M 1998 *JHEP* **07** 007
[35] Garkusha A V and Kataev A L 2011 *Phys. Lett.* **B705** 400
[36] Baikov P A, Chetyrkin K G and Kühn J H 2010 *Phys. Rev. Lett.* **104** 132004
[37] Kataev A L, Parente G and Sidorov A V 2003 *Phys. Part. Nucl.* **34** 20
[38] Blumlein J 2013 *Prog. Part. Nucl. Phys.* **69** 28
[39] Khandramai V L, Pasechnik R S, Shirkov D V, Solovtsova O P and Teryaev O V 2012 *Phys. Lett.* **B706** 340

Journal of Physics: Conference Series **523** (2014) 012063 doi:10.1088/1742-6596/523/1/012063

One loop integration with hypergeometric series by using recursion relations

Norihisa Watanabe, Toshiaki Kaneko

High Energy Accelerator Research Organization (KEK), Computing Research Center, 1-1, O-ho, Tsukuba, Ibaraki 305-0801, Japan

E-mail: norihisa@post.kek.jp,toshiaki.kaneko@kek.jp

Abstract. General one-loop integrals with arbitrary mass and kinematical parameters in d-dimensional space-time are studied. By using Bernstein theorem, a recursion relation is obtained which connects $(n+1)$-point to n-point functions. In solving this recursion relation, we have shown that one-loop integrals are expressed by a newly defined hypergeometric function, which is a special case of Aomoto-Gelfand hypergeometric functions.

We have also obtained coefficients of power series expansion around 4-dimensional space-time for two-, three- and four-point functions. The numerical results are compared with "LoopTools" for the case of two- and three-point functions as examples.

1. Introduction

For discovery of the beyond standard model, we need to know the precise theoretical prediction of standard model. For Large Hadron Collider at CERN and the international linear collider, at least next-to-leading order(NLO) electroweak corrections are necessary. However, it is not easy to calculate Feynman integrations with highly accuracy even for one-loop level. There appear many kinematical parameters including masses, momenta of particles, and space-time dimension. This means that the loop integrals are analytic functions with singularities on a multiple dimensional complex vector space. It is difficult to obtain the numerical accuracy in all kinematical region by a simple numerical approach. Since the numerical stability of an expression is connected to its analytic properties, suitable analytic expressions of one-loop integral are still important.

It is known that any loop integrals are expressed by GKZ-hypergeometric functions[1]. However, theses functions are so general extension of hypergeometric function that it is not easy to obtain numerical values. It is desirable to find a subset of GKZ-hypergeometric functions which corresponds to specific loop integrals to be calculated.

The analytic properties of hypergeometric functions, such as position of singularities, have been investigated for many hypergeometric functions. Since these singularities correspond to physical singularities or large cancellations in numerical calculations, information about singularities helps us to obtain accurate numerical results.

There are various methods to express one-loop integrals by hypergeometric functions[2, 3, 4, 5, 6]. In this article, we show a method to obtain analytic expressions of n-point functions with arbitrary kinematical parameters. Our method is based on Bernstein theorem [7] (see also [8]). This theorem implies that for given polynomial $\mathcal{D}$ of variables $x = (x_1, ..., x_n)$, there exist differential operator $\mathcal{P}$ and polynomial $b(s)$ of parameter s such that

$$\mathcal{P}(\partial, x, s)\mathcal{D}^{s+1}(x) = b(s)\mathcal{D}^{s}(x), \tag{1}$$

where $\mathcal{P}$ is a polynomial of x, differential operator $\partial = (\partial_1, ..., \partial_n)$, and parameter s. Applying this theorem to the integrands of one-loop integrals, we obtain a recursion relation. In solving this relation, we show that one-loop integrals are expressed by newly introduced hypergeometric function G_n. This function is found to be one of Aomoto-Gelfand hypergeometric functions (or general hypergeometric functions on complex Grassmannians) [9, 10, 11] which make a subset of GKZ-hypergeometric functions.

Starting from our general expression of one-loop integral, two- and three-point function are re-expressed in a linear combination of Gauss hypergeometric function F and Appell's function F_3[12], respectively. For the case of scalar integral of four-point function, we have expanded the analytic expression around 4-dimensional space-time. The result is expressed by Lauricella's function F_D[12] up to the finite order of space-time dimension $d = 4 - 2\epsilon$.

We have numerically calculated two- and three-point function as examples and have compared the results with `LoopTools`[13].

2. Formulation

Let us consider the one-loop $(n+1)$-point function denoted by I_{n+1}. After performing integration of δ-function, the Feynman parameter integration of I_{n+1} is written as:

$$I_{n+1}(s) = \int_{\Delta_n} d^n x\, \mathcal{D}_n^s, \tag{2}$$

$$\begin{aligned} \mathcal{D}_n(x_1, ..., x_n) &= -\frac{1}{2}\sum_{j,k=0}^{n} q_{jk}^2 x_j x_k + \frac{1}{2}\sum_{j,k=0}^{n}(m_j^2 + m_k^2)x_j x_k, \\ &= \frac{1}{2}\sum_{i,j=1}^{n} A_{ij}x_i x_j + \sum_i B_i x_i + C \\ &= \frac{1}{2}(Ax, x) + (B, x) + C, \end{aligned} \tag{3}$$

where m_i are masses of the propagators, $q_{jk} \equiv -\sum_{i=j+1}^{k} p_i$ with external momenta p_i, x_i are Feynman parameters with $x_0 = 1 - \sum_{j\geq 0} x_j$. Here, Δ_n refers to n-dimensional simplex and $A_{ij} \equiv \partial_i\partial_j\mathcal{D}_n$, $B_i \equiv \partial_i\mathcal{D}_n(0)$ and $C \equiv \mathcal{D}_n(0)$. Parameter s depends on the space-time dimension d. If $d = 4 - 2\epsilon$ are chosen, we find $s = 1 - n - \epsilon$ for the standard scalar model.

Let us define operator $\mathcal{P}$ by

$$\mathcal{P} \equiv -(s+2)\frac{1}{E_n} + \frac{1}{2E_n}(A^{-1}\partial\mathcal{D}_n, \partial), \tag{4}$$

where $E_n = (A^{-1}B, B)/2 - C$. We can find the following relation:

$$\mathcal{P}\mathcal{D}_n^{s+1} = (s+1)\mathcal{D}_n^s. \tag{5}$$

This is the explicit expression of Bernstein theorem for scalar one-loop integral and polynomial b-function is found to be $s+1$. Applying Eq.(5) to Eq.(2) iteratively with partial integrations, we obtain

$$I_{n+1}(s) = J_{n+1,m}(s) + K_{n+1,m}(s), \quad (6)$$

$$J_{n+1,m}(s) = \frac{1}{2(s+1)E_n} \sum_{j=0}^{m-1} \frac{(s+n/2+1)_j}{(s+2)_j} \times \int_{\Delta_n} d^n x \sum_k \partial_k \left\{ (A^{-1}\partial \mathcal{D}_n)_k \mathcal{D}_n^{s+1} \left(-\frac{\mathcal{D}_n}{E_n} \right)^j \right\}, \quad (7)$$

$$K_{n+1,m}(s) = \frac{(s+n/2+1)_m}{(s+1)_m} \int_{\Delta_n} d^n x \, \mathcal{D}_n^s \left(-\frac{\mathcal{D}_n}{E_n} \right)^m, \quad (8)$$

where $(a)_j \equiv a(a+1)\cdots(a+j-1)$ is Pochhammer's symbol.

If we choose the appropriate parameter region, the second term $K_{n+1,m}(s)$ goes to be zero in the limit $m \to \infty$. The first term is surface integration and can be integrated once easily. The remaining integrations are expressed by n-point functions. We obtain a recursion relation between $(n+1)$- and n-point functions:

$$I_{n+1}(s) = \frac{1}{2} \sum_{j=0}^{\infty} \frac{(s+n/2+1)_j}{(s+1)_{j+1}} \left(-\frac{1}{E_n} \right)^{j+1} \sum_{k=0}^{n} h_{\rho(n),k} I_{n,\rho(n;k)}(s+j+1), \quad (9)$$

where coefficients $h_{\rho(n),k}$ are rational functions of kinematical variables. The suffix $\rho(n) = \Delta_n$ and $\rho(n;k)$ represents a $(n-1)$-dimensional simplex which is obtained by eliminating k-th vertex and faces attaching to the vertex from the original n-dimensional simplex, which appears as a part of the boundary of the original integration domain. In a similar way, we define $\rho(n;k_1,k_2) \equiv \rho(\rho(n;k_1),k_2)$. Using Eq.(9) repetitively, Eq.(2) eventually depend only on the remaining vertex, in which no integration is left. The final formula of $(n+1)$-point function is

$$\begin{aligned} I_{n+1}(s) = \; & \frac{1}{2^n} \sum_{k_n=0}^{n} \sum_{k_{n-1}=0}^{n-1} \cdots \sum_{k_1=0}^{1} \mathcal{D}(n;k_n,k_{n-1},...,k_1)^s h_{\rho(n),k_n} h_{\rho(n;k_n),k_{n-1}} \cdots h_{\rho(n;k_n,...,k_2),k_1} \\ & \times \sum_{j_n=0}^{\infty} \sum_{j_{n-1}=0}^{\infty} \cdots \sum_{j_1=0}^{\infty} \frac{(s+n/2+1)_{j_n}}{(s+1)_{j_n+1}} \frac{(s+j_n+(n-1)/2+2)_{j_{n-1}}}{(s+j_n+2)_{j_{n-1}+1}} \cdots \\ & \frac{(s+j_n+\cdots+j_2+1/2+n)_{j_1}}{(s+j_n+\cdots+j_2+n)_{j_1+1}} \\ & \times \left(-\frac{\mathcal{D}(n;k_n,k_{n-1},...,k_1)}{E_{\rho(n)}} \right)^{j_n+1} \left(-\frac{\mathcal{D}(n;k_n,k_{n-1},...,k_1)}{E_{\rho(n;k_n)}} \right)^{j_{n-1}+1} \cdots \\ & \left(-\frac{\mathcal{D}(n;k_n,k_{n-1},...,k_1)}{E_{\rho(n;k_n,...,k_2)}} \right)^{j_1+1}. \end{aligned} \quad (10)$$

Here, $\mathcal{D}(n;k_n,k_{n-1},...,k_1)$ is the value at the vertex k_{n+1} which does not appear in the list $(k_n,k_{n-1},\cdots,k_1)$. The right-hand side of Eq.(10) shows that it is expressed by a kind of hypergeometric series. We call this hypergeometric series function G_n, which is defined by Eq.(13) in the next section. Using this function, one-loop scalar integral becomes

$$\begin{aligned} I_{n+1}(s) = \; & \frac{1}{2^n(s+1)_n} \sum_{k_n=0}^{n} \sum_{k_{n-1}=0}^{n-1} \cdots \sum_{k_1=0}^{1} \frac{\mathcal{D}(n;k_n,k_{n-1},...,k_1)^{s+n}}{E_{\rho(n)} \cdots E_{\rho(n;k_n,...,k_2)}} \\ & \times h_{\rho(n),k_n} h_{\rho(n;k_n),k_{n-1}} \cdots h_{\rho(n;k_n,...,k_2),k_1} \\ & \times G_n \left(\alpha, \beta; \gamma; \left(-\frac{\mathcal{D}(n;k_n,k_{n-1},...,k_1)}{E_{\rho(n)}} \right), \cdots, \left(-\frac{\mathcal{D}(n;k_n,k_{n-1},...,k_1)}{E_{\rho(n;k_n,...,k_2)}} \right) \right), \end{aligned} \quad (11)$$

where

$$\alpha = (\underbrace{1, \cdots, 1}_{n}), \quad \beta = (\underbrace{1/2, \cdots, 1/2}_{n-1}, s + n/2 + 1), \quad \gamma = s + n + 1. \tag{12}$$

3. G_n-functions

In this section, we discuss the properties of G_n function. This function is defined by:

$$G_n(\alpha, \beta; \gamma; x) = \sum_{j_n=0}^{\infty} \sum_{j_{n-1}=0}^{\infty} \cdots \sum_{j_1=0}^{\infty} \frac{\prod_{i=1}^n (\alpha_i)_{j_i} \prod_{k=1}^n (\sum_{\ell=k}^n \beta_\ell)_{\sum_{\ell=k}^n j_\ell}}{(\gamma)_{\sum_{i=1}^n j_i} \prod_{k=1}^n (\sum_{\ell=k}^n \beta_\ell)_{\sum_{\ell=k+1}^n j_\ell} \prod_{i=1}^n j_i!} \prod_{i=1}^n x_i^{j_i}, \tag{13}$$

where $x = (x_1, ..., x_n) \in \mathbb{C}^n$ are variables, $\alpha = (\alpha_1, ..., \alpha_n)$ and $\beta = (\beta_1, ..., \beta_n)$ are complex vectors, and γ is a complex parameter.

Euler type integral representation of G_n is obtained as:

$$\begin{aligned} G_n(\alpha, \beta; \gamma; x) &= \frac{\Gamma(\gamma)}{\prod_{j=1}^n \Gamma(\alpha_j)\Gamma(\gamma - \sum_{k=1}^n \alpha_k)} \int_{\Delta_n} d^n u \prod_{k=1}^n u_k^{\alpha_k - 1} \\ &\quad \times \left(1 - \sum_{j=1}^n u_j\right)^{\gamma - \sum_{k=1}^n \alpha_k - 1} \prod_{k=1}^n \left(1 - \sum_{j=1}^k x_j u_j\right)^{-\beta_k}. \end{aligned} \tag{14}$$

The integrand is a product of powers of linear factors of integration variables, while the original integrand of I_{n+1} is a power of quadratic term. This representation shows that this function is a member of a class of hypergeometric functions, which are called Aomoto-Gelfand hypergeometric functions or hypergeometric functions on complex Grassmannians. This fact and their analytic properties give us important information for numerical calculation. One can easily show the following formulae:

- differentiation (with k-th unit vector e_k)

$$\frac{\partial}{\partial x_\ell} G_n(\alpha, \beta; \gamma; x) = \sum_{k=\ell}^n \frac{\alpha_\ell \beta_k}{\gamma} G_n(\alpha + e_\ell, \beta + e_k; \gamma + 1; x). \tag{15}$$

- recursion relation

$$G_n(\alpha, \beta; \gamma; x) = \sum_{j_n=0}^{\infty} \frac{(\alpha_n)_{j_n} (\beta_n)_{j_n}}{(\gamma)_{j_n} j_n!} x_n^{j_n} G_{n-1}(\alpha', \beta'; \gamma'; x') \tag{16}$$

$$\begin{aligned} &= \frac{\Gamma(\gamma)}{\Gamma(\alpha_1)\Gamma(\gamma - \alpha_1)} \\ &\times \int_0^1 dw\, w^{\alpha_1 - 1}(1 - w)^{\gamma - \alpha_1 - 1}(1 - x_1 w)^{-\sum_{j=1}^n \beta_j} G_{n-1}(\alpha', \beta'; \gamma'; x'). \end{aligned} \tag{17}$$

Tensor integrals are obtained by differentiating I_{n+1} in terms of mass parameters. Eq.(15) shows that tensor integrals are also expressed by G_n.

Based on the above results, the problem of calculating one-loop integral is converted to one of establishing methods of expansion around 4 space-time dimension and numerical evaluations.

We show some samples of scalar case in the next section.

4. Calculations of n-point functions

In this section, we will discuss how to evaluate and expand G_n functions. Let us discuss in detail for 2-, 3- and 4-point functions separately.

4.1. Two-point function

Let's first consider the two-point function. This is a good example of the understanding of how to evaluate G_n. From the formula Eq.(10), the scalar two-point function is expressed by G_1 functions with $\alpha = 1$, $\beta = s + 3/2$ and $\gamma = s + 2$. Function G_1 is nothing but Gaussian hypergeometric function:

$$G_1(1, s+3/2; s+2; x) = \sum_{j=0}^{\infty} \frac{(1)_j (s+3/2)_j}{(s+2)_j} \frac{x^j}{j!} = F(1, s+3/2, s+2; x). \tag{18}$$

Combining the kinematical factor, we obtain

$$\begin{aligned} I_2(s) &= \frac{(p^2 + m_1^2 - m_2^2)(m_1^2)^{s+1}}{(s+1)E_1} F\left(1, s+\frac{3}{2}, s+2; -\frac{4p^2 m_1^2}{E_1}\right) \\ &\quad + \frac{(p^2 + m_2^2 - m_1^2)(m_2^2)^{s+1}}{(s+1)E_1} F\left(1, s+\frac{3}{2}, s+2; -\frac{4p^2 m_2^2}{E_1}\right), \end{aligned} \tag{19}$$

where $E_1 \equiv (p^2 - (m_1 + m_2)^2)(p^2 - (m_1 - m_2)^2)$. The case of $s = -\epsilon$ corresponds to the usual dimensional regularization $d = 4 - 2\epsilon$. It is more convenient for the expansion when half-integer parameter is converted to integer. By using identities of F, we obtain

$$I_2(s) = \frac{\xi_-(m_1^2)^s}{s+1} F\left(1, -s; s+2; \frac{\xi_-}{\xi_+}\right) + \frac{(1-\xi_-)(m_2^2)^s}{s+1} F\left(1, -s; s+2; \frac{1-\xi_-}{1-\xi_+}\right), \tag{20}$$

where $\xi_\pm = (p^2 - m_1^2 + m_2^2 \pm \sqrt{E_1})/(2p^2)$. After the conversion, G_1 can be expanded in arbitrary order of ϵ with multiple polylogarithmic functions[14, 15].

With the known analytic properties of F, it is found that two-point function may only be singular when $\xi_\pm = 0$, 1, ∞ and $\xi_1 = \xi_2$. These cases correspond to massless or on-shell limit. Let us investigate the case both of masses are taken massless limit as an example, where $\xi_- \to 0$ and $\xi_+ \to 1$. The first term in Eq.(20) goes to zero, but second term is not well-defined at this limit. However, with using identities of F, we can transform the expression into the well-defined form at this limit under the condition $\mathrm{Re}(s) > 0$:

$$\lim_{m_1, m_2 \to 0} I_2(s) = \lim_{m_1, m_2 \to 0} \frac{(-p^2 \xi_+)^s}{s+1} F\left(s+1, -s; s+2; \frac{1}{\xi_+}\right) = (-p^2)^s B(s+1, s+1), \tag{21}$$

where B is beta-function. This means that we can select appropriate representations in terms of kinematical conditions.

4.2. Three-point function

Eq.(11) shows that scalar three-point function is obtained as a linear combination of G_2. It is expressed as

$$I_3(s) = \frac{1}{(s+1)(s+2)} \sum_{k_1=1}^{2} \sum_{k_2=1}^{3} h_{k_1,k_2} G_2\left(\{1,1\}, \{1/2, s+2\}; s+3; x_{1,(k_1,k_2)}, x_{2,(k_1,k_2)}\right). \tag{22}$$

Function G_2 is equivalent to Appell's function F_3:

$$G_2((\alpha_1, \alpha_2), (\beta_1, \beta_2), \gamma; x_1, x_2) = (1 - x_1)^{-\alpha_1} F_3\Big(\alpha_1, \alpha_2, \gamma - \beta_1 - \beta_2, \beta_2, \gamma; \frac{x_1}{x_1 - 1}, x_2\Big). \tag{23}$$

When $\alpha_1 = \alpha_2 = 1$, G_2 reduces to Appell's function F_1. It is also convenient when the half-integer parameter is transformed to integer as same as the case of two-point function. We apply nontrivial identity:

$$F_1(\alpha,\beta,\beta',\gamma;x,y) \\ = (1+z)^\alpha F_D\Big(\alpha;\alpha-\gamma+1,1-\alpha,2\beta-1,\beta',\beta';\gamma;z,\frac{z}{1+z},\frac{2z}{1+z},\frac{1}{v_+},\frac{1}{v_-}\Big), \quad (24)$$

where $z = (1-\sqrt{1-x})/(1+\sqrt{1-x})$, $v_\pm = (1+\sqrt{1-x})/(y \mp \sqrt{y(y-x)})$ and F_D is on of Lauricella's functions[12]. For the case $s = -1-\epsilon$, Eq.(22) reduces to

$$G_2 \to F_1(\text{half-integer}) \to F_D(\text{integer}) \to (\text{expansion}) \to F_1(\text{integer}) \\ \to \text{multiple polylogarithmic functions in arbitrary order of } \epsilon.$$

Investigating the limit $\epsilon \to 0$, $1/\epsilon$ pole appears from $1/(s+1)$ of Eq.(22) for both massive and massless cases. However, this poles canceled out when all contributions are summed up for the massive case. So we can obtain the value of integration in this limit.

4.3. Four-point function

After performing δ-function integration, three Feynman parameters remain on the four-point function.

In this case, G_3 appears in Eq.(11).

$$G_3(\alpha,\beta;\gamma;x) = \frac{\Gamma(\gamma)}{\Gamma(\alpha_1)\Gamma(\alpha_2)\Gamma(\alpha_3)\Gamma(\gamma-\alpha_1-\alpha_2-\alpha_3)} \\ \times \int_{\Delta_3} d^3u\, u_1^{\alpha_1-1}u_2^{\alpha_2-1}u_3^{\alpha_3-1}(1-u_1-u_2-u_3)^{\gamma-\alpha_1-\alpha_2-\alpha_3-1} \\ \times(1-x_1u_1)^{-\beta_1}(1-x_1u_1-x_2u_2)^{-\beta_2}(1-x_1u_1-x_2u_2-x_3u_3)^{-\beta_3}. \quad (25)$$

For scalar integral case, parameters take the values $\alpha = \{1,1,1\}$, $\beta = \{1/2, 1/2, s+5/2\}$, $\gamma = s+4$, $x = \{x_1,x_2,x_3\}$, and $s = -2-\epsilon$ for $d = 4-2\epsilon$. This function can be written in a linear combination of F_D up to $\mathcal{O}(\epsilon)$, which corresponds to finite order of I_4, since

$$G_3\left(\{1,1,1\},\left\{\frac{1}{2},\frac{1}{2},\frac{1}{2}-\epsilon\right\};2-\epsilon,x_1,x_2,x_3\right) \\ = C_1F_1\left(1,1,\frac{1}{2};2;\frac{x_2-x_1}{x_2-1},\frac{x_3-x_1}{x_3-1}\right) + C_2F_D\left(1,2\epsilon,1,1,\frac{1}{2},\frac{1}{2};2;\frac{1-\sqrt{1-x_3}}{2},\frac{1}{\eta_1},\frac{1}{\eta_2},\frac{1}{\eta_3},\frac{1}{\eta_4}\right) \\ +C_3F_D\left(1,-\epsilon,1,\frac{1}{2};2-\epsilon;x_1,\frac{x_1-x_2}{1-x_2},\frac{x_1-x_3}{1-x_3}\right) + \mathcal{O}(\epsilon^2), \quad (26)$$

where the coefficients C_i's and η_i's are algebraic functions of x_1, x_2, and x_3. The half-integer parameters are converted to integers by using extended identities from Eq.(24).

4.4. Summary numerical calculation

We have compared the numerical results between our method and `LoopTools`[13]. We show the compared results of two- and three-point function in Figs.1 and 2, respectively. The results are consistent in satisfactory accuracy.

5. Conclusion and discussion

In the discussion of Sec.2, it is necessary to select appropriate kinematical region in order to make Eq.(8) vanishes at the limit $m \to \infty$. However, we can show the following identity of

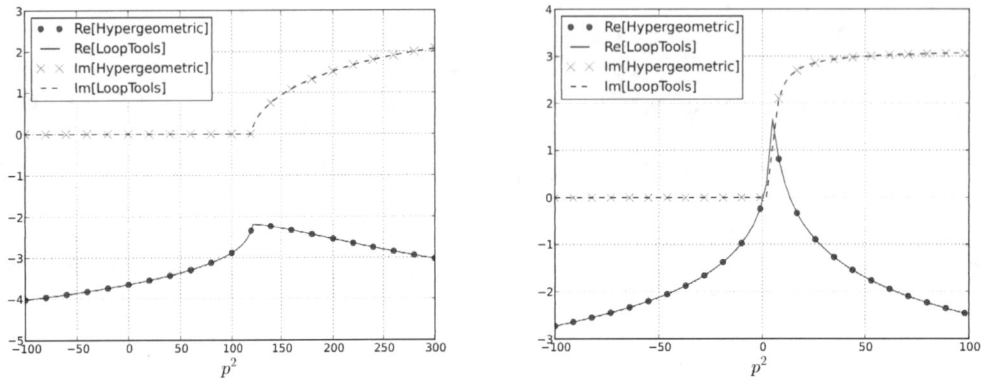

(a) $m_1^2 = 10$, $m_2^2 = 1$ (b) $m_1^2 = 1$, $m_2^2 = 1.1$

Figure 1. Comparison of numerical results with `LoopTools` for two-point functions in two parameter sets (a) and (b). Circles and crosses are the numerical results of real and imaginary part of our calculation, respectively. Solid and dashed lines are the results of real and imaginary part which are obtained from `LoopTools`, respectively.

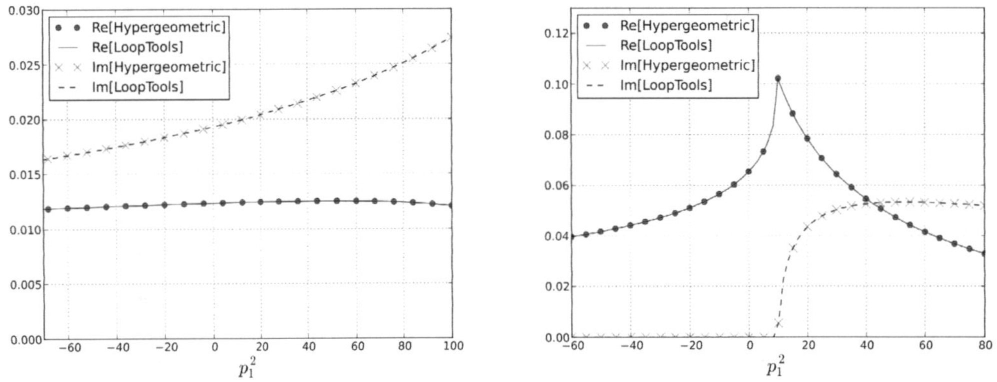

(a) $p_2^2 = -10$, $p_3^2 = 100$, $m_1^2 = 1$, $m_2^2 = 10$, $m_3^2 = 100$ (b) $p_2^2 = -10$, $p_3^2 = -100$, $m_1^2 = 1$, $m_2^2 = 2$, $m_3^2 = 3$

Figure 2. Comparison of numerical results of three-point functions for two parameter sets (a) and (b). Circles and crosses are the numerical results of real and imaginary part of our calculation, respectively. Solid and dashed lines are the results of real and imaginary part which are obtained from `LoopTools`, respectively.

Gauss hypergeometric function:

$$\mathcal{D}_n^s = \frac{1}{2(s+1)E_n} \sum_k \partial_k \left[\left(A^{-1}\partial\mathcal{D}_n\right)_k F\left(1, s+\frac{n}{2}+1; s+2; -\frac{\mathcal{D}_n}{E_n}\right)\right]. \tag{27}$$

From this identity, one can derive recursion relation Eq.(9) and confirm it holds in all kinematical region[16].

We have shown that general one-loop integral is expressed by G_n, one of hypergeometric functions on complex Grassmannian. Especially, scalar two- and three-point functions are expressed in terms of Gaussian and Appell's functions, respectively for any kinematics variables and space-time dimension. Four-point function is expressed in Lauricella's functions up to finite order for arbitrary kinematical parameters. We have also shown the sample numerical calculation in terms of two-, and three-point functions and results are consistent with `LoopTools` package.

Acknowledgment

The authors wish to thanks to the members of Minami-tateya group for useful discussions. Especially, we would like to thank to Y. Shimizu and J. Fujimoto for their focus on Bernstein theorem and suggestions.

References

[1] Gel'fand I M, Zelevinsky A V and Kapranov M M 1990 Soviet Math. Dokl. **40** 239–243
I.M. Gel'fand, A.V. Zelevinsky and M,M, Kapranov, 1990 Adv. Math. **84** 255–271
[2] Davydychev A I 1991 J. Math. Phys. **32** 1052–1060
Davydychev A I 1992 J. Math. Phys. **33** 385–369
[3] Tarasov O V 1996 Phys. Rev. D. **54** 6479–6490
Fleischer J, Jegerlehner F and Tarasov O V 2003 Nucl. Phys. B **672** 303
[4] Duplancic G and Nizic B 2001 Eur. Phys. J. C **20** 357
[5] Kurihara Y 2006 Eur. Phys. J. C **45** 427
[6] Kaneko T 2010 PoS CPP **2010** 010
[7] See for example, Coutinho S C 1995 *A Primer of Algebraic D-modules* Cambridge University Press
[8] Tkachov F V 1997 Nucl. Instrum. Meth. A **389** 309
[9] Aomoto K 1977 J. Fac. Sci. Univ. Tokyo Sect. IA Math. **27** 49–61
Aomoto K and Kita M 2011 *Theory of Hypergeometric Functions* Springer
[10] Gelfand I M 1986 Soviet Math. Dokl. **33** 573–577
[11] Vasil'ev V A, Gel'fand I M and Zelevinskii A V 1987 Funk. Anal, Appl. 21 19–3
[12] 1953–1955 Erdéley A, Magnus W, Oberhettinger F, and Tricomi F G, *Higher transcendental functions, I–III* Bateman manuscript Project McGraw-Hill
[13] Hahn T `LoopTools` http://www.feynarts.de/looptools/
[14] Moch S and Uwer P 2006 Comput. Phys. Commun. **174** 759
[15] Kalmykov M Y, . Ward B F L and Yost S 2007 JHEP **0702** 040
Kalmykov M Y, . Ward B F L and Yost S 2007 JHEP **0711** 009
[16] Kaneko T and Watanabe N in prepararion

The NLO Calculations of Heavy Quarkonium Production at B Factories

Yi-Jie Li[1], Guang-Zhi Xu[1], Kui-Yong Liu[2], Yu-Jie Zhang[1,3]

[1] Key Laboratory of Micro-nano Measurement-Manipulation and Physics (Ministry of Education) and School of Physics, Beihang University, Beijing 100191, China
[2] Department of Physics, Liaoning University, Shenyang 110036 , China
[3] International Research Center for Nuclei and Particles in the Cosmos, Beihang University, Beijing 100191, China

E-mail: [1] yijiegood@gmail.com

E-mail: [2] still200@gmail.com

E-mail: [3] liukuiyong@lnu.edu.cn

E-mail: [4] nophy0@gmail.com

Abstract.
We calculated the next to leading order quantum chromodynamics corrections of J/ψ production at B factories. The next to leading order corrections are very important for heavy quarkonium production at B factories. The next to leading order cross sections can be compared with the B factories data. Some detail of the next to leading order calculations within the non-relativistic quantum chromodynamics factorization framework are given here.

1. Introduction

It is widely believed that the production and annihilation of the heavy quarkonium can be described by non-relativistic quantum chromodynamics (NRQCD)[1]. Defferring from the case in the color-singlet model, in a color-octet production process, the intermediate pair of quark and antiquark can be created at short distances with different colors and then forms the non color quarkonium at long distances by emitting or absorbing soft gluons. The process at short distances called short-distance coefficients which can be calculated perturbatively, and the process at long distances dependents on non-perturbative NRQCD matrix elements which are universal, process-independence and must be determined by experimental extraction, potential model or lattice calculations.

The NRQCD factorization scheme seems to acquire some significant successes in describing heavy quarkonium decay and production. But recently, several next-to-leading order (NLO) QCD corrections for the inclusive and exclusive heavy quarkonium production in the color-singlet piece are found to be large and significantly, which relieve the conflicts between the color-singlet model predictions and experiments. It may imply, though inconclusively, that the color-octet contributions in the production processes are not as big as previously expected, and the color-octet mechanism should be studied more carefully. Lots of work have been done to investigate the color-octet mechanism in NRQCD for heavy quarkonium production[2, 3, 4, 5].

The charmonium production in e^+e^- annihilation at B factories has provided an important test ground for NRQCD and color-octet mechanism. The production of double charmonium in e^+e^- annihilation at B factories [6, 7, 8, 9, 10, 11, 12, 13, 14, 15, 16, 17, 18, 19] is one of the challenging problems in heavy quarkonium physics and NRQCD[1]. The large discrepancies of $J/\psi + \eta_c$ production in e^+e^- annihilation at B factories between LO theoretical predictions[9, 10, 11, 20, 21] and experimental results [7, 13] were challenging issues once but now are largely resolved by higher order corrections: NLO QCD[22, 23, 24, 25, 26, 27, 28, 29] and relativistic [19, 30, 18, 31, 32, 33, 34] corrections, and the results show that the color-singlet NLO corrections (both in α_s and v)[35, 36] may increase the cross section of double charmonium production e.g. $e^+e^- \to J/\psi\eta_c$ by an order of magnitude. However, the cross sections of other processes, i.e., $e^+e^- \to J/\psi(\psi(2S))\chi_{c0}$ measured by Belle[6] $\sigma[J/\psi + \chi_{c0}] \times B^{\chi_{c0}}[> 2] = (16 \pm 5 \pm 4)$ fb are also larger than LO NRQCD predictions by about an order of magnitude or at least a factor of 5. Here $B^{\chi_{c0}}[> 2]$ is the branching fraction for the χ_{c0} decay into more than 2 charged tracks. Theoretically, two studies in NRQCD by Braaten and Lee[9] and by Liu, He, and Chao[10] showed that, at LO in the QCD coupling constant α_s and the charm quark relative velocity v, the cross-section of $e^+e^- \to J/\psi(\psi(2S))\chi_{c0}$ at $\sqrt{s} = 10.6$GeV is only about $2.4 \sim 6.7(1.0 \sim 4.4)$fb (depending on the used parameters, e.g., the long-distance matrix elements, m_c and α_s). The QCD radiative correction can also greatly enhance $\sigma[J/\psi(\psi(2S))\chi_{c0}]$[24, 37, 38, 39, 37, 40].

On the inclusive J/ψ production, the total prompt J/ψ cross sections in e^+e^- annihilation were measured to be $\sigma_{tot} = 2.52 \pm 0.21 \pm 0.21$ pb by BaBar [41], whereas Belle gave a much smaller value $\sigma_{tot} = 1.47 \pm 0.10 \pm 0.13$ pb [42]. Obviously, the large discrepancy between the two measurements should be further clarified. Belle reported new measurements[43]

$$\sigma(e^+e^- \to J/\psi + X) = (1.17 \pm 0.02 \pm 0.07)pb, \quad (1)$$
$$\sigma(e^+e^- \to J/\psi + c\bar{c}) = (0.74 \pm 0.08^{+0.09}_{-0.08})pb, \quad (2)$$
$$\sigma(e^+e^- \to J/\psi + X_{non\ c\bar{c}}) = (0.43 \pm 0.09 \pm 0.09)pb. \quad (3)$$

The J/ψ inclusive production cross section given in Eq.(1) is significantly smaller than that given by BaBar[41] and Belle[42] previously. The double charm production cross section given in Eq.(2) also becomes smaller accordingly. However, the ratio of J/ψ production rate through double $c\bar{c}$ to that of J/ψ inclusive production measured by Belle[7, 43]

$$R_{c\bar{c}} = \frac{\sigma[e^+e^- \to J/\psi + c\bar{c}]}{\sigma[e^+e^- \to J/\psi + X]} = 0.59^{+0.15}_{-0.13} \pm 0.12, \quad (4)$$

are much larger than LO NRQCD predictions. If only including the color-singlet contribution at LO in α_s, the ratio is about $0.2 \sim 0.4$[44].

The cross section of $J/\psi + X_{non\ c\bar{c}}$ includes the color-singlet contribution of $e^+e^- \to J/\psi + gg$ and the color-octet one of $e^+e^- \to c\bar{c}(^3P_0^{(8)}$ or $^3P_0^{(8)})g$ at leading order of α_s. The color-singlet piece has been investigated by including the NLO $O(\alpha_s)$ correction[29, 28] and $O(v^2)$ relativistic correction[32, 33], of which each contributes an enhancement factor of $1.2 - 1.3$ to the cross section of $e^+e^- \to J/\psi + gg$. As a result, the color-singlet contribution has saturated the observed value given in Eq.(3), leaving little room for the color-octet contribution.

In the color octet J/ψ inclusive production $e^+e^- \to J/\psi + X$ at B factories, there are two processes. One is $e^+e^- \to q\bar{q} + J/\psi + X(q = u, d, s)$, which was found to be negligible at $\sqrt{s} = 10.6$ GeV and can only be important at much higher energies than $\sqrt{s} = 10.6$ GeV[44]. The other color-octet process $e^+e^- \to c\bar{c}(^1S_0^{(8)}$ or $^3P_J^{(8)})g$ was studied by Braaten and Chen[45]. Based on the LO NRQCD calculation, they predicted that the J/ψ production is dominated in the region near the upper endpoint in the J/ψ energy distribution, and the width of the peak near the endpoint is of the order of 150 MeV. But the measured J/ψ spectra in e^+e^- annihilation by BaBar [41] and Belle[42] do not exhibit any enhancement near the endpoint. And a large color-octet contribution to the J/ψ inclusive production would enhance the denominator and then further decrease this ratio. So, this became a very puzzling issue. Some theoretical studies have been suggested in resolving this problem. Fleming, Leibovich and Mehen use the Soft-Collinear Effective Theory (SCET) to resum the color-octet contribution[46]. Lin and Zhu use SCET to analyze the color-singlet contribution to $e^+e^- \to J/\psi gg$ [47]. Leibovich and Liu sum the leading and next-to-leading logarithms in the color-singlet contribution to the J/ψ production cross section[48]. As a new step, Zhang and Chao find the NLO QCD corrections to $e^+e^- \to J/\psi + c\bar{c}$[23] to be large, and increase the cross sections by a factor of about 2 (using the same matrix elements as LO), making the ratio R larger than the LO results.

In order to further clarify this problem, it is certainly useful to study the the higher order corrections of J/ψ production at B factories. We have calculated $e^+e^- \to J/\psi + \eta_c$[22], $e^+e^- \to J/\psi + \chi_{c0}$[24], $e^+e^- \to J/\psi + c\bar{c}$[23], $e^+e^- \to J/\psi + gg$[29], and $e^+e^- \to J/\psi + g$[49]. The paper is organized as follows. In Section II, we will give the frame of the calculation. In Section III , we will give the detail of the loop integrate. In section IV, we will give the numerical results and relations to the color-octet matrix elements. A summary will be given in Section V.

2. The frame of the calculation

In leading order in strong coupling constant α_s, for example, $J/\psi + \eta_c$ can be produced at order $\alpha^2\alpha_s^2$. We refer to Ref [10]. There are four feynman diagrams, two of which are shown in Fig. 1. The other two diagrams are gotten through inversing quark lines. Momenta for the involving particles are assigned as $e^-(k_1)e^+(k_2) \to J/\psi(2p_1) + \eta_c(2p_2)$. We now write down the scattering amplitude in the nonrelativistic approximation to describe the creation of two color-singlet $c\bar{c}$ pairs which subsequently hadronize to two charmonium states in the e^+e^- annihilation process in Fig. 1

$$\begin{aligned}
\mathcal{A}(a+b \to Q\bar{Q}(^{2S_\psi+1}L_{J_\psi})(p_3) + Q\bar{Q}(^{2S+1}L_J)(p_4)) &= \sqrt{C_{L_\psi}}\sqrt{C_L} \sum_{L_{\psi z}S_{\psi z}} \sum_{s_1s_2} \sum_{jk} \sum_{L_zS_z} \sum_{s_3s_4} \sum_{il} \\
&\times \langle s_1; s_2 \mid S_\psi S_{\psi z}\rangle \langle L_\psi L_{\psi z}; S_\psi S_{\psi z} \mid J_\psi J_{\psi z}\rangle \langle 3j; \bar{3}k \mid 1\rangle \\
&\times \langle s_3; s_4 \mid SS_z\rangle \langle LL_z; SS_z \mid JJ_z\rangle \langle 3l; \bar{3}i \mid 1\rangle \\
&\times \begin{cases} \mathcal{A}(a+b \to Q_j(\frac{p_3}{2}) + \bar{Q}_k(\frac{p_3}{2}) + Q_l(\frac{p_4}{2}) + \bar{Q}_i(\frac{p_4}{2})) & (L=S), \\ \epsilon^*_\alpha(L_Z)\mathcal{A}^\alpha(a+b \to Q_j(\frac{p_3}{2}) + \bar{Q}_k(\frac{p_3}{2}) + Q_l(\frac{p_4}{2}) + \bar{Q}_i(\frac{p_4}{2})) & (L=P), \end{cases}
\end{aligned} \tag{5}$$

where $\langle 3j; \bar{3}k \mid 1\rangle = \delta_{jk}/\sqrt{N_c}$, $\langle 3l; \bar{3}i \mid 1\rangle = \delta_{li}/\sqrt{N_c}$, $\langle s_1; s_2 \mid S_\psi S_{\psi z}\rangle$, $\langle s_3; s_4 \mid SS_z\rangle$, $\langle L_\psi L_{\psi z}; S_\psi S_{\psi z} \mid J_\psi J_{\psi z}\rangle$ and $\langle LL_z; SS_z \mid JJ_z\rangle$ are respectively the color-SU(3), spin-SU(2), and angular momentum Clebsch-Gordan coefficients for $Q\bar{Q}$ pairs projecting out appropriate bound states. $\mathcal{A}(a+b \to Q_j(\frac{p_3}{2}) + \bar{Q}_k(\frac{p_3}{2}) + Q_l(\frac{p_4}{2}) + \bar{Q}_i(\frac{p_4}{2}))$ is the scattering amplitude for double $Q\bar{Q}$ production and $\mathcal{A}^\alpha$ is the derivative of the amplitude with respect to the relative momentum between the quark and anti-quark in the bound state. The coefficients C_{L_ψ} and C_L can be related to the radial wave function of the bound states or its derivative with respect to the relative spacing as

$$\begin{aligned}
C_s &= \frac{1}{4\pi}|R_s(0)|^2, \\
C_p &= \frac{3}{4\pi}|R'_p(0)|^2
\end{aligned} \tag{6}$$

We introduce the spin projection operators $P_{SS_z}(p,q)$ as[10]

$$P_{SS_z}(2p,q) \equiv \sum_{s_1s_2} \langle s_1; s_2|SS_z\rangle c(p-q; s_1)\bar{c}(p+q; s_2). \tag{7}$$

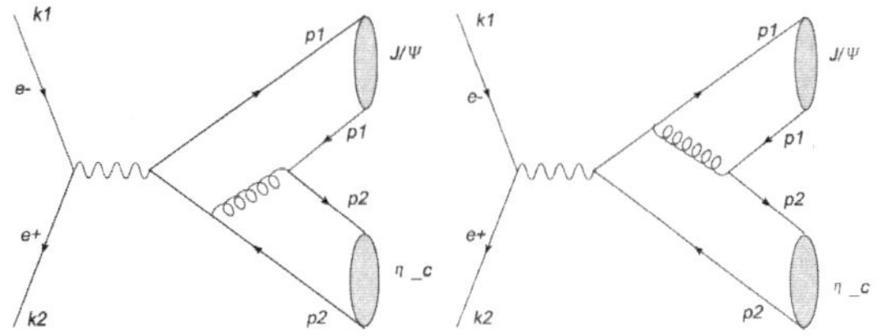

Figure 1. Two of the four born diagrams for $e^-(k_1)e^+(k_2) \to J/\psi(2p_1)\eta_c(2p_2)(\chi_{c0}(2p_2))$.

Expanding $P_{SS_z}(P, q)$ in terms of the relative momentum q, we get the projection operators and their derivatives, which will be used in our calculation, as follows

$$\begin{aligned} P_{00}(2p, q) &= \sum_{s\bar{s}} v(s)\bar{u}(\bar{s})\langle \frac{1}{2}, s; \frac{1}{2}, \bar{s}|0, 0\rangle \\ &= \frac{1}{2\sqrt{2}(E_q + m)}(-\not{p}_{\bar{c}} + m_c)\gamma_5 \frac{\not{P} + 2E_q}{2E_q}(\not{p}_c + m_c). \end{aligned} \quad (8)$$

For spin-triplet case, the expression is defined as

$$\begin{aligned} P_{1S_z}(2p, q) &= \sum_{s\bar{s}} v(s)\bar{u}(\bar{s})\langle \frac{1}{2}, s; \frac{1}{2}, \bar{s}|1, S_z\rangle \\ &= \frac{1}{2\sqrt{2}(E_q + m)}(-\not{p}_{\bar{c}} + m_c)\not{\epsilon} \frac{\not{P} + 2E_q}{2E_q}(\not{p}_c + m_c), \end{aligned} \quad (9)$$

In next-to-leading order in strong coupling constant α_s,the cross section is

$$\sigma = |\mathcal{M}_{Born} + \mathcal{M}_{NLO}|^2 = |\mathcal{M}_{Born}|^2 + 2\mathrm{Re}(\mathcal{M}_{Born}\mathcal{M}^*_{NLO}) + \mathcal{O}(\alpha^2\alpha_s^4) \quad (10)$$

Then we need to calculate $\mathcal{M}_{NLO}$. We use `FeynArts` [50, 51] to generate feynman diagrams, `FeynCalc` [52] for the tensor reduction, and `LoopTools` [53] for the numerical evaluation of the IR-safe integrals. One-loop diagrams contain $\gamma \to gg, g \to J/\psi, g \to \eta_c, gg \to J/\psi$ will vanish for color and charge conjugation. Then the self-energy, triangle diagrams are all correspond to propagators and vertexes of born diagrams. And it remains twenty-four box and pentagon diagrams. Twelve diagrams of them are shown in Fig. 2. The upper $c\bar{c}$ hadronize to J/ψ which momentum is $2p_1$,and the lower $c\bar{c}$ hadronize to η_c which momentum is $2p_2$. The other twelve diagrams are gotten through inversing quark lines. Specially, the associated diagram of Pentagon N12 is only inversing the lower quark line which hardonize to η_c, and inverse all quark lines get itself.

The self-energy and triangle diagrams are in general UV divergent; the triangle, box, and pentagon diagrams are in general IR divergent. Box N5 and N8 and Pentagon N10 ,which have a virtual gluon line connecting the $c\bar{c}$ in a meson, also contain Coulomb singularities, which are cancelled after taking into account the corresponding corrections to the operators $\left\langle \mathcal{O}^{J/\psi}\left[{}^3S_1^{(1)}\right]\right\rangle$ and $\left\langle \mathcal{O}^{\eta_c}\left[{}^1S_0^{(1)}\right]\right\rangle$. In the practical calculation, the IR and UV singularities are regularized with $D = 4 - 2\epsilon$ space-time dimension, and the Coulomb singularities are first regularized by a small relative velocity v between the $c\bar{c}$ [54]. $v = |\overrightarrow{p_{1c}} - \overrightarrow{p_{1\bar{c}}}|/m$, defined in meson c.m. frame. For the

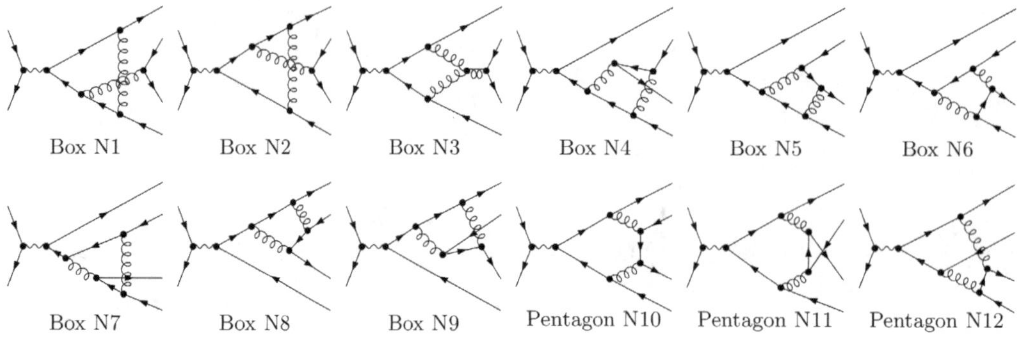

Figure 2. Twelve of the twenty-four box and pentagon diagrams.

Coulomb-singularity part of the virtual cross section, we find

$$\sigma \;=\; |R_S(0)|^4 \hat{\sigma}^{(0)} \left(1 + \frac{2\pi\alpha_s C_F}{v} + \frac{\alpha_s \hat{C}}{\pi} + \mathcal{O}(\alpha_s^2)\right), \tag{11}$$

where the color factor is given by $C_F = 4/3$ in $SU(3)_c$. Leading order of operators $\left\langle \mathcal{O}^{J/\psi}\left[{}^3S_1^{(1)}\right]\right\rangle$ and $\left\langle \mathcal{O}^{\eta_c}\left[{}^1S_0^{(1)}\right]\right\rangle$ are associated with $R_S(0)$, and next-to-leading order are $\pi\alpha_s C_F/v$ [1]. And two operators give a factor of 2 at $\mathcal{O}(\alpha_s)$, just the Coulomb-singularity term in Eq. (11). Then the corresponding contribution of Coulomb-singularity has to be factored out and mapped into the wave functions of J/ψ and η_c:

$$\begin{aligned} \sigma \;&=\; |R_S(0)|^4 \left(1 + \frac{2\alpha_s}{\pi}\frac{C_F \pi^2}{v}\right) \hat{\sigma}^{(0)} \left[1 + \frac{\alpha_s}{\pi}\hat{C} + \mathcal{O}(\alpha_s^2)\right] \\ &\Rightarrow\; |R_S(0)|^4\, \hat{\sigma}^{(0)} \left[1 + \frac{\alpha_s}{\pi}\hat{C} + \mathcal{O}(\alpha_s^2)\right]. \end{aligned} \tag{12}$$

Selecting feynman gauge, then we get the amplitude of born diagrams

$$i\mathcal{M}_{Born} = \frac{4096\pi e_c \alpha \alpha_s m |R_s(0)|^2}{3s^3}\, \epsilon_{\alpha\beta\nu\rho} p_1^\alpha p_2^\beta \varepsilon^{*\nu} \bar{v}_e(k_2)\gamma^\rho u_e(k_1), \tag{13}$$

where $s = (k_1 + k_2)^2$, $e_c = \frac{2}{3}$ is the fractional electric charge of the charm quark. The vector indices ρ is for the virtual photon. ε is the polarization vector of J/ψ. p_1 and p_2 are half momenta of J/ψ and η_c respectively. The coefficients $R_S(0)$ is the radial wave function at origin of the bound states.

3. Detail of the calculation

There are only two independent momentum p_1, p_2 in the process of $\gamma^* \to J/\psi + \eta_c(\chi_{c0})$, but the five points loop integral will be

$$\int d^D q \frac{1}{N_0^2 N_1 N_2 N_3 N_4}, \tag{14}$$

For the number of independent momentum is less than 4, the $GramDeterminant = 0$[55]. And the traditional Passarino-Veltman reduction does not work here. The first steps is separation the IR divergence[56]. Then we can introduce the mass of gluon and solve the equation

$$\sum_{i=0}^{4} a_i N_i / C = 1$$
$$\int \frac{\mathrm{d}^D q}{N_0 N_1 N_2 N_3 N_4} = \sum_{i=0}^{4} \frac{a_i}{C} \int \frac{N_i \mathrm{d}^D q}{N_0 N_1 N_2 N_3 N_4}. \tag{15}$$

where $C \neq 0$ is independent on integral momentum q and a_i is constant. They become four points integral and can be reduced again. Then we can get a set of three points integral and calculate directly.

If we do not introduce the mass of gluon, $C = 0$ when the gluon connect with the both legs of J/ψ. Then we can reduce the integral to a set of basic three, four, and five points function. The five points integral is the integral which appears in N10 of Fig.2

$$\begin{aligned}
& E_0^{fin}[p_1, 2p_1, -p_2, -2p_2, m, 0, m, 0, m] \\
&= E_0 - \frac{2}{s} D_0[-p_1, -p_1 - p_2, p_1, 0, m, 0, m] - \frac{2}{s} D_0[p_1 \leftrightarrow p_2] \\
&= \int \frac{\mathrm{d}^D q/(2\pi)^D \left(s/2 - 2(q^2 - m^2) - 4q \cdot p_1 + 4q \cdot p_2 - 8m^2\right) 2/s}{(q^2 - m^2)(q + p_1)^2((q + 2p_1)^2 - m^2)(q - p_2)^2((q - 2p_2)^2 - m^2)} \\
&= \frac{-4}{s} D_0[p_1 + p_2, p_1 + 2p_2, -p_1, 0, 0, m, m] + \int \frac{\mathrm{d}^D q}{(2\pi)^D} \\
& \frac{2/s(s/2 - 4q \cdot p_1 + 4q \cdot p_2 - 8m^2)}{(q^2 - m^2)(q + p_1)^2((q + 2p_1)^2 - m^2)(q - p_2)^2((q - 2p_2)^2 - m^2)} \\
&= \text{First Term} + \int \frac{\mathrm{d}^D q}{(2\pi)^D} \int_0^1 \\
& \frac{\Pi_{i=1}^5 \mathrm{d}x_i \delta(\sum_{j=1}^5 x_j - 1) 4!(1 - 16m^2/s)(1 - X - Y)}{\left[(q + Xp_1 - Yp_2)^2 - m^2(1 - X - Y)^2 + XYs/4\right]^5}
\end{aligned}$$

where $X = x_1 + 2x_2, Y = x_3 + 2x_4$. The `First Term` is IR- and Coulomb-finite. It can be calculated in $D = 4$ space-time dimension and $v = 0$, it is

$$\frac{2\sqrt{4m^2 - s}\, \tan^{-1} \frac{\sqrt{s}}{\sqrt{4m^2 - s}} - \sqrt{s}\, \ln \frac{-s}{m^2}}{-i\pi^2 m^2 s^{5/2}} \tag{16}$$

The second term is IR- and Coulomb-finite too. Choose $\{x_1, x_2, x_3, x_4, x_5\} = \{1–a, ab(1–c), a(1–b), abcd, abc(1–d)\}$ and integral a, b, d, c step by step in `Mathematica`,

$$\frac{2(4m^2 - s)^{3/2} \tan^{-1} \frac{\sqrt{s}}{\sqrt{4m^2 - s}} + \sqrt{s}\left(i\pi(3m^2 - s) + (s - 4m^2)\ln\frac{-s}{m^2}\right)}{8im^4\pi^2(4m^2 - s)s^{5/2}(16m^2 - s)^{-1}} \tag{17}$$

and $\ln(-s/m^2) = \ln(-(s + i0)/m^2) = \ln(s/m^2) - i\pi$. $D_0[-p_1, -p_1 - p_2, p_1, 0, m, 0, m]$ term in Eq. (16) is,

$$D_0[-p_1, -p_1 - p_2, p_1, 0, m, 0, m] = \frac{4}{s} C_0[-p_1, p_1, 0, m, m] + \frac{i}{(4\pi)^2} \frac{2i\pi - 2\ln 4}{m^2 s}. \tag{18}$$

This term will appear in Box N5, N8.

$$C_0[p_{1c}, -p_{1\bar{c}}, 0, m, m] = \frac{-i}{2m^2(4\pi)^2} \left(\frac{4\pi\mu^2}{m^2}\right)^{\epsilon} \Gamma(1 + \epsilon) \left[\frac{1}{\epsilon} + \frac{\pi^2}{v} - 2\right] \tag{19}$$

where $v = |\overrightarrow{p_{1c}} - \overrightarrow{p_{1\bar{c}}}|/m$, defined in meson c.m. frame. For the $J/\psi + \chi_{c0}$, the corresponded integral is

$$\int \frac{\mathrm{d}^D q}{N_0^2 N_1 N_2 N_3 N_4}. \tag{20}$$

It can be integrald analysis too[24].

In another way, we can introduce the equation

$$\sum_{i=1}^{4} a_i \frac{N_i}{N_0} = 1, \tag{21}$$

where a_i is constant. Then we get

$$\int \frac{\mathrm{d}^D q}{N_0 N_1 N_2 N_3 N_4} = \sum_{i=1}^{4} a_i \int \frac{N_i \mathrm{d}^D q}{N_0^2 N_1 N_2 N_3 N_4}. \tag{22}$$

They become four points integral but N_0^2. We can repeat the process, and get a set of three points integral such as $\int \frac{\mathrm{d}^D q}{N_0^3 N_1^2 N_2}$, $\int \frac{\mathrm{d}^D q}{N_0^2 N_1^2 N_2^2}$, and so on. This integral can be calculated by Integration-By-Parts (IBP) method [57, 58, 59, 60].

4. Numerical result

In the numerical result, we select $m_{J/\psi} = m_{\eta_c} = 2m$ $\Lambda_{\overline{\rm MS}}^{(4)} = 338\text{MeV}$, $m = 1.5\text{GeV}$, $\mu = 2m$, the NLO cross section is

$$\sigma(e^+ + e^- \to J/\psi + \eta_c) = 15.7\text{fb}, \tag{23}$$

which is larger than LO cross section 8.0 fb by a factor of 1.96. It is can be compared with the B factories data 17 $\sim$ 25 fb. And

$$\sigma(e^+ + e^- \to J/\psi + \chi_{c0}) = 10.1\text{fb}, \tag{24}$$

which are a factor of 1.6 larger than the LO cross sections 6.35 fb.

If we select $m_c = 1.4$ GeV and μ=$2m$, the cross section at for $J/\psi + c\bar{c}$ is

$$\sigma(e^+ + e^- \to J/\psi + c\bar{c} + X) = 0.47 \text{ pb}. \tag{25}$$

It is about a factor of 1.7 larger than leading order cross section 0.27 pb. Including the contributions of the higher charmonium transition of other contributions, then the prompt cross section is

$$\sigma_{prompt}(e^+ + e^- \to J/\psi + c\bar{c} + X) = 0.71 \text{ pb}. \tag{26}$$

It is about 96% of the new Belle date 0.74 pb. The cross sections of prompt (feeddown included) $J/\psi gg$ and $J/\psi c\bar{c}$ production in e^+e^- annihilation at B factories are shown in Tab.1.

For the numerical calculations of the color octet cross sections, we select $m_c = 1.55$ GeV and μ=$2m$. The cross section at LO in α_s is

$$\sigma(e^+ + e^- \to J/\psi + X) = \left[11\frac{\left\langle 0\left|\mathcal{O}^{J/\psi}[^1S_0^{(8)}]\right|0\right\rangle}{\text{GeV}^3} + 18\frac{\left\langle 0\left|\mathcal{O}^{J/\psi}[^3P_0^{(8)}]\right|0\right\rangle}{\text{GeV}^5}\right]\text{pb}, \tag{27}$$

while the cross section at NLO in α_s becomes

$$\sigma(e^+ + e^- \to J/\psi + X) = \left[21\frac{\left\langle 0\left|\mathcal{O}^{J/\psi}[^1S_0^{(8)}]\right|0\right\rangle}{\text{GeV}^3} + 35\frac{\left\langle 0\left|\mathcal{O}^{J/\psi}[^3P_0^{(8)}]\right|0\right\rangle}{\text{GeV}^5}\right]\text{pb}. \tag{28}$$

Table 1. Cross sections of prompt (feeddown included) $J/\psi gg$ and $J/\psi c\bar{c}$ production in e^+e^- annihilation at B factories in units of pb.

	Belle Data	$\mu = 2.8$ GeV LO	$\mu = 2.8$ GeV NLO	$\mu = 5.3$ GeV LO	$\mu = 5.3$ GeV NLO
$\sigma(gg)$	0.43	0.57	0.67	0.36	0.53
$\sigma(c\bar{c})$	0.74	0.38	0.71	0.24	0.53
$R_{c\bar{c}}$	0.63	0.40	0.51	0.40	0.50

The NLO short-distance coefficients are larger than the LO coefficients by a factor of about 1.9. The color-octet matrix element

$$M_k = \langle 0|\mathcal{O}^{J/\psi}[^1S_0^{(8)}]|0\rangle + k\,\langle 0|\mathcal{O}^{J/\psi}[^3P_0^{(8)}]|0\rangle/m_c^2 \tag{29}$$

In fact, for the $e^+e^- \to J/\psi + X_{non\ c\bar{c}}$ production, the color-singlet process $e^+e^- \to J/\psi + gg$ has been found to make a dominant contribution to the cross section: $\sigma(e^+e^- \to J/\psi + gg) = 0.4 \sim 0.7$ pb at NLO in α_s and v^2 [29, 28, 32, 33], thus leaves little room to the color-octet contributions. This gives a very stringent constraint on the color-octet contribution, and may imply that the values of color-octet matrix elements are much smaller than expected earlier by using the naive velocity scaling rules or extracted from fitting experimental data with the leading-order calculations. Even if we disregard the dominant contribution of $e^+e^- \to J/\psi + gg$ to $e^+e^- \to [J/\psi + X_{non\ c\bar{c}}]$ by setting the color-singlet contribution $\sigma(e^+e^- \to J/\psi + gg)$ to be zero, and combining Eq.(3) with Eq.(28), we can get an upper bound of the color-octet matrix element:

$$M_{4.0}^{NLO} \quad < \quad (2.0 \pm 0.6) \times 10^{-2}\ \mathrm{GeV}^3. \tag{30}$$

5. Summary

We calculated the NLO QCD corrections of J/ψ production at B factories. The NLO corrections are very important for heavy quarkonium production at B factories. The NLO cross sections can be compared with the B factories data. Some detail of the NLO calculations within the NRQCD factorization framework are given here.

Acknowledgments

The authors would like to thank Professor K.T. Chao for useful discussion. This work was supported by the National Natural Science Foundation of China (Grants No.11075011 and No. 11375021), the Foundation for the Author of National Excellent Doctoral Dissertation of China (Grants No. 2007B18 and No. 201020), the Fundamental Research Funds for the Central Universities, and the Education Ministry of LiaoNing Province.

References

[1] Bodwin G T, Braaten E and Lepage G P 1995 *Phys.Rev.* **D51** 1125–1171 (*Preprint* hep-ph/9407339)
[2] Brambilla N *et al.* (Quarkonium Working Group) 2004 (*Preprint* hep-ph/0412158)
[3] Lansberg J 2006 *Int.J.Mod.Phys.* **A21** 3857–3916 (*Preprint* hep-ph/0602091)
[4] Lansberg J P *et al.* 2008 *AIP Conf. Proc.* **1038** 15–44 (*Preprint* 0807.3666)
[5] Brambilla N, Eidelman S, Heltsley B, Vogt R, Bodwin G *et al.* 2011 *Eur.Phys.J.* **C71** 1534 (*Preprint* 1010.5827)
[6] Uglov T 2004 *Eur.Phys.J.* **C33** S235–S237
[7] Abe K *et al.* (Belle) 2002 *Phys. Rev. Lett.* **89** 142001 (*Preprint* hep-ex/0205104)
[8] Abe K *et al.* (Belle) 2004 *Phys. Rev.* **D70** 071102 (*Preprint* hep-ex/0407009)
[9] Braaten E and Lee J 2003 *Phys. Rev.* **D67** 054007 (*Preprint* hep-ph/0211085)
[10] Liu K Y, He Z G and Chao K T 2003 *Phys. Lett.* **B557** 45–54 (*Preprint* hep-ph/0211181)

[11] Hagiwara K, Kou E and Qiao C F 2003 *Phys. Lett.* **B570** 39–45 (*Preprint* hep-ph/0305102)
[12] Bodwin G T, Lee J and Braaten E 2003 *Phys. Rev. Lett.* **90** 162001 (*Preprint* hep-ph/0212181)
[13] Aubert B *et al.* (BABAR) 2005 *Phys. Rev.* **D72** 031101 (*Preprint* hep-ex/0506062)
[14] Braguta V V, Likhoded A K and Luchinsky A V 2006 *Phys. Lett.* **B635** 299–304 (*Preprint* hep-ph/0602047)
[15] Ma J P and Si Z G 2004 *Phys. Rev.* **D70** 074007 (*Preprint* hep-ph/0405111)
[16] Bodwin G T, Kang D and Lee J 2006 *Phys. Rev.* **D74** 014014 (*Preprint* hep-ph/0603186)
[17] Bodwin G T, Kang D and Lee J 2006 *Phys. Rev.* **D74** 114028 (*Preprint* hep-ph/0603185)
[18] Bodwin G T, Lee J and Yu C 2008 *Phys.Rev.* **D77** 094018 (*Preprint* 0710.0995)
[19] He Z G, Fan Y and Chao K T 2007 *Phys.Rev.* **D75** 074011 (*Preprint* hep-ph/0702239)
[20] Liu K Y, He Z G and Chao K T 2003 *Phys. Rev.* **D68** 031501 (*Preprint* hep-ph/0305084)
[21] Liu K Y, He Z G and Chao K T 2004 *Phys. Rev.* **D69** 094027 (*Preprint* hep-ph/0301218)
[22] Zhang Y J, Gao Y J and Chao K T 2006 *Phys.Rev.Lett.* **96** 092001 (*Preprint* hep-ph/0506076)
[23] Zhang Y J and Chao K T 2007 *Phys.Rev.Lett.* **98** 092003 (*Preprint* hep-ph/0611086)
[24] Zhang Y J, Ma Y Q and Chao K T 2008 *Phys. Rev.* **D78** 054006 (*Preprint* 0802.3655)
[25] Gong B and Wang J X 2008 *Phys. Rev. Lett.* **100** 181803 (*Preprint* 0801.0648)
[26] Gong B and Wang J X 2008 *Phys. Rev.* **D77** 054028 (*Preprint* 0712.4220)
[27] Gong B and Wang J X 2009 (*Preprint* 0904.1103)
[28] Gong B and Wang J X 2009 *Phys. Rev. Lett.* **102** 162003 (*Preprint* 0901.0117)
[29] Ma Y Q, Zhang Y J and Chao K T 2009 *Phys. Rev. Lett.* **102** 162002 (*Preprint* 0812.5106)
[30] Bodwin G T, Kang D, Kim T, Lee J and Yu C 2006 (*Preprint* hep-ph/0611002)
[31] Elekina E and Martynenko A 2010 *Phys.Rev.* **D81** 054006 (*Preprint* 0910.0394)
[32] He Z G, Fan Y and Chao K T 2010 *Phys.Rev.* **D81** 054036 (*Preprint* 0910.3636)
[33] Jia Y 2010 *Phys.Rev.* **D82** 034017 (*Preprint* 0912.5498)
[34] Martynenko A and Trunin A 2013 (*Preprint* 1308.3998)
[35] Dong H R, Feng F and Jia Y 2012 *Phys.Rev.* **D85** 114018 (*Preprint* 1204.4128)
[36] Li X H and Wang J X 2013 (*Preprint* 1301.0376)
[37] Wang K, Ma Y Q and Chao K T 2011 *Phys.Rev.* **D84** 034022 (*Preprint* 1107.2646)
[38] Dong H R, Feng F and Jia Y 2011 *JHEP* **1110** 141 (*Preprint* 1107.4351)
[39] Dong H R, Feng F and Jia Y 2013 (*Preprint* 1301.1946)
[40] Wang S Q, Wu X G, Zheng X C, Shen J M and Zhang Q L 2013 (*Preprint* 1301.2992)
[41] Aubert B *et al.* (BABAR) 2001 *Phys. Rev. Lett.* **87** 162002 (*Preprint* hep-ex/0106044)
[42] Abe K *et al.* (BELLE) 2002 *Phys. Rev. Lett.* **88** 052001 (*Preprint* hep-ex/0110012)
[43] Pakhlov P *et al.* (Belle Collaboration) 2009 *Phys.Rev.* **D79** 071101 (*Preprint* 0901.2775)
[44] Yuan F, Qiao C F and Chao K T 1997 *Phys. Rev.* **D56** 321–328 (*Preprint* hep-ph/9703438)
[45] Braaten E and Chen Y Q 1996 *Phys. Rev. Lett.* **76** 730–733 (*Preprint* hep-ph/9508373)
[46] Fleming S, Leibovich A K and Mehen T 2003 *Phys. Rev.* **D68** 094011 (*Preprint* hep-ph/0306139)
[47] Lin Z H and Zhu G h 2004 *Phys.Lett.* **B597** 382–390 (*Preprint* hep-ph/0406121)
[48] Leibovich A K and Liu X 2007 *Phys.Rev.* **D76** 034005 (*Preprint* 0705.3230)
[49] Zhang Y J, Ma Y Q, Wang K and Chao K T 2010 *Phys.Rev.* **D81** 034015 (*Preprint* 0911.2166)
[50] Kublbeck J, Bohm M and Denner A 1990 *Comput. Phys. Commun.* **60** 165–180
[51] Hahn T 2001 *Comput. Phys. Commun.* **140** 418–431 (*Preprint* hep-ph/0012260)
[52] Mertig R, Bohm M and Denner A 1991 *Comput. Phys. Commun.* **64** 345–359
[53] Hahn T and Perez-Victoria M 1999 *Comput. Phys. Commun.* **118** 153–165 (*Preprint* hep-ph/9807565)
[54] Kramer M 1996 *Nucl. Phys.* **B459** 3–50 (*Preprint* hep-ph/9508409)
[55] Fleischer J and Riemann T 2011 *Phys.Rev.* **D83** 073004 (*Preprint* 1009.4436)
[56] Dittmaier S 2003 *Nucl. Phys.* **B675** 447–466 (*Preprint* hep-ph/0308246)
[57] Smirnov A 2008 *JHEP* **0810** 107 (*Preprint* 0807.3243)
[58] Feng F 2013 (*Preprint* 1307.5587)
[59] Feng F and Mertig R 2012 (*Preprint* 1212.3522)
[60] Feng F 2012 *Comput.Phys.Commun.* **183** 2158–2164 (*Preprint* 1204.2314)

From Black Holes to Cosmology : The Universe in the Computer

J.-P. Luminet
Laboratoire Univers et Théories (LUTH)
Observatoire de Paris, CNRS, Université Paris Diderot
5 Place Jules Janssen, 92190 Meudon, France

E-mail: jean-pierre.luminet @obspm.fr

Abstract. I discuss some aspects of the use of computers in Relativity, Astrophysics and Cosmology. For each section I provide two examples representative of the field, including gravitational collapse, black hole imagery, supernovae explosions, star-black hole tidal interactions, N-body cosmological simulations and detection of cosmic topology.

1. Introduction

In this plenary lecture I shall discuss various aspects of the use of computers in relativity, astrophysics and cosmology, by presenting a few representative examples of what is been done in the field. These examples are by no means exhaustive, but reflect a personal choice on subjects that I know better than others, such as black holes and cosmology, and I apologize for not discussing all the excellent work in numerical astrophysics that could not find place in this short presentation.

Of course the use of computers is present in all the activities of astronomers and astrophysicists, but it is important to distinguish between numerical modelisation (involving computer simulations) from data treatment and analysis – although both may consume a lot of computing time and memory. As a theoretician, I shall concentrate mainly on numerical modelisation.

Before discussing the physics, let us begin with a brief survey of the computing facilities available to the astronomical community, from the local scale of a laboratory to the global scale of international organizations.

• At the local scale, e.g. a typical laboratory comprising about fifty researchers in theoretical astrophysics, such as mine (Laboratory Universe and Theories), most of calculations are performed on sequential machines – personal laptops and computers - and small clusters (10 to 100 nodes) owned by the laboratory. For heavier calculations, use is done of mesocenters (around 200 nodes) located in Paris Observatory, grids and supercomputers (more than 10,000 nodes) owned by national institutions. Numerical tools are developed, including codes for hydrodynamics, shock capture methods, N-body simulations, ray-tracing, etc., as well as numerical libraries, which provide a set of tools in the form of free public softwares for building various codes. For instance the publicly available library Kadath [http://luth.obspm.fr/~luthier/grandclement/kadath.html] implements spectral methods for solving partial differential equations in general relativity, astrophysics and theoretical physics, written in C++ or Fortran 77. Fully parallel and running on several hundreds of processors, Kadath has successfully recovered known solutions in various areas of physics and is the main tool for the study of quasi-periodic solutions in gauge field theory [1].

• At the scale of a country such as France, computing facilities are provided by governmental organizations. For instance, the GENCI (Grand Equipement National de Calcul Intensif [http://www.genci.fr/en]) coordinates the principal French equipments in high performance computing. This legal entity takes the form of a «société civile» under French law, owned by the French State represented by the Ministry for Higher Education and Research, by CEA (Commission for alternative Energies and Atomic energy), CNRS (National Centre for Scientific Research) and Universities. Created in 2007, GENCI implements and ensures the coordination of the major equipments of the national High Performance Computing (HPC) centres by providing funding and by assuming ownership, promotes the organization of an European HPC area and participates to its achievements, and sets up R&D collaborations in order to optimize HPC.

• At a more global scale such as European Union, the PRACE infrastructure (Partnership foR Advanced Computing in Europe, [http://www.prace-ri.eu]) creates a pan-European supercomputing infrastructure for large scale scientific and engineering applications. The computer systems and their operations accessible through PRACE are provided by 4 country members (BSC representing Spain, CINECA representing Italy, GCS representing Germany and GENCI representing France) who committed a total funding of €400 million for the initial PRACE systems and operations. In pace with the needs of the scientific communities and technical developments, systems deployed by PRACE are continuously updated and upgraded to be at the apex of HPC technology.

The Curie supercomputer, owned by GENCI and operated by CEA, is the first French Tier0 system open to scientists through the French participation into the PRACE infrastructure. It offers a peak performance of 1,7 Petaflops but at the present date (september 2013) it ranks only #15 in the list of the most powerful computers.

For the sake of comparison, table 1 below lists some of the TOP 20 supercomputer sites for June 2013 (keeping in mind that the list evolves rapidly).

Table 1.

Rank	Site	Name	Cores	Rpeak *(Petaflops)*
1	National University of Defense Technology (China)	Tianhe 2	3,120,000	55
2	DOE/SC/Oak Ridge National Lab (United States)	Titan	560,640	27
3	DOE/NNSA/LLNL (UnitedStaes)	Sequoia	1,572,864	20
4	RIKEN/AICS (Japan)	K computer	705,024	11
5	DOE/SC/Argonne National Lab (United States)	Mira	786,432	10
...	...	...	...	...
7	Forschungszentrum Juelich (Germany)	JUQUEEN	458,752	5,8
...	...	...	...	...
12	CINECA (Italy)	Fermi	163,840	2,1
...	...	...	...	...
15	CEA/TGCC-GENCI (France)	Curie	77,184	1,7

We conclude this introduction by recalling the common definition of what is called a "numerical grand challenge" : a fundamental problem in science or engineering, with broad applications, whose solution would be enabled by the application of HPC resources that could become available in the near future. Each grand challenge requires simulations on supercomputers with more than 10,000 nodes. The specificites and requirements of such programs are massive parallelization, very large fluxes of I/O data, and full teams of researchers, engineers and technicians to correctly operate them. In the last section of this article we briefly describe the numerical grand challenge in cosmology.

2. Numerical Relativity

Numerical relativity aims to obtain solutions to Einstein's equations with computers. As it provides a powerful tool to explore fundamental problems in physics and astrophysics, it has been a constant field of research since the 1990's, but spectacular progress has been made in the last decade due to

- larger computational facilities
- more advanced and accurate numerical techniques
- new formulation of Einstein's and magnetohydrodynamics (MHD) equations well-suited for numerical evolution.

Indeed a very important step is the choice of formulation and gauge for Einstein's equations : as these are covariant, one must specify a "good" gauge in order to be able to write them as a well-posed system of partial differential equations, which can also be numerically integrated without the appearance of instabilities. In this section we discuss two representative examples, related to gravitational collapse and black hole imagery.

2.1. From gravitational collapse to gamma-ray bursts

Gamma-ray bursts (GRBs) are flashes of gamma rays associated with extremely energetic explosions that have been observed in distant galaxies. They can last from ten milliseconds to several minutes, and their light curves vary considerably from one event to another. Two relatively distinct groups can nevertheless be identified: short bursts (SGRB), representing about one-third of those observed, which have a mean duration of 0.3 seconds, and long bursts, which last more than 2 seconds, with a mean of 30 seconds. It is tempting to explain this apparent dichotomy by different physical origins, especially as short bursts generally emit harder radiation (i.e higher frequency) than long bursts. Most investigators consider that GRBs are linked to the formation of stellar mass black holes, whose stronger gravitational fields can result in the release of more energy. Stellar black holes could be formed in at least two ways: either by the catastrophic collapse of a very massive star, or by the fusion of two compact stars.

In the model of coalescence, bursts would be formed from a pair of neutron stars orbiting around each other or from a pair consisting of a neutron star and a black hole. The theory of general relativity indicates that in such a situation the two compact stars would rapidly lose orbital energy in the form of gravitational waves. Over time, the decrease in the energy of the pair will inexorably shorten the distance between them. The ballet will end when the two bodies collide and fuse, thus giving birth to a black hole and an accretion disc, accompanied by a spurt of ultrahot matter orthogonal to the disc, which is responsible for a short burst.

This qualitative scenario is supported to a good extent by fully general-relativistic simulations. I refer the reader to the excellent review by Rezzolla [2]. The numerical investigation of the inspiral and merger of binary neutron stars in full general relativity has seen enormous progress made in recent years. Crucial improvements in the formulation of the equations and numerical methods, along with increased computational resources, have extended the scope of early simulations. These developments have made it possible to compute the full evolution, from large binary-separations up to black-hole formation, without and with magnetic fields and with idealised or realistic equations-of-state. Numerical simulations show that the formation of a torus around a rapidly rotating black hole is inevitable (figure 1). They also provide the first evidence that the merger of a binary of modestly magnetised neutron stars naturally forms many of the conditions needed to produce a jet of ultrastrong magnetic field, with properties that are consistent with SGRB observations. This missing link between

Journal of Physics: Conference Series **523** (2014) 012065 doi:10.1088/1742-6596/523/1/012065

the astrophysical phenomenology of GRBs and the theoretical expectations is a genuine example of the potential of numerical relativity.

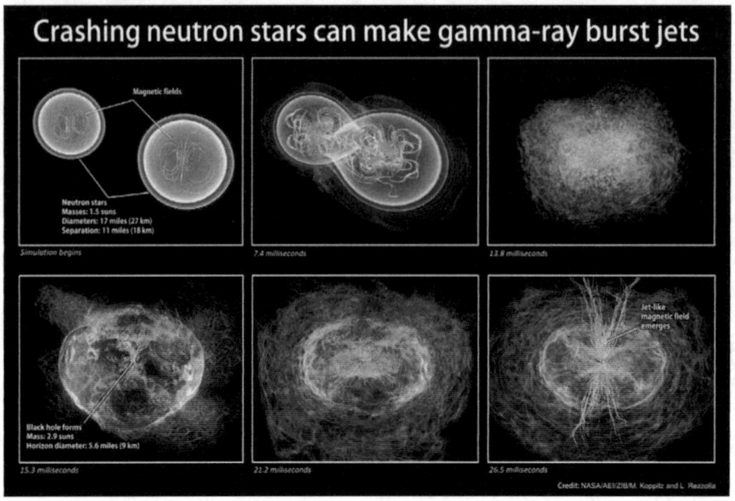

Figure 1. These images simulate the merger of two neutron stars using a new supercomputer model. Ribbons and lines represent magnetic fields. The orbiting neutron stars rapidly lose energy by emitting gravitational waves and merge in less than 8 milliseconds. The merger amplifies and scrambles the merged magnetic field. A black hole forms and the magnetic field becomes more organized, eventually producing structures capable of supporting the jets that power SGRB. Credit: NASA/AEI/ZIB/M. Koppitz and L. Rezzolla.

This remarkable advancement also provides information about the entire gravitational waveform, from the early inspiral up to the ringing of the black hole. Advanced interferometric detectors starting from 2014 (advanced Virgo /advanced LIGO) are expected to observe such sources at a rate of 40 – 400 events per year.

2.2. Black hole imagery

Since black holes cause extreme deformations of spacetime, they also create the strongest possible deflections of light rays passing in their vicinity and give rise to spectacular optical illusions. Today, it is possible to observe black holes only indirectly, through the effects that they have on their environment; for instance, their powerful gravitational field sucks in the neighboring gaz into accretion disks.

The first computer images of the appearance of a black hole surrounded by an accretion disk were obtained by myself in 1979 [3]. Calculations were done on an IBM 7040 machine of Paris Observatory; it was the time of punched cards, and no visualization device was available, so that I had to produce the final image by hand from numerical data (figure 2) ! The system is observed from a great distance at an angle of 10° above the plane of the disk. The light rays are received on a photographic plate (rather a bolometer in order to capture all wavelengths). A realistic image, e.g. taking account of the spacetime curvature, of the blue- and redshift effects, of the physical properties of the disk and so on, can be precisely calculated at any point of spacetime. Because of the curvature of spacetime in the neighborhood of the black hole, the image of the system is very different from the ellipses which would be observed if an ordinary celestial body (like the planet Saturn) replaced the black hole. The light emitted from the upper side of the disk forms a direct image and is considerably distorted, so that it is completely visible. There is no hidden part. The lower side of the disk is also visible as an indirect image, caused by highly curved light rays. In theory, there are multiples images of higher orders that give extremely distorted views of the top and of the bottom that are even more squashed, and so on to infinity.

With the improvement of computers, color and animated calculations were later performed by Marck [4] on a DEC-VAX 8600 computer, see fig. 3. At the time, black hole imagery was done mainly for pedagogical purpose, but the situation has changed very recently. Indeed black hole physics will develop considerably in the coming years, driven by observations at very high angular resolution (micro-arcseconds), as well in infrared (GRAVITY instrument installed at the VLT in 2014) as at submillimeter wavelength (Event Horizon Telescope, circa 2020). For the first time, we shall see the immediate environment of the event horizon of a black hole, especially the central black hole of our Galaxy, Sgr A*. Thus it is now essential to prepare the observations by numerical simulations which compute images and spectra of black holes and their close environment. Figure 4 depicts the result recently obtained by a new ray-tracing code, GYOTO (General relativitY Orbit Tracer of Paris Observatory) [5], publicly released at http://gyoto.obspm.fr. With respect to existing codes, GYOTO has the distinctive capability to compute geodesics in any metric, in particular in metrics that are known only numerically. A first application has been the modelling of the spectrum of Sgr A* by a magnetized ion torus [6], see figure 5.

Figure 2. First simulation of a spherical black hole with thin accretion disk, 1979. Credit: J.-P. Luminet/CNRS

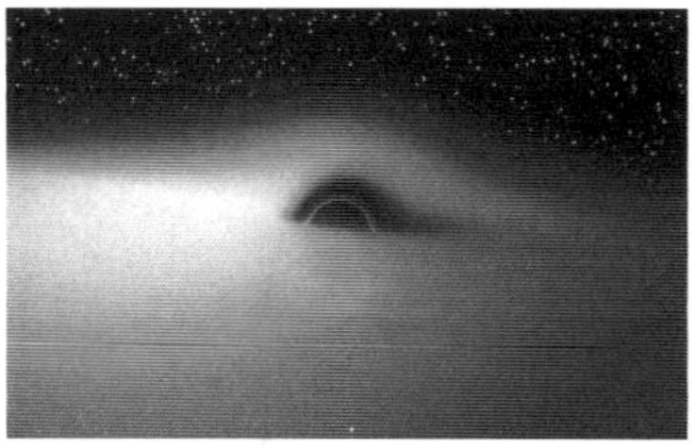

Figure 3. Image of a spherical black hole with thin accretion disk, 1996. Credit: J.-A. Marck/CNRS

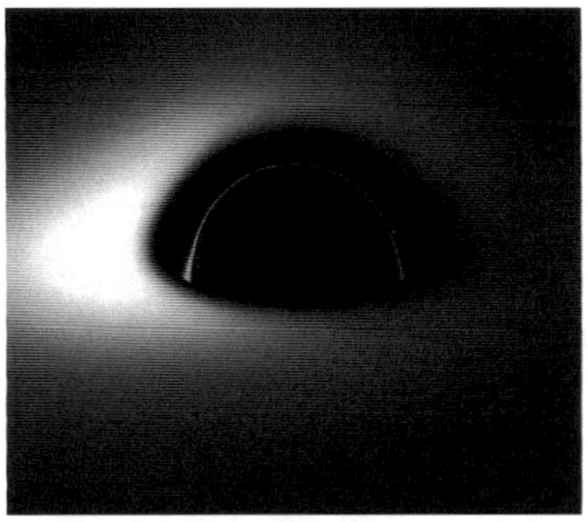

Figure 4. Same image calculated by a ray-tracing code, 2011. Credit: Vincent et al. /CNRS

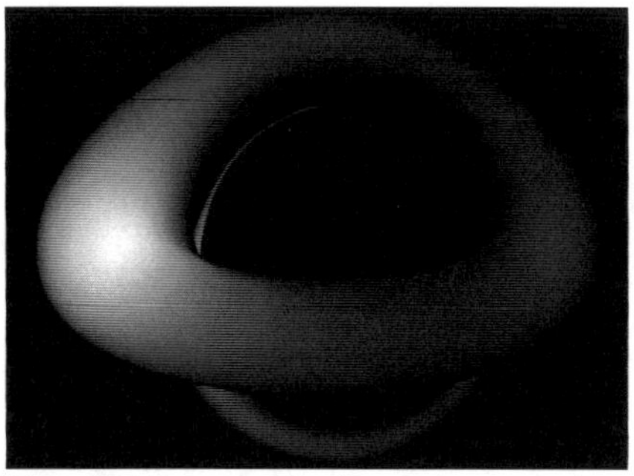

Figure 5. Image of an ion torus around a Kerr black hole, with radiative transfer, 2011. Credit: Vincent et al./CNRS

Another spectacular example of black hole imagery is the simulation of gravitational lensing. The spacetime curvature created by a massive object along the line of sight of a starry sky causes optical illusions for all objects in the background, because the curvature increases the number of distinct trajectories taken by light rays. Powerful modern telescopes can detect gravitational mirages created by intervening ordinary bodies such as stars (micro-lensing), galaxies or galaxy clusters which function as a gravitational lens, but not yet those created by black holes, due to lack of resolution. Computer simulations [7] recreate the effect of a large black hole acting as a gravitational lens to distort a starry landscape in the background and multiply the images of distant objects to create a mirage. Figures 6-7 show gravitational lensing and optical illusions produced by a black hole in the line of sight of the Milky Way center in the Southern Hemisphere (left) and in front of the Magellanic Clouds (right).

Figure 6. Simulation of gravitational lensing by a black hole in the Southern Hemisphere. The full Milky Way is seen as a mirror image close to the black hole. Credit A. Riazuelo IAP/UPMC/CNRS

Figure 7. Gravitational lensing by a black hole in the line of sight of the Magellanic Clouds. The large rings are two mirror images of the Large Magellanic Cloud. The brightest star is Canopus, seen twice. Credit A. Riazuelo IAP/UPMC/CNRS

3. Numerical Astrophysics

Numerical simulations play a vital role in all the fields of astrophysics, partly because many astronomical phenomena are results of nonlinearly coupled multiple factors. Thus they are powerful tools to elucidate the properties and rules of such nonlinear systems. They also serve as numerical experiments, since real experiments are impossible in astrophysics. Thanks to rapid increase in computational power, numerical simulations have become sophisticated enough to achieve high spatial resolution, wide dynamic range and inclusion of many factors and effects. Such a sophistication enables us to compare the simulations with observations in wide areas of astrophysics. We give below two examples representative of the field.

3.1. Simulations of type II supernovae explosions

Type II supernovae result from the rapid collapse and violent explosion of stars more massive than 8 solar masses. They are among the most powerful cosmic phenomena, releasing more energy within seconds than stars like the sun produce in billions of years. They terminate the lives of massive stars, giving birth to neutron stars or stellar-mass black holes, and are responsible for the production of about half of the chemical elements heavier than iron. By far most of the energy release occurs in the form of elusive neutrinos, elementary particles that are abundantly produced when the hot, compact remnant settles and cools.

The major unsolved problem with Type II supernovae is that it is not understood how the burst of neutrinos transfers its energy to the rest of the star, producing the shock wave which causes the star to explode. Only one percent of the energy needs to be transferred to produce an explosion, but

explaining how that one percent of transfer occurs has proven very difficult, even though the particle interactions involved are believed to be well understood.
A number of problems have plagued numerical simulations of type II supernovae for years. Supernova equations don't explode in 1D. One of the reasons is that stellar explosions are generically multi-dimensional phenomena. All implications of this fact were recognized only as computational models of supernovae explosions became increasingly more sophisticated and began to be advanced to the third dimension and to longer evolution periods. The other crucial area of investigation is the hydrodynamics of the plasma that makes up the dying star; how it behaves during the core collapse determines when and how the shock wave forms and when and how it stalls and is reenergized. Computer models are now successful at calculating the behavior of Type II supernovae [8][9]. They have led to a better understanding of the role of neutrinos for driving the explosion, to new insights into the characteristics of the neutrino and gravitational-wave emission of supernovae, and to the discovery of links between pulsar kicks and spins, supernova nucleosynthesis, and the origin of asymmetries seen in the gaseous remnants of stellar explosions.

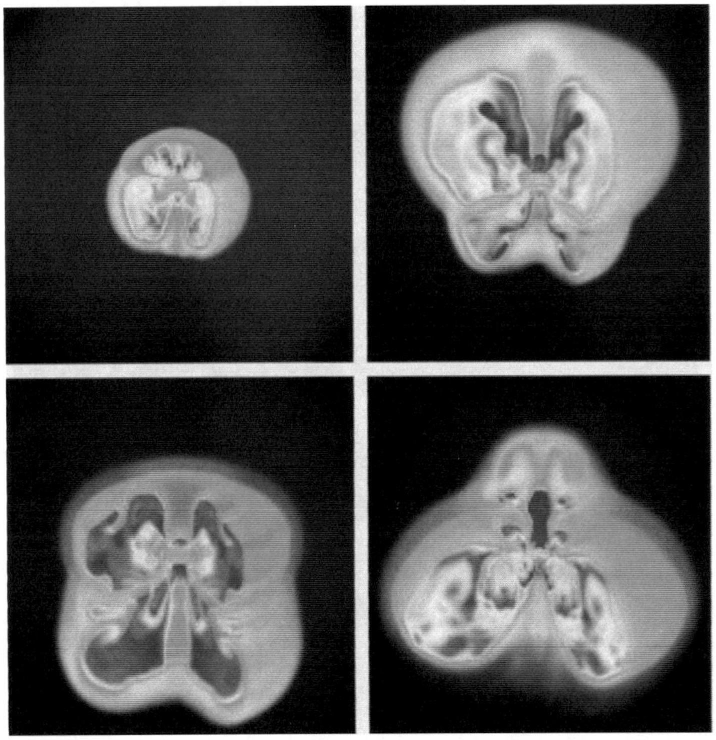

Figure 8. Onset of the type II supernova explosion of a star with 15 solar masses. The images show a sequence of moments from the computer simulation at 0.53, 0.61, 0.65, and 0.7 seconds (from top left to bottom right) after the stellar core has collapsed to a neutron star.
Full movie and more detailed explanations are available on http://www.youtube.com/watch?v=_ljcM4ftWnc (Visualization M. Rampp, RZG).

3.2. Tidal destruction of stars by giant black holes

Astrophysicists have long understood that massive black holes lurking in galactic nuclei and weighing millions of Suns can disrupt stars that come too close. Due to intense tidal forces, the black hole's gravity pulls harder on the nearest part of the star, an imbalance that pulls the star apart over a period of hours, once it gets inside the so-called "tidal radius". First calculations were performed by myself and collaborators in the 1980's [10]. We first used linearized equations to describe the final moments of a star as it penetrates deeply into the tidal field of a massive black hole, and we found that the tidal forces flatten it into a pancake shape; this flattening would increase the density and temperature inside the star enough to trigger intense nuclear reactions that would tear it apart.

But the real picture would be complicated by shock waves generated during the flattening process, so that other studies claimed that no nuclear explosion should occur. The controversy could be solved only by more sophisticated computer simulations, since no observational data were available. Various groups [11] [12] have recently investigated the effects of shock waves in detail; they confirmed that high compression and heating factors in the stellar core are able to trigger a thermonuclear explosion, and that the shock waves carry a brief but very high peak of temperature from the star's center outwards (figures 9-10), that could give rise to a new type of X-ray or gamma-ray bursts.

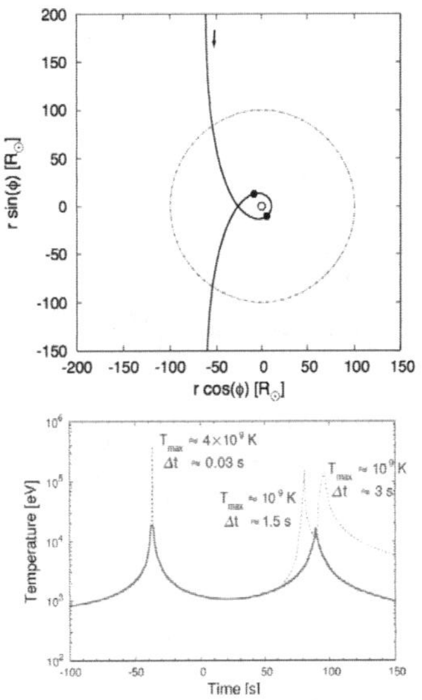

Figure 9. Top: Relativistic orbit of a solar-type star deeply plunging within the tidal radius of a 10^6 M_S black hole. The points on the orbit show the instants of maximum compression at the centre of the star. Bottom : as the stellar orbit winds up inside the tidal radius, the tidal field induces two successive compressions and heatings of the stellar core.
1D simulation. Credit: Luminet

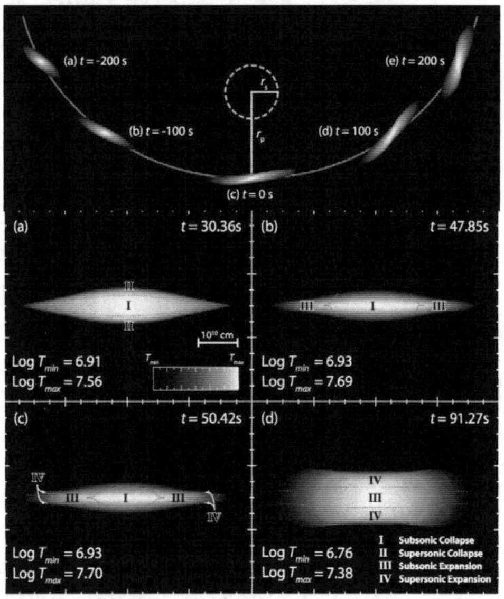

Figure 10. Top : Tidal deformation of a star along a non-relativistic parabolic orbit around a 10^6 M_S black hole. Bottom : Four snapshots showing a slice through the center of mass perpendicular to the orbital plane and parallel to the minor axis of the remnant during the bounce phase of the passage. Regimes of subsonic/supersonic collapse/expansion are indicated with roman numerals. The 3D adaptive mesh grid code recovers the results of the 1D approximation.
Credit : Guillochon et al.

Explosive disruption of stars by black holes will be an important subject in the next years. Recently observed X-ray, UV and gamma-ray bursts at the centers of galaxies are currently interpreted as the disruption of a star by the tidal forces exerted by the central massive black hole and place new constraints on the scenario that did not exist ten years ago. The search for X-ray flares, which would be due to stellar pancakes are now advancing to the next step, with new X-ray and gamma-ray observatories acquiring data. After the pioneering studies, the models have now to be improved with the inclusion of realistic equations of state for different stellar progenitors (solar-type, white dwarfs, red giant cores, ...) interacting with massive black holes in galaxies or intermediate-mass black holes in globular clusters, relativistic effects with Schwarzschild or Kerr metric (orbital precession induces multiple compressions), coupling with a nuclear reaction network, and computation of the energy transfer across the upper layers of the star in order to characterize the spectra of the resulting X-ray or gamma-ray flares. The aim is to predict observational signatures of these "tidally-induced supernovae" and to compare them to new data coming from telescopes.

4. Numerical Cosmology

In order to account for the observable Universe, any comprehensive model of cosmology must draw from many disciplines of physics, including general relativity, gauge theories of strong and weak interactions, the hydrodynamics and microphysics of baryonic matter, electromagnetic fields, spacetime curvature and topology, and so on. Although it is difficult to incorporate all these physical elements into a single complete model of our Universe, advances in computing methods and technologies have contributed significantly to our understanding of cosmological models.

4.1. Dark Energy and Formation of Structures

Here I closely follow [13]. 95 % of the cosmos is in the form of two invisible components, ~ 25 % is in Cold Dark Matter (CDM) particles, which are primarily responsible for the formation of the visible structures in the universe, ~ 70 % is in an unknown exotic form dubbed dark energy (DE), which is responsible for the present phase of cosmic accelerated expansion. The quest for DE has become one of the main topics of cosmological research. DE may have different theoretical explanations, but to date a coherent physical understanding is still missing. The study of DE through its imprints on cosmic structure formation is one of the core topics of investigation for numerical cosmology. One has to follow the gravitational collapse of dark matter throughout the history of the universe and across several orders of magnitude length scales for different models of DE. These calculations rely on N-body simulations. Many computational difficulties arise from the extreme number of particles involved, such as the great number of operations needed and the huge size of the generated data.

During the past 10 years several groups have pushed to the limits both size and resolution of cosmological N-body simulations. The Millenium Simulation in 2005 [14] has run a 2.2 billion light-years simulation box with 10 billion particles. Since then, the performance of cosmological simulations has increased more rapidly than the Moore's law (figure 11).

The Dark Energy Universe (DEUS) simulations [15] have performed the first-ever N-body simulations of the full observable universe (~95 billion light-years box) for three different DE models that are currently indistinguishable using cosmological observations. DEUSS have evolved 550 billion particles on an Adaptive Mesh Refinement grid with more than two and half trillion computing points along the entire evolutionary history of the universe and across 6 orders of magnitudes length scales, from the size of the Milky Way (mass of one particle) to that of the whole observable Universe. Such runs provide unique information on the formation and evolution of the largest structures in the universe and a support to future observational programs dedicated to the mapping of the distribution of matter and galaxies in the universe, such as the EUCLID mission (http://sci.esa.int/euclid). Each simulation has run on 4752 (of 5040) thin nodes of GENCI's supercomputer CURIE, using more than 300 TB of memory for 5 million hours of computing time. About 50 PBytes of rough data were generated throughout each run. Using an advanced reduction workflow, the amount of useful stored data has been reduced almost one the fly for each to 500 TBytes. Overall the realization of such large simulations required the development of a global application which integrated all aspects of the physical computational problem: initial conditions, dynamical computing, data validation, reduction and storage. Hence, it required optimizing not only the efficiency of the numerical dynamical solver, but also the memory usage, communications and I/O at the same time. Previous cosmological "grand-challenge" simulations could limit the use of efficient parallel HPC schemes to the dynamical computational solver only. Instead the DEUS experiment and similar projects clearly show the need for the use of MPI instructions in all parts of the application, scalable to more than 10,000 to 100 000 processes and optimized to match the specificities of supercomputing machine in which these have been run.

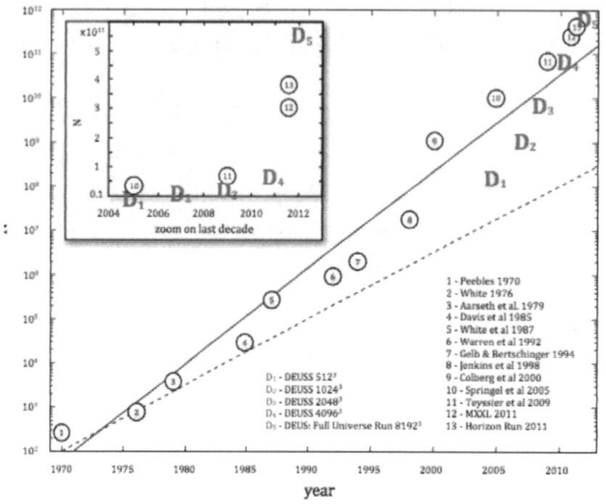

Figure 11. Evolution of cosmological N-body simulations over the last 40 years. A zoom shows the acceleration in performance occurred in the last decade, especially for DEUS collaboration. The solid line is the mean evolution of the simulation size, the dashed line is "Moore's Law", which corresponds to a factor 2 increase every 18 months. Courtesy J.-M. Alimi.

Computing time has been granted on a number of supercomputing facilities in France by GENCI. The realization of these simulations has had a great impact on the national supercomputing community in France : in fact to perform such project the supercomputing machines have been used at their full capacities. This has allowed testing and optimizing their nominal performances.

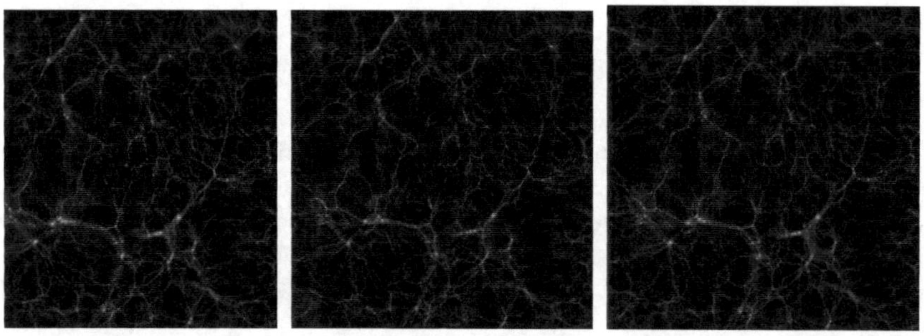

Figure 12. Formation of structures at the scale of 162 Mpc from DEUS simulations, in three different models of DE. Left : LCDM; center : RPCDM; right : SUCDM. Credit : DEUS-Consortium.org

Figure 12 depicts the formation of structures obtained by DEUS for three models of DE. Full movies are available on http://www.deus-consortium.org. The numerical data are publicly accessible through the Dark Energy Universe Virtual Observatory (DEUVO) database (http://roxxor.obspm.fr/deuvo-ui/).

4.2. Cosmic Topology

It is presently believed that our Universe is correctly described at large scale by a Friedmann-Lemaître (hereafter FL) model. The FL models are homogeneous and isotropic solutions of Einstein's equations, of which the spatial sections have constant curvature ; thus they fall into 3 general classes according to the sign of their spatial curvature, resp. k= +1, 0, or -1.

The spatial topology is usually assumed to be the same as that of the corresponding simply connected, universal covering space, resp. the hypersphere, Euclidean space or the 3D-hyperboloid, the first being finite and the other two infinite. However, there is no particular reason for space to have a simply connected topology: to a given FL metric element there are several possible topologies, and thus different models for the physical Universe [16]. For example, the hypertorus and familiar Euclidean space are locally identical, and relativistic cosmological models describe them with the same FL equations, even though the former is finite and the latter infinite ; in fact the multi-connected cosmological models share exactly the same kinematics and dynamics as the corresponding simply connected ones.

The global topology of the universe can be tested by studying the 3D distribution of discrete sources and the 2D fluctuations in the Cosmic Microwave Background (CMB). The methods are all based on the search for topological images of a same celestial object such as a galaxy, a cluster or a spot in the CMB. Such topological images can appear in a multi-connected space a characteristic length scale of which is smaller than the size of the observable space, because light emitted by a distant source can reach an observer along several null geodesics [17]. Figure 13 shows a computer visualization of the structure of the Poincaré Dodecahedral Space (PDS), a positively curved space which has been proposed [18] to account for the peculiarities of the observed CMB angular power spectrum.

If space has a non trivial topology, there must be particular correlations in the CMB, namely pairs of "matched circles" along which temperature fluctuations should be the same (figure 14). The PDS model predicts 6 pairs of antipodal circles with an angular radius less than 35°. Such circles have been searched in WMAP and PLANCK data by several teams, using various statistical indicators and massive computer calculations. The simplest multi-connected models, in which the matched circles are antipodal or nearly antipodal, such as the hypertorus and PDS, seem to be excluded by recent analyses [19].

However in generic topologies the circles are not antipodal, and their position in the sky depend on the observer's position in the fundamental domain. The corresponding larger number of degrees of freedom for the circle search in the data generates a dramatic increase of the computer time, up to values which are out-of-reach of the present facilities. To be more precise, a full search requires 6 free parameters : 2 for the location of first circle center, 2 for the location of second circle center, 1 for the radius of the circle and 1 for the relative phase of the two circles. A full search naively costs about $n^{5/2}$ operations, where n is the number of sky pixels. The recently released Planck CMB map has 5.10^7 pixels, which implies 10^{27} operations, i.e. ~3,000 years of computing time on petaflop supercomputers! Various computational tricks can reduce the cost, but not much.

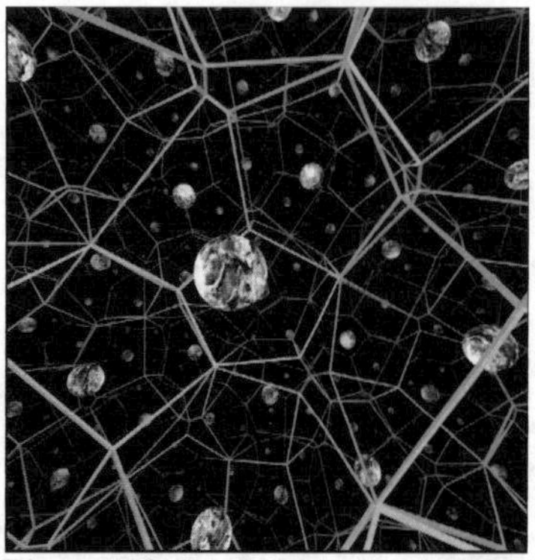

Figure 13. View from inside PDS along an arbitrary direction, calculated by the CurvedSpaces program [20], with multiple images of the Earth. The observer has the illusion to live in a space 120 times vaster, made of tiled dodecahedra which duplicate like in a mirror hall. Credit : J Weeks.

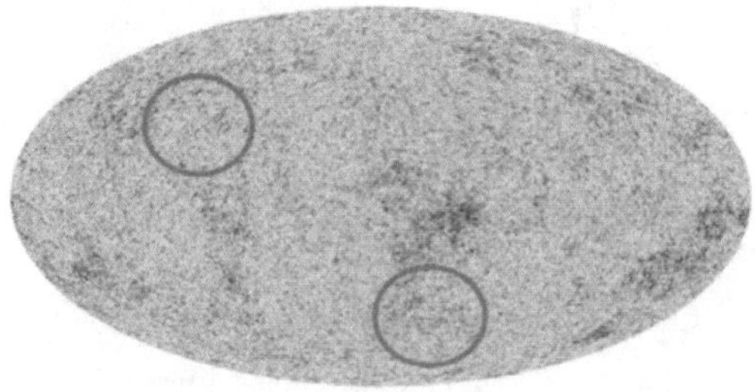

Figure 14. A pair of matched circles arbitrarily superimposed on the Planck CMB map. The number, location and size of matched circles depend on space topology and characteristic lengths of the fundamental domain compared to the size of the observable universe. A positive detection would be a signature of a multi-connected space at a sub-horizon scale.
Credit : Planck Collaboration.

5. References

[1] Grandclément P 2010 *J. Comput. Phys.* **229** 3334
[2] Rezzolla L 2013 Three little pieces for computer and relativity *Preprint* arXiv 1303.6464
[3] Luminet J P, 1979 *Astron. Astrophys.* **75** 228
[4] Marck J A 1996 *Class. Quantum Grav.* **13** 393
[5] Vincent F H, Paumard T, Gourgoulhon E and Perrin G 2011 *Class. Quantum Grav.* **28** 225011
[6] Straub O, Vincent F H, Abramowicz M A, Gourgoulhon E and Paumard T 2012 *Astron. Astrophys.* **543** A83
[7] Riazuelo A 2010 http://www2.iap.fr/users/riazuelo/bh/APOD.php
[8] Marek A and Janka H Th 2009 *Astrophys.J.* **694** 664
[9] Rampp M, Buras R, Janka H-Th and Raffelt G 2002 *Proceedings of the 11th Workshop on "Nuclear Astrophysics"*; eds.: W. Hillebrandt and E. Müller; *Preprint* astro-ph/0203493
[10] Luminet J-P and Carter B 1986 *Astrophys.J. Suppl.* **61**, 219
[11] Brassart M and Luminet JP 2008 *Astron. Astrophys.* **481** 259
[12] Guillochon J, Ramirez-Ruiz E, Rosswog S and Kasen D 2009 *Astrophys. J. Lett.* **705** 844
[13] Alimi J M 2013 The Challenges of the Next Decade in Numerical Cosmology: Big Data and Extreme-Scale Computing *Preprint* http://www.exascale.org/bdec/
[14] Springel V, White S, Jenkins A 2005, *Nature* **435**, 629
[15] Alimi J A et al. 2012 DEUS Full Observable CDM Universe Simulation: the numerical challenge *Preprint* arXiv 1206.2838
[16] Lachièze-Rey M and Luminet J-P 1995 *Physics Reports* **254** 135
[17] Luminet J-P 2008 *The Wraparound Universe* (NewYork: AK Peters)
[18] Luminet J-P, Weeks J, Riazuelo A, Lehoucq R and Uzan J-P 2003 *Nature* **425** 593
[19] Planck Collaboration 2013 *Preprint* arXiv 1303.5086
[20] Weeks J 2013 http://www.geometrygames.org/CurvedSpaces/

Summary of the ACAT Round Table Discussion: Open-source, knowledge sharing and scientific collaboration

Federico Carminati
CERN-PH, 1211 Geneva 23, Switzerland
E-mail: Federico.Carminati@cern.ch

Denis Perret-Gallix
LAPP/IN2P3, CNRS, 74941 Annecy-le-Vieux, France
E-mail: Denis.Perret-Gallix@in2p3.fr

Tord Riemann
DESY, 15738 Zeuthen, Germany
E-mail: Tord.Riemann@desy.de

Abstract. Round table discussions are in the tradition of ACAT. This year's plenary round table discussion was devoted to questions related to the use of scientific software in High Energy Physics and beyond. The 90 minutes of discussion were lively, and quite a lot of diverse opinions were spelled out. Although the discussion was, in part, controversial, the participants agreed unanimously on several basic issues in software sharing:

- The importance of having various licensing models in academic research;
- The basic value of proper recognition and attribution of intellectual property, including scientific software;
- The user respect for the conditions of use, including licence statements, as formulated by the author.

The need of a similar discussion on the issues of data sharing was emphasized and it was recommended to cover this subject at the conference round table discussion of next ACAT. In this contribution, we summarise selected topics that were covered in the introductory talks and in the following discussion.

1. Introduction

ACAT, the "Workshop on Advanced Computing and Analysis Techniques in Physics Research", is with us since more than 20 years now. It began with a software workshop in Lyon in 1990, went on under the name AIHENP since 1992 in La Londe Les Maures, and the name ACAT was first coined for the meeting at Fermilab in 2000. Following this nomenclature, we have now the 15th ACAT, see [1]. High Energy Physics is at the heart of ACAT, which traditionally includes three tracks: "Computing technology for physics research", "Data analysis - algorithms and tools", and "Techniques and methods in theoretical physics".

At ACAT 2008, there was a plenary talk on "Aspects of intellectual property law for HEP software developers" [2]; see also [3]. At ACAT 2009, we had a round table discussion devoted to "Event generation – are we ready for LHC?" [4]. In one section, "The quest for publicly available code" was discussed. In fact, software intellectual property related questions are getting more and more important in our research activity

This time, at ACAT 2013, we discussed "High Energy Physics software, knowledge sharing and scientific collaboration" not from the point of view of law, strongly affected by national regulations but, our field being vastly international, focusing on ethics and fairness of use There is a lot of literature available on the subject, but the complexity of legal regulations makes it of little practical help; see several Wikipedia articles on the topics, e.g. [5].

During the round table we specifically addressed the problems related to software sharing when "Good Scientific Practice" refers, in general, to "data, text or theories". For a summary on good practice see e.g. "Springer Policy on Publishing Integrity - Guidelines for Journal Editors" [6].

In recent decades, developing scientific software has become a full-fledged scientific activity of its own, see e.g. the discussions in [7, 8], or the famous CERNLIB history [9] with several licence regulations [10] for its different parts, see also the historical overview on the ZFITTER package [11] with a CPC licence, and many other long-term software projects. The very existence of conference series like ACAT together with dedicated scientific journals like Elseviers "Computer Physics Communications" and its companion "Computer Physics Communications Program Library - Programs in Physics & Physical Chemistry" at Belfast University, issuing the well- known CPC software licence statements [12], are clear signs of this evolution.

From our scientific experience, we have learned that there are specific rules to follow when dealing with software in knowledge sharing, even if, before the conference, controversial positions were expressed on the need or usefulness of discussing these issues.

Actually, High Energy Physics and, to some extent, even theoretical particle physics has become more and more a kind of industry. Due to this trend, the contacts between cooperating scientists tend to become more anonymous and virtually driven by Internet search. At the same time, the competition has become harsher and less open.

Knowledge sharing in early times was simple:

- Typically an author wrote software for his/her project to get numerical results. Sometimes he/she could hand over the Fortran code to a more or less close colleague. Sometimes the name of the author and a date was simply written in the code, sometimes not even that. Often people just copied what they needed into their Fortran codes without specific mention of the origin.[1]
- The other extreme was the way CERNLIB [9] was managed; its software was nearly free but better protected; see the 2004 licence statement for a typical function in CERNLIB [15].

Even if no rules or "conditions of use" were formulated, there was a common consensus on how to cite the software used in the scientific publication. Today we call this "proper attribution".

Actually, knowledge sharing today is much more complex than in the old days:

- There are many means of distribution, most of them are more or less anonymous.
- The software is now, primarily, written with the purpose of being used by other people: experimentalists or theorists and, not only, by the original authors.

[1] We again like to use as an example the ZFITTER project [13]. The package exists since 1985, and is distributed among users since 1992. For the first time, in the 1999 version of its description [14], the software pieces written by others were listed completely.

- The complexity of the software becomes higher by orders of magnitude, and it often results in many man-years involvement. This has consequences. The main one, for our discussion here, is the fact that software itself is the final product of sometimes highly complex scientific studies, and not only a by-product of a research activity.

Facing to these difficulties, people react differently, and during the round-table, we had the following reactions:

- When I create software, I want to get cited for its use in a way which I define. Sometimes this is achieved by applying a GPL-type licence (see http://en.wikipedia.org/wiki/GNU_General-_Public_License), sometimes by references to published articles.
- Our software has not to be touched by the user, because we guarantee for its high standards or because it is a standard candle [an etalon] for others, and it was created as such. So, please link my software to yours, or refrain from using it. Or, please write your interface to my package as a whole. Or, please contact me directly.

The following rather unethical statements, some of them often heard, express contradicting opinions of the use of software:

- Any software on Internet with anonymous download is, for me, open-source software.
- And I define by myself what the term "open-source" means: just free for any use.
- I need open-source software of this type because I have to adapt and develop your software for my purposes, in the interest of scientific progress.
- If you do not like misuse, you should not publish your software on Internet. This is the only safe way. I myself go this way.
- As there are no commonly accepted rules for the use and citation of open-source software, of course, I need not cite the software when using it. Why should I? And I can do what I like with the software I got, even if the authors claim to have a "licence".

We do not have the space to go into details of the discussion on definitions of "open-source software" or "licence". Two positions were expressed, though, in the discussion:

- It is the contradiction of the opinions, all with some justifications, which made the round table discussion lively. We discussed mainly what one might call source-open software, to mark the fact that the source code is open for read-only. While, the word open-source refers to some specific conditions of use. To get a flavour of the current confusion, one may consult the various articles in various national editions of Wikipedia.
- We agreed that the software licences mentioned here are not licences in the usual, legal meaning. They are, for us, standardized formulations of "conditions of use", enforced by authors or by their organizations. This approach reflects scientific practices and makes us independent of national law - as long as the rules are followed.

We discussed exclusively about academic, basic research, and in particular only about academic software, to be specific. This excluded e.g. any reference to commercial software.

- We live in an international community. As a consequence, national law, national licences, institutional regulations are not automatically valid. See the Wikipedia article on the Bern convention for the protection of literary and scientific works [16].
- We researchers at ACAT work on long-term projects, often in teams, sometimes really huge teams of several thousand people (e.g. ATLAS and CMS collaborations), and, in addition, with a variable membership. As a consequence, the definition of an "author" as well as his practical rights in a project are difficult to define with precision.

- Academic researchers career and funding depend, in many respects, on the recognition of the contributions to scientific progress. And this recognition has influence on:
  - the project budget
  - the resources that can be used, e.g. large disc spaces and clusters of computers for large-scale calculations
  - the possibilities to hire PhD and postdoc.
  - the permanent position that can be obtained
  - the recognition in the science "hall of fame" [non-monetary recognition]

For all these reasons, since the "Renaissance" the tradition of citation - or attribution - of the work of others [works = creations] became more and more an essential part of scientific ethics in European and world-wide basic research.[2]

Attribution applies either because the work relies on others' work, or because the work of others is directly integrated. This need for "attribution" in basic research is beyond commercial arguments, but not beyond material interests.

The usual mean of scientific attribution in basic academic research is citation as it is usually done in well-written articles.[3] But the balance between Competition and Cooperation gets destroyed when researchers use others' achievements without attribution.

In practice, there are additional expectations of the "creators of scientific work" toward their users that are well accepted by the society (in Germany, they are called "personal copyrights"). See for example, the regulations in the Internet for photographs, videos, music, texts, etc. It is well known by everyone that one has to care about the "conditions of use" when downloading anything.

For software, there may be certain very specific regulations. E.g. in German law this applies.[4]

Because national law often cannot be applied in practice, it is of high importance that researchers feel an ethical need to respect the conditions of use of any work, formulated by the authors of software. Licences are integral part of these "conditions of work usage".

Examples of popular licence statements:

- Gnu General Public licence = GPL and the derivatives, like e.g. the lesserGPL [17].
  It is often used and/or recommended. But it is not quite appropriate for academic software because it does not enforce proper citation.
- Creative Commons Licences = CC licences, with a complete tree of derivatives [19].
  This type of licences seems to be appropriate for scientific work, because it defines how to deal with attributions. Nevertheless, it is not often recommended for software licencing.
- Computer Physics Communications software deposit licence [12].
  This licence is in use since decades by many high energy physics software packages, e.g. GEANT, MINUIT, ZFITTER, FF, etc. Since recently, the users of the CPC software deposit have a choice of licence.
- FLUKA licence [20].
  The licence is rather strict.
- POWHEG BOX licence [21, 22].
  Example of guidelines that you may find short, simple and useful.

[2] One of the first holders of a kind of licence police, released by the emperor Maximilian, was the German painter Albrecht Dürer (1471-1528). The perhaps most famous scientific struggle about copyright was put forward by Issac Newton (1642-1727) and Gottfried Wilhelm Leibniz (1646-1716).

[3] The licence GPL [17] does not demand attribution in an article based on the use of GPL-licenced software. For this reason, the GPL is not truly appropriate for scientific purposes.

[4] For German speaking colleagues: German copyright law, see [18]; it is easy to read and understand. Although, some practiced regulations of that law deviate heavily from a naive understanding of the text.

- MCnet guidelines [23].
  They recommend GPLv2.
- HepForge, a development environment for High Energy Physics software projects [22, 24].
  No licence statement has been released.

These examples are shown in order to demonstrate one conclusion from the discussion: There is not a single widely appropriate licence model for all academic software projects and their purposes and we must keep at hand a variety of licence agreements fitting each one particular needs.

Finally, the decision to select a licence is taken jointly by the authors of the software, the employer and the funding agencies. The last two, the "third parties" often will not get involved explicitly. But, in other cases, like for the huge competing collaborations (ATLAS, ALICE, or CMS), the "licencing rules" can get much stronger and it might happen that the usage of software be restricted, exclusively, to the collaboration members.

In the view of some experts of law, the software licences of High Energy Physics are "so-called licences", i.e. no true licences, because domestic law might have different definitions on what is a licence. But, rejecting these licences is not a good approach as they are, at least for the best of them, a commonly accepted frame for the use of others' work. They are, in practice, the rules of the game.

Finally, the world of research is largely embedded in the society. Researchers in academic basic research are supported by the society and, as such, they do their best to:

- Show honesty in their research;
- Make the scientific achievements publicly available.

To summarize:

- We, software developers and authors are taking an active part in the design of the ethical regulations on the creation of knowledge.
- Conditions of use, including licences, are formulated by the authors, their employers and the funding agencies. Every efforts should be made to have them as convenient as possible to the users.
  Publish whatever work you have been producing: ASAP = As Source-open As Possible.
- Institutions should have compliance statements. This helps to regulate the relations with the researchers, in case of conflicting situations.
  Definitions of plagiarism, ombudsperson, and lists of sanctions are perhaps also needed, but the best approach is to prevent conflicts.

The summary of the Round Table Session on the last day's final session made evident that there is a broad understanding on the problems related to the collaborative use of software in the community. From time to time, problematic situations may arise, but it is usually clear how to handle them.

Concerning the collaborative handling of data, the situation is less clear. This is mainly due to the fact that such a large amount of data as produced by the LHC collaborations has never been collected in the past, so that there is less experience.

It might be a good idea to devote a round table discussion at the next ACAT conference to the problems of data sharing, data security, data property and data long term integrity. It would also fit well into this year's round table title:

Open-source, knowledge sharing and scientific collaboration

References

[1] For former ACAT meetings, see webpage http://acat2013.ihep.ac.cn/previous.htm
[2] Pinsky L S 2008 PoS ACAT08 003
[3] L. Pinsky, Aspects of Intellectual Property Law for HEP Software Developers, in: R. Brun, F. Carminati, G. Galli Carminati (Eds.), From the Web to the Grid and Beyond: Computing Paradigms Driven by High-Energy Physics, pp. 201-224, Springer Publisher 2012
[4] Bartalini P, Binoth T, Blümlein J, Buckley A, Hahn T, Kataev A, Perkins W, Riemann T and Sherstnev A, 2008 PoS ACAT08 129 section 7 by T. Riemann, The quest for publicly available code
[5] Wikipedia webpage, https://en.wikipedia.org/wiki/Legal_aspects_of_computing(12 July 2013)
[6] Springer Policy on Publishing Integrity - Guidelines for Journal Editors, text available from http://static. springer.com/sgw/documents/784498/application/pdf/Policy_on_Publishing_Integrity2010.pdf (7 January 2014)
[7] Brun R 2011 Eur. Phys. J. Plus 126 14
[8] Brun R 2012 J. Phys. Conf. Ser. 396 052016
[9] Webpage of the CERN Program Library - CERNLIB, http://cernlib.web.cern.ch/cernlib/(7 January 2014), and references therein
[10] Webpage http://cernlib.web.cern.ch/cernlib/conditions.html(dated 18 March 2004)
[11] Akhundov A, Arbuzov A, Riemann S and Riemann T 2013, http://arxivorg/abs/13021395v2, to appear in Particles & Nuclei, JINR, Dubna, Russia; ZFITTER was used by the LEPEWWG at CERN for the prediction of the Higgs boson mass in the Standard Model
[12] Webpage of CPC, licence statement, http://cpc.cs.qub.ac.uk/licence/licence.html
[13] The ZFITTER project, webpage http://zfitter.education (12 May 2014)
[14] Bardin D, Bilenky M, Christova P, Jack M, Kalinovskaya L, Olchevski A, Riemann S and Riemann T 2001 Comput. Phys. Commun. 133 229–395 (Preprint hep-ph/9908433)
[15] An item of CERNLIB, http://wwwasdoc.web.cern.ch/wwwasdoc/psdir/shortwrups.dir/c304.ps.gz
[16] Wikipedia webpage, http://en.wikipedia.org/wiki/Berne_Convention_for_the_Protection_of_Literary_and_Artistic_Works
[17] The GNU general public license, http://www.gnu.org/licenses/gpl-3.0.en.html
[18] Webpage http://www.gesetze-im-internet.de/bundesrecht/urhg/gesamt.pdf(7 January 2014)
[19] Creative Commons licences are formulated in many languages. English version: http://creativecommons. org/licenses/by-sa/3.0/, Chinese version: http://creativecommons.org/licenses/by-sa/3.0/deed.
[20] FLUKA license, http://www.fluka.org/fluka.php?id=license
[21] POWHEG box licence, http://powhegbox.mib.infn.it/
[22] HEPFORGE, project page http://www.hepforge.org/
[23] MCnet guidelines webpage http://www.montecarlonet.org/index.php?p=Publications/Guidelines
[24] A. Buckley, HepForge and HEP software engineering, talk at CHEP06, Mumbay, Feb. 2006, http://indico. cern.ch/getFile.py/access?contribId=120&sessionId=5&resId=0&materialId=slides&confId=048

AUTHOR INDEX

Abdurachmanov, D....54
Ablinger, J....464
Adzhemyan, L Ts....373
Alloul, A....336
Al-Turany, M....239
Ammendola, R....42, 80
Aoki, S....350
Aoyama, T....350
Apostolakis, J....22
Arbuzov, A....328
Arya, K....97
Badger, S....440, 474
Bagnasco, S....73
Balewski, J....194
Bardin, D....328
Bednyakov, A V....342
Bern, Z....388
Bertini, D....239
Berzano, D....73
Betev, L....88
Biagioni, A....42, 80
Biedermann, B....440
Blümlein, J....464
Bondarenko, S....328
Borowka, S....364
Braguta, V V....279
Brock, I C....205
Brun, R....22
Brunetti, R....73
Carena, F....31
Carminati, F....22, 113, 519
Cascioli, F....448
Christensen, N D....336
Christova, P....328
Cicero, F Lo....42, 80
Conte, E....226
Cooperman, G....97
Cordero, F Febres....388
Cullen, G....431
De Doncker, E....393
Degrande, C....336
Dell'Asta, L....121
Deri, L....42
Dixon, L J....388
Dotti, A....97
Duhr, C....336
Elmer, P....54, 97
Erdmann, M....148
Eulisse, G....54
Feng, F....313
Fiorini, M....42
Fischer, R....148
Frellesvig, H....474
Frezza, O....42, 80
Fuks, B....226, 336
Galster, G....221
Gheata, A....22, 88
Gheata, M....88
Glaser, C....148
Gong, B....287
Greiner, N....431
Grigoras, A G....60
Grigoras, C....60, 88
Hahn, T....381
Heinrich, G....364, 431
Hoang, S....186
Höche, S....388, 448
Hristov, P....88
Idarraga, J....186
Ionita, C....31
Ita, H....388
Kalinovskaya, L....328
Kama, S....255
Kanaya, K....350
Kaneko, T....490
Karabowicz, R....239
Khandramai, V....482
Kirpichnikov, D....426
Kisel, I....153
Klingebiel, D....148
Kolesnikov, V....328
Komm, M....148
Kompaniets, M V....373
Kosower, D A....388
Kövesárki, P....205
Krasznahorkay, A....129
Krauss, F....448
Kresan, D....239
Kroupa, M....186
Lamanna, G....42
Lampén, T....173
Lang, J-N....381
Lauret, J....37, 105
Lee, R N....456
Lei, G....414
Leney, K J C....164
Li, Y-J....498
Likhoded, A K....279
Liu, B....50
Liu, K-Y....498
Liu, Y B....16
Lonardo, A....42, 80
Luchinsky, A V....279
Luisoni, G....431
Luminet, J-P....507
Lusso, S....73
Luzzi, C....113
Maierhöfer, P....448
Maître, D....388
Makatun, D....105

AUTHOR INDEX

Manafov, A 239
Mastrolia, P 431
Matsufuru, H 350
Mertens, A 199
Messina, A 42
Mirabella, E 381, 431
Motoki, S 350
Müller, G 148
Muzaffar, S 54
Nairz, A 137
Namekawa, Y 350
Nejad, B C 381
Nemura, H 350
Nowak, A 7
Ossola, G 300, 431
Oyanagi, Y 1
Ozeren, K J 388
Pantaleo, F 42
Paolucci, P S 80
Paolucci, Ps 42
Pedreira, M M 60
Peraro, T 431
Perret-Gallix, D 519
Pikelner, A F 342
Pinsky, L 186
Poslavsky, S V 279
Pozzorini, S 448
Reichel, J 431
Ren-You, Z 414
Rieger, M 148
Riemann, T 519
Ritsch, E 246
Rossetti, D 42, 80
Rybalchenko, A 239
Sadykov, R 328
Saiz, P 60
Sangaline, E 37
Sapronov, A 328
Schlenk, J 431
Schneider, C 262, 464
Schreiner, S 60
Selvaggi, M 234
Seuster, R 255
Siegert, F 448
Simula, F 42, 80
Sozzi, M 42
Steggemann, J 148
Stelzer, J 221
Stewart, G A 255
Stoffle, N 186
Šumbera, M 105
Taniguchi, Y 350
Tkachov, F 214
Tkachov, F V 67
Tosoratto, L 42, 80
Tramontano, F 431
Ueda, S 350
Ueda, T 358
Uglov, E 328
Uhlig, F 239
Ukita, N 350
Urban, M 148
Uwer, P 440
Vallero, S 73
Van Deurzen, H 431
Velizhanin, V N 342
Vermaseren, J 358
Vicini, P 42, 80
Vilalta, R 186
Vitillo, R A 255
Von Soden-Fraunhofen, J F 431
Wan, L-P 287
Wang, J-X 287
Watanabe, N 490
Wen-Gan, M 414
Wenzel, S 22
Wiedenmann, W 221
Winchen, T 148
Wu, X-G 321
Xia, D 181
Xu, G-Z 498
Yuasa, F 393
Yundin, V 440
Zhang, H-F 287
Zhang, Y 474
Zhang, Y-J 498
Zykunov, V A 405